Urban Computing

Information Systems

Eric Yu, Florian Matthes, Michael P. Papazoglou, editors

Foundations of Neural Networks, Fuzzy Systems, and Knowledge Engineering, Nikola K. Kasabov

Advances in Object-Oriented Data Modeling, Michael P. Papazoglou, Stefano Spaccapietra, and Zahir Tari, editors

Workflow Management: Models, Methods, and Systems, Wil van der Aalst and Kees van Hee

A Semantic Web Primer, Grigoris Antoniou and Frank van Harmelen

Aligning Modern Business Processes and Legacy Systems: A Component-Based Perspective, Willem-Jan van den Heuvel

A Semantic Web Primer, second edition, Grigoris Antoniou and Frank van Harmelen

Service-Oriented Computing, Dimitrios Georgakopoulos and Michael P. Papazoglou, editors

At Your Service, Elisabetta di Nitto, Anne-Marie Sassen, Paolo Traverso, and Arian Zwegers, editors

Metamodeling for Method Engineering, Manfred A. Jeusfeld, Matthias Jarke, and John Mylopoulos, editors

Social Modeling for Requirements Engineering, Eric Yu, Paolo Giorgini, Neil Maiden, and John Mylopoulos, editors

Modeling Business Processes: A Petri Net-Oriented Approach, Wil van der Aalst and Christian Stahl

A Semantic Web Primer, third edition, Grigoris Antoniou, Paul Groth, Frank van Harmelen, and Rinke Hoekstra

Workflow Patterns: The Definitive Guide, Nick Russell, Wil van der Aalst, and Arthur ter Hofstede

Security Requirements Engineering: Designing Secure Socio-Technical Systems, Fabiano Dalpiaz, Elda Paja, and Paolo Giorgini

Urban Computing, Yu Zheng

Urban Computing

Yu Zheng

The MIT Press
Cambridge, Massachusetts
London, England

This book was set in Stone Serif by Westchester Publishing Services. Printed and bound in the United States of America.

Library of Congress Cataloging-in-Publication Data

Names: Zheng, Yu (Data scientist), author.
Title: Urban computing / Yu Zheng.
Description: Cambridge, MA : The MIT Press, [2018] | Series: Information systems | Includes bibliographical references and index.
Identifiers: LCCN 2018011906 | ISBN 9780262039086 (hardcover : alk. paper)
Subjects: LCSH: Smart cities. | Municipal engineering—Data processing. | Municipal services—Data processing. | Cities and towns—Technological innovations. | Big data.
Classification: LCC TD159.4 .Z43 2018 | DDC 628.0285—dc23 LC record available at https://lccn.loc.gov/2018011906

10 9 8 7 6 5 4 3 2 1

Contents

Preface

Rapid urbanization has led to the expansion of numerous large cities, not only modernizing many people's lives but also posing big challenges, such as air pollution, energy consumption, and traffic congestion. Tackling these challenges seemed nearly impossible only a few years ago given the complex and dynamic settings of cities. Nowadays, sensing technologies and large-scale computing infrastructure have produced a variety of big data, such as human-mobility, meteorology, traffic pattern, and geographical data. The corresponding big data implies a rich knowledge about a city and can help tackle these challenges when used correctly. In addition, the rise of computing technology, such as cloud computing and artificial intelligence (AI), has provided us with unprecedented data-processing capabilities.

Under such a circumstance, urban computing has proliferated as an interdisciplinary field, where computer science meets conventional related urban fields, such as urban planning, transportation, environmental science, energy engineering, economics, and sociology. Urban computing aims to unlock the power of knowledge from urban data to solve major issues in cities, resulting in a win-win-win situation among people, city operation systems, and the environment. In short, urban computing is about to tackle cities' challenges by using big data and advanced computing technology, such as cloud computing and AI.

People have been spending many years discussing a vision of smart cities, which integrate multiple sources of information and communication technology to improve quality of life. However, it is not yet clear how such a broad vision can be substantialized and then achieved. Instead of getting lost in the endless discussion of the vision of smart cities, urban computing tackles specific urban challenges with concrete methodologies in a data-centric computing framework comprising urban sensing, urban data management, urban data analytics, and services provided.

Although there are a few books on urban informatics, this is the first book dedicated to urban computing, with a broad spectrum of coverage and an authoritative overview.

This book introduces a general framework, key research problems, methodologies, and applications of urban computing from a computer science perspective. More specifically, this book focuses on data and computing, distinguishing urban computing from tradition urban science based on classical models and empirical assumptions.

Aimed at advanced undergraduates, graduate students, researchers, and professionals, this book covers the major fundamentals and key advanced topics that shape the field. Each chapter is a tutorial that provides readers with an introduction to one important aspect of urban computing, with many valuable references to relevant research studies. This book provides researchers and application developers with a comprehensive overview of the general concepts, techniques, and applications of urban sensing, urban data management, urban data analytics, and services provided, helping them to explore this exciting field and develop new methods and applications that can eventually lead to greener and smarter cities. It also offers graduate students and other interested readers a general introduction to the most recent developments in this promising field of research.

The book is organized according to the framework of urban computing, as shown in figure 1, consisting of four parts: concepts and framework, urban sensing and data acquisition, urban data management, and urban data analytics.

The first part of this book provides an overview of urban computing in chapters 1 and 2.

Chapter 1 introduces the key concepts and framework of urban computing, discussing the main challenges in each layer of the framework from a computer science perspective. Sources of data in cities are introduced, classifying urban big data into six categories in terms of data structures and spatiotemporal properties. Some public urban datasets are listed at chapter's end.

Chapter 2 presents typical applications of urban computing in different domains, consisting of transportation, urban planning, environmental protection, energy, economics, public security, and social and entertainment. These applications broaden our horizons, suggesting new research topics and inspiring new ideas.

The second part of this book introduces where data comes from and how it is collected in chapter 3.

Chapter 3 introduces four paradigms of urban sensing, consisting of static sensing, mobile sensing, passive crowd sensing, and active crowd sensing. The first two paradigms belong in the sensor-centric sensing category, and the rest pertain to human-centric sensing. For the sensor-centric sensing paradigm, four sensor deployment models have been introduced. For a human-centric sensing paradigm, the technology for participant recruitment and task design has been presented. Finally, this chapter introduces

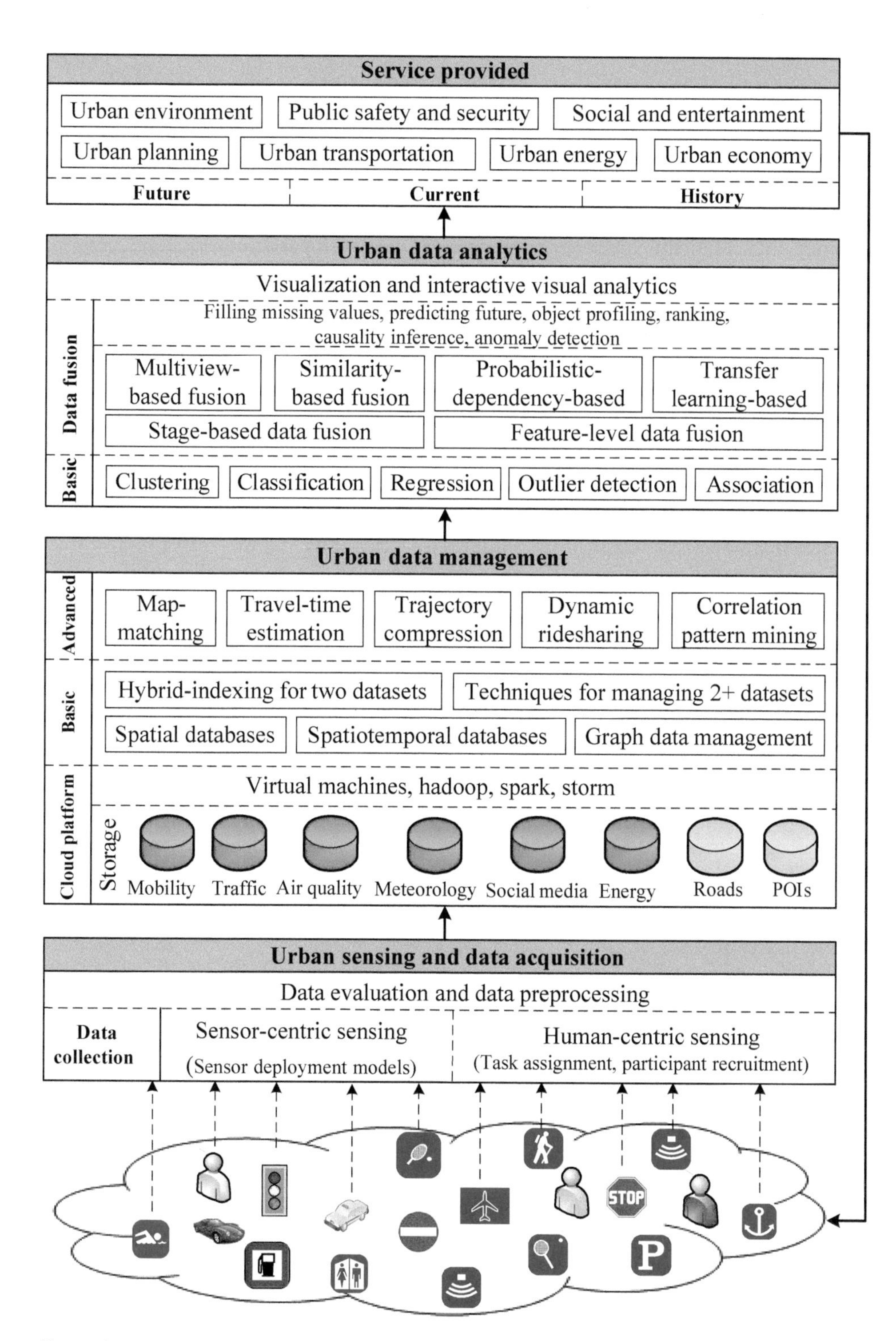

Figure 1
Urban computing framework.

three categories of models, consisting of spatial models, temporal models, and spatio-temporal models, for filling in missing values in geosensory data.

The third part of this book, composed of chapters 4, 5, and 6, introduces data management for spatial and spatiotemporal data. It starts with basic indexing and retrieval algorithms and then discusses the technology that uses cloud-computing platforms to manage spatial and spatiotemporal data.

Chapter 4 first introduces four widely used indexing structures, consisting of the grid-based index, quadtree-based index, K-d trees, and R-trees, for spatial data from four perspectives: building an index, serving spatiotemporal range queries, serving nearest neighbor queries, and updating the index. This chapter then presents techniques for managing spatiotemporal data, consisting of moving-object databases and trajectory data management. The former is concerned more with the specific location of a moving object at a (usually recent) time stamp. The latter concerns the continuous movement (e.g., the path) a moving object has traversed at a given time interval. Regarding moving object databases, three types of queries and two indexing approaches have been described. One indexing approach builds a spatial index at each time stamp and reuses the unchanged substructures of an index across consecutive time intervals. The other indexing approach treats time as the third dimension, extending spatial indexing structures from managing two-dimensional spatial data to three-dimensional spatiotemporal data. Regarding trajectory data management, three types of queries, consisting of range queries, *k*-nearest neighbor queries, and path queries, as well as different distance metrics designed for trajectory data, are presented. Finally, we introduce hybrid indexing structures for managing multiple datasets.

Chapter 5 introduces the main components in cloud-computing platforms from the perspective of storage, computing, and application interfaces. The framework of each component and the general procedure for using them are described using Microsoft Azure as an exemplary platform. Storage in Microsoft Azure is further composed of SQL Server, Azure Storage, and Redis. Microsoft Azure computing resources consist of Virtual Machine, Cloud Services, and HDInsight. HDInsight is a distributed-computing component in Microsoft Azure to perform large-scale data preprocessing, management, and mining and containing the widely used Hadoop, Spark, and Storm. The application interfaces provided by Azure are composed of Web Apps, Mobile Apps, and API Apps. These components have ensured the smooth, reliable implementation of urban-computing applications.

Chapter 6 introduces data management schemes respectively designed for six types of spatiotemporal data that enable current cloud-computing platforms to manage massive and dynamic spatiotemporal data with minimal effort. For each type of data, four data management schemes are presented, based on whether a spatial or spatiotemporal index is employed and whether it is deployed on a distributed system. Instead of fundamentally rebuilding a new platform, existing resources and architecture, such as cloud storage and HDInsight, on current clouds are leveraged to create an enhanced data management platform for spatial and spatiotemporal data. The most advanced data management scheme (among the four schemes) integrates spatial and spatiotemporal indexes—for example, grid-based indexes, R-trees, and 3D R-trees—into distributed-computing systems, such as Spark and Storm in HDInsight. This advanced scheme combines the strength from both sides, empowering us to process a larger scale of spatiotemporal data more efficiently while using fewer computing resources.

The fourth part of the book, consisting of chapters 7, 8, 9, and 10, introduces the basic technology of and advanced topics on mining knowledge from urban big data. It starts with fundamental data-mining algorithms and then introduces advanced machine-learning techniques designed for spatiotemporal data as well as cross-domain knowledge fusion methods. Some advanced topics, such as selecting relevant datasets, trajectory data mining, combining database techniques with machine-learning models, and interactive visual data analytics, are also discussed for urban computing.

Chapter 7 introduces the general framework of data mining, which comprises two main parts: data preprocessing and data analytics. The data preprocessing part further consists of data cleaning, data transformation, and data integration. Data analytics is further composed of a variety of data-mining models, results presentation, and evaluation. Based on the task that a model is designed to complete, data-mining models can be divided into five main categories: frequent pattern mining, clustering, classification, regression analysis, and outlier detection. For each category of model, this chapter presents its general idea and concrete examples within the context of mining knowledge from spatial and spatiotemporal data.

Chapter 8 first discusses the unique properties of spatiotemporal data, as compared to images and text data. Spatial properties consist of spatial distance and spatial hierarchy. Temporal properties are composed of temporal closeness, period, and trend. These unique properties call for advanced machine-learning algorithms that are specifically designed for spatiotemporal data. This chapter then introduces the principle of six categories of machine-learning algorithms, consisting of collaborative filtering, matrix factorization, tensor decomposition, probabilistic graphical models, deep learning, and reinforcement learning, within the context of spatiotemporal data. Rich examples are presented to show how these machine-learning algorithms should be adapted to handle spatiotemporal data. For example, coupled matrix factorization is designed to enable location recommendations and traffic condition estimations. Bayesian networks are respectively employed to infer traffic volumes, do map matching, and discover the latent function of a region. Specific Markov random fields are designed to predict the transportation modes of a user and the air quality of a location. A unique deep-learning model that is dedicatedly designed to predict the flow of crowds in each and every region throughout a city is also presented.

Traditional data mining usually deals with data from a single domain. In the big-data era, we face myriad datasets from different sources in different domains. These datasets consist of multiple modalities, each of which has a different representation, distribution, scale, and density. The goals of fusing multiple datasets include filling in missing values, predicting the future, inferring causality, profiling objects, ranking, and detecting anomalies. The ability to unlock the power of knowledge from multiple disparate (but potentially connected) datasets is paramount in big-data research, essentially distinguishing big data from traditional data-mining tasks. This calls for advanced techniques that can fuse the knowledge from various datasets organically in a machine-learning and data-mining task. Chapter 9 introduces three categories of knowledge fusion methods, consisting of stage-based, feature level–based, and semantic meaning–based methods. The last category of fusion methods is further divided into four groups: multi-view learning–based, similarity-based, probabilistic dependency–based, and transfer learning–based methods. These methods focus on knowledge fusion rather than schema mapping and data merging, significantly distinguishing between cross-domain data fusion and the traditional data fusion studied in the database community. This chapter introduces not only high-level principles of each category of method but also valuable examples in which these techniques are used to handle real big-data problems. In addition, this chapter positions existing works within a framework, exploring the relationships and differences between different knowledge fusion methods.

Chapter 10 discusses a few advanced urban data analytics topics based on the fundamental techniques introduced in previous chapters. First, given an urban-computing problem, we usually need to answer the question of which datasets should be selected to solve a given problem. By choosing the right datasets, we are more likely to solve a problem efficiently and effectively. Second, trajectory data has a complex data model and contains rich knowledge about moving objects, calling for unique data-mining techniques. Third, extracting (profound) knowledge from large-scale datasets requires both efficient data management techniques and effective machine-learning models. An organic integration of the two techniques is imperative to completing an urban-computing task. Finally, solving an urban-computing problem requires both data science and domain knowledge. How to incorporate human intelligence with machine intelligence is an advanced topic worthy of discussion. Interactive visual data analytics may be an approach to addressing this issue.

I hope you find this book to be a useful overview of and a practical tutorial on the young and evolving field of urban computing.

Yu Zheng

Acknowledgments

Many thanks to my colleagues, Dr. Jie Bao and Dr. Junbo Zhang, for collaborating with me over the past few years, including on many of the research projects introduced in this book. We have also coauthored quite a few other publications referenced in this work.

Thank you to my students, Ruiyuan Li, Shenggong Ji, Yexin Li, Zheyi Pan, and Yuxuan Liang, for assisting me in writing some parts of this book. We generated concrete content for a few sections after having several rounds of constructive discussions. The same appreciation is also extended to Xiuwen Yi, Huichu Zhang, Sijie Ruan, Junkai Sun, and Tianfu He for contributing to the research projects I introduce here.

There are many more—for example, Ye Liu, Chuishi Meng, Xianyuan Zhan, Yixuan Zhu, Yuhong Li, Chao Zhang, Xuxu Chen, Yubiao Chen, Jingbo Shang, Wenzhu Tong, Yilun Wang, Tong Liu, Yexiang Xue, Hsun-Ping Hsie, Ka Wai Yung, Furui Liu, David Wilkie, Bei Pan, Yanjie Fu, Shuo Ma, Kai Zheng, Lu-An Tang, Ling-Ying Wei, Wei Liu, Wenlei Xie, Hyoseok Yoon, Vincent Wenchen Zheng, Jing Yuan, Zaiben Chen, Xiangye Xiao, Chengyang Zhang, Yin Lou, Ye Yang, Lizhu Zhang, Yukun Chen, Quannan Li, Like Liu, and Longhao Wang, who worked with me as interns at Microsoft Research. Though I cannot list all their names here, I am very grateful for having them on my team. Most of them coauthored publications with me. Some of these publications are referenced or introduced in this book.

Thanks to Prof. Liang Hong for contributing useful content about graph pattern mining in section 7.3.4 and to Prof. Licia Capra for contributing a survey on mining subway data in section 2.3.4. This part has been revised from its original version, which was a subsection in a survey article about urban computing published in *Transactions on Intelligent Systems and Technology*.

I would also like to express my gratitude to Prof. Xiaofang Zhou, who coedited the book *Computing with Spatial Trajectories*, and to that book's chapter contributors, including Dr. John Krumm, Prof. Wang-Chien Lee, Dr. Ke Deng, Prof. Goce Trajcevski,

Prof. Chi-Yin Chow, Prof. Mohamed Mokbel, Dr. Hoyoung Jeung, Prof. Christian Jensen, Dr. Yin Zhu, and Prof. Qiang Yang. The main content in section 10.2 (of this book) comes from my survey paper on trajectory data mining, a part of which is derived from *Computing with Spatial Trajectories*.

Thanks go to Prof. Jaiwei Han, Prof. Micheline Kaber, and Prof. Jian Pei for their book *Data Mining: Concepts and Techniques*, which was an important reference book when I was writing sections 7.2, 7.3, and 7.4. I am very grateful for Prof. Jiawei Han's long-term support of my research and career. I was also inspired by Dr. Charu C. Aggarwal's book *Data Mining: The Text Book* when organizing the content of chapter 7.

I Concepts and Framework

1 Overview

Abstract: This chapter defines urban computing, presenting a four-layer framework and discussing the key challenges of each layer from a computer science perspective. Afterward, sources of data in cities are introduced, classifying urban big data into six categories in terms of data structures and spatiotemporal properties. Some public urban datasets are presented at chapter's end.

1.1 Introduction

Urbanization's rapid progress has led to the development and expansion of many big cities, not only modernizing many people's lives but also exacerbating big challenges, such as air pollution, increased energy consumption, and traffic congestion. Tackling these challenges seemed nearly impossible just a few years ago, given the complex and dynamic settings of cities. Nowadays, sensing technologies and large-scale computing infrastructures have produced a variety of big data in urban spaces, such as human mobility, air quality, meteorology, traffic patterns, and geographical data. Big data implies rich knowledge about a city and can help tackle these challenges when used correctly. For instance, we can detect the underlying problems in a city's road network by analyzing citywide human mobility data. This can help cities to formulate and implement better urban planning initiatives in the future [74]. Another example is the exploration of the root causes of urban air pollution by studying the correlation between air quality and other data sources, such as traffic flow and points of interest (POIs) [51, 54].

Urban computing has been proliferating as an interdisciplinary field in which computer science meets conventional city-related areas, such as urban planning, transportation, environmental science, energy engineering, economics, and sociology. As illustrated in figure 1.1, urban computing aims to unlock the power of knowledge from urban data to solve major issues in cities, resulting in a win-win-win situation among people, city operation systems, and the environment [66].

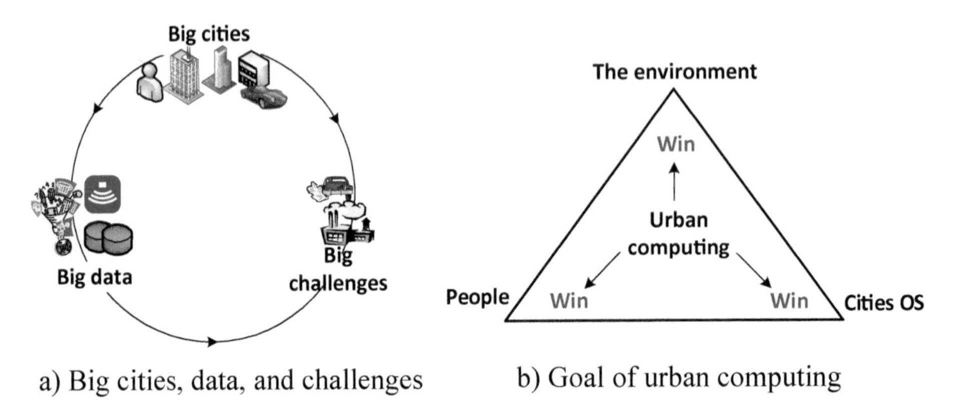

Figure 1.1
The motivation and goal of urban computing.

Although there are a few articles discussing urban computing [22], it is still a vague concept with many open questions. For example, what are the core research problems for urban computing? What are the challenges posed in this research field? What are the key methodologies for urban computing? What are the representative applications, and how does an urban-computing system work? To address these issues, this chapter formally defines the main concepts of urban computing, introducing the general framework and key research challenges of urban computing from a computer science perspective.

1.2 Definition of Urban Computing

Urban computing is a process of acquisition, integration, and analysis of big and heterogeneous data generated by a variety of sources in urban spaces, such as sensors, devices, vehicles, buildings, and humans, to tackle the major issues that cities face (e.g., air pollution, increased energy consumption, and traffic congestion). Urban computing connects unobtrusive and ubiquitous sensing technologies, advanced data management and analytics models, and novel visualization methods to create win-win-win solutions that improve urban environments, quality of life, and city operation systems. Urban computing also helps us understand the nature of urban phenomena and even predict the future of cities. Urban computing is an interdisciplinary field fusing computer science and information technology with traditional city-related fields, such as urban planning, transportation, civil engineering, economy, ecology, and sociology, in the context of urban spaces.

1.3 General Framework

1.3.1 Brief and Example

Figure 1.2 depicts a general urban-computing framework comprised of four layers: urban sensing, urban data management, urban data analytics, and services provided. Using urban anomaly detection as an example [36], we first briefly outline the framework and then introduce the function of each layer, respectively, in section 1.3.2.

In urban sensing, we constantly probe people's mobility (e.g., routing behavior in a city's road network) using GPS sensors or their mobile phone signals. We also continuously collect people's social media posts (in this city) from the Internet. In data management, human mobility data and social media are organized by some indexing structures that simultaneously incorporate spatiotemporal information and texts for supporting efficient data analytics. In data analytics, once an anomaly occurs, we are able to identify the locations where people's mobility differs significantly from its origin patterns, based on human mobility data. We can then describe the anomaly by mining representative terms from social media relating to these locations and the time span in which the anomaly has been detected. In services provided, the locations and descriptions of the anomaly will be sent to nearby drivers, so they can choose a bypass. The information is delivered to transportation authorities for dispersing traffic congestion and diagnosing the anomaly. The system continuously repeats the aforementioned four steps in a loop for an instantaneous and unobtrusive detection of urban anomalies, improving people's driving experiences and reducing traffic congestion.

Compared with other information systems (e.g., language translation or image recognition engines), which are usually based on a single (modal)-data, single-task framework, urban computing holds a multi (modal)-data, multitask framework. The tasks of urban computing include improving urban planning, easing traffic congestion, saving energy, reducing air pollution, and so on. Additionally, we usually need to harness a wide array of datasets in a single task. For instance, the aforementioned anomaly detection uses human mobility data, road networks, and social media. Different tasks can be fulfilled by combining data from different sources with different data acquisition, management, and analytics techniques from different layers of the framework.

1.3.2 Functions of Each Layer

This section discusses functions and key components of each layer in the urban-computing framework.

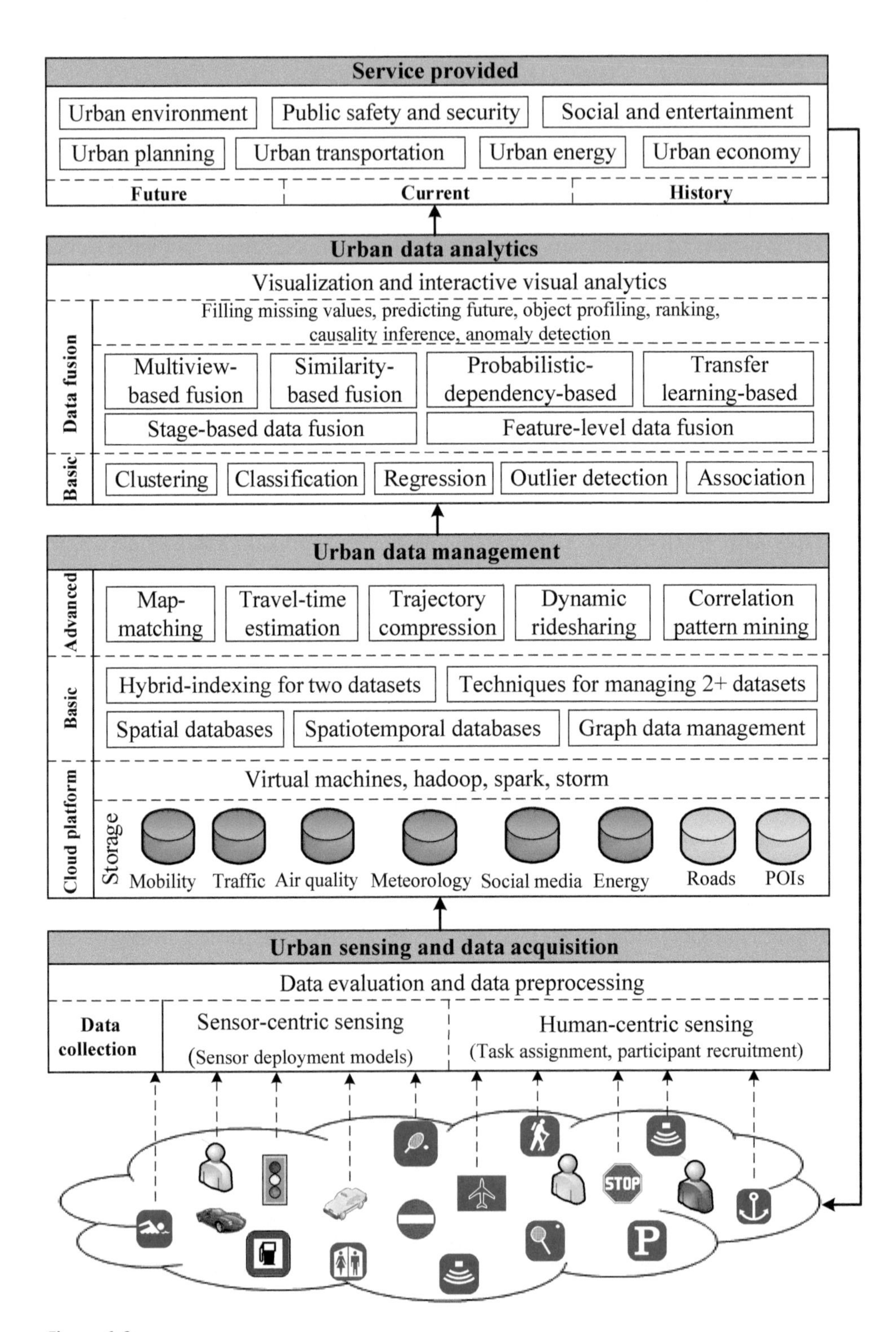

Figure 1.2
The general framework of urban computing.

1.3.2.1 Urban sensing Urban sensing collects data from different sources through sensors or people in a city. There are two main urban-sensing modes consisting of sensor-centric sensing and human-centric sensing. The former mode deploys a collection of sensors in fixed locations (e.g., at meteorological stations), as shown in figure 1.3a, also called *static sensing*, or with moving objects, such as buses or taxis, as illustrated in figure 1.3b, also called *mobile sensing*. Those sensors continuously send readings to a back end system without people involved in the loop once they have been deployed.

The human-centric sensing mode leverages humans as sensors to probe urban dynamics when they move around in cities. The information collected by individuals is then used to solve a problem collectively. This sensing mode can be further divided into two categories: passive and active crowd sensing.

A *passive crowd-sensing* program collects individual users' data passively when they use existing urban infrastructures, such as wireless communication systems and public transportation systems, as illustrated in figure 1.3c. People do not even know they are contributing data in a passive crowd-sensing program. For instance, while wireless cellular networks are built for mobile communication between individuals, large groups of people's mobile phone signals can help in understanding citywide commuting patterns, thereby improving urban planning [5]. Likewise, passengers' swipe-in and swipe-out data at subway stations can also describe people's commuting patterns in a city, though such a ticketing system was originally created for charging people's travel expenses. Commuters have no idea they are performing a sensing task when traveling through an automatic fare gate.

Active crowd sensing can be regarded as a combination of crowd sourcing [18] and participatory sensing [4]. In such a sensing mode, as illustrated in figure 1.3d, people

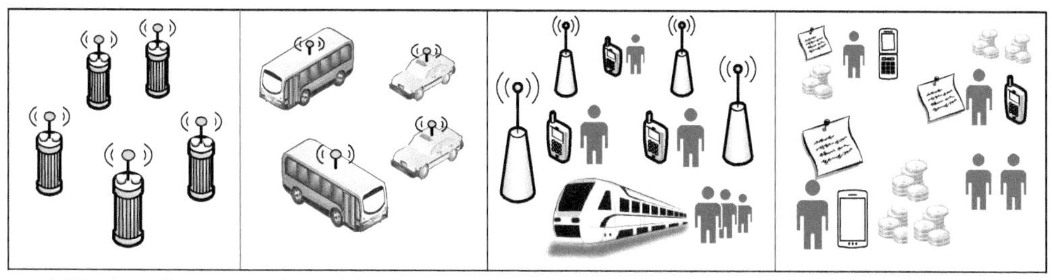

a) Fixed sensing b) Mobile sensing c) Passive crowd sensing d) Active crowd sensing
Sensor-centric sensing Human-centric sensing: urban crowd sensing

Figure 1.3
Different modes of urban sensing.

actively obtain information from around them and contribute their own data to formulate collective knowledge that can solve a problem. People explicitly know the purpose of sharing and what they have contributed in a participatory-sensing program. They can also control where and when to participate in such a sensing program based on their availability and the incentives offered. When there are many participants and a limited budget, an active crowd-sensing program also involves participant recruitment and task assignment processes [6, 14].

1.3.2.2 The urban data management layer The urban data management layer manages large-scale and dynamic urban data from different domains, such as traffic, meteorology, human mobility, and POIs, using cloud-computing platforms, indexing structures, and retrieval algorithms.

First, this layer devises different storage mechanisms on the cloud for different types of urban data. According to data structures, urban data is classified into two categories: point data and network data. Alternatively, urban data can be divided into three categories by its spatiotemporal dynamics, consisting of spatiotemporal static data, spatial static-temporal dynamic data, and spatiotemporal dynamic data.

Second, this layer designs unique indexing structures and retrieval algorithms for spatial and spatiotemporal data, as most urban data is associated with spatial and temporal properties. Furthermore, to support upper-level cross-domain data-mining tasks, hybrid indexing structures are needed for organizing multimodality data from different domains. Those indexing and retrieval techniques are foundations of upper-level data-mining and machine-learning tasks.

Third, this layer also enables some advanced data management functions, including map matching [31, 56], trajectory compression [7, 42], finding the maximum k-coverage [27], and dynamic dispatching [32, 33, 62], which can solve many urban computing problems by themselves.

1.3.2.3 The urban data analytics layer This layer applies a diversity of data-mining models and machine-learning algorithms to unlock the power of knowledge from data across different domains. This layer adapts basic data-mining and machine-learning models, such as clustering, classification, regression, and anomaly detection algorithms, to handle spatiotemporal data. This layer also fuses knowledge from multiple disparate datasets based on cross-domain data fusion methods [64], such as deep learning based [61], multi-view based, probabilistic dependency based, and transfer learning based [48]. As many urban-computing applications need instantaneous services, it is also important to combine database techniques with machine-learning algorithms in data mining. Based on the aforementioned components, advanced topics on this layer include filling missing values in spatiotemporal data, predictive models, object profiling, and causality inference. It is also imperative to enable interactive visual data analytics [28], which combine human wisdom with machine intelligence by involving domain experts in a data-mining loop.

1.3.2.4 The services layer The services layer offers an interface that allows domain systems to draw on knowledge from an urban-computing application through cloud-computing platforms. As urban computing is an interdisciplinary field, the knowledge from data must be integrated into existing domain systems to inform their decision-making. For example, through a set of application programming interfaces (APIs), as shown in figure 1.4, air quality forecasts from an urban-computing application can be integrated into existing mobile apps to help people's travel planning or used by environmental protection agencies' systems to inform their decision-making on pollution control.

In terms of the timing that a service is created for, this layer provides three categories of services consisting of understanding the current situation, predicting the future, and diagnosing history. For instance, inferring real-time and fine-grained air quality throughout a city based on big data belongs in the first category [71], while forecasting future air quality is an example of the second category [77]; diagnosing the root cause of air pollution based on data accumulated over a long period belongs in the last category [80]. Based on the domain that a service is created for, the services provided

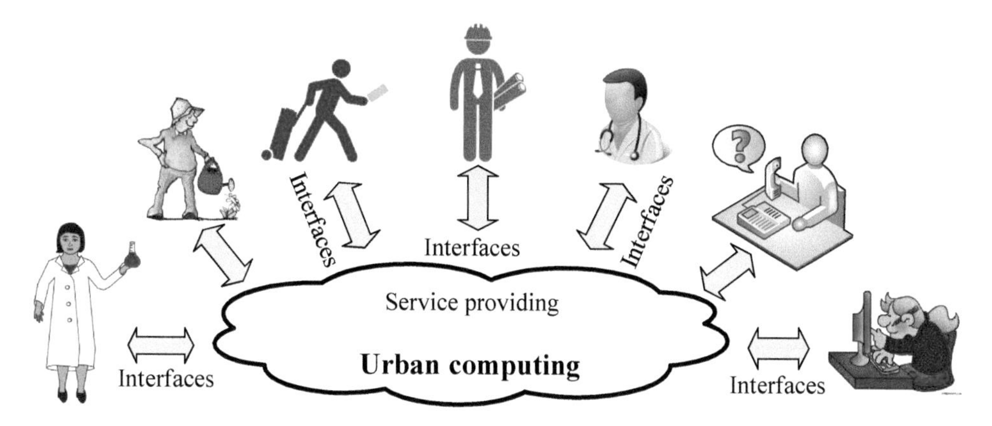

Figure 1.4
Urban-computing services.

by this layer range from transportation to environmental protection, urban planning, energy saving, social functions and entertainment, and public security.

1.4 Key Urban-Computing Challenges

This section discusses the challenges posed by each layer of urban computing from a computing perspective.

1.4.1 Urban-Sensing Challenges

Many urban computing applications need data acquisition techniques that can *unobtrusively* and *continually* collect *city-scale* data. This is a nontrivial problem given the three italicized terms. While monitoring the traffic flow on an individual road segment is easy, continuously probing citywide traffic is challenging, as we do not have sensors deployed on every road segment. Building new sensing infrastructures could accomplish such a task but would in turn increase the burden on cities. How to leverage what we already have in urban spaces creatively to enable such data acquisition techniques has yet to be explored.

More specifically, the challenges of urban sensing are fourfold: (1) skewed sample data, (2) missing data and sparsity, (3) implicit and noisy data, and (4) resource deployment. Table 1.1 presents the challenges that different urban-sensing modes face. We will detail each challenge in later parts of this subsection.

1.4.1.1 Skewed sample data The dataset we can sense in an urban-computing scenario is usually a sample, which may not accurately represent the entire dataset, as depicted in figure 1.5a. The distribution of some properties in a sample may be skewed from the entire dataset. To derive true knowledge about the entire dataset from a sample of data remains a challenge.

Table 1.1
Challenges for different urban-sensing modes.

Urban-sensing modes		Skewed sample data	Missing data and sparsity	Implicit and noisy	Resource allocation
Sensor-centric	Fixed sensing		√		√
	Mobile sensing	√	√		√
Human-centric	Passive crowd sensing	√	√	√	
	Active crowd sensing	√	√		√

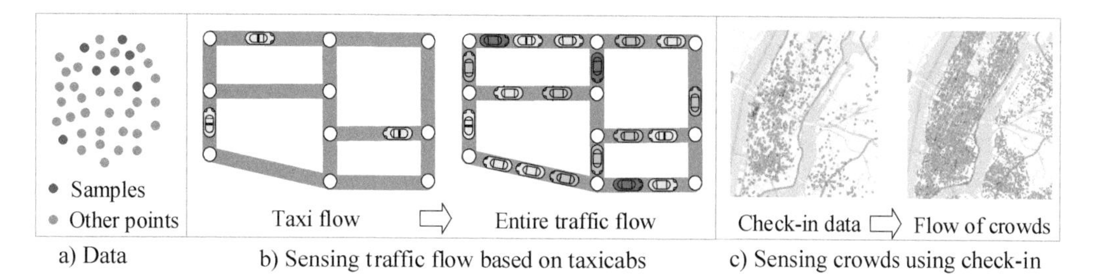

Figure 1.5
Skewed distribution of sampled data.

For instance, it is almost impossible to track each and every vehicle traveling in a city, though we can collect the GPS trajectories of taxicabs. Those taxicabs are only a sample of vehicles, with a geographical distribution quite different from others. As illustrated in figure 1.5b, we may find many private cars but few (or zero) taxicabs on some road segments, whereas many taxicabs but few private cars are on other segments. Thus, to estimate a road's traffic volume based on taxicabs' trajectories, we cannot simply multiply the number of taxicabs on the road by a specific factor. To derive the true traffic flow of a road from sampled data (i.e., taxicabs' trajectories) remains a challenge.

Likewise, as illustrated in figure 1.5c, users' check-in data on an online social networking service denotes the human mobility of a portion of people in a city. As there are many people who do not check in at a place they have visited, the geographical distribution of the check-in data may be quite different from the true crowd distribution in a city. Thus, to estimate the flow of crowds based on the check-in data is challenging.

To tackle the aforementioned challenge, we need to know what kind of knowledge derived from a sample of data can represent the entire dataset—and what cannot. For instance, a road segment's travel speed can be derived from the GPS trajectories of the taxicabs traversing it. As vehicles traveling on the same road segment usually have a similar speed, the speed information derived from taxicabs, which is a sample of vehicles, can represent the speed information for all vehicles. However, the volume of taxicabs is not directly transferrable to the entire volume of vehicles. Under such a circumstance, we need to incorporate knowledge from other datasets, such as POIs, the structure of road networks, and weather conditions, to infer the total traffic volume on a road [41].

1.4.1.2 Data sparsity and missing data

Data sparsity Many sensing systems have only a limited number of sensors deployed in a few locations of a city. It is challenging to collect detailed information throughout a city based on those sparse sensors in a geographical space.

For instance, as shown in the left part of figure 1.6a, there are only thirty-five air quality–monitoring stations deployed in the urban area of Beijing whose size is about 40 km×50 km. While those sensors are sparsely distributed in the city, we hope to sense the fine-grained air quality throughout a city [71], as demonstrated in the right part of figure 1.6a.

Figure 1.6b presents another example that aims to sense the urban noise of New York City based on people's complaints about noise [73]. As it is almost impossible to have people reporting ambient noise conditions everywhere and at every time interval, the collected data is very sparse in a spatiotemporal space. To diagnose the noise conditions throughout New York City based on sparse data remains a challenge.

Missing data *Missing data* is a different concept from data sparsity, denoting the absence of data that should be obtained. For example, as illustrated in figure 1.7, an air quality–monitoring station s_1 is supposed to generate a reading about air pollution every hour. When experiencing communication or device errors, however, we lose some sensor readings—for example, the reading of s_1 at t_2 and that of s_3 at t_{i+1}, called *missing data* [50]. While supplementing such missing values is important for supporting monitoring and further data analysis, the task is challenging for two reasons.

First, readings can be absent at arbitrary sensors and time stamps. In some extreme cases, we may lose readings from a sensor at consecutive time stamps (e.g., readings of s_2 from t_1 to t_i) or lose readings of all sensors in one (or more) time stamp(s) simultaneously (e.g., at t_2). We call these extreme cases *missing blocks*. It is very difficult for existing models to handle the missing block problem, as we may not be able to find stable inputs for a model.

Second, affected by multiple complex factors, sensor readings change over location and time significantly and nonlinearly. The readings of sensors with a shorter distance may not always be more similar than those with a farther distance. In addition, sensor readings fluctuate tremendously over time, sometimes coming with a sudden change.

1.4.1.3 Implicit and noisy data The data generated by traditional sensors is well structured, explicit, clean, and easy to understand. However, the data contributed by users in a passive crowd-sensing program, for example, is usually in a free format, such

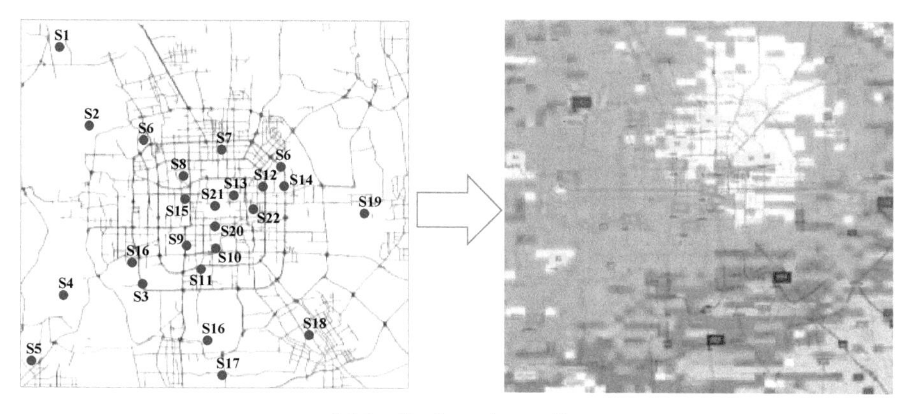

a) Monitoring air quality

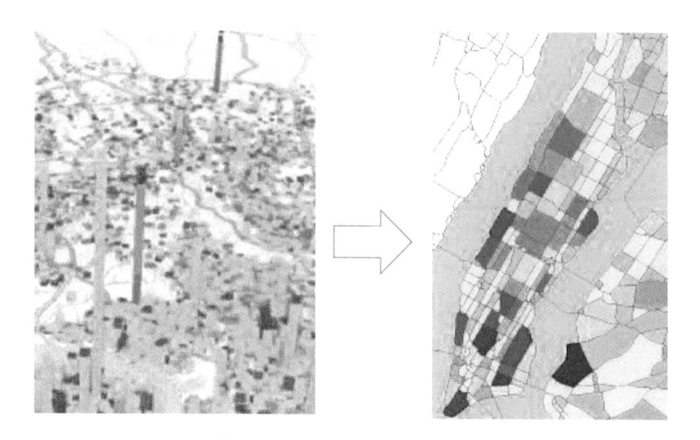

b) Diagnosing urban noises

Figure 1.6
Data sparsity and missing in urban sensing.

as texts and images, or cannot explicitly lead us to the final goal as can traditional sensors. Sometimes, the information is also quite noisy because people do not collect data with an intention.

For example, Zhang et al. [59, 60] aim to use GPS-equipped taxi drivers as sensors to detect the queuing time in a gas station (when they are refueling taxis) and further infer the number of people who are also refueling their vehicles there. The goal is to estimate the gas consumption of a station and finally the citywide gas consumption in a given time span. In this application, what we can collect are taxi drivers' GPS trajectories, which do not tell us the result of gas consumption explicitly. In the meantime, taxi

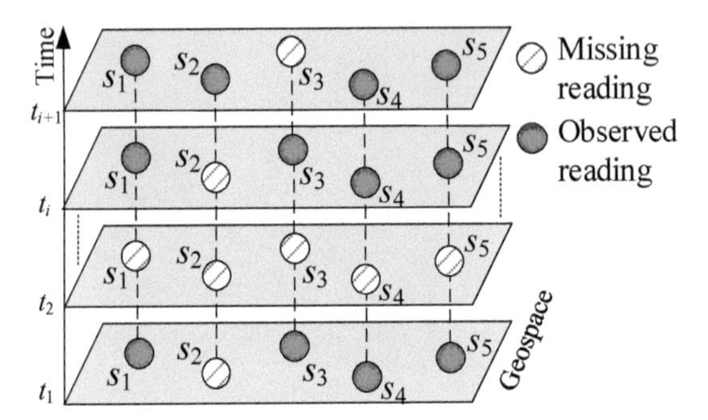

Figure 1.7
Missing data in spatiotemporal datasets.

drivers may park taxis somewhere close to a gas station just to rest or wait for a traffic light. These observations from GPS trajectory data are noise, as they are not really a refueling behavior.

1.4.1.4 Resource allocation While expecting to maximize the coverage and quality of data collection in an urban-sensing program, we usually face resource constraints, such as money, sensors, and workforce. This is challenging for two specific reasons: measurement of data quality and candidate selection.

Measurement of data quality We need a clear measurement for evaluating the data collected by a sensing program. The measurement changes over different applications, including amount, coverage, balance, redundancy, and stability. In some applications, the measurement is easy to quantify—for example, covering the maximum number of unique trajectories. In other applications, however, defining the measurement is a nontrivial task. For instance, as shown in figure 1.8a, we hope to build four new monitoring stations in order to best monitor the air quality in the entire city. To define "to best monitor" is challenging, as we do not have the ground truth on air quality throughout an entire city [71].

Likewise, as illustrated in figure 1.8b, to define the coverage of data collected by an active crowd-sensing program is nontrivial in a spatiotemporal space, where spatial and temporal dimensions can have different granularities [20]. Partitioned by different geographical sizes and different lengths of temporal intervals, collected data will present different distributions. The evaluation of data coverage will significantly affect the task design and participant recruitment in an active crowd-sensing program.

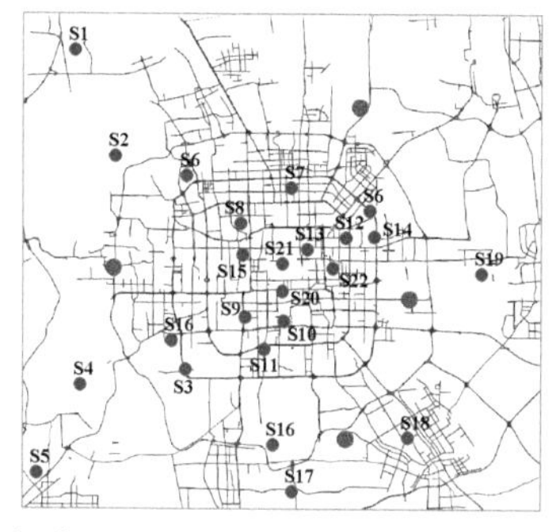

a) Deploying new sensors in fixed sensing mode

b) Task assignment for active crowd sensing mode

Figure 1.8
Illustrations of challenges in resource deployment.

Candidate selection Based on the aforementioned measurement, we need to select some candidates, such as locations, vehicles, and people, from a large pool for an urban-sensing program. This is a very complex and sometimes NP-hard problem. For instance, finding a *k*-location set (from a road network) with the maximum number of unique trajectories traversing them [27] is a typical fixed sensor deployment problem, which can be converted into a submodular approximation problem with an NP computing complexity. Another example in mobile sensing is to select a few buses on which to put commercial advertisements to maximize the number of people seeing the ads.

In an active crowd-sensing program, as illustrated in figure 1.8b, given a limited budget, a sensing scheme needs to select proper participants based on their mobility and provide them with unobstructed tasks that do not break their origin commuting plans. This is difficult because human mobility is highly skewed in a city. Without an effective participant recruitment and task assignment mechanism, some locations will lack participant-contributing data, whereas a few places (e.g., popular tourist attractions) may have overly sufficient and even redundant data, thus wasting resources. As a result, we cannot ensure the best coverage of data to better support upper-level applications.

1.4.2 Urban Data Management Challenges

There are many types of data, such as text, image, traffic, and meteorology, all of which are continuously generated by different sources in urban spaces. For better supporting upper-level data mining and services, *large-scale* and *dynamic* datasets need to be managed *across different domains*. Cloud-computing platforms, such as Microsoft Azure and Amazon Web Services, are regarded as ideal infrastructures for large-scale and dynamic data. As most urban data is associated with spatial information and temporal properties (called *spatiotemporal data*), urban computing needs cloud-computing technology for managing spatiotemporal data. However, current cloud-computing platforms are not specifically designed for spatiotemporal data. The following are the primary challenges in handling spatiotemporal data.

1.4.2.1 Unique data structures Spatiotemporal data has unique data structures that are different from texts and images. For example, the size of a photo is fixed once taken, while the length of a trajectory keeps increasing as a taxi travels around a city. The trajectory data of one moving object continues streaming into the cloud with a sequential order that cannot be changed. We do not know when the trajectory terminates and how large it will be in advance. In addition, to support different queries, the storage mechanism for spatiotemporal data is different.

1.4.2.2 Different queries We usually query documents by key words, that is, by matching a few key words against a collection of documents precisely or approximately. However, in urban computing, we usually need to handle spatiotemporal range queries or the *k*-nearest neighboring queries over spatiotemporal data. For example, finding nearby vacant taxis over a sixty-second time span is a spatiotemporal range query over taxi trajectory data [65]. Likewise, searching for the nearest gas station while driving is a continuous 1-nearest neighbor query over POIs. Existing components in the cloud cannot handle spatiotemporal queries directly and efficiently, as indexing and retrieval algorithms are not found in most cloud storage systems.

1.4.2.3 Hybrid indexing structures Traditional indexing and query algorithms are usually proposed to handle a single type of data. For instance, R-tree is proposed to index spatial point data, and inverted indexing is designed for handling text documents. In urban computing, however, we need to harness multiple datasets with different formats and updating frequencies across different domains. This calls for (hybrid) indexing structures that can organize multimodality data (e.g., managing trajectories, POIs, and air quality data together). Without such hybrid indexing structures, online feature extraction processes of upper-level machine-learning models will take a very long time, consequently failing to provide instantaneous services. Such research is very rare at the present time.

1.4.2.4 Integrating spatiotemporal indexing with the cloud

When processing texts and images, a cloud-computing platform typically relies on the distributed computing environment, such as Spark, Storm, and Hadoop, because the functions of indexing structures designed for texts and images are quite limited. For example, an inverted index is typically employed to maintain the relationship between the words and documents that contain them. However, for spatiotemporal data, an effective indexing structure can improve the efficiency of a retrieval algorithm by order of magnitude or several orders of magnitude. Combining the strength of spatiotemporal indexes and the distributed computing system in a cloud can empower the cloud to process a larger scale of data more efficiently while using fewer computing resources.

Integrating spatiotemporal indexes into the in-parallel computing environment in a cloud is a nontrivial task, particularly when there are massive trajectories streaming into a cloud. For example, to find the vehicles traversing a sequence of road segments over a few minutes requires an organic integration of trajectory indexing structures into a Storm-based computing framework. Managing the trade-off between memory and input/output (I/O) throughput, conditioned on the in-parallel computing framework and indexing structures, is challenging, calling for knowledge from distributed systems, cloud storage, and indexing algorithms. Some indexing structures can work very well in a stand-alone machine but may not be a good choice for a distributed environment, considering the potential problems of index partitioning and updating.

1.4.3 Urban Data Analytics Challenges

1.4.3.1 Adapting machine-learning algorithms to spatiotemporal data

Spatiotemporal data has unique properties compared to image and text data. Adapting existing machine-learning algorithms to deal with spatiotemporal properties poses the following challenges.

Spatial properties Spatial properties consist of geographical distance and hierarchy. For example, according to the first law of geography, which states that "everything is related to everything else, but near things are more related than distant things" [45], machine-learning algorithms should be able to differentiate between the similarities of objects with a short distance between them and those with a larger one. The distance refers not only to a geographical distance but also to the distance in a semantic space such as a feature space.

For example, as illustrated in figure 1.9a, there are four air-quality-monitoring stations (s_1, s_2, s_3, and s_4) deployed in a city. According to geographical Euclidean distance, s_4 is the closest station to s_2, with air quality more similar to s_2 than the rest of the

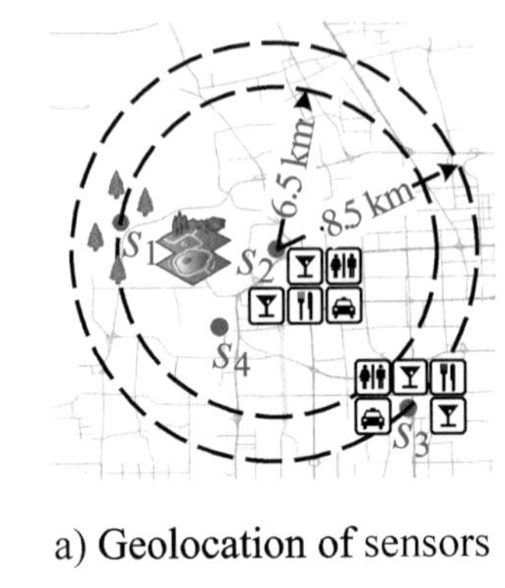

a) Geolocation of sensors

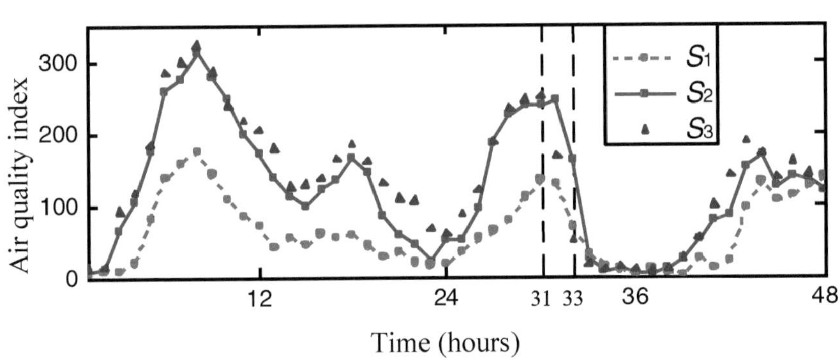

Time (hours)

b) Air quality index over time

Figure 1.9
The spatial distance of spatiotemporal data.

stations. However, this does not hold for s_1 and s_3. Although s_1 is closer to s_2 than s_3 in a geographical space, the air quality readings of s_2 are more similar to s_3 than s_1, as shown in figure 1.9b. The reason is that s_2 and s_3 are located in two regions with a similar geographical context, such as POIs and traffic patterns, while s_1 is deployed in a forest and with a lake between it and s_2. As a result, in a feature space, s_3 is actually closer to s_1 than s_2.

In addition, different granularities of locations result in a natural hierarchy; for example, a state is comprised of many counties, each of which further consists of many cities. The data representation at different levels of the hierarchy implies different levels of knowledge. It is a challenging task for a machine-learning algorithm to capture the information of spatiotemporal data with different granularities.

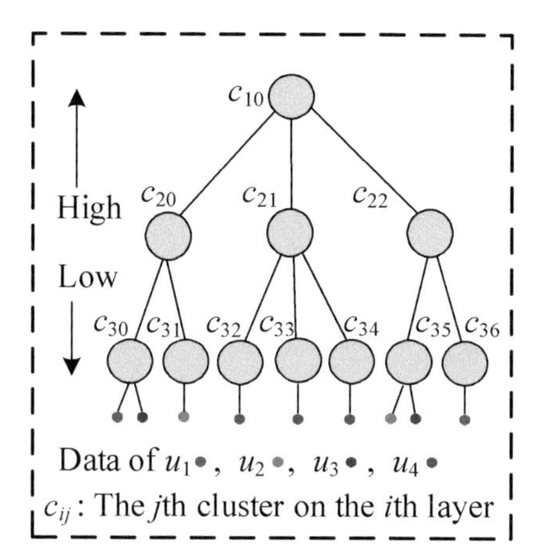

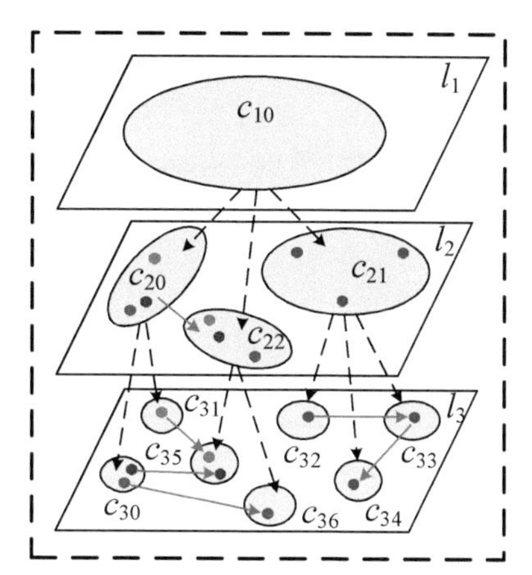

Figure 1.10
Hierarchical properties of spatial data.

For instance, as illustrated in figure 1.10, different individuals' location histories form a geographical hierarchy where nodes on an upper level denote location clusters with a coarser granularity. We want to estimate the similarity between four users (u_1, u_2, u_3, u_4) based on their location histories, with a general insight that similar users should share more location histories [25, 79]. If only checking their location histories on the second layer l_2, we cannot differentiate among u_1, u_2, and u_3, as all of them have traveled through $c_{20} \rightarrow c_{22}$. However, if checking the third layer, we find that both u_2 and u_3 have visited c_{35}. Consequently, u_3 is more similar to u_2 compared to u_1. Likewise, u_3 is more similar to u_1 compared to u_2, as u_1 and u_3 have both visited c_{30}. If we only explore these individuals' data on l_3, u_1, u_2, and u_4 become hard to compare, as they do not share any locations on this layer. By checking their location histories on l_2, we can distinguish u_2 from u_4 because u_1 and u_2 both travel through $c_{20} \rightarrow c_{22}$, whereas u_1 and u_2 do not.

Temporal properties Temporal properties are comprised of three parts: temporal closeness, periodic patterns, and trend patterns. The first one is very similar to spatial distances, denoting that datasets generated at two close time stamps are usually more similar than those from two distant time stamps [50]. However, this is not always true given the periodic patterns of spatiotemporal data. For example, as demonstrated in figure 1.11a, traffic speeds on a road at 8 a.m. are almost the same on consecutive workdays. This could be quite different from 11 a.m. on the same days, although 11 a.m. is closer

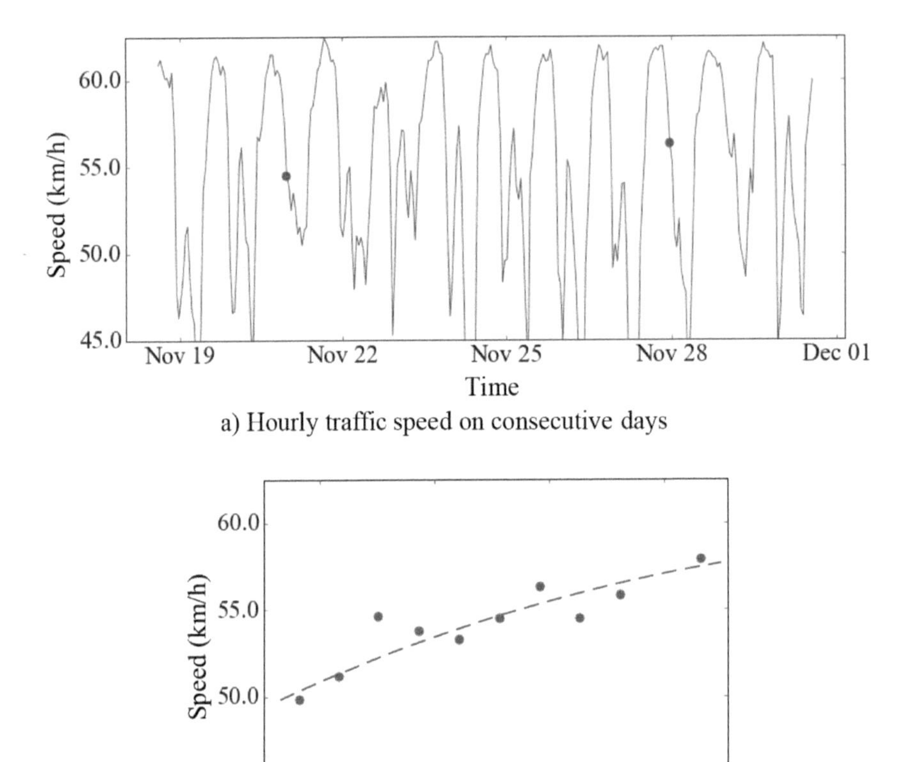

Figure 1.11
Temporal properties of spatiotemporal data: an example of hourly traffic speeds on the north segment of the fourth ring road of Beijing.

to 8 a.m. than 8 a.m. the day before. Furthermore, the period evolves as time goes by. As illustrated in figure 1.11b, the travel speed from 9 a.m. to 10 a.m. on weekends keeps increasing as winter approaches. When the temperature drops, people postpone the time they start their weekend activities. Thus, traffic conditions on this road become better and better at 9 a.m. In short, machine-learning algorithms should be able to model the temporal closeness, periodic patterns, and trend properties of spatiotemporal data.

1.4.3.2 Combining machine-learning algorithms with database techniques Machine learning and databases are two distinct fields in computer science with their own communities and conferences. While work from these two communities rarely intersects, knowledge from both areas is needed when designing data analytic methods for urban computing, as well as various other big-data projects. It is a challenging task for people from both communities to design effective and efficient data analytics methods that seamlessly and organically integrate the knowledge of databases and machine learning. There are three main approaches to the combination.

First, we can design spatial and spatiotemporal indexing structures to expedite the information retrieval and feature extraction processes for machine-learning algorithms. For example, to infer the air quality of a location, we need to extract traffic features from the GPS trajectories of taxicabs traversing each region. Without a spatiotemporal indexing structure, the process of feature extraction will last for a couple of hours [71].

Second, we can use database techniques to generate candidates for machine-learning algorithms, scaling down searching spaces of a complex learning algorithm. For example, causality inference algorithms are usually very complex and inefficient and thereby hard to apply to big data directly. Using pattern-mining techniques from database fields, we can find some correlation patterns from a large amount of data. Then, causality inference algorithms can take those correlation patterns, which are much smaller than the original data, as inputs to derive the causality between different objects or datasets.

Third, we can use upper and lower bounds derived from database techniques to prune search spaces. For instance, in [37, 78], the authors derive an upper bound for a log-likelihood ratio test to expedite the anomaly detection process.

1.4.3.3 Cross-domain knowledge-fusion methods While fusing knowledge from multiple disparate datasets is crucial to big-data projects, cross-domain data fusion is a nontrivial task for a few reasons.

First, simply concatenating features extracted from different datasets into a single feature vector may compromise the performance of a task, as different data sources may have very different feature spaces, distributions, and levels of significance [35].

Second, the more types of data that are involved in a task, the more likely it is that data scarcity will occur. For example, in [71], five data sources, consisting of traffic, meteorology, POIs, road networks, and air quality readings, are used to predict fine-grained air quality throughout a city. When trying to apply this method to other cities, however, we find that many cities cannot find enough data in each domain (e.g., do not have enough monitoring stations to generate air quality data) or may not even have the data from a given domain (like traffic data) at all [48].

1.4.3.4 Interactive visual data analytics

Data visualization is not solely about displaying raw data and presenting results, though the two are common motivations for using visualizations. Interactive visual data analytics becomes even more important in urban computing, detecting and describing patterns, trends, and relations in data, motivated by certain investigation directions [1, 34]. As something relevant is detected in data, new questions arise, causing specific parts to be viewed in more detail. A visualization also helps to tune and refine the parameters of data-mining models. Interactive visual data analytics provides an approach to the combination of human intelligence with machine intelligence. It also empowers people to integrate domain knowledge (such as urban planning) with data science, enabling domain experts (such as urban planners or environmentalists) to work with data scientists on solving problems in a specific domain.

Interactive visual data analytics calls for the following two aspects of technology: One is a seamless integration of visualization methods with data-mining algorithms as well as a deployment of the integration on a cloud-computing platform. The other is to combine the strengths of human and digital data processing in an interactive way, involving hypothesis generation rather than mere hypothesis testing [1].

1.4.4 Urban Service Challenges

This layer bridges the gap between urban computing and existing city-related fields, such as urban planning, environmental theory, and transportation, and faces the following challenges: One is the fusion of domain knowledge with data science. The other is an integration of an urban-computing system with existing domain systems.

1.4.4.1 Fusion domain knowledge with data science

When enabling an urban-computing application for these fields, we need a certain level of domain knowledge. For instance, in order to predict traffic conditions in a city, we need to know what would affect the traffic flow in a given location. However, domain experts and data scientists are usually two separate groups of people who barely share common knowledge. The former have rich domain knowledge and experiences but usually know little about data science. The latter are equipped with a diversity of data science techniques but lack domain knowledge. Moreover, the domain knowledge may be too sophisticated to specify explicitly and thus cannot be precisely modeled by an intelligent algorithm. There are two approaches to addressing this issue.

First, data scientists need to gain a certain level of domain knowledge by communicating with domain experts or learning from existing literature published in a domain. Collaborating with domain experts, data scientists need to identify the key problems (in the domain) that are mission critical and can be solved by data science. Data scientists should know the factors that could cause a problem, selecting the corresponding datasets to solve it. They need to understand the principle of existing methods proposed in the domain, leveraging good insights and making up for a deficiency in these methods.

Second, it is better to design some visual data analytics tools that allow domain experts to interact with the intelligent techniques created by data scientists. This is a way of combining machine intelligence and human intelligence. It is also an approach to integrating domain knowledge with data sciences. For example, using intelligent techniques, data scientists can generate some preliminary results based on a few simple criteria or some preliminary settings. The results are presented to domain experts, who will then refine the results based on their domain knowledge. The refinement might involve removing a few illegible candidates from the results or adjusting a few meaningful parameters. With the refined results, the intelligent techniques continue to generate another round of results, receiving feedback from domain experts until a satisfactory result is obtained. Through the interaction, domain knowledge and human intelligence have been integrated into intelligent techniques powered by data science. With such interactive visual data analytics tools, domain experts are more likely to contribute more knowledge to an urban-computing project.

1.4.4.2 System integration An ideal situation is to be able to fundamentally build an urban-computing system for a domain application following the framework shown in figure 1.12. In reality, however, many domain applications may already have their own systems that receive data from sensors and generate decisions for the applications. Domain experts may want to test a new urban-computing system before completely switching. Such circumstances are challenging for two reasons.

First, it is almost impossible to deploy an urban-computing system's key components, such as data analytics models, into an existing domain system because these components may need to be updated (i.e., rebuilt) based on recent data. The training process is not completely automatic, involving data scientists in the parameter-tuning and visualization processes. For example, we may retrain the air quality prediction model every few months, as the traffic and weather conditions may change significantly in a city over a few months. However, it is not within the majority of domain experts' capacity to retrain the machine-learning model.

Simply assigning a data scientist to each existing domain system may address this issue, but this raises another concern. Given that the number of data scientists is far less than the number of domain applications requiring data science, it will be very laborious to try to handle more domain requests. Tightly integrating key components into existing domain systems increases the work for data scientists. For example, they need to learn more about how a specific domain system works, besides the domain knowledge mentioned in the first challenge. In addition, they need to maintain multiple similar components deployed in different domain systems.

Second, to prevent data from being disclosed to the public, these systems are sometimes built based on private clouds. Thus, an urban-computing system may not be able to fully access data from the domain systems.

To tackle these challenges, figure 1.12 presents a possible loose integration strategy. In this strategy, an existing domain system continues to receive original data. It then provides processed data (e.g., features extracted from the original data) to an urban-computing system if there is a data security issue. As the feature extraction function is not dynamic and does not involve data scientists, it can be easily deployed into an existing domain system. Data scientists work on the urban-computing system side, training new models when necessary (without knowledge of how the domain system works). These models running in the urban-computing system consume the data from the domain system, continuously generating results (e.g., traffic predictions). The urban-computing system then provides results as a service to the domain system through cloud APIs. Based on the results from the urban-computing system, the domain system can derive the final decision to operate domain applications. The same set of APIs can be provided to many other domain systems that need traffic predictions for a given city. Thus, one model can serve many domain applications, and one data scientist can handle many similar requests.

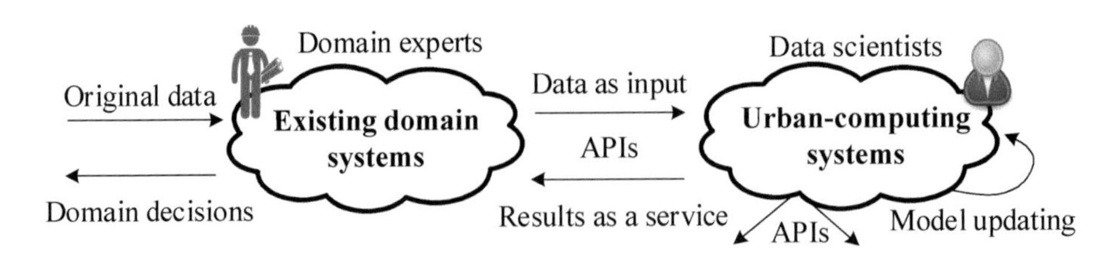

Figure 1.12
A system integration strategy for urban-computing services.

1.4.4.3 Training data scientists Although data scientists play a paramount role in many urban-computing projects, the number of data scientists is very limited because data science is a nascent field. In addition, training a data scientist is very challenging—much more difficult than training a data analyst. Table 1.2 presents the differences between data analysts and data scientists.

Generally speaking, a data analyst can employ existing data analytics tools to solve a well-formulated (data mining or machine learning) problem, using given datasets and generating results with predefined schemas. For example, to determine if a credit card application should be approved or rejected, a data analyst can train a binary classification (e.g., decision trees) using previous applicants' forms and repayment histories. A set of features, such as age, job, and income, can be extracted from an applicant's form. A corresponding label (either yes or no) can be derived based on the applicant's repayment history after a credit card has been issued. If the repayment is always on time, the label is set to yes; otherwise, no. Once the model has been trained, it can predict the label of a new application based on the features from the new form. If the label is yes, the application is approved; otherwise, it is rejected. In short, this is a well-formulated

Table 1.2
A comparison between a data analyst and a data scientist.

		Problem	Data	Method	Results
	Examples	Well formulated	Given	Existing tools	Predefined
Data analyst	"Shall a bank issue a credit card for a user based on his or her application form?"	Binary classification	1. Previous application forms, including a user's information, such as income, age, and job 2. Labels (Y or N) derived from these users' repayment histories	Classification models such as decision trees or random forest	Y: approve an application N: reject an application
		Not formulated	Unknown	Customized	Undefined
Data scientist	"What percentage of PM2.5 in the environment is generated by vehicles?"	Clustering? Classification? Regression? ???	Vehicles' emission data? Trajectories? Road networks? Meteorology? ???	Context-aware matrix factorization? Graphical models?	???

binary classification problem, with given datasets (i.e., previous applicants' forms and repayment histories) and predefined results (i.e., reject or approve).

In comparison, a data scientist may face the following questions—for example, What percentage of PM2.5 in the environment is generated by vehicles? What is the impact to Beijing's transportation and economy if we move the Beijing government to its rural part? How do we reduce a city's noise pollution? Such kind of questions are not well-formulated data analytics problems, as they are not a simple clustering, classification, regression, or causality analysis task. In addition, what datasets are relevant to the problem is unknown. No existing tools can solve the problem, and the schema of results cannot be easily predefined as some labels like yes or no. Instead, data scientists need to analyze such a problem, identify the datasets that are relevant to the problem and available in the real world, design customized data analytics models for the problem, and derive final results.

In many cases, our customers (e.g., governmental officials) cannot even propose a problem. Under such a circumstance, data scientists need to identify valuable problems (on their own) that are mission critical to a domain and fit data science better than traditional solutions. This is even harder than solving a given problem.

More specifically, as depicted in figure 1.13, a data scientist should be equipped with the following four aspects of skills.

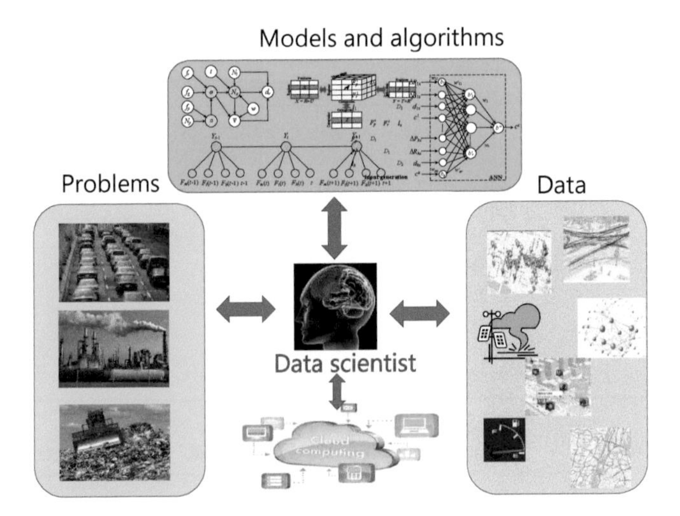

Figure 1.13
The definition of a data scientist.

Understanding problems The problems that data scientists need to solve in urban computing are usually from other domains, such as transportation, energy, and environment, rather than computer science. It is not necessary for a data scientist to become a domain expert, but the scientist must understand the following issues: what a problem is, why this problem is challenging, what factors could cause this problem, how traditional solutions solve this problem, and why these methods cannot solve this problem thoroughly. Answers to these questions are derived from commonsense knowledge, existing literature published in the domain, and simple data visualization.

For example, to infer a location's air quality, data scientists need to know the factors that could cause air pollution, such as emissions from factories and vehicular traffic, meteorology, and dispersion conditions. Knowing only these factors, data scientists can select proper datasets that represent or indicate corresponding factors. By learning from existing literature, we know that air pollution is a result of multiple complex processes consisting of local emissions, external propagation, and chemical reactions. Thus, on the one hand, we know traditional physical dispersion models cannot solve this problem, as they only consider the first two processes. In addition, it is difficult to accurately model the first two processes, as capturing all pollution sources is impractical in the real world. On the other hand, data scientists can learn from existing methods the features to extract and the principle inspiring the design of data analytics models.

Understanding the insights that data implies Data scientists need to understand the insight behind data besides its formats and properties. For example, GPS trajectories of taxicabs not only denote traffic conditions on roads but also imply people's commuting patterns because each taxi trajectory contains a taxi rider's pickup and drop-off points. Massive pickup and drop-off pairs represent people's origins and destinations as well as departure and arrival times, which provide key information about commuting patterns. Further, people's commuting patterns indicate a region's functions and economy as well as natural environment. With such insights, we can employ datasets from one domain to solve problems in another domain, enabling cross-domain data fusion and tackling the data sparsity challenges. For instance, we can combine taxi trajectories, which indicate traffic conditions and people's commuting patterns, with other datasets like POIs and road networks to infer a region's functions. We can even rank real estate in terms of its potential value, using taxi trajectories as one of the inputs.

Proficiency in different types of data analytics models Data scientists need to harness a variety of models and algorithms in data science, including data management, data mining, machine learning, and visualization. To solve a real problem with an end-to-end solution, data scientists need to organically integrate algorithms from different divisions of data science. In some cases, designing an algorithm in one step of a solution depends on the algorithms of its predecessor and successor steps. For example, when designing data management algorithms, we need to consider the properties of cloud-computing platforms and upper-level machine-learning algorithms.

Using cloud-computing platforms In the big-data era, data can no longer be held in a single machine. Cloud-computing platforms are becoming a common infrastructure for much big-data research, including urban computing. Knowing how to use such a platform is imperative for data scientists to deploy their solutions. The unique design of the cloud-computing platform affects the design of algorithms. Furthermore, it is better to know how to improve a cloud-computing platform by adding new components or intermedia layers. An enhanced cloud-computing platform can support an urban-computing system more effectively (see chapter 6 for details).

1.5 Urban Data

1.5.1 Taxonomy of Urban Data

1.5.1.1 Based on data structures and spatiotemporal properties As illustrated in figure 1.14, the form of these datasets can be classified into six categories in terms of their structures and spatiotemporal properties. With respect to data structures, there are point-based and network-based datasets, shown at the bottom and top, respectively. For spatiotemporal properties, there are three types of datasets: spatiotemporal static

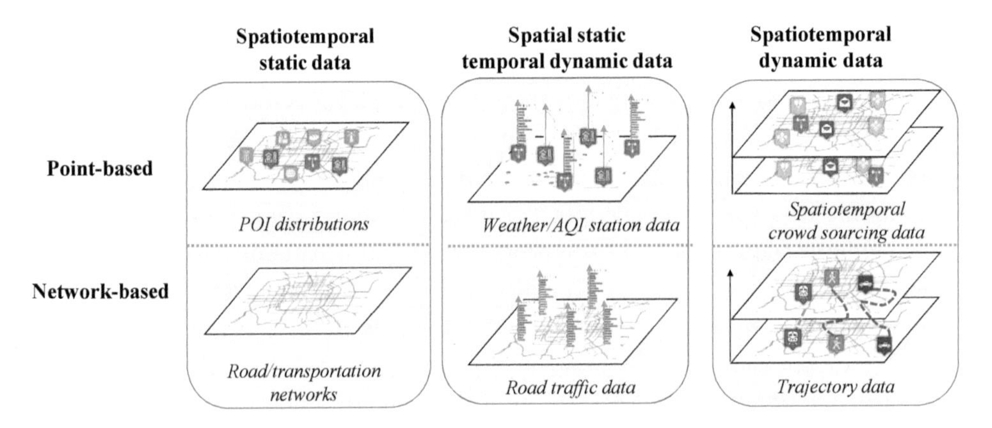

Figure 1.14
Six types of data forms for urban big data.

data, spatial static but temporal dynamic data, and spatiotemporal dynamic data, illustrated by the three columns.

For instance, as depicted in the first column of figure 1.14, represented by a point data structure, POI data has a static geoposition and fixed properties, such as name, address, and categories, which do not change over time.

Represented by a network structure, roads in a city also have static spatiotemporal properties, such as location, name, number of lanes, and speed limits. As shown in the bottom part of the middle column in figure 1.14, when associated with traffic, which changes over time, road traffic data becomes network-based spatially static but temporally dynamic data.

As illustrated in the top part of the middle column, most geosensory data, such as meteorological data and air quality, has a static point location where fixed sensors are deployed but generates dynamic readings continuously.

As demonstrated in the bottom part of the last column in figure 1.14, the most complex form of data is trajectory data, which usually represents the movement of moving objects like vehicles or people. Both spatial information, like the geoposition of a moving object, and temporal properties, such as speed and heading direction, change over time.

Spatiotemporal crowd-sensing data can be regarded as a point-based spatiotemporal dynamic dataset. The difference between the top and bottom parts of the last column is that the sequential property between consecutive points in the former is much weaker than the latter. For instance, people may post a geotagged tweet at one location and some other tweets at another location a few days later. The connection between the two locations becomes very weak, as people could have visited many other places between the two locations.

1.5.1.2 Based on data sources Another way to classify urban data is based on its sources—for example, geographical data, traffic data, commuting data, environmental-monitoring data, social network data, economy, energy consumption, and health-care data. Each source may be further composed of a few subcategories. For instance, commuting data includes people's ticketing data for buses, subways, and bike-sharing systems.

In addition, data from a source can formulate different data structures when used in different ways and viewed from different perspectives. For example, from a bike-sharing station's point of view, people's bike-renting data in an individual station is a point-based spatially static and temporally dynamic dataset. By putting together massive bike-renting data from multiple stations, however, we can formulate a network among bike

Table 1.3
Data sources mapping to different categories in a taxonomy.

	Spatiotemporal properties		
Data structures	Spatiotemporal static data	Spatially static and temporally dynamic data	Spatiotemporal dynamic data
Point-based	POI, land use	Meteorological data, air quality, loop detector data, surveillance camera data, data in a bike-sharing station, housing price, electricity consumption	Check-in data in location-based social networks, crowd sensing data, credit card transactions
Network-based	Road networks and structures of rivers, railways, and subway systems	Traffic on road networks, wireless communication networks, railways, subways, and bike-sharing systems	Personal mobile phone signals, GPS trajectories of vehicles, movements of animals, and hurricanes

stations as people check out a bike from one station and return it to another. Moreover, the bike traffic between different stations changes over time. Consequently, massive bike-sharing data from many stations can be regarded as a network-based spatially static and temporally dynamic dataset. If equipped with a GPS sensor, a bike's movement formulates a trajectory, which belongs to the spatiotemporal dynamic network data.

Given these two factors, data from one source may belong to multiple categories in the taxonomy based on data structures and spatiotemporal properties, as shown in table 1.3.

The following subsections introduce each type of urban data based on its sources, discussing potential applications in urban computing and possible challenges.

1.5.2 Geographical Data

1.5.2.1 Road network data Road network data may be the most frequently used geographical data in urban computing (e.g., traffic monitoring and prediction, urban planning, routing, and energy consumption analysis). It is usually represented by a graph that comprises a set of edges (denoting road segments) and a collection of nodes (standing for road intersections). Each node has a unique identity and geospatial coordinates; each edge is described by two nodes (sometimes called *terminals*) and a sequence of intermediate geospatial points (in case a road segment is not a straight line). Other properties, such as length, speed constraint, level of road (highway, avenue, or street), one-way or bidirectional, and number of lanes, are associated with an edge.

1.5.2.2 POI data A POI, such as a restaurant or a shopping mall, is usually described by a name, address, category, and set of geospatial coordinates. Once a POI is built, its properties rarely change over time, though a restaurant may occasionally change its name or move to a new location. As there are massive POIs in a city, collecting POI data is not an easy task. Generally, there are two approaches to the generation of POI data.

One is to obtain POIs through existing yellow page data using a geocoding algorithm to derive a POI's geospatial coordinates from its text address. The other approach is to manually collect POI information in the real world—for instance, carrying a GPS logger to record the geospatial coordinates of a POI. Major data providers have devoted substantial efforts to the second approach. Some recent location-based social networking services, like Foursquare, allow end users to create a new POI if it has not been included in the system.

In order to have a large coverage of POIs, widely used online map services, such as Bing and Google Maps, usually collect POI data through both of these approaches. As a result, quite a few issues have arisen, such as how to verify whether the information of a POI is correct. Sometimes, a POI's geospatial coordinates may be inaccurate, leading people to the wrong place. Another major issue is how to merge POIs generated from different sources or approaches [69].

1.5.2.3 Land use data Land use data describes the function of a region, such as residential, suburban, and nature areas, originally planned by urban planners and roughly measured by satellite images in practice [23]. For example, the United States Geological Survey categorizes each 30 m × 30 m square of the United States into twenty-one types of ground cover, such as grassland, water, and commercial. In many developing countries where cities change over time and many new infrastructure works are built and old buildings are removed, the reality of a city may be different from its original planning. As most satellite imagery cannot differentiate between fine-grained land use categories, such as educational, commercial, and residential areas, obtaining the current land use data for a big city requires a certain level of inference based on other datasets, such as human mobility and POIs [51, 54].

1.5.3 Traffic Data on Road Networks

1.5.3.1 Loop detector data There are multiple ways to collect traffic data, such as using loop detectors, surveillance cameras, and floating cars. Loop detectors are usually embedded in a pair in major roads (e.g., highways), detecting the time interval that a vehicle takes to travel across them. Dividing the distance between a pair of loop detectors by the time interval, we can estimate the travel speed on the road. Meanwhile, the traffic volume on a road can be obtained by counting the number of vehicles traversing these loop detectors in a time slot. As deploying and maintaining loop detectors is very expensive in terms of money and workforce, such traffic-monitoring technology is usually employed for major roads rather than low-level streets. As a result, the coverage of loop detectors is quite limited in a city. Additionally, loop detector data does not tell us how a vehicle travels on a road and between two roads. Consequently, the travel time that a vehicle spends at an intersection (e.g., waiting for traffic lights and direction turns) cannot be derived from this kind of sensor data.

1.5.3.2 Surveillance cameras Surveillance cameras are widely deployed in urban areas, generating a huge volume of images and videos. This data provides a visual ground truth of traffic conditions to people. However, it is still a challenging task to automatically turn these images and videos into a specific traffic volume and travel speed. It is difficult to apply a machine-learning model trained for a location to other places, as it is subject to the structure of roads in a location and a camera's settings, such as height (to the ground), angle, and focus. As a result, monitoring citywide traffic conditions with this approach still relies on human efforts.

1.5.3.3 Floating car data Floating car data [40] is generated by vehicles traveling around a city with a GPS sensor. The trajectories of these vehicles are sent to a central system and matched onto a road network for deriving speeds on road segments. As many cities have already installed GPS sensors in taxicabs, buses, and cargo trucks for different purposes, floating car data is already widely available.

In contrast to loop detectors and surveillance camera–based approaches, floating car–based traffic monitoring methods have a higher level of flexibility and a lower deployment cost. However, the coverage of floating car data depends on the distribution of the probing vehicles, which may change over time and be skewed in a city at various time intervals. This calls for advanced knowledge discovery technology that can recover citywide traffic conditions based on limited and skewed data [41].

Besides using floating car data to determine traffic conditions, we can turn taxis' GPS trajectories into social and community dynamics [6]. For example, knowing the pickup and drop-off points of taxi passengers, which can be regarded as a kind of human mobility data, we can study city-scale commuting patterns, thereby helping to improve urban planning [74]. In addition, human mobility data indicates the function of a region [51, 54], which is related to business and commerce [10, 11] and environmental protection [71].

1.5.4 Mobile Phone Data

There are three types of mobile phone data that can contribute to urban computing: call detail records (CDR), mobile phone location data, and mobile app logs.

1.5.4.1 Call detail records A CDR is a data record, produced by a telephone exchange, containing attributes specific to a single instance of a phone call, such as the phone numbers of both the calling and receiving parties, the originating and terminating stations, the start time, and the call duration [17, 38]. CDRs serve a variety of functions. For telephone service providers, they are the critical basis for generating telephone bills. For law enforcement, CDRs provide a wealth of information that can help identify suspects, as the originating and terminating stations can derive the whereabouts of an individual during a call. CDRs also reveal details about an individual's relationships with associates and his or her communication and behavior patterns. Thus, they can build a network between users and estimate the similarity and correlation between different users.

1.5.4.2 Mobile phone location data Mobile phone location data is a category of mobile phone signals that identify the location of a mobile phone rather than the communication between users. There are two ways to obtain the location of a mobile phone: One is to roughly estimate a mobile phone's location based on signals received by three or more base stations using a triangle-positioning algorithm. The other is to extract GPS coordinates from the data flow of mobile apps running on users' smartphones. The first way is widely available in that the location can be estimated as long as a mobile phone is connected to a wireless communication network, but it has a low positioning accuracy depending on the density of base stations. The second way has a low availability; that is, the location is not available if users do not run any mobile apps that acquire GPS readings, but it has a very high positioning accuracy (generated by a GPS sensor rather than the station-based triangle-positioning algorithm).

Mobile phone location data can provide a wealth of location information much better than CDRs, as the latter can only derive the location of a mobile phone based on the originating and terminating stations of a phone call. If people are not on phone calls, their locations cannot be derived. In addition, a base station usually covers a certain-sized area that may not be precise enough to localize a user. The mobile phone location data denotes citywide human mobility, which can be used for detecting urban anomalies or, in the long run, for studying a city's functional regions and urban planning. Sometimes, the two kinds of mobile phone data are integrated; that is, transaction records between phones and regarding the location of each phone are kept.

1.5.4.3 Mobile apps' logs Many mobile apps log user behaviors when people interact with a smartphone. These user logs imply a user's preferences and profile, helping to improve the design of an app and enable personalized advertisements. When used aggregately, massive user logs reveal people's life patterns and living styles in a given region. The logs may even help predict the commercial trend of a region. For example, if many people search for a certain piece of real estate using a mobile app, the price of the real estimate could rise. In principle, the log data of an application can only be accessed by the app operator. However, the data may be accessed by the wireless network operators and mobile phone manufactures if the data is not encrypted. Protecting users' privacy while leveraging these mobile app logs poses a challenge.

1.5.5 Commuting Data

People traveling in cities generate a huge volume of commuting data, such as card-swiping data on subway and bus systems, bike-sharing data, taxi fare data, and ticketing data from parking lots.

1.5.5.1 Card-swiping data This data is widely available in a city's public transportation systems, where people swipe a radio-frequency identification (RFID) card when entering a subway station or boarding a bus. Some systems also require people to swipe their cards again when leaving a station or getting off a bus. Each transaction record consists of the station's ID and a time stamp of entering/leaving the station, as well as the trip's fare. While the card-swiping data was originally created for the generation of transportation bills, massive card-swiping data can improve existing public transportation systems—for example, optimizing the schedules of existing buses and subways or planning new bus and subway lines.

1.5.5.2 Bike-sharing data Bike-sharing systems are widely deployed in many major cities, including New York, Paris, and Beijing, providing a convenient transportation mode for people's commutes. A user can rent (i.e., check out) a bike at a nearby station and return it (i.e., check in) to a station close to his or her destination. Users are required to swipe an RFID card when checking out/in a bike. A record consisting of the bike ID, time stamp, and station ID is generated for each card swipe. Bike-sharing systems face challenges in bike rebalancing between stations. Intrinsically, bike usages are skewed, changing over time and location. Consequently, some stations may be jammed, without enough docks for returned bikes, while some may lack available bikes for interested users [26]. The data can be used not only to monitor the current number of bikes at each station but also to predict future bike demands so that operators can reallocate bikes in advance. The data can also help plan a better deployment of stations.

Recently, some stationless bicycle-sharing systems allow users to conveniently park and pick up bikes from arbitrary locations. Such systems record users' riding trajectories, which can help plan bike lanes in a city effectively [2].

1.5.5.3 Taxi records A taxi fare record consists of two types of information: taxi fare data and trip data. The trip data includes pickup and drop-off locations and times, the duration and distance of each trip, the taxi ID, and the number of passengers. The fare data records the taxi fare, the tips, and the tax for each trip.

1.5.5.4 Parking data Street-side parking is usually paid through a parking meter. The payment information for parking spots may include the parking fare and the time the ticket is issued. The data indicates the vehicle traffic around a place, which can be used not only to improve a city's parking infrastructure but also to analyze people's travel patterns. The latter can support geo-ads and location choosing for a business.

1.5.6 Environmental-Monitoring Data

1.5.6.1 Meteorological data Meteorological data includes humidity; temperature; barometer pressure; wind speed; wind direction; precipitation; and weather conditions such as sunny, overcast, cloudy, and rainy. Meteorological data is generated by on-the-ground meteorological monitoring stations and published on public websites. The temporal granularity of meteorological data ranges from minutes to hours, changing over cities and countries. Its spatial granularity ranges from a monitoring station to a city. Forecasting meteorology is always an important task that matters to many fields, including the aviation, marine, and agricultural industries. Current weather forecasts are derived from the results of a set of classic models and human intervention.

1.5.6.2 Air quality data Air quality data, such as the concentration of PM2.5 (particulate matter with a diameter smaller than 2.5 μm), NO_2, and SO_2, can be obtained from air quality–monitoring stations. Though some gasses like CO_2 and CO can be detected by portable sensors, a device needs to absorb enough air to derive a reasonably accurate reading for PM2.5 and PM10. Thus, such a monitoring station is usually very big and expensive, requiring a certain area of land for deployment and a workforce for maintenance.

Readings from a monitoring station are concentrations of different air pollutants, for example, 0.0014 $\mu g/m^3$. When communicating with people, the concentration of an

Table 1.4
AQI values, descriptors, and color codes.

AQI	Values levels of health concern	Colors
0–50	Good (G)	Green
51–100	Moderate (M)	Yellow
101–150	Unhealthy for sensitive groups (U-S)	Orange
151–200	Unhealthy (U)	Red
201–300	Very unhealthy (VU)	Purple
301–500	Hazardous (H)	Maroon

air pollutant is converted into an individual air quality index (AQI) ranging from zero to five hundred. Different countries have their own standards for this conversion (refer to [81] for details). Later, the maximum individual AQI of all air pollutants during a time interval is selected to represent the AQI of the interval. The AQI range is then divided into six levels of air pollution, denoted by different colors. Table 1.4 presents the AQI standard of the United States. For example, an AQI falling between zero and fifty denotes good air quality, represented by green.

Influenced by multiple complex factors, such as traffic flow and land use, urban air quality varies significantly by location and changes frequently over time. As a consequence, a limited number of monitoring stations cannot reveal fine-grained air quality throughout a city [71]. Furthermore, there has been a recent rise in demand for advanced air quality forecasting.

1.5.6.3 Noise data Noise data is another type of environmental data that has a direct impact on people's mental and physical health [15]. Many cities have deployed sensors to measure the sound levels, with most outdoor noise worldwide caused by machines, transportation systems, motor vehicles, aircraft, and trains [16]. However, the level of noise pollution depends on both the intensity of noises and people's tolerance for noise [15]; the latter changes over time and can vary greatly from person to person. In addition, noise is a mixture of different types of sound. Sound sensors cannot tell us the composition of noise in a location, let alone the fact that sound changes over time and location significantly.

In recent years, there has been a series of research efforts that uses humans as sensors to collect noise data. For example, in cities like New York City, there is a 311 platform that allows people to register nonurgent complaints over the phone. Each complaint is associated with a time stamp, a location, and a category. Noise is the third-largest category in the data, which can be used to diagnose a city's noise pollution [73]. Other research projects leverage users' mobile phones to collect the noise level of a location and ask those users to tag the type of noise they heard.

1.5.6.4 Urban water quality Urban water quality refers to the physical, chemical, and biological characteristics of a body of water. It serves as "a powerful environmental determinant" and "a foundation for the prevention and control of waterborne diseases" [49]. Several indexes, such as residual chlorine, turbidity, and pH, are often used to measure the chemical properties of urban water in its distribution systems [39]. Other types of sensors detecting the physical properties of water, such as pressure, temperature, and flow, have also been deployed. Sensors detecting the physical and chemical properties of urban water may not be installed in the same location. The data generated by these sensors updates every few minutes. As very few such sensors are installed in a water distribution system, which usually contains hundreds of thousands of nodes and pipes, effectively monitoring water quality is still very difficult [30].

1.5.6.5 Satellite remote sensing Satellite remote sensing scans the surface of the earth with rays of different lengths to generate images representing the ecology and meteorology of a wide region. These images can be used to calibrate urban planning, control environmental pollution, and deal with catastrophic disasters.

1.5.7 Social Network Data

Social network data consists of three parts. The first is the user profile, which is composed of a user's personal properties, such as gender, home location, and age. This information is usually scarce and incomplete, as people with a privacy concern would not fill out all information. The information can help with different kinds of recommendations (such as advertisements).

The second is the social structure, represented by a graph, denoting the relationship, interdependency, or interaction between users. The social structure can help us detect communities of people, understand information diffusion in crowds, and even predict missing values in a user's profile.

The third is user-generated social media, such as texts, photos, and videos, that contain rich information about a user's behavior/interests. When adding a location to social media [63]—for example, check-in data from Foursquare and geotagged tweets—we can model people's mobility in urban areas, which is helpful for urban planning and anomaly detection [36].

1.5.8 Energy

1.5.8.1 Energy consumption of vehicles The gas consumption of vehicles on road surfaces and at gas stations reflects a city's energy consumption. The corresponding data can be obtained directly from sensors, such as those used by insurance companies to collect myriad types of sensor data from a vehicle. Alternatively, the data can be inferred from other sources implicitly—for example, from GPS trajectories of a vehicle [59, 60]. The data can be used to evaluate a city's energy infrastructure, such as its distribution of gas stations; to calculate pollution emissions from vehicles on road surfaces; or to find the most efficient route.

Many electric vehicles have recently been produced and are charged frequently. The charging data, consisting of where and when an electric vehicle has been charged and the corresponding electricity consumption, can inform the decision-making on deploying charging stations. The data is also helpful for improving the battery design in electric vehicles.

1.5.8.2 Smart grid technologies Smart grid technologies [9] have emerged from earlier attempts to use electronic controlling, metering, and monitoring in electricity infrastructure. In recent years, many smart meters and sensors have been installed in electrical grids, generating data about electricity consumption, transmission, and distribution [13]. The electricity consumption of an apartment or a building can be used to optimize residential energy usage, shifting peak loads to periods of low demand. The sensor data from electrical networks can help optimize energy transmission and distribution.

1.5.8.3 Household energy consumption Many smart meters have been installed to remotely monitor the electricity, water, and gas consumption in a household. The data can help understand a household's living patterns and estimate a family's economic ability, which is useful in precision marketing. When used aggregately, the data can infer the economic prosperity of a neighborhood, which can be used to predict the future value of real estate in the neighborhood.

1.5.8.4 Power plants A variety of datasets have been continuously generated by thermal power plant components, such as coal-fired boilers, blowing machines, and tail gas purification machines. The data can be used to improve energy efficiency, producing more electricity with less coal.

1.5.9 Economy

There are a variety of data representing a city's economic dynamics, such as transaction records of credit cards, stock prices, housing prices, and personal income. Each record of this type of data is associated with a location, a time stamp, and a value. When used aggregately, these datasets can capture the economic rhythm of a city, thereby predicting its future economy.

1.5.10 Health Care

There is already an abundance of health-care and disease data generated by hospitals and clinics, including datasets on medical treatments and reports of medical examinations. The latter may contain a variety of numbers, images (e.g., chest x-rays), time series, and charts such as electrocardiograms.

In addition, advances in wearable computing enable people to monitor their own health conditions, such as heart rate, pulse, and sleep time, with wearable devices such as wristbands. The data can even be sent to a cloud for diagnosing a disease and performing a remote medical examination. In urban computing, we can use these datasets aggregately to study the impact of the environmental change to people's health, such as analyzing how air pollution is related to the asthma situation in a city or how urban noises affect people's mental health in New York City.

1.6 Public Datasets

Quite a few cities, including New York City and Chicago, have opened their datasets to the public. Here are some links to the open datasets:

- New York City open data: https://data.cityofnewyork.us/.
- Chicago open data: https://data.cityofchicago.org/.
- Urban computing in Microsoft research: https://www.microsoft.com/en-us/research/project/urban-computing/ [66].
- Urban noise: 311 complaint data with social media, POIs, and road networks in New York City. https://www.microsoft.com/en-us/research/publication/diagnosing-new-york-citys-noises-with-ubiquitous-data/ [73].
- Urban air: Air quality data with meteorological data and weather forecasts in five Chinese cities [19, 71, 77]. https://www.microsoft.com/en-us/research/publication/forecasting-fine-grained-air-quality-based-on-big-data/.
- Traffic speed plus POIs plus road network: Features extracted from three datasets in Beijing have been accommodated in three matrices, used in [41].

https://www.microsoft.com/en-us/research/publication/travel-time-estimation-of-a-path-using-sparse-trajectories/. By adding a user dimension into the data, a tensor is built to describe the travel time of a particular user on a particular road in a specific time slot. The data was used in [46] and can be downloaded from the following URL: https://www.microsoft.com/en-us/research/publication/travel-time-estimation-of-a-path-using-sparse-trajectories/.

- GeoLife trajectory dataset [82]: A GPS trajectory dataset from Microsoft Research GeoLife project [76], collected by 182 users from April 2007 to August 2012. The dataset has been used to estimate the similarity between users [25], which enables friend and location recommendations [75, 79]. It was also used by [8] for studying the problem of finding the nearest trajectory to a sequence of query points.
- T-Drive taxi trajectories [83]: A sample of trajectories from the Microsoft Research T-Drive Project [52, 53, 55] generated by over ten thousand Beijing taxicabs over a week in 2008. The full dataset was used to suggest the fastest practical driving directions to normal drivers [53], recommend passenger pickup locations for taxi drivers [55, 57], enable dynamic taxi ride sharing [32, 33], glean problematic design in a city's transportation network [74], and identify urban functional regions [51, 54].
- GPS trajectory with transportation labels [84]: Each trajectory has a set of transportation mode labels, such as driving, taking a bus, riding a bike, and walking. The dataset can be used to evaluate trajectory classification and activity recognition [67, 70, 72].
- Check-in data from location-based social networks [85]: This dataset consists of the check-in data generated by over forty-nine thousand users in New York City and thirty-one thousand users in Los Angeles, as well as the social structure of the users. Each check-in includes a venue ID, the category of the venue, a time stamp, and a user ID. As the check-in data of a user can be regarded as a low-sampling-rate trajectory, this dataset has been used to study the uncertainty of trajectories [47] and evaluate location recommendations [3].
- Hurricane trajectories [86]: This dataset provided by the National Hurricane Center (NHC) contains 1,740 trajectories of Atlantic hurricanes (formally defined as tropical cyclones) from 1851 to 2012. The NHC also provides annotations of typical hurricane tracks for each month throughout the annual hurricane season, which spans from June to November. The dataset can be used to test trajectory clustering and uncertainty.
- Greek truck trajectories [87]: This dataset contains 1,100 trajectories from fifty different trucks delivering concrete around Athens, Greece. It was used to evaluate trajectory pattern mining in [12].
- Movebank animal tracking data [88]: Movebank is a free online database that helps animal-tracking researchers manage, share, protect, analyze, and archive their data.

References

[1] Andrienko, N., G. Andrienko, and P. Gatalsky. 2003. "Exploratory Spatio-Temporal Visualization: An Analytical Review." *Journal of Visual Languages and Computing* 14 (6): 503–541.

[2] Bao, J., T. He, S. Ruan, Y. Li, and Y. Zheng. 2017. "Planning Bike Lanes Based on Sharing-Bike's Trajectories." In *Proceedings of the 23rd SIGKDD Conference on Knowledge Discovery and Data Mining*. New York: Association for Computing Machinery (ACM).

[3] Bao, J., Y. Zheng, and M. F. Mokbel. 2012. "Location-Based and Preference-Aware Recommendation Using Sparse Geo-Social Networking Data." In *Proceedings of the 20th ACM SIGSPATIAL International Conference on Advances in Geographic Information Systems*. New York: ACM, 199–208.

[4] Burke, J. A., D. Estrin, M. Hansen, A. Parker, N. Ramanathan, S. Reddy, and M. B. Srivastava. 2006. *Participatory Sensing*. Los Angeles: Center for Embedded Network Sensing.

[5] Candia, J., M. C. González, P. Wang, T. Schoenharl, G. Madey, and A. L. Barabási. 2012. "Uncovering Individual and Collective Human Dynamics from Mobile Phone Records." *Journal of Physics A: Mathematical and Theoretical* 41 (22): 224015.

[6] Castro, P. S., D. Zhang, C. Chen, S. Li, and G. Pan. 2013. "From Taxi GPS Traces to Social and Community Dynamics: A Survey." *ACM Computer Survey* 46 (2), article no. 17.

[7] Chen, Y., Kai Jiang, Yu Zheng, Chunping Li, and Nenghai Yu. 2009. "Trajectory Simplification Method for Location-Based Social Networking Services." In *Proceedings of the ACM GIS Workshop on Location-Based Social Networking Services*. New York: ACM.

[8] Chen, Zaiben, Heng Tao Shen, Xiaofang Zhou, Yu Zheng, and Xing Xie. 2010. "Searching Trajectories by Locations: An Efficiency Study." In *Proceedings of the ACM SIGMOD International Conference on Management of Data*. New York: ACM.

[9] Farhangi, H. 2010. "The Path of the Smart Grid." *IEEE Power and Energy Magazine* 8 (1): 18–28.

[10] Fu, Y., Yong Ge, Yu Zheng, Zijun Yao, Yanchi Liu, Hui Xiong, and Nicholas Jing Yuan. 2014. "Sparse Real Estate Ranking with Online User Reviews and Offline Moving Behaviors." Washington, DC: Institute of Electrical and Electronics Engineers (IEEE) Computer Society Press.

[11] Fu, Y., H. Xiong, Yong Ge, Zijun Yao, and Y. Zheng. 2014. "Exploiting Geographic Dependencies for Real Estate Appraisal: A Mutual Perspective of Ranking and Clustering." In *Proceedings of the 20th SIGKDD Conference on Knowledge Discovery and Data Mining*. New York: ACM.

[12] Giannotti, F., M. Nanni, F. Pinelli, and D. Pedreschi. 2007. "Trajectory Pattern Mining." In *Proceedings of the 13th ACM SIGKDD International Conference on Knowledge Discovery and Data Mining*. New York: ACM, 330–339.

[13] Gungor, V. C., D. Sahin, T. Kocak, S. Ergut, C. Buccella, C. Cecati, and G. P. Hancke. 2011. "Smart Grid Technologies: Communication Technologies and Standards." *IEEE Transactions on Industrial Informatics* 7 (4): 529–539.

[14] Guo, B., Z. Yu, X. Zhou, and D. Zhang. 2014. "From Participatory Sensing to Mobile Crowd Sensing." In *Pervasive Computing and Communications Workshops (PERCOM Workshops), 2014 IEEE International Conference*. Washington, DC: IEEE Computer Society Press, 593–598.

[15] Hoffmann, B., S. Moebus, A. Stang, E. M. Beck, N. Dragano, S. Möhlenkamp, A. Schmermund, M. Memmesheimer, K. Mann, R. Erbel, and K. H. Jöckel. 2006. "Residence Close to High Traffic and Prevalence of Coronary Heart Disease." *European Heart Journal* 27 (22): 2696–2702.

[16] Hogan, C.M., and G. L. Latshaw. 1973. "The Relationship between Highway Planning and Urban Noise." In *Proceedings of the ASCE Urban Transportation Division Environment Impact Specialty Conference*. New York: American Society of Civil Engineers.

[17] Horak, R. 2007. *Telecommunications and Data Communications Handbook*. Hoboken, NJ: Wiley-Interscience, 110–111.

[18] Howe, J. 2006. "The Rise of Crowdsourcing." *Wired Magazine* 14 (6): 1–4.

[19] Hsieh, H.-P., S.-D. Lin, and Y. Zheng. 2015. "Inferring Air Quality for Station Location Recommendation Based on Big Data." In *Proceedings of the 21st SIGKDD Conference on Knowledge Discovery and Data Mining*. New York: ACM.

[20] Ji, S., Y. Zheng, and Tianrui Li. 2016. "Urban Sensing Based on Human Mobility." In *Proceedings of the 18th ACM International Conference on Ubiquitous Computing*. New York: ACM.

[21] Kindberg, T., M. Chalmers, and E. Paulos. 2007. "Guest Editors' Introduction: Urban Computing." *Pervasive Computing* 6 (3): 18–20.

[22] Kostakos, V., and E. O'Neill. 2008. "Cityware: Urban Computing to Bridge Online and Real-World Social Networks." *Handbook of Research on Urban Informatics*. Hershey, PA: IGI Global.

[23] Krumm, J., and Eric Horvitz. 2006. "Predestination: Inferring Destinations from Partial Trajectories." In *Proceedings of the 8th International Conference on Ubiquitous Computing*. New York: ACM.

[24] Lee, R., and K. Sumiya. 2010. "Measuring Geographical Regularities of Crowd Behaviors for Twitter-Based Geo-social Event Detection." In *Proceedings of the 2nd ACM SIGSPATIAL GIS Workshop on Location Based Social Networks*. New York: ACM, 1–10.

[25] Li, Q., Yu Zheng, Xing Xie, Yukun Chen, Wenyu Liu, and Wei-Ying Ma. 2008. "Mining User Similarity Based on Location History." In *Proceedings of the 16th ACM SIGSPATIAL Conference on Advances in Geographical Information Systems*. New York: ACM, 1–10.

[26] Li, Yexin, Yu Zheng, Huichu Zhang, and Lei Chen. 2015. "Traffic Prediction in a Bike-Sharing System." In *Proceedings of the 23rd ACM International Conference on Advances in Geographical Information Systems*. New York: ACM.

[27] Li, Y., Jie Bao, Yanhua Li, Zhiguo Gong, and Yu Zheng. 2017. "Mining the Most Influential k-Location Set from Massive Trajectories." *IEEE Transactions on Big Data*. doi:10.1109/TBDATA.2017.2717978.

[28] Liu, D., D. Weng, Y. Li, J. Bao, Y. Zheng, H. Qu, and Y. Wu. 2017. "SmartAdP: Visual Analytics of Large-Scale Taxi Trajectories for Selecting Billboard Locations." *IEEE Transactions on Visualization and Computer Graphics* 1:1–10.

[29] Liu, W., Yu Zheng, Sanjay Chawla, Jing Yuan, and Xing Xie. 2011. "Discovering Spatio-temporal Causal Interactions in Traffic Data Streams." In *Proceedings of the 17th SIGKDD Conference on Knowledge Discovery and Data Mining*. New York: ACM.

[30] Liu, Y., Yu Zheng, Yuxuan Liang, Shuming Liu, and David S. Rosenblum. 2016. "Urban Water Quality Prediction Based on Multi-task Multi-view Learning." In *Proceedings of the 25th International Joint Conference on Artificial Intelligence*. Pasadena, CA: International Joint Conferences on Artificial Intelligence Organization (IJCAI).

[31] Lou, Y., Chengyang Zhang, Yu Zheng, Xing Xie, Wei Wang, and Yan Huang. 2009. "Map-Matching for Low-Sampling-Rate GPS Trajectories." In *Proceedings of the 17th ACM SIGSPATIAL Conference on Geographical Information Systems*. New York: ACM.

[32] Ma, Shuo, Yu Zheng, and Ouri Wolfson. 2013. "T-Share: A Large-Scale Dynamic Taxi Ridesharing Service." In *Proceedings of the 29th IEEE International Conference on Data Engineering*. Washington, DC: IEEE Computer Society Press.

[33] Ma, Shuo, Yu Zheng, and Ouri Wolfson. 2015. "Real-Time City-Scale Taxi Ridesharing." *IEEE Transactions on Knowledge and Data Engineering* 27, no. 7 (July): 1782–1785.

[34] Martinoc, D., S. M. Bertolottoa, F. Ferruccic, T. Kechadi, and P. Compieta. 2007. "Exploratory Spatio-temporal Data Mining and Visualization." *Journal of Visual Languages and Computing* 18 (3): 255–279.

[35] Ngiam, J., A. Khosla, M. Kim, J. Nam, H. Lee, and A. Y. Ng. 2011. "Multimodal Deep Learning." In *Proceedings of the 28th International Conference on Machine Learning*. Pittsburgh, PA: International Machine Learning Society, 689–696.

[36] Pan, B., Y. Zheng, D. Wilkie, and C. Shahabi. 2013. "Crowd Sensing of Traffic Anomalies Based on Human Mobility and Social Media." In *Proceedings of the 21st ACM SIGSPATIAL Conference Advances in Geographic Information Systems*. New York: ACM, 334–343.

[37] Pang, L. X., Sanjay Chawla, Wei Liu, and Yu Zheng. 2013. "On Detection of Emerging Anomalous Traffic Patterns Using GPS Data." *Data and Knowledge Engineering* 87 (September): 357–373.

[38] Peterson, K. 2000. *Business Telecom Systems: A Guide to Choosing the Best Technologies and Services*. New York: CMP Books.

[39] Rossman, Lewis A., Robert M. Clark, and Walter M. Grayman. 1994. "Modeling Chlorine Residuals in Drinking Water Distribution Systems." *Journal of Environmental Engineering* 120 (4): 803–820.

[40] Schäfer, R. P., K. U. Thiessenhusen, and P. Wagner. 2002. "A Traffic Information System by Means of Real-Time Floating-Car Data." *ITS World Congress* 11 (October): 14.

[41] Shang, J., Yu Zheng, Wenzhu Tong, Eric Chang, and Yong Yu. 2014. "Inferring Gas Consumption and Pollution Emission of Vehicles throughout a City." In *Proceedings of the 20th SIGKDD Conference on Knowledge Discovery and Data Mining*. New York: ACM.

[42] Song, R., Weiwei Sun, Baihua Zheng, and Yu Zheng. 2014. "Press: A Novel Framework of Trajectory Compression in Road Networks." In *Proceedings of the 40th International Conference on Very Large Data Bases*. San Jose, CA: Very Large Data Bases Endowment (VLDB).

[43] Srivastava, N., and R. R. Salakhutdinov. 2012. "Multimodal Learning with Deep Boltzmann Machines." *Neural Information Processing Systems* (NIPS), 2222–2230.

[44] Sun, Yu, Jianzhong Qi, Yu Zheng, and Rui Zhang. 2015. "K-Nearest Neighbor Temporal Aggregate Queries." In *Proceedings of the 18th International Conference on Extending Database Technology*. Konstanz, Germany: Extending Database Technology.

[45] Tobler W. 1970. "A Computer Movie Simulating Urban Growth in the Detroit Region." *Economic Geography* 46 (2): 234–240.

[46] Wang, Y., Yu Zheng, and Yexiang Xue. 2014. "Travel Time Estimation of a Path Using Sparse Trajectories." In *Proceedings of the 20th SIGKDD Conference on Knowledge Discovery and Data Mining*. New York: ACM.

[47] Wei, L., Y. Zheng, and W. Peng. 2012. "Constructing Popular Routes from Uncertain Trajectories." In *Proceedings of the 18th ACM SIGKDD International Conference on Knowledge Discovery and Data Mining*. New York: ACM, 195–203.

[48] Wei, Ying, Yu Zheng, and Qiang Yang. 2016. "Transfer Knowledge between Cities." In *Proceedings of the 22nd SIGKDD Conference on Knowledge Discovery and Data Mining*. New York: ACM.

[49] World Health Organization. 2004. "Guidelines for Drinking-Water Quality." Volume 3.

[50] Yi, X., Yu Zheng, Junbo Zhang, and Tianrui Li. 2016. "ST-MVL: Filling Missing Values in Geo-sensory Time Series Data." In *Proceedings of the 25th International Joint Conference on Artificial Intelligence*. Pasadena, CA: IJCAI.

[51] Yuan, J., Yu Zheng, and Xing Xie. 2012. "Discovering Regions of Different Functions in a City Using Human Mobility and POIs." In *Proceedings of the 18th SIGKDD Conference on Knowledge Discovery and Data Mining*. New York: ACM.

[52] Yuan, J., Y. Zheng, X. Xie, and G. Sun. 2011. "Driving with Knowledge from the Physical World." In *Proceedings of the 17th SIGKDD Conference on Knowledge Discovery and Data Mining*. New York: ACM, 316–324.

[53] Yuan, J., Y. Zheng, X. Xie, and G. Sun. 2013. "T-Drive: Enhancing Driving Directions with Taxi Drivers' Intelligence." *IEEE Transactions on Knowledge and Data Engineering* 25 (1): 220–232.

[54] Yuan, N. J., Yu Zheng, Xing Xie, Y. Wang, Kai Zheng, and Hui Xiong. 2015. "Discovering Urban Functional Zones Using Latent Activity Trajectories." *IEEE Transactions on Knowledge and Data Engineering* 27 (3): 1041–4347.

[55] Yuan, N. J., Y. Zheng, L. Zhang, and X. Xie. 2013. "T-Finder: A Recommender System for Finding Passengers and Vacant Taxis." *IEEE Transactions on Knowledge and Data Engineering* 25 (10): 2390–2403.

[56] Yuan, J., Y. Zheng, C. Zhang, X. Xie, and Guangzhong Sun. 2010. "An Interactive-Voting Based Map Matching Algorithm." In *Proceedings of the Eleventh International Conference on Mobile Data Management*. Washington, DC: IEEE Computer Society Press.

[57] Yuan, J., Y. Zheng, L. Zhang, X. Xie, and G. Sun. 2011. "Where to Find My Next Passenger?" In *Proceedings of the 13th ACM International Conference on Ubiquitous Computing*. New York: ACM, 109–118.

[58] Yuan, J., Y. Zheng, C. Zhang, W. Xie, X. Xie, G. Sun, and Y. Huang. 2010. "T-Drive: Driving Directions Based on Taxi Trajectories." In *Proceedings of the 18th ACM SIGSPATIAL Conference on Advances in Geographical Information Systems*. New York: ACM, 99–108.

[59] Zhang, F., David Wilkie, Yu Zheng, and Xing Xie. 2013. "Sensing the Pulse of Urban Refueling Behavior." In *Proceedings of the 15th ACM International Conference on Ubiquitous Computing*. New York: ACM.

[60] Zhang, F., Nicholas Jing Yuan, David Wilkie, Yu Zheng, and Xing Xie. 2015. "Sensing the Pulse of Urban Refueling Behavior: A Perspective from Taxi Mobility." *ACM Transaction on Intelligent Systems and Technology* 6 (3): 16–34.

[61] Zhang, Junbo, Yu Zheng, and Dekang Qi. 2017. "Deep Spatio-Temporal Residual Networks for Citywide Crowd Flows Prediction." In *Proceedings of the 31st AAAI Conference*. Menlo Park, CA: AAAI Press.

[62] Zhang, Siyuan, Lu Qin, Yu Zheng, and Hong Cheng. 2016. "Effective and Efficient: Large-Scale Dynamic City Express." *IEEE Transactions on Data Engineering* 28, no. 12 (December): 3203–3217.

[63] Zheng, Y. 2011. "Location-Based Social Networks: Users." In *Computing with Spatial Trajectories*, edited by Y. Zheng and X. Zhou, 243–276. Berlin: Springer.

[64] Zheng, Yu. 2015. "Methodologies for Cross-Domain Data Fusion: An Overview." *IEEE Transactions on Big Data* 1 (1): 16–34.

[65] Zheng, Yu. 2015. "Trajectory Data Mining: An Overview." *ACM Transactions on Intelligent Systems and Technology* 6 (3), article no. 29.

[66] Zheng, Y., L. Capra, O. Wolfson, and H. Yang. 2014. "Urban Computing: Concepts, Methodologies, and Applications." *ACM Transactions on Intelligent Systems and Technology* 5 (3): 38–55.

[67] Zheng, Y., Y. Chen, Q. Li, X. Xie, and W.-Y. Ma. 2010. "Understanding Transportation Modes Based on GPS Data for Web Applications." *ACM Transactions on the Web* 4 (1): 1–36.

[68] Zheng, Y., Y. Chen, X. Xie, and Wei-Ying Ma. 2009. "GeoLife2.0: A Location-Based Social Networking Service." In *Proceedings of the 10th International Conference on Mobile Data Management.* Washington, DC: IEEE Computer Society Press.

[69] Zheng, Y., X. Feng, Xing Xie, Shuang Peng, and James Fu. 2010. "Detecting Nearly Duplicated Records in Location Datasets." In *Proceedings of the 18th ACM SIGSPATIAL Conference on Advances in Geographical Information Systems.* New York: ACM.

[70] Zheng, Y., Q. Li, Y. Chen, and X. Xie. 2008. "Understanding Mobility Based on GPS Data." In *Proceedings of the 11th International Conference on Ubiquitous Computing.* New York: ACM, 312–321.

[71] Zheng, Y., F. Liu, and H. P. Hsieh. 2013. "U-Air: When Urban Air Quality Inference Meets Big Data." In *Proceedings of the 19th SIGKDD Conference on Knowledge Discovery and Data Mining.* New York: ACM, 1436–1444.

[72] Zheng, Y., L. Liu, L. Wang, and X. Xie. 2008. "Learning Transportation Mode from Raw GPS Data for Geographic Application on the Web." In *Proceedings of the 17th International Conference on World Wide Web.* New York: ACM, 247–256.

[73] Zheng, Y., T. Liu, Yilun Wang, Yanchi Liu, Yanmin Zhu, and Eric Chang. 2014. "Diagnosing New York City's Noises with Ubiquitous Data." In *Proceedings of the 16th ACM International Joint Conference on Pervasive and Ubiquitous Computing.* New York: ACM.

[74] Zheng, Y., Y. Liu, J. Yuan, and X. Xie. 2011. "Urban Computing with Taxicabs." In *Proceedings of the 13th ACM Conference on Ubiquitous Computing.* New York: ACM, 89–98.

[75] Zheng, Y., and X. Xie. 2011. "Learning Travel Recommendations from User-Generated GPS Traces." *ACM Transactions on Intelligent Systems and Technology* 2 (1): 2–19.

[76] Yu Zheng, X. Xie, and Wei-Ying Ma. 2010. "GeoLife: A Collaborative Social Networking Service among User, Location and Trajectory." *IEEE Data Engineering Bulletin* 33 (2): 32–40.

[77] Zheng, Y., X. Yi, M. Li, R. Li, Z. Shan, E. Chang, and T. Li. 2015. "Forecasting Fine-Grained Air Quality Based on Big Data." In *Proceedings of the 21st SIGKDD Conference on Knowledge Discovery and Data Mining.* New York: ACM.

[78] Zheng, Y., H. Zhang, and Y. Yu. 2015. "Detecting Collective Anomalies from Multiple Spatio-temporal Datasets across Different Domains." In *Proceedings of the 23rd ACM International Conference on Advances in Geographical Information Systems.* New York: ACM.

[79] Zheng, Y., L. Zhang, Z. Ma, X. Xie, and W.-Y. Ma. 2011. "Recommending Friends and Locations Based on Individual Location History." *ACM Transactions on the Web* 5 (1), article no. 5.

[80] Zhu, J. Y., Y. Zheng, Xiuwen Yi, and Victor O. K. Li. 2016. "A Gaussian Bayesian Model to Identify Spatiotemporal Causalities for Air Pollution Based on Urban Big Data." Paper presented at Workshop on Smart Cities at InforCom, San Francisco, CA.

[81] *Wikipedia.* https://en.wikipedia.org/wiki/Air_quality_index.

[82] GeoLife GPS Trajectories. http://research.microsoft.com/en-us/downloads/b16d359d-d164-469e-9fd4-daa38f2b2e13/default.aspx.

[83] T-Drive Trajectory Data Sample. http://research.microsoft.com/apps/pubs/?id=152883.

[84] GPS Trajectories with Transportation Mode Labels. http://research.microsoft.com/apps/pubs/?id=141896.

[85] User check-in data. https://www.dropbox.com/s/4nwb7zpsj25ibyh/check-in%20data.zip.

[86] National Hurricane Center Data Archive. HURDAT. http://www.nhc.noaa.gov/data/hurdat.

[87] The Greek Trucks Dataset. http://www.chorochronos.org.

[88] Movebank. https://www.movebank.org/.

2 Urban-Computing Applications

Abstract: This chapter presents some typical applications of urban computing in different domains consisting of transportation, urban planning, environmental protection, energy, economy, public security, and social and entertainment.

2.1 Introduction

Before introducing the technology involved in urban computing, as illustrated in figure 2.1, we list seven categories of application scenarios: urban planning, transportation, environment, public safety and security, energy, economy, and social and entertainment. In each category, we first briefly mention the research progress in traditional fields and then present some representative urban-computing applications. In this chapter, we focus on the goal, motivation, results, and data used in an application rather than the methodology details, which will be discussed in later chapters.

2.2 Urban Computing for Urban Planning

2.2.1 Gleaning Underlying Problems in Transportation Networks

Effective planning is of great importance in building an intelligent city. Formulating urban planning requires evaluating a broad range of factors, such as traffic flow, human mobility, points of interest (POIs), and road network structures. These complex and quickly evolving factors turn urban planning into a very challenging task.

Traditionally, urban planners rely on labor-intensive surveys to inform their decision-making. For example, to understand urban-commuting patterns, a body of research has been performed based on travel survey data [7, 61, 71], but the information obtained through the surveys may not be sufficient and timely enough.

Recent widely available human mobility data generated in urban spaces actually reflects the underlying problems of a city, providing urban planners with opportunities to better formulate future planning [185].

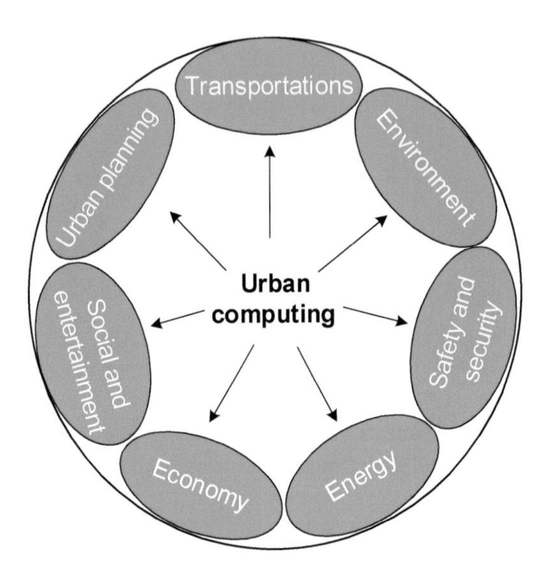

Figure 2.1
Main categories of applications in urban computing.

Zheng et al. [185] glean the underlying problems in Beijing's transportation network by analyzing the GPS trajectories generated by thirty-three thousand taxicabs over a period of three years. They first partition the urban areas of Beijing into disjointed regions using major roads, such as highways and arterial roads, and by employing the map segmentation method [155], as illustrated in figure 2.2a. The pickup and drop-off points of passengers are extracted from each taxi trajectory to formulate the origin-destination (OD) transitions between these regions. A region graph is then built based on the OD transitions, where a node is a region, and an edge denotes the aggregation of the transitions between two regions, as depicted in figure 2.2b.

Using a data-driven method, a day is divided into a few time spans that correspond to morning rush hours, the evening peak hours, and the rest. For each time span, a region graph is built based on taxi trajectories falling into the time span.

As demonstrated in figure 2.2c, three features consisting of the volume of taxis $|S|$, the average speed of these taxis $E(V)$, and a detour ratio θ are extracted for each edge based on the associated taxi trajectories. Representing an edge with a point in the three-feature dimension space, points with a large $|S|$, small $E(V)$, and big θ could be underlying problems. That is, the connection between two regions is not effective enough to support the traffic traveling between them, resulting in large volumes, low speeds, and big detour ratios.

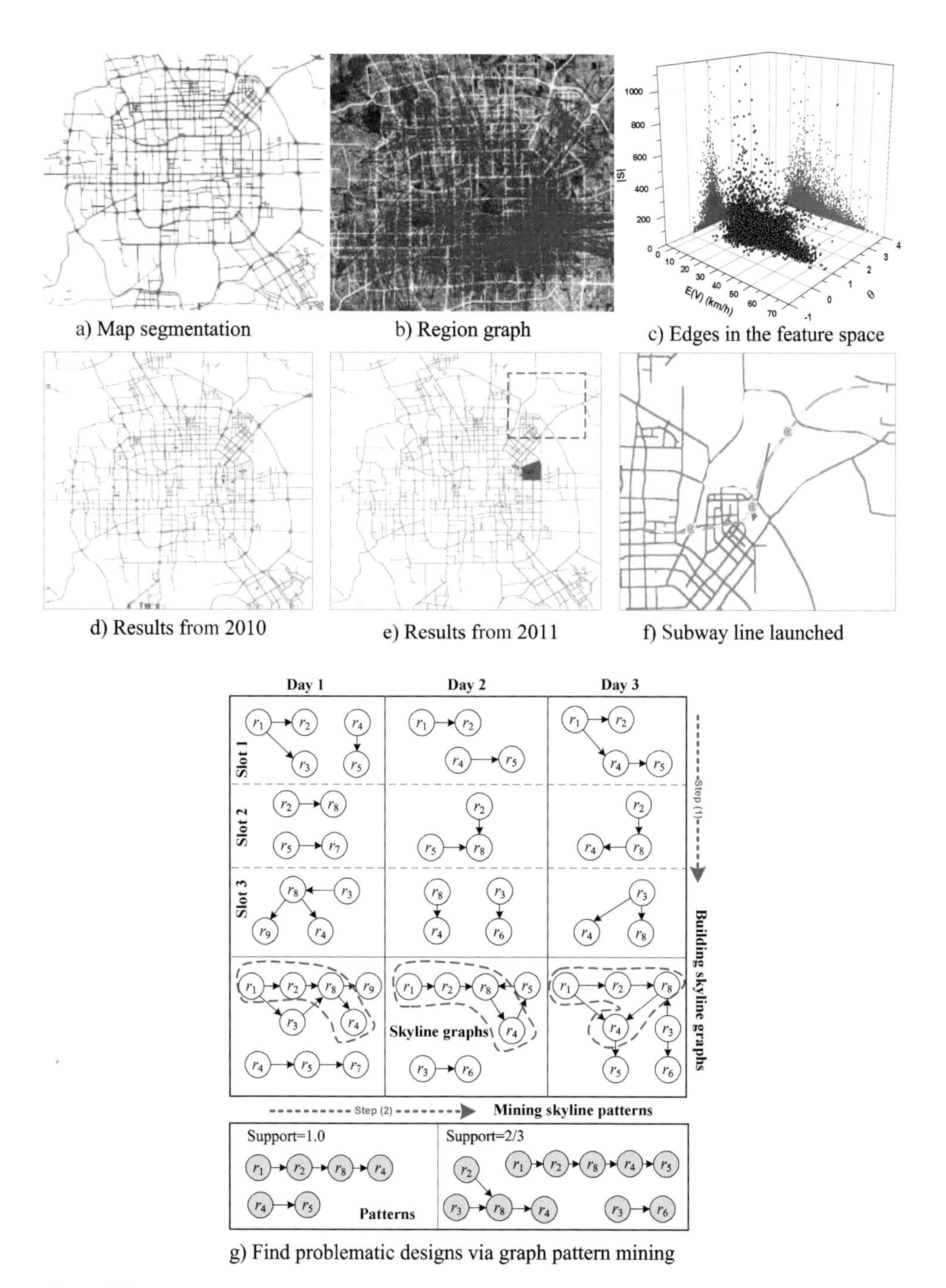

Figure 2.2

Finding underlying problems with Beijing's road network using taxi trajectories.

Using a skyline detection algorithm [20], a set of points (called *skyline edges*) can be detected from the data of each time slot. As illustrated in figure 2.2g, the skyline edges from different time slots of the same day are connected to formulate skyline graphs if they are spatially overlapped by some nodes and temporally adjacent.

Finally, some subgraph patterns [146] are obtained by mining skyline graphs across multiple days, such as $r_1 \rightarrow r_2 \rightarrow r_8 \rightarrow r_4$ occurring on all three days. Such patterns represent the underlying problems in a road network, showing the correlations between individual regions and avoiding the false alerts that could be caused by some traffic accidents.

By comparing the results detected from two consecutive years, the research can even evaluate if a newly built transportation facility works well. As demonstrated in figure 2.2d, e, and f, the underlying problem detected in 2010 disappeared in 2011 because of a newly launched subway line. In short, the subway line played a major role in resolving the problem.

2.2.2 Discovering Functional Regions

The development of a city gradually fosters different functional regions [7], such as educational areas, residential neighborhoods, and business districts, that support different people's needs and serve as a valuable organizing technique for framing detailed knowledge of a metropolitan area. Artificially designed by urban planners or naturally derived from people's actual demands, these regions may change their functions and borders over time. The understanding of functional regions in a city can calibrate urban planning and facilitate other applications such as site selection for commerce and resource allocation.

Functional regions have been studied in geographical information systems (GIS) and urban planning for years. Methods in these fields usually employ clustering algorithms to identify functional regions [77]. For instance, some network-based clustering algorithms (e.g., spectral clustering) are employed to identify functional regions based on interaction data, such as economic transactions and people's movements, between regions. There is also a body of research on using classification algorithms to identify the land use of a region based on satellite remote-sensing data [132].

Recently, user-generated content, such as social media and human mobility, has been used to study the topics of a region. For example, Yin et al. [150] studied the distributions of some geographical topics (such as beaches, hiking, and sunsets) in the United States using geotagged photos acquired from Flickr. Pozdnoukhov et al. [114] explored the space-time structure of topical content based on a large number of geotweets. Qi et al. [115] observed that the getting on/off records from taxi passengers in a region can depict the dynamics of social activities in that region.

Yuan et al. [154] propose a framework that *discovers regions of different functions* in a city (called DRoF) using human mobility between regions and POIs located in a region. For example, the red regions shown in figure 2.3a denote the educational and scientific areas of Beijing. However, the function of a region is compound, represented by a distribution across multiple functions. Regions with the same color actually share a similar distribution of functions.

On the other hand, even if a region is recognized as an educational area, that does not mean every part of the region serves this function. For instance, there could be some shopping centers around a university. As a result, given a function, Yuan et al. [162] further identify its kernel density distribution. Figure 2.3b shows the density distribution of commercial areas in Beijing; the darker the color, the higher probability the location could be a commercial area.

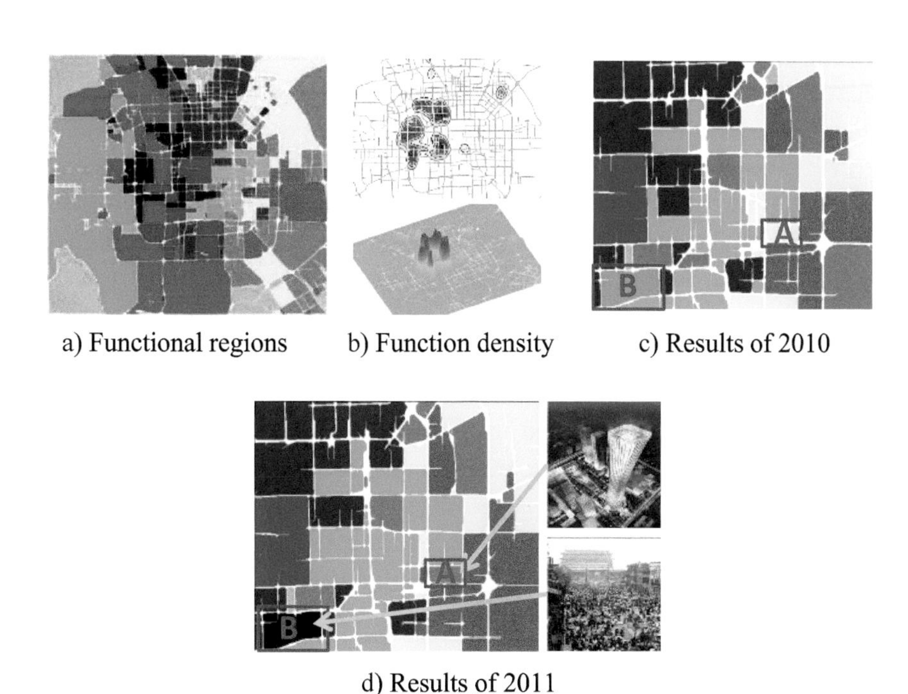

a) Functional regions b) Function density c) Results of 2010

d) Results of 2011

Figure 2.3
Identifying functional regions in a city using human mobility and POIs.

In their methodology, a city is first segmented into disjointed regions by major roads, such as highways and urban expressways. The functions of each region are inferred using a topic-based inference model [100], which regards a region as a document, a function as a topic, categories of POIs (e.g., restaurants and shopping malls) as metadata (e.g., authors, affiliations, and key words), and human mobility patterns (when people reach/leave a region and where people come from and leave for) as words. As a result, a region is represented by a distribution of functions, each of which is further denoted by a distribution of mobility patterns.

Here, human mobility can differentiate between the popularities of POIs belonging to the same category. It also indicates the function of a region (e.g., people leave residential areas in the morning and return in the evening). Specifically, the human mobility data was extracted from the GPS trajectories generated by over thirty-three thousand taxis over a period of three months in 2010 and 2012, respectively. Finally, nine kinds of functional regions are identified based on the clustering results and human labeling.

2.2.3 Detecting a City's Boundaries

The regional boundaries defined by governments may not respect the natural ways that people interact across space. The discovery of the *real* borders of regions according to the interaction between people can provide decision-support tools for policy makers, suggesting optimal administrative borders for a city [118, 119]. The discovery also helps government understand the evolution of a city's territory. The general idea of this category of research is to first build a network between locations based on human interaction (e.g., GPS tracking or phone call records) and then partition the network using some community discovery method that finds location clusters with denser interaction between locations in the cluster than between clusters.

Ratti et al. [118] propose a fine-grained approach to regional delineation, analyzing the human network inferred from a large telecommunications database in Great Britain. Given a geographical area and some measuring of the strength of links between its inhabitants, they partition the area into smaller, nonoverlapping regions while minimizing the disruption to each person's links. The algorithm yields geographically cohesive regions that correspond with administrative regions while unveiling unexpected spatial structures that have previously only been hypothesized in literature.

Rinzivillo et al. [119] address the problem of finding the borders of human mobility at the lower spatial resolution of municipalities or counties. They map vehicle GPS tracks onto regions to formulate a complex network in Pisa. A community discovery algorithm, namely Infomap [122], is then used to partition the network into nonoverlapped subgraphs.

2.2.4 Facility and Resource Deployment

To meet the increasing demands of urban life, we usually need to build new infrastructure and facilities, such as ambulance centers, bus stops, and charging stations for electrical vehicles. Considering many factors such as population, weather, and traffic conditions, where to place these new facilities to maximize their function remains a challenge to urban planners. To this end, we introduce four categories of models (originating from data science) that can be employed for resource deployment: (1) finding optimal meeting points, (2) maximizing coverage, (3) learning to rank candidates, and (4) minimizing uncertainty. In this section, we focus on their concepts and application scenarios. Detailed techniques of each model will be introduced in chapter 3.

2.2.4.1 Finding optimal meeting points This category of models aims to find a set of meeting points (from many candidates) that a group of objects can arrive at, or be reached from, with the (overall) minimum cost (e.g., travel time). For example, figure 2.4a shows the best meeting point that the seven objects can reach with the minimal travel time or distance in total. Figure 2.4b illustrates two optimal meeting points for these seven objects. Figure 2.4c shows an equivalent example in which those seven objects can be reached from the two optimal meeting points.

Now, we can see a more concrete example about the deployment of ambulance stations: emergency medical service, also known as ambulance services, provides a variety of services dedicated to out-of-hospital acute medical care, transport to definitive care, and other medical transport of patients with illnesses and injuries that prevent them from transporting themselves [138]. Given a certain number of ambulances, one of the major challenges that emergency service providers face is how to select appropriate locations for ambulance stations in order to maximize the ability to serve more patients.

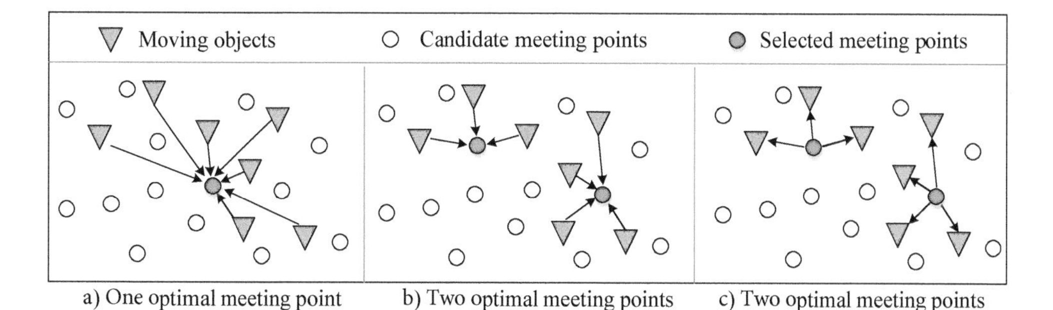

Figure 2.4
Finding optimal meeting points.

Aiming to minimize the average travel time to reach emergency requests in a given time interval, Li et al. [85] suggest nearly optimal locations for ambulance stations based on historical emergency requests and real traffic conditions. Evaluations based on real data collected in Tianjin show that the travel time to reach emergency requests can be reduced by 30 percent if ambulance stations are moved to the places suggested by the proposed algorithm. Prior research solely measured spatial proximity in a Euclidean space or static road network when placing a station. Other similar applications are possible. For example, by analyzing the trajectory data of massive electric taxis over a long period in Shenzhen city, Li et al. [84] propose a framework that could result in a reduction of about 26 percent in the average time required to find a charging station than current settings in Shenzhen.

2.2.4.2 Maximizing coverage This category of models aims to select a set of locations (from many candidates) that can cover the maximum number of objects. For example, as illustrated in figure 2.5a, based on vehicles' GPS trajectories, we want to deploy a charging station at two road intersections, respectively, so that the two intersections can cover the maximum number of electric vehicles. It turns out that n_1 and n_3 is the best combination, covering five vehicles in total. Likewise, as illustrated in figure 2.5b, according to users' check-in data, we want to place billboards at two regions that jointly have the maximum number of nonoverlapping users. As depicted in figure 2.5c, we expect to build two monitoring stations to observe the maximum number of migratory birds, based on their flying traces.

Enabling those applications is equivalent to solving a maximum coverage problem over trajectory data [82], which is NP-hard. By integrating advanced spatiotemporal indexing structures into a cloud-computing platform, Li et al. [82] efficiently find a *k*-location set with almost the maximum coverage of trajectories. Based on Li's method and platform, Liu et al. [90] search for a *k*-location set for placing billboards. They create an interactive visual data analytics system that allows urban planners to refine the locations suggested by the searching algorithm iteratively. Keeping humans in a data-mining loop, this system combines human knowledge with machine intelligence and fuses domain knowledge with data science.

2.2.4.3 Learning to rank This category of models was originally proposed in the community of information retrieval, learning from history to rank a set of candidates based on their properties. In urban planning, we can rank a set of locations so that the top-*k* best candidates are selected to deploy resources or facilities. For example, to open a profitable shopping mall, we select the best place by ranking a set of candidate

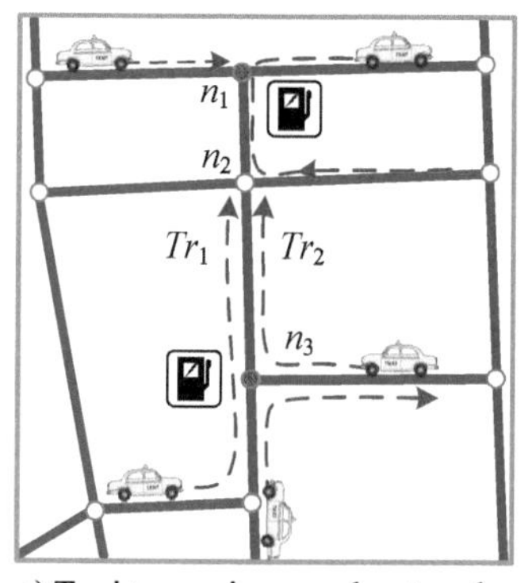

a) Taxi traces in a road network

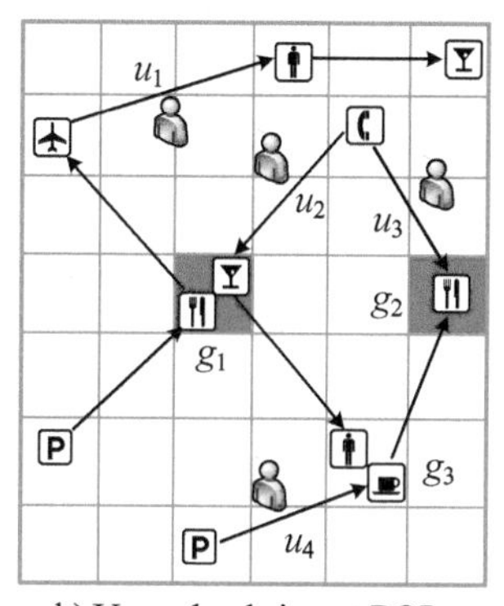

b) User check-ins at POIs

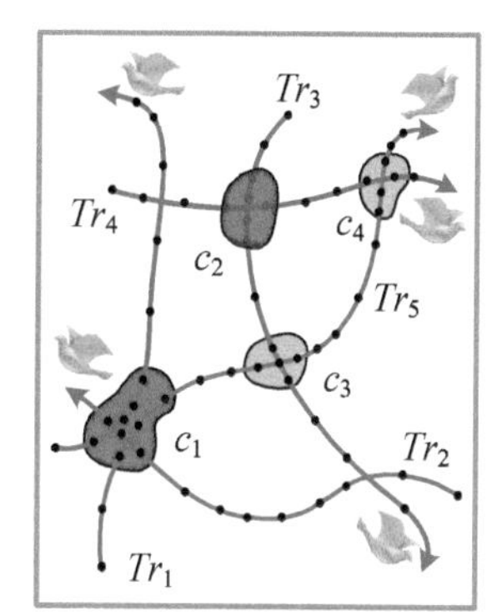

c) Bird trajectories

Figure 2.5
The maximum coverage problem.

locations based on multiple factors, such as their surrounding POIs, transportation facilities, traffic conditions, and neighborhood popularity. The ranking function of different factors can be learned from examining existing shopping malls and their historical revenues [47, 48].

2.2.4.4 Minimizing uncertainty This category of models aims to reduce the uncertainty of an urban-sensing system. Here is an example of fixed sensor-centric urban sensing. Many cities have deployed air quality–monitoring stations to inform people about the ambient air quality every hour. Because those stations are expensive in terms of deployment cost and maintenance, the number of stations that can be placed in a city is very limited. However, air quality is highly skewed in a city, changing over location and time nonlinearly. To address this issue, some machine-learning models [65, 182] have been proposed to infer the fine-grained air quality of those locations without a station. When dealing with a set budget for deploying new stations, determining where to place them to maximize monitoring efficacy remains a challenge. A key idea proposed in [65] is to minimize the uncertainty of air quality inference. That is, if the air quality of a location can be confidently inferred by a machine-learning model, it is not necessary to deploy a station there anymore. On the contrary, we should place new stations at those locations with air quality that cannot be inferred for certain—for example, if the inferred probability of air quality across five different categories is <0.20, 0.21, 0.19, 0.22, 0.18>.

Another example concerns active crowd sensing, which attempts to collect urban noise through users' mobile phones. In this urban-sensing scenario, Ji et al. [70] select proper participants and plan data-collecting routes for them according to their schedule, origins, and destinations. The general principle of choosing a participant and designing a route is to minimize the uncertainty of collected data in spatiotemporal spaces. More specifically, they prefer to select those users who can collect data in some locations with scarce data rather than those traveling to locations with sufficient data. The uncertainty of collected data is measured by a hierarchical information entropy [70] that calculates the probability of a location being sensed by a user in different granularities of a geographical space.

2.3 Urban Computing for Transportation Systems

By 2050 it is expected that 70 percent of the world's population will be living in cities. Municipal planners will face an increasingly urbanized and polluted world, with cities everywhere suffering from overly stressed road transportation networks. To tackle this challenge, different efforts have focused on improving people's driving experiences; enhancing the operation of existing taxi systems; and building more effective public transportation systems, including buses, subways, and shared bicycle schemes. The following subsections will review the literature of the aforementioned effort, respectively.

2.3.1 Improving Driving Experiences

A fast driving route saves both travel time and energy [68, 75]. Extensive studies have been done to learn historical traffic patterns [16, 63], estimate real-time traffic flows [63], and forecast future traffic conditions [24] on individual roads based on floating car data [110, 111], such as the GPS trajectories of vehicles, Wi-Fi, and global system for mobile communications (GSM) signals. However, research modeling and predicting citywide traffic conditions are still rare.

2.3.1.1 Suggesting the practically quickest driving directions T-Drive [156, 157, 159] is a system that provides personalized driving directions that adapt to weather, traffic conditions, and one's own driving habits. The first version of this system [159] only suggests the practically fastest path based on historical trajectories of taxicabs. The key insights consist of two parts: (1) GPS-equipped taxicabs can be regarded as mobile sensors continually probing traffic patterns on road surfaces, and (2) taxi drivers are experienced drivers who can find a really quick route based on their own knowledge, which incorporates not only the route's distance but also traffic conditions and the probability of accidents, meaning taxi trajectories imply traffic patterns and human intelligence. To deal with data sparsity (i.e., many road segments would not have taxis traversing), citywide traffic patterns are modeled as a landmark graph, as

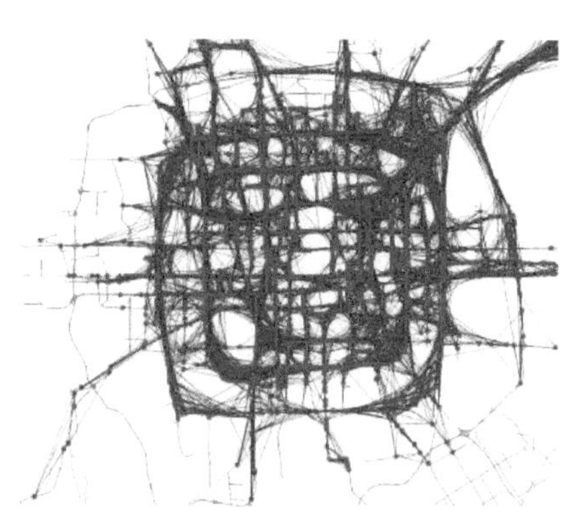

a) Landmark graph of Beijing ($k = 4000$)

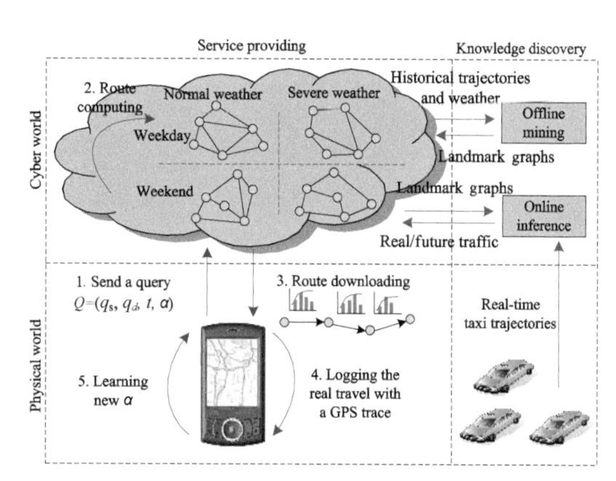

b) Framework of T-drive system

Figure 2.6
T-Drive: driving directions based on taxi trajectories.

shown in figure 2.6a, where red nodes are top-k road segments (entitled landmarks) frequently traveled by taxis, and each blue edge denotes the aggregation of taxis' commutes between two landmarks. The travel time of each landmark edge is estimated based on the taxi data using a VE (variance and entropy)-clustering algorithm. T-Drive uses a two-stage routing algorithm that first searches the landmark graph for a rough route (represented by a sequence of landmarks) and then connects these landmarks with a detailed route.

The second version of T-Drive [156, 157] mines historical taxi trajectories and weather condition records to build four landmark graphs corresponding to different weather and days, as shown in figure 2.6b. The system also calculates real-time traffic according to the recently received taxi trajectories and predicts future traffic conditions based on the real-time traffic and the corresponding landmark graph. A user submits a query consisting of a source q_s, a destination q_d, a departure time t, and a custom factor α from a GPS-enabled mobile phone. Here, α is a vector representing how fast the user typically drives on different landmark edges. α is set by a default value at the very beginning and is gradually updated based on the trajectories the user has actually driven. T-Drive gives a much more accurate estimate for each user and will adjust its suggestions if a person's driving habits change over time. As a result, the system saves five minutes per thirty-minute drive.

2.3.1.2 Travel-time estimation of a driving path VTrack [129] is a system for travel-time estimation based on Wi-Fi signals, measuring and localizing time delays. The system uses a hidden Markov model (HMM)–based map-matching scheme that interpolates sparse data to identify the most probable road segments driven by a user. A

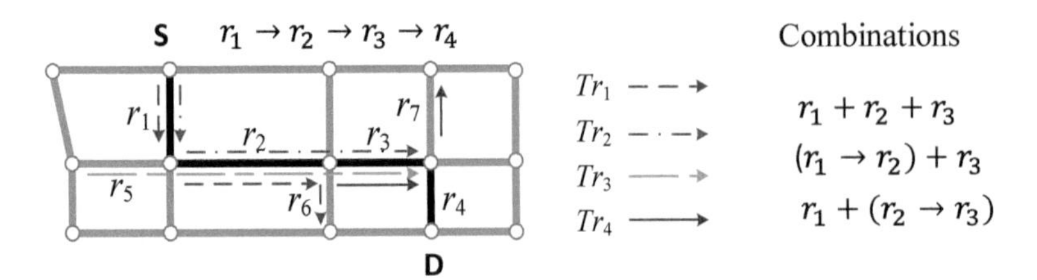

Figure 2.7
Travel time estimation–based sparse trajectories.

travel-time estimation method is then proposed to attribute travel times to those segments. The experiments show that VTrack can tolerate significant noise and outages in these location estimates and still successfully identify delay-prone segments.

Wang et al. [135] propose a citywide and real-time model for estimating the travel time of any path at a given time in a city, based on the GPS trajectories of vehicles received in present time slots and over a period of time as well as map data. Figure 2.7 illustrates an example in which we estimate the travel time of a path $r_1 \rightarrow r_2 \rightarrow r_3 \rightarrow r_4$ based on four trajectories: Tr_1, Tr_2, Tr_3, and Tr_3. There are three challenges if we want to solve the problem.

The first is data sparsity—that is, many road segments (e.g., r_4) may not have been traveled on by any GPS-equipped vehicles in the present time slot. In most cases, we cannot find a trajectory exactly traversing a query path (i.e., $r_1 \rightarrow r_2 \rightarrow r_3 \rightarrow r_4$) either.

Second, for the fragment of a path with trajectories (e.g., $r_1 \rightarrow r_2 \rightarrow r_3$), there are multiple ways of using (or combining) the trajectories (Tr_1, Tr_2, and Tr_3) to estimate the corresponding travel time. For instance, we can compute the travel time for r_1, r_2, and r_3, respectively, and then add the corresponding time costs to estimate the travel time for $r_1 \rightarrow r_2 \rightarrow r_3$. Alternatively, we can calculate the travel time of $r_1 \rightarrow r_2$ based on Tr_1 and Tr_2 and that for r_3 based on Tr_1, Tr_2, and Tr_3. Then, we can combine the travel times of the two parts to estimate that of $r_1 \rightarrow r_2 \rightarrow r_3$. Finding an optimal combination is a challenging problem, subject to a trade-off between the length of a path and the number of trajectories traversing the path (i.e., support). The ideal situation is to estimate the travel time of $r_1 \rightarrow r_2 \rightarrow r_3$ using many trajectories like Tr_2, covering the entire path. Such trajectories reflect the traffic conditions of an entire path, including intersections, traffic lights, and direction turns, hence there is no need to model these complex factors separately and explicitly.

However, as the length of a path increases, the number of trajectories traveling on the path decreases. Consequently, the confidence of the travel time (derived from few drivers) decreases. For example, Tr_2 might be generated by an uncommon driver or in an unusual situation, such as pedestrians crossing a street. On the other hand, using the concatenation of shorter subpaths can have more occurrences of trajectories on each subpath (i.e., having a high level of confidence in the derived travel time for each subpath). But this results in more fragments, across which the aforementioned complex factors are difficult to model. The more fragments a concatenation contains, the more inaccuracy a path's travel time could involve.

Third, we need to instantly answer users' queries, which may occur in any part of a city. This calls for an efficient, scalable, and effective solution that can enable a citywide and real-time travel-time estimation.

To address these challenges, Wang et al. model different drivers' travel times on different road segments in different time slots with a three-dimensional tensor. Combining them with geospatial, temporal, and historical contexts learned from trajectories and map data, they fill in the tensor's missing values through a context-aware tensor decomposition approach. They then devise and prove an object function to model the aforementioned trade-off, resulting in the most optimal concatenation of trajectories for an estimate through a dynamic programming solution. In addition, they propose using frequent trajectory patterns (mined from historical trajectories) to scale down the candidates of concatenation and a suffix tree–based index to manage the trajectories received in the current time slot. The proposed solution is evaluated based on extensive experiments using GPS trajectories generated by more than thirty-two thousand taxis over a period of two months. The results demonstrate the effectiveness, efficiency, and scalability of the method beyond baseline approaches, such as a simple summation of each individual road segment's travel time.

2.3.2 Improving Taxi Services

Taxi is an important commuting mode between public and private transportation, providing almost door-to-door traveling services. In major cities like New York City and Beijing, people usually wait for a nontrivial time before taking a vacant taxi, while taxi drivers are eager to find passengers. Effectively connecting passengers with vacant taxis is of great importance to reducing people's wait times, increasing taxi drivers' profits, and reducing unnecessary traffic and energy consumption. To address this issue, three categories of research have been performed.

2.3.2.1 Taxi dispatching systems These kinds of systems [10, 80, 123, 145] accept a user's booking requests and assign taxis to pick up the user. Most systems request people to book a taxi in advance, thereby reducing the flexibility of taxi services. Some real-time dispatching systems like Uber search for vehicles around a user based on the nearest neighbor principle of distance and time. The main challenge the system faces is the uncertainty of taxis' movements when searching for them [112, 145]. As shown in figure 2.8a, taxi *K* might be a better candidate than *X*, *Y*, or *Z* to pick up the user if we can determine that taxi *K* is moving toward the user while others are leaving the spatial range. The traffic conditions on routes should also be considered when estimating the travel time to pick up the user [46].

2.3.2.2 Taxi recommendation systems This category of systems solves the aforementioned problem from the perspective of recommendation [140, 163]. Ge et al. [55] have developed a mobile recommender system that can recommend a sequence of pickup points for taxi drivers or a sequence of potential parking positions. The goal of the system is to maximize the probability of business success and reduce energy consumption.

T-Finder [160, 163] provides taxi drivers with some locations and the routes to these locations, toward which they are more likely to pick up passengers quickly (along the routes or in these locations) and maximize the profit of the next trip. T-Finder also suggests to people some locations (within walking distance) where they can easily find vacant taxis. As illustrated in figure 2.8b, the probabilities of finding a vacant taxi on different road segments are visualized with different colors, in which *red* means very difficult, and *blue* means very likely. The parking places of taxis are also detected from the GPS trajectories of taxis, with an estimate of the number of taxis that will be arriving in the next half hour. The major challenge of this category of system is dealing with the data sparsity problem, such as calculating the probability of finding a vacant taxi on road segments without sufficient data.

2.3.2.3 Taxi ridesharing services Ridesharing is of great importance to saving energy and easing traffic congestion while satisfying people's commuting demands. Furuhata et al. [50] summarize three major challenges for ridesharing services: the design of attractive pricing and incentive mechanisms, proper ride arrangement, and the building of trust among travelers in online systems. There are two types of ridesharing: *static* ridesharing and *dynamic* ridesharing.

Static ridesharing Static ridesharing, often referred to as *carpooling*, has long been studied in operation research. Static ridesharing requests passengers to register their trip with identity information prior to a trip. Given a small group of people, researchers are able to solve static ridesharing optimally by using linear programming techniques [11, 22].

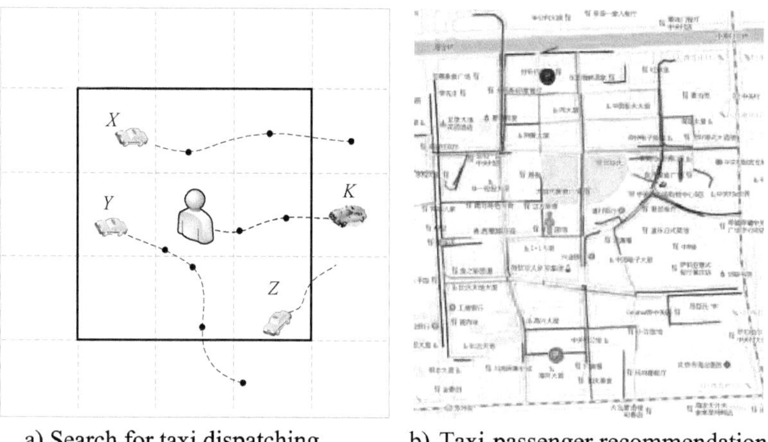

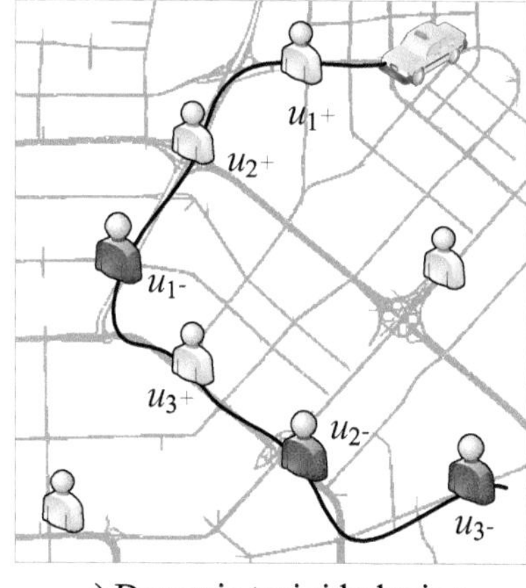

Figure 2.8
Three categories of systems for improving taxi services.

Dynamic ridesharing Unlike carpooling, where ride requests are known in advance, dynamic ridesharing is more challenging because ride requests are generated on the fly, and the routes of vehicles change continuously. Agatz et al. [3] review the optimization challenges of dynamic ridesharing systems. As a type of dynamic ridesharing, real-time taxi sharing is even more challenging, as the number of taxis and ride requests are much larger than general carpool services. In addition, there are other constraints, such as monetary ones, to consider for ridesharing. As illustrated in figure 2.8c, in a taxi-sharing service, a taxi is scheduled to sequentially pick up u_1 and u_2, drop off u_1, pick up u_3, and drop off u_2 and u_3, where + means a pickup, and − denotes a drop-off.

The taxi-sharing problem can be viewed as a special member of the general class of dial-a-ride problems (DARP). DARP originates from and has been studied in various transport scenarios, notably goods transport [42] and paratransit for handicapped and elderly personnel [15]. Existing works on DARP have primarily focused on static DARP, where all customer ride requests are known in advance. Since general DARP is NP-hard, only small instances (involving only a few cars and dozens of ride requests) can be solved optimally (often by resorting to integer programming techniques [34, 69]).

Large static DARP instances are usually solved by using the two-phase scheduling strategy [9, 35, 36, 141] with heuristics. Specifically, phase I partitions ride requests into some group and computes an initial schedule for delivering the riders in each group. In phase II, ride requests are swapped between different groups, aiming to find new schedules optimizing a predefined objective function. However, the two-phase strategy is not feasible for real-time taxi sharing. If the strategy is applied, the cloud will not serve a new request immediately. Instead, it needs to wait for more requests to make the second phase possible, which prolongs the response time of a request. In addition, the heavy computation load of phase II will further increase response time, resulting in many requests being unfulfilled.

Real-time ridesharing Some recent research efforts explore the real-time taxi-sharing problem. Early research, such as [22, 41, 56, 95], does not consider time and monetary constraints in a rideshare. T-Share [96, 97] is a large-scale dynamic taxi-sharing system that accepts passengers' real-time ride requests sent from smartphones and schedules taxis to pick up passengers via ridesharing subject to time, capacity, and monetary constraints. T-Share maintains a spatiotemporal index that stores the status of each taxi, consisting of the current location, the number of passengers on board, and the planned route to deliver said passengers. When receiving a ride request, T-Share first searches the index for a set of candidate taxis that are likely to satisfy a user's query based on some temporal constraints. A scheduling algorithm is then proposed to insert the query's trip into the existing schedule of each candidate taxi, finding the taxi that satisfies the query with the minimum increase of travel distance.

The system creates a win-win-win scenario, yielding significant social and environmental benefits. According to a simulation based on the taxi trajectories generated by over thirty thousand taxis in Beijing, compared with traditional nonridesharing, the technology is able to save 120 million liters of gasoline per year in Beijing, which can support 1 million cars for 1.5 months, save 150 million U.S. dollars, and reduce 246 million kilograms of CO_2 emissions. In addition, passengers save 7 percent in taxi fare and have a 300 percent higher chance of being served, while the income of taxi drivers increases by 10 percent.

There are two challenges in achieving such a taxi-sharing system. One is modeling the time, capacity, and monetary constraints for taxi trips. The other is the heavy computational load caused by the dynamics and large scale of passengers and taxis, which calls for efficient search and scheduling algorithms.

Santi et al. [124] introduce the notion of a shareability network to model the collective benefits of sharing as a function of passenger inconvenience. They apply this framework to a dataset of millions of taxi trips taken in New York City, showing that with a relatively low level of passenger discomfort, the cumulative trip length can be cut by 40 percent or more. There is also a branch of research considering user privacy [45, 58] and the social context of a user [83] when performing ridesharing.

2.3.3 Improving Bus Services

Public mass transit systems, coupled with integrated fare management and advanced traveler information systems, are considered key enablers to better mobility management.

2.3.3.1 Bus arrival time estimation In order to attract more riders, bus services need to be not only more frequent but also more reliable. Watkins et al. [136] have studied the impact of providing real-time bus arrival information directly to riders' mobile phones and have found that it reduces not only the perceived wait time of those already at a bus stop but also the actual wait time experienced by customers who plan their journey using such information. In other words, mobile real-time information has the ability to improve the experience of transit riders by making information available to them before they reach the stop.

In cases where GPS receivers have not been deployed on buses themselves, alternative solutions have been explored to gather the same information but in a cheaper and less intrusive manner. Zimmerman et al. [200] were the first to develop, deploy, and evaluate a system called Tiramisu, where commuters share GPS traces collected from the GPS receivers on their mobile phones. Tiramisu then processes incoming traces and generates real-time arrival time predictions for buses. As the GPS trajectories may be a mixture of different transportation modes—for example, first taking a bus and then walking—Zheng et al. [177, 181, 183] propose a method to infer a user's transportation modes (consisting of driving, walking, riding a bike, and taking a bus) in each segment of a trajectory. Once the trajectories have been classified by transportation modes, a more accurate estimate can be made for bus travel time or driving time prediction.

2.3.3.2 Bus route planning As urbanization continuously changes cities, it is essential for bus transit services to adapt their routes over time to keep meeting the mobility demands of their citizens. However, the pace at which bus routes are updated is much slower than the pace at which citizens' needs change. Bastani et al. [14] propose a data-centric approach to tackling the issue: they have developed a new minishuttle transit system called flexi, for which routes are flexibly derived from actual demand by analyzing passenger trip data from a large set of taxi trajectories.

In a similar vein, Berlingerio et al. [18] present an analysis of the anonymized and aggregated call detail records (CDRs) from Abidjan, in the Ivory Coast, with the aim of informing the planning of the city's public transit network using mobile phones. In this context, the resource-intensive transportation-planning processes prevalent in the West are not affordable; using mobile phone data to perform transit analysis and optimization represents a new frontier for transport planning in developing countries, where mobile phones have deep penetration, so that the anonymized flow data can be readily mined.

Chen et al. [28] aim to plan night bus routes by using taxi GPS traces. A two-phase approach for bidirectional night bus route planning has been proposed. In the first phase, taxi passengers' pickups and drop-offs are clustered into groups of a certain size, in each of which a candidate bus stop is selected. In the second phase, given the bus route origin, destination, candidate bus stops, and bus operation time constraints, a bus route graph is built and pruned iteratively. Finally, the best bidirectional bus route with the maximum number of passengers under the given conditions is selected.

2.3.4 Subway Services

Automated fare collection (AFC) systems (e.g., London's Oyster card, Seattle's Orca, Beijing's Yikatong, Hong Kong's Octopus, etc.) have been introduced and are now widely adopted in many metropolitan cities around the world. Apart from simplifying access to the city subway network of train services, these smart cards create a digital record every time a trip is made, which can be linked back to the individual traveler. Mining the travel data created as travelers enter and exit stations can give vast insight into the travelers themselves: their implicit preferences, travel times, and commuting habits.

Lathia et al. [78] mined AFC data with the aim of building more accurate travel route planners. They used data collected from the London Underground (tube) system, which implements electronic ticketing in the form of radio frequency identification (RFID)–based contactless smart cards (Oyster cards). Unlike some AFC systems, Oyster cards must be used when both entering and exiting stations. An in-depth analysis of two large datasets of London's tube usage demonstrates that there are substantial differences between travelers. Based on the insights, they automatically extracted features from the AFC data that implicitly capture information about a user's familiarity with a journey, a user's similarity to other travelers, and a user's journey context. Finally, they used these features to develop personalized travel tools with aims that can be formalized as prediction problems: (1) predicting personalized travel times between any origin and destination pair to provide users with accurate estimates of their transit time and (2) predicting and ranking the interest that individual travelers will have in receiving alert notifications about particular stations based on their past travel histories.

In a follow-up work, Ceapa et al. [25] perform a spatiotemporal analysis of the same historical Oyster card traces and discover that crowdedness is a highly regular phenomenon during the work week, with spikes occurring for rather short time intervals. They go on to build predictors of crowding levels, which can then be incorporated in advanced traveler information systems, to offer travelers more personalized and high-quality planning services.

Xue et al. [144] try to differentiate between tourists and normal commuters in subway systems based on trip data from smart cards. Besides that, Wu et al. [139] further extract each tourist's trajectory from his or her smart card data. These trajectories imply tourists' travel patterns and personal interests, which can enable intelligent travel recommendations.

2.3.5 Bike-Sharing Systems

As the world population grows and an ever-increasing proportion of people live in cities, designing, maintaining, and promoting sustainable urban mobility modes is becoming of paramount importance. Shared bike schemes [125] are one such example: their proliferation throughout the world's metropolises clearly reflects the belief that providing easy access to healthy (and quick) modes of transport will lead cities away from the congestion and pollution problems they currently face. Detailed records are often available about shared bikes' movements (from where/when a bike was taken to where/when a bike was returned), allowing researchers to analyze these digital traces to help end users, who may benefit from both understanding and forecasting how the system will be used when planning their own trips; transport operators, who may benefit from more accurate models of bike flows in order to appropriately load-balance the stations throughout the day; and urban planners, who can leverage flow data when designing social spaces and policy interventions.

2.3.5.1 Bike system planning The planning of a bike-sharing system usually consists of three steps: a feasibility study, a detailed design, and a business plan [54]. Dell'Olio et al. [38] present a feasibility study to estimate citizens' demand for bike sharing and the willingness to pay for using such a system. To build a new system, a location model is then proposed to estimate the optimal locations for bike-sharing stations. Lin et al. [87] propose a systematical method to estimate the number of bike stations needed and their locations. They also suggest paths that should be built between bike stations and recommend a particular path for a user given the origin and destination.

Bao et al. [12] suggest bike lane planning based on massive users' bike-riding trajectories in a stationless bike sharing system, considering the following three constraints. First, there is a budget constraint on the government side and a space constraint in current road networks. That is, we cannot build bike lanes on each and every road segment. Second, there is a trade-off between the number of riders who can be served and the continuous length of bike trips for each individual rider. If we try to serve an individual rider as long as possible, we may not be able to simultaneously serve as many people as possible. Third, considering riding experiences, we hope the planned bike lanes can be connected locally (in some areas) rather than brokenly distributed in different parts of a city. Of course, we cannot require that all bike lanes be globally connected throughout an entire city either. Thus, a more reasonable setting is to build bike lanes with a total length shorter than x km (i.e., a budget) in k areas, each of which has a locally connected bike network. Given the three constraints, this problem becomes NP-hard. Using a greedy expansion strategy and an objective function, the nearly optimal solution can be found within a reasonable time interval.

Chen et al. [29] predict the potential bike demands in different locations of a city following a regression-and-ranking methodology. In addition, they propose a semi-supervised feature selection method to select features extracted from heterogeneous urban datasets. García-Palomares et al. [53] estimate the spatial distribution of potential bike demands, determining the location and capacity of a station based on a location-allocation model. Research on business plans is currently very rare, focusing mainly on the trade-off between customer charges and operating costs, including system maintenance, reallocation, labor payment, and so on [54].

2.3.5.2 Bike usage patterns Froehlich et al. [46] were among the first to take a data-centric approach to shared bike systems by applying a host of data-mining techniques to uncover spatiotemporal trends in a city's data. They performed an in-depth analysis of thirteen weeks of Barcelona's Bicing system (Spain), clearly demonstrating the relationships between time of day, geography (especially clusters of stations within geographic areas of the city), and usage.

Kaltenbrunner et al. [74] performed a similar study of bike-sharing systems in Barcelona and Borgnat et al. in Lyon, France [19]. In these studies, the authors focus on temporal properties of bike station data in order to train and test classifiers that predict the state (availability of bikes) of each station. Nair et al. [104] analyze data from Paris' (France) Vélib', relating usage to rail station proximity: they uncover the relationship between bike usage and multimodal trips, thus providing key insights into station placement policy. Lathia et al. [79] analyze London's Cycle Hire scheme over two different three-month periods, deriving quantitative evidence of how an access policy change impacts bike usage across the whole city.

2.3.5.3 Bike usage prediction Affected by multiple complex factors, such as events, weather, and the bike demands of nearby stations, the bike usage in an individual station (e.g., S_1 and S_2 shown in figure 2.9a) is usually small and fluctuates over time almost randomly (see figure 2.9b and c). As a result, it is very difficult to accurately predict bike usage at an individual station level.

To address this issue, Li et al. [86] propose a bipartite clustering algorithm to cluster bike stations into groups (e.g., C_1, C_2, and C_3 depicted in figure 2.9a) based on their geographical locations and bike usage patterns. The bike usage in a cluster like C_1 becomes quite stable, showing a certain level of periodicity (see figure 2.9d). The total number of bikes that will be rented in a city is predicted by a gradient boosting regression tree (GBRT), as shown in figure 2.9e. A multisimilarity-based inference model is then proposed to predict the rent proportion across clusters and the intercluster transition, as illustrated in figure 2.9f, based on which the number of bikes rented from/returned to each cluster can be easily inferred. This model is evaluated by the bike-sharing systems in New York City and Washington, DC, respectively, resulting in a better performance beyond baseline approaches.

Following Li and Zheng's research [86], Yang et al. [147] propose a spatiotemporal bike mobility model and a traffic prediction mechanism to forecast the bike usage at each individual station over a half-hour period. The proposed system is evaluated based on bike-sharing data from the city of Hangzhou, with over twenty-eight hundred stations.

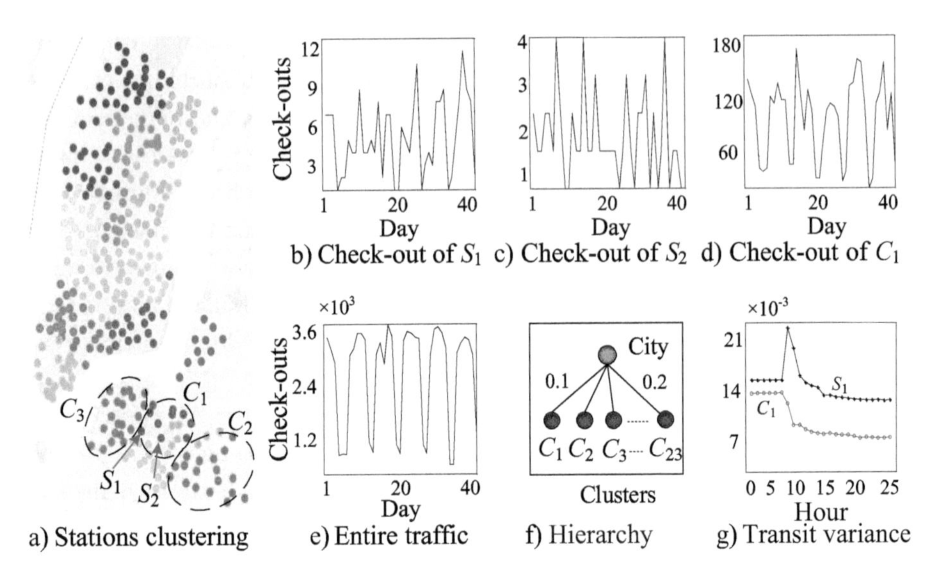

Figure 2.9
Traffic prediction in a bike-sharing system.

2.3.5.4 System operation Because of the spatially and temporally skewed distribution of bike usage, bike reallocation between stations is necessary to meet customers' bike demands. Currently, operators usually send trucks to reallocate bikes between different stations by monitoring real-time bike usage at each station. A body of research (e.g., [17, 27, 33, 91]) formulates this problem as a constrained optimization, designing reallocation routes for trucks based on the capacity of a truck and the unbalanced distributions of each station. Of note is the proposal by Liu et al. [91] to cluster stations into groups and then design routes that minimize the total travel distance of trucks in a cluster using mixed-integer nonlinear programming.

2.4 Urban Computing for the Environment

Without effective and adaptive planning, urbanization's rapid progress will become a potential threat to cities' environments. We have witnessed an increase in different pollutants in urban environments, such as air quality issues, noise, and rubbish, around the world. Protecting the environment while modernizing people's lives is of paramount importance in urban computing. This section will introduce some technical works on tackling air pollution and urban noise, as well as protecting water quality.

2.4.1 Air Quality

2.4.1.1 Outdoor air quality Information about urban air quality, such as concentrations of PM2.5, is of great importance to protecting human health and controlling air pollution. Many cities are monitoring PM2.5 by building ground-based air quality measurement stations. While covering a large spatial area, cities have only a limited number of measuring stations due to the high cost of building and maintaining such stations. Unfortunately, air quality varies by location nonlinearly and significantly, impacted by multiple complex factors including meteorology, traffic flow, and land use. As a result, we do not really know the air quality of a location without a measuring station.

Monitoring air quality Advances in mobile communication and sensing technologies have led to the proliferation of crowdsourcing-based applications, which decompose a complex problem into small tasks and distribute them to a network of users. The returns from individual users formulate collective knowledge that can solve the problem.

The Copenhagen Wheel [128] is a project that installs environmental sensors in bike wheels to sense the fine-grained environmental data of a city, including temperature, humidity, and the concentration of CO_2. The human energy required to ride a bike is converted into power to support the operation of on-bike sensors. In addition, the wheel can communicate with a user's mobile phone, through which the collected information is sent to a back-end system.

Likewise, Devarakonda et al. [39] present a vehicular-based approach for measuring fine-grained air quality in real time. They have devised a mobile device consisting of a GPS receiver, a CO_2 sensor, and a cellular modem. By installing such a device on multiple vehicles, they can monitor the concentration of CO_2 throughout a city.

Though it has tremendous potential, monitoring the environment through crowdsourcing only works well for a few gases, such as CO_2 and NO. Devices that measure aerosols, like PM2.5 and PM10, are not easily portable for an individual. Moreover, these devices need a relatively long period of sensing time, often at least one hour, before generating an accurate measurement.

Another branch of research [59] estimates the traffic flow on road surfaces based on floating car data and then calculates the emission of vehicles based on some empirical equations formulated by environmentalists. This is a promising approach to estimating air pollution near roads, but it cannot reveal the air quality of an entire city, as the emissions from vehicles are only one source of pollution.

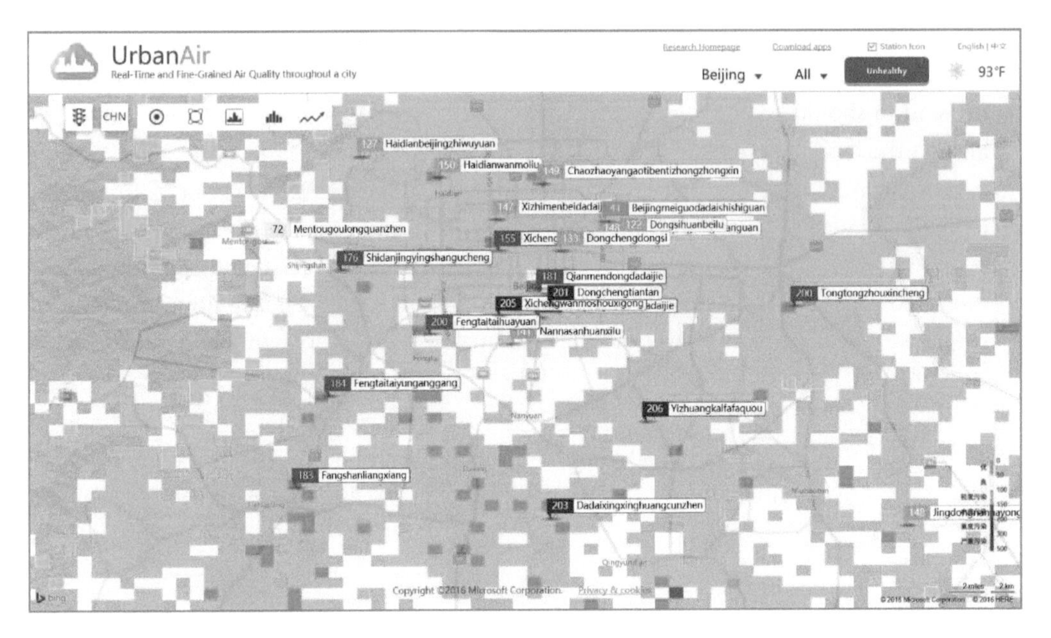

Figure 2.10
Monitoring real-time and fine-grained air quality using big data.

Air quality inference Different from existing solutions, Zheng et al. [176, 182] infer real-time and fine-grained air quality information throughout a city (as demonstrated in figure 2.10) based on (historical and real-time) air quality data reported by existing monitoring stations and a variety of datasets collected in a city, such as meteorology, traffic flow, human mobility, road network structure, and POIs. Instead of using classical physical models that explicitly combine factors in a formula based on empirical assumptions, they approach this problem from a big-data perspective—that is, using data mining and machine-learning techniques to build a network between diverse data and air quality indexes (more details in section 9.4.1). The fine-grained air quality can inform governments' decision-making on pollution control. It can also help people figure out when and where to go jogging—or when they should shut the window or put on a face mask. A system called Urban Air that uses this sort of technology has been deployed in over three hundred Chinese cities, with publicly available data at http://urbanair.msra.cn/.

Monitoring station deployment Following this research theme, Hsieh et al. [65] propose a probabilistic graphical model–based method to infer fine-grained air quality. They further suggest locations for building new monitoring stations based on the entropy of inferred air quality in a given location. The bigger the entropy of the inferred air quality in a location, the higher the uncertainty for prediction. As a result, a monitoring station should be deployed in such a location rather than using predictive models to infer the corresponding air quality.

Forecasting air quality The Urban Air system can also forecast air quality at each monitoring station forty-eight hours in advance [191] by using a data-driven method that considers current meteorological data, weather forecasts, and the air quality data of the station and of other stations within a few hundred kilometers. Figure 2.11 presents the user interface of the prediction function, where an hourly prediction is generated for each of the next six hours and a max-min range of air quality for the next seven to twelve, thirteen to twenty-four, and twenty-five to forty-eight hours, respectively. The number shown on the top of each time interval denotes the accuracy of predictions over forty-eight hours.

Diagnosing root causes of air pollution Zhang et al. [164] examine the problem of mining spatial coevolving patterns (SCPs) from air quality data. An SCP denotes a group of sensors that are spatially correlated and coevolve frequently in their readings.

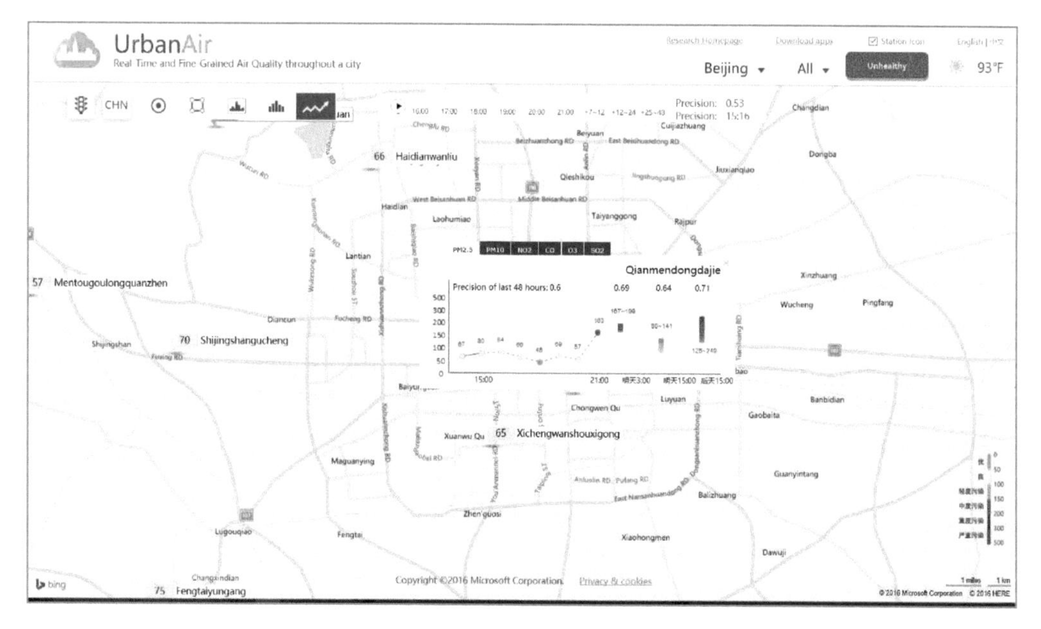

Figure 2.11
Forecasted air quality based on big data.

SCPs of air quality can help infer the propagation path of air pollution in geographical spaces.

Zhu et al. [199] apply a causality inference model to the problem of identifying the root cause of air pollution. Later, Zhu et al. [198] combine the coevolving pattern-mining techniques with a Gaussian Bayesian network to infer the spatiotemporal causality of air pollutants among different locations. For instance, when the wind speed is less than 5 m/s, the high concentration of PM10 is mainly caused by the high concentration of SO_2 in Zhangjiakou and that of PM2.5 in Baoding. Further, the high concentration of PM2.5 in Baoding is mainly caused by NO_2 from Hengshui and Cangzhou. The method first finds SCPs between the air pollutants of each pair of locations, calculating the correlation between them. Then the top-N most correlated factors (to an air pollutant) are selected to construct a Bayesian network. The combination of the two techniques reduces the complexity of a Bayesian network while enhancing the accuracy of the inference results.

2.4.1.2 Indoor air quality Indoor air quality has been monitored by a wide array of sensors [72]. There is also a body of research on predicting indoor air quality. For example, [73] present an approach to predicting indoor air pollution generated by stove emissions using a Monte Carlo model. Other mathematical models [107] are proposed for predicting indoor air quality based on smoking activity.

The sensing and prediction of indoor air quality have been used to control the operation of heating, ventilation, and air conditioning (HVAC) in a building in order to minimize energy consumption [2, 31]. Chen et al. [31] introduce an indoor air quality–monitoring system deployed on four Microsoft campuses in China. The system is comprised of sensors deployed on different floors of a building, a cloud collecting and analyzing the data from the sensors, public air pollution information, and clients that display the real-time air quality data of both outdoor and indoor environments to end users. The system provides users with indoor air quality information that can inform people's decision-making, such as when to work out in a gym or turn on an additional air filter in an office. The gap between the concentration of PM2.5 in outdoor and indoor environments can measure the effectiveness of an HVAC system in filtering PM2.5.

The system also integrates outdoor air quality information with indoor measurements to adaptively control HVAC settings with a view toward optimizing runtimes with regard to energy efficiency and air quality conservation. Using a neural network–based approach, the system can even predict the purification time that an HVAC needs to reduce the concentration of indoor PM2.5 to below a healthy threshold based on six factors, such as the concentration of outdoor/indoor PM2.5, barometer pressure, and humidity. Given the purification time and the time that people start working in a building, the number of hours that an HVAC system should be turned on ahead of its original schedule can be recommended, thus saving a great deal of energy. Extensive experiments using three-month data demonstrate the advantage of the approach beyond baseline methods—for example, linear regression.

2.4.2 Noise Pollution

The compound functions of a city and its complex settings that incorporate different infrastructures and millions of people inevitably generate a lot of environmental noise. As a result, a large number of people around the world are exposed to high levels of noise pollution, which can cause serious illnesses ranging from hearing impairment to negatively influencing productivity and social behavior [116].

2.4.2.1 Monitoring urban noises As an abatement strategy, a number of countries, such as the United States, the United Kingdom, and Germany, have started monitoring noise pollution. They typically use a noise map (a visual representation of the noise level of an area) to assess noise pollution levels. The noise map is computed using simulations based on inputs, such as traffic flow data, road or rail type, and vehicle type. Since the collection of such input data is very expensive, updating these maps usually takes a long period of time. Silvia et al. [127] assess environmental noise pollution in urban areas using wireless sensor networks. To deploy and maintain a citywide sensor network, however, especially in major cities like New York, is very expensive in terms of money and human resources.

Another solution is to take advantage of crowdsourcing, where people collect and share their ambient environmental information using a mobile device. For example, NoiseTube [105] presents a person-centric approach that leverages the noise measurements shared by mobile phone users to paint a noise map in a city. Based on NoiseTube, D'Hondt and Stevens [40] conducted a citizen science experiment for noise mapping a 1 km^2 area in the city of Antwerp. Extensive calibration experiments were also carried out, investigating both frequency-dependent and white noise behavior. The main objective of this experiment was to investigate the quality of the obtained noise map by participatory sensing, compared with official simulation-based noise maps.

In [116] an end-to-end, context-aware, noise-mapping system called Ear-Phone is designed and implemented. Different from phone users actively uploading their measurements in [40, 105], an opportunistic sensing approach is proposed in which noise level data is collected without informing smartphone users. One major problem solved in this paper is the classification of phone-sensing contexts (i.e., in a pocket [bag] or in a hand), which is related to the accuracy of the sensed data. To recover a noise map from incomplete and random samples, Rana et al. [117] further study a number of different interpolation and regularization methods, including linear interpolation, nearest neighbor interpolation, Gaussian process interpolation, and L1-norm minimization methods.

2.4.2.2 Urban noise inference Modeling citywide noise pollution is actually much more than just measuring the intensity of noises, as the measurement of noise pollution also depends on people's tolerance to noise, which changes with the time of day and from person to person. For example, people's tolerance to noise at night is much lower than during the daytime. A less loud noise at night may nevertheless be considered heavier noise pollution. Consequently, even if we could deploy sound sensors everywhere, diagnosing urban noise pollution solely based on sensor data is insufficient. Furthermore, urban noises are usually a mixture of multiple sources. Understanding the composition of urban noises (e.g., that during evening rush hour, 40 percent of noise is from pub music, 30 percent from vehicular traffic, and 10 percent from construction) is vital to reducing noise pollution.

Since 2001, New York City has operated the 311 platform to allow people to register nonemergency city disturbances by using a mobile app or making a phone call; noise is the system's third-largest category of complaints. Each complaint about noise is associated with a location, a time stamp, and a fine-grained noise category, such as loud music or construction noises. Figure 2.12 paints the 236-day 311 complaints about noises on a digital map, where the height of a bar denotes the number of complaints in a location. For example, we can see that Lower Manhattan was suffering from construction and loud music/party noise.

This 311 data is actually the result of "human as a sensor" and "crowd sensing," where each individual contributes her or his own information about ambient noises, helping to diagnose the noise pollution throughout a city when used aggregately. More specifically, the more 311 calls are made in a location, the louder the real noise could be in the location. In addition, the noise categories tagged by a complainer can help analyze the composition of noises in a location. For example, if one hundred 311 complaints have been received in a location over the recent month, consisting of fifty complaints about traffic, thirty about construction, and twenty about loud music, the composition of noise in the location could be traffic: 50 percent, construction: 30 percent, and loud music: 20 percent.

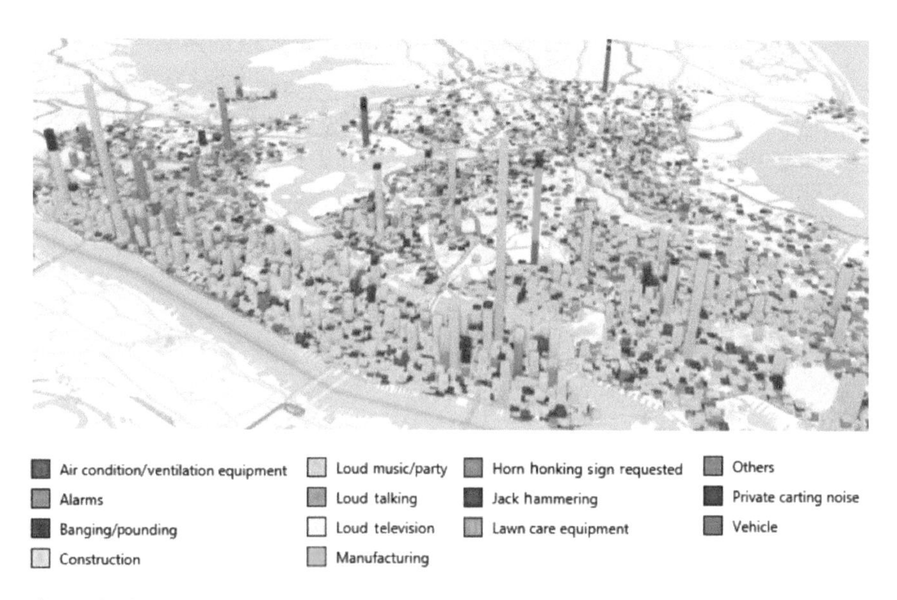

Figure 2.12
Complaints of noise in New York City (May 23, 2012–January 13, 2014).

However, the 311 data is rather sparse, as we cannot expect people to report the noise situation around them anytime and anywhere. Sometimes, they may be too busy or too lazy to make a phone call, even if they have been disturbed by noise. To address this issue, Zheng et al. [184] infer the fine-grained noise situation (consisting of a noise pollution indicator and the composition of noises) at different times of day for each region of New York City using the 311 complaint data together with social media, road network data, and POIs. Based on the overall noise pollution indicator, we can rank locations at different time intervals such as 0 a.m. to 5 a.m. on weekdays and 7 p.m. to 11 p.m. on weekends, as illustrated in figure 2.13a; the darker the color of a region, the heavier the noise pollution the region suffers from. Alternatively, we can rank locations by a particular noise category like construction, as depicted in figure 2.13b. We can also check the noise composition of a particular location changing over time, like Times Square, as shown in figure 2.13c.

They model the noise situation in New York City with a three-dimensional tensor, where the three dimensions stand for regions, noise categories, and time slots, respectively. By filling in the missing entries of the tensor through a context-aware tensor decomposition approach, they recover the noise situation throughout New York City. The noise information can not only facilitate an individual's lifestyle (e.g., to help find a

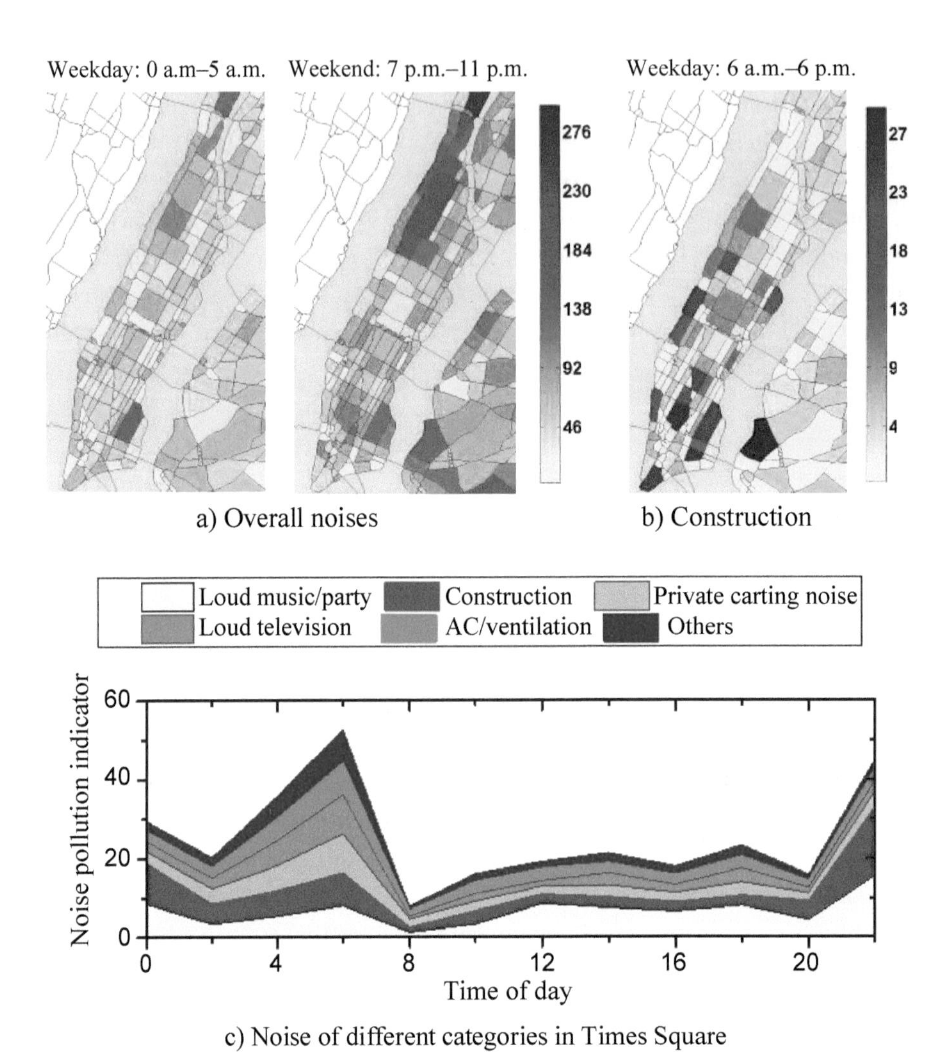

Figure 2.13
Diagnosing noise pollution in New York City.

quiet place to settle down), but can also inform government officials' decision-making on reducing noise pollution.

2.4.3 Urban Water

Urban water is a vital resource that affects various aspects of human health and security. People living in cities are increasingly concerned about water properties (e.g., water quality, flow, and pressure) and water-related events (e.g., urban flooding and sudden rainstorms), calling for technologies that can model or predict these issues. The research in this area mainly deals with the issues relating to surface water, underground water, and pipe water, respectively.

2.4.3.1 Surface water–related issues Surface water is water on planet surfaces such as that found in rivers, lakes, wetlands, or rain runoff, which can be contrasted with groundwater and atmospheric water. In surface water research, there is significant interest in modeling rainfall-runoff processes and tackling hydrological issues, such as water flow prediction and flood forecasting.

Surface runoff (also known as *rainfall runoff*) is the flow of water that occurs when excess storm water or other sources flow over the earth's surface. Generally, the rainfall-runoff process is believed to be highly nonlinear over time and location and thus is not easily described by simple models [66]. Amorocho et al. [6] evaluate two distinct approaches to the problem of establishing the relationship between rainfall and streamflow. One is referred to as *physical hydrology*, and the other is called *systems investigation*. Duan et al. [42] present an effective and efficient global optimization for conceptual rainfall-runoff models. Over the last decade, many physical-based hydrological models have been developed that focus on flow forecasting and are both physically based and spatially distributed, such as the European Hydrological System (SHE) [1].

Conversely, subject to constraints posed by the availability of data instead of physical considerations [101], several data-driven approaches have been widely applied to model these dynamic processes. For instance, Hsu et al. [66] present a three-layer feed-forward artificial neural network (ANN) model for simulating and predicting the rainfall. Toth et al. [130] prove that rainfall forecasting based on ANN achieves better performance than *k*-nearest neighbors (KNN) and auto regressive moving average (ARMA). Bray et al. [21] employ a support vector machine (SVM) to forecast rainfall runoff, generating better performance than ANN-based methods. They also explore the relationship among various model structures and kernel functions. Yu et al. [153] also model the rainfall-runoff process by SVM and demonstrate that the SVM model outperforms the neural network model.

2.4.3.2 Groundwater-related works Groundwater is the water present beneath the earth's surface in soil pore spaces and in the fractures of rock formations, constituting an important component of many water resource systems by supplying water for domestic, industrial, and agricultural uses [8].

Groundwater hydraulic management models enable the determination of optimal locations and pumping rates for numerous wells under a variety of restrictions placed on local drawdown, hydraulic gradients, and water production targets [99]. Aguado et al. [4] suggest that the physical behavior of the groundwater system is an integral part of an optimization model. They then reformulate the problem as a "fixed charge problem" where costs due to pumping plus fixed costs due to well installation are considered [5, 57].

To model the groundwater flow, Arnold et al. [8] add a comprehensive groundwater flow and height model to an existing basin-scale surface water model and validate this model on a 471 km^2 watershed near Waco, Texas. Based on a hybrid multiobjective paradigm of evolutionary polynomial regression, Giustolisi et al. [57] introduce a modeling approach aimed at the management of groundwater resources, including testing rainfall depth and water table depth. Without focusing on the hydraulic properties of groundwater, Liu et al. [89] apply a factor analysis method to explore the correlation between groundwater quality and thirteen hydrochemical parameters.

2.4.3.3 Pipe water quality Since urban water is closely related to people's lives, research on pipe water quality (e.g., residual chlorine) is of great importance to managing disinfectant concentrations and protecting public health. However, modeling the pipe water quality throughout a pipe network is challenging due to its spatial and temporal variability.

Several experiments indicate that chlorine decay can be characterized as a combination of first-order reactions in the bulk liquid and mass transfer–limited reactions at the pipe wall [131]. In general, the physical-based models have been well employed in many water distribution systems by utilizing the first- or higher-order decay kinetics to describe twofold chlorine consumption: bulk water and pipe wall [120, 121]. There are many applications of first-order kinetics using EPANET software on a real network to model chlorine decay [23, 103]. To extend these models, Clark et al. [32] propose two second-order terms to predict chlorine loss or decay since it travels throughout the network. According to different functional dependencies between bulk water and wall reaction, an intuitive idea is to model free chlorine rates separately. Pipe materials are classified as having either high reactivity or low reactivity. Hallam et al. [60] show that wall decay rates for the former are limited by chlorine transport and for the latter by pipe material characteristics. Despite the wall reaction, Powell et al. [113] investigate many factors that influence bulk decay, based on over two hundred determinations of bulk chlorine decay against time.

In terms of data-driven approaches, Liu et al. [94] propose a novel multi-view multi-task learning framework to forecast water quality at about thirty monitoring points in a pipe network over a couple of hours. In this framework, different sources of urban data, such as road networks, POIs, and meteorology, are regarded as spatial or temporal views of water quality, respectively. In addition, the prediction at each monitoring point is deemed a task. The correlation between different tasks is measured by the connectivity between the two monitoring points in the pipe network. Experiments show that the multitask multi-view framework outperforms classical prediction models and other machine-learning algorithms.

Besides modeling chlorine decay, Ostfeld et al. [106] combine a "hurdle-race" and a hybrid genetic-*k*-nearest neighbor algorithm (GA-KNN) into a calibration model for a two-dimensional hydrodynamic and water quality (e.g., nitrogen, phosphorus, excluding chlorine) model.

2.5 Urban Computing for Urban Energy Consumption

The rapid progress of urbanization is resulting in increasingly more energy consumption, calling for technologies that can sense city-scale energy cost, improve energy infrastructure, and save energy.

2.5.1 Gas Consumption

Zhang et al. [166, 167] propose estimating the number of vehicles in each gas station in a timely manner using taxicabs as sensors. Such information can not only suggest gas stations with a shorter waiting time to drivers but can also help plan gas station locations more efficiently. Intuitively, in the case that gas stations have been overbuilt in a region with few people refueling cars, we can reduce the operation time of some stations (e.g., by only opening them for half a day). Furthermore, this information can estimate how many liters of gas have been put into vehicles over a certain time interval given the statistics on the capacity of the fuel tanks of vehicles in a city.

In the proposed method, taxicabs' refueling events at gas stations are first detected from their GPS trajectories, as illustrated in figure 2.14. The time that a taxicab spends on refueling at a station is used to estimate the length of a queue (waiting for refueling) and then the number of vehicles at the station. Later, a context-aware tensor decomposition method is proposed to estimate the number of vehicles for those stations that have not been visited by taxicabs recently.

Shang et al. [126] instantly infer the gas consumption and emissions of vehicles traveling on a city's road network in a certain time slot using GPS trajectories from a sample of vehicles (e.g., taxicabs), as illustrated in figure 2.15a. The knowledge can be used not only to suggest cost-efficient driving routes but to identify road segments where gas has been significantly wasted. In the meantime, the instant estimation of the

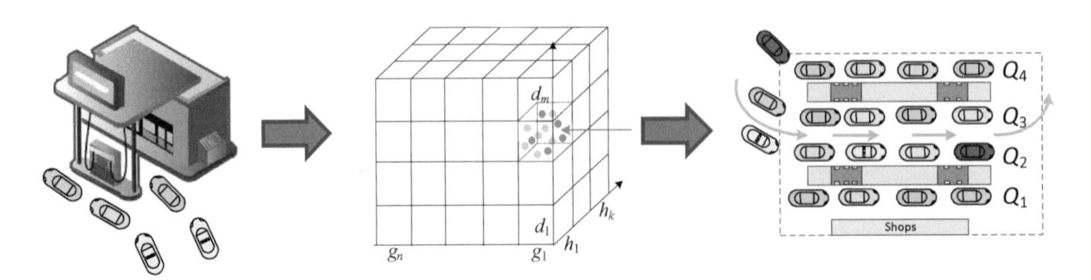

Figure 2.14
Crowd sensing urban refueling behavior with GPS-equipped taxis.

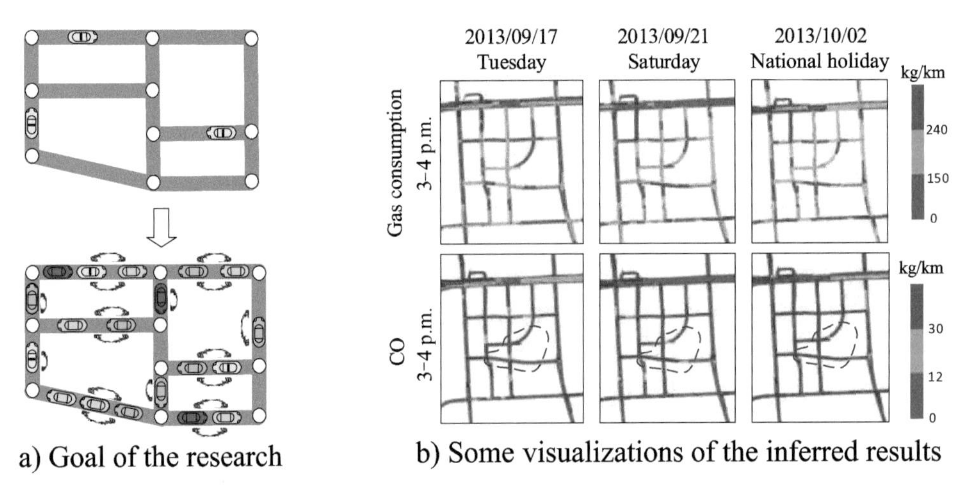

Figure 2.15
Inferring gas consumption and pollution emissions from vehicles based on sparse trajectories.

polluting emissions from vehicles can enable pollution alerts and, in the long run, help diagnose the root cause of air pollution.

To achieve this goal, they first compute the travel speed of each road segment using recently received GPS trajectories. As many road segments are not traversed by trajectories (i.e., data sparsity), a travel speed estimation model is proposed, based on a context-aware matrix factorization approach. The model leverages features learned from other data sources (e.g., map data and historical trajectories) to deal with the data sparsity problem. A traffic volume inference model (TVI) is then proposed to infer the number of vehicles passing each road segment per minute. TVI is an unsupervised dynamic Bayesian network that incorporates multiple factors, such as the travel speed, weather conditions, and geographical features of a road. Given the travel speed and traffic volume of a road segment, the gas consumption and emissions can be calculated based on existing environmental theories.

Figure 2.15b demonstrates the gas consumption and emissions of NO_x around the Zhongguancun neighborhood, a mixed area with many companies and entertainment venues, on three different days, respectively. In the 3 p.m. to 4 p.m. time slot, before evening rush hour, this area has less gas consumption on the workday than the weekend and holidays because people are still working indoors. During weekends and holidays, many people travel to this region for entertainment purposes (e.g., shopping or seeing a movie), leading to more energy consumption and emission of CO, denoted by the red segments. A movie theater, a supermarket, and two shopping centers are located in the region marked by the broken curve.

2.5.2 Electricity Consumption

The efficient integration of energy from renewable sources and meeting the increased demand resulting from an increase in the electrification of vehicles and heating are key to the sustainable use of the electricity supply. In order to optimize residential energy usage, intelligent demand response mechanisms are needed to shift energy usage to periods of low demand or to periods of high availability of renewable energy. Intelligent algorithms, implemented at either the device level or at a community/transformer level, enable devices to meet individual device and user policies as well as stay within community-assigned energy usage limits.

In [44], each electric vehicle (EV) within a community is controlled by a reinforcement-learning agent further supported by a short-term load prediction algorithm [98]. Each agent's local goals are to minimize the charging price (which is dynamic and directly proportional to current energy demand) and to meet desired user utility (e.g., have an EV's battery 80 percent charged in time for departure). Each agent also wants to keep the community transformer level under a target limit as a goal (by minimizing, for example, the number of vehicles charging during peak periods). Demand is dynamically repredicted if real-time monitoring shows actual deviations from predicted demand.

Galvan-Lopez et al. [51] propose an alternative approach, where instead of each vehicle agent making its own decisions, a globally optimal charging schedule adapts to demand using genetic algorithms and communicates with EVs. In [62], intelligent set point control algorithms at the transformer level send out signals to controllable devices (e.g., EVs or water heaters) indicating either a probability they should use to determine whether they should be charging/on at any particular point or the degree to which each device's variable power chargers should be turned on. This enables the fine-grained control of device demand to fill up the gaps between uncontrollable electric load and the target transformer load to even out the overall energy demand.

Momtazpour et al. [102] present a framework to support charging and storage infrastructure design for EVs. A coordinated clustering technique is proposed to work with network models of urban environments to aid in the placement of charging stations for EV deployment support. Issues that have been taken into account include (1) the prediction of EV charging needs based on their owners' activities, (2) the prediction of EV charging demands at different urban locations and the available charge of EV batteries, (3) the design of distributed mechanisms that manage the movements of EVs to different charging stations, and (4) the optimization of the charging cycles of EVs to satisfy users' requirements while maximizing vehicle-to-grid profits.

2.6 Urban Computing for Social Applications

2.6.1 Concepts of Location-Based Social Networks

Though there are already many social networking services on the Internet, in this section we focus on introducing location-based social networks (LBSNs), which are formally defined as follows in [171, 172]:

> A location-based social network (LBSN) does not only mean adding a location to an existing social network so that people in the social structure can share location-embedded information, but also consists of the new social structure made up of individuals connected by the interdependency derived from their locations in the physical world as well as their location-tagged media content, such as photos, video, and texts. Here, the physical location consists of the instant location of an individual at a given timestamp and the location history that an individual has accumulated in a certain period. Further, the interdependency includes not only that two persons co-occur in the same physical location or share similar location histories but also the knowledge, e.g., common interests, behavior, and activities, inferred from an individual's location (history) and location-tagged data.

LBSNs bridge the gap between users' behaviors in the digital and physical worlds [37], which sufficiently matches the nature of urban computing. In a location-based social network, people can not only track and share the location-related information of an individual but also leverage collaborative social knowledge learned from user-generated and location-related content, such as check-ins, GPS trajectories, and geo-tagged photos [180, 186]. Examples of LBSNs include the widely used Foursquare and a research prototype called GeoLife [178, 190]. With LBSNs, we can understand users and locations, respectively, and explore the relationship between them. More details about LBSNs can be found in [171, 189], as well as a survey on recommendations in LBSNs [13].

Here, we discuss the research on LBSNs from the perspectives of users and locations, respectively.

2.6.2 Understanding Users in Location-Based Social Networks

2.6.2.1 Estimating user similarity An individual's location history in the real world implies, to some extent, her interests and behaviors. Accordingly, people who share similar location histories are likely to have common interests and behaviors. The similarity between users inferred from their location histories can enable friend recommendations [81], which connect users with similar interests even when they may not have known each other previously, and community discovery, which identifies a group of people sharing common interests.

To better estimate the similarity between users, more information, such as the visiting sequences between locations, the geospatial granularity of a location, and the popularity of a location, is considered in [193]. In addition, in order to be able to calculate the similarity of users living in different cities (i.e., having little geospatial overlaps in users' location histories), Xiao et al. [142, 143] extend Zheng's research from physical locations into the semantic space of locations by considering the categories of POIs in the location visited by a user.

2.6.2.2 Finding local experts in a region With users' locations, we are able to identify local experts who have richer knowledge about a region (or a topic like shopping) than others. Their travel experiences (e.g., the locations where they have been) are more accountable and valuable for travel recommendation. For instance, local experts are more likely to know about high-quality restaurants than some tourists [194].

2.6.2.3 Understanding life patterns and styles Social media data, especially geotagged tweets, photos, and check-ins, can help to understand not only an individual's life patterns [149] but also a city's dynamics [37], topics [150], behavior patterns [133], or lifestyles [161] when used aggregately. We can also compute the similarity between two cities according to the social media generated in both locations.

2.6.3 Location Recommendations

2.6.3.1 Generic point location recommendations Finding the most interesting locations in a city is a general task that a tourist wants to fulfill when traveling to an unfamiliar city [194]. However, the interest level of a location depends not only on the number of people who have visited the location but also on these people's travel knowledge. For example, the most frequently visited location in a city could be its railway stations

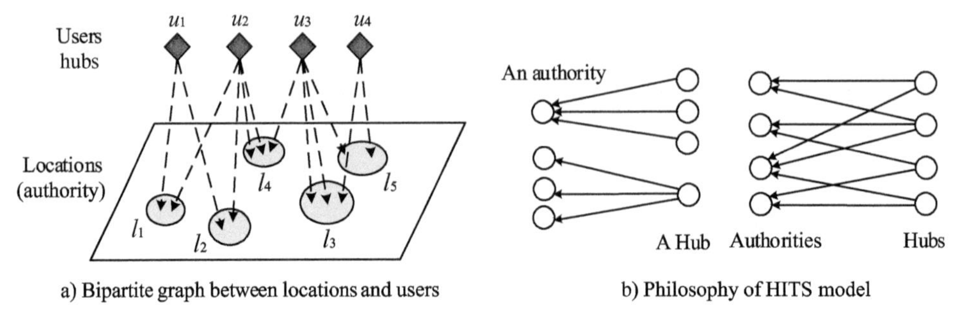

Figure 2.16
Inferring the most interesting places and most experienced users.

or airports, which might not be interesting location recommendations. On the contrary, some locations that attract experienced people (i.e., with rich travel knowledge) may be truly interesting. The problem is then how to determine an individual's travel experiences.

As shown in figure 2.16, Zheng et al. [194] formulate a bipartite graph between users and locations and employ a hypertext induced topic search (HITS)-based model to infer the interest level of a location and the travel knowledge of a user. The general idea is that users' travel experiences and the interest level of a location have a mutual reinforcement relationship. More specifically, a user's knowledge can be represented by the summation of the interests of the locations the user has visited; in turn, the interest of a location is represented by the summation of the knowledge of the users who have visited this location.

2.6.3.2 Personalized point location recommendations In some scenarios, we can consider a user's preferences (e.g., enjoys Italian food and watching movies) and contexts (like current location and time) when suggesting location recommendations [88, 148]. One simple method is to formulate a user-location matrix where each row denotes a user, each column denotes a location, and each entry stands for the number of visits by a particular user in a particular location. Then some collaborative filtering methods can be used to fill in the entries without a value. This kind of method calculates the similarity between users based solely on the two rows denoting the two users' location histories, not counting useful information such as the aforementioned visiting sequences between locations.

Considering this rich information, Zheng et al. incorporate the user similarity they infer in paper [81] into a user-based collaborative-filtering (CF) model to infer the missing values in the user-location matrix. Though it has a deeper understanding of user similarity, the method suffers from the increasing scale of users since the model needs to calculate the similarity between each pair of users. To address this issue, location-based collaborative filtering is proposed in [188]. This model computes the correlation between locations based on the location histories of the users visiting these locations [187]. The correlation is then used as a kind of similarity between locations in an item-based CF model. Given the limited geographical space (i.e., the number of locations is limited), this location-based model is more practical for a real system.

As a user can only visit a limited number of locations, the user-locations matrix is very sparse, leading to a big challenge to traditional CF-based location recommender systems. The problem becomes even more challenging when people travel to a new city they have not visited. To this end, Bao et al. [13] present a location-based and preference-aware recommender system that offers a particular user a set of venues (such as restaurants and shopping malls) within a geospatial range with the consideration of both (1) user personal preferences, which are automatically learned from a user's location history and (2) social opinions, which are mined from the location histories of the local experts. This recommender system can facilitate people's travel not only near their living areas but also to a city that is new to them.

2.6.3.3 Itinerary planning Sometimes, a user needs a sophisticated itinerary conditioned by the user's travel duration and departure place. The itinerary could include not only stand-alone locations but also detailed routes connecting these locations and a proper schedule—for example, the best time to reach a location and an appropriate time length to stay there. Yoon et al. [151, 152] plan a trip in terms of the collective knowledge learned from people's massive GPS trajectories. Wei et al. [137] determine the most likely route traveling between two query points by learning from many check-in data points.

2.6.3.4 Location-activity recommender This recommender provides a user with two types of recommendations: (1) the most popular activities that can be performed in a given location and (2) the most popular locations for conducting a given activity, such as shopping. These two categories of recommendations can be mined from a large number of users' location histories with activity tags. To successfully generate these two types of recommendations, Zheng et al. [196] propose a context-aware collaborative filtering model, which is solved with a matrix factorization method.

Furthermore, Zheng et al. [195, 197] extend the location-activity matrix into a tensor, considering users as the third dimension. A personalized location-activity recommendation is proposed, based on a context-aware tensor decomposition method.

2.7 Urban Computing for the Economy

The dynamics of a city (e.g., human mobility and the number of changes in a POI category) may indicate trends in a city's economy. For instance, the number of movie theaters in Beijing kept increasing from 2008 to 2012, reaching 260. This could mean that more and more people living in Beijing want to watch a movie in a theater. On the contrary, some categories of POIs might get smaller or disappear, denoting a downturn in business for that sector. Likewise, human mobility data could indicate the unemployment rate of some major cities, helping to predict business and stock market trends.

2.7.1 Location Selection for Businesses

Human mobility, combined with POIs, can also help with the placement of some businesses. Karamshuk et al. [76] study the problem of optimal retail store placement in the context of location-based social networks. They collect human mobility data from Foursquare and analyze it to understand how the popularity levels of three retail store chains in New York City are shaped in terms of number of check-ins. A diverse set of data-mining features are evaluated, modeling spatial and semantic information about places and patterns of user movements in the surrounding area. As a result, among those features, the presence of user attractors (i.e., train stations or airports), as well as retail stores of the same type as the target chain (i.e., coffee shops or restaurants), encoding the local commercial competition of an area are the strongest indicators of popularity.

Combining multiple data sources, we can even predict the ranking of properties. Fu et al. [47–49] present research on predicting the future ranking of residential real estate in a city according to the potential value inferred from various datasets, such as human mobility and urban geography, currently observed around the properties. Here, *value* means the ability to increase faster in a rising market and decrease more slowly than others in a falling market, quantified by discretizing the increasing or decreasing ratio over its previous price into five levels (R1–R5), where R1 stands for the best, and R5 denotes the worst. The rank is of great importance to people when settling down or allocating capital investment.

As illustrated in figure 2.17, they considered three categories of factors consisting of geographical utility, neighborhood popularity, and the business zone's prosperity. These factors correspond with the common adage "The value of real estate is determined by its location, location, and location." More specifically, they identified a set of discriminative features for each property by mining the surrounding geographic data (e.g., road networks and POIs), traffic data (such as taxi traces and card-swiping records in public transportation systems), and social media. They then trained a pairwise

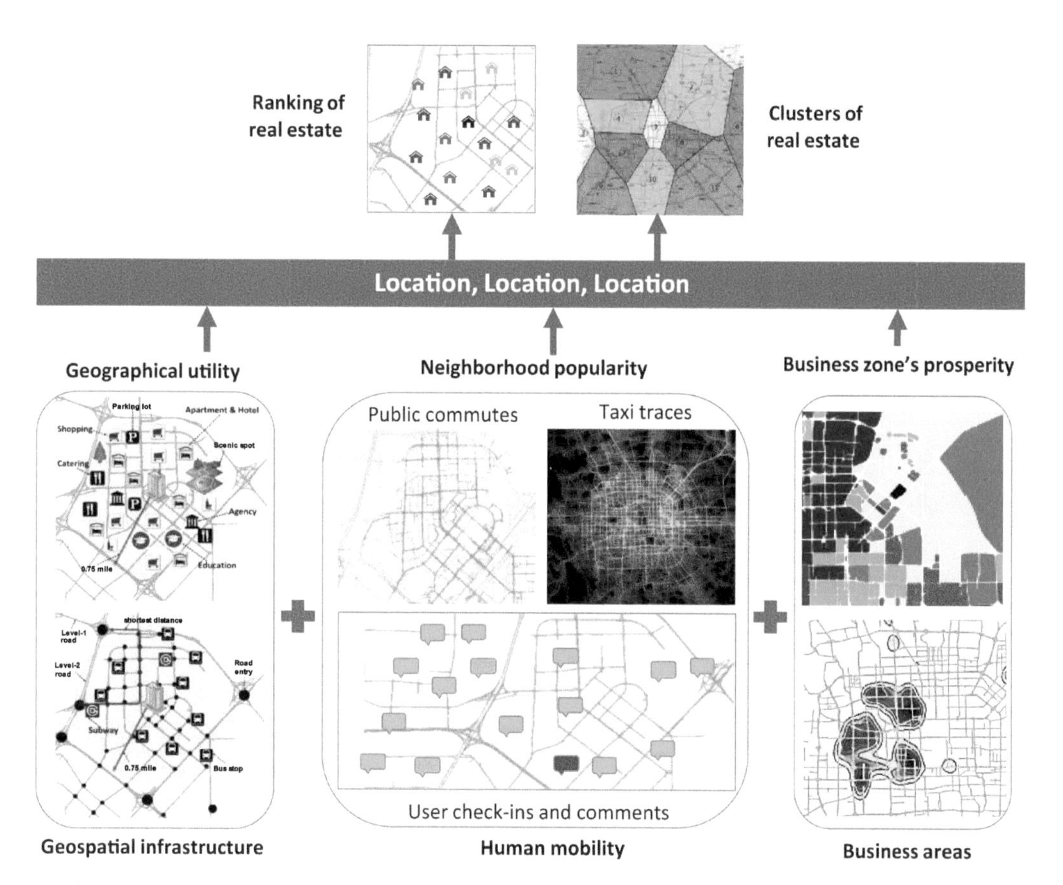

Figure 2.17
Ranking residential real estate based on big data.

learning-to-rank model [92], feeding a list of features-ranking pairs into an ANN. A metric-learning algorithm was also applied to identify the top ten most influential features affecting the ranking, implicitly revealing the important factors determining the value of a piece of real estate.

2.7.2 Optimizing Urban Logistics

The increasing population and developing businesses in a city result in a tremendous demand on urban logistics. It is of great importance to the sustainability of a city if we can enhance the throughput of existing urban logistics systems.

Figure 2.18 illustrates the working process of current city express systems, which are a major type of urban logistics. A city is divided into several regions (e.g., R_1 and R_2), each

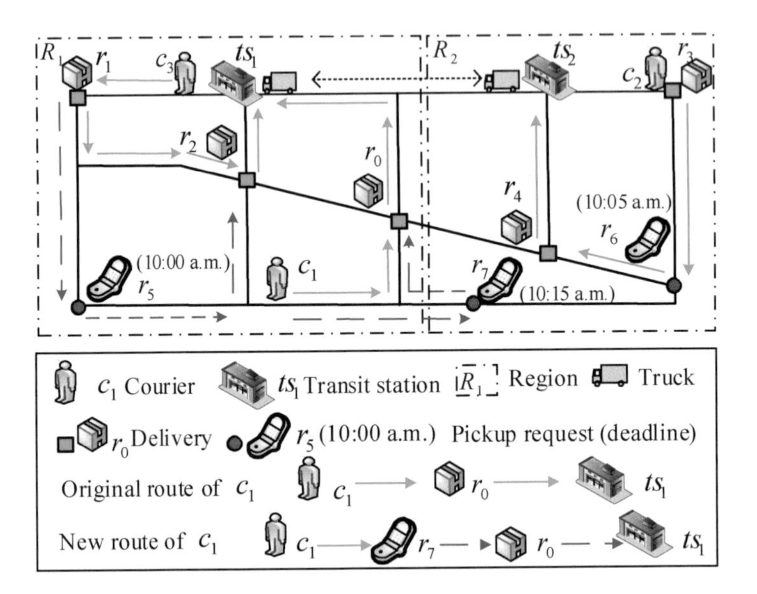

Figure 2.18
City express systems.

of which covers some streets and neighborhoods. A transit station is built in a region to temporarily store the parcels received in the region (e.g., ts_1 in R_1). The received parcels at a transit station are further organized into groups according to their destinations. Each group of parcels will regularly be sent to a corresponding transit station by trucks (e.g., from ts_1 to ts_2). In each region, there are a team of couriers (e.g., c_1 and c_3 in R_1) delivering parcels to and receiving parcels from specific locations in the region. When a truck carrying parcels arrives at a transit station, each courier will send a portion of these parcels to their final destinations by a small delivery van, bike, or motorcycle, which has a fixed capacity limit. Before departing from the transit station, they will precompute the delivery routes (e.g., the blue lines), usually based on their own knowledge.

During the delivery, each courier might receive pickup requests (e.g., r_5, r_6, and r_7) from a central dispatch system or directly from end users. Each pickup request is associated with a location and a pickup time deadline. A courier may change the originally planned route to fetch the new parcels or decline the pickup request due to constraints in the courier's schedule or vehicle capacity. All couriers are required to return to their own transit station by a specific time (to fit the schedule of trucks that travel between transit stations regularly) or when fully loaded.

The service quality and operational efficiency of current express services are unsatisfactory for three reasons:

First, current central dispatch systems process each pickup request individually without global optimization. For example, a new pickup request (e.g., r_5) is usually assigned to the nearest courier (e.g., c_1) to the pickup location. In the meantime, each courier makes a decision about whether to pick up a new parcel based solely on his or her own situation, without knowing the status of other couriers in the same region (e.g., c_3 instead of c_1 can pick up a parcel at r_5).

Second, dispatch systems do not know the current status of a courier (e.g., the remaining capacity, the number of parcels that have not been delivered, and the following pickup-delivery route) before assigning a new pickup request. Note that these statuses would change dynamically due to the new pickup requests.

Third, requests near the boundary of a region (e.g., r_7) are ignored by couriers (e.g., c_1) in other regions because couriers only pick up parcels in their own regions.

To address these issues, Zhang et al. [170] have designed a central dispatch system with an effective scheduling algorithm for couriers to deliver and pick up parcels in real time. Each courier in the system carries a handheld device that records their location, uploads their status after delivering or picking up a parcel, and receives new pickup requests. The central dispatch system receives information on couriers from their handheld devices and manages their schedules consisting of pickup and delivery times and routes. After collecting pickup requests from customers during a short period, the system processes requests in a batch according to the minimum incurred distance. Finally, the system sends the updated schedules to all couriers and confirmation or declining messages to customers. A dynamic dispatching system can increase the throughput of current city express services by 30 percent.

2.8 Urban Computing for Public Safety and Security

Large events, pandemics, severe accidents, environmental disasters, and terrorist attacks pose significant hindrances to public security and order. The wide availability of urban data facilitates the ability to learn from past events how to handle such threats correctly. It also enables us to detect threats or even predict them in advance.

2.8.1 Detecting Urban Anomalies

Urban anomalies might be caused by car accidents, traffic control, protests, sports, celebrations, disasters, and other events. The detection of urban anomalies can help disperse congestion, diagnose unexpected events, and facilitate people's movement around them.

2.8.1.1 Anomaly detection based on single datasets Liu et al. [93] partition a city into disjointed regions with major roads [155] and glean anomalous links between two regions based on the traffic of vehicles traveling between the two regions. They divide the time of day into time bins and identify three features for each link consisting of the number of vehicles traveling the link in a time bin (#Obj), the proportion of vehicles moving toward a given destination region (Pct_d), and those moving out of a given origin region (Pct_o). As shown in figure 2.19a, regarding link $a \rightarrow b$, #Obj = 5, $Pct_d = 5/14$, and $Pct_o = 5/9$. The three features are respectively compared with those in the equivalent time bins of previous days to calculate the minimum distortion of each feature (i.e., minDistort #Obj, minDistort Pct_d, and minDistort Pct_o). Then, the link of the time bin can be represented in a three-dimensional space, with each dimension denoting the minimum distort of a feature, as depicted in figure 2.19b. To normalize the effect of variances along different directions, the Mahalanobis distance is used to measure the most extreme points, which are regarded as outliers.

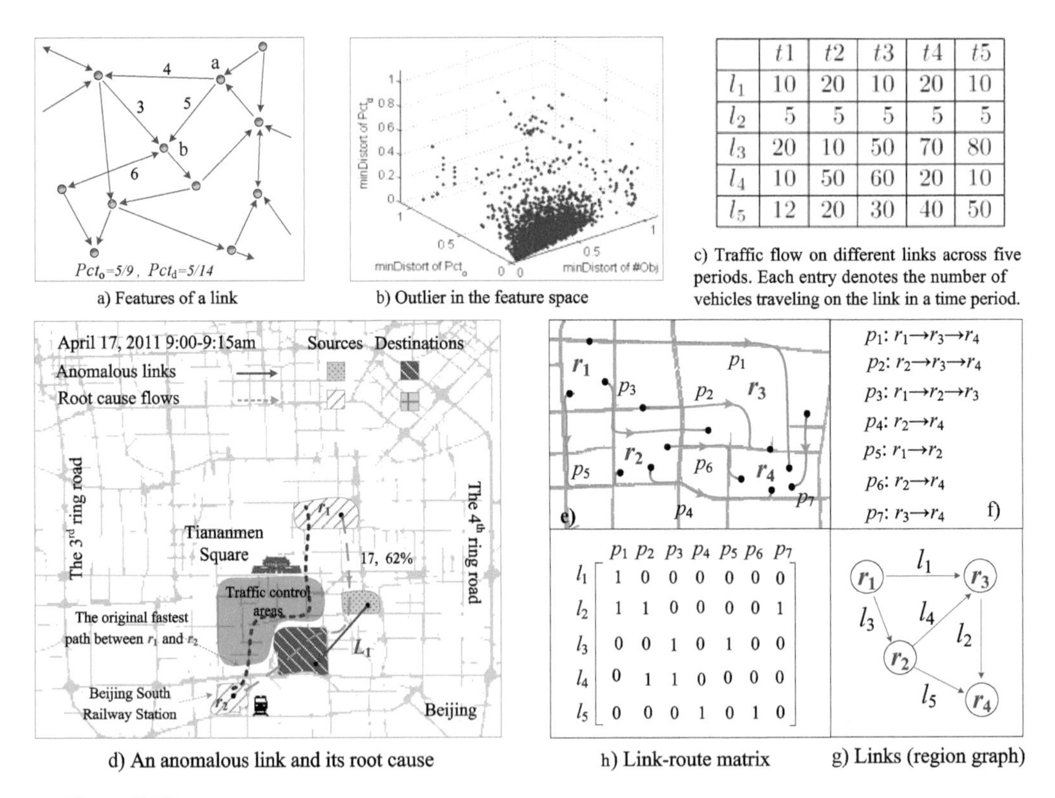

	$t1$	$t2$	$t3$	$t4$	$t5$
l_1	10	20	10	20	10
l_2	5	5	5	5	5
l_3	20	10	50	70	80
l_4	10	50	60	20	10
l_5	12	20	30	40	50

	p_1	p_2	p_3	p_4	p_5	p_6	p_7
l_1	1	0	0	0	0	0	0
l_2	1	1	0	0	0	0	1
l_3	0	0	1	0	1	0	0
l_4	0	1	1	0	0	0	0
l_5	0	0	0	1	0	1	0

Figure 2.19
Detecting anomalies in urban traffic based on distance.

Following the aforementioned research, Chawla et al. [26] propose a two-step mining and optimization framework to detect traffic anomalies between two regions and explain an anomaly with traffic flows passing the two regions. As illustrated in figure 2.19d, an anomalies link L_1 is found between two regions. However, the problem may not lie in the two regions. On April 17, 2011, traffic in Beijing was diverted away from Tiananmen Square because of the Beijing marathon. Thus, the normal traffic route (shown as a dotted path) from region r_1 to the Beijing South Railway Station in r_2 was diverted; the dashed (green) path witnessed excess traffic. In short, the traffic flow on the green path leads to the anomaly.

In the methodology, given a link matrix like that shown in figure 2.19c, they first use a principal component analysis (PCA) algorithm to detect some anomalous links, which are represented by a column vector b with 1 denoting an anomaly detected on the link. An adjacent link-route matrix A is formulated based on the trajectories of vehicles, as illustrated from figure 2.19d to g. Each entry of the matrix denotes whether a route passes a link; 1 denotes yes and 0 denotes no. For instance, route p_1 passes l_1 and l_2. The relationship between anomalous links and routes is then captured by solving the equation, $Ax=b$, where x is a column vector denoting which paths contribute to the emergency of these anomalies shown in b. Using L_1 optimization techniques, the x can be inferred.

Pang et al. [109] adopt a likelihood ratio test (LRT), which has previously been used in epidemiological studies to describe traffic patterns. They partition a city into uniform grids and count the number of vehicles arriving in the grid over a time period. The objective is to identify contiguous sets of cells and time intervals that have the largest statistically significant deviation from the expected behavior (i.e., the number of vehicles). The regions whose log-likelihood ratio statistic value drops in the tail of χ^2 distribution are likely anomalous.

2.8.1.2 Detecting urban anomalies based on multiple datasets Pan et al. [108] identify traffic anomalies according to drivers' routing behavior on an urban road network. Here, a detected anomaly is represented by a subgraph of a road network where drivers' routing behaviors differ significantly from their original patterns. They then try to describe a detected anomaly by mining representative terms from the social media that people posted when the anomaly happened. The system for detecting such traffic anomalies can benefit both drivers and transportation authorities (e.g., by notifying drivers approaching an anomaly and suggesting alternative routes, as well as supporting traffic jam diagnosis and dispersal).

Zheng et al. [192] detect a collective anomaly that denotes a collection of nearby locations that are anomalous during a few consecutive time intervals in terms of phenomena collectively witnessed by multiple datasets. Here, *collective* has two types of meanings:

1. One is that an anomaly might not be that anomalous in terms of a single dataset but is considered an anomaly when checking multiple datasets simultaneously. As illustrated in figure 2.20, an unusual event has just happened at location r_1, affecting its surrounding locations (e.g., from r_2 to r_6). As a result, the traffic flow entering r_1 from its surrounding locations increases 10 percent. Meanwhile, social media posts and bike rental flow around these locations change slightly. The deviation in each single dataset against its common pattern is not significant enough to be considered anomalous. However, when putting them together, we might be able to identify the anomaly, as the three datasets barely change simultaneously to that extent.
2. The other denotes spatiotemporal collectiveness. That is, a collection of nearby locations is anomalous during a few consecutive time intervals, as illustrated in figure 2.20d, while a single location in the collection may not be anomalous at a single time interval if checked individually. For example, locations from r_1 to r_6 formulate a collective anomaly in a few consecutive time intervals (e.g., from 2 p.m. to 4 p.m.). If we check location r_2 individually at 2 p.m., it might not be considered an anomaly. It also associates individual locations and time intervals, formulating a panoramic view of an event. Such collective anomalies could denote an early stage of an epidemic disease, the beginning of a natural disaster, an underlying problem, or a potentially catastrophic accident.

To detect a collective anomaly is very challenging, however, as different datasets have different densities, distributions, and scales. Additionally, to find the spatiotemporal

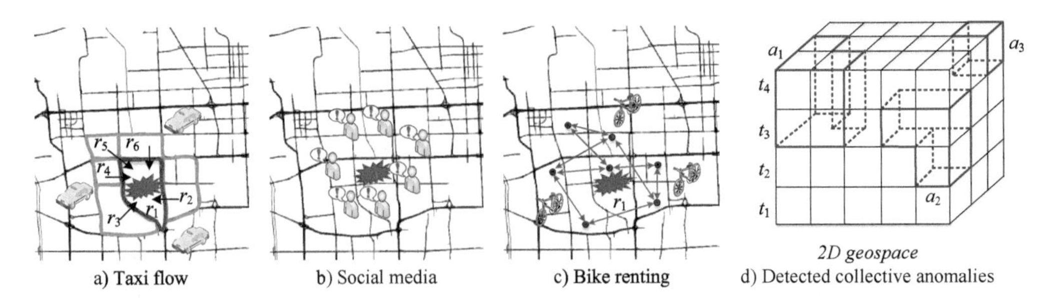

Figure 2.20
Detecting collective anomalies based on multiple datasets.

scope of a collective anomaly is time consuming because there are many ways to combine regions and time slots. Zheng's method consists of three components: a multiple-source latent-topic (MSLT) model, a spatiotemporal likelihood ratio test (ST_LRT) model, and a candidate generation algorithm [192]. MSLT combines multiple datasets to infer the latent functions of a geographic region in the framework of a topic model. In turn, a region's latent functions help estimate the underlying distribution of a sparse dataset generated in the region. ST_LRT learns a proper underlying distribution for different datasets and calculates an anomalous degree for each dataset based on an LRT. It then aggregates the anomalous degrees of different datasets using a skyline detection algorithm.

2.8.2 Predicting the Flow of Crowds

Predicting the movement of crowds in a city is strategically important for traffic management, risk assessment, and public safety. For example, thirty-six people died, and forty-seven others were injured in the Shanghai Bund stampede of 2015, turning a New Year's celebration into a catastrophic accident. Massive flows of people streamed into a strip region that could not really hold them to watch the New Year's Eve Light Show, making the region overloaded and out of police control. Similar stampedes happened at the 2010 German Love Parade. If we can predict the arrival of crowds in a region and know that crowd flows will exceed the region's safe capacity, we can launch emergency mechanisms (e.g., sending warnings to people and conducting traffic controls) or evacuate people in advance.

Hoang et al. [64] propose a novel approach to predicting two types of flows of crowds in every region of a city based on big data, including human mobility data, weather conditions, and road networks. As shown in figure 2.21a, a region (such as r_1) is bound by major roads, and the two flows are (1) inflow, or the traffic of crowds originating from a region at a given time interval (e.g., people start driving from a parking spot); and (2) outflow, or the traffic of crowds that is terminated in a region (e.g., people stop driving and park their cars). Intuitively, inflow and outflow track the origins and the destinations of crowds, thus summarizing the movements of crowds for traffic management and risk assessment. The problem is different from the predictions of each individual's movements and each road segment's traffic conditions, which are computationally costly and not necessary from the perspective of public safety on a citywide scale.

To build a practical solution for citywide traffic prediction, they first partition the map of a city into regions using both its road network and historical records of human mobility. To model the multiple complex factors affecting the crowd flows, they decompose the flows into three components: seasonal (periodic patterns), trend (changes in

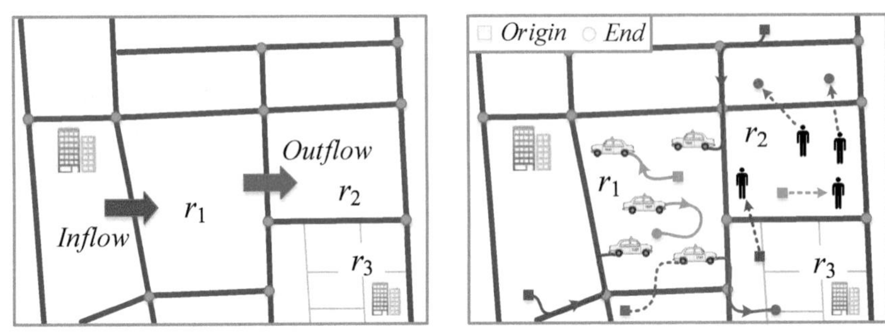

a) Two types of flow to be predicted b) Illustration of measurement of flow

Figure 2.21
Detecting collective anomalies based on multiple datasets.

periodic patterns), and residual flows (instantaneous changes). The seasonal and trend models are built as intrinsic Gaussian Markov random fields, which can cope with noisy and missing data, whereas the residual model exploits the spatiotemporal dependence among different flows and regions as well as the effect of weather.

Following this idea, Zhang et al. [168, 169] propose a deep learning-based prediction model, called ST-ResNet, to collectively forecast the flow of crowds in each and every region throughout a city. They base the architecture of ST-ResNet on the spatial and temporal properties of flow data. It is comprised of two main components: spatiotemporal modeling and global factor modeling. The spatiotemporal component employs the framework of convolutional neural networks to simultaneously model near and distant spatial dependencies, temporal closeness, periods, and trends. The global component is used to capture external factors, such as weather conditions, time of day, and day of the week. ST-ResNet is evaluated using taxi data from Beijing, loop detector data from Guiyang, and bike-sharing data from New York City, showing its advantages beyond four baseline methods. More details about this technology will be presented in chapter 8, section 8.7.

2.9 Summary

This chapter introduces the applications of urban computing in seven different domains, consisting of urban planning, transportation, environment, energy, social and entertainment, economy, and public safety and security. Using different data science techniques, these examples have revolutionized traditional approaches to solving urban challenges.

References

[1] Abbott, M. B., J. C. Bathurst, J. A. Cunge, P. E. O'Connell, and J. Rasmussen. 1986. "An Introduction to the European Hydrological System—Systeme Hydrologique Europeen, 'SHE,' 2: Structure of a Physically-Based, Distributed Modelling System." *Journal of Hydrology* 87 (1–2): 61–77.

[2] Agarwal, Y., B. Balaji, S. Dutta, R. K. Gupta, and T. Weng. 2011. "Duty-Cycling Buildings Aggressively: The Next Frontier in HVAC Control." In *Proceedings of the 10th ACM/IEEE International Conference on Information Processing in Sensor Networks*. Washington, DC: Institute of Electrical and Electronics Engineers (IEEE) Computer Society Press, 246–257.

[3] Agatz, N., A. Erera, M. Savelsbergh, and X. Wang. 2012. "Optimization for Dynamic Ride-Sharing: A Review." *European Journal of Operational Research* 223 (2): 295–303.

[4] Aguado, E., and I. Remson. 1974. "Ground-Water Hydraulics in Aquifer Management." *Journal of the Hydraulics Division* 100 (1): 103–118.

[5] Aguado, E., and I. Remson. 1980. "Ground-Water Management with Fixed Charges." *Journal of the Water Resources Planning and Management Division* 106 (2): 375–382.

[6] Amorocho, J., and W. E. Hart. 1964. "A Critique of Current Methods in Hydrologic Systems Investigation." *Eos, Transactions American Geophysical Union* 45 (2): 307–321.

[7] Antikainen, J. 2005. "The Concept of Functional Urban Area." *Findings of the Espon Project* 1 (1): 447–452.

[8] Arnold, J. G., P. M. Allen, and G. Bernhardt. 1993. "A Comprehensive Surface-Groundwater Flow Model." *Journal of Hydrology* 142 (1–4): 47–69.

[9] Attanasio, A., J.-F. Cordeau, G. Ghiani, and G. Laporte. 2004. "Parallel Tabu Search Heuristics for the Dynamic Multi-Vehicle Dial-a-Ride Problem." *Parallel Computing* 30, no. 3 (March): 377–387.

[10] Balan, R., K. Nguyen, and L. Jiang. 2011. "Real-Time Trip Information Service for a Large Taxi Fleet." In *Mobisys '11. Proceedings of the 9th International Conference on Mobile Systems, Applications, and Services*. New York: Association for Computing Machinery (ACM), 99–112.

[11] Baldacci, R., V. Maniezzo, and A. Mingozzi. 2004. "An Exact Method for the Carpooling Problem Based on Lagrangean Column Generation." *Operations Research* 52 (3): 422–439.

[12] Bao, J., T. He, S. Ruan, Y. Li, and Y. Zheng. 2017. "Planning Bike Lanes Based on Sharing-Bikes' Trajectories." In *Proceedings of the 23rd SIGKDD Conference on Knowledge Discovery and Data Mining*. New York: ACM.

[13] Bao, J., Y. Zheng, D. Wilkie, and M. Mokbel. 2015. "Recommendations in Location-Based Social Networks: A Survey." *Geoinformatica* 19 (3): 525–565.

[14] Bastani, F., Y. Huang, X. Xie, and J. W. Powell. 2011. "A Greener Transportation Mode: Flexible Routes Discovery from GPS Trajectory Data." In *Proceedings of the 19th ACM SIGSPATIAL International Conference on Advances in Geographic Information Systems*. New York: ACM, 405–408.

[15] Beaudry, A., G. Laporte, T. Melo, and S. Nickel. 2010. "Dynamic Transportation of Patients in Hospitals." *OR Spectrum* 32 (1): 77–107.

[16] Bejan, A. I., R. J. Gibbens, D. Evans, A. R. Beresford, J. Bacon, and A. Friday. 2010. "Statistical Modelling and Analysis of Sparse Bus Probe Data in Urban Areas." In *Proceedings of the 13th IEEE International Conference on Intelligent Transportation Systems*. Washington, DC: IEEE Computer Society Press, 1256–1263.

[17] Benchimol, M., P. Benchimol, B. Chappert, A. De La Taille, F. Laroche, F. Meunier, and L. Robinet. 2011. "Balancing the Stations of a Self-Service 'Bike Hire' System." *RAIRO-Operations Research* 45 (1): 37–61.

[18] Berlingerio, M., F. Calabrese, Giusy Di Lorenzo, R. Nair, F. Pinelli, and M. L. Sbodio. 2013. "AllAboard: A System for Exploring Urban Mobility and Optimizing Public Transport Using Cellphone Data." In *Proceedings of the 12th European Conference on Machine Learning and Principles and Practice of Knowledge Discovery in Databases*. Berlin: Springer, 663–666.

[19] Borgnat, P., E. Fleury, C. Robardet, and A. Scherrer. 2009. "Spatial Analysis of Dynamic Movements of Vlov, Lyon's Shared Bicycle Program." In *Proceedings of the European Conference on Complex Systems*. Coventry, UK: Warwick University.

[20] Borzsony, S., D. Kossmann, and K. Stocker. 2001. "The Skyline Operator." In *Proceedings of the 17th International Conference on Data Engineering*. Washington, DC: IEEE Computer Society Press, 421–430.

[21] Bray, M., and H. Dawei. 2004. "Identification of Support Vector Machines for Runoff Modelling." *Journal of Hydroinformatics* 6 (4): 265–280.

[22] Calvo, R. W., F. de Luigi, P. Haastrup, and V. Maniezzo. 2004. "A Distributed Geographic Information System for the Daily Carpooling Problem." *Computers and Operations Research* 31: 2263–2278.

[23] Castro, P., and M. Neves. 2003. "Chlorine Decay in Water Distribution Systems Case Study—Lousada Network." *Electronic Journal of Environmental, Agricultural and Food Chemistry* 2 (2): 261–266.

[24] Castro-Neto, M., Y. S. Jeong, M. K. Jeong, and L. D. Han. 2009. "Online-SVR for Short-Term Traffic Prediction under Typical and Atypical Traffic Conditions." *Expert Systems with Applications* 36 (3): 6164–6173.

[25] Ceapa, I., C. Smith, and L Capra. 2012. "Avoiding the Crowds: Understanding Tube Station Congestion Patterns from Trip Data." In *Proceedings of the 1st ACM SIGKDD International Workshop on Urban Computing*. New York: ACM, 134–141.

[26] Chawla, S., Y. Zheng, and J. Hu. 2012. "Inferring the Root Cause in Road Traffic Anomalies." In *Proceedings of the 12th International Conference on Data Mining*. Washington, DC: IEEE Computer Society Press, 141–150.

[27] Chemla, D., F. Meunier, and R. Wolfler-Calvo. 2011. "Balancing a Bike-Sharing System with Multiple Vehicles." In *Proceedings of Congress annual de la société Française de recherche opérationelle et d'aidea la décision*. Saint-Etienne, France: Roadef.

[28] Chen, C., D. Zhang, N. Li, and Z. H. Zhou. 2014. "B-Planner: Planning Bidirectional Night Bus Routes Using Large-Scale Taxi GPS Traces." *IEEE Transactions on Intelligent Transportation Systems* 15 (4): 1451–1465.

[29] Chen, L., D. Zhang, G. Pan, X. Ma, D. Yang, K. Kushlev, W. Zhang, and S. Li. 2015. "Bike Sharing Station Placement Leveraging Heterogeneous Urban Open Data." In *Proceedings of the 2015 ACM International Joint Conference on Pervasive and Ubiquitous Computing*. New York: ACM, 571–575.

[30] Chen, P.-Y, J.-W. Liu, and W.-T. Chen. 2010. "A Fuel-Saving and Pollution-Reducing Dynamic Taxi-Sharing Protocol in VANETs." In *Proceedings of the IEEE 72nd Vehicular Technology Conference*. Washington, DC: IEEE Computer Society Press, 1–5.

[31] Chen, X., Y. Zheng, Y. Chen, Q. Jin, W. Sun, E. Chang, and W. Y. Ma. 2014. "Indoor Air Quality Monitoring System for Smart Buildings." In *Proceedings of the 16th ACM International Conference on Ubiquitous Computing*. New York: ACM.

[32] Clark, R. M., and M. Sivaganesan. 2002. "Predicting Chlorine Residuals in Drinking Water: Second Order Model." *Journal of Water Resources Planning and Management* 128 (2): 152–161.

[33] Contardo, C., C. Morency, and L. M. Rousseau. 2012. "Balancing a Dynamic Public Bike-Sharing System." Volume 4. Montreal: Cirrelt.

[34] Cordeau, J. 2003. "A Branch-and-Cut Algorithm for the Dial-a-Ride Problem." *Operations Research* 54 (3): 573–586.

[35] Cordeau, J. F., and G. Laporte. 2003. "A Tabu Search Heuristic for the Static Multi-Vehicle Dial-a-Ride Problem." *Transportation Research Part B: Methodological* 37 (6): 579–594.

[36] Cordeau, J. F., and G. Laporte. 2007. "The Dial-a-Ride Problem: Models and Algorithms." *Annals of Operations Research* 153 (1): 29–46.

[37] Cranshaw, J., E. Toch, J. Hong, A. Kittur, and N. Sadeh. 2010. "Bridging the Gap between Physical Location and Online Social Networks." In *Proceedings of the 12th ACM International Conference on Ubiquitous Computing*. New York: ACM, 119–128.

[38] Dell'Olio, L., A. Ibeas, and J. L. Moura. 2011. "Implementing Bike-Sharing Systems." *Proceedings of the Institution of Civil Engineers-Municipal Engineer* 164 (2): 89–101.

[39] Devarakonda, S., P. Sevusu, H. Liu, R. Liu, L. Iftode, and B. Nath. 2013. "Real-Time Air Quality Monitoring through Mobile Sensing in Metropolitan Areas." In *Proceedings of the 2nd ACM SIGKDD International Workshop on Urban Computing*. New York: ACM.

[40] D'Hondt, E., and M. Stevens. 2011. "Participatory Noise Mapping." In *Proceedings of the 9th International Conference on Pervasive Computing*. Berlin: Springer, 33–36.

[41] d'Orey, P. M., R. Fernandes, and M. Ferreira. 2012. "Empirical Evaluation of a Dynamic and Distributed Taxi-Sharing System." In *Proceedings of the 15th International IEEE Conference on Intelligent Transportation Systems*. Washington, DC: IEEE Computer Society Press, 140–146.

[42] Duan, Q., S. Sorooshian, and V. Gupta. 1992. "Effective and Efficient Global Optimization for Conceptual Rainfall-Runoff Models." *Water Resources Research* 28 (4): 1015–1031.

[43] Dumas, Y., J. Desrosiers, and F. Soumis. 1991. "The Pickup and Delivery Problem with Time Windows." *European Journal of Operational Research* 54, no. 1 (September): 7–22.

[44] Dusparic, I., C. Harris, A. Marinescu, V. Cahill, and S. Clarke. 2013. "Multi-Agent Residential Demand Response Based on Load Forecasting." In *Proceedings of the 1st IEEE Conference on Technologies for Sustainability—Engineering and the Environment*. Washington, DC: IEEE Computer Society Press.

[45] Friginal, J., S. Gambs, J. Guiochet, and M. O. Killijian. 2014. "Towards Privacy-Driven Design of a Dynamic Carpooling System." *Pervasive and Mobile Computing* 14:71–82.

[46] Froehlich, J., J. Neumann, and N. Oliver. 2009. "Sensing and Predicting the Pulse of the City through Shared Bicycling." In *Proceedings of the 21st International Joint Conference on Artificial Intelligence*. Pasadena, CA: International Joint Conferences on Artificial Intelligence Organization (IJCAI), 1420–1426.

[47] Fu, Y., Y. Ge, Y. Zheng, Z. Yao, Y. Liu, H. Xiong, and J. Yuan. 2014. "Sparse Real Estate Ranking with Online User Reviews and Offline Moving Behaviors." In *Proceedings of the 2014 IEEE International Conference on Data Mining*. Washington, DC: IEEE Computer Society Press, 120–129.

[48] Fu, Y., H. Xiong, Y. Ge, Z. Yao, and Y. Zheng. 2014. "Exploiting Geographic Dependencies for Real Estate Appraisal: A Mutual Perspective of Ranking and Clustering." In *Proceedings of the 20th SIGKDD Conference on Knowledge Discovery and Data Mining*. New York: ACM.

[49] Fu, Y., H. Xiong, Y. Ge, Y. Zheng, Z. Yao, and Z. H. Zhou. 2016. "Modeling of Geographic Dependencies for Real Estate Ranking." *ACM Transactions on Knowledge Discovery from Data* 11 (1): 11.

[50] Furuhata, M., M. Dessouky, Fernando Ordóñez, Marc-Etienne Brunet, Xiaoqing Wang, and Sven Koenig. 2013. "Ridesharing: The State-of-the-Art and Future Directions." *Transportation Research Part B: Methodological* 57 (November): 28–46.

[51] Galvan-Lopez, E., A. Taylor, S. Clarke, and V. Cahill. 2014. "Design of an Automatic Demand-Side Management System Based on Evolutionary Algorithms." In *Proceedings of the 29th Annual ACM Symposium on Applied Computing*. New York: ACM, 24–28.

[52] Gandia, R. 2015. "City Outlines Travel Diary Plan to Determine Future Transportation Needs." *Calgary Sun*, May 7.

[53] García-Palomares, J. C., J. Gutiérrez, and M. Latorre. 2012. "Optimizing the Location of Stations in Bike-Sharing Programs: A GIS Approach." *Applied Geography* 35 (1): 235–246.

[54] Gauthier A., C. Hughes, C. Kost, S. Li, C. Linke, S. Lotshaw, J. Mason, C. Pardo, C. Rasore, B. Schroeder, and X. Treviño. *The Bike-Share Planning Guide*. New York: Institute for Transportation and Development Policy.

[55] Ge, Y., H. Xiong, A. Tuzhilin, K. Xiao, M. Gruteser, and M. Pazzani. 2010. "An Energy-Efficient Mobile Recommender System." In *Proceedings of the 16th SIGKDD Conference on Knowledge Discovery and Data Mining*. New York: ACM, 899–908.

[56] Gidofalvi, G., T. B. Pedersen, T. Risch, and E. Zeitler. 2008. "Highly Scalable Trip Grouping for Large-Scale Collective Transportation Systems." In *Proceedings of the 11th International Conference on Extending Database Technology: Advanced Database Technology*. New York: ACM, 678–689.

[57] Giustolisi, O., A. Doglioni, D. A. Savic, and F. Di Pierro. 2008. "An Evolutionary Multiobjective Strategy for the Effective Management of Groundwater Resources." *Water Resources Research* 44 (1). doi:10.1029/2006WR005359.

[58] Goel, P., L. Kulik, and K. Ramamohanarao. 2016. "Privacy-Aware Dynamic Ride Sharing." *ACM Transactions on Spatial Algorithms and Systems* 2 (1): 4.

[59] Guehnemann, A., R. P. Schaefer, K. U. Thiessenhusen, and P. Wagner. 2004. *Monitoring Traffic and Emissions by Floating Car Data*. Clayton, Australia: Institute of Transport Studies.

[60] Hallam, N. B., J. R. West, C. F. Forster, J. C. Powell, and I. Spencer. 2002. "The Decay of Chlorine Associated with the Pipe Wall in Water Distribution Systems." *Water Research* 36 (14): 3479–3488.

[61] Hanson, S., and P. Hanson. 1980. "Gender and Urban Activity Patterns in Uppsala, Sweden." *Geographical Review* 70 (3): 291–299.

[62] Harris, C., R. Doolan, I. Dusparic, A. Marinescu, V. Cahill, and S. Clarke. 2014. "A Distributed Agent Based Mechanism for Shaping of Aggregate Demand." Paper presented at Energycon, Dubrovnik, Croatia.

[63] Herrera, J. C., D. Work, X. Ban, R. Herring, Q. Jacobson, and A. Bayen. 2010. "Evaluation of Traffic Data Obtained via GPS-Enabled Mobile Phones: The Mobile Century Field Experiment." *Transportation Research C* 18 (4): 568–583.

[64] Hoang, M. X., Y. Zheng, and A. K. Singh. "FCCF: Forecasting Citywide Crowd Flows Based on Big Data." In *Proceedings of the 24th ACM International Conference on Advances in Geographical Information Systems*. New York: ACM.

[65] Hsieh, H. P., S. D. Lin, and Y. Zheng. 2015. "Inferring Air Quality for Station Location Recommendation Based on Urban Big Data." In *Proceedings of the 21st ACM SIGKDD International Conference on Knowledge Discovery and Data Mining*. New York: ACM, 437–446.

[66] Hsu, Kuo-lin, Hoshin Vijai Gupta, and Soroosh Sorooshian. 1995. "Artificial Neural Network Modeling of the Rainfall-Runoff Process." *Water Resources Research* 31 (10): 2517–2530.

[67] Hung, C. C., C. W. Chang, and W. C. Peng. 2009. "Mining Trajectory Profiles for Discovering User Communities." In *Proceedings of the 1st ACM SIGSPATIAL GIS Workshop on Location Based Social Networks*. New York: ACM, 1–8.

[68] Hunter, T., R. Herring, P. Abbeel, and A. Bayen. 2009. "Path and Travel Time Inference from GPS Probe Vehicle Data." In *Proceedings of the International Workshop on Analyzing Networks and Learning with Graphs*. Vancouver, Canada: Neural Information Processing Systems Foundation.

[69] Hvattum, L. M., A. Løkketangen, and G. Laporte. 2007. "A Branch-Andregret Heuristic for Stochastic and Dynamic Vehicle Routing Problems." *Networks* 49, no. 4 (July): 330–340.

[70] Ji, S., Y. Zheng, and T. Li. 2016. "Urban Sensing Based on Human Mobility." In *Proceedings of the 2016 ACM International Joint Conference on Pervasive and Ubiquitous Computing*. New York: ACM, 1040–1051.

[71] Jiang, S., J. Ferreira, and M. C. Gonzalez. 2012. "Discovering Urban Spatial-Temporal Structure from Human Activity Patterns." In *Proceedings of the 1st ACM SIGKDD International Workshop on Urban Computing*. New York: ACM, 95–102.

[72] Jiang, Y., K. Li, L. Tian, R. Piedrahita, X. Yun, L. Q. Mansata, L. Shang. 2011. "MAQS: A Personalized Mobile Sensing System for Indoor Air Quality Monitoring." In *Proceedings of the 13th International Conference on Ubiquitous Computing*. New York: ACM, 271–280.

[73] Johnson, M., N. Lam, S. Brant, C. Gray, and D. Pennise. 2011. "Modeling Indoor Air Pollution from Cookstove Emissions in Developing Countries Using a Monte Carlo Single-Box Model." *Atmospheric Environment* 45 (19): 3237–3243.

[74] Kaltenbrunner, A., R. Meza, J. Grivolla, J. Codina, and R. Banchs. 2010. "Urban Cycles and Mobility Patterns: Exploring and Predicting Trends in a Bicycle-Based Public Transport System." *IEEE Pervasive and Mobile Computing* 6:455–466.

[75] Kanoulas, E., Y. Du, T. Xia, and D. Zhang. 2006. "Finding Fastest Paths on a Road Network with Speed Patterns." In *Proceedings of the 22nd International Conference on Data Engineering*. Washington, DC: IEEE Computer Society Press.

[76] Karamshuk, D., A. Noulas, S. Scellato, V. Nicosia, and M. Cecilia. 2013. "Geo-Spotting: Mining Online Location-Based Services for Optimal Retail Store Placement." In *Proceedings of the 19th ACM International Conference on Knowledge Discovery and Data Mining*. New York: ACM, 793–801.

[77] Karlsson, C. 2007. "Clusters, Functional Regions and Cluster Policies." JIBS and CESIS Electronic Working Paper Series 84.

[78] Lathia, N., J. Froehlich, and L. Capra. 2010. "Mining Public Transport Usage for Personalised Intelligent Transport Systems." In *Proceedings of the 10th IEEE International Conference on Data Mining*. Washington, DC: IEEE Computer Society Press, 887–892.

[79] Lathia, S. A., and L. Capra. 2012. "Measuring the Impact of Opening the London Shared Bicycle Scheme to Casual Users." *Transportation Research Part C* 22:88–102.

[80] Lee, D., H. Wang, R. Cheu, and S. Teo. 2004. "Taxi Dispatch System Based on Current Demands and Real-Time Traffic Conditions." *Transportation Research Record: Journal of the Transportation Research Board* 1882 (1): 193–200.

[81] Li, Q., Y. Zheng, X. Xie, Y. Chen, W. Liu, and W. Y. Ma. 2008. "Mining User Similarity Based on Location History." In *Proceedings of the 16th ACM SIGSPATIAL International Conference on Advances in Geographic Information Systems*. New York: ACM, 34.

[82] Li, Y., J. Bao, Y. Li, Y. Wu, Z. Gong, and Y. Zheng. 2016. "Mining the Most Influential *k*-Location Set from Massive Trajectories." *IEEE Transactions on Big Data*. doi:10.1109/TBDATA.2017.2717978.

[83] Li, Y., R. Chen, L. Chen, and J. Xu. 2015. "Towards Social-Aware Ridesharing Group Query Services." *IEEE Transactions on Services Computing* 10 (4): 646–659.

[84] Li, Y., J. Luo, C. Y. Chow, K. L. Chan, Y. Ding, and F. Zhang. 2015. "Growing the Charging Station Network for Electric Vehicles with Trajectory Data Analytics." In *Proceedings of the 2015 IEEE 31st International Conference on Data Engineering*. Washington, DC: IEEE Computer Society Press, 1376–1387.

[85] Li, Y., Y. Zheng, S. Ji, W. Wang, and Z. Gong. 2015. "Location Selection for Ambulance Stations: A Data-Driven Approach." In *Proceedings of the 23rd SIGSPATIAL International Conference on Advances in Geographic Information Systems*. New York: ACM, 85.

[86] Li, Y., Y. Zheng, H. Zhang, and L. Chen. 2015. "Traffic Prediction in a Bike-Sharing System." In *Proceedings of the 23rd SIGSPATIAL International Conference on Advances in Geographic Information Systems*. New York: ACM, 33.

[87] Lin, J. R., and T. H. Yang. 2011. "Strategic Design of Public Bicycle Sharing Systems with Service Level Constraints." *Transportation Research Part E: Logistics and Transportation Review* 47 (2): 284–294.

[88] Liu, B., Y. Fu, Z. Yao, and H. Xiong. 2013. "Learning Geographical Preferences for Point-of-Interest Recommendation." In *Proceedings of the 19th ACM SIGKDD International Conference on Knowledge Discovery and Data Mining*. New York: ACM.

[89] Liu, C. W., K. H. Lin, and Y. M. Kuo. 2003. "Application of Factor Analysis in the Assessment of Groundwater Quality in a Blackfoot Disease Area in Taiwan." *Science of the Total Environment* 313 (1): 77–89.

[90] Liu, D., Di Weng, Yuhong Li, Yingcai Wu, Jie Bao, Yu Zheng, and Huaming Qu. 2016. "SmartAdP: Visual Analytics of Large-Scale Taxi Trajectories for Selecting Billboard Locations." *IEEE Transactions on Visualization and Computer Graphics* 23, no. 1 (January): 1–10.

[91] Liu, J., L. Sun, W. Chen, and H. Xiong. 2016. "Rebalancing Bike Sharing Systems: A Multi-source Data Smart Optimization." In *Proceedings of the 22nd SIGKDD Conference on Knowledge Discovery and Data Mining*. New York: ACM.

[92] Liu, T. Y. 2009. "Learning to Rank for Information Retrieval." *Foundations and Trends in Information Retrieval* 3 (3): 225–331.

[93] Liu, W., Y. Zheng, S. Chawla, J. Yuan, and X. Xing. 2011. "Discovering Spatio-temporal Causal Interactions in Traffic Data Streams." In *Proceedings of the 17th ACM SIGKDD International Conference on Knowledge Discovery and Data Mining*. New York: ACM, 1010–1018.

[94] Liu, Y., Y. Zheng, Y. Liang, S. Liu, and D. S. Rosenblum. 2016. "Urban Water Quality Prediction Based on Multi-Task Multi-View Learning." In *Proceedings of the Twenty-Fifth International Joint Conference on Artificial Intelligence*. Pasadena, CA: IJCAI.

[95] Ma, S., and O. Wolfson. 2013. "Analysis and Evaluation of the Slugging Form of Ridesharing." In *Proceedings of the 21st ACM SIGSPATIAL International Conference on Advances in Geographic Information Systems*. New York: ACM, 64–73.

[96] Ma, Shuo, Yu Zheng, and Ouri Wolfson. 2013. "T-Share: A Large-Scale Dynamic Taxi Ridesharing Service." In *Proceedings of the 29th IEEE International Conference on Data Engineering*. Washington, DC: IEEE Computer Society Press.

[97] Ma, Shuo, Yu Zheng, and Ouri Wolfson. 2015. "Real-Time City-Scale Taxi Ridesharing." *IEEE Transactions on Knowledge and Data Engineering* 27, no. 7 (July): 1782–1795.

[98] Marinescu, A., I. Dusparic, C. Harris, S. Clarke, and V. Cahill. 2014. "A Dynamic Forecasting Method for Small Scale Residential Electrical Demand." In *Proceedings of the 2014 International Joint Conference on Neural Networks*. Washington, DC: IEEE Computer Society Press.

[99] McNeill, J. D. 1990. "Use of Electromagnetic Methods for Groundwater Studies." *Geotechnical and Environmental Geophysics* 1:191–218.

[100] Mimno, D., and A. McCallum. 2008. "Topic Models Conditioned on Arbitrary Features with Dirichlet-Multinomial Regression." In *Proceedings of the Twenty-Fourth Conference on Uncertainty in Artificial Intelligence*. Arlington, VA: AUAI Press, 411–418.

[101] Minns, A. W., and M. J. Hall. 1996. "Artificial Neural Networks as Rainfall-Runoff Models." *Hydrological Sciences Journal* 41 (3): 399–417.

[102] Momtazpour, M., P. Butler, N. Ramakrishnan, M. S. Hossain, M. C. Bozchalui, and R. Sharma. 2014. "Charging and Storage Infrastructure Design for Electric Vehicles." *ACM Transactions on Intelligent Systems and Technology* 5 (3): 42.

[103] Monteiro, L., D. Figueiredo, S. Dias, R. Freitas, D. Covas, J. Menaia, and S. T. Coelho. 2014. "Modeling of Chlorine Decay in Drinking Water Supply Systems Using EPANET MSX." *Procedia Engineering* 70:1192–1200.

[104] Nair, R., E. Miller-Hooks, R. Hampshire, and A. Busic. 2012. "Large-Scale Bicycle Sharing Systems: Analysis of V'Elib." *International Journal of Sustainable Transportation* 7 (1): 85–106.

[105] Nicolas, M., M. Stevens, M. E. Niessen, and L. Steels. 2009. "NoiseTube: Measuring and Mapping Noise Pollution with Mobile Phones." *Information Technologies in Environmental Engineering*. Berlin: Springer, 215–228.

[106] Ostfeld, A., and S. Salomons. 2005. "A Hybrid Genetic—Instance Based Learning Algorithm for CE-QUAL-W2 Calibration." *Journal of Hydrology* 310 (1): 122–142.

[107] Ott, W. R. 1999. "Mathematical Models for Predicting Indoor Air Quality from Smoking Activity." *Environmental Health Perspectives* 107 (Suppl 2): 375.

[108] Pan, B., Y. Zheng, D. Wilkie, and C. Shahabi. 2013. "Crowd Sensing of Traffic Anomalies Based on Human Mobility and Social Media." In *Proceedings of the 21st ACM SIGSPATIAL International Conference on Advances in Geographic Information Systems*. New York: ACM, 344–353.

[109] Pang, L. X., S. Chawla, W. Liu, and Y. Zheng. 2013. "On Detection of Emerging Anomalous Traffic Patterns Using GPS Data." *Data and Knowledge Engineering* 87:357–373.

[110] Pfoser, D. 2008. "Floating Car Data." *Encyclopedia of GIS*. Berlin: Springer.

[111] Pfoser, D., S. Brakatsoulas, P. Brosch, M. Umlauft, N. Tryfona, and G. Tsironis. 2008. "Dynamic Travel Time Provision for Road Networks." In *Proceedings of the 16th International Conference on Advances in Geographic Information Systems*. New York: ACM.

[112] Phithakkitnukoon, S., M. Veloso, C. Bento, A. Biderman, and C. Ratti. 2010. "Taxi-Aware Map: Identifying and Predicting Vacant Taxis in the City." In *Proceedings of the 1st International Joint Conference on Ambient Intelligence*. Berlin: Springer, 86.

[113] Powell, J. C., N. B. Hallam, J. R. West, C. F. Forster, and J. Simms. 2000. "Factors Which Control Bulk Chlorine Decay Rates." *Water Research* 34 (1): 117–126.

[114] Pozdnoukhov, A., and C. Kaiser. 2011. "Space-Time Dynamics of Topics in Streaming Text." In *Proceedings of the 3rd ACM SIGSPATIAL GIS Workshop on Location Based Social Networks*. New York: ACM, 8:1–8:8.

[115] Qi, G., X. Li, S. Li, G. Pan, Z. Wang, and D. Zhang. 2011. "Measuring Social Functions of City Regions from Large-Scale Taxi Behaviors." In *Proceedings of Pervasive Computing and Communications Workshops (PERCOM Workshops), 2011 IEEE International Conference*. Washington, DC: IEEE Computer Society Press, 384–388.

[116] Rana, R. K., C. T. Chou, S. S. Kanhere, N. Bulusu, and W. Hu. 2010. "Ear-Phone: An End-to-End Participatory Urban Noise Mapping System." In *Proceedings of the 9th ACM/IEEE International Conference on Information Processing in Sensor Networks*. New York: ACM, 105–116.

[117] Rana, R. K., C. T. Chou, S. S. Kanhere, N. Bulusu, and W. Hu. 2013. "Ear-Phone: A Context-Aware Noise Mapping Using Smart Phones." doi:arXiv:1310.4270.

[118] Ratti, C., S. Sobolevsky, F. Calabrese, C. Andris, J. Reades, M. Martino, R. Claxton, and S. H. Strogatz. 2010. "Redrawing the Map of Great Britain from a Network of Human Interactions." *PLoS ONE* 5 (12): e14248.

[119] Rinzivillo, S., S. Mainardi, F. Pezzoni, M. Coscia, D. Pedreschi, and F. Giannotti. 2012. "Discovering the Geographical Borders of Human Mobility." *Künstl intell* 26:253–260.

[120] Rossman, L. A., and P. F. Boulos. 1996. "Numerical Methods for Modeling Water Quality in Distribution Systems: A Comparison." *Journal of Water Resources Planning and Management* 122 (2): 137–146.

[121] Rossman, L. A., R. M. Clark, and W. M. Grayman. 1994. "Modeling Chlorine Residuals in Drinking-Water Distribution Systems." *Journal of Environmental Engineering* 120 (4): 803–820.

[122] Rosvall, M., and C. T. Bergstrom. 2008. "Maps of Random Walks on Complex Networks Reveal Community Structure." *Proceedings of the National Academy of Sciences* 105 (4): 1118–1123.

[123] Santani, D., R. K. Balan, and C. J. Woodard. 2008. "Spatio-Temporal Efficiency in a Taxi Dispatch System." Research Collection. School of Information Systems, Singapore Management University, October.

[124] Santi, P., G. Resta, M. Szell, S. Sobolevsky, S. H. Strogatz, and C. Ratti. 2014. "Quantifying the Benefits of Vehicle Pooling with Shareability Networks." *Proceedings of the National Academy of Sciences* 111 (37): 13290–13294.

[125] Shaheen, S., S. Guzman, and H. Zhang. 2010. "Bikesharing in Europe, the Americas, and Asia: Past, Present, and Future." Paper presented at the 2010 Transportation Research Board Annual Meeting, Washington, DC.

[126] Shang, J., Y. Zheng, W. Tong, and E. Chang. 2014. "Inferring Gas Consumption and Pollution Emission of Vehicles throughout a City." In *Proceedings of the 20th SIGKDD Conference on Knowledge Discovery and Data Mining*. New York: ACM.

[127] Silvia, S., B. Ostermaier, and A. Vitaletti. 2008. "First Experiences Using Wireless Sensor Networks for Noise Pollution Monitoring." In *Proceedings of the Workshop on Real-World Wireless Sensor Networks*. New York: ACM, 61–65.

[128] The Copenhagen Wheel. https://www.superpedestrian.com/.

[129] Thiagarajan A., L. Ravindranath, K. Lacurts, S. Madden, H. Balakrishnan, S. Toledo, and J. Eriksson. 2009. "VTrack: Accurate, Energy-Aware Road Traffic Delay Estimation Using Mobile Phones." In *Proceedings of the 7th ACM Conference on Embedded Networked Sensor Systems*. New York: ACM.

[130] Toth, E., A. Brath, and A. Montanari. 2000. "Comparison of Short-Term Rainfall Prediction Models for Real-Time Flood Forecasting." *Journal of Hydrology* 239 (1): 132–147.

[131] Vasconcelos, J. J., L. A. Rossman, W. M. Grayman, P. F. Boulos, and R. M. Clark. 1997. "Kinetics of Chlorine Decay." *American Water Works Association Journal* 89 (7): 54.

[132] Vatsavai, R. R., E. Bright, C. Varun, B. Budhendra, A. Cheriyadat, and J. Grasser. 2011. "Machine Learning Approaches for High-Resolution Urban Land Cover Classification: A Comparative Study." In *Proceedings of the 2nd International Conference on Computing for Geospatial Research and Applications*. New York: ACM, 11:1–11:10.

[133] Wakamiya, S., R. Lee, and K. Sumiya. 2012. "Crowd-Sourced Urban Life Monitoring: Urban Area Characterization Based Crowd Behavioral Patterns from Twitter." In *Proceedings of the 6th International Conference on Ubiquitous Information Management and Communication*. New York: ACM, article no. 26.

[134] Wand, M., and M. Jones. 1995. *Kernel Smoothing*. Volume 60. London: Chapman and Hall.

[135] Wang, Y., Yu Zheng, and Yexiang Xue. 2014. "Travel Time Estimation of a Path Using Sparse Trajectories." In *Proceedings of the 20th SIGKDD Conference on Knowledge Discovery and Data Mining*. New York: ACM.

[136] Watkins, K., B. Ferris, A. Borning, S. Rutherford, and D. Layton. 2011. "Where Is My Bus? Impact of Mobile Real-Time Information on the Perceived and Actual Wait Time of Transit Riders." *Transportation Research Part A* 45 (8): 839–848.

[137] Wei, L. Y., Y. Zheng, and W. C. Peng. 2012. "Constructing Popular Routes from Uncertain Trajectories." In *Proceedings of the 18th SIGKDD Conference on Knowledge Discovery and Data Mining*. New York: ACM, 195–203.

[138] "What is EMS?" 2008. Washington, DC: National Highway Traffic Safety Administration (NHTSA). Accessed August 9, 2008. https://www.ems.gov/whatisems.html.

[139] Wu, H., J. A. Tan, and W. S. Ng et al. 2015. "FTT: A System for Finding and Tracking Tourists in Public Transport Services." In *Proceedings of the 2015 ACM SIGMOD International Conference on Management of Data*. New York: ACM, 1093–1098.

[140] Wu, W., W. S. Ng, S. Krishnaswamy, and A. Sinha. 2012. "To Taxi or Not to Taxi?—Enabling Personalised and Real-Time Transportation Decisions for Mobile Users." In *Proceedings of the IEEE 13th International Conference on Mobile Data Management*. Washington, DC: IEEE Computer Society Press, 320–323.

[141] Xiang, Z., C. Chu, and H. Chen. 2006. "A Fast Heuristic for Solving a Large-Scale Static Dial-a-Ride Problem under Complex Constraints." *European Journal of Operational Research* 174 (2): 1117–1139.

[142] Xiao, X., Y. Zheng, Q. Luo, and X. Xie. 2010. "Finding Similar Users Using Category-Based Location History." In *Proceedings of the 18th ACM SIGSPATIAL Conference on Advances in Geographical Information Systems*. New York: ACM, 442–445.

[143] Xiao, X., Y. Zheng, Q. Luo, and X. Xie. 2014. "Inferring Social Ties between Users with Human Location History." *Journal of Ambient Intelligence and Humanized Computing* 5 (1): 3–19.

[144] Xue, M., H. Wu, W. Chen, W. Siong Ng, and G. Howe Goh. 2014. "Identifying Tourists from Public Transport Commuters." In *Proceedings of the 20th ACM SIGKDD International Conference On Knowledge Discovery and Data Mining*. New York: ACM, 779–1788.

[145] Yamamoto, K., K. Uesugi, and T. Watanabe. 2010. "Adaptive Routing of Cruising Taxis by Mutual Exchange of Pathways." *Knowledge-Based Intelligent Information and Engineering Systems* 5178:559–566.

[146] Yan, X., and J. Han. 2002. "Gspan: Graph-Based Substructure Pattern Mining." In proceedings, *2002 IEEE International Conference on Data Mining*. Washington, DC: IEEE Computer Society Press, 721–724.

[147] Yang, Z., J. Hu, Y. Shu, P. Cheng, J. Chen, and T. Moscibroda. "Mobility Modeling and Prediction in Bike-Sharing Systems." In *ACM International Conference on Mobile Systems, Applications, and Services*. New York: ACM.

[148] Ye, M., Y. Yin, W. Q. Lee, and D. L. Lee. 2011. "Exploiting Geographical Influence for Collaborative Point-of-Interest Recommendation." In *Proceedings of the 34th International ACM SIGIR Conference on Research and Development in Information Retrieval*. New York: ACM.

[149] Ye, Y., Y. Zheng, Y. Chen, J. Feng, and X. Xie. 2009. "Mining Individual Life Pattern Based on Location History." In *Proceedings, 2009 Tenth International Conference on Mobile Data Management: Systems, Services and Middleware*. Washington, DC: IEEE Computer Society Press, 1–10.

[150] Yin, Z., L. Cao, J. Han, C. Zhai, and T. Huang. 2011. "Geographical Topic Discovery and Comparison." In *Proceedings of the 20th International Conference on World Wide Web*. New York: ACM, 247–256.

[151] Yoon, H., Y. Zheng, X. Xie, and W. Woo. 2010. "Smart Itinerary Recommendation Based on User-Generated GPS Trajectories." In *Proceedings of the 7th International Conference on Ubiquitous Intelligence and Computing*. Berlin: Springer-Verlag, 19–34.

[152] Yoon, H., Y. Zheng, X. Xie, and W. Woo. 2011. "Social Itinerary Recommendation from User-Generated Digital Trails." *Journal of Personal and Ubiquitous Computing* 16 (5): 469–484.

[153] Yu, P. S., S. T. Chen, and I. F. Chang. 2006. "Support Vector Regression for Real-Time Flood Stage Forecasting." *Journal of Hydrology* 328 (3): 704–716.

[154] Yuan, J., Yu Zheng, and Xing Xie. 2012. "Discovering Regions of Different Functions in a City Using Human Mobility and POIs." In *Proceedings of the 18th SIGKDD Conference on Knowledge Discovery and Data Mining*. New York: ACM.

[155] Yuan, J., Yu Zheng, and Xing Xie. 2012. "Segmentation of Urban Areas Using Road Networks." *Microsoft Technical Report*, July 1.

[156] Yuan, J., Y. Zheng, X. Xie, and G. Sun. 2011. "Driving with Knowledge from the Physical World." In *Proceedings of the 17th SIGKDD Conference on Knowledge Discovery and Data Mining*. New York: ACM, 316–324.

[157] Yuan, J., Y. Zheng, X. Xie, and G. Sun. 2013. "T-Drive: Enhancing Driving Directions with Taxi Drivers' Intelligence." *IEEE Transactions on Knowledge and Data Engineering* 25 (1): 220–232.

[158] Yuan, J., Y. Zheng, C. Zhang, X. Xie, and Guangzhong Sun. 2010. "An Interactive-Voting Based Map Matching Algorithm." In *Proceedings of the Eleventh International Conference on Mobile Data Management*. Washington, DC: IEEE Computer Society Press.

[159] Yuan, J., Y. Zheng, C. Zhang, W. Xie, X. Xie, G. Sun, and Y. Huang. 2010. "T-Drive: Driving Directions Based on Taxi Trajectories." In *Proceedings of the ACM SIGSPATIAL Conference on Advances in Geographical Information Systems*. New York: ACM, 99–108.

[160] Yuan, J., Y. Zheng, L. Zhang, X. Xie, and G. Sun. 2011. "Where to Find My Next Passenger?" In *Proceedings of the 13th ACM International Conference on Ubiquitous Computing*. New York: ACM, 109–118.

[161] Yuan, N. J., F. Zhang, D. Lian, K. Zheng, S. Yu, and X. Xie. 2013. "We Know How You Live: Exploring the Spectrum of Urban Lifestyles." In *Proceedings of the First ACM Conference on Online Social Networks*. New York: ACM, 3–14.

[162] Yuan, N. J., Y. Zheng, X. Xie, Y. Wang, K. Zheng, and H. Xiong. 2015. "Discovering Urban Functional Zones Using Latent Activity Trajectories." *IEEE Transactions on Knowledge and Data Engineering* 27 (3): 1041–4347.

[163] Yuan, N. J., Y. Zheng, L. Zhang, and X. Xie. 2014. "T-Finder: A Recommender System for Finding Passengers and Vacant Taxis." *IEEE Transactions on Knowledge and Data Engineering* 25 (10): 2390–2403.

[164] Zhang, C., Y. Zheng, X. Ma, and J. Han. 2015. "Assembler: Efficient Discovery of Spatial Co-evolving Patterns in Massive Geo-Sensory Data." In *Proceedings of the 21st ACM SIGKDD International Conference on Knowledge Discovery and Data Mining*. New York: ACM, 1415–1424.

[165] Zhang, D., and T. He. 2013. "CallCab: A Unified Recommendation System for Carpooling and Regular Taxicab Services." In *Proceedings of the IEEE International Conference on Big Data*. Washington, DC: IEEE Computer Society Press, 439–447.

[166] Zhang, F., David Wilkie, Yu Zheng, and Xing Xie. 2013. "Sensing the Pulse of Urban Refueling Behavior." In *Proceedings of the 15th ACM International Conference on Ubiquitous Computing*. New York: ACM.

[167] Zhang, F., Nicholas Jing Yuan, David Wilkie, Yu Zheng, and Xing Xie. 2015. "Sensing the Pulse of Urban Refueling Behavior: A Perspective from Taxi Mobility." *ACM Transactions on Intelligent Systems and Technology* 6 (3), article no. 37.

[168] Zhang, J., Y. Zheng, and D. Qi. 2017. "Deep Spatio-temporal Residual Networks for Citywide Crowd Flows Prediction." In *Proceedings of the 31st AAAI Conference*. Menlo Park, CA: AAAI Press.

[169] Zhang, J., Y. Zheng, D. Qi, R. Li, and X. Yi. 2016. "DNN-Based Prediction Model for Spatial-Temporal Data." In *Proceedings of the 24th ACM International Conference on Advances in Geographical Information Systems*. New York: ACM.

[170] Zhang, Siyuan, Lu Qin, Yu Zheng, and Hong Cheng. 2016. "Effective and Efficient: Large-Scale Dynamic City Express." *IEEE Transactions on Data Engineering* 28 (12): 3203–3217.

[171] Zheng, Y. 2011. "Location-Based Social Networks: Users." In *Computing with Spatial Trajectories*, edited by Y. Zheng and X. Zhou, 243–276. Berlin: Springer.

[172] Zheng, Y. 2012. "Tutorial on Location-Based Social Networks." In *Proceedings of the International Conference on World Wide Web*. New York: ACM.

[173] Zheng, Y. 2015. "Methodologies for Cross-Domain Data Fusion: An Overview." *IEEE Transactions on Big Data* 1 (1): 16–34.

[174] Zheng, Y. 2015. "Trajectory Data Mining: An Overview." *ACM Transactions on Intelligent Systems and Technology* 6 (3), article no. 29.

[175] Zheng, Y., L. Capra, O. Wolfson, and H. Yang. 2014. "Urban Computing: Concepts, Methodologies, and Applications." *ACM Transactions on Intelligent Systems and Technology* 5 (3): 38–55.

[176] Zheng, Y., Xuxu Chen, Qiwei Jin, Yubiao Chen, Xiangyun Qu, Xin Liu, Eric Chang, Wei-Ying Ma, Yong Rui, and Weiwei Sun. 2014. "A Cloud-Based Knowledge Discovery System for Monitoring Fine-Grained Air Quality." *Microsoft Technical Report*, MSR-TR-2014-40, March 1.

[177] Zheng, Y., Y. Chen, Q. Li, X. Xie, and W.-Y. Ma. 2010. "Understanding Transportation Modes Based on GPS Data for Web Applications." *ACM Transactions on the Web* 4 (1): 1–36.

[178] Zheng, Y., Y. Chen, X. Xie, and Wei-Ying Ma. 2009. "GeoLife2.0: A Location-Based Social Networking Service." In *Proceedings of the 10th International Conference on Mobile Data Management*. Washington, DC: IEEE Computer Society Press.

[179] Zheng, Y., X. Feng, Xing Xie, Shuang Peng, and James Fu. 2010. "Detecting Nearly Duplicated Records in Location Datasets." In *Proceedings of the ACM SIGSPATIAL Conference on Advances in Geographical Information Systems*. New York: ACM.

[180] Zheng, Y., and J. Hong. 2012. *Proceedings of the 4th International Workshop on Location-Based Social Networks*. New York: ACM.

[181] Zheng, Y., Q. Li, Y. Chen, and X. Xie. 2008. "Understanding Mobility Based on GPS Data." In *Proceedings of the 11th International Conference on Ubiquitous Computing*. New York: ACM, 312–321.

[182] Zheng, Y., F. Liu, and H.P. Hsieh. 2013. "U-Air: When Urban Air Quality Inference Meets Big Data." In *Proceedings of the 19th SIGKDD Conference on Knowledge Discovery and Data Mining*. New York: ACM, 1436–1444.

[183] Zheng, Y., L. Liu, L. Wang, and X. Xie. 2008. "Learning Transportation Mode from Raw GPS Data for Geographic Application on the Web." In *Proceedings of the 17th International Conference on World Wide Web*. New York: ACM, 247–256.

[184] Zheng, Y., T. Liu, Yilun Wang, Yanchi Liu, Yanmin Zhu, and Eric Chang. 2014. "Diagnosing New York City's Noises with Ubiquitous Data." In *Proceedings of the 16th ACM International Joint Conference on Pervasive and Ubiquitous Computing*. New York: ACM.

[185] Zheng, Y., Y. Liu, J. Yuan, and X. Xie. 2011. "Urban Computing with Taxicabs." In *Proceedings of the 13th International ACM Conference on Ubiquitous Computing*. New York: ACM, 89–98.

[186] Zheng, Y., and M. F. Mokbel. 2011. *Proceedings of the 3rd ACM SIGSPATIAL International Workshop on Location-Based Social Networks*. New York: ACM.

[187] Zheng, Y., and X. Xie. 2009. "Learning Location Correlation from GPS Trajectories." In *Proceedings of the Eleventh International Conference on Mobile Data Management*. Washington, DC: IEEE Computer Society Press, 27–32.

[188] Zheng, Y., and X. Xie. 2011. "Learning Travel Recommendations from User-Generated GPS Traces." *ACM Transactions on Intelligent Systems and Technology* 2 (1): 2–19.

[189] Zheng, Y., and X. Xie. 2011. "Location-Based Social Networks: Locations." In *Computing with Spatial Trajectories*, edited by Y. Zheng and X. Zhou, 277–308. Berlin: Springer.

[190] Zheng, Y., X. Xie, and W.-Y. Ma. 2010. "GeoLife: A Collaborative Social Networking Service among User, Location and Trajectory." *IEEE Data Engineering Bulletin* 33 (2): 32–40.

[191] Zheng, Y., X. Yi, M. Li, R. Li, Z. Shan, E. Chang, and T. Li. 2015. "Forecasting Fine-Grained Air Quality Based on Big Data." In *Proceedings of the 21st SIGKDD Conference on Knowledge Discovery and Data Mining*. New York: ACM.

[192] Zheng, Y., H. Zhang, and Y. Yu. 2015. "Detecting Collective Anomalies from Multiple Spatio-temporal Datasets across Different Domains." In *Proceedings of the 23rd ACM International Conference on Advances in Geographical Information Systems*. New York: ACM, 2.

[193] Zheng, Y., L. Zhang, Z. Ma, X. Xie, and W.-Y. Ma. 2011. "Recommending Friends and Locations Based on Individual Location History." *ACM Transactions on the Web* 5 (1), article no. 5.

[194] Zheng, Y., L. Zhang, X. Xie, and W.-Y. Ma. 2009. "Mining Interesting Locations and Travel Sequences from GPS Trajectories." In *Proceedings of the 18th International Conference on World Wide Web*. New York: ACM, 791–800.

[195] Zheng, V. W., B. Cao, Y. Zheng, X. Xie, and Q. Yang. 2010, July. "Collaborative Filtering Meets Mobile Recommendation: A User-Centered Approach." In *Proceedings of the Twenty-Fourth AAAI Conference on Artificial Intelligence*. Volume 10. Menlo Park, CA: AAAI Press, 236–241.

[196] Zheng, V. W., Y. Zheng, X. Xie, and Q. Yang. 2010. "Collaborative Location and Activity Recommendations with GPS History Data." In *Proceedings of the 19th International Conference on World Wide Web*. New York: ACM, 1029–1038.

[197] Zheng, V. W., Y. Zheng, X. Xie, and Q. Yang. 2012. "Towards Mobile Intelligence: Learning from GPS History Data for Collaborative Recommendation." *Artificial Intelligence* 184:17–37.

[198] Zhu, J. Y., C. Zhang, H. Zhang, S. Zhi, Victor O. K. Li, J. Han, and Y. Zheng. 2017. "pg-Causality: Identifying Spatiotemporal Causal Pathways for Air Pollutants with Urban Big Data." *IEEE Transactions on Big Data*. doi:10.1109/TBDATA.2017.2723899.

[199] Zhu, J. Y., Y. Zheng, Xiuwen Yi, and Victor O. K. Li. 2016. "A Gaussian Bayesian Model to Identify Spatiotemporal Causalities for Air Pollution Based on Urban Big Data." Paper presented at the IEEE InfoCom Workshop on Smart Cities and Urban Computing, San Francisco, CA.

[200] Zimmerman, J., A. Tomasic, C. Garrod, D. Yoo, C. Hiruncharoenvate, R. Aziz, N. R. Thiruvengadam, Y. Huang, and A. Steinfeld. 2011. "Field Trial of Tiramisu: Crowd-Sourcing Bus Arrival Times to Spur Co-Design." In *Proceedings of the 2011 Annual Conference on Human Factors in Computing Systems*. New York: ACM, 1677–1686.

II Urban Sensing and Data Acquisition

3 Urban Sensing

Abstract: This chapter introduces four paradigms of urban sensing, discussing the challenges of and the techniques for each paradigm. It also shows a general framework of urban sensing as well as rich examples for each paradigm of urban sensing.

3.1 Introduction

Technological advances in sensing, computation, and communications have provided an unprecedented opportunity to collect a wide spectrum of data in a metropolitan area. Urban sensing, as the foundation of urban computing, probes the pulse and rhythm of a city by collecting citywide data from sensors and people. The data is then managed and analyzed in the upper layers of the urban-computing framework. Though the concept of urban sensing has been previously discussed in [7, 8, 29] in 2008, the evolution of urban science and computing technology has generated new functions and meanings that enrich the intension and extension of urban sensing. This section introduces each paradigm of urban sensing, presenting its representative applications and discussing its key challenges and technology.

3.1.1 Four Paradigms of Urban Sensing

There are two main urban-sensing paradigms, consisting of sensor-centric and human-centric sensing, as shown in figure 3.1. The former paradigm further comprises two subcategories: static sensing and mobile sensing. The latter consists of two subcategories: passive and active crowd sensing.

3.1.1.1 Static sensor-centric sensing This sensing paradigm deploys a collection of static sensors in (fixed) locations (e.g., at meteorological monitoring stations), as shown in figure 3.1a. Once a sensor has been deployed, its location will not change

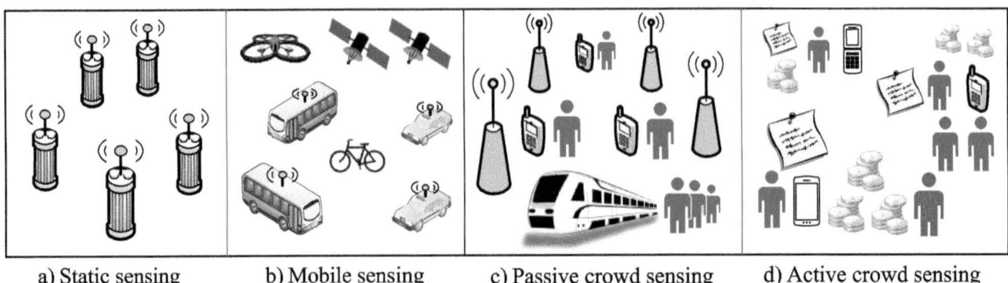

Figure 3.1
Four urban-sensing paradigms.

over time. These sensors then send data to a back-end system automatically with a certain frequency (e.g., every hour). Each sensor can send data respectively, alternatively first formulating a sensor network with its neighbors and then uploading data collectively through some gateways in the network. The majority of sensing systems deployed in current cities, such as loop detectors for sensing traffic conditions and air quality stations for sensing the concentrations of air pollutants, belong to this category. Such an urban-sensing paradigm faces two main challenges.

First, because of resource constraints like budgets, land use, and workforce, the number of sensors we can deploy in a city is usually very limited, resulting in sparse data coverage that poses challenges for citywide monitoring and further data analytics. For instance, while hoping to monitor citywide traffic conditions, we can usually deploy loop detectors on some main roads due to different kinds of constraints. As traffic flow varies by location significantly and nonlinearly, traffic conditions on roads without a loop detector cannot be linearly interpolated based on others with a sensor.

Second, there are significant challenges to selecting the proper locations for deploying sensors. One is measuring the quality of a location to deploy a sensor before the sensor has been deployed—for example, where to deploy four additional air quality–monitoring stations to maximize improvement in the air quality monitoring of an entire city. Unfortunately, one does not know the ground truth of air quality at a location before a station is deployed. Additionally, there are usually many candidates to select, leading to high computing complexity for candidate selection. For instance, selecting three road intersections at which to deploy gas stations so these stations cover the maximum number of vehicles is equivalent to a submodular maximization problem, which is NP-hard.

3.1.1.2 Mobile sensor-centric sensing This paradigm deploys sensors with moving objects, such as bikes, buses, taxies, drones, and satellites, as illustrated in figure 3.1b. A sensor's location changes along the movement of a moving object, continuously collecting and sending data to a back-end system at different locations. Once finished deploying these sensors, people are no longer actively involved with the sensing loop. For example, we can deploy a sensor with many buses to probe the air quality of a city. The collected air quality data is automatically sent to a back-end system (e.g., a cloud-computing platform) every ten minutes through a wireless communication module. The power need for a sensor and communication module can be supplied by the battery of a bus. As a sensor moves around a city, the coverage of data collected by this paradigm should be larger than using static sensing, given the same number of sensors. In addition, it is more flexible and nonobtrusive to deploy sensors in a mobile sensor-centric paradigm than in the static paradigm.

Such a sensing paradigm also faces its own challenges. Because the movement of moving objects is highly skewed in a city, this sensing paradigm may result in imbalanced data distribution. The collected data is very sparse in those locations with few moving objects traversing them. On the contrary, we may have redundant data in other places with massive moving objects passing by. As a coin has two sides, while the noninvolvement of people in this sensing paradigm reduces human effort, we lose control over what data can be collected.

3.1.1.3 Passive crowd sensing This sensing paradigm collects the data people passively generate when using urban infrastructure, such as wireless communication systems and public transportation systems, as illustrated in figure 3.1c. In such a sensing program, people do not even know they are contributing data, let alone its purpose. For instance, while wireless cellular networks are built for mobile communication between individuals, massive quantities of mobile phone signals can help understand citywide commuting patterns, thereby improving urban planning. Likewise, passengers' card-swiping data in subway stations or at bus stops can describe people's commuting patterns in a city, though such a ticketing system was originally created for charging people's travel expenses. People have no idea that they are performing a sensing task when traveling through an automatic fare gate. Other examples include people's check-in/out records in a bike-sharing system and the taxi fare data of passengers. The challenge this sensing paradigm faces is the scale of urban infrastructure, which constrains the geographical scope of a sensing task. How to protect people's privacy while utilizing their data is a major concern.

3.1.1.4 Active crowd sensing This sensing paradigm is a combination of crowd sourcing [24] and participatory sensing [6]. As illustrated in figure 3.1d, it usually consists of a sensing program owner and a group of participants. The program owner defines a sensing task and provides a budget to achieve the task. A task could be comprised of the type of data to be collected, a geographical region and a time interval in which this type of data is expected to be collected, and an incentive mechanism to encourage people to contribute data. People can decide where and when to join such a sensing program based on their availability and the incentives a sensing program offers. While participating in a sensing program, people actively collect the information around them and contribute their own data to formulate collective knowledge that can solve a problem. People explicitly know the purpose of a collection task and what they have contributed to a participatory sensing program.

When there are many participants and a limited budget, an active crowd-sensing program would involve participant recruitment and task assignment processes. The former process selects quality participants according to their historical performance and the potential contribution of the data (they can collect) to the entire sensing program. The task design process can customize a task (e.g., where and when to collect a type of data) for a particular participant in terms of location and time.

The challenges that an active sensing paradigm faces are fourfold. First, the distribution (and mobility) of people is intrinsically skewed in a city, resulting in imbalanced data coverage. While data is redundantly received in some locations, there may be a lack of data in other places. Second, it is difficult to measure the quality (concerned with balance and amount) of data that has already been collected. Apparently, the data collected in a location where it is lacking should be valued more than those with sufficient data. Third, we do not want to recruit a participant who would contribute noisy or fake data, or who cannot fulfill their commitment. However, it is not easy to evaluate a participant's performance, as in most cases there is no ground truth for the data we want to collect. Fourth, the task design process is usually very consuming of computational resources, meaning sometimes it is an NP-hard problem.

To advance participatory sensing in an active crowd-sensing program, Campbell [7, 8] and Lane et al. [32] propose an opportunistic sensing concept, in which a participant's device (e.g., cell phone) is automatically utilized whenever its state (e.g., geographic location or body location) matches the requirements of an application. To maintain transparency, opportunistic use of a device should not noticeably impact the normal user experience of a participant as it is used for the participant's own needs [7, 8]. According to Campbell's statement, the primary challenges in opportunistic sensing are determining when the state of the sensing device matches the requirements of applications and sampling when the device state and participants' requirements (i.e., privacy and transparency) are met [8].

3.1.2 General Framework of Urban Sensing

Figure 3.2 presents a general framework for urban sensing, which comprises five layers.

3.1.2.1 The first layer The first layer consists of four elements defining a sensing program: the geographical location, the time interval in which a sensing program is to be held, a budget for the program, and a specific task, such as sensing urban noise or traffic conditions. The four elements are defined by a sensing program owner before a program starts.

3.1.2.2 The second layer The second layer aims to deploy sensors or assign tasks to people so they can collect data for a sensing program. In a sensor-centric sensing program, a sensor deployment process is called to install a number of sensors at proper locations or with some moving objects, constrained by the budget given in the first layer. This process aims to optimize sensing results according to different measurements through some deploying models. Section 3.3 will introduce four types of models.

In active crowd sensing, we need to first select some quality participants according to some measurements. Sometimes, a task design process is utilized to customize a task for a particular participant according to the participant's preference or constraints. For

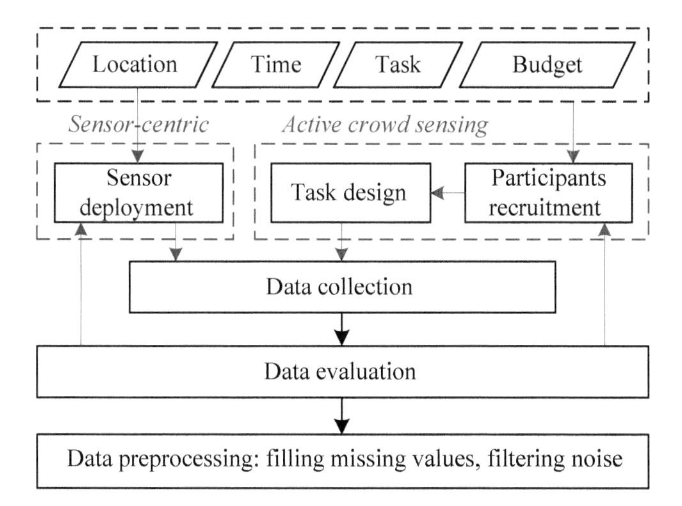

Figure 3.2
The general framework for urban sensing.

example, Ji et al. [29] design a task, composed of a sequence of geographical locations (and visiting time intervals), to sense urban noise for a particular participant based on the participant's origin, destination, departure times, and arrival times. As a passive sensing program leverages existing infrastructures that have already been built for other purposes, it is not necessary to select participants or design tasks or deploy sensors in such a program.

3.1.2.3 The third layer The third layer collects data from sensors or participants through different kinds of approaches. Typically, in a static sensor-centric sensing program, each sensor can send data respectively or first formulate a sensor network with its neighbors and then upload data collectively through gateways in the network. Regarding mobile sensor-centric sensing, each sensor usually sends data respectively, as moving objects may travel individually and randomly. The most frequently used devices for active crowd sensing programs are mobile phones, each of which is equipped with quite a few sensors (such as three-dimensional accelerators, gyroscopes, microphones, cameras, and GPS devices) and communication modules like GSM, Wi-Fi, Bluetooth, and near field communication (NFC). The collected data then streams to a cloud-computing platform.

3.1.2.4 Quality evaluation After data has been collected, we need to evaluate its quality. For a sensor-centric sensing program, the data evaluation process can help determine whether additional sensors need to be deployed and where they should be placed. For example, we have deployed thirty-six air quality–monitoring stations in Beijing. How effective is the data from these stations in monitoring air quality for the entire city? This is a very challenging task as we do not know the ground truth of air quality in a place before a station has been deployed there. Zheng et al. [53] solve this problem through a machine-learning inference approach. The general idea is to train an inference model based on data from existing stations and apply the model to infer the air quality of locations without a station. If the confidence of an inference on a location's air quality is very high, it is not necessary to deploy a station there. On the contrary, top-*k* locations with the most uncertain inference should be considered as a place of deployment. Based on this idea, Hsieh et al. [25] suggest the locations for deploying additional stations when a new budget is available.

For an active crowd-sensing program, a good data evaluation metric can help determine if more participants should be recruited. It can even help select participants before an active crowd-sensing program truly takes place. For example, Ji et al. [29] propose a hierarchical entropy-based metric to evaluate the quality of data that has been collected or is planned to be collected, considering both the balance and amount of data. This metric is then employed to select candidate participants who could collect data to improve the metric significantly.

3.1.2.5 Data preprocessing After a sensing program closes, we need to do some data preprocessing, such as filling in missing values and filtering noisy data. Because of communication or device errors, we may lose the data from some sensors at some intervals. This poses challenges to real-time monitoring and further data analytics. To address this issue, Yi et al. [51] propose collectively filling missing values in geosensory data through a multi-view learning approach, which considers the temporal correlation between readings at different time stamps in the same series and the spatial correlation between different time series.

3.2 Sensor and Facility Deployment

In an urban-sensing program, we need to deploy sensors and facilities at static locations or with moving objects before collecting data. Where to deploy these sensors and facilities becomes the first problem we need to solve. This section introduces four types of sensor and facility deployment models, which can be used to address different applications' issues.

3.2.1 Finding Optimal Meeting Points

Finding optimal meeting points, also known as the *facility location problem*, has been studied extensively in operations research and database communities. Based on the objective function, models proposed for solving this problem can be divided into two categories: the minsum [3] and minmax [11] models. The minsum model aims to locate k facilities such that the average cost to reach all clients can be minimized, while the minmax model expects to minimize the maximum cost to reach those clients.

Given two datasets $C=\{c_1, c_2, \ldots, c_n\}$ denoting clients and $F=\{f_1, f_2, \ldots, f_m\}$ representing facilities, the facility location problem is to find a k-location set R from F, such that

$$\text{minsum model: } R = arg\ Min_{R' \subset F} \sum_i cost(c_i, R'),\ where\ |R'| = k; \quad (3.1)$$

$$\text{minmax model: } R = arg\ Min_{R' \subset F}\ Max_i\ cost(c_i, R'),\ where\ |R'| = k; \quad (3.2)$$

where $cost(c_i, R')$ denotes the minimum cost between c_i and any facility $f \in R'$. The cost can be travel time, distances, or expenses. When using distance as a cost, there are two main measures: the Euclidean distance and the network distance.

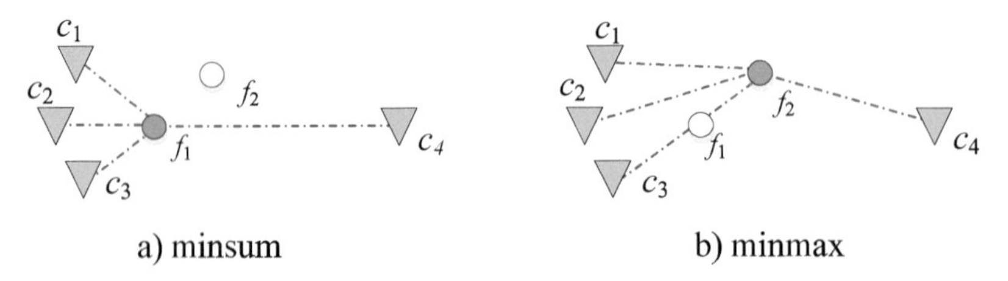

Figure 3.3
Examples of minsum and minmax models.

Figure 3.3 presents an example for the two models, respectively, using the Euclidean distance as a cost measure. When set $k=1$ (i.e., finding the best place), the minsum model will select f_1 as the optimal meeting point for the four clients c_1, c_2, c_3, and c_4. However, the minmax model will select f_2 because the distance between f_1 and c_4 is bigger than the distances between f_2 and clients. The two models have different application scenarios. For example, governments want to set up k polling stations for an election such that voters' average travel distance to their closest stations can be minimized. This problem should be solved by the minsum model. Another example is to use the minmax model to set up fire stations in a city so that the maximum travel time to reach a fire is minimized.

The facility location problem is NP-hard no matter which model is used to solve this problem. To this end, a branch of research—for example, the local search heuristic [3], the pseudo-approximation [36], and the farthest-point clustering heuristic [21]—has been conducted to solve this problem approximately with a performance guarantee.

3.2.1.1 Local search heuristic and pseudo-approximation Regarding the minsum model, the local search heuristic [3] is a widely used solution, which provides an approximation within five times the optimum value. Given an initial location set R_{ini}, the local search heuristic tentatively replaces each current location in R_{ini} by every candidate $f \in \{F - R_{ini}\}$ and estimates the distance reduction of this replacement. Among all these tentative replacements, the one with the maximum distance reduction is carried out. A well-known implementation of the local search heuristic is the partitioning around medoids (PAM) [14].

Figure 3.4 illustrates the PAM with an example that aims to select two optimal locations out of three candidates $\{f_1, f_2, f_3\}$ for four clients $\{c_1, c_2, c_3, c_4\}$. The initial location set is $\{f_1, f_2\}$ with an overall distance of 17. If we replace f_1 by f_3, as shown in figure 3.4b, the overall distance is reduced to 13. Likewise, the overall distance is reduced to 12 if

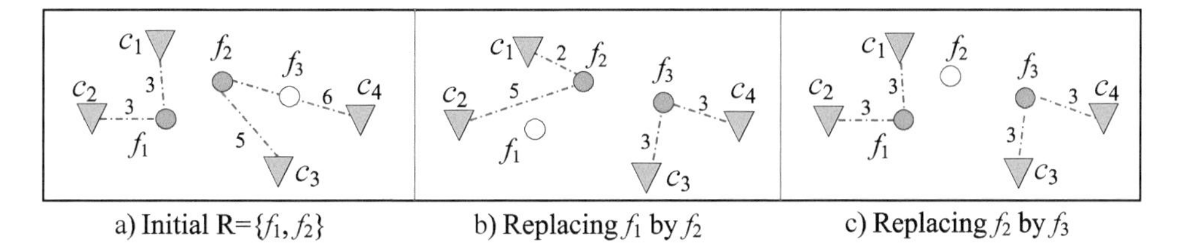

Figure 3.4
An example of PAM (a local search heuristic solution for the minsum problem).

we replace f_2 by f_3, as depicted in figure 3.4c. That is, the latter replacement is more effective. As a result, $\{f_1, f_3\}$ are selected as the optimal 2-location set. The approximation guarantee can be further improved to $3+2/p$ if p locations can be swapped simultaneously [3]. In practice, the actual gap is usually much smaller than these upper bounds. Except for the local search heuristic, the pseudo-approximation has been proposed to improve the approximation guarantee from $3+2/p$ to $1+\sqrt{3}+\varepsilon$, with a longer running time [36].

3.2.1.2 Farthest-point clustering heuristic With respect to the minmax model, the farthest-point clustering heuristic [21] is a widely used algorithm that can provide an approximation within two times the optimal value. Using this algorithm to find the k-optimal location set is also known as the *k-center problem*. Figure 3.5 presents an example of the farthest-point clustering heuristic, which finds the three-optimal location set from four location candidates for five clients. The algorithm first selects f_1 to the result set R arbitrarily. Afterward, f_4 is added to the result set, as it has the farthest distance to $R=\{f_1\}$. Finally, f_3 is added to R, as $d(f_3, R)$ is maximal when $R=\{f_1, f_4\}$, where $d(f_3, R)$ denotes the minimum distance between f_3 and any facility in R. Apparently, $dist(f_3, f_4)$ is bigger than $dist(f_1, f_2)$. As a consequence, the final result set $R=\{f_1, f_3, f_4\}$.
Formally, the procedure of the algorithm can be defined as follows:

1. Select a location $f \in F$ arbitrarily and add it to the result set R.
2. Calculate the distance between f and each facility $f' \subset F-R$ and select the minimum distance to represent $d(f', R)$.
3. Add $f = arg\ max_{f' \subset F-R}\ d(f', R)$ to R until $|R|=k$.

Though the aforementioned approximation solutions significantly reduce running time for solving the facility location problem, they are still very time-consuming when facing big data. To expedite the local search heuristic in the Euclidean space, [14] proposes several database techniques (e.g., candidates grouping and best-first search) to

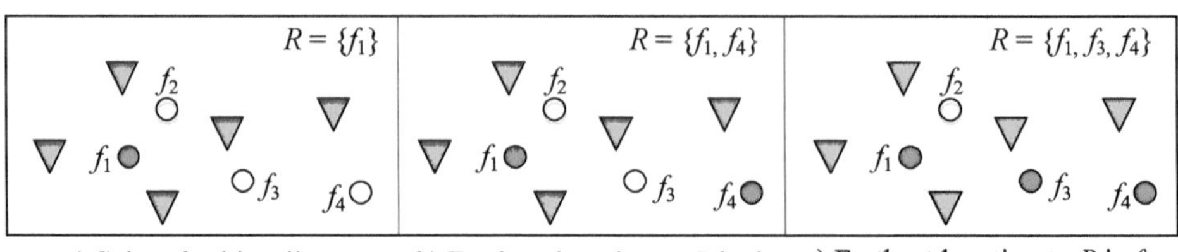

Figure 3.5
An example of the minmax model.

prune unpromising location candidates efficiently. [50] proposes an efficient framework that considers both Euclidean and network distances. However, this framework can only solve the facility location problem with $k=1$.

In some application scenarios, where some facilities have already been deployed, we try to add additional facilities or move some existing ones to new locations in order to minimize the cost function [11, 43, 49]. One can simply use the adaption of the PAM and farthest-point clustering heuristic to solve these two problems with the minsum and minmax model, respectively.

Recently, Li et al. [38] suggested relocating ambulance stations in a city, trying to minimize the average travel time for reaching emergency requests of the entire city, based on a three-step method. Different from previous research only when considering the Euclidean and network distances, Li and Zheng's solution uses ambulances' real travel times as a cost measure. Figure 3.6 illustrates this method using an example with ten clients and six location candidates $\{f_1, f_2, f_3, f_4, f_5, f_6\}$:

1. As shown in figure 3.6a, this method first applies a *k-medoids clustering* algorithm to the clients ($k=2$ in this example), selecting the location candidates close to the centers of the k clusters (i.e., f_1 and f_6) as initial points for the minsum model. Good initial points can significantly reduce computational loads of further replacements.
2. In the second step, as depicted in figure 3.6b, this method groups candidates together, calculating the lower bound of the travel time between clients and a group of candidates (e.g., f_2 and f_3). The travel time between each client and a candidate location can actually be precomputed, as it will be reused multiple times during the replacement.
3. The third step conducts many iterations, each of which aims to replace a location selected in previous rounds, like f_1, with other candidates. Each iteration is further composed of many replacements, each of which tries to find a group of candidates to replace f_1. For example, we try to replace f_1 with a group of candidates (f_2 and f_3).

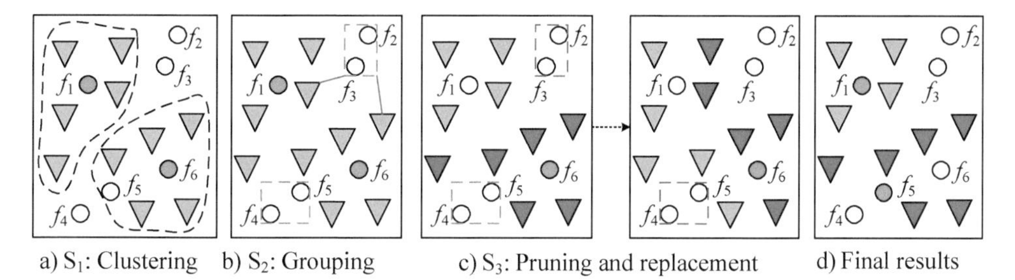

Figure 3.6
Selecting locations for ambulance stations with a minsum model.

The candidate group can be pruned; that is, it is not necessary to replace f_1 using each individual element in the group if its travel-time lower bound is bigger than that of f_1. Such a grouping and pruning strategy can reduce the number of replacements in each iteration, particularly when there are many candidates in a group. When a group of candidates (e.g., f_4 and f_5) cannot be pruned, we will further check the possibility of replacement for each element in the group. If f_1 cannot be replaced, we will then try to replace f_6 in another iteration.

3.2.2 Maximizing Coverage

3.2.2.1 The maximum coverage problem Also called the MAX-k-COVER problem, this has been studied extensively in recent decades. One typical example is to select k nodes from a graph so that these nodes' one-hop neighbors are collectively maximized. Another example is to find at most k articles aggregately containing the maximum number of topics. Formally, this problem can be defined as follows: Given an integer k, a universal set E, and a collection of subsets $S=\{s_1, s_2, \ldots, s_m\}$, select k subsets from S to formulate a union set s' so that $\sum_{ej\in s'} y_j$ is maximized, where y_j denotes if an element $e_j \in E$ is covered by s'; $y_j=1$ if e_j is covered, otherwise $y_j=0$.

The problem can also be formulated as an integer linear programming (ILP) as follows:

$$\max: \sum\nolimits_{ej\in E} y_j, \quad s.t. \sum x_i \le k, \sum\nolimits_{ej\in s_i} x_i \ge y_j, x_i \in \{0,1\}, y_j \in \{0,1\}, \tag{3.3}$$

where x_i denotes if s_i is selected; $x_i=1$ if selected in the result set, and $x_i=0$ otherwise.

For example, as shown in figure 3.7, there are three subsets $\{s_1, s_2, s_3\}$ covering eight elements in $E=\{e_1, e_2, e_3, \ldots, e_8\}$. $\{s_2, s_3\}$ is the max-2-cover set, as it contains all elements, while $\{s_1, s_3\}$ misses e_6 and $\{s_1, s_2\}$ lacks $\{e_7, e_8\}$.

The maximum coverage problem is NP-hard, thus computational intensive. The greedy heuristic algorithm proposed in [15] has the best polynomial time solution with

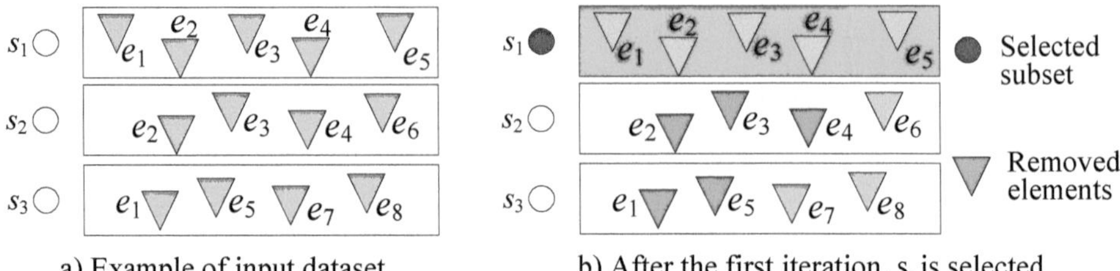

Figure 3.7

a (1 – 1/e) approximation guarantee. The greedy algorithm executes k iteration of the following two phases:

1. *Selection phase*. In this phase, the greedy heuristic algorithm inserts the subset s_i with the maximum increase of coverage to the current iteration into the result set.
2. *Updating phase*. In this phase, it removes the elements that have already been covered by s_i from unselected subsets.

Using figure 3.7b as an example, we illustrate the greedy heuristic algorithm. In the first iteration, the greedy algorithm adds subset s_1 to the result set as it covers the maximum number of elements (i.e., 5). The updating phase then removes the elements $\{e_1, e_2, e_3, e_4, e_5\}$ covered by s_1 from the unselected subsets $\{s_2, s_3\}$. Supposing $k=2$, in the second iteration, the greedy algorithm will add s_3, covering two more elements, to the result set, as s_2 only covers one more new element. As a result, the final solution is $\{s_1, s_3\}$ with seven elements covered. Apparently, this is not the optimal solution, which should be $\{s_2, s_3\}$.

3.2.2.2 The weighted maximum coverage problem In the weighted version, each element e_j in the universal set E is associated with a weight $w(e_j)$. The objective is to select k subsets that have the maximum covered weight. The weighted version can be formally defined as follows:

$$\max: \sum_{e_j \in E} w(e_j) \cdot y_j, \quad s.t. \sum x_i \le k, \sum_{e_j \in s_i} x_i \ge y_j, x_i \in \{0,1\}, y_j \in \{0,1\}. \tag{3.4}$$

The greedy heuristic algorithm can also be adopted to solve the weighted maximum coverage problem, choosing the subset with the maximum weight instead of the maximum covered elements in the selection phase. With a minor modification, the greedy algorithm also achieves an approximation ratio of (1 – 1/e).

3.2.2.3 The budget constraint maximum coverage problem In some application scenarios, a budget constraint B is specified when each subset s_i is associated with a cost $c(s_i)$. For example, different locations may need different setup costs in the facility

1. $H_1 \leftarrow argmax\{w\{G\}, s.t. G \subset S, |G| < k, and\ c(G) \leq B\}$
2. $H_2 \leftarrow \emptyset$
3. For each $R \subset S$, whose $|R| = k\ and\ c(R) \leq B$ do
4. $\quad U \leftarrow S \backslash R$
5. $\quad$ Repeat
6. $\quad\quad$ select $s_i \in U$ that maximizes $\frac{w\prime(s_i)}{c(s_i)}$
7. $\quad\quad$ if $c(R) + c(s_i) \leq B$ then
8. $\quad\quad\quad R \leftarrow R \cup s_i$
9. $\quad\quad U \leftarrow U \backslash s_i$
10. $\quad$ Until $U = \emptyset$
11. $\quad$ if $w(R) > w(H_2)$ then $H_2 \leftarrow R$
12. If $w(H_1) > w(H_2)$, output H_1, otherwise, output H_2.

Figure 3.8
Algorithm for the budget constraint maximum coverage problem.

location problem. We formally define the budget constraint maximum coverage problem as below:

$$\max: \sum_{ej \in E} w(e_j) \cdot y_j, \quad s.t. \sum c(s_i) \cdot x_i \leq B, \sum_{ej \in S_i} x_i \geq y_j, x_i \in \{0,1\}, y_j \in \{0,1\}. \tag{3.5}$$

To solve the budget constraint maximum coverage problem, [31] proposes an algorithm with an approximation ratio of $(1-1/e)$ when the fixed integer $k \geq 3$. Though k denotes the number of subsets to select, it is not a constraint for the problem. The pseudo code of this algorithm [31] is illustrated in figure 3.8.

As shown in line 1, this solution first explores all subsets of S whose cardinality is smaller than k and whose cost is less than B, finding the subset G with the biggest weight. The algorithm then searches for each subset (of S) whose cardinality equals δ and whose cost is smaller than B, denoted by R, as shown in line 3. For each R, the algorithm selects an element s_i from the rest of S (i.e., U), adding s_i to R if the updated cost is still smaller than B. A heuristic is proposed to select s_i, calculated by $w'(s_i)/c(s_i)$, where $w'(s_i)$ is the incremental weight after adding s_i. The aforementioned operation (i.e., line 6–9) repeats iteratively until all elements in U have been tested. After the test iterations, we find an approximately optimal solution for the corresponding R. The algorithm tries all possible R satisfying the criteria presented in line 3, selecting the best H_2. Finally, the one between H_1 and H_2 with a larger weight is returned as the final result (i.e., line 12).

3.2.2.4 The most influential *k*-location set problem Recently, Li et al. [37] have proposed mining the most influential k-location set traversed by the maximum number of vehicles. The problem is formally defined as follows: given a user-specified spatial

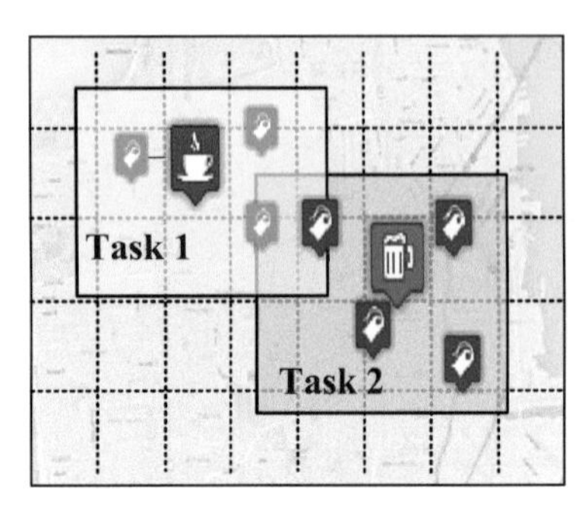

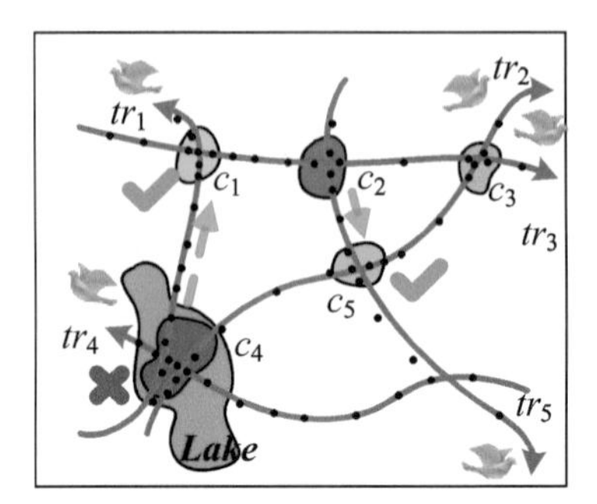

a) Location awareness b) Domain knowledge and interactions

Figure 3.9
The challenges of mining the most influential k-location set.

region R, a k value and a set of trajectories Tr, and a spatial network in R denoted as $G_s=(V_s, E_s)$, the most influential k-location set in R finds k locations (vertices) in V_s such that the total number of unique trajectories covered by the k locations is maximized.

Mining the most influential k-location set can be mapped to the MAX-k-COVER problem with two extra challenges. The first challenge is that different users may be interested in mining k locations in different spatial areas. For instance, as shown in figure 3.9a, two local business owners want to place different numbers of billboards in two different areas, calling for the computing of the most influential k-location set in the two areas, respectively. The change of a spatial area of interest leads to a new round of computing. The second challenge is that users—for example, domain experts—may need to interact with a mining system several times, adding their domain knowledge into the mining process. For example, as depicted in figure 3.9b, $\{c_2, c_4\}$ is the 2-location set that covers the highest number of migratory birds' trajectories. But, as c_4 is located in a lake where we cannot find land to place an observation station, we need to remove it from the result set. Finally, $\{c_2, c_4\}$ becomes the alternative result.

Figure 3.10 presents the framework for mining the most influential k-location set, which comprises two major modules: preprocessing and location set mining. As shown in the left portion of figure 3.10, the first module processes a trajectory dataset through three steps:

1. *Spatial network mapping* projects raw trajectories onto a spatial network, which can be a road network, a grid partition, or clusters of stay points.
2. *Inverted trajectory indexing* generates two types of indexes for each vertex. One is a vertex-trajectory index to manage the trajectories traversing a vertex; the other is a vertex-vertex index to manage trajectory IDs shared between two vertexes.

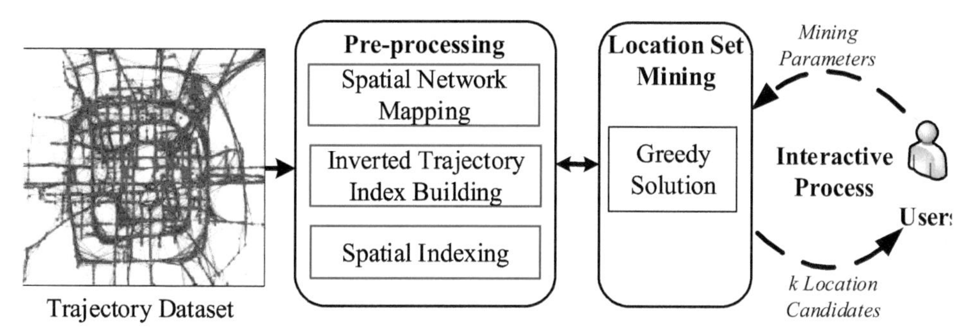

Figure 3.10
The framework of mining the most influential k-location set.

3. *Spatial indexing* organizes vertexes based on their spatial coordinates using spatial indexing structures like quadtrees or R-trees.

As presented in the right part of figure 3.10, the location set mining module takes a user's query parameters, consisting of a spatial range R, a k, and a set of premarked vertexes as an input, and finds the k locations through iterations of the following two steps: selecting and updating. In the selection phase, the algorithm selects the vertex with the maximum trajectory coverage at the current iteration and puts it in the result set. In the updating phase, the algorithm updates the coverage values for all the unselected vertices by removing the newly covered trajectories from their coverage. When a trajectory dataset is huge, however, the removing of newly covered trajectories is time-consuming. Based on the indexes built in the preprocessing module, the efficiency of the updating phase improves significantly.

3.2.3 Learning-to-Rank Candidates

In urban planning, we can rank a set of locations so that the top-k best candidates can be selected to deploy resources or facilities. For example, to open a profitable shopping mall, we usually need to select the best place by ranking a set of candidate locations based on multiple factors, such as their surrounding POIs, transportation facilities, traffic conditions, and neighborhood popularity. The ranking function of different factors can be learned from existing shopping malls and their historical revenues.

This category of models was originally proposed for information retrieval, learning from history to rank a set of candidates based on their properties. Recently, machine-learning technologies have been successfully applied to ranking, resulting in three categories of "learning-to-rank" algorithms: the pointwise, pairwise, and listwise approaches [9, 48]. Hereafter, we introduce the three approaches using the example shown in figure 3.11.

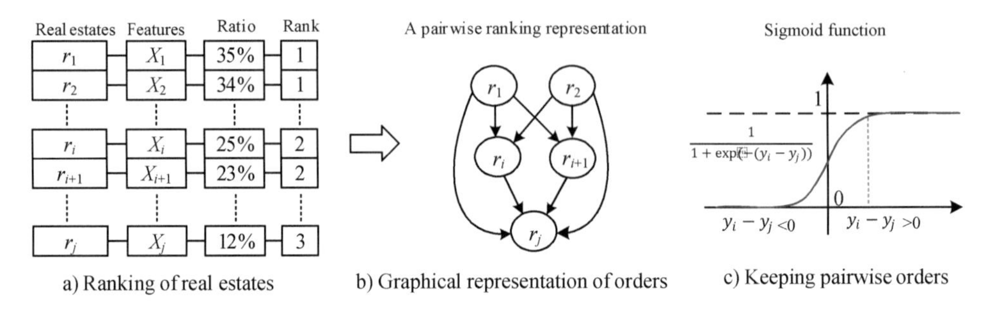

Figure 3.11
An example of ranking real estate using learning-to-rank approaches.

In this example, we aim to rank real estate locations $\{r_1, r_2, \ldots, r_i, \ldots, r_j\}$ based on a diversity of features. X_i denotes the features of real estate r_i, such as the number of bus stops and shopping malls around r_i. The third column stands for the price-increasing ratio of a piece of real estate in the past year. We can sort these locations according to their increasing ratios and further discretize the ranking into three levels, where level one is the highest rank and level three is the lowest. Sometimes, differentiating between orders caused by some small gaps (e.g., 0.35 and 0.34) is not really meaningful. Additionally, such a fine-grained ranking leads to heavy computational loads and even compromises the performance of a ranking model.

3.2.3.1 The pointwise approach This category of approach (e.g., [41]) transforms ranking into a regression or classification problem on single objects. With respect to the example, we can train a linear regression model as shown in equation (3.6) to predict the increasing ratio of a piece of real estate individually. Once the increasing ratio of each property is predicted, we can rank them accordingly.

$$y_i = \boldsymbol{\omega} \cdot X_i + \varepsilon = \sum\nolimits_m \omega_m x_{im} + \varepsilon. \tag{3.6}$$

Another pointwise approach for this example is to train a classification model (e.g., using a decision tree) to predict the discretized rank {1, 2, 3} of each individual piece of real estate. Once the rank value is predicted, we can sort these properties accordingly. As the regression and classification models only consider information from an individual property, they are pointwise approaches.

3.2.3.2 The pairwise approach This category of approach (e.g., [5, 23]) transforms ranking into classification on object pairs. That is, it classifies object pairs into two categories: correctly ranked and incorrectly ranked. Following the aforementioned example, the pairwise approach first groups every two properties into a pair, such as

(X_i, X_j), and labels each pair with either 1 or −1, where 1 denoting X_i is ranked before X_j, and −1 denoting X_i is ranked after X_j. Using such a <feature pair, label>, such as $f(X_i, X_j) \rightarrow 1$; $f(X_i, X_1) \rightarrow -1$; $f(X_j, X_1) \rightarrow -1$; $f(X_2, X_i) \rightarrow 1$, we can train a classification model to predict the rank between two properties based on their features. Once the rank between any two objects is determined, the final rank of those objects can be derived.

Toward this research direction, Herbrich et al. [23] propose Ranking SVM, which employs SVM techniques to build a classification model. Burges et al. [5] suggest RankNet, using cross entropy as a loss function to train a neural network model. RankNet has been applied to the Bing search engine. Fu et al. [16–18] propose a learning-to-rank approach with an objective function considering both the individual prediction of instances (i.e., the increasing ratio of a property) and the pairwise order between two instances. Figure 3.11b presents a graphical representation of the properties shown in figure 3.11a, using a direct edge denoting a descending order between two instances. For example, $r_i \rightarrow r_j$ means r_i is ranked higher than r_j. The objective function is presented in equation (3.7). The first part of this equation $\prod_i^N P(y_i|X_i)$ is to maximize the individual prediction accuracy of each instance. The second part ensures correct orders (between each pair of instances), represented by the direct edges shown in figure 3.11b.

$$Obj = \prod_i^N P(y_i|X_i) \prod_i^{N-1} \prod_{i+1}^N P(r_i \rightarrow r_j). \tag{3.7}$$

More specifically, the pairwise order is modeled by a sigmoid function that turns the gap between two properties' individual predictions (i.e., $y_i - y_j$) into a real value within (0, 1).

$$P(r_i \rightarrow r_j) = Sigmoid\,(y_i - y_j) = \frac{1}{1 + \exp(-(y_i - y_j))}. \tag{3.8}$$

Here is the insight of equation (3.8): If r_i is truly ranked ahead of r_j and its predicted increasing ratio is higher than r_j (i.e., a correct inference keeping the right order), $P(r_i \rightarrow r_j)$ tends to be a relatively big value. According to the distribution of the sigmoid curve, the value of $Sigmoid\,(y_i - y_j)$ approaches 1 quickly when $y_i - y_j > 0$. This gives a positive reward to correct individual predictions that keep the right order. Conversely, if r_i is truly ranked ahead of r_j but r_i's predicted increasing ratio is smaller than r_j—that is, the predictions for the two properties could be inaccurate—$P(r_i \rightarrow r_j)$ will be very small. $Sigmoid(y_i - y_j)$ approaches 0 quickly when $y_i - y_j > 0$, penalizing such an inaccurate prediction.

3.2.3.3 The listwise approach This category of approach (e.g., ListNet [9] and RankCosine [44]) takes ranked lists of objects (e.g., ranked lists of documents) as instances and trains a ranking function through the minimization of a listwise loss function defined from the predicted list and the ground truth list. The listwise approach is claimed to capture ranking problems in a more conceptually natural way than previous work [9, 44]. With a higher computational load than other approaches, the listwise approach demonstrates its own advantages based on some dataset tests.

3.2.4 Minimizing Uncertainty

In some types of sensor deployment, it is hard to explicitly model the criteria for choosing the best location for deployment. For example, we hope to find some places to deploy a few more air quality sensors to significantly improve the air quality–monitoring capabilities of an existing system. However, it is difficult to model such a criterion as we do not know the air quality of a location before deploying a sensor there. We face the same challenge when deploying sensors for monitoring water quality, meteorology, minerals, and so on.

A general approach to solving this problem is to first train an inference model based on data generated by existing sensors and then apply the model to infer the value of locations without a sensor. Intuitively, it is not necessary to deploy sensors at those locations that can always be inferred by the model confidently. On the contrary, we should deploy sensors at those locations that cannot be handled by the model. The confidence (alternatively, we can call it uncertainty) of an inference can be measured by entropy or probability. For instance, if an inference model can predict the air quality of a location with a probability distribution across different categories as <good: 0.85, moderate: 0.1, unhealthy: 0.05>, the uncertainty is very low and confidence is very high. Note that the confidence is not an accuracy, derived by an inference model without needing the ground truth of air quality in this location.

Figure 3.12 presents an example of deploying air quality sensors through minimizing the uncertainty of inference [25]. In this example, an inference model is first trained based on data from existing sensors (e.g., G, H, M, and N) and applied to predict the air quality of locations without a sensor, such as A, B, and C. The model can be a decision tree or a Bayesian network. The inference result is a probabilistic distribution across different label categories such as good, moderate, unhealthy, and so on. We can then rank locations based on the entropy of these inference results in a descent manner. The most straightforward way is to select the location with the highest entropy—that is, the most uncertain inference. However, there are many rounds of inference (e.g., hourly). The ranking of locations may change in different rounds. In addition, these uncertain locations may be correlated and near to each other. It is rather wasteful to deploy multiple sensors in a small geographical area.

To this end, the method proposed in [65] selects the location (e.g., B) with the minimum entropy—that is, the most certain inference—as a labeled location, adding such

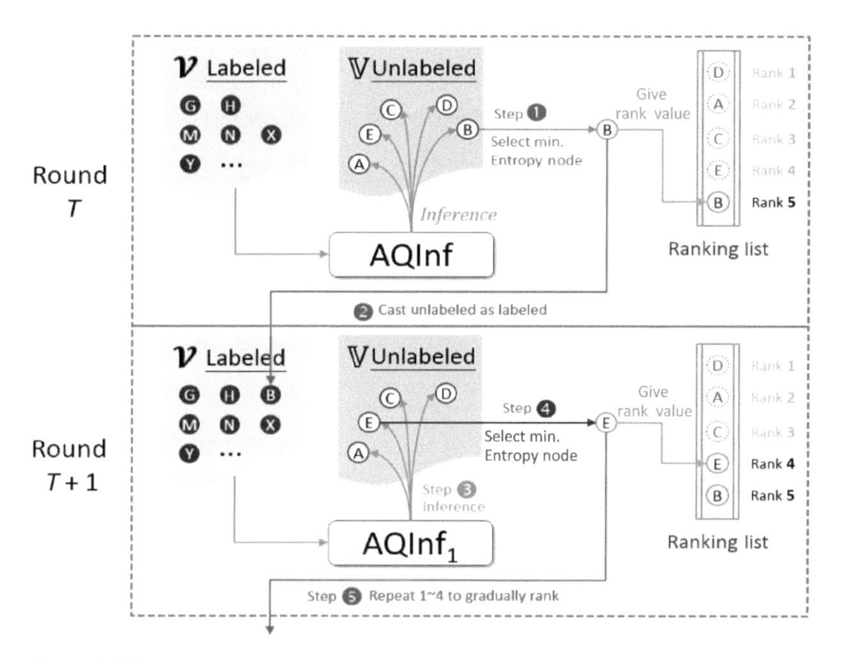

Figure 3.12
Selecting locations for deploying air quality sensors.

locations into the labeled dataset. In the next round of prediction, the inference model is retrained based on the augmented label dataset and then applied to infer the air quality of the rest of the locations, such as A, C, D, and E. Likewise, the location (e.g., E) with the minimum entropy is then added into the labeled dataset. After performing many rounds of the inference iteratively, we aggregate the rank of each location, selecting the top-*k* ranked locations to deploy new sensors.

3.3 Human-Centric Urban Sensing

Human-centric urban sensing consists of two paradigms: passive crowd sensing and active crowd sensing. Though the passive crowd-sensing paradigm may contribute more data than active crowd sensing, in reality, given the scale of existing infrastructure, we have very little control over people and sensors in such a sensing paradigm. The main challenges of the passive crowd-sensing paradigm can be solved by the upper layers of the urban-computing framework. Thus, in this section, we focus on discussing the active crowd-sensing paradigm.

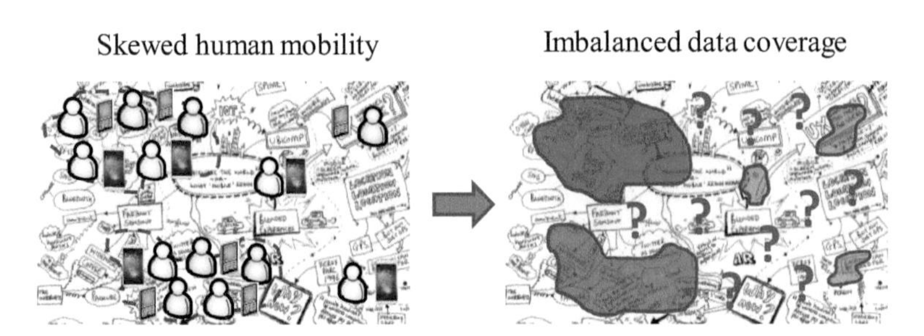

Figure 3.13
Imbalanced data coverage caused by skewed human mobility.

The first issue that we face with the active crowd-sensing paradigm is skewed human mobility, as some places like central business districts are intrinsically more crowded than other regions, such as rural areas. As illustrated in figure 3.13, this results in an imbalanced data coverage that poses challenges to a real-time monitoring program and further data analytics. On the one hand, we may receive redundant data from regions where many people cluster. Since the active crowd sensing usually encourages people to contribute data by incentive, redundant data could mean a waste of resources like money. On the other hand, we do not know what was going on in these locations with a question mark. To address this issue, we need to evaluate the quality of the data that has been collected; for example, how imbalanced is the data? Further, we need to sophistically recruit participants and creatively design proper tasks for them in order to collect a quality dataset.

3.3.1 Data Evaluation

Given a limited budget, skewed human mobility, and a large geographical space for sensing, we cannot guarantee data can be collected at any location and time interval. Under such a circumstance, it is important to have a metric with which to evaluate the quality of data that has been collected. This evaluation metric can not only measure the success of a sensing program but can also be involved in the objective function for participant recruitment and task design. Currently, there are two data evaluation approaches: coverage based and inference based.

3.3.1.1 Coverage-based data evaluation The data coverage ratio [1, 22, 54]—that is, the ratio of data collected across the whole sensing region and the entire sensing time span—is commonly used to measure the value (i.e., utility) of collected data. In general, a coverage ratio is predefined before an urban-sensing program starts. The sensing

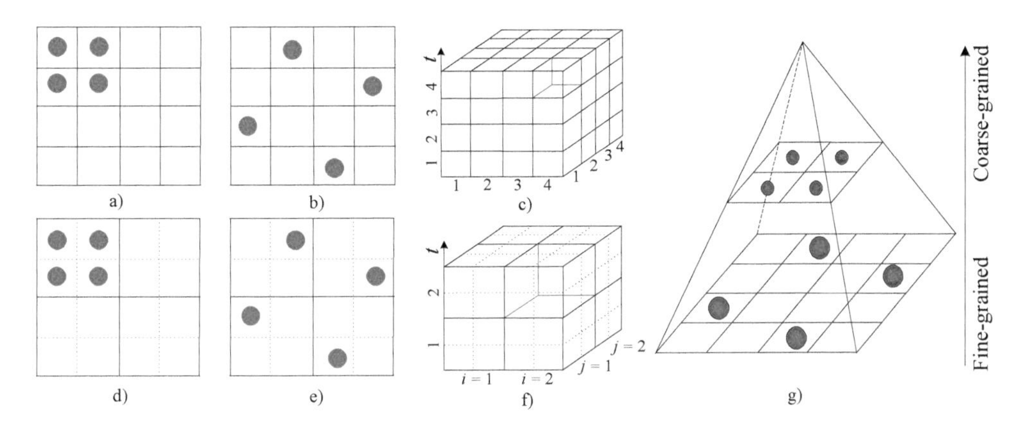

Figure 3.14
Data coverage of active crowd sensing.

program then recruits a minimum number of participants, ensuring the collected data can meet the predefined ratio [1, 22, 54]. Such a ratio ignores the distribution of data in spatial and temporal spaces, failing to reveal the balance of data. For instance, there are four locations with data collected in figure 3.14a and b. Although b seems to be having better coverage than a, they have the same coverage ratio of 0.25 and are therefore not differentiable. However, if we look at the two datasets with a coarse granularity, as depicted in figure 3.14d and e, the coverage of e becomes more balanced than d.

To address this issue, Ji et al. [29] propose a hierarchical entropy-based data coverage evaluation method that measures both the balance and amount of data in a spatiotemporal space. As a spatiotemporal space has different granularities of partitions with different geographical sizes and different lengths of temporal intervals, as shown in figure 3.14c and f, this method aggregates the entropy of data distributions at different spatial and temporal granularities in a hierarchical structure, as illustrated in figure 3.14g. More specifically, as shown in equation (3.9), the quality of data is represented by a weighted sum of two parts: data balance $E(\mathcal{A})$ and amount $Q(\mathcal{A})$.

$$\varphi(\mathcal{A}) = \alpha \cdot E(\mathcal{A}) + (1-\alpha) \cdot Q(\mathcal{A}); \tag{3.9}$$

$Q(\mathcal{A})$ is the total amount of data collected in all spatial partitions and time intervals. $\mathcal{A}_k(i,j,t)$ denotes the number of pieces of an entry at the k-th layer. For instance, there are four pieces of data collected in figure 3.14a and b, respectively.

$$Q(A) = \sum_{i,j,t} \mathcal{A}_k(i,j,t); \tag{3.10}$$

$E(\mathcal{A}_k)$ is the entropy at the k-th layer of the hierarchy shown in figure 3.14g. ω_k is the weight normalizing the scale of $E(\mathcal{A}_k)$ at different layers, with an insight that the entropies on different layers are equally important. For instance, without normalization, $E(\mathcal{A}_1)$—that is, the entropy of the bottom layer—could be dramatically larger than that of $E(\mathcal{A}_2)$. As a result, the latter will be dominated by the former, losing its contribution to $E(\mathcal{A})$.

$$E(\mathcal{A}) = \frac{\sum_k \omega_k E(\mathcal{A}_k)}{k_{max}}, \tag{3.11}$$

$$E(\mathcal{A}_k) = -\sum_{i,j,t} p(i,j,t|k) \log_2 p(i,j,t|k), \tag{3.12}$$

$$p(i,j,t|k) = \mathcal{A}_k(i,j,t) / Q(A). \tag{3.13}$$

3.3.1.2 The inference-based evaluation method As there is a spatial and temporal correlation between different location-time entries, this type of method infers the value of locations without being sensed based on partially collected data. The value of collected data can be defined as the capability of inferring missing data [47, 52]. Collected data is deemed to be valuable if missing data can be well inferred using the data collected.

3.3.2 Participant Recruitment and Task Design

The participant recruitment and task design in an urban-sensing program are usually concerned with human mobility, as people need to reach a sensing location before collecting data. Based on whether to change participants' original traveling routes, there are two approaches to participant recruitment and task assignment.

The first approach selects participants to collect data without changing their own travel routes. For example, Jaimes et al. [28] assume participants can collect data within the circles of their current locations. Zhang et al. [54] recruit participants by predicting participants' future trips using some mathematical models, such as the truncated Levy walk model [22] and the Markov chain model [1], and their historical trajectories.

The second approach is to ask participants to change their original traveling routes to collect data with higher values/utilities, as human mobility in a city is highly skewed by nature [13]. Without changing participants' original travel routes, areas with sparse human coverage cannot be sensed intrinsically. To tackle this challenge, Kawajiri et al. [30] design a reward mechanism that sets discriminative rewards for data in different areas, steering participants to collect data in areas with sparse human coverage. It is critical in this framework to model participants' reactions to the discriminative

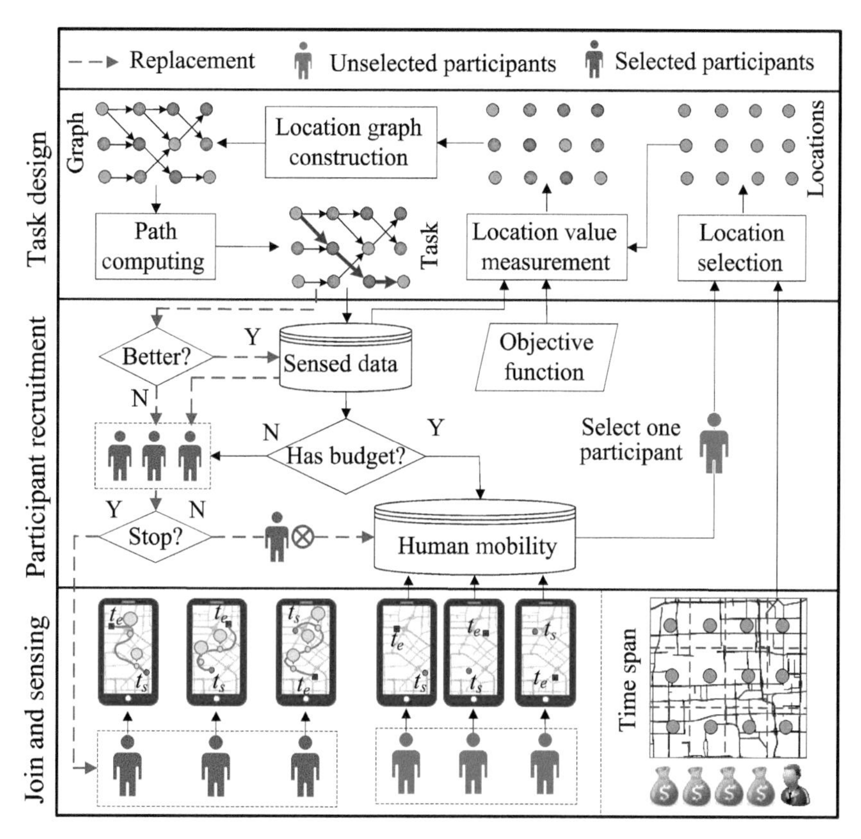

Figure 3.15
The framework of urban sensing, considering human mobility.

rewards. Such a model, however, is very difficult to build due to its many complex factors, including the participants' psychologies, the distribution of rewards, the type of data to collect, the weather conditions, and the traffic conditions.

More recently, Ji et al. [29] propose a novel task design mechanism that selects participants according to their commuting plans, consisting of the origin, destination, departure times, and arrival times, aiming to collect data with good coverage concerning both the total amount and data balance. With some incentives (e.g., money), many participants would willingly submit their commuting plans and travel on new routes passing areas with valuable data as long as they can reach their destinations before the predefined arrival times. Figure 3.15 presents the framework of the urban-sensing approach proposed in [29], which is composed of three main components: join and sensing, task design, and participant recruitment.

In the first component, as illustrated in the bottom part of figure 3.15, people can create an urban-sensing program by submitting a geographical region and a time span where data will be collected together with the type of data to be collected, a budget (e.g., the amount of money they would like to pay), and the minimum reward that a participant can obtain each hour. If interested, participants (*denoted by gray human icons*) can submit information on their mobility, consisting of an origin, a destination, a departure time, and an arrival time, through their mobile devices. If being recruited, a participant (*denoted by red human icons*) will receive a task that consists of a sequence of collecting points and corresponding time intervals where the participant should collect data. Afterward, those selected participants go collect data in the real world, following the assigned task, and return the collected data to our system through their mobile devices.

In the task design component, as depicted at the top of figure 3.15, a task is designed for each participant, based on his or her mobility and the data that is expected to be collected, through four steps. First, we check each location individually in the geographical region, finding the locations that can be reached by a participant between the departure time and arrival time. Each selected location is associated with a time interval in which data can be collected (called *collecting time interval*). Second, we measure the value of each selected location (for improving data coverage, denoted by different colors in figure 3.15) based on an objective function (e.g., equation [3.9]) and the data that is supposed to be collected at the location. In the third step, we connect two selected locations L_1 and L_2 if a participant can reach L_2 at its collecting time interval after collecting data in L_1. In this way, we can construct a location graph with each node denoting a location and with a collecting time interval and a coverage value. In the location graph, each path from the participant's origin to destination is an unobstructed task candidate. Finally, we search the location graph for a near-optimal path with near-maximum coverage value.

The participant recruitment component comprises two steps: participant selection and participant replacement [11], as demonstrated in the middle part of figure 3.15. We first select participants from a candidate pool, randomly and one by one. With the task design component, we assign each participant a task and update the data that is expected to be collected (denoted by the sensed data). At this moment, the data has not really been collected yet. The total budget is then reduced by the reward that will be given to the participant. We repeat the participant selection process until the budget has been used up. Afterward, we start the participant replacement process, which randomly replaces one participant from the selected group (*denoted by the red human icons*) with another participant from the candidate pool. If the data coverage is improved by the replacement, we keep the change; otherwise, we drop the replacement and continue to find another pair of participants to do the replacement. We repeat the replacement process until the data coverage is not improved at all after a certain number (e.g., one hundred) of consecutive attempts.

3.4 Filling Missing Values

3.4.1 Problem and Challenges

Many sensors have been deployed in the physical world, generating massive amounts of geotagged time series data. In general, as illustrated in figure 3.16a, each sensor is associated with a location where the sensor has been deployed, generating a reading at every time interval (e.g., an hour). In reality, readings are often lost at various unexpected

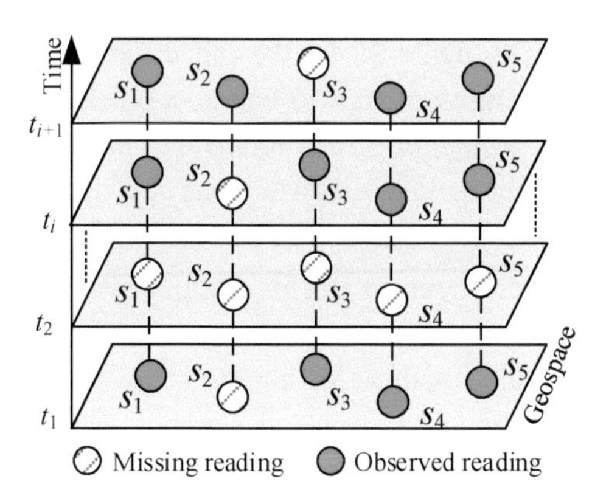

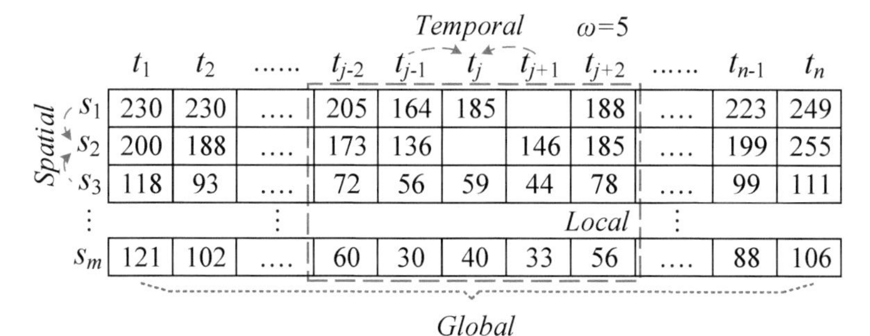

Figure 3.16
Problem statement of filling missing values in geosensory data.

moments because of sensor or communication errors. Those missing readings not only affect real-time monitoring but also compromise the performance of further data analysis.

Figure 3.16b presents m sensors' readings at n consecutive time stamps, which are stored in a form of matrix, where a row stands for a sensor, and a column denotes a time stamp. An entry v_{ij} refers to the reading of the ith sensor at the jth time stamp. If the reading of a sensor is not successfully received at a time interval, a blank entry arises. Now, the problem of filling sensors' missing values can be converted into imputing blank entries in the matrix.

Filling missing readings in a collection of geosensory time series data, however, is challenging for two reasons:

1. Readings can be absent at arbitrary sensors and time stamps. In some extreme cases, we may lose readings from a sensor at consecutive time stamps—for example, s_2 shown in figure 3.16a—or lose readings of all sensors at one (or more) time stamp(s) simultaneously, as t_2 illustrates in figure 3.16a. We call these extreme cases *block missing*. It is very difficult for existing models to handle a block missing problem, as we may not be able to find stable inputs for a model. For instance, nonnegative matrix factorization (NMF) cannot handle cases where the data of a column or a row are completely missing in a matrix.
2. Affected by multiple complex factors, sensor readings change over location and time significantly and nonlinearly. First, the readings of sensors with a shorter distance may not always be more similar than those with a farther distance. As illustrated in figure 3.17a, s_1 is closer to s_2 than s_3 in terms of geographical Euclidian distance. As shown in figure 3.17b, however, the air quality readings of s_2 are more similar to s_3 than s_1. The reason is that s_2 and s_3 are located in two regions with a similar geographical context, such as POIs and traffic patterns, while s_1 is deployed in a forest and with a lake between it and s_2. These cases violate the first law of geography, devaluing some interpolation-based models. Second, sensor readings fluctuate tremendously over time, sometimes coming with a sudden change. As illustrated in figure 3.16b, the reading of s_2 at the thirty-first time stamp drops over two hundred in two hours. Such a sudden change is actually very important to real-time monitoring and further data analysis but cannot be well handled by existing smoothing or interpolation methods.

Many solutions have been proposed to solve this problem from different perspectives. For instance, in the matrix shown in figure 3.16b, v_{2j} and $v_{1,j+1}$ are missing. We can estimate the reading of v_{2j} based on its spatial neighborhoods, such as s_1 and s_3; we call this a *spatial view*. v_{2j} can also be estimated based on readings of adjacent time

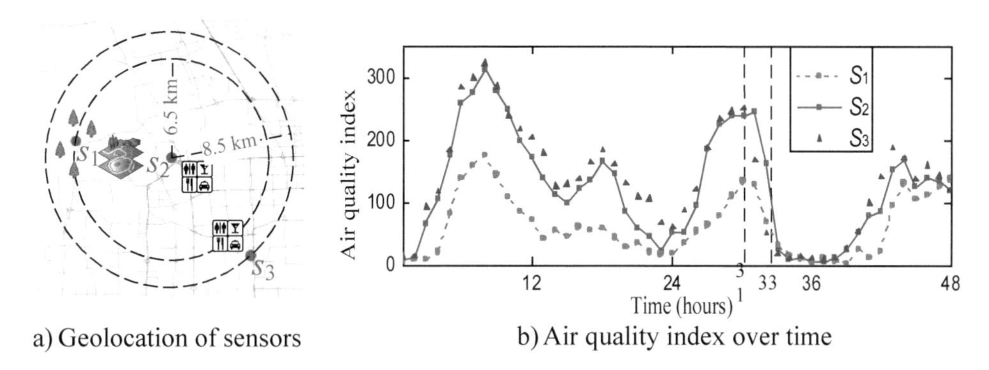

a) Geolocation of sensors b) Air quality index over time

Figure 3.17
The challenges of filling missing values in geosensory data.

stamps, such as t_{j-1} and t_{j+1}, and so on; we call this a *temporal view*. We can also use data of different time lengths for an estimation, enabling local and global views. For example, we consider adjacent readings of v_{2j} from t_{j-2} to t_{j+2} in a local data matrix, which is regarded as a *local view*. Alternatively, we can take into account readings over a very long time period (e.g., from t_1 to t_n), which is regarded as a *global view*. Local view captures instantaneous changes, whereas global view represents long-term patterns.

3.4.2 Spatial Models

This category of models considers a sensor's neighborhoods to impute its missing value at a time interval. The inverse distance weighting (IDW) [39], linear estimation, and kriging are three widely used methods.

3.4.2.1 Inverse distance weighting This model assigns a weight to each available reading of geospatially adjacent sensors according to their distance to the target sensor and aggregates these weights by equation (3.14) to make prediction $\hat{v}_{gs}$.

$$\hat{v}_{gs} = \frac{\sum_{i=1}^{m} v_i * d_i^{-\alpha}}{\sum_{i=1}^{m} d_i^{-\alpha}}, \tag{3.14}$$

where d_i is the spatial distance between a candidate sensor s_i and the target sensor, and α is a positive power parameter that controls the decay rate of a sensor's weight by $d_i^{-\alpha}$. $d_i^{-\alpha}$ assigns a bigger weight to closer sensors' readings, for which a bigger α denotes a faster decay of weight by distance.

Figure 3.18 presents the insight of the IDW using the statistics on two datasets: air quality data and meteorological data in Beijing from May 2014 to May 2015. Here, we

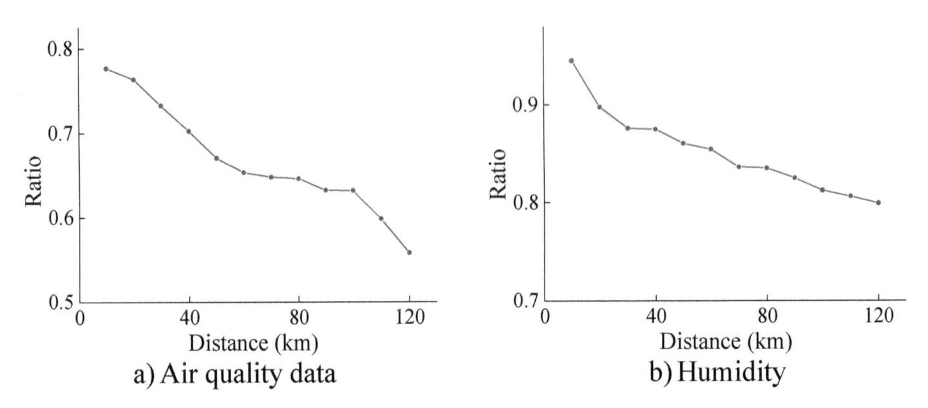

Figure 3.18
Empirical spatial correlation in different sensors' data.

calculate the ratio between an arbitrary two sensors' readings at the same time stamp. The ratio decreases as the distance between two sensors increases in both datasets. This follows the first law of geography [46]: everything is related to everything else, but near things are more related than distant things. This can be regarded as an empirical spatial correlation in geosensory data.

Here, we demonstrate the IDW using the running example shown in figure 3.16. Suppose two sensors s_1 and s_3 are near s_2 with a spatial distance of 6.5 km and 8.5 km, respectively. We aim to fill the missing value of s_2 at time interval t_j with the reading of s_1 and s_3 at t_j (i.e., 185 and 59). If setting $\alpha=1$, the weights for the two sensors are 1/6.5 and 1/8.5, respectively. According to equation (3.14), we compute the prediction $\hat{v}_{gs} = (185/6.5 + 59/8.5)/(1/6.5 + 1/8.5) = 130.4$.

3.4.2.2 Linear estimation The spatial inference, or estimation, of a quantity $Z: R^n \rightarrow R$ at an unobserved location x_0 is calculated from a linear combination of the observed values $z_i = Z(x_i)$ and weights $\omega_i(x_0)$, $i=0, 1, 2, \ldots, N$:

$$\hat{Z}(x_0) = \sum_{i=1}^{N} \omega_i(x_0) \cdot Z(x_i). \tag{3.15}$$

The weight $\omega_i(x_0)$ can be learned by minimizing the mean square error of the estimation and its true value $[\hat{Z}(x_0) - Z(x_i)]^2$. This is a kind of linear regression model in machine learning. We will introduce more about linear regression in chapter 7.

Pan and Li [42] propose a k-nearest neighbor–based algorithm, titled AKE, to estimate the missing values of a sensor. AKE first adopts a linear regression model to calculate a time-dependent spatial correlation between each pair of sensors based on recently received data. A determination coefficient is generated after the learning process, describing the fitness between a learned linear function and the received data. The determination coefficient is then employed as a weight $\omega_i(x_0)$ of corresponding sensors $Z(x_i)$ in a weighted average function like equation (3.11). Though the weight is updated dynamically in different time intervals, the dependency between different time intervals is not considered.

3.4.2.3 Kriging In statistics, *kriging* is a method of interpolation for which an interpolated value is modeled by a Gaussian process with a prior covariance. Based on reasonable assumptions on the priors, kriging provides the best linear unbiased prediction for a missing value.

Suppose there are $n+1$ points whose indexes are from 0 to n. The ith point is denoted as p_i, whose value z_i is generated by a variable Z_i. We can then estimate an unobserved point p_0's value $\hat{z}_0$ by a linear combination of the observed values $z_1, z_2, \ldots, z_n$ as follows:

$$\hat{z}_0 = \sum_{i=1}^{n} \lambda_i \times z_i, \tag{3.16}$$

where $\lambda_1, \lambda_2, \ldots, \lambda_n$ are weights corresponding to different points. When determining these weights, kriging considers not only the pairwise distance between two points but also the correlation between the two points' random variables. The correlation between a pair of variables (Z_i, Z_j) is defined by a covariance matrix, denoted by $Cov(Z_i, Z_j)$. Based on assumptions on those variables, different types of kriging have been proposed, such as ordinary kriging [27], universal kriging [2], IRFk-kriging [40], indicator kriging [45], and so on.

In this section, we focus on introducing ordinary kriging, which assumes the expected values of all points' variables share the same constant c, though the value of c may be unknown, denoted by equation (3.17):

$$E(Z_i) = c,\ i = 0, 1, 2, \ldots, n. \tag{3.17}$$

The estimated error for $\hat{z}_0$ is defined as equation (3.18):

$$\varepsilon_0 = \hat{z}_0 - z_0 = \sum_{i=1}^{n} \lambda_i \times z_i - z_0. \tag{3.18}$$

Then the algorithm finds parameters $\lambda_1, \lambda_2, \ldots, \lambda_n$ that minimize $Var(\varepsilon_0)$ subject to $E(\varepsilon_0) = 0$, defined as equation (3.19) and (3.20):

$$\begin{aligned} E(\varepsilon_0) = 0 &\Leftrightarrow \sum_{i=1}^{n} \lambda_i \times E(Z_i) - E(Z_0) = 0 \\ &\Leftrightarrow \sum_{i=1}^{n} \lambda_i \times E(Z_i) = E(Z_0) \Leftrightarrow c\sum_{i=1}^{n} \lambda_i = c \Leftrightarrow \sum_{i=1}^{n} \lambda_i = 1 \end{aligned} \tag{3.19}$$

$$\begin{aligned}Var(\varepsilon_0) &= Var\left(\sum_{i=1}^{n} \lambda_i \times Z_i - Z_0\right) \\ &= Var\left(\sum_{i=1}^{n} \lambda_i \times Z_i\right) - 2Cov\left(\sum_{i=1}^{n} \lambda_i \times Z_i, Z_0\right) + Cov(Z_0, Z_0) \\ &= \sum_{i=1}^{n} \sum_{j=1}^{n} \lambda_i \lambda_j Cov(Z_i, Z_j) - 2\sum_{i=1}^{n} \lambda_i Cov(Z_i, Z_0) + Cov(Z_0, Z_0).\end{aligned} \tag{3.20}$$

Then the optimization problem becomes

$$\begin{aligned}&argmin_{\lambda} \sum_{i=1}^{n} \sum_{j=1}^{n} \lambda_i \lambda_j Cov(Z_i, Z_j) - 2\sum_{i=1}^{n} \lambda_i Cov(Z_i, Z_0) + Cov(Z_0, Z_0) \\ &\quad subject\ to \sum_{i=1}^{n} \lambda_i = 1.\end{aligned} \tag{3.21}$$

By adding a Lagrange multiplier ϕ, it can be solved by

$$\begin{bmatrix} \lambda_1 \\ \cdots \\ \lambda_n \\ \phi \end{bmatrix} = \begin{bmatrix} Cov(Z_1, Z_1) & \cdots & Cov(Z_1, Z_n) & 1 \\ \vdots & \ddots & \vdots & \vdots \\ Cov(Z_n, Z_1) & \cdots & Cov(Z_n, Z_n) & 1 \\ 1 & \cdots & 1 & 0 \end{bmatrix}^{-1} \begin{bmatrix} Cov(Z_1, Z_0) \\ \vdots \\ Cov(Z_n, Z_0) \\ 1 \end{bmatrix}. \tag{3.22}$$

In reality $Cov\,(Z_i, Z_j)$ is not easy to compute. An alternative way is to compute a variogram $\gamma(Z_i, Z_j)$, defined as equation (3.23):

$$\gamma(Z_i, Z_j) = \frac{1}{2} E\left((Z_i - Z_j)^2\right). \tag{3.23}$$

Assuming all variables share the same variance σ^2, we derive the following relationship:

$$\gamma(Z_i, Z_j) = \sigma^2 - Cov(Z_i, Z_j),$$

thus,

$$Cov(Z_i, Z_j) = \sigma^2 - \gamma(Z_i, Z_j). \tag{3.24}$$

By replacing $Cov(Z_i, Z_j)$ with $\sigma^2 - \gamma(Z_i, Z_j)$ in equation (3.21), the ordinary kriging is further defined as follows:

$$\begin{aligned}&argmin_{\lambda} \sum_{i=1}^{n} \sum_{j=1}^{n} \lambda_i \lambda_j (\sigma^2 - \gamma(Z_i, Z_j)) - 2\sum_{i=1}^{n} \lambda_i (\sigma^2 - \gamma(Z_i, Z_0)) + \sigma^2 - \gamma(Z_0, Z_0), \\ &\quad subject\ to \sum_{i=1}^{n} \lambda_i = 1.\end{aligned} \tag{3.25}$$

Finally, by adding a Lagrange multiplier ϕ, it can be solved by:

$$\begin{bmatrix} \lambda_1 \\ \cdots \\ \lambda_n \\ \phi \end{bmatrix} = \begin{bmatrix} \gamma(Z_1, Z_1) & \cdots & \gamma(Z_1, Z_n) & 1 \\ \vdots & \ddots & \vdots & \vdots \\ \gamma(Z_n, Z_1) & \cdots & \gamma(Z_n, Z_n) & 1 \\ 1 & \cdots & 1 & 0 \end{bmatrix}^{-1} \begin{bmatrix} \gamma(Z_1, Z_0) \\ \vdots \\ \gamma(Z_n, Z_0) \\ 1 \end{bmatrix}. \quad (3.26)$$

Once parameters $\lambda_1, \ldots, \lambda_n$ have been determined, we can calculate $\hat{z}_0$ based on equation (3.16).

3.4.3 Temporal Models

This category of approach imputes the missing values of a sensor at a time interval based on the sensor's own readings at other time intervals. Methods in this category can be classified into two groups: nonfeature based and feature based. The first group of methods, such as simple exponential smoothing (SES) [19], autoregressive moving average (ARMA) [4], and SARIMA [26], solely considers a sensor's readings when imputing a missing value. Ceylan et al. [10] conduct a comparison of these methods in filling missing values in time series. The second group of methods, such as graphical models and regression models [19, 34], exploits a function of features, considering the temporal correlation of readings over time. As the second group of methods will be elaborated on in chapter 7, hereafter we focus on introducing the first category of methods.

3.4.3.1 Simple exponential smoothing SES is frequently used in the time series domain as an exponential moving–average model, formally defined as:

$$\hat{v}_{gt} = \beta v_j + \beta(1-\beta)v_{j-1} + \cdots + \beta(1-\beta)^{t-1}v_1, \quad (3.27)$$

where t is a time interval between a candidate reading v_j and a target reading; β is a smoothing parameter within (0, 1). In general, $\beta * (1-\beta)^{t-1}$ gives a bigger weight to recent readings than distant ones, and a smaller β denotes a slower decay of weight over the time interval.

Traditional SES only uses predecessors of the target time stamp as input. Yi et al. [51] extend SES by using both predecessors and successors of a target time stamp. Given a target time stamp, the improved SES gives a weight $\beta * (1-\beta)^{t-1}$ to each reading of the same sensor, calculating $\hat{v}_{gt}$ by normalizing the weight according to equation (3.28):

$$\hat{v}_{gt} = \frac{\sum_{j=1}^{n} v_j * \beta * (1-\beta)^{t-1}}{\sum_{j=1}^{n} \beta * (1-\beta)^{t-1}}. \quad (3.28)$$

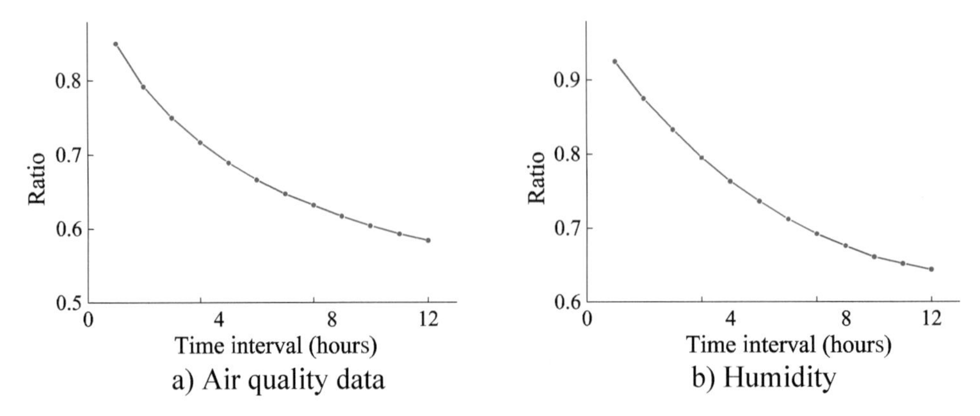

Figure 3.19
Empirical temporal correlation in a sensor's data.

In implementation, readings within a temporal threshold (e.g., twelve hours) are selected, as distant readings are not very useful. The SES model is inspired by the observations from time series data. Figure 3.19 presents the ratio between an arbitrary two readings at two different time stamps of the same sensor using the same air quality data and meteorological data presented in figure 3.18. Both curves in figure 3.19 decrease as the time interval increases, showing an empirical temporal correlation in time series; that is, readings of recent time stamps are more relevant than readings of distant time stamps.

We demonstrate SES continually using the example shown in figure 3.16. Suppose we aim to fill the missing value of s_2 at time interval t_j by using the readings of four adjacent time stamps (t_{j-2}, t_{j-1}, t_{j+1}, t_{j+2}) whose values are (173, 136, 146, 185). If setting $\beta=0.5$, the weights for the four time stamps are (0.25, 0.5, 0.5, 0.25), respectively. Thus, the final result is

$$\hat{v}_{gt} = \frac{173 \times 0.25 + 136 \times 0.5 + 146 \times 0.5 + 185 \times 0.25}{0.25 + 0.5 + 0.5 + 0.25} = 230.5.$$

3.4.3.2 ARMA In the statistical analysis of time series, ARMA models provide a description of a (weakly) stationary stochastic process in terms of two polynomials: one for the autoregression AR(p) and the other for the moving average MA(q) [4].

The notation AR(p) refers to the autoregressive model of order p, written as

$$X_t = c + \sum_{i=1}^{p} \varphi_i X_{t-i} + \varepsilon_t, \tag{3.29}$$

where X_t denotes the reading of a time series at the t-th time interval; $\varphi_1, \varphi_2, \ldots, \varphi_i$ are parameters; c is a constant; and the random variable ε_t is white noise.

The notation MA(q) refers to the moving-average model of order q:

$$X_t = \mu + \varepsilon_t + \sum_{i=1}^{q} \theta_i \varepsilon_{t-i}, \tag{3.30}$$

where $\theta_1, \theta_2, \ldots, \theta_i$ are parameters of the model; μ is the expectation of X_t (often assumed to equal 0); and $\varepsilon_t, \varepsilon_{t-1}, \ldots$ are white noise error terms.

The notation ARMA(p, q) refers to the model with p autoregressive terms and q moving-average terms. This model contains the AR(p) and MA(q) models,

$$X_t = c + \varepsilon_t + \sum_{i=1}^{p} \varphi_i X_{t-i} + \sum_{i=1}^{q} \theta_i \varepsilon_{t-i}. \tag{3.31}$$

In lag operator polynomial notation $L^i X_t = X_{t-i}$, defining the degree p AR lag operator polynomial $\varphi(L) = (1 - \varphi_1 L - \varphi_2 L^2 - \cdots - \varphi_p L^p)$ and defining the degree q MA lag operator polynomial $\theta(L) = (1 + \theta_1 L + \theta_2 L^2 + \cdots + \theta_q L^q)$, we can write the ARMA($p$, q) model as:

$$\varphi(L) X_t = c + \theta(L) \varepsilon_t. \tag{3.32}$$

When implementing ARMA, we can either use default parameters for φ_i and θ_i or predefine these parameters. ε_t are generally assumed to be independent, identically distributed variables sampled from a normal distribution with a zero mean.

3.4.3.3 ARIMA and SARIMA In order to include more realistic dynamics, in particular nonstationarity in mean and seasonal behaviors, quite a few variants of ARMA models, including an autoregressive integrated–moving average (ARIMA) model and a seasonal autoregressive integrated–moving average (SARIMA) model, have been proposed. The AR part of ARIMA indicates that the evolving variable of interest is regressed on its own prior values. The MA part indicates that the regression error is actually a linear combination of error terms whose values occurred contemporaneously and at various times in the past. The I (for "integrated") indicates that the data values have been replaced with the difference between their values and the previous values (and this differencing process may have been performed more than once). The purpose of each of these features is to make the model fit the data as well as possible.

Nonseasonal ARIMA models are generally denoted ARIMA(p, d, q), where parameters p, d, and q are nonnegative integers; p is the order (number of time lags) of the autoregressive model; d is the degree of differencing (the number of times the data have had past values subtracted); and q is the order of the moving-average model. SARIMA models are usually denoted by ARIMA(p, d, q)(P, D, Q,)$_m$, where m refers to the number of periods in each season, and the uppercase P, D, Q refer to the autoregressive, differencing, and moving-average terms for the seasonal part of the ARIMA model [26].

3.4.4 Spatiotemporal Models

This category of models simultaneously considers spatial correlations between different locations and the temporal dependencies between different time intervals when imputing a missing value. There are three different types of combinations.

3.4.4.1 Collaborative filtering-based methods This category of methods accommodates sensor data over a period of time with a matrix, as illustrated in figure 3.16b, where a row denotes a sensor, and a column stands for a time interval; an entry stores the value of a particular sensor at a specific time interval. Now, the sensor data imputation problem can be converted into a recommendation problem that infers the values of missing entries. The similarity between two rows represents the spatial correlation between two sensors, and the similarity between two columns denotes the temporal correlation between two time intervals. Classical models for solving a recommendation problem are called *collaborative filtering*, which will be detailed in chapter 8.

Li et al. [35] propose two matrix factorization–based models, consisting of STR-MF and MTR-MF, to fill the missing values of a sensor, considering both the correlation between different sensors and that between different time intervals. Specifically, STR-MF incorporates spatiotemporal similarities into a matrix factorization model by adding a temporal regularized term and a spatial proximity term to its loss function. With respect to a location with multiple sensors (e.g., temperature and humidity), MTR-MF accommodates data from each type of sensor with a matrix, respectively. Those matrices are then factorized collectively, sharing the same latent temporal space in a loss function. Further, Li et al. extend MRT-MF to a three-dimensional tensor, where the three dimensions are location, time, and sensors. A tensor completion method with a temporal regularized term is then proposed to estimate entries missing from the tensor.

3.4.4.2 Multi-view-based methods Yi et al. [51] propose a spatiotemporal multi-view–based learning (ST-MVL) method to collectively fill missing readings in a collection of geosensory time series data, considering (1) the temporal correlation between readings at different time stamps in the same series and (2) the spatial correlation between different time series. As shown in figure 3.20, the ST-MVL consists of four views: IDW, SES, user-based collaborative filtering (UCF), and item-based collaborative filtering (ICF). The four views are then aggregated to generate a final estimate for missing readings.

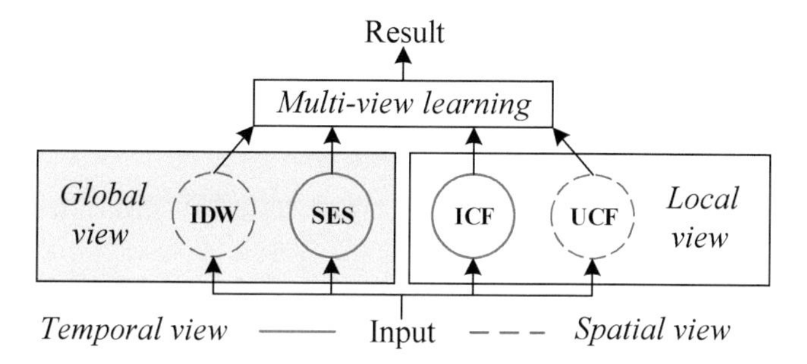

Figure 3.20
The framework of our method.

IDW computes an estimate for a missing reading of a sensor based on the readings of the sensor's spatial neighborhoods. SES estimates the missing reading of a sensor based on readings at other time stamps of the same sensor. As IDW and SES are actually empirical models derived from data over a long period of time, they denote a global spatial view and a global temporal view on the missing reading, respectively.

On the contrary, UCF estimates a missing reading based only on the local similarity between a sensor's recent readings and those of the sensor's spatial neighbors, where a sensor is regarded as a user. Likewise, ICF estimates a missing reading based on the local similarity between recent readings of different time stamps, where a time stamp denotes an item. As UCF and ICF only consider local similarity from a spatial and temporal perspective, they stand for local spatial and local temporal views, respectively.

To leverage advantages of different views, we propose a multi-view learning algorithm that finds a linear combination of different views' predictions with minimal square errors, according to equation (3.33):

$$\hat{v}_{mvl} = w_1 * \hat{v}_{gs} + w_2 * \hat{v}_{gt} + w_3 * \hat{v}_{ls} + w_4 * \hat{v}_{lt} + b, \tag{3.33}$$

where b is a residual, and w_i ($i=1, 2, 3, 4$) is a weight assigned to each view. Algorithm 1 presents the procedure of ST-MVL. When a dataset encounters a block missing problem, in which ICF and UCF do not work very well, ST-MVL leverages IDW and SES to generate an initial value for those missing entries (see line 3 in figure 3.21). Then ST-MVL predicts each missing entry using ICF, UCF, IDW, and SES, respectively (line 4–9), combining the four predictions based on a linear kernel–based multi-view learning framework (see line 10 and equation [3.29]). The model is trained for each sensor, respectively, minimizing the linear least square error [33] between predictions and ground truth.

Input: Original data matrix M, ω, α, β;
Output: Final data matrix;
1. $O \leftarrow$ Get_All_Missing_Values(M);
2. **If** there are block missing problem
3. $M \leftarrow$ Initialization(M, α, β); //using IDW or SES
4. **For each** target t in O
5. $\hat{v}_{ls} \leftarrow$ UCF(M, t, ω);
6. $\hat{v}_{lt} \leftarrow$ ICF(M, t, ω);
7. $\hat{v}gs \leftarrow$ IDW(M, t, α);
8. $\hat{v}gt \leftarrow$ SES(M, t, β);
9. $\hat{v}mvl \leftarrow$ Mutiview_Learning($\hat{v}_{ls}, \hat{v}_{lt}, \hat{v}gs, \hat{v}gt$);
10. Add $\hat{v}mvl$ into M;
11. **Return** M;

Figure 3.21
The procedure of the ST-MVL algorithm.

3.5 Summary

This chapter has introduced four paradigms of urban sensing consisting of static sensing, mobile sensing, passive crowd sensing, and active crowd sensing. A general framework for urban sensing is presented, which comprises five layers: (1) a sensing program's definition, (2) sensor deployment for a sensor-centric sensing paradigm or task design for a crowd-sensing paradigm, (3) data collection, (4) data evaluation, and (5) preprocessing.

The functions and challenges of each layer have been discussed, with some key technologies presented in detail. For example, four models for sensor deployment have been introduced in section 3.3. In addition, the technology for participant recruitment and task design has been presented for a human-centric sensing paradigm in section 3.4.

Finally, three categories of models for filling missing values in geosensory data have been introduced in section 3.5, consisting of spatial models, temporal models, and spatiotemporal models.

References

[1] Ahmed, A., K. Yasumoto, Y. Yamauchi, and M. Ito. 2011. "Distance and Time Based Node Selection for Probabilistic Coverage in People-Centric Sensing." In *Proceedings: 2011 8th Annual IEEE Communications Society Conference on Sensor, Mesh and Ad Hoc Communications and Networks (SECON)*. Washington, DC: Institute of Electrical and Electronics Engineers (IEEE) Computer Society Press, 134–142.

[2] Armstrong, M. 1984. "Problems with Universal Kriging." *Mathematical Geology* 16 (1): 101–108.

[3] Arya, V., N. Garg, R. Khandekar, A. Meyerson, K. Munagala, and V. Pandit. 2004. "Local Search Heuristics for K-Median and Facility Location Problems." *SIAM Journal on Computing* 33 (3): 544–562.

[4] Box, G., G. M. Jenkins, and G. C. Reinsel. 1994. *Time Series Analysis: Forecasting and Control.* 3rd edition. Upper Saddle River, NJ: Prentice Hall.

[5] Burges, C., T. Shaked, E. Renshaw, A. Lazier, M. Deeds, N. Hamilton, and G. Hullender. 2005. "Learning to Rank Using Gradient Descent." In *Proceedings of the 22nd International Conference on Machine Learning*. New York: Association for Computing Machinery (ACM), 89–96.

[6] Burke, J. A., D. Estrin, M. Hansen, A. Parker, N. Ramanathan, S. Reddy, and M. B. Srivastava. 2006. *Participatory Sensing*. Los Angeles: University of California Center for Embedded Network Sensing.

[7] Campbell, A. T., S. B. Eisenman, N. D. Lane, E. Miluzzo, and R. A. Peterson. 2006. "People-Centric Urban Sensing." In *Proceedings of the 2nd Annual International Workshop on Wireless Internet*. New York: ACM, 18.

[8] Campbell, A. T., S. B. Eisenman, N. D. Lane, E. Miluzzo, R. A. Peterson, H. Lu, X. Zheng, M. Musolesi, K. Fodor, and G. S. Ahn. 2008. "The Rise of People-Centric Sensing." *IEEE Internet Computing* 12 (4): 12–21.

[9] Cao, Z., T. Qin, T. Y. Liu, M. F. Tsai, and H. Li. 2007. "Learning to Rank: From Pairwise Approach to Listwise Approach." In *Proceedings of the 24th International Conference on Machine Learning*. New York: ACM, 129–136.

[10] Ceylan, Y., S. Aslan, C. Iyigun, and I. Batmaz. 2013. "Comparison of Missing Value Imputation Methods in Time Series: The Case of Turkish Meteorological Data." *Theoretical and Applied Climatology* 112 (1): 143–167.

[11] Chen, N., N. Gravin, and P. Lu. 2011. "On the Approximability of Budget Feasible Mechanisms." In *Proceedings of the Twenty-Second Annual ACM-SIAM Symposium on Discrete Algorithms*. Philadelphia: Society for Industrial and Applied Mathematics (SIAM), 685–699.

[12] Chen, Z., Y. Liu, R. C. W. Wong, J. Xiong, G. Mai, and C. Long. 2014. "Efficient Algorithms for Optimal Location Queries in Road Networks." In *Proceedings of the 2014 ACM SIGMOD International Conference on Management of Data*. New York: ACM, 123–134.

[13] Chon, Y., N. D. Lane, Y. Kim, F., Zhao, and H. Cha. 2013. "Understanding the Coverage and Scalability of Place-Centric Crowdsensing." In *Proceedings of the 2013 ACM International Joint Conference on Pervasive and Ubiquitous Computing*. New York: ACM, 3–12.

[14] Deng, K., S. W. Sadiq, X. Zhou, H. Xu, G. P. C. Fung, and Y. Lu. 2012. "On Group Nearest Group Query Processing." *IEEE Transactions on Knowledge and Data Engineering* 24 (2): 295–308.

[15] Feige, U. 1996. "A Threshold of Ln N for Approximating Set Cover (Preliminary Version)." In *Proceedings of the Twenty-Eighth Annual ACM Symposium on Theory of Computing*. New York: ACM, 314–318.

[16] Fu, Y., Y. Ge, Y. Zheng, Z. Yao, Y. Liu, H. Xiong, and J. Yuan. 2014. "Sparse Real Estate Ranking with Online User Reviews and Offline Moving Behaviors." In *2014 IEEE International Conference on Data Mining Workshop*. Washington, DC: IEEE Computer Society Press.

[17] Fu, Y., H. Xiong, Y. Ge. Z. Yao, and Y. Zheng. 2014. "Exploiting Geographic Dependencies for Real Estate Appraisal: A Mutual Perspective of Ranking and Clustering." In *Proceedings of the 20th SIGKDD Conference on Knowledge Discovery and Data Mining*. New York: ACM.

[18] Fu, Y., Hui Xiong, Yong Ge, Yu Zheng, Zijun Yao, and Zhi-Hua Zhou. "Modeling of Geographic Dependencies for Real Estate Ranking." *ACM Transactions on Knowledge Discovery from Data* 11.

[19] Fung, David S. C. 2006. "Methods for the Estimation of Missing Values in Time Series." PhD diss., Edith Cowan University, Perth, Australia.

[20] Gardner, Everette S. 2006. "Exponential Smoothing: The State of the Art—Part II." *International Journal of Forecasting*, 22 (4): 637–666.

[21] Gonzalez, T. F. 1985. "Clustering to Minimize the Maximum Intercluster Distance." *Theoretical Computer Science* 38:293–306.

[22] Hachem, S., A. Pathak, and V. Issarny. 2013. "Probabilistic Registration for Large-Scale Mobile Participatory Sensing." In *Proceedings: 2013 IEEE International Conference on Pervasive Computing and Communications (PerCom)*. Washington, DC: IEEE Computer Society Press, 132–140.

[23] Herbrich, R., T. Graepel, and K. Obermayer. 1999. "Support Vector Learning for Ordinal Regression." In *Ninth International Conference on Artificial Neural Networks, 1999. ICANN 1999*. Conference publication no. 470. Washington, DC: IEEE Computer Society Press, 97–102.

[24] Howe, J. 2006. "The Rise of Crowdsourcing." *Wired Magazine* 14 (6): 1–4.

[25] Hsieh, H. P., S. D. Lin, and Y. Zheng. 2015. "Inferring Air Quality for Station Location Recommendation Based on Urban Big Data." In *Proceedings of the 21st ACM SIGKDD International Conference on Knowledge Discovery and Data Mining*. New York: ACM, 437–446.

[26] Hyndman, Rob J., and George Athanasopoulos. 2015. "8.9 Seasonal ARIMA Models." *Forecasting: Principles and Practice*. Accessed May 19, 2015. https://www.otexts.org/fpp/8/9.

[27] Isaaks, Edward H. 1989. "Applied Geostatistics." No. 551.72 I86. Oxford: Oxford University Press, 278–290.

[28] Jaimes, L. G., I. Vergara-Laurens, and M. A. Labrador. 2012. "A Location-Based Incentive Mechanism for Participatory Sensing Systems with Budget Constraints." In *Proceedings, 2012 IEEE International Conference on Pervasive Computing and Communications*. Washington, DC: IEEE Computer Society Press, 103–108.

[29] Ji, S., Y. Zheng, and T. Li. 2016. "Urban Sensing Based on Human Mobility." In *Proceedings of the 2016 ACM International Joint Conference on Pervasive and Ubiquitous Computing*. New York: ACM, 1040–1051.

[30] Kawajiri, R., M. Shimosaka, and H. Kashima. 2014. "Steered Crowdsensing: Incentive Design towards Quality-Oriented Place-Centric Crowdsensing." In *Proceedings of the 2014 ACM International Joint Conference on Pervasive and Ubiquitous Computing*. New York: ACM, 691–701.

[31] Khuller, S., A. Moss, and J. S. Naor. 1999. "The Budgeted Maximum Coverage Problem." *Information Processing Letters* 70 (1): 39–45.

[32] Lane, N. D., S. B. Eisenman, M. Musolesi, E. Miluzzo, and A. T. Campbell. 2008. "Urban Sensing Systems: Opportunistic or Participatory?" In *Proceedings of the 9th Workshop on Mobile Computing Systems and Applications*. New York: ACM, 11–16.

[33] Lawson, Charles L., and Richard J. Hanson. 1974. *Solving Least Squares Problems*. Englewood Cliffs, NJ: Prentice Hall, 161.

[34] Lee, D., Dana Kulic, and Yoshihiko Nakamura. 2008. "Missing Motion Data Recovery Using Factorial Hidden Markov Models." In *Proceedings, 12th International Conference on Robotics and Automation*. Washington, DC: IEEE Computer Society Press, 1722–1728.

[35] Li, Chung-Yi, Wei-Lun Su, Todd G. McKenzie, Fu-Chun Hsu, Shou-De Lin, Jane Yung-Jen Hsu, and Phillip B. Gibbons. 2015. "Recommending Missing Sensor Values." In *Proceedings, 2015 IEEE International Conference on Big Data*. Washington, DC: IEEE Computer Society Press, 381–390.

[36] Li, S., and O. Svensson. 2016. "Approximating k-Median via Pseudo-Approximation." *SIAM Journal on Computing* 45 (2): 530–547.

[37] Li, Y., J. Bao, Y. Li, Y. Wu, Z. Gong, and Y. Zheng. 2016. "Mining the Most Influential k-Location Set from Massive Trajectories." *IEEE Transactions on Big Data*. doi:10.1109/TBDATA.2017.2717978.

[38] Li, Y., Y. Zheng, S. Ji, W. Wang, and Z. Gong. 2015. "Location Selection for Ambulance Stations: A Data-Driven Approach." In *Proceedings of the 23rd SIGSPATIAL International Conference on Advances in Geographic Information Systems*. New York: ACM, 85.

[39] Lu, George Y., and David W. Wong. 2008. "An Adaptive Inverse-Distance Weighting Spatial Interpolation Technique." *Computers and Geosciences* 34 (9): 1044–1055.

[40] Marcotte, D., and M. David. 1988. "Trend Surface Analysis as a Special Case of IRF-K Kriging." *Mathematical Geology* 20 (7): 821–824.

[41] Nallapati, R. 2004. "Discriminative Models for Information Retrieval." In *Proceedings of the 27th Annual International ACM SIGIR Conference on Research and Development in Information Retrieval*. New York: ACM, 64–71.

[42] Pan, Liqiang, and Jianzhong Li. 2010. "K-Nearest Neighbor Based Missing Data Estimation Algorithm in Wireless Sensor Networks." *Wireless Sensor Network* 2 (2): 115.

[43] Qi, J., R. Zhang, L. Kulik, D. Lin, and Y. Xue. 2012. "The Min-dist Location Selection Query." In *Proceedings, 2012 IEEE 28th International Conference on Data Engineering*. Washington, DC: IEEE Computer Society Press, 366–377.

[44] Qin, T., X.-D. Zhang, M.-F Tsai, D.-S. Wang, T.-Y. Liu, and H. Li. 2007. "Query-Level Loss Functions for Information Retrieval." *Information Processing and Management* 44 (2): 838–855.

[45] Solow, Andrew R. 1986. "Mapping by Simple Indicator Kriging." *Mathematical Geology* 18 (3): 335–352.

[46] Tobler, Waldo R. 1970. "A Computer Movie Simulating Urban Growth in the Detroit Region." *Economic Geography* 46 (2): 234–240.

[47] Wang, L., D. Zhang, A. Pathak, C. Chen, H. Xiong, D. Yang, and Y. Wang. 2015. "CCS-TA: Quality-Guaranteed Online Task Allocation in Compressive Crowdsensing." In *Proceedings of the 2015 ACM International Joint Conference on Pervasive and Ubiquitous Computing*. New York: ACM, 683–694.

[48] Xia, F., T. Y. Liu, J. Wang, W. Zhang, and H. Li. 2008. "Listwise Approach to Learning to Rank: Theory and Algorithm." In *Proceedings of the 25th International Conference on Machine Learning*. New York: ACM, 1192–1199.

[49] Xiao, X., B. Yao, and F. Li. 2011. "Optimal Location Queries in Road Network Databases." In *Proceedings, 2011 IEEE 27th International Conference on Data Engineering*. Washington, DC: IEEE Computer Society Press, 804–815.

[50] Yan, D., Z. Zhao, and W. Ng. 2015. "Efficient Processing of Optimal Meeting Point Queries in Euclidean Space and Road Networks." *Knowledge and Information Systems* 42 (2): 319–351.

[51] Yi, X., Y. Zheng, J. Zhang, and T. Li. 2016. "ST-MVL: Filling Missing Values in Geo-Sensory Time Series Data." In *Proceedings of the 25th International Joint Conference on Artificial Intelligence*. Pasadena, CA: International Joint Conferences on Artificial Intelligence Organization (IJCAI).

[52] Zhang, Y., M. Roughan, W. Willinger, and L. Qiu. 2009. "Spatio-Temporal Compressive Sensing and Internet Traffic Matrices." *ACM SIGCOMM Computer Communication Review* 39 (4): 267–278.

[53] Zheng, Y., F. Liu, and H. P. Hsieh. 2013. "U-air: When Urban Air Quality Inference Meets Big Data." In *Proceedings of the 19th SIGKDD Conference on Knowledge Discovery and Data Mining*. New York: ACM, 1436–1444.

[54] Zhang, D., H. Xiong, L. Wang, and G. Chen. 2014. "CrowdRecruiter: Selecting Participants for Piggyback Crowdsensing under Probabilistic Coverage Constraint." In *Proceedings of the 2014 ACM International Joint Conference on Pervasive and Ubiquitous Computing*. New York: ACM, 703–714.

III Urban Data Management

4 Spatiotemporal Data Management

Abstract: The management of spatial and spatiotemporal data is of great importance to urban computing, providing upper-level machine-learning tasks with efficient access to and queries on data. This chapter first introduces data management techniques, such as indexing and retrieval algorithms, for spatial data. It then presents techniques for managing spatiotemporal data, consisting of moving object databases and trajectory data management. The former is concerned more with the specific location of a moving object at a (usually recent) time stamp. The latter concerns the continuous movement (e.g., the path) a moving object has traversed at a given time interval. Finally, we introduce hybrid indexing structures for managing multiple datasets.

4.1 Introduction

Cities receive huge and heterogeneous data from different domains every second. An effective management of such datasets is imperative for supporting the real-time monitoring of a city and the implementation of data analytics. In order to design an effective data management system for urban big data, we need to pay attention to four aspects: data structures, queries, indexes, and retrieval algorithms.

4.1.1 Data Structures

The majority of data generated in a city is associated with unique spatiotemporal properties (i.e., a spatial coordinate and temporally dynamic readings). As shown in figure 4.1, there are six forms of urban big data in terms of structure and spatiotemporal dynamics. According to data structures, we can categorize urban data into two groups consisting of point-based and network-based data, denoted by the two rows in figure 4.1. Based on its spatiotemporal dynamics, we can partition the data into three categories consisting of spatiotemporal static, spatial static and temporal dynamic, and spatiotemporal dynamic data, represented by the three columns. For instance, a point of interest (POI) is associated with a static point location. Its properties, such as size and name, do not change over time. Thus, it belongs to spatiotemporal static data (refer to

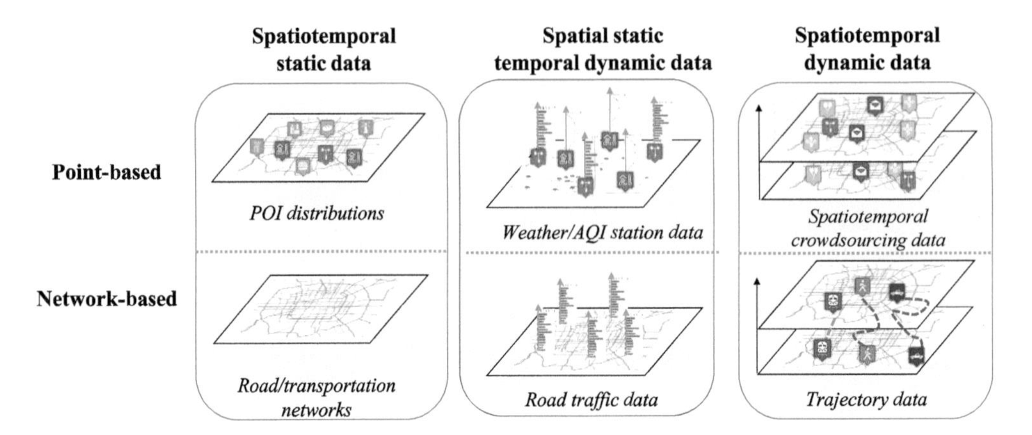

Figure 4.1
Six types of data forms for urban big data.

section 1.5.1 for details). We need to define proper data structures to contain different types of data. Afterward, we can design different storage mechanisms, indexing, and retrieval algorithms for different data structures.

4.1.2 Queries

Typical queries on spatiotemporal data can be classified into two categories: range queries and nearest neighbor queries [26]. As illustrated in figure 4.2a, the first category of queries searches for objects that lie partially or fully inside a specific range, which can be a geographical region or a spatiotemporal range. For example, finding buildings located in a geographical region is a spatial range query. Searching for vacant taxis traversing a plaza in the past two minutes is a spatiotemporal range query. In the second category of queries, given an object or a point, we search for *k*-nearest neighbor (KNN) objects satisfying the given conditions. This category of queries usually involves a distance metric—for instance, querying the nearest gas station around a driver. Here, a driver's location is a query point, and gas stations are objects to search. The distance here should be a network distance. The two categories of queries on spatiotemporal data are very different from searching documents for key words, calling for unique indexing and retrieval algorithms.

4.1.3 Indexes

In order to expedite the accessing and retrieval processes of data, we need to design indexing structures to sufficiently organize data in advance. Typical spatial indexes can be divided into two groups: space partition–based and data-driven indexes.

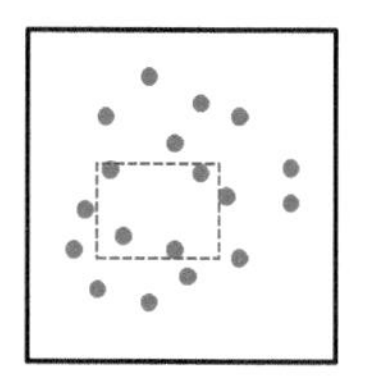

a) Range (region) queries

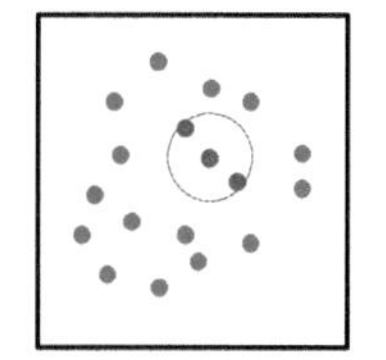

b) Nearest neighbor queries

Figure 4.2
Queries for spatial and spatiotemporal data.

The first category of indexes, such as grid based and quadtree based [29], as illustrated in figure 4.3a and b, partitions a spatial space into grids (with an equal or non-uniform size) according to some rules regardless of the distribution of data. It then builds the relation between a grid and the data falling into the grid. Such grids help to prune the search space of a retrieval algorithm tremendously, as data falling into grids that cannot satisfy a query condition can be ignored.

The second category of indexes (e.g., k-d tree [1] and R-tree [16]) builds an index based on the distribution of data. For instance, k-d tree partitions a dataset by using a few points from the dataset. R-tree formulates many rectangles based on data and further organizes small rectangles into big ones. We will detail this further in section 4.2.1.

Spatial indexes can be extended to a spatiotemporal index through one of the following three approaches.

First, as shown in figure 4.4a, we can treat time as the third dimension, directly applying spatial indexes to spatiotemporal data. For instance, 3D R-tree [39] is an extension of R-tree, formulating three-dimensional cuboids (in contrast to rectangles) based on spatiotemporal points.

As illustrated in figure 4.4b, the second approach, such as multiple-version R-tree (MVR-tree) [37], historical R-tree (HR-tree) [26], and HR+−tree [36], is to build a spatial index for each time stamp. To enhance the efficiency of such an index, unchanged substructures of the index across different time stamps can be reused.

The third approach first divides a spatial space into regions using a spatial index (e.g., quadtree or R-tree), as depicted in figure 4.4c. A temporal index is then built for data falling into each spatial region. SETI (scalable and efficient trajectory index) [6], SEB (start-end time B-tree) [32], and CSE-tree (compressed start-end tree) [40] belong in this category.

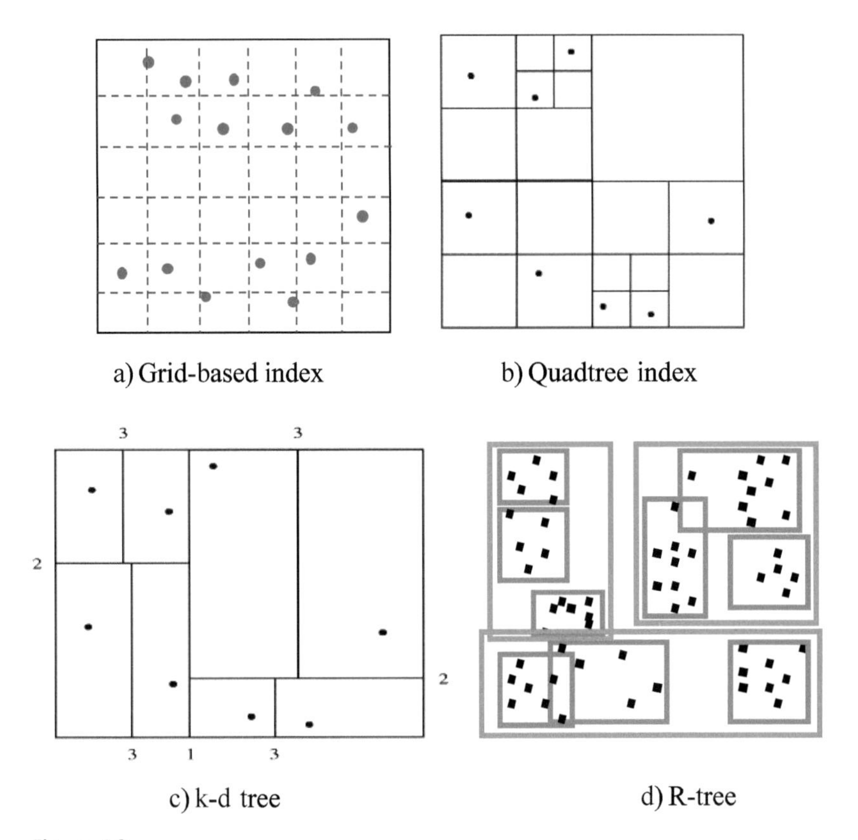

a) Grid-based index
b) Quadtree index
c) k-d tree
d) R-tree

Figure 4.3
Some frequently used spatial indexes.

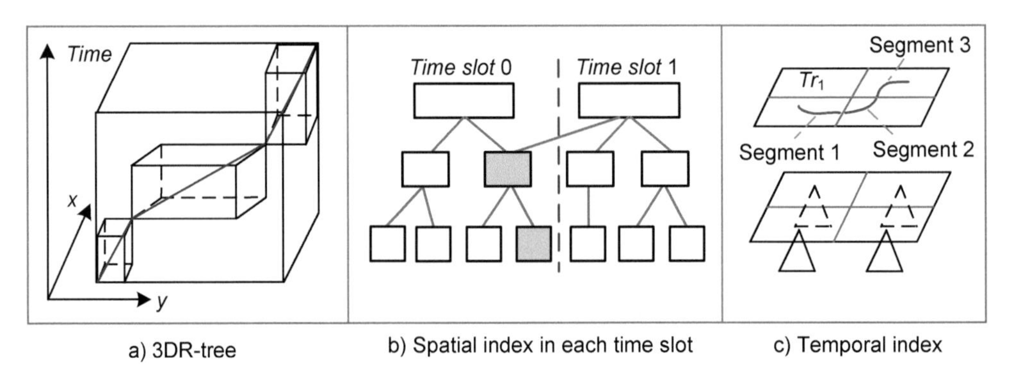

a) 3DR-tree
b) Spatial index in each time slot
c) Temporal index

Figure 4.4
Spatiotemporal indexing approaches.

4.1.4 Retrieval Algorithms

Given a query and an index, a retrieval algorithm searches for objects that satisfy the criteria of a query. To expedite retrieval efficiency, such algorithms usually prune searching spaces significantly based on spatial or spatiotemporal indexes. For example, when answering a spatial range query based on a grid-based index as shown in figure 4.3a, we can only retrieve the grids that lie partially or fully inside a query rectangle for a further refinement; other grids are filtered out. Sometimes a retrieval algorithm derives some upper or lower bounds to prune searching spaces. For instance, given a query point and an R-tree index, we can quickly estimate the shortest Euclidean distance d_s and the longest distance d_l to a group of points falling in a rectangle. If another object has a distance shorter than d_s, none of the points in this rectangle can be the query point's nearest neighbor. That is, they can be pruned without calculating the distance between the query and them. We explain this part further in section 4.3.

4.2 Data Structures

In order to accommodate the six forms of data presented in figure 4.1, we design six types of data structures, as shown in figure 4.5. Each column stands for one type of data structure, which is further composed of some components. The left three columns denote data structures for point-based data, consisting of point-based spatial static data, point-based spatial time series data, and point-based spatiotemporal data. The right three columns stand for network-based data, which comprises network-based spatial static data, network-based spatial time series data, and network-based temporal data. The six types of data structures share some common components, such as spatial point, time series, road network, metadata, and property readings. We will detail each type of data structure in the following paragraphs.

4.2.1 Point-Based Spatial Static Data

This category of data structure is devised for storing point-based data (e.g., POIs) whose location and readings do not change over time. For example, once a gas station is built, its location does not change over time. Its size and category do not change over time either. As shown in figure 4.6a, the data structure is composed of three parts: an ID, a spatial point, and metadata. A spatial point further consists of four fields: an ID, latitude, longitude, and altitude. Metadata contains a summary of an object, including name, category, size, textual descriptions, and others. As the component of a spatial point is employed by several data structures (e.g., trajectory data, which may contain a list of spatial points), an identity for a spatial point is necessary to differentiate between

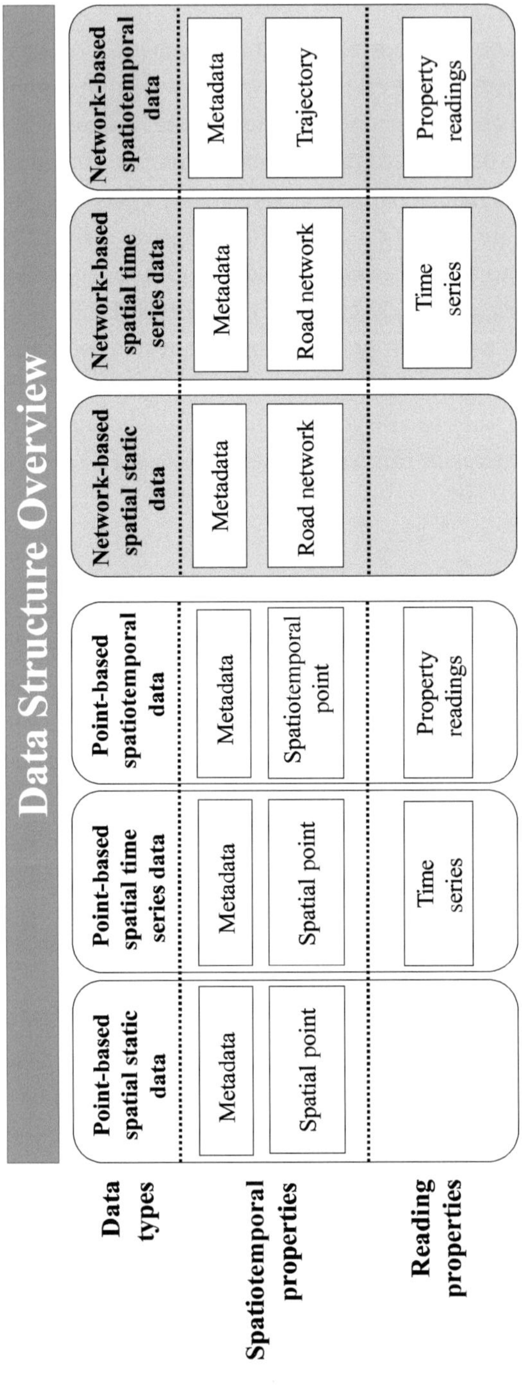

Data Structure Overview

Data types	**Point-based spatial static data**	**Point-based spatial time series data**	**Point-based spatiotemporal data**	**Network-based spatial static data**	**Network-based spatial time series data**	**Network-based spatiotemporal data**
Spatiotemporal properties	Metadata	Metadata	Metadata	Metadata	Metadata	Metadata
	Spatial point	Spatial point	Spatiotemporal point	Road network	Road network	Trajectory
Reading properties		Time series	Property readings		Time series	Property readings

Figure 4.5
Data structure overview on urban big data.

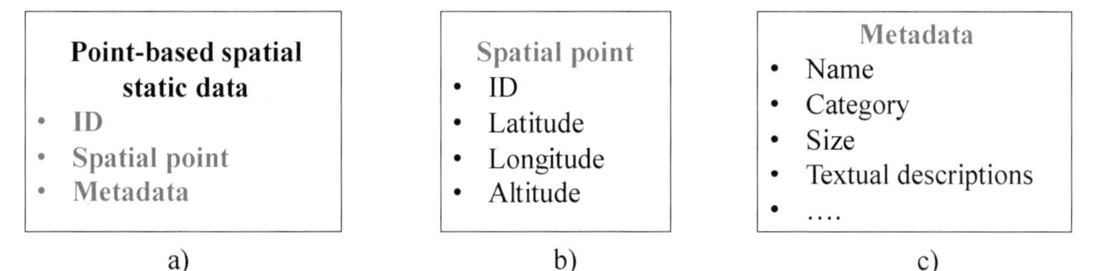

Figure 4.6
Data structure for point-based spatial static data.

different points. When we store a POI, the ID shown in figure 4.6a and b is the same thing.

Such point-based spatial static data can also be used to accommodate static polygons representing geographical regions, such as a plaza with a specific shape and tiles of satellite images. In such a case, the point-based spatial static data can contain a list of spatial points describing a polygon, and the metadata can contain the size of the region formed by the polygon. Regarding satellite images with a regular shape, the spatial point component can be the center of a tile, and the metadata can be used to record the tile's size (e.g., width and height). The pixel level content of a tile image can be stored in a disk file. An alternative way is to store the upper-left and bottom-right points of a tile by two spatial points, respectively.

4.2.2 Point-Based Spatial Time Series Data

This category of data structure is designed for storing point-based data, such as geosensory readings, which is associated with a static location but with readings changing over time constantly. For instance, once a meteorological sensor has been deployed, its location is fixed, but the readings from this sensor change over time every hour, formulating a time series. Thus, a data structure is composed of four parts: an identity, metadata, a spatial point, and a time series, as shown in figure 4.7. More specifically, a time series is a sequence of pairs, each of which consists of a time stamp and a property reading. A property reading can be a categorical or numerical value. The structure of the metadata and spatial point is shown in figure 4.7.

4.2.3 Point-Based Spatiotemporal Data

This category of data structure is designed for point-based data whose location and reading change over time. For example, in a crowd-sensing program, participants collect data in different locations and at different time intervals. Likewise, in a taxi-dispatching

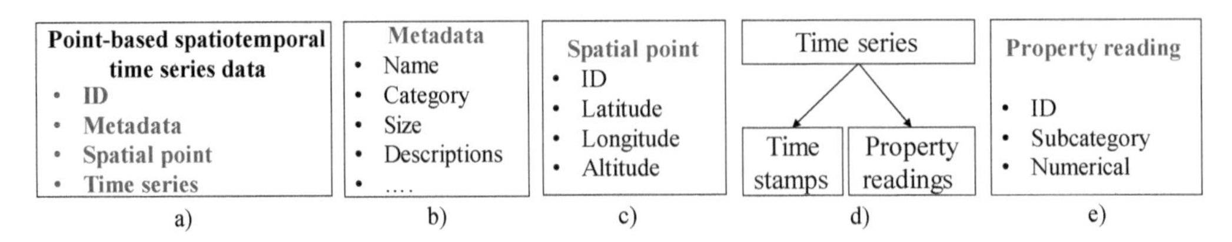

Figure 4.7
Data structure for point-based time series data.

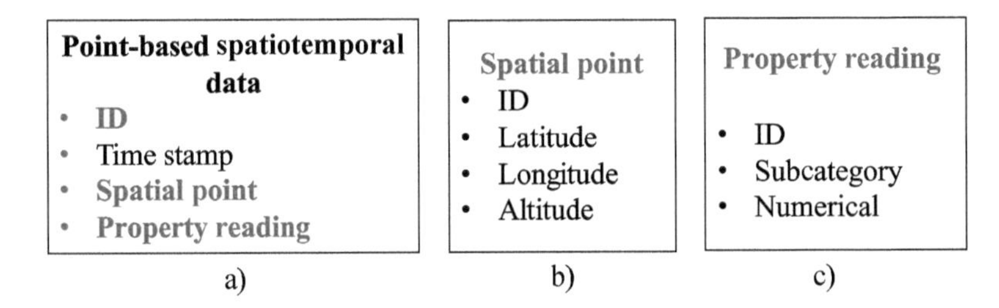

Figure 4.8
Data structure for point-based spatiotemporal data.

system, passengers submit ride requests from different places and at different time stamps. Thus, in the aforementioned scenarios, each instance of data is associated with a location and a time stamp. Different instances are independent, having different locations and time stamps. As presented in figure 4.8a, the data structure is composed of four parts: an identity, a time stamp, a spatial point, and metadata, all of which have been introduced in previous paragraphs.

4.2.4 Network-Based Spatial Static Data

This category of data structure stores spatial network-based data, such as road networks, that is represented by three types of substructures: nodes, edges, and an adjacent list. Once such a spatial network is built, its properties do not change over time (i.e., with static properties). As illustrated in figure 4.9a, an edge can be further represented by a combination of spatial points and metadata, and a node is actually a kind of spatial point. More specifically, as shown in figure 4.9c, an edge consists of an identity, two nodes denoting its terminals, a list of spatial points describing its shape, and metadata. As presented in figure 4.9d, the metadata could include the name, number of lanes, direction (bidirectional or one-way), level, and a bounding box of an edge. In addition, in order to facilitate some applications, such as the maximum k-coverage problem, we can

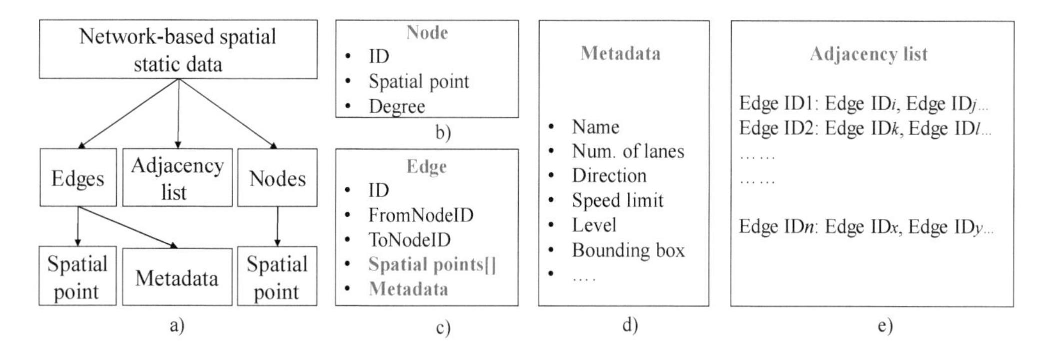

Figure 4.9
Data structure of network-based spatial static data.

count the number of edges connecting to a node, called *degree*, for each node in advance and store the degree in its data structure, as shown in figure 4.9b. Finally, an adjacency list [12] is maintained to represent the structure of a network, showing the neighboring edges of a given edge. As nodes in a spatial network are not densely connected (e.g., a road segment usually has two or three neighbors), an adjacent list is a better choice than an adjacency matrix [4].

4.2.5 Network-Based Spatial Time Series Data

This category of data structure stores time series data over a spatial network, such as the traffic conditions on a road network and the water flow in a pipe network. As illustrated in figure 4.10, the data structure comprises two main parts: one is a network-based spatial static data structure, which has been introduced in figure 4.9. The other is a content structure consisting of a list of edge IDs, each of which is associated with a time series. The design of a time series is the same as in figure 4.7.

4.2.6 Network-Based Spatiotemporal Data

This category of data structure stores network-based data whose location and temporal information change over time constantly. As illustrated in figure 4.11, there are two major subcategories: One is titled *trajectory data*, which records traces of a moving object, such as vehicles, people, and animals [43]. The other is called *spatiotemporal graphs*, denoting dynamic connections and interactions between different moving objects. For example, in a vehicle-to-vehicle network, vehicles move constantly, connecting to different neighbors at different time intervals. Likewise, soldiers communicate with the nearest tank in the middle of a battle. The connections between them and the tanks change over time as they move forward. Different from the aforementioned

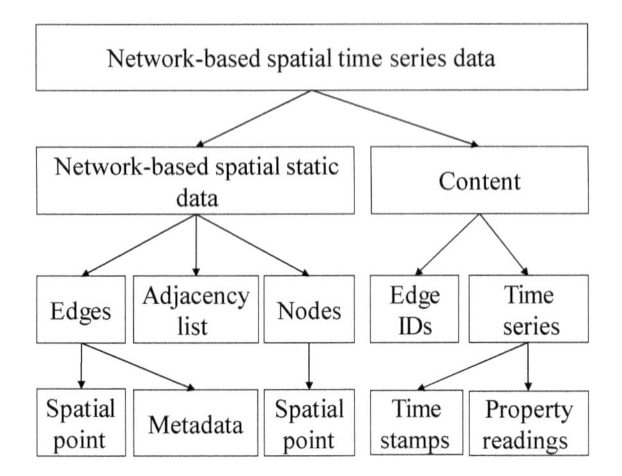

Figure 4.10
Data structure of network-based spatial time series.

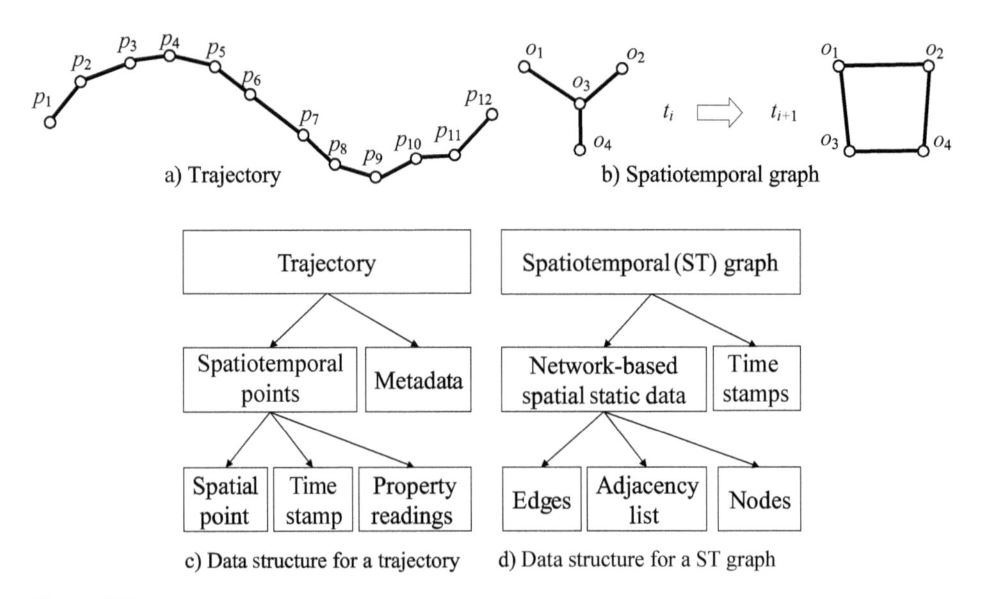

Figure 4.11
Structures for network-based spatiotemporal data.

point-based spatiotemporal data, in which points are independent, there is a clear connection and a strong relationship between different points in the network-based spatiotemporal data.

Figure 4.11c presents the data structure designed for trajectory data, consisting of metadata and a list of spatiotemporal points (defined in figure 4.8). Figure 4.11d shows the data structure devised for a spatiotemporal graph, building a series of network-based spatial static data that corresponds to different time stamps. The definition of network-based spatial static data is the same as shown in figure 4.10.

4.3 Spatial Data Management

In this section, we introduce indexing structures and retrieval algorithms designed for managing spatial data, such as POIs and road networks, regardless of temporal dynamics. As shown in figure 4.3, spatial indexes can be classified into two categories: space partition–based and data-driven indexes. The grid-based indexes and quadtree [29] are frequently used space partition–based spatial indexes. This category of indexes divides geographical spaces into regions regardless of the distribution of data, creating the relationship between regions and data instances. R-tree [16] and k-d tree [1] belong to the data-driven category, building indexes according to the distribution of data.

The requirements for a spatial index are twofold. One is to build a spatial index before query, calling for scanning a given dataset at least one time. The other is to update an index if there are new data instances incoming or removed. Updating consists of inserting and deleting processes, which may require the partial or entire rebuilding of an index. Though we need additional effort to build and maintain an index, the index will be reused for serving the same type of query many times, significantly reducing computational cost for each search. Table 4.1 presents a comparison among four main spatial indexing structures, which will be detailed in the following subsections.

Table 4.1
Comparison of spatial indexing structures.

	Deal with unbalanced data	Range query	Nearest neighbor	Construction	Balanced structure	Size of index
Grid-based	Poor	Good	Normal	Easy	Yes	Big
Quadtree	Good	Best	Poor	Easy	No	Median
K-d tree	Good	Good	Good	Easy	Almost	Median
R-tree	Good	Good	Best	Difficult	Yes	Small

4.3.1 Grid-Based Spatial Index

The grid-based spatial index is the simplest approach to the management of spatial data, organizing data with grids. Hereafter, we introduce the processes for building indexes, serving spatial range queries and nearest neighbor queries, and updating indexes, respectively.

4.3.1.1 Building indexes As illustrated in figure 4.12a, we first partition a given geographical region into uniform and disjoint grids. We then project a given dataset (e.g., the POIs of a city) into these grids, building a grid-points index, as shown in figure 4.12b, that stores a list of identities of points falling in each grid. For example, a point p_1 falls in a grid g_1. Likewise, p_3 and p_4 fall in g_2.

4.3.1.2 Serving spatial range queries When a spatial range query comes, such as the broken rectangle depicted in figure 4.12c, we quickly find the grids that fall partially or fully in the given spatial region. As the region is uniformly partitioned, it is not necessary to match every grid against the query range. In the example shown in figure 4.12, according to its upper-left and bottom-right coordinates, we find the query range is within the second and fourth horizontal partition lines and that of the vertical lines. Thus, we can retrieve g_1, g_2, g_3, and g_4 quickly, ignoring other grids outside the range. We further check if the points from the four grids truly fall in the given spatial range. This is usually called *refinement*. As a result, p_1, p_2, and p_3 are returned as an answer, whereas p_4 is filtered out.

4.3.1.3 Serving nearest neighbor queries As illustrated in figure 4.12d, we aim to retrieve the nearest data instance close to a given query point q based on the grid-based instance. The most intuitive idea is to retrieve the instances falling in the same

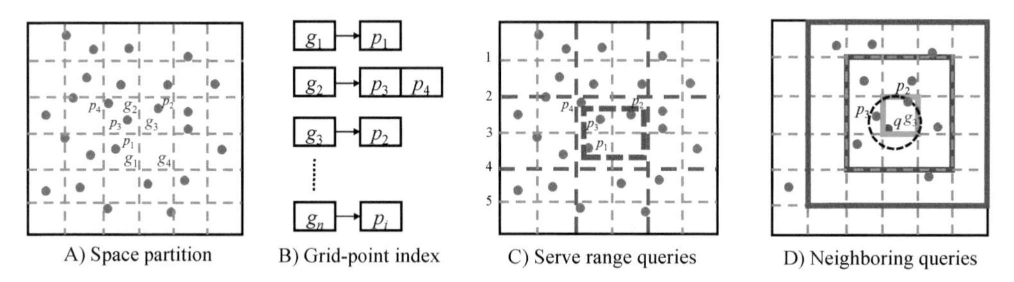

Figure 4.12
Grid-based spatial index.

grid (e.g., g_3 here) where q stands and then to find the one (e.g., p_2) with the shortest distance to q. However, the nearest instance in g_3 may not be the nearest one in the entire dataset because of the shape of the grids. In this example, while p_3 is the nearest neighbor of q, it is not located in g_3. Thus, a safe method is to further explore the eight grids surrounding g_3, checking if there is an instance (like p_3) with a distance to q smaller than p_2. If yes, instances in the neighboring grids are returned. Otherwise, p_2 is returned. In case there are no instances in the grid where a query point is located, we need to explore its neighboring grids outward until at least one instance is found. Then, we can apply the refinement process to these instances, just like dealing with p_2 and p_3.

The way to answer the k-nearest neighbor query is the same as the aforementioned process. In each round of search and expansion, we keep the distance of the first k candidates, respectively, and use the k-th candidate's distance as an upper bound to prune other instances during the later search. That is, if the distance between an instance and a query point is larger than that of the k-th candidate, it cannot be one of the top k-nearest neighbors of the query.

4.3.1.4 Updating indexes The updating process of a grid-based index is fairly simple, not changing the structure of the index. When new data instances come, we can project them onto the grids according to their spatial coordinates and then append the identities of new instances in each grid to the corresponding grid-point list shown in figure 4.12b. The deletion process is similar—removing the identities of instances from the corresponding grid-point indexes.

4.3.1.5 Advantages and disadvantages The grid-based indexes are easy to understand and simple to implement. The effort for updating a grid-based index is minor. However, when handling data with an imbalanced distribution in geographical spaces (e.g., a majority of data instances are concentrated in a few grids while most grids are almost empty), such an index will lead to a grid storing many instances. As a result, the refinement workload increases dramatically, compromising the space-pruning effect of an index.

4.3.2 Quadtree-Based Spatial Index

To deal with the data imbalance problem mentioned above, a quadtree-based indexing structure [29] has been proposed, as illustrated in figure 4.13.

4.3.2.1 Building indexes This indexing structure keeps dividing a given region into four equal subregions, as depicted in figure 4.13a, until the number of instances in a region (or subregion) is less than a given threshold. As depicted in figure 4.13b, a tree structure is built to represent the hierarchical space partitions, where each parent node (standing for a region) has four children nodes (denoting four subregions). Each non-leaf node stores a spatial bounding box it covers and points to its four offspring, and data instances are stored in leaf nodes. In this example, each leaf node only contains one data instance. This partition strategy can tackle the data imbalance problem, providing a dense region with a deep tree structure.

Typically, nodes in a quadtree are named after a number from 0 to 3. For instance, 0 denotes the first node in the first partition layer, and 3 stands for the last node on the layer. Further, 00, 01, 02, and 03 mean the first to fourth offspring nodes of node 0. Because node 1 does not contain any data instances, we do not need to partition it

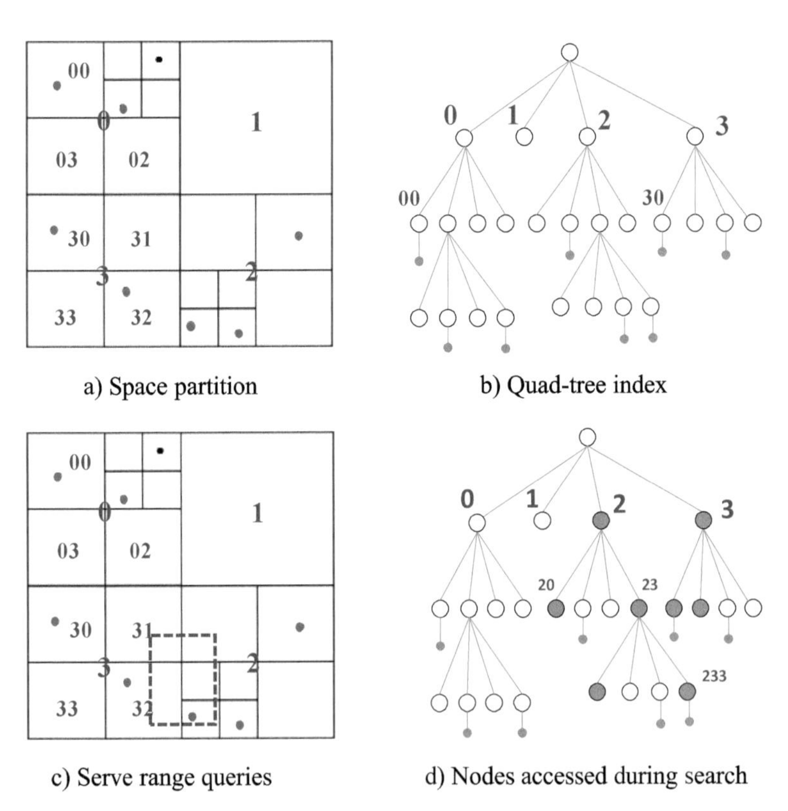

Figure 4.13
Quadtree-based indexing structure.

further. On the contrary, as nodes 0 and 2 have more instances, we need to further partition nodes 01 and 22.

4.3.2.2 Serving spatial range queries Given a spatial range query, denoted by a broken rectangle in figure 4.13c, a retrieval algorithm first finds the nodes that fall partially or fully inside the query range on the first layer of a quadtree. As illustrated in figure 4.13d, the spatial region of both node 2 and 3 intersects the range query in this case. Other nodes consisting of nodes 0 and 1 will be ignored (sometimes we say pruned) in a further searching process. The retrieval algorithm then drills down into nodes 2 and 3, searching for their offspring nodes that fall partially or fully inside the query range iteratively, and so on until they reach data instances stored in the leaf node 233. Finally, a refinement process checks whether each instance in the leaf node truly falls into the given range query.

The gray points denote nodes accessed by the retrieval algorithm during the search process. Because they are a small subset of nodes in the quadtree, the computational cost has been significantly reduced. In the meantime, there are only four children with the same rectangular shape under each parent node. Thus, we can judge if an offspring node intersects a spatial range with minimal effort. Finally, the imbalanced tree structure ensures that the number of instances in each leaf node is less than a threshold. Thus, the workload of the final refinement is minor.

4.3.2.3 Serving nearest neighbor queries Quadtree-based indexes may not be good at serving nearest neighbor queries, but we cannot say they cannot do this at all. Because of its imbalanced tree structure, the neighboring relation between different regions on the same layer of a tree is not straightforward. In addition, given a node, its neighboring nodes' shapes could be irregular. Some nodes that lack data might have a region with a very big size, whereas others with many data instances might correspond to a very small region.

4.3.2.4 Updating indexes When new instances arrive, they will be inserted into a corresponding leaf node of a quadtree through a searching process. Once the number of instances in the leaf node exceeds a given threshold, a split process is called to further partition the spatial region of the leaf node into four equal subregions. When some data instances are removed from a leaf node, a merging process is called to merge the remaining data instances in the four offspring leaf nodes into their parent node if the sum of these instances is smaller than a given threshold. Otherwise, the structure of a quadtree remains unchanged.

4.3.2.5 Advantages and disadvantages The quadtree-based indexes are easy to understand, simple to implement, and capable of handling imbalanced spatial data. The effort for updating such an index is also minor. However, this index cannot easily deal with the nearest neighbor queries.

4.3.3 K-D Tree–Based Spatial Index

A k-dimensional tree (k-d tree) is a binary tree in which every node is a k-dimensional point. Every nonleaf node can be regarded as a splitting hyperplane that divides the space into two parts. Points to the left of this hyperplane are represented by the left subtree of that node, and points to the right of the hyperplane are represented by the right subtree. The hyperplane direction is chosen in the following way: every node in the tree is associated with one of the k-dimensions, with the hyperplane perpendicular to that dimension's axis. So, for example, if for a particular split the x axis is chosen, all points in the subtree with a smaller x value than the node will appear in the left subtree, and all points with a larger x value will be in the right subtree [1].

4.3.3.1 Building indexes Figure 4.14a shows an example of building a k-d tree based on two dimensional points. Each line in the figure (other than the outside box) corresponds to a node in the k-d tree. We first sort all points based on their values on the x dimension, selecting the median point (i.e., $x=5$) as a splitting hyperplane to partition the space into two parts. Points with an x value smaller than 5 are thrown into the left subtree of the node, and those with an x value larger than 5 are placed in the right subtree, as depicted in figure 4.14b. Afterward, we sort nodes of the left subtree based on their value on the y dimension, selecting a median (i.e., $y=6$) point to further partition these nodes into two parts. The left subtree accommodates nodes whose y value is smaller than 6, and the right one stores nodes with a y value larger than 6. We apply the same strategy to each subtree recursively, choosing the median of x and y dimensions alternatively to partition nodes in the subtree until the number of nodes in a subtree is smaller than a given threshold. The maximum number of points in a leaf node has been set to 1 in this example.

This partition strategy results in a balanced tree in which each leaf node is approximately the same distance from the root. The balanced tree structure may have optimal performances in many applications. Note that it is not required to select the median point. In the case where median points are not selected, there is no guarantee that the tree will be balanced. K-d tree is generally regarded as a space partition–based indexing structure. However, as splitting points of a k-d tree are selected from the data to be organized based on the distribution of the data (e.g., median), we deem it a data-driven

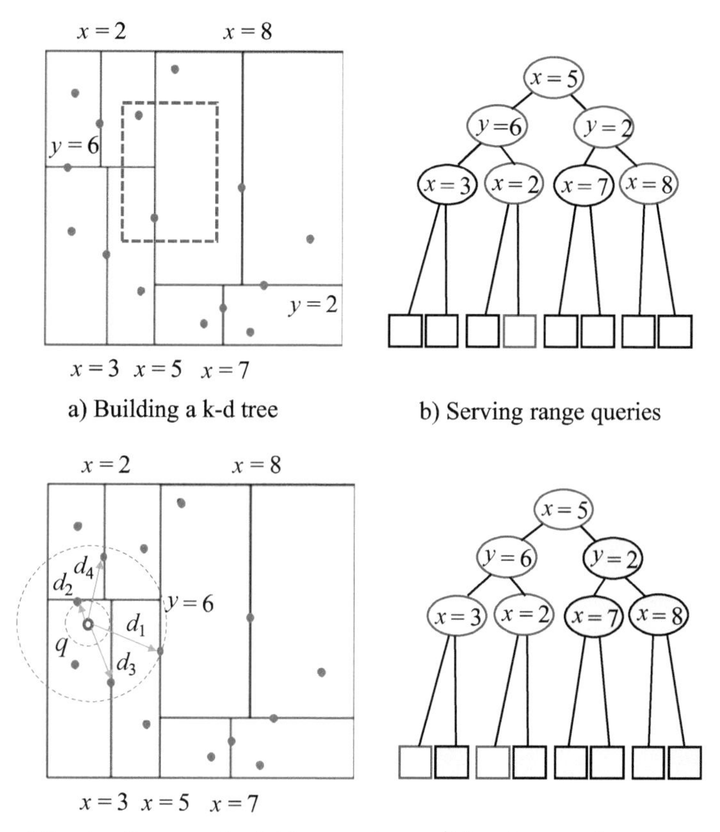

Figure 4.14
K-d tree indexing structure.

indexing structure in this book. That is, given slightly different datasets in the same space, the key indexing structures (e.g., the nodes in a tree) would be dramatically different, relying more on data itself than on a fixed space partition.

4.3.3.2 Serving spatial range queries To answer a range query, a retrieval algorithm traverses a k-d tree, starting from the root and moving to either the left or the right offspring depending on whether the range of the query is on the "left" or "right" side of the splitting plane. Figure 4.14b presents an example of answering spatial range queries based on a k-d tree. A retrieval algorithm first matches a given range query (e.g., the broken rectangle shown in figure 4.14a) against the root of the k-d tree. The spatial range is denoted by its upper-left point (4.5, 8) and bottom-right point (6.7, 4). If the query's spatial range falls only to one side of the root-splitting plane ($x=5$)—that is, the farthest right x value of the rectangle is smaller than 5 or the farthest left x value of the rectangle is larger than 5—the retrieval algorithm will solely check the corresponding subtree of the root and ignore the other subtree. In this example, as the range of the query's x value (4.5, 6.7) covers both sides, we need to further check both subtrees of the root.

In the second round of checking, we compare the range of the query's y value (4, 8) against the splitting plane denoted by $y=6$, finding the query range intersects both sides of $y=6$. As a result, we need to further check the two subtrees of the node denoted by $y=6$. Finally, we find that the query's x range (4.5, 6.7) is on the right side of the splitting plane ($x=2$). Thus, the left side of $x=2$ can be ignored. The searching process on the right subtree of the root is the same as that of the left, pruning the node $x=7$ and the right subtree of node $x=8$. Note that the nodes of a k-d tree can also be a result falling into a range query. For example, the root point is actually inside the range query.

4.3.3.3 Serving nearest neighbor queries The nearest neighbor search algorithm aims to find the point in a k-d tree that is nearest to a given input query point. This search can be done efficiently by using the tree properties to quickly prune large portions of the search space. Starting with the root node, the algorithm moves down the tree recursively, in the same way as if the query point were to be inserted. Once the algorithm reaches the leaf node, it saves that node point as the "current best" and calculates the distance d_s between the query point q and the node. The algorithm then unwinds the recursion of the tree, performing the following steps at each node from the root to the leaf node.

First, the algorithm calculates the distance d between the node and q. If d is shorter than d_s, it notes this node as the current best and updates d_s with d.

Second, the algorithm checks whether there are any points on the other side of the splitting plane that are closer to q than the current best. In concept, this is done by intersecting the splitting hyperplane with a hypersphere around the query point that has a radius equal to the current nearest distance. If the hypersphere does not intersect the splitting plane, the entire branch on the other side of that node is eliminated. Otherwise, there could be nearer points on the other side of the plane. Thus, the algorithm must move down the other branch of the tree from the current node to look for closer points, following the same recursive process as the entire search.

For example, as illustrated in figure 4.14c, after being inserted into the k-d tree, the query point q is located in the left subtree of node $x=3$. The current best is node $x=3$ with $d_s=d_3$. The recursion is a path of three nodes: $x=5$, $y=6$, $x=3$. Starting from the root $x=5$, we can draw a circle (i.e., a hyperplane in a two-dimensional space) with q

as a center and d_1 as a radius. As this circle intersects both sides of the splitting plane, we cannot ignore any of the two subtrees. As $d_1 > d_s$, no replacement happens. Here, we can first select node $y=6$ to check, as it is in the recursion path (and thus more likely to result in space elimination). We calculate the distance d_2 between q and node $y=6$, drawing a corresponding circle with d_2 and q. As this circle does not intersect the splitting plane of $x=5$ (i.e., it is impossible to find other nodes on the right side of the plane with a shorter distance to q than d_2), the right subtree of the root now can be eliminated. Because $d_2 < d_s$, we replace d_s with d_2 and set node $y=6$ as the current best. Because this circle intersects both sides of the plane $y=6$, we need to further check the two subtrees under this node. Because d_3 and d_4 are larger than d_s, no replacement occurs. As the (current best) circle of d_2 does not cross the plane of $x=2$ and $x=3$, the right subtrees of the two planes can be pruned. After checking the instances in the leaf node of the two left subtrees, we find no one has a shorter distance to q than d_s. Consequently, node $y=6$ is the nearest neighbor of q.

4.3.3.4 Updating indexes To insert a new instance into a k-d tree, we traverse the tree, starting from the root and moving to either the left or the right offspring depending on whether the new instance is on the "left" or "right" side of the splitting plane, until reaching a leaf node. Adding points in this manner can cause the tree to become unbalanced, leading to decreased tree performance. The rate of tree performance degradation is dependent upon the spatial distribution of tree points being added and the number of points added in relation to the tree size. If a tree becomes too unbalanced, it may need to be rebalanced to restore the performance of queries that rely on the tree balancing, such as nearest neighbor searching.

4.3.3.5 Advantages and disadvantages K-d trees are easy to understand, simple to implement, and capable of handling range queries and nearest neighbor queries. As it is a binary tree, a k-d tree is deeper than a quadtree, given the same dataset and threshold on instances in a leaf node. Thus, to answer a range query, the number of nodes accessed in a k-d tree is much more than a quadtree.

4.3.4 R-Tree-Based Spatial Index

4.3.4.1 Basic R-trees R-trees were proposed by Antonin Guttman in 1984 [16], using tree structures and minimal bounding rectangles (MBR) for indexing multidimensional information, such as geographical points, rectangles, or polygons. The key idea of the data structure is to group nearby objects and represent them with their MBR in the next higher level of the tree; the *R* in R-tree is for *rectangle*. Since all objects lie within this bounding rectangle, a query that does not intersect the bounding rectangle cannot intersect any of the contained objects. Similar to the B-tree, the R-tree is a balanced search tree (so all leaf nodes are at the same height) organizing the data in pages, each of which can contain a maximum number of entries.

Building indexes The general idea for building an R-tree is to group data instances by MBRs, as illustrated in figure 4.15a, and further aggregate these MBRs into bigger MBRs recursively until all data instances are merged into one MBR (i.e., the root node). Each MBR is represented by its bottom-left ($L.x$, $L.y$) and upper-right points ($U.x$, $U.y$). Figure 4.15b presents an R-tree built for storing the data instances shown in figure 4.15a, where bigger MBRs consisting of R_{10}, R_{11}, and R_{12} are represented by nodes in the second level of the R-tree, and each of them is further composed of a few small MBRs (e.g., R_1, R_2, and R_3 belong to R_{10}). Each entry within a nonleaf node stores two pieces of data: a way of identifying an offspring node and the bounding box of all

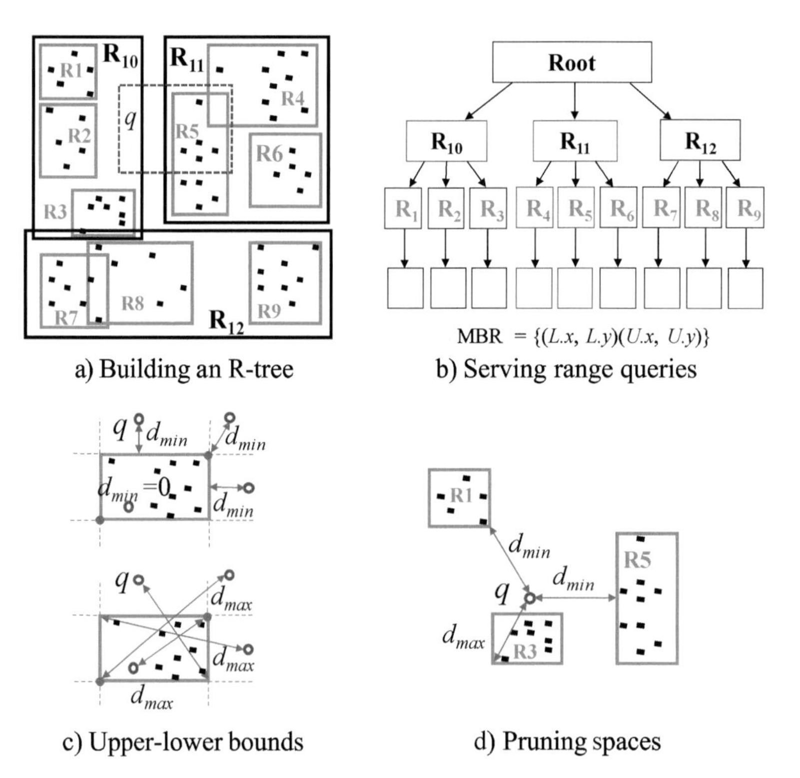

Figure 4.15
R-tree indexing structure.

entries within this offspring node. Leaf nodes store the data required for each offspring: for point data, the leaf entries can be just the points themselves; for polygon data (that often requires more storage), the common setup is to store only the MBR of the polygon along with a unique identifier in the tree.

The key difficulty of R-trees is to build an efficient tree satisfying the following three criteria: First, the tree must be balanced (so the leaf nodes are the same height). Second, rectangles in the R-tree should cover empty spaces to make them as small as possible (i.e., the density of data instances in each MBR should be as high as possible). Third, overlaps between these rectangles should be as small as possible (so that during a search, fewer subtrees need to be processed). Given these three criteria, there is no polynomial-time algorithm that can find the optimal solution. Alternatively, some density-based clustering algorithms, such as OPTICS [21] and DBSCAN [15], can be employed to first generate some preliminary instance groups, which will be used as candidates to formulate MBRs with further adjustments.

Serving spatial range queries Given a spatial range query, the searching process is quite similar to that of quadtree and k-d tree. The search starts from the root node of the tree, checking if a node's MBR overlaps the query rectangle or not. If yes, the corresponding offspring node has to be further searched. Searching is done this way in a recursive manner until all overlapping nodes have been traversed. When a leaf node is reached, the contained points or bounding boxes (for polygons) are tested against the query rectangle. The objects that lie within the query rectangle are returned as results.

Figure 4.15b presents the nodes (*in red*) that have been accessed for answering the range query q shown in figure 4.15a. As the query q intersects both R_{10} and R_{11} at the second layer, the two nodes' offspring have to be further searched. After further matching the MBRs of R_{10}'s offspring against q, we find no children actually intersect the query rectangle. Thus, we stop the search originating from R_{10}. Among R_{11}'s children, R_4 and R_5's MBRs intersect q. Thus, we need to retrieve the data instances stored in the two leaf nodes, searching for instances lying within the query rectangle. Although no instance from R_4 truly lies within the query rectangle, we still need to match each instance stored in it against q. This is caused by the overlap between R_4 and R_5's MBRs. To reduce the searching workload, we need to minimize the overlap between different nodes' MBR when building an R-tree.

Serving nearest neighbor queries The shape of an MBR provides an effective way to calculate the minimum and maximum (also called lower bound d_{min} and upper bound d_{max}) distances between a query point and a group of instances. These upper and lower bounds can help prune search spaces significantly when answering the nearest neighbor queries. As illustrated in figure 4.15c, when a query point lies within an MBR, d_{min} is zero, and d_{max} is the maximum distance to one of the four vertexes of the MBR.

If $L.x \leq q.x \leq U.x$ and $q.y > U.y$, (4.1)

$d_{min} = q.y - U.y$, and $d_{max} = \max\{dist(q.x, q.y.; L.x, L.y), dist(q.x, q.y; U.x, L.y)\}$.

If $q.x > U.x$ and $q.y > U.y$, (4.2)

$d_{min} = dist(q.x, q.y; U.x, U.y)$, and $d_{max} = dist(q.x, q.y; L.x, L.y)$.

If $q.x > U.x$ and $L.y \leq q.y \leq U.y$, (4.3)

$d_{min} = q.x - U.x$, $d_{max} = \max\{dist(q.x, q.y; L.x, L.y), dist(q.x, q.y; L.x, U.y)\}$.

d_{min} and d_{max} in other situations can be easily derived in a way similar to that mentioned above.

Figure 4.15d shows the benefit of calculating such bounds through an example. If the maximum distance between q and R_3's instances is smaller than the minimum distance of R_1 and R_5, all instances in R_1 and R_5 cannot have a shorter distance to q than those in R_3 (i.e., they cannot be the nearest neighbor). Consequently, they can be pruned from the search process without being further checked. In case d_{min} and d_{max} cannot help prune any rectangles at the level of an R-tree, we need to further check the bounds of their offspring.

Updating indexes The original idea for inserting elements into an R-tree is to insert those into the subtree that require the least enlargement of their bounding boxes. Once that page is full, the data instances are split into two sets that should cover each minimal area. More specifically, to insert an instance, an R-tree is traversed recursively from its root node. At each step, all rectangles in the current directory node are examined, and a candidate is chosen using a heuristic, such as choosing the rectangle that requires the least enlargement. The search then descends into this page until reaching a leaf node. If the leaf node is full, it must be split before the insertion is made. A heuristic is employed to split the node into two, as an exhaustive search is very costly. By adding the newly created node to the previous level, this level can again overflow, propagating up to the root node. If the root node also overflows, a new root node is created, and the tree has increased in height.

Advantages and disadvantages R-trees are built based on the distribution of data, which may result in an efficient index with a precise structure. R-tree is a balanced tree, having an optimal performance in many search scenarios, such as the nearest neighbor search. The MBRs of an R-tree can be used to derive upper and lower bounds, helping to prune search spaces significantly during a nearest neighbor search. Once an R-tree is built, the search process is easy to understand and simple to implement.

However, as compared to the grid-based indexes, quadtrees, and k-d trees, building an effective R-tree is complex. Considering the aforementioned three criteria—a balanced tree, minimal overlaps, and a dense representation in each MBR—there is no polynomial-time algorithm that can find the optimal solution under the three criteria.

When there are new instances, updating an R-tree is also computationally expensive and has to employ some heuristics to find a compromised solution. As R-tree structures are highly susceptible to the order in which their entries are inserted, an insertion-built (rather than bulk-loaded) structure is likely to be suboptimal.

4.3.4.2 R*-trees and R+ trees R*-trees [2], a variant of R-trees, attempt to address this issue using a combination of a revised node split algorithm and the concept of forced reinsertion at node overflow. The deletion and reinsertion of entries allows them to find a place in the tree that may be more appropriate than their original location. When a node overflows, a portion of its entries are removed from the node and reinserted into the tree. In order to avoid an indefinite cascade of reinsertions caused by subsequent node overflow, the reinsertion routine may be called only once in each level of the tree when inserting any new entry. This has the effect of producing more well-clustered groups of entries in nodes, reducing node coverage. Furthermore, actual node splits are often postponed, causing average node occupancy to rise. Reinsertion can be seen as a method of incremental tree optimization triggered on node overflow. R*-trees have slightly higher construction costs than standard R-trees but result in better query performance.

When overlaps between MBRs become large, the search efficiency of an R-tree decreases significantly. To tackle this issue, R+ trees [30] avoid the overlapping of internal nodes' MBR by inserting an object into multiple leaves if necessary. As illustrated in figure 4.16, MBR E causes an overlap between MBR A and B in R-trees. To address this issue, R+ trees reconstruct the MBR for A and B, respectively, so that they have no overlap. As a result, MBR E belongs to both A and B. R+ trees differ from R-trees in that nodes are not guaranteed to be at least half filled, the entries of any internal node do not overlap, and an object ID may be stored in more than one leaf node. Because nodes do not overlap with each other, point query can be handled by traversing a single path in an R+ tree (i.e., fewer nodes are visited than with the R-tree). Since an MBR may belong to multiple parents, an R+ tree can be larger than an R-tree built on the same

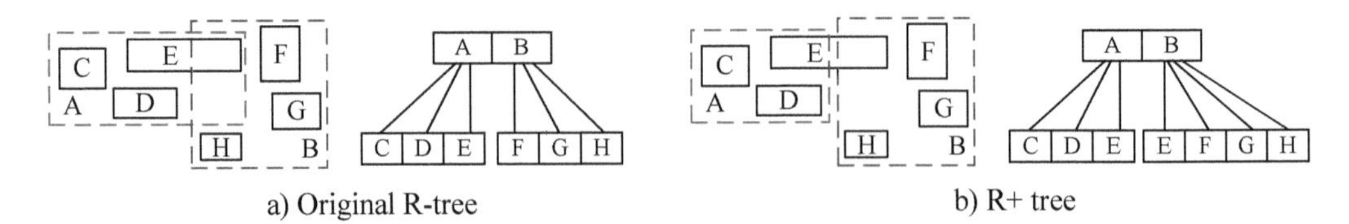

Figure 4.16
The differences between R+ tree and R-tree.

dataset. The construction and maintenance of R+ trees are more complex than that of R-trees and other variants of R-trees.

4.4 Spatiotemporal Data Management

As illustrated by the three columns in figure 4.1, there are three types of data in cities according to data's spatiotemporal dynamics. The first type of data—that is, spatial (and temporal static) data—has been discussed in section 4.3. In this section, we will discuss the other types of data.

One is the spatial static and temporal dynamic data (i.e., the location of an object is static, but readings from the object constantly change over time). Most sensory data belongs to this category. The main challenge of managing this type of data is to effectively handle the continuously increasing time series at each location so that we can answer spatiotemporal queries efficiently.

The other is the spatiotemporal dynamic data (i.e., both the location and readings of an object constantly change over time). Trajectory data and crowd-sensing data belong in this category. Research on this category of data has two perspectives. One thread of research, also known as moving object databases [5, 23, 43, 44], focuses on querying moving objects' current locations. The other branch of research, such as [28, 40, 48, 49], pays more attention to the management of objects' historical trajectories.

Range queries on spatiotemporal data are concerned with a spatial range and a temporal span, or a spatial range plus a value range of readings. A nearest neighbor query on spatiotemporal data is associated with a location and a time stamp.

4.4.1 Managing Spatial Static and Temporal Dynamic Data

This category of data is associated with a static location and dynamic readings. Typical datasets in this category include geosensory readings and traffic flows on road segments, as illustrated in the second column of figure 4.1. Figure 4.4c presents a general approach to managing such data. In this approach, we can manage objects' locations, which do not change over time, using a spatial indexing structure introduced in section 4.3. For each object, its dynamic readings are further organized by a temporal index, such as sorted array and B+ trees [27]. When answering a spatiotemporal range query, we first search for objects that lie in the given spatial range based on the spatial index and then find the elements that fall within the temporal range in terms of said objects' temporal index. Since spatial indexes have been introduced in section 4.3, we focus on discussing some possible temporal indexing structures in this section.

4.4.1.1 Queries There are two typical range queries on this category of data. One is to retrieve the data (from objects) that lie within a given spatial region and time span. For example, retrieving the humidity data from 8 a.m. to 1 p.m. on this Monday from Central Park in New York City. The other is to search for data whose spatial locations are within a given spatial range and whose temporal readings are within a range—for instance, retrieving air quality data whose value is larger than 200 but smaller than 300 in Beijing.

4.4.1.2 Temporal indexing structures Sorted array and B+ tree are two indexing structures that can be employed to manage dynamic readings generated by each object.

Sorted array As readings of an object are generated chronologically, there is a natural order between them. In other words, we can save each object's readings with a sorted array in which each element stores a time stamp t_i and the corresponding value v_i. New readings will be naturally appended to the end of the array without calling for a sorting process. Thus, finding the readings generated within a temporal span (e.g., $[t_3, t_k]$, as shown in figure 4.17a) can be easily handled by two binary search processes that search for t_3 and t_k, respectively, within the sorted array. The elements between $[t_3, t_k]$ are retrieved as search results. In computer science, binary search, also known as *half-interval search* [42] or *logarithmic search* [20], is a search algorithm that finds the position of a target value within a sorted array. Binary search compares the target value to the middle element of the array; if they are unequal, the half in which the target cannot lie is eliminated. The search continues on the remaining half until it is successful.

Such an indexing structure can only be employed to handle temporal queries concerned with a time span. If aiming to find elements within a value range $[v_k, v_i]$, we need to sort the array by each element's value and insert a new reading into a proper position of the array; each insert calls for a searching process and results in a movement of elements after the insert position. This is a very time-consuming process.

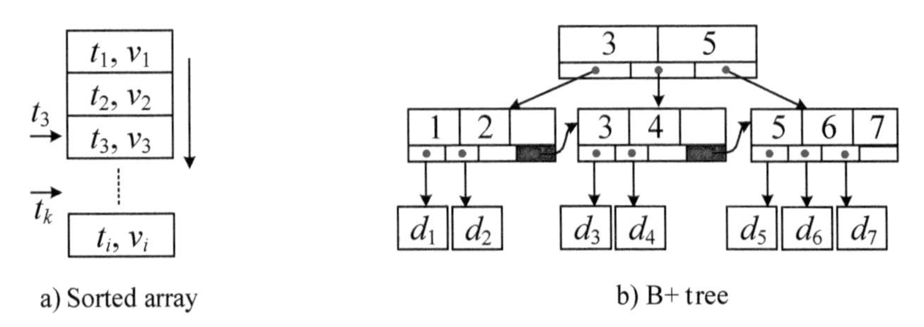

a) Sorted array b) B+ tree

Figure 4.17
Indexing structures for temporal dynamics.

B+ tree To address the aforementioned issue, a B+ tree is employed to store dynamic readings based on their values. Consisting of a root, internal nodes, and leaves, a B+ tree is a tree with a variable but often large number of children per node. A B+ tree can be viewed as a B-tree in which each node contains only keys (not key-value pairs) and to which an additional level is added at the bottom with linked leaves. The primary value of a B+ tree is in storing data for efficient retrieval in a block-oriented storage context.

Figure 4.17b presents an example that manages seven data instances ($d_1, d_2, \ldots, d_7$) with a B+ tree. The root node splits the seven instances into three segments. Data instances with a value smaller than 3 (i.e., $v<3$) are placed into the left subtree; instances with $3 \le v \le 5$ are stored in the middle subtree; instances with $v>5$ are thrown into the right subtree. The left subtree is further split into three segments by two values (1, 2) with two links pointing to two data instances d_1 and d_2, respectively, where d_1's value v is smaller than 1, and d_2's value $1 \le v \le 2$. The same strategy is applied to the middle and right subtrees, with data instances assigned to corresponding range segments. The (*red*) linked list between leaf nodes allows rapid in-order traversal, which expedites the retrieval process after finding the two boundary data instances. For example, if trying to search for data instances with $1 \le v \le 5$, we can first find the instance with the smallest value (i.e., d_2) in the first round of search and then the instance with the largest value (i.e., d_5) in another round of search. Finally, we can quickly retrieve d_2, d_3, d_4, and d_5 through the red links between leaf nodes.

When a new instance comes, a B+ tree can insert it into corresponding leaf nodes through a search process without calling for a resort process like a sorted array. In most cases, an insert is very efficient. When the bucket of a leaf node is full, a split process is called to allocate a new leaf and move half the bucket's elements to the new bucket.

4.4.2 Moving-Object Databases

A moving-object database is composed of a collection of queries, indexes, and retrieval algorithms that are concerned with the current locations (or the location at a particular historical time) of moving objects, such as people, vehicles, and animals.

4.4.2.1 Queries A typical range query in moving-object databases is to retrieve objects (e.g., free taxicabs) that are currently within a spatial range (e.g., one mile of 33 North Michigan Avenue, Chicago) and a time span (e.g., in the past three minutes) or at a time stamp, as illustrated in figure 4.18a.

Regarding the nearest neighbor queries, there are three scenarios, depending on whether the location of a query point and that of objects are dynamic or not.

1. In the first scenario, the location of a query point is constantly moving, while the location of objects to be searched is static. A representative example is to search for the nearest gas stations around a vehicle when the vehicle is being driven, as shown in figure 4.18b.
2. In the second scenario, the location of a query point is static while objects are moving. A typical example is to find the nearest vacant taxi around a person, as depicted in figure 4.18c.
3. In the third scenario, the location of a query point and that of objects are both dynamic. For instance, in a battle, a soldier wants to find the nearest tank around him to access more information about the entire battle. As presented in figure 4.18d, both tanks and soldiers are moving in the field when a search is performed.

4.4.2.2 Indexing structure overview Figure 4.19 presents the evolvement of representative indexing structures that have been proposed (or can be employed) for managing spatial and spatiotemporal data. Since there have been many indexes proposed in the past decades, this road map provides a panorama showing the connection and differences between different indexing structures for spatial and spatiotemporal data.

The vertical axis stands for the years that different indexes are proposed. The horizontal axis denotes the dimension of data an index is concerned with. For example, B-trees and B+ trees are proposed to handle one-dimensional (1D) data (like time), and R-trees are proposed to manage two-dimensional (2D) spatial data. Multiple-version B-trees is an indexing structure that builds a B-tree for each time stamp. Thus, it is between 1D and 2D indexes. Indexes that are 2.5D, such as multiple-version R-tree (MVR-tree) [37] and historical R-tree (HR-tree) [26], build a 2D spatial index for each time stamp. They are not real 3D indexes, like 3D R-tree [39], which treat time as the third dimension when

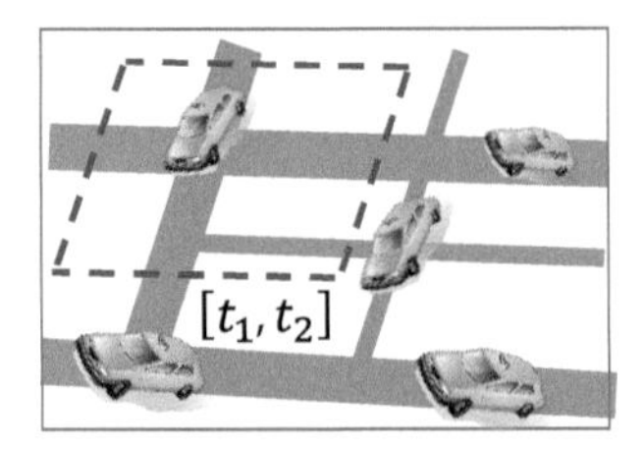

a) Spatiotemporal range query

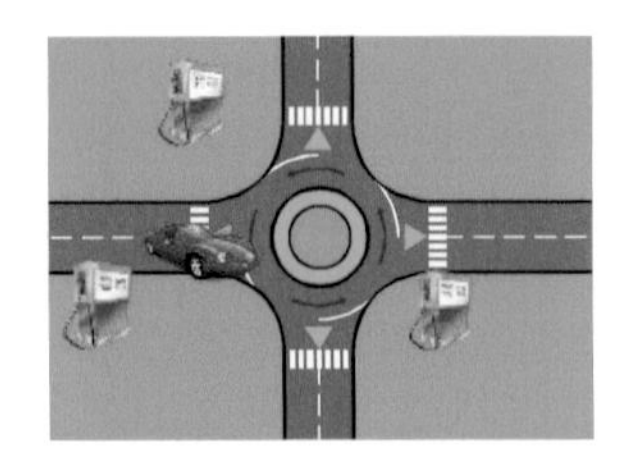

b) KNN query: moving query point and static objects

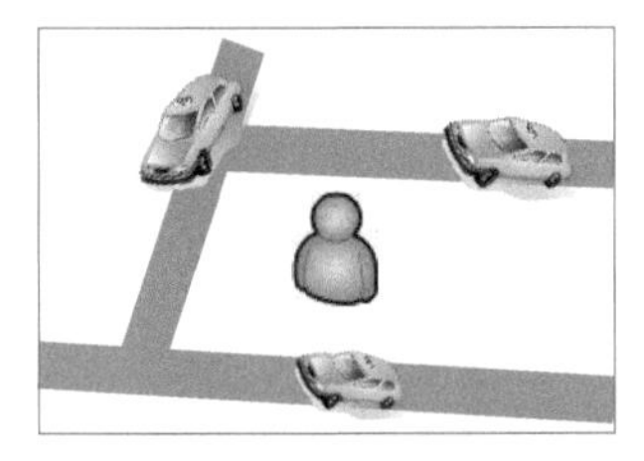

c) KNN query: static query point and moving objects

d) KNN query: moving query point and moving object

Figure 4.18
The nearest neighbor queries in moving-object databases.

dealing with spatiotemporal data. The arrows pointing from index A to index B denote that B is derived from or based on A. For instance, both R+ trees and R*-trees are derived from R-trees. Likewise, compressed start-end time (CSE)-trees are based on grid-based indexes and B+ trees. Indexing structures that lie in the same broken box, such as MVR-trees and MV3R-trees, are proposed in the same paper [37].

As 2D indexes have been introduced in section 4.3 and 1D indexes have been discussed in section 4.4.1, we focus on discussing 2.5D, 3D, and beyond indexing structures for moving-object databases.

4.4.2.3 Multiple version–based spatiotemporal indexes The first category of indexes, such as historical R-tree (HR-tree) [26], HR+ tree [36], and multiple-version R-tree (MVR-tree) [37], builds a spatial index at each time stamp. To enhance the efficiency of such indexes, unchanged substructures of an index across consecutive time intervals can be reused. A temporal indexing structure can be employed to manage time stamps, expediting the search for a given time stamp.

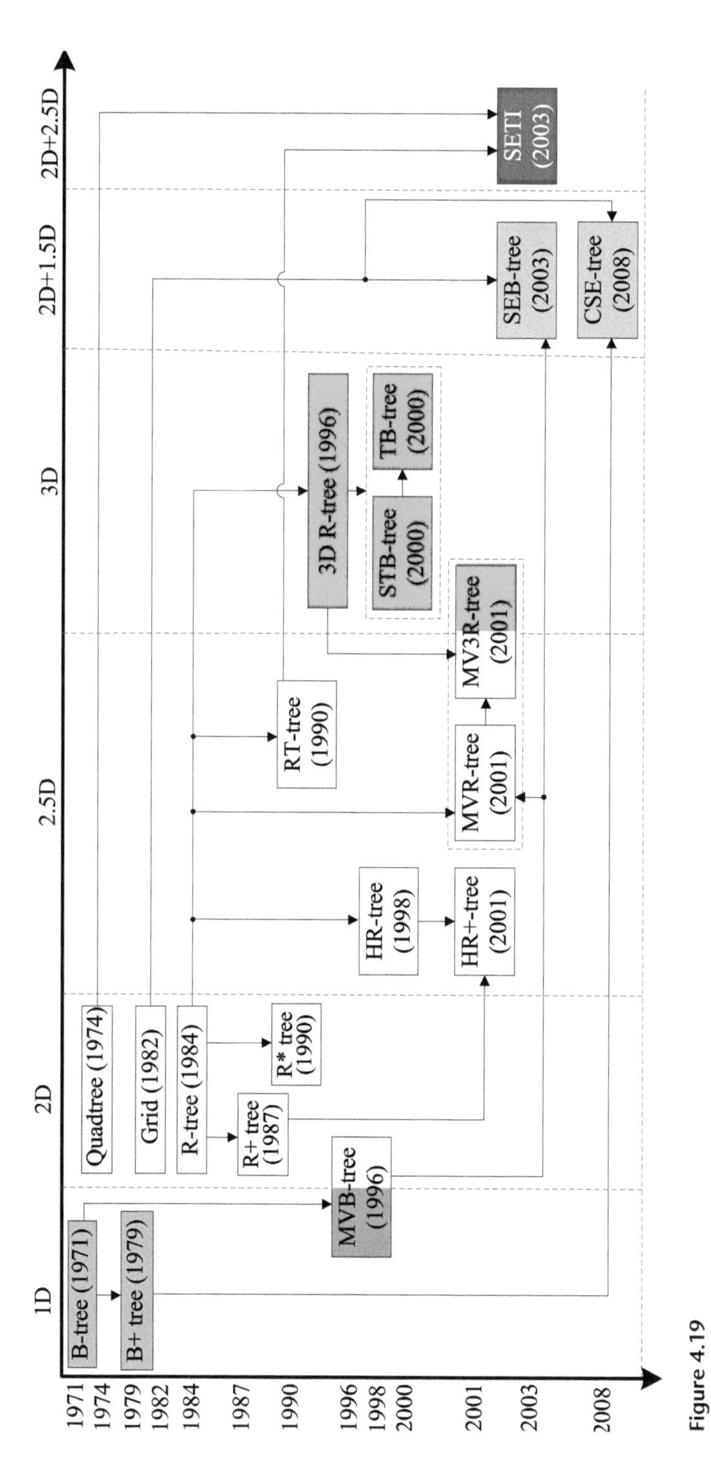

Figure 4.19
Indexing structure for spatiotemporal data.

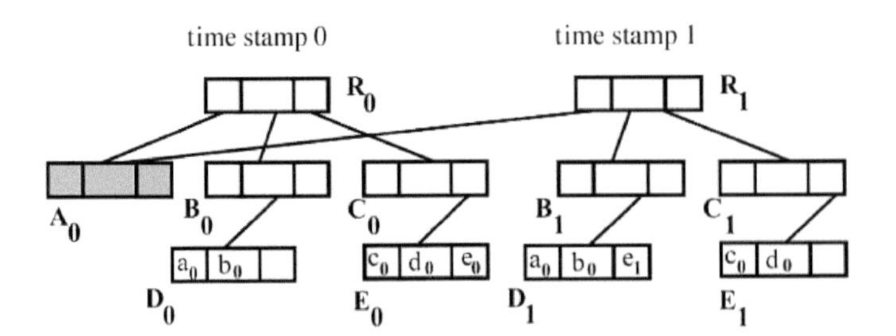

Figure 4.20
The structure of HR-tree [26].

Figure 4.20 illustrates a part of HR-tree with two time stamps. At time stamp 1, object e changes its location. Thus, its old version e_0 should be deleted from the R-tree for time stamp 0, while its new version e_0 should be inserted into the R-tree for time stamp 1. This causes the creation of two leaf nodes: D_1, which contains the entries of D_0 plus e_1, and E_1, which contains the entries of E_0 after the deletion of e_0. The changes are propagated to the root (causing the creation of B_1 and C_1). Even if only one object changes its position, the entire path may need to be duplicated. Trees of previous time stamps are never modified. Notice that node A_0 is shared by both trees, indicating that no object in its subtree has changed its location at time stamp 1 [26].

The spatiotemporal queries this category of indexes can handle are usually time stamp queries, such as finding objects that lie in a given spatial range at a particular time stamp (i.e., time stamp + spatial range) or finding the spatial k-nearest neighbors of a given object at a particular time stamp (time stamp + KNN). When a time stamp spatial range query comes, we first search for the time stamp based on the temporal index (refer to section 4.4.1 for details). We then check the spatial index of the time stamp, finding the objects that lie in the query's spatial range. Likewise, when a time-stamped KNN query comes, a search algorithm first finds the corresponding time stamp and then searches for the k-nearest neighbors of the query based on the spatial index of the time stamp. The disadvantages of such multiple version–based indexes are threefold.

First, such indexes are not very effective in handling temporal range (a.k.a. interval) queries (e.g., finding free taxis around a plaza over an hour). An interval query calls for searching in many time stamps (most of which may be redundant) and an operation to merge search results across different time stamps.

Second, it is subtle to set the granularity of time (i.e., the interval between two time stamps). If we build such an index with a very fine granularity on time (e.g., every five seconds), there are extensive duplications of objects that lead to many nearly redundant (but not exactly the same) nodes in each time stamp's spatial index. As the temporal granularity increases, the total size of the index becomes very large, thereby compromising search efficiency. On the other hand, if we choose a very coarse temporal granularity, we may not be able to accurately answer a time-stamp query if the query does not have a corresponding spatial index built in advance. Moving objects' locations become uncertain between two time stamps. The bigger the time interval, the higher uncertainty that exists, thus less search accuracy is guaranteed.

Third, such indexes are vulnerable to different objects' sampling rates. In a real-world setting, different moving objects usually have different sampling rates, reporting on their location at different time stamps. That is, moving objects maintain the same position until the new position is updated. As a consequence, objects without reports at a given time stamp will be lost in its spatial index and thus cannot be found for the query.

To tackle the first challenge, a hybrid structure, titled MV3R-tree [37], is proposed, using a multiversion R-tree for time-stamp queries and a small 3D R-tree for time-interval queries, as illustrated in figure 4.21. Thus, it is a hybrid structure fusing 2.5D and 3D indexes. MVR-tree is an extension of multiversion B-tree. A 3D R-tree is built on the leaf nodes of the MVR-tree. An MV3R-tree can handle both time-stamp queries and time-interval queries. For a discrete event, it outperforms other indexing structures, such as 3D R-tree and HR-tree. However, the two indices share the same leaf nodes, which leads to a rather complex insert algorithm. In addition, the MVR-tree models the temporal change as a discrete event. Thus, it cannot handle the third challenge mentioned above.

Before the multiple version–based indexes have been proposed, RT-tree [45] couples time intervals with spatial ranges in each node of the tree so that only one index tree is maintained, as opposed to the multi-index trees maintained by MVR-tree or HR-tree. When inserting a record (MBR, t_i) into the index tree, if there is a leaf node with

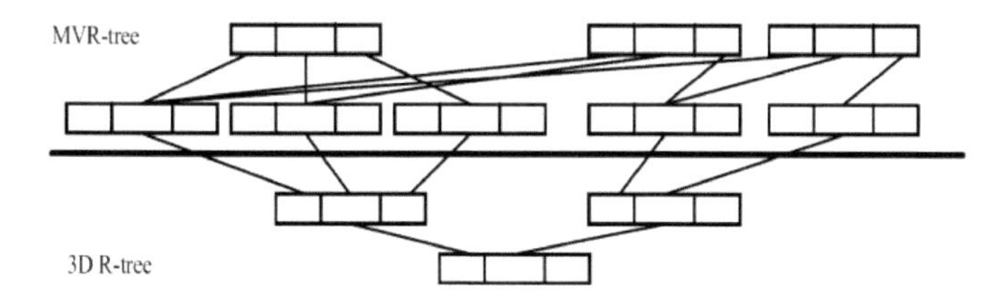

Figure 4.21
The structure of MV3R-tree [37].

the same MBR and the same date as this record, the time interval of the entry will be expanded to t_i. Otherwise, a new entry is created based on the minimal time interval and/or spatial covering MBR. RT-tree needs much fewer nodes than its corresponding MVR-tree because it does not create duplicating paths. However, any temporal query has to search the entire RT-tree, as there is no discrimination in the index search along the temporal dimension.

4.4.2.4 Three-dimensional spatiotemporal indexes The second category of indexes, such as 3D R-tree [39], treats time as the third dimension, extending spatial indexing structures from managing two-dimensional spatial data to three-dimensional spatiotemporal data, as illustrated in figure 4.4a. Thus, there are real 3D indexes.

For example, 3D R-tree records the state of each moving object with a $<x, y, t>$, where $<x, y>$ are two-dimensional coordinates in the geographical space, and t is a time stamp. 3D R-tree groups the data generated by moving objects at recent time intervals with 3D minimal bounding boxes, which are further aggregated into bigger bounding boxes to formulate a tree. Similar to R-trees, a 3D R-tree is a balanced tree, keeping each bounding box as dense as possible and minimizing the overlaps between bounding boxes. Each node in a 3D R-tree stores the coordinates of the minimal bounding box, covering its children and the links pointing to its children nodes. Each leaf node can store a number of data instances smaller than a given threshold.

Answering a spatiotemporal range query based on a 3D R-tree is the equivalent to finding objects that lie within a query 3D box. The search process is quite similar to that of R-trees. The search starts from the root node of the tree, checking if a node's 3D bounding box overlaps the query box or not. If yes, the corresponding child node has to be further searched. Searching is done this way in a recursive manner until all overlapping nodes have been traversed. When a leaf node is reached, the contained data instances are tested against the query box. Data instances that lie within the query box are returned as results.

Likewise, answering the KNN query can be regarded as searching for the k-nearest neighbor of a 3D query point in a 3D space. We can efficiently calculate the upper bound and lower bound of the distance between a query point and a group of data instances within a 3D bounding box, following the way shown in figure 4.15c and d. If a data instance has a distance to a query point smaller than a bounding box's lower bound of distance to the query point, all data instances within the bounding box cannot be the nearest neighbor of the query point and are thereby filtered without being checked one by one.

The process of updating a 3D R-tree is the same as that of R-trees. It inserts a new data instance into the corresponding leaf node through a search process, which starts from the root node and matches the coordinates of the new instance against the bounding box of intermedia nodes until reaching a leaf node. If the number of instances in a leaf node does not exceed a given threshold, the new instance is stored there. Otherwise, a splitting process is called to partition a leaf node into several smaller leaf nodes. If the leaf node's parent is also full, the splitting process needs to be performed at a higher level until the new instance can be accommodated.

When using 3D R-trees to organize data of long history, the overlap between different nodes' minimum bounding boxes inevitably becomes large. This compromises the performance of 3D R-trees in answering spatiotemporal queries. Thus, 3D R-trees are usually employed to manage the moving objects' recent status rather than a long history. Though the spatiotemporal R-tree (STR-tree) and trajectory bundle (TB)-tree [28] have been proposed to address this issue, the overlap among different 3D boxes still keeps increasing as time goes by. As STR-tree and TB-tree are proposed for managing trajectory data, we will introduce them in section 4.4.3.

4.4.3 Trajectory Data Management

Different from moving-object databases that are concerned with the current location (or the location of a particular time stamp) of a moving object, the trajectory data management introduced in this section deals with the traveling history of a moving object, particularly the path of a moving object at a given time interval [14].

4.4.3.1 Query overview There are three major types of queries: range queries, KNN queries, and path queries. Other advanced queries can be derived from the combination of these three fundamental queries.

Range queries retrieve the trajectories falling into (or intersecting) a spatial (or spatiotemporal) range. For example, as shown in figure 4.22a, a range query can help us retrieve the trajectories of vehicles passing a given rectangular region between 2 p.m. and 4 p.m. in the past month. The retrieved trajectories (or segments) can then be used to derive features, such as the travel speed and traffic flow, for data-mining tasks like classification and prediction.

KNN queries retrieve the top-*k* trajectories with the minimum aggregated distance to a few points (called the *KNN point queries* [9, 35, 38]) or a specific trajectory (called the *KNN trajectory queries* [1, 46]). As depicted in figure 4.22b, an example of the KNN point query is to retrieve the trajectories of vehicles that are close to two given restaurants

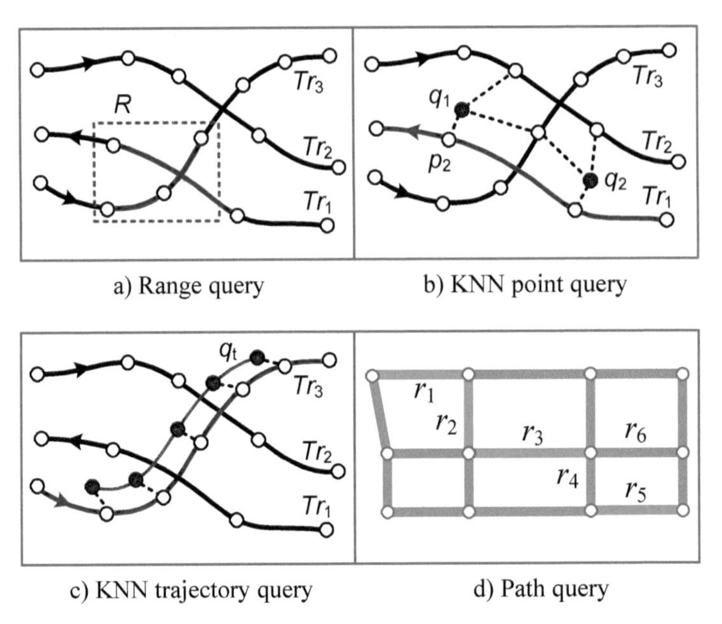

Figure 4.22
Typical queries on trajectories.

(e.g., q_1 and q_2). Sometimes, the order between the query points is also considered [9] (e.g., finding the top-k nearest trajectories first passing q_1 and then q_2). Without the order, Tr_1 is the nearest trajectory to the two points. However, Tr_2 becomes the nearest after considering the order.

As illustrated in figure 4.22c, the most similar trajectory (i.e., with the minimum aggregated distance) to the query trajectory q_t is Tr_3. Such a query can help a person find other people's hiking trails that are most similar to the path the user is going to explore. Thus, the person can learn from other people's hiking experiences before a hiking activity really takes place. The query can also help us find people or animals traveling together. For example, given the GPS track of a tiger, zoologists can identify similar trajectories generated by other tigers, which might be accompanying it or are its family members, based on this query.

The differences between the aforementioned two KNN queries are twofold. First, the KNN point queries are concerned with whether a trajectory provides a good connection to query locations rather than whether the trajectory is similar to the query in shape. It is not important for the KNN point query to consider the movement of a moving object traveling between two query points. Second, the number of query points is usually very small and can be far away from each other in applications. As a result, we cannot connect these query points sequentially to formulate a trajectory and then call solutions designed for the KNN trajectory query to solve it.

Path queries retrieve trajectories exactly traversing a given path. This type of query is usually performed in a spatial graph, such as road networks or airline networks. The path comprises a sequence of graph edges, which can be consecutive or disjointed. As illustrated in figure 4.22d, a path query retrieves trajectories of vehicles sequentially passing the following four road segments: r_1, r_2, r_3, r_5. Note that r_3 and r_5 are not connected. The path query can also be associated with a time interval, such as finding vehicles' GPS trajectories traversing $r_1 \rightarrow r_2 \rightarrow r_3$ over the past hour. Trajectories retrieved by such a query can be used to estimate the travel time of a path at the current time interval. When associated with a long interval (e.g., the last month), retrieved trajectories can be employed to calculate the congestion index of the path. Before performing such a query, a map-matching algorithm [24, 47] is usually called to project a trajectory onto a spatial graph.

4.4.3.2 Distances of trajectories When answering KNN queries or clustering trajectories, we need to calculate the distance (alternatively we can say the similarity) between a trajectory and a few points, or between two trajectories.

The distance between a point q and a trajectory A is usually measured by the distance from q to its nearest point in A, denoted as $D(q, A) = min_{p \in A} D(p, q)$ (e.g., q_1 and p_2 shown in figure 4.22b). An approach extending the distance from a single point q to multiple query points Q is

$$D(Q, A) = \sum_{q \in Q} e^{D(q, A)}, \text{ or } S(Q, A) = \sum_{q \in Q} e^{-D(q, A)}, \tag{4.4}$$

written in a similarity fashion [9]. The intuition of using the exponential function is to assign a larger contribution to a closer matched pair of points while giving a much lower value to those faraway pairs.

The distance between two trajectories is usually measured by the aggregation of distances between the two trajectories' points. Closest-pair distance uses the minimal distance between the points in two trajectories (A, B) to represent the similarity of trajectories as $CPD(A, B) = min_{p \in A, p' \in B} D(p, p')$. Assuming that two trajectories are of the same length, sum-of-pairs distance uses the sum of corresponding points from the two trajectories to denote the distance as $SPD(A, B) = \sum_{i=1}^{n} D(p_i, p_i')$.

As the assumption may not hold in reality, dynamic time–wrapping (DTW) distance has been proposed to allow the "repeating" of some points as many times as needed in order to get the best alignment [1]. As some noise points from a trajectory may cause a big distance between two trajectories, the concept of the longest common subsequence (LCS) is employed to address this issue. The LCS-based distance allows for the skipping of some noise points when calculating the distance of trajectories using a threshold δ to control how far in time we can go in order to match one point from a trajectory to a point in another trajectory. Another threshold ε is used to determine whether two points (from two different trajectories) are matched. Chen et al. [7] propose the EDR distance, which is similar to LCS in using a threshold ε to determine a match while assigning penalties to the gaps between two matched subtrajectories. In [8], Chen et al. propose the ERP distance, aiming to combine the merits of DTW and EDR by using a constant reference point for computing distance. Note that DTW is not a metric, as it does not satisfy the triangle inequality. EDR is a metric that can be used to prune unnecessary trajectories.

Basically, LCS and edit distance were proposed for matching strings. When used to match two trajectories, there is a threshold ε to set; this is not easy. To address this issue, Chen et al. [9] define the best connect distance (BCT), which is a parameter-free similarity metric for trajectories combining the merits of DTW and LCS. During the matching process, K-BCT can repeat some trajectory points and skip unmatched trajectory points, including outliers.

One type of distance measure for trajectory segments is based on the MBR of segments [19]. As demonstrated in figure 4.23a, the MBRs of two segments (L_1, L_2) are (B_1, B_2), each of which is described by the coordinates of the low bound point (x_l, y_l) and upper-bound point (x_u, y_u). The MBR-based distance D_{min} (B_1, B_2) is defined as the minimum distance between any two points from (B_1, B_2), calculated as

$$\sqrt{\left(\Delta([x_l, x_u],[x_l', x_u'])\right)^2 + \left(\Delta([y_l, y_u],[y_l', y_u'])\right)^2}, \tag{4.5}$$

where the distance between two intervals is defined as

$$\Delta([x_l, x_u],[x_l', x_u']) = \begin{cases} 0 & [x_l, x_u] \cap [x_l', x_u'] \neq \varnothing \\ x_l' - x_u & x_l' > x_u \\ x_l - x_u' & x_l > x_u' \end{cases}. \tag{4.6}$$

In the two examples shown in figure 4.23a, the distance between L_1 and L_2 is 0 and $y_l' - y_u$, respectively.

Lee et al. [22] propose a distance function, titled the trajectory-Hausdorff distance (D_{Haus}), as depicted in figure 4.23b, which is a weighted sum of three terms: (1) the aggregate perpendicular distance ($d_{\perp}$) that measures the separation between two trajectories, (2) the aggregate parallel distance ($d_{//}$) that captures the difference in length between two trajectories, and (3) the angular distance (d_{θ}) that reflects the orientation difference between two trajectories. Formally,

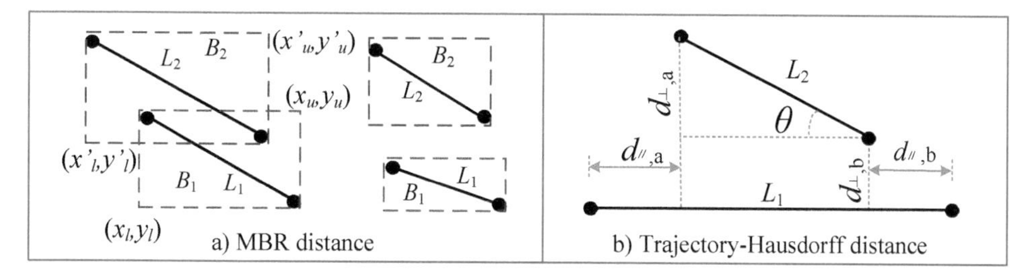

Figure 4.23
Distance metrics for trajectory segments.

$$D_{Haus} = w_1\, d_{\perp} + w_2\, d_{//} + w_3\, d_{\theta}, \tag{4.7}$$

where $d_{\perp} = \dfrac{d_{\perp,a}^2 + d_{\perp,b}^2}{d_{\perp,a} + d_{\perp,b}}$, $d_{//} = \min\,(d_{//,a}, d_{//,b})$, $d_{\theta} = ||L_2|| \cdot sin\theta$, and w_1, w_2, and w_3 are weights depending on applications.

4.4.3.3 Algorithms for range queries There are three main approaches to answering the spatiotemporal queries.

The first is to employ multiple version–based indexes, such as HR-tree, to manage trajectory data. However, as mentioned previously, such kinds of indexes are not very efficient in dealing with temporal interval queries, as they need to search for objects in the spatial indexes of many time stamps and merge search results across different time stamps.

The second approach is to employ the 3D R-tree like indexes, which bound segments of trajectories with 3D boxes, to manage trajectories. For example, the STR-tree [28] is an extension of R-trees to support the efficient query processing of the trajectories of moving objects. Different from R-trees, the STR-tree organizes line segments not only according to spatial properties but also by attempting to group the segments according to the trajectories they belong to. This property is termed *trajectory preservation*. The TB-tree [28] aims only for trajectory preservation and leaves other spatial properties aside. Figure 4.24 presents a part of a TB-tree structure, where a trajectory is represented by a gray band rather than a line. The trajectory is fragmented across six nodes (e.g., c_1 and c_2). In the TB-tree, a leaf node only contains segments belonging to the same trajectory, thus the index is best understood as a trajectory bundle. These leaf nodes are connected through a linked list, which allows us to retrieve the (partial) trajectory with minimal effort. As a drawback, line segments (from different trajectories) that lie spatially close will be stored in different nodes. As the overlap increases, the space discrimination decreases. Thus, the cost of answering a range query increases significantly [28].

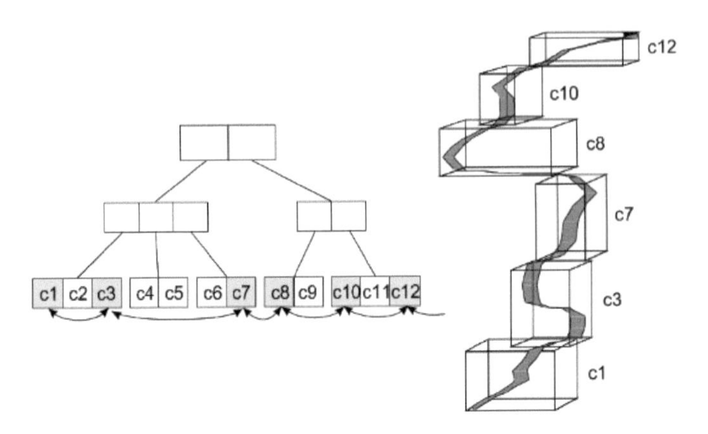

Figure 4.24
A partial structure of TB-trees [28].

The third approach partitions a geographical space into grids and then builds a temporal index for trajectories falling in each grid. As shown in figure 4.25a, CSE-tree [40] divides a trajectory into several segments by grids, which can be a result of grid-based indexes or quadtree indexes. Each grid maintains a temporal index to organize the segments lying within it. Each segment in a grid is represented by a 2D point whose coordinates are the starting time t_s and ending time t_e of the segment. The trajectory segments in a grid are then represented by points in a two-dimensional space, where the horizontal axis denotes t_s, and the vertical axis stands for t_e, as illustrated in figure 4.25b. A temporal range query seeks the segments whose time spans $[t_s, t_e]$ intersect, lie within, or contain a given temporal interval $[T_{min}, T_{max}]$. Figure 4.25b presents the four situations in which a segment should be retrieved. The four kinds of situations can be converted to the following two criteria: $t_s \leq T_{max}$ and $t_e \geq T_{min}$. If a segment satisfies the two criteria simultaneously, it should be retrieved. A visual representation of the temporal range query shows the retrieval of the points falling in the blue area.

To find such an area quickly, points in the space are divided into groups, each of which has the same number of points. As illustrated in figure 4.25c, once the number of points in a group reaches its limit, a splitting line (e.g., $t_1, t_2, \ldots, t_i$) is generated, and a new group is created. A start-time index S_i is built for each collection of points, indexing these points' start times with a B+ tree. An end-time index is then built over these splitting time lines based on a B+ tree.

When retrieving trajectories satisfying a spatiotemporal query, CSE-tree first finds the grids intersecting the spatial range of the query and then searches the temporal indexes of these grids for the segments of trajectories falling in the temporal range of

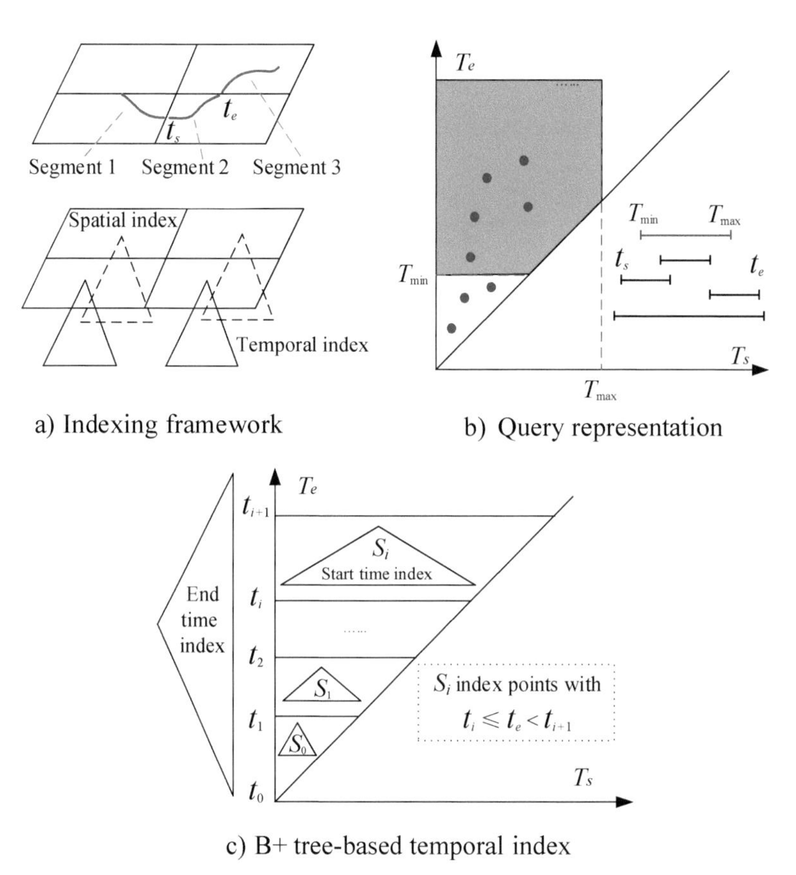

Figure 4.25
The indexing structure of CSE-tree.

the query. During the temporal search, the algorithm first searches the end time B+ tree for time lines with $t_e \geq T_{min}$. It then drills down into the start-time indexes of the group above T_{min}, seeking the points (i.e., trajectory segments) with $t_s \leq T_{max}$. Finally, CSE-tree merges the IDs of trajectory segments (and their starting and ending times) retrieved from different grids.

The reason for using two B+ trees rather than two sorted arrays is to deal with frequent inserts of new trajectories. For example, in a trajectory-sharing website, people can upload trajectories of any time that may not follow a chronological order. That is, a trajectory, generated at an earlier time in the real world, may be uploaded to the system after another trajectory is generated at a later time in the real world. This phenomenon causes random inserts of new records in the temporal indexes, which result in an intensive workload for a sorted array (refer to section 4.4.1 for details). If indexing a collection of historical trajectories, which will not be updated, we can simply concert the two B+ trees into two sorted arrays to organize the start and end times of each segment.

4.4.3.4 Algorithms for KNN queries Given a set of trajectories $T = \{R_1, R_2, \ldots, R_n\}$ and a small set of query locations $Q = \{q_1, q_2, \ldots, q_m\}$, the k-BCT query aims to find k trajectories that have the highest similarity (defined in equation [4.4]) to Q. Though the original k-BCT query does not consider the temporal dimension, it can be extended to a spatiotemporal query by adding a time stamp to each query point. The distance is then calculated in a 3D space.

The original k-BCT query [9] employs an R-tree to organize the trajectory set T. Based on the spatial index, the algorithm finds the point from each trajectory with the shortest distance to a query point. For example, as depicted in figure 4.26a, p_1 is the closest point from R_1 to q_1, p_2 is that to q_1, and p_5 is that to q_3. The algorithm then searches for the λ-nearest neighbor (λ-NN) points (from T) for each query point ($\lambda \geq k$). Note that these λ-NN points for a query point may come from the same trajectory or different ones. The circle denotes the distance between a query and the λ-th neighbor. Though p_5 is the closest point from R_1 to q_3, it is not a member of q_3's λ-NN points. In other words, there are other trajectories with at least λ points closer than p_5. The algorithm merges the trajectories these λ-NN points belong to, formulating a candidate trajectory set C for further checking.

The algorithm then estimates the lower bound of the similarity between a trajectory in C and Q. As illustrated in figure 4.26a, the similarity between Q and R_1 ($R_1 \in C$) is defined as:

$$Sim(Q, R_1) = e^{-D(q_1, p_1)} + e^{-D(q_2, p_2)} + e^{-D(q_3, p_5)}. \tag{4.8}$$

Thus, the lower bound of $Sim(Q, R_1)$ can be estimated as

$$Sim(Q, R_1) \geq e^{-D(q_1, p_1)} + e^{-D(q_2, p_2)}. \tag{4.9}$$

That is, the lower bound of similarity between a query set Q and a trajectory in C can be calculated by only considering the points satisfying two criteria: (1) they are the nearest neighbor points to Q in a trajectory and (2) they belong to Q's λ-NN points.

Likewise, as shown in figure 4.26b, if a trajectory $R_5 \notin C$ (i.e., it does not contain any λ-NN points of Q), the upper bound of the similarity between R_5 and Q should be

$$Sim(Q, R_1) = e^{-D(q_1, R_5)} + e^{-D(q_2, R_5)} + e^{-D(q_3, R_5)} \leq e^{-radius1} + e^{-radius2} + e^{-radius3}. \tag{4.10}$$

If we can find k trajectories (from C) whose lower bounds are not less than the upper bound of similarity for all unscanned trajectories, then the k best-connected

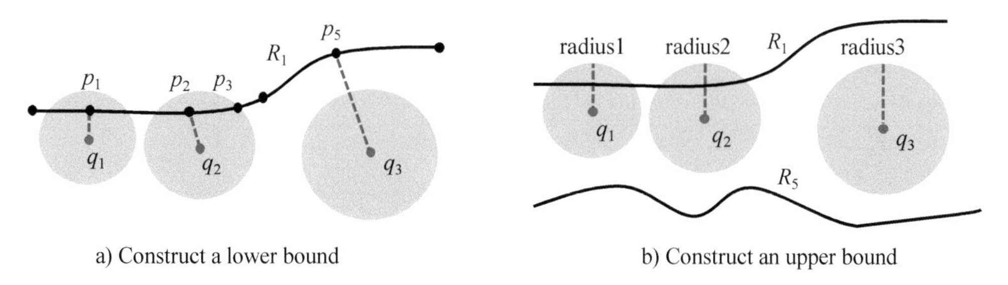

Figure 4.26
Answering the *k*-BCT query.

trajectories must be included in C. Then, we can calculate the exact similarity between Q and each trajectory in C, selecting the top-k most similar trajectories. Otherwise, we increase λ gradually to include more trajectories into C and repeat calculating the lower and upper bounds of similarity until we find the top-k most similar trajectories.

4.4.3.5 Algorithms for path queries Before answering such a query, we need to convert a trajectory into a sequence of road segments by using a map-matching algorithm. That is, a trajectory Tr is represented by $r_1 \rightarrow r_2 \rightarrow r_3$. Afterward, there are two approaches to answering a path query.

One straightforward approach is to build an inverted index for each road segment, recording the ID (and time) of moving objects passing a road segment. When a path (plus time interval) query comes, we first retrieve the moving objects passing each road segment contained in the path, respectively. Some moving objects that do not travel on a road segment at the given time interval can be filtered. As a moving object may traverse several road segments in a path, its ID is recorded in these segments redundantly. Thus, the moving objects' IDs retrieved from each road segment are merged together through a joint operation. When a path is long (i.e., the number of road segments contained in a path is big), the joint operation does not work very efficiently.

To address this issue, another approach, titled suffix tree–based indexing structure [33, 41], is proposed. As illustrated in figure 4.27, four trajectories Tr_1, Tr_2, Tr_3, and Tr_4 traverse a road network. After a map-matching process, a trajectory is converted into a sequence of road segments (e.g., Tr_1: $r_1 \rightarrow r_2 \rightarrow r_6$). Regarding each trajectory as a string and each road segment as a character, we can build a suffix tree for the four trajectories. Here, each node in the indexing tree stands for a road segment; each path on the tree corresponds to a route in the road network. Each node stores the IDs and travel times of the trajectories that traverse the path from the root to the node. For example, $t_{r_1 \rightarrow r_2 \rightarrow r_3}$ stands for the time for traveling path $r_1 \rightarrow r_2 \rightarrow r_3$. In order to find any existing paths, a

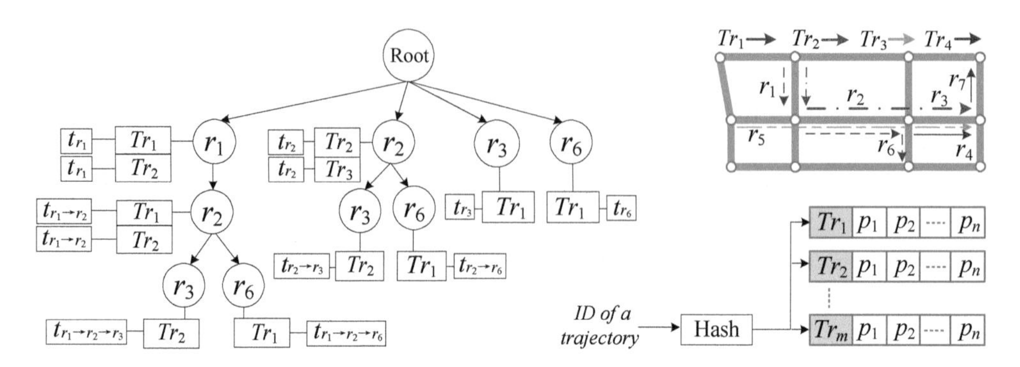

Figure 4.27
A suffix-tree-based index for answering path queries.

suffix tree [25] finds all suffixes of a string (i.e., a map-matched trajectory) and inserts them into the tree. For instance, the suffix of $r_1 \rightarrow r_2 \rightarrow r_6$ consists of r_6 and $r_2 \rightarrow r_6$. To reduce the size of the index, points (like p_1 and p_2) of a trajectory are not stored in the tree.

During a search, we can easily find a query path in such a suffix tree, starting from the root and ending at the node denoting the last road segment of the query path. Then we can retrieve the trajectory IDs and their corresponding travel times from the ending node. Based on a trajectory's ID, we can retrieve its points through a hash table (as shown in the bottom-right part of figure 4.27). If there is no trajectory passing a given path query, the path does not exist in the suffix tree.

The suffix tree is easy to build and efficient for answering a path query. However, its size is several times larger than that of the original trajectories it is going to manage. As the number of trajectories increases, the size of the index grows quickly and thus may not be able to be stored in a computer's memory. Such an index is only suitable for managing trajectories generated in a short period (e.g., at recent time intervals). Using such an indexing structure to manage trajectory data over a long time calls for a sophisticated indexing design and the usage of cloud computing.

4.5 Hybrid Indexes for Managing Multiple Datasets

4.5.1 Queries and Motivations

This section discusses indexing structures that can simultaneously manage multiple datasets across different domains. The motivations for studying such kinds of hybrid indexes are twofold.

First, we usually need to explore the correlation between different types of datasets. For instance, given three datasets consisting of weather, traffic conditions, and air quality, we find the following correlation patterns: when it is foggy and the traffic flow is congested in a given region, the region's air quality tends to be unhealthy. Those correlation patterns are valuable for diagnosing the root causes of air pollution. To discover such patterns, we need to issue many spatiotemporal range queries that find co-occurrences of different types of data within a spatial distance and a temporal span. As illustrated in figure 4.28, where different shapes denote different types of data, we find squares always occur within a spatial distance d to triangles. In addition, the time span between their occurrences is smaller than t. We call them *co-occurs* in a spatial distance d and time span t. Such queries call for a search across different datasets, such as using each square instance as a query to search for the objects of a triangle and a circle within (d, t), respectively. Likewise, we also need to search for squares and triangles within (d, t) using each circle instance as a query.

Without a hybrid index, such a search process is very time-consuming and sometimes redundant. If there is a hybrid index showing the high-level relationship between different types of datasets, we can reduce a lot of unnecessary query processes. For example, knowing there are no circle objects around the region that square objects s_1 fall in, we can avoid searching circle objects by using s_1 as a query point. In addition, with a good hybrid index, we can even avoid the search for a group of objects, such as s_1 and s_2. Alternatively, we can guarantee a group of squares is within (d, t) of a group of another dataset such as triangles without a further check. We will elaborate on these types of hybrid indexes in section 4.5.2.

Second, an object may generate or be associated with multiple datasets, while we may want to query objects using different criteria for different datasets. For instance, a restaurant is associated with a location and some textual information, such as menus and user comments. Users want to search for the nearest restaurant (to them) serving lobster. This query concerns the location and textual information of a restaurant,

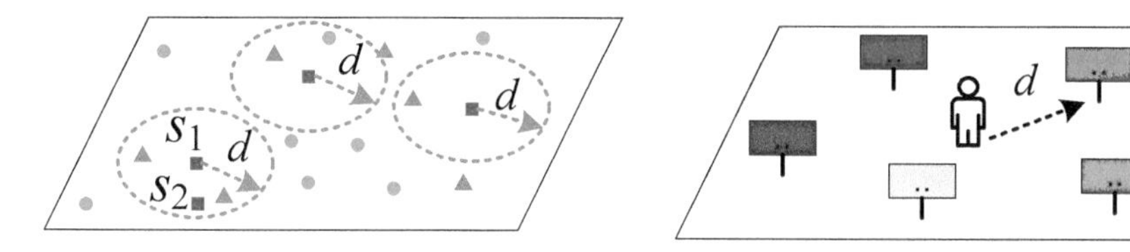

Figure 4.28
Queries on multiple datasets.

aiming to rank restaurants for a given query point and return the top-k objects as a result. Without a hybrid index, a straightforward way to answer this query is to first find all restaurants with textual information "lobster" and then search for the nearest one to the query point from those restaurants. The search process is clearly inefficient. If we hope to find restaurants offering lobster and with an average expense per person lower than forty U.S. dollars, the search becomes even more time-consuming. A body of research on solving this problem is known as *spatial key words* [10, 13, 17], which focuses on integrating spatial/spatiotemporal information with text information in a hybrid index in order to reduce the search effort. In the following sections, we will discuss the aforementioned two queries, respectively.

4.5.2 Spatial Key Words

Figure 4.29 presents the evolvement of representative research on spatial key words. The horizontal axis denotes the year when a method was proposed, and methods within a broken box were proposed in the same publication. Generally speaking, the evolution of the research can be divided into three stages: (1) spatial key word search with a prespecified geographical region and a Boolean key word query, (2) top-k spatial key word search ranked by an aggregated score considering both distance and semantic relevance, and (3) spatial key word on trajectories.

4.5.2.1 Separate search methods This category of research aims to retrieve documents relevant to one or several query key words (Boolean query) within a prespecified spatial region. Each document is associated with a spatial location. For example, user-generated comments about a restaurant can be regarded as a document. Likewise, geotagged social media, such as tweets and articles, are also associated with a location.

Zhou et al. [50] propose three schemes to integrate spatial and textual indexes together. As illustrated in figure 4.30, the first scheme builds two separate indexes over location and textual information, respectively. It employs an R*-tree to organize the locations of documents. In the meantime, it creates an inverted list that maintains the relationship between a key word and the documents containing it. To answer a spatial key word query with a geographical region and several key words, the scheme searches the R*-tree for the documents within the given spatial region and searches the inverted list for the documents containing the given key words. The retrieved documents from the two indexes are then merged as the final result through a joint operation. Considering that the candidate set obtained based on one feature may be extremely huge, this method is seldom used in practice.

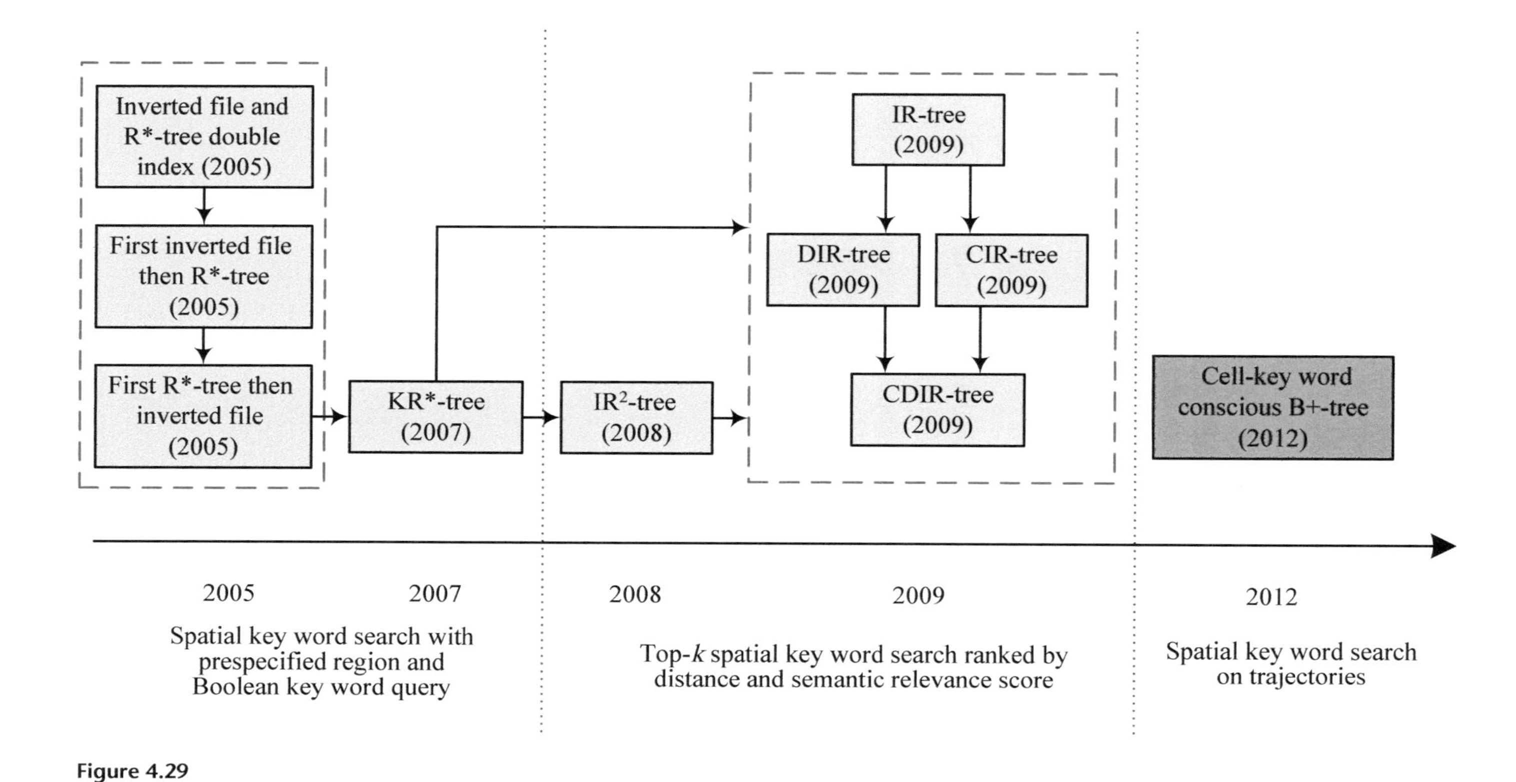

Figure 4.29
Evolvement of representative research on spatial key words.

To address this issue, as shown in figure 4.30b, the second scheme proposes building an R*-tree for each distinct key word over the spatial objects whose textual description contains the key word. Given a spatial key word query with a set of key words and a specific spatial range, the R*-tree corresponding to each query key word is used to filter the spatial part of the query. The final answer set can be obtained by merging the object IDs from the multiple R*-trees according to the designated Boolean relationship. However, this index does not take advantage of the spatial correlation between key words. When query key words are closely correlated in the geographical space, this approach suffers from paying extra disk costs from accessing different R*-trees and high overhead in the subsequent merging process. When a query contains multiple key words, the cost for searching and merging results becomes heavy. In addition, building a separate R*-tree for each distinct key word requires substantial storage.

Contrary to the second scheme, as illustrated in figure 4.30c, the third one first builds an R*-tree for all objects without considering textual information. Then it creates an inverted list file for key words that appear in each leaf node of the tree. Each key word in this inverted file index points to a list of object IDs whose textual information contains the key word. When a query is issued, a set of leaf nodes intersecting with the query rectangle is retrieved first, based on the R*-tree. Then we find the objects satisfying the query key words using these nodes' inverted file indexes. Because the number of objects in a leaf node is generally small, the inverted file indexes of leaf nodes are usually small. Thus, this approach speeds up the key word–filtering process. However, when a query covers a large area in geographical space, many candidates will be retrieved for further key word filtering; this is time-consuming.

To better leverage the association of key words in geographical space, Hariharan proposes a KR*-tree [17], enhancing Zhou et al.'s work in two ways. First, instead of filtering objects by text and space separately, or one followed by the other, KR*-tree filters objects by space and text simultaneously. Second, instead of merging the separate results for multiple key words, KR*-tree exploits the joint distribution of key words. As depicted in figure 4.31, this method augments an R*-tree with a KR*-tree list that stores the node IDs (instead of object IDs) that a key word occurs in. As the number of nodes is much smaller than the number of objects, the size of the index has been significantly reduced. At query time, the KR*-tree–based algorithm finds the nodes that spatially intersect with the query region and checks the KR*-tree list if the query key words are contained in these nodes. If a node ID does not occur in the list of a key word, the search can stop at this node. It is not necessary to search for its children nodes anymore.

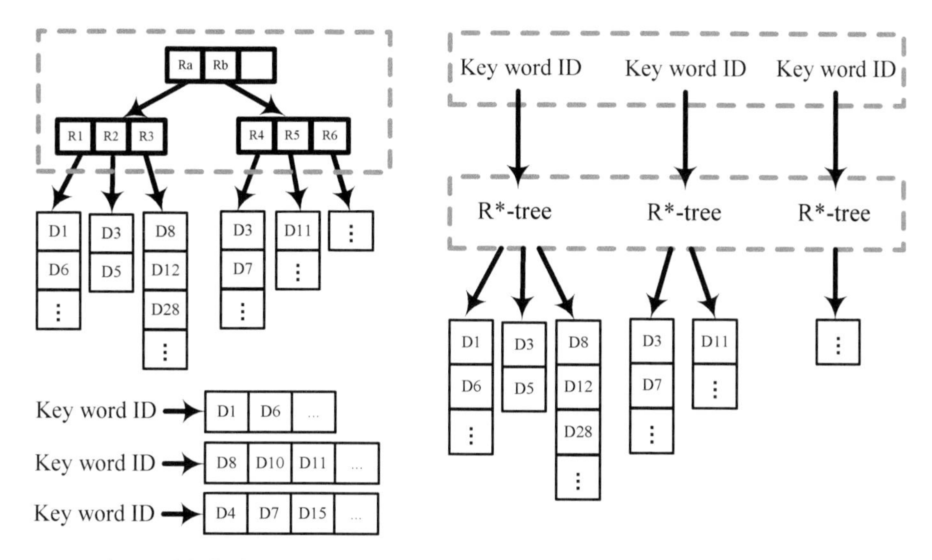

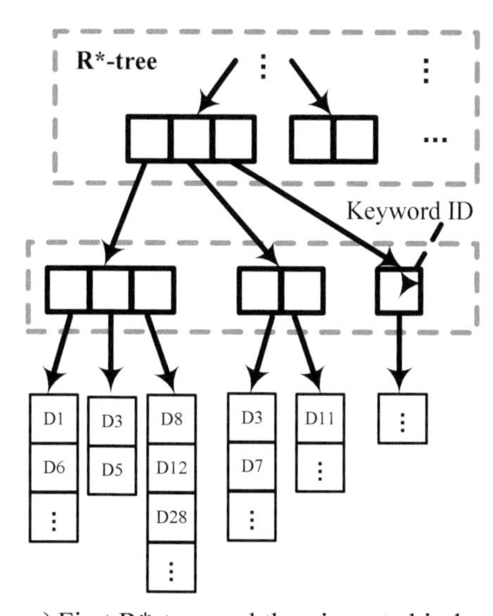

Figure 4.30
Three schemes for the first stage of research on spatial key words.

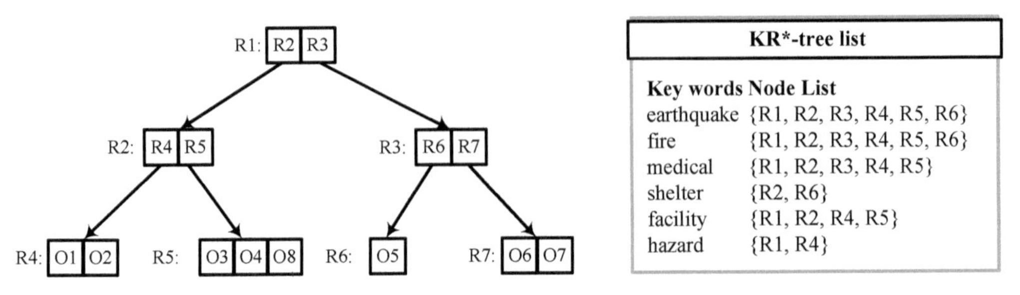

Figure 4.31
A partial structure of KR*-tree [14].

4.5.2.2 Aggregated relevance This category of research aims to retrieve the top-*k* objects based on a metric that considers both distance and key word relevance. It typically is designed as a function *f (dist, key)* that aggregates the distance to a query point and the relevance between query key words and an object's textual information in a weighted manner.

IR^2-tree (information retrieval R-tree) [13] augments each R-tree node with a specific signature file, as illustrated in figure 4.32. Each key word is represented by an m-bit vector (i.e., signature). Each node has a signature that superimposes (i.e., binary ORs) the signatures of its children.

For instance, as shown in figure 4.32, each key word is represented by a sixteen-bit vector. A leaf node R_4 contains two objects: O_1 and O_2. The key words associated with O_1 are aggregated into a two-byte signature s_1=<10001011 00000010> through an OR operation. Likewise, O_2's signature is represented by another two-byte vector s_2=<00001110 00100011>. Then, the two-byte signature of internal node R_2 is a binary OR of O_1 and O_3's signatures. During query time, query key words are represented by an m-bit vector, following the same way of encoding objects' key words and then superimposing them onto a single signature through a binary OR operation. Starting from the root node, the search algorithm matches q against each node's signature s, checking if $q=q\&s$. If the condition does not hold, the query cannot be contained in the node and its children nodes. For example, a query is represented by q=<10000011 10000110>. Obviously, it is impossible for R_3 to contain the query, as its third bit is 0. As a consequence, we can prune R_3 and its children nodes in the following search process. Finally, we find O_3's signature satisfies the condition; therefore, it could contain the query key words. After checking each key word associated with O_3, we will find if O_3 truly contains the query key words.

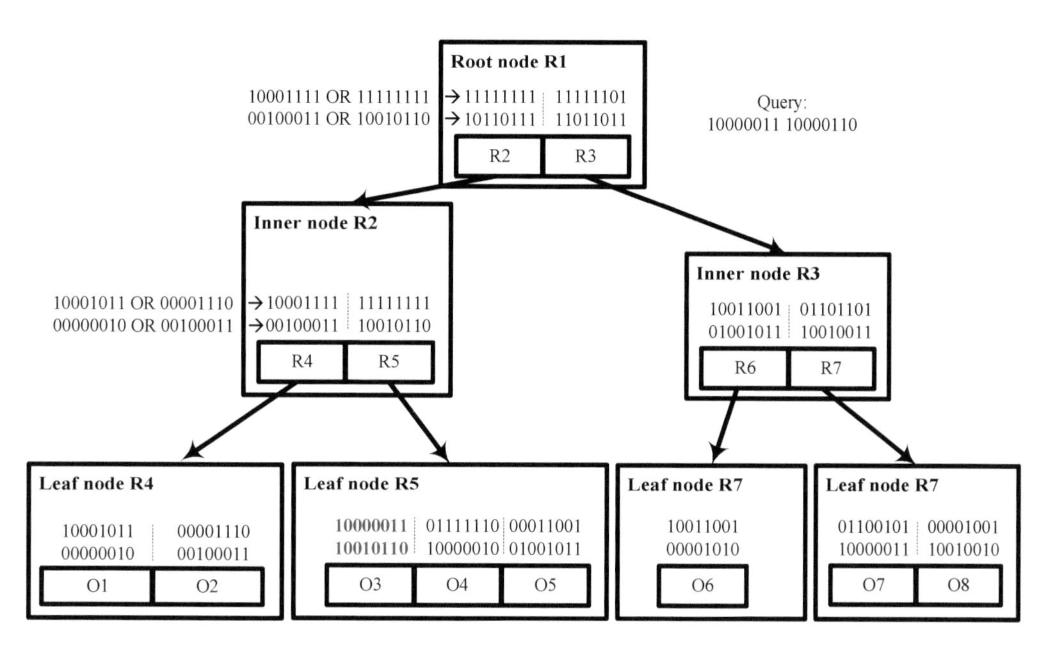

Figure 4.32
A partial structure of an IR^2-tree.

In order to deal with the case in which query key words occur in multiple objects separately (but no object contains the entire set of key words), we need to turn each query key word into an m-bit vector and match them individually against the IR^2-tree. The number of key words that a node could contain can be employed as a priority score to guide the search strategy. For instance, given a query consisting of three key words, we find R_4 could contain one and R_5 could have two. We will check R_5 first during the search process. The priority can also be combined with the distance (between an object and the query) to rank the final search results. However, an IR^2-tree does not consider the frequency with which a key word occurs in an object's textual information. The relevance of an object to a query is based solely on a binary occurrence of a key word in the object's textual information. Intuitively, the more frequently a query key word occurs in an object's textual information, the more relevant this object could be to the query.

To address this issue, Cong et al. [10] propose IR-tree (inverted file R-tree), which augments each node with an inverted file rather than a signature file. As illustrated in figure 4.33, each leaf node is comprised of some objects (e.g., object O_1 and O_2 belong to leaf node R_4), and each object's textual information is regarded as a document. An

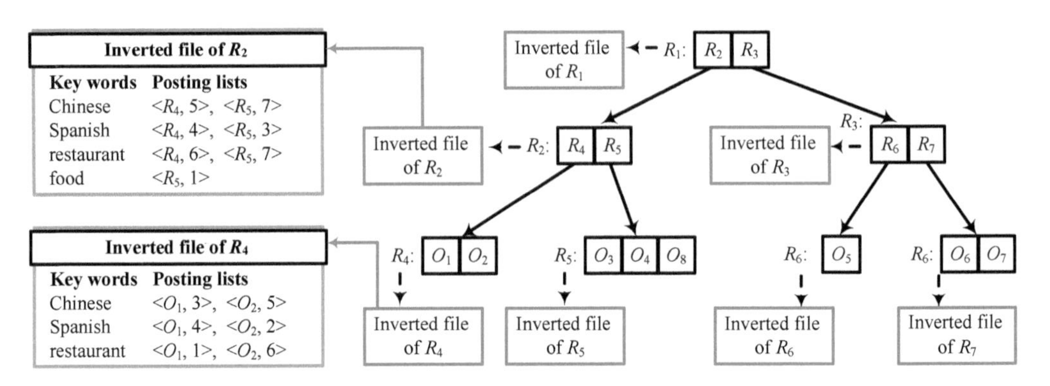

Figure 4.33
A partial structure of an IR-tree.

inverted file of a leaf node consists of a vocabulary of distinct terms and their corresponding posting lists in the documents (the leaf node contains). The posting list of a term t is a sequence of pairs $<d, w_{d,t}>$, where d is a document containing t, and $w_{d,t}$ is the weight of t in d. Commonly used term-weighting approaches, such as TF (term frequency) or TF-IDF (term frequency-inverse document frequency), can be adopted here. As illustrated in figure 4.33, the word *Chinese* occurs three times in O_1's textual information and five times in that of O_2. Regarding a nonleaf node like R_2, which consists of two leaf nodes R_4 and R_5, its documents are an aggregation of its children nodes' documents. The weight of a term is the maximum $w_{d,t}$ in its children nodes' documents. For instance, the documents of object O_1 and O_2 are merged into a virtual document denoting that of R_4. The word *Chinese* is considered to be *occurring* five times in R_4's document, as five is the maximum number of occurrences of the word in its two children's documents.

The aforementioned inverted files are used to estimate the relevance between a query and the objects contained in a node. The more relevant an object is to a query, the smaller the textual distance between them. Combined with the geographical distance between a query and an object, these inverted files derive a spatial-textual distance for each node through a sort of aggregation function $f(dist, key)$.

IR-tree employs a best-first traversal algorithm for retrieving the top-k nearest objects. There is a priority queue keeping track of the nodes and objects that have yet to be visited. At each iteration, the algorithm picks the node with the smallest minimum spatial-textual distance from the queue to visit. Given a query q and a node R in an IR-tree, the so-called minimum spatial-textual distance offers a lower bound on the actual spatial-textual distance between q and the objects within the rectangle of node R. If the lower bound is bigger than the k-th nearest candidate (to q) that has already been retrieved, all objects in R can be pruned without a further check. Retrieval terminates when k-nearest objects have been found.

A variant of IR-trees is CIR-tree (cluster-enhanced IR-tree). The main idea of CIR-tree is to cluster objects into groups according to their corresponding documents. Instead of constructing a single virtual document for each node, a virtual document is constructed for each cluster in each node. Since objects within the same cluster are more similar than objects in other clusters, the bounds estimated using clusters in a node will be tighter than those estimated for an entire node, thereby further improving query performance.

Sun et al. [34] propose the k-nearest neighbor temporal aggregate (kNNTA) query. Given a query point and a time interval, it returns the top-k locations that have the smallest weighted sums of (1) the spatial distance to the query point and (2) a temporal aggregate on a certain attribute over the time interval—for example, finding a nearby club that has the largest number of people visiting in the last hour. This type of query has emerging applications in location-based social networks, location-based mobile advertising, and social event recommendations. However, it is very challenging to efficiently answer such a query because of the large volume of the data and queries as well as the highly dynamic nature of the applications.

To tackle this challenge, an index, named TAR-tree, was proposed to organize locations by integrating an R-tree with temporal indexes. As illustrated in figure 4.34, TAR-tree first organizes objects according to their spatial location, based on an R-tree. Each node of the TAR-tree contains several entries, each of which is associated with an MBR and a pointer to a temporal index. The temporal index stores the nonzero aggregate over each epoch and keeps each record as a triple $<t_s, t_e, agg>$, where t_s is the start time, and t_e is the end time of the epoch, and agg is the aggregate value during the epoch (e.g., the number of people visiting a POI). The temporal index of a leaf entry stores the temporal aggregation of POIs it contains. The temporal index of a nonleaf entry (a.k.a. internal entry) stores the largest aggregate value of its child nodes for each epoch. For example, c, g, and b are entries contained in a leaf node. Their temporal aggregates are as follows:

c: $<t_0, t_1, 2>$, $<t_1, t_2, 2>$, and $<t_2, *, 2>$;

g: $<t_0, t_1, 2>$, $<t_1, t_2, 3>$, and $<t_2, *, 1>$;

b: $<t_0, t_1, 1>$, and $<t_2, *, 1>$.

R_2 is an internal entry, consisting of an MBR and three leaf entries c, g, and b. Its temporal aggregation is

$<t_0, t_1, \max(2, 2, 1)>$, $<t_1, t_2, \max(2, 3)>$, $<t_2, *, \max(2, 1, 1)>$.

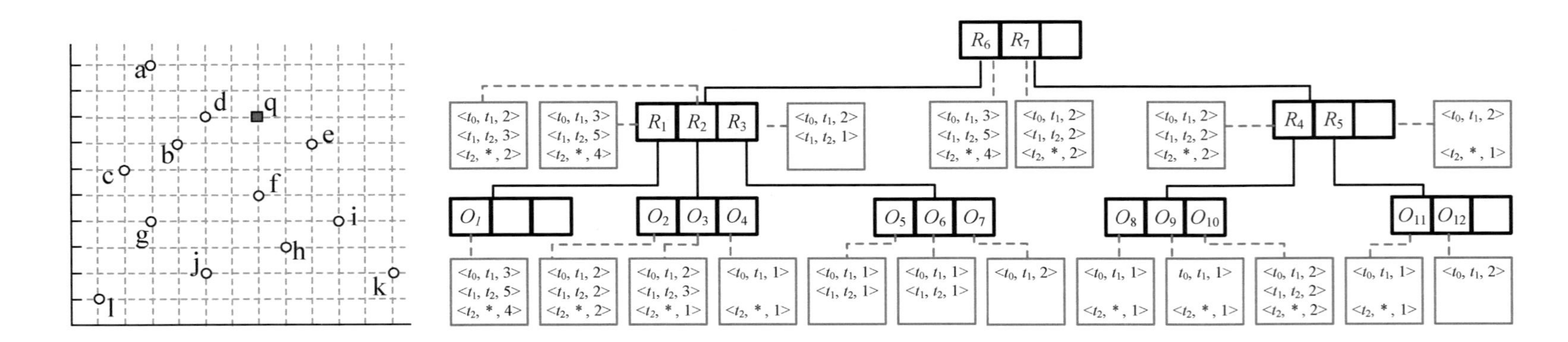

Figure 4.34
A partial structure of a TAR-tree.

Any temporal indexing structures (e.g., B+ trees introduced in section 4.4.1) can be employed to implement the temporal aggregate index. For example, we can partition time by equal-length epoch, counting the aggregate of each epoch and then indexing these triples using a B+ tree. When a temporal interval query I comes, we can quickly retrieve the epochs intersecting I and then sum the aggregates of these epochs.

The best-first search (BFS) [18] can be used for query processing, which works as follows:

1. The entries in the root node are first inserted into a priority queue in which the priority is determined by an entry's score $f(p)$. The smaller the score, the higher priority an entry has. The score is defined as equation (4.11).

$$f(p) = \alpha \cdot dist(q, p) + (1 - \alpha) \cdot (1 - g(I)), \tag{4.11}$$

   where $0 < \alpha < 1$ is a weight to merge the distance between a query point and a POI—*dist* (q, p)—and the temporal aggregate that a time interval I covers. For example, as shown in figure 4.33, $g(I)$ of entry f in the time interval from t_0 to now is $3+5+4=12$, and $g(I)$ of entry g is $2+3+1=6$. To ensure $f(p)$'s value is within [0, 1], $dist(q,p)$ is normalized by the maximum distance between POIs and the query point (i.e., $dist(q,p)/\max Dist$). Likewise, $g(I)$ is normalized by the maximum temporal aggregate among all points (i.e., $g(I)/\max g(I)$). In this example, entry f's $g(I)=12$ is the maximum. Thus, after normalization, $g(I)$ of entry g becomes $6/12=0.5$.
2. The front entry of the queue is ejected. If the entry is a leaf entry, the POI it contains is added to the result list; otherwise, each of its offspring entries is inserted into the queue.
3. Step 2 is repeated until k POIs are obtained.

4.5.2.3 Spatial key words over trajectories The aforementioned hybrid indexes are basically intended for spatial key word queries on static objects with geospatial location and textual description. Cong et al. [11] extend the problem to trajectories. Given a dataset of trajectories contain text descriptions, users may want to retrieve a set of trajectories that cover several query key words and have the shortest match distance to a query location. The match distance is measured by the sum of two distances: the length of a subtrajectory covering all query key words and the distance from the query location to the start location of the subtrajectory.

They propose cell key word–conscious B+ trees to serve such queries, which consist of two individual components. The first component is used to locate the IDs of trajectories that are close to the query location and contain all the key words. This component contains two separate index structures that are loosely coupled using a quadtree to divide the spatial space and then building a B+ tree to index trajectories together with their text descriptions. The second component computes the minimum match distance of a selected trajectory to query q. Top-k trajectories with the minimum distance to q are retrieved by iteratively performing range queries with an incrementally expanded search region.

4.5.3 Indexes for Managing Multiple Datasets

There are intrinsic connections between different urban phenomena, such as traffic conditions, the environment, human behavior, and economy. Sometimes, these connections are invisible but can be revealed by the correlation between the datasets these phenomena have generated. For example, when the humidity is within [70, 90], the air quality is typically within [150, 200]. When the average travel speed in a region is lower than 20 km/h, the concentration of NO_2 is most likely to be above 0.03 $\mu g/m^3$. When the number of user check-ins around a coffee shop is larger than one thousand times per week and the density of POIs is more than 100/km^2, the weekly revenue of the shop would be higher than ten thousand U.S. dollars. Such correlation patterns reveal the insight that can explain complex urban phenomena and can be used as inputs to diagnose the root cause of a problem.

To discover the correlation patterns across multiple datasets, we first need to define the co-occurrence of two datasets. Afterward, we can count the frequency of the co-occurrence. As datasets are generated by different sources independently and respectively, however, there is no clear transaction, like "beer" and "diaper" bought together in one transaction record, that naturally contains instances from two different datasets. Instead, we need to define the co-occurrence using a spatial distance threshold δ and a time interval threshold t. If two instances respectively from two datasets are with a spatial distance smaller than δ and a time interval smaller than t, they are regarded as a co-occurrence. Counting the co-occurrences for two datasets calls for many times of spatiotemporal searches, which are very time-consuming.

As illustrated in figure 4.35a, there are three datasets represented by three different shapes: blue diamond (c_1), red circle (c_2), and yellow square (c_3). A naïve method to count the occurrences of (c_1 and c_2) is using each and every instance of c_1 as a spatiotemporal query point to search c_2 for the instances that are within threshold δ and t to the query point. When the datasets are huge, it is impossible to achieve this goal within a reasonable time interval. Another straightforward method is to build a grid-based spatial index for all the datasets. Each grid stores all the instances falling in the grid. Ideally, the index can prune the search space significantly when answering spatiotemporal

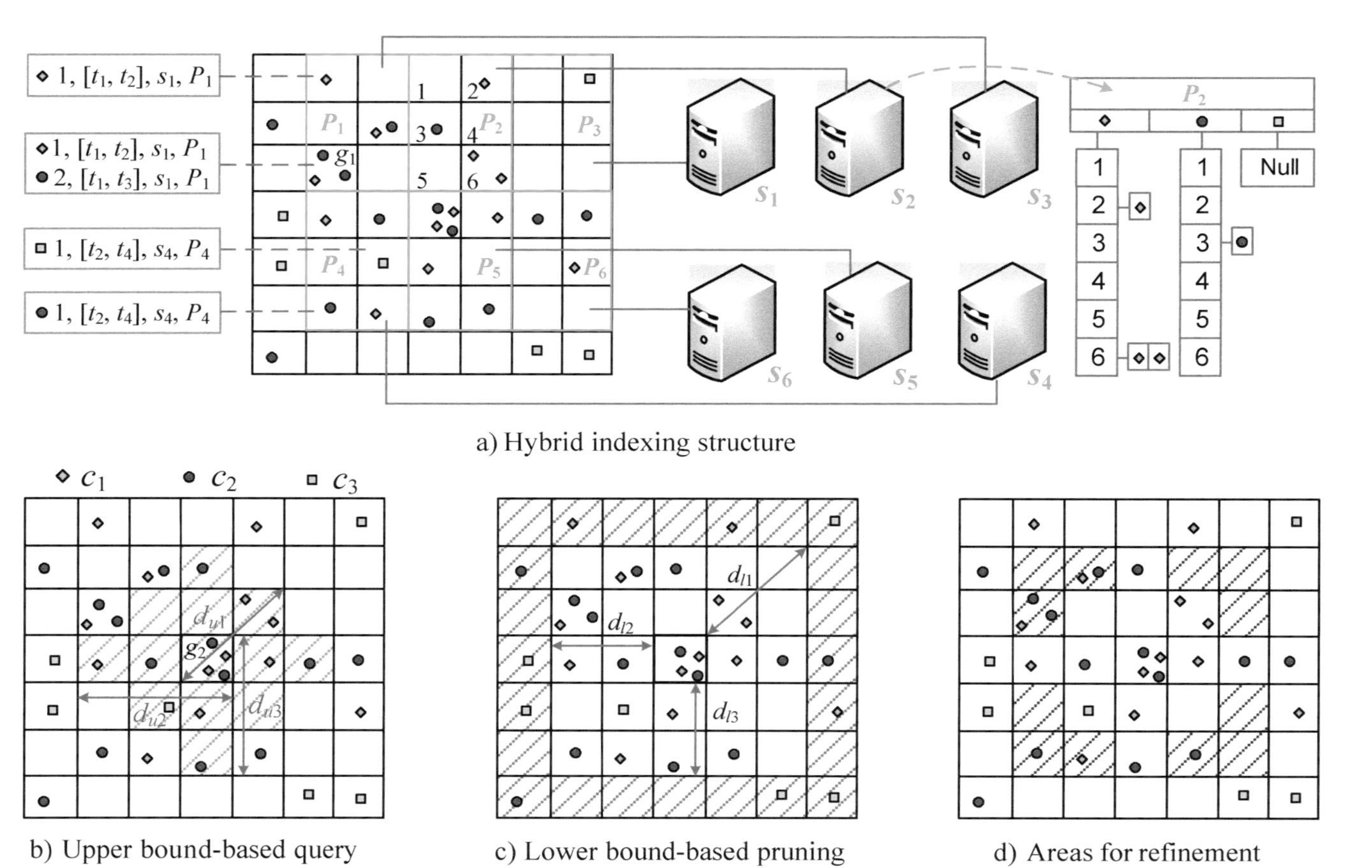

Figure 4.35
Hybrid index for cross-domain correlation pattern mining.

range queries. However, when datasets are very big, we cannot hold such a big indexing structure in a single machine's memory. In reality, a big dataset is partitioned into many portions that are respectively stored by multiple machines or nodes in a cloud-computing platform. Without an overview of datasets, we need to send each query to all machines to process. This not only wastes computing resources but also generates heavy communicational load between machines.

To address this issue, we design a two-layer hybrid index to efficiently manage multiple datasets. As illustrated in figure 4.35a, for simplicity, we use a grid-based index as an example, dividing a space into uniform partitions, each of which is further partitioned into uniform grids.

The first layer of the index only stores the high-level information of each grid, consisting of the number of instances from each dataset, the time span in which these instances are generated, and the server and partition ID where the detailed instances are stored. Thus, the size of the first-layer index is small and can be stored in the memory of a single machine like the spout node in an Azure Storm. As shown in the left part of figure 4.35a, grid g_1 contains one instance generated at time interval $[t_1, t_2]$ from c_1 and two instances from c_2 generated at $[t_1, t_3]$. The three instances are stored in machine s_1 and partition P_1.

The second layer of the index stores the detailed information of each instance in each partition. A machine (or a node in a cloud-computing platform) can store one or more partitions. In this example, each partition is stored in one machine. For instance, partition P_2, which is comprised of six grids, is stored in the memory of machine s_2. There are three lists saving the detailed information of the instances that occur in each of the six grids belonging to P_2. For example, there is one blue diamond in grid 2 and two in grid 6. As there is no yellow square occurring in partition P_2, the list can be null. As the second-layer index only stores detailed information for one or more partitions, its size is not too big and thus can be held in a machine's memory.

The first layer of the hybrid indexing structure enables fast pruning of search space. For instance, as illustrated in figure 4.35b, given the two red circles in grid g_2 and a distance threshold δ, we can quickly find a set of grids whose instances are within δ to the two circles without computing the distance between each circle and each instance from other datasets. d_{u1}, d_{u2}, and d_{u3} are upper bounds of the distances between instances in grid g_2 and those in other grids. If $d_{u1} \leq \delta$, $d_{u2} \leq \delta$, and $d_{u3} \leq \delta$, all the instances falling in the shadow grids must have a distance smaller than δ to any of the two circles. Thus, we can simply sum up the number of instances of different datasets in these shadow grids' first-layer indexes.

Likewise, we can derive the lower bound of distances between instances in g_2 and those in other grids (i.e., d_{l1}, d_{l2}, and d_{l3}). As illustrated in figure 4.35c, if $d_{l1} > \delta$, $d_{l2} > \delta$, and $d_{l3} > \delta$, all instances in the shadow grids must have a distance larger than δ to the two circles in g_2. Thus, these (and farther) grids can be pruned without individually checking instances in them.

By combining the two shadow areas from figure 4.35b and c, we only need to thoroughly check the shadow grids shown in figure 4.35d. By checking the first-layer index, we find only three grids containing instances of another dataset (i.e., blue diamonds). The three grids belong to partition P_1 and P_4, which are stored in machine s_1 and s_4, respectively. As a consequence, we only need to send the two red circles as queries to s_1 and s_4. Other machines will not be involved in the query. This reduces many unnecessary communications between machines, which is one of the main costs in distributed-computing systems.

Thus, the advantages of the two-layer hybrid index are twofold:

1. The first-layer index derives upper and lower bounds that prune search spaces significantly, enabling query and prune instances in a group manner. This reduces not only unnecessary query processes but also communications between machines, such as between the spout and bolts in Azure Storm. The latter is usually a bottleneck in distributed-computing systems.
2. The two-layer index allows us to load summary and detail information of datasets into different machines' memory, expediting the search process significantly. Such an indexing structure with grids and partitions can be easily employed in a parallel-computing framework.

4.6 Summary

This chapter introduces indexing and retrieval algorithms for spatial data. Four widely used indexing structures, consisting of grid-based indexes, quadtree-based indexes, k-d trees, and R-trees, have been introduced from the following four perspectives: building indexes, serving spatiotemporal range queries, serving nearest neighbor queries, and updating indexes. The advantages and disadvantages of these indexing structures are compared.

This chapter then presents techniques for managing spatiotemporal data, consisting of moving-object databases and trajectory data management. The former is concerned more with the specific location of a moving object at a (usually recent) time stamp.

The latter concerns the continuous movement (e.g., the path) a moving object has traversed at a given time interval.

Regarding the moving-object databases, three types of queries are presented, and two indexing approaches have been introduced. One indexing approach, such as historical R-tree (HR-tree), HR+ tree, and multiple-version R-tree (MVR-tree), builds a spatial index at each time stamp and reuses unchanged substructures of an index across consecutive time intervals. The other indexing approach, such as 3D R-tree, treats time as the third dimension, extending spatial indexing structures from managing two-dimensional spatial data to three-dimensional spatiotemporal data.

Regarding trajectory data management, there are three major types of queries: range queries, KNN queries, and path queries. Other advanced queries can be derived from the combination of these three fundamental queries. Different distance metrics designed for trajectory data are also introduced.

Finally, we introduce hybrid indexing structures for managing multiple datasets. Quite a few data management techniques have been introduced to deal with spatial key word queries, which are concerned with location and text simultaneously.

An advanced two-layer hybrid indexing structure is then proposed to mine correlation patterns across three types of datasets. The two-layer indexing structure reduces computational effort in machines and communication loads between machines significantly, using upper- and lower-distance bounds between grids as well as the summary of instances in a grid in the first layer. The indexing structure can be easily employed by a distributed-computing system, dealing with large-scale datasets efficiently.

References

[1] Agrawal, R., C. Faloutsos, and A. Swami. 1993. *Efficient Similarity Search in Sequence Databases*. Berlin: Springer, 69–84.

[2] Beckmann, N., H. P. Kriegel, R. Schneider, and B. Seeger. 1990. "The R*-tree: An Efficient and Robust Access Method for Points and Rectangles." In *Proceedings of the 1990 ACM SIGMOD International Conference on Management of Data*. New York: Association for Computing Machinery (ACM). doi:10.1145/93597.98741.

[3] Bentley, J. L. 1975. "Multidimensional Binary Search Trees Used for Associative Searching." *Communications of the ACM* 18 (9): 509. doi:10.1145/361002.361007.

[4] Biggs, Norman. 1993. *Algebraic Graph Theory*. 2nd edition. Cambridge: Cambridge University Press, 7.

[5] Brakatsoulas, S., D. Pfoser, and N. Tryfona. 2004. "Modeling, Storing and Mining Moving Object Databases." In *Database Engineering and Applications Symposium, 2004. IDEAS'04. Proceedings.*

International. Washington, DC: Institute of Electrical and Electronics Engineers (IEEE) Computer Society Press, 68–77.

[6] Chakka, V. P., A. Everspaugh, and J. M. Patel. 2003. "Indexing Large Trajectory Data Sets with SETI." In *Proceedings of the Conference on Innovative Data Systems Research*. Asilomar, CA: Conference on Innovative Data Research.

[7] Chen, L., and R. Ng. 2004. "On the Marriage of Lp-Norms and Edit Distance." In *Proceedings of the 30th International Conference on Very Large Data Bases*. Burlington, MA: Morgan Kaufmann, 792–803.

[8] Chen, L., M. T. Özsu, and V. Oria. 2005. "Robust and Fast Similarity Search for Moving Object Trajectories." In *Proceedings of the 2005 ACM SIGMOD International Conference on Management of Data*. New York: ACM, 491–502.

[9] Chen, Z., H. T. Shen, X. Zhou, Y. Zheng, and X. Xie. 2010. "Searching Trajectories by Locations—An Efficient Study." In *Proceedings of the 29th ACM SIGMOD International Conference on Management of Data*. New York: ACM, 255–266.

[10] Cong, G., C. S. Jensen, and D. Wu. 2009. "Efficient Retrieval of the Top-k Most Relevant Spatial Web Objects." *PVLDB* 2 (1): 337–348.

[11] Cong, G., H. Lu, B. C. Ooi, D. Zhang, and M. Zhang. 2012. "Efficient Spatial Keyword Search in Trajectory Databases." Cornell University Library Computer Science Database (arXiv preprint arXiv:1205.2880).

[12] Cormen, Thomas H., Charles E. Leiserson, Ronald L. Rivest, and Clifford Stein. 2001. *Introduction to Algorithms*. 2nd edition. Cambridge, MA: MIT Press, 527–529.

[13] De Felipe, I., V. Hristidis, and N. Rishe. 2008. "Keyword Search on Spatial Databases." In *Proceedings of the 2008 IEEE 24th International Conference on Data Engineering*. Washington, DC: IEEE Computer Society Press, 656–665.

[14] Deng, K., K. Xie, K. Zheng, and X. Zhou. 2011. "Trajectory Indexing and Retrieval." In *Computing with Spatial Trajectories*, edited by Y. Zheng and X. Zhou, 35–60. Berlin: Springer.

[15] Ester, M., Hans-Peter Kriegel, Jörg Sander, and Xiaowei Xu. 1996. Evangelos Simoudis, Jiawei Han, Usama M. Fayyad, eds. "A Density-Based Algorithm for Discovering Clusters in Large Spatial Databases with Noise." In *Proceedings of the Second International Conference on Knowledge Discovery and Data Mining*. Menlo Park, CA: AAAI Press, 226–231.

[16] Guttman, A. 1984. "R-trees: A Dynamic Index Structure for Spatial Searching." *ACM SIGMOD Record* 14 (2): 47–57.

[17] Hariharan, R., B. Hore, C. Li, and S. Mehrotra. 2007. "Processing Spatial-Keyword (SK) Queries in Geographic Information Retrieval (GIR) Systems." In *Proceedings, 19th International Conference on Scientific and Statistical Database Management*. Washington, DC: IEEE Computer Society Press, 16.

[18] Hjaltason, G. R., and H. Samet. 1999. "Distance Browsing in Spatial Databases." *ACM Transactions on Database Systems* 24 (2): 265–318.

[19] Jeung, H., M. L. Yiu, and C. S. Jensen. 2011. "Trajectory Pattern Mining." In *Computing with Spatial Trajectories*, edited by Y. Zheng and X. Zhou. Berlin: Springer, 143–177.

[20] Knuth. 1998. §6.2.1 "Searching an ordered table." Subsection "Binary search."

[21] Kriegel, Hans-Peter, Peer Kröger, Jörg Sander, and Arthur Zimek. 2011. "Density-Based Clustering." *Wiley Interdisciplinary Reviews: Data Mining and Knowledge Discovery* 1, no. 3 (May): 231–240. doi:10.1002/widm.30.

[22] Lee, J. G., J. Han, and K. Y. Whang. 2007. "Trajectory Clustering: A Partition-and-Group Framework." In *Proceedings of the 2007 ACM SIGMOD International Conference on Management of Data*. New York: ACM, 593–604.

[23] Li, Z., M. Ji, J. G. Lee, L. A. Tang, Y. Yu, J. Han, and R. Kays. 2010. "MoveMine: Mining Moving Object Databases." In *Proceedings of the 2010 ACM SIGMOD International Conference on Management of Data*. New York: ACM, 1203–1206.

[24] Lou, Y., C. Zhang, Y. Zheng, X. Xie, W. Wang, and Y. Huang. 2009. "Map-Matching for Low-Sampling-Rate GPS Trajectories." In *Proceedings of the 17th ACM SIGSPATIAL International Conference on Advances in Geographic Information Systems*. New York: ACM, 352–361.

[25] McCreight, E. M. 1976. "A Space-Economical Suffix Tree Construction Algorithm." *Journal of the ACM* 23 (2): 262–272.

[26] Nascimento, M., and J. Silva. 1998. "Towards Historical R-Trees." In *Proceedings of the 1998 ACM Symposium on Applied Computing*. New York: ACM, 235–240.

[27] Navathe, Ramez Elmasri, and B. Shamkant. 2010. *Fundamentals of Database Systems*. 6th edition. Upper Saddle River, NJ: Pearson Education, 652–660.

[28] Pfoser, D., C. S. Jensen, and Y. Theodoridis. 2000. "Novel Approaches to the Indexing of Moving Object Trajectories." In *Proceedings of the International Conference on Very Large Data Bases*. Burlington, MA: Morgan Kaufmann, 395–406.

[29] Samet, H. 1984. "The Quad-Tree and Related Hierarchical Data Structures." *ACM Computing Surveys* 16 (2): 187–260.

[30] Sellis, T., N. Roussopoulos, and C. Faloutsos. 1987. "The R+-Tree: A Dynamic Index for Multi-Dimensional Objects." In *Proceedings of the 13th International Conference on Very Large Data Bases*. Burlington, MA: Morgan Kaufmann.

[31] Shekhar, S., and S. Chawla. 2003. *Spatial Databases: A Tour*. Upper Saddle River, NJ: Prentice Hall.

[32] Song, Z., and N. Roussopoulos. 2003. "SEB-Tree: An Approach to Index Continuously Moving Objects." In *Mobile Data Management*. Berlin: Springer.

[33] Song, R., W. Sun, B. Zheng, and Y. Zheng. 2014. "PRESS: A Novel Framework of Trajectory Compression in Road Networks." *Proceedings of the VLDB Endowment* 7 (9): 661–672.

[34] Sun, Y., J. Qi, Yu Zheng, and Rui Zhang. 2015. "K-nearest Neighbor Temporal Aggregate Queries." In *Proceedings of the 18th International Conference on Extending Database Technology*. Konstanz, Germany: Extending Database Technology.

[35] Tang, L. A., Y. Zheng, X. Xie, J. Yuan, X. Yu, and J. Han. 2011. "Retrieving k-nearest Neighboring Trajectories by a Set of Point Locations." In *Proceedings of the 12th Symposium on Spatial and Temporal Databases*. Berlin: Springer, 223–241.

[36] Tao, Y., and D. Papadias. 2001. "Efficient Historical R-trees." In *Proceedings of the 13th International Conference on Scientific and Statistical Database Management*. Washington, DC: IEEE Computer Society Press, 223–232.

[37] Tao, Y., and D. Papadias. 2001. "MV3R-tree: A Spatio-Temporal Access Method for Timestamp and Interval Queries." In *Proceedings of the 27th International Conference on Very Large Data Bases*. Burlington, MA: Morgan Kaufmann, 431–440.

[38] Tao, Y., D. Papadias, and Q. Shen. 2002. "Continuous Nearest Neighbour Search." In *Proceedings of the 28th International Conference on Very Large Data Bases*. Burlington, MA: Morgan Kaufmann, 287–298.

[39] Theodoridis, Y., M. Vazirgiannis, and T. Sellis. 1996. "Spatio-Temporal Indexing for Large Multimedia Applications." In *Proceedings of the 3rd IEEE Conference on Multimedia Computing and Systems*. Washington, DC: IEEE Computer Society Press, 441–448.

[40] Wang, L., Y. Zheng, X. Xie, and W. Y. Ma. 2008. "A Flexible Spatio-Temporal Indexing Scheme for Large-Scale GPS Track Retrieval." In *Proceedings of the Ninth International Conference on Mobile Data Management*. Washington, DC: IEEE Computer Society Press, 1–8.

[41] Wang, Y., Y. Zheng, and Y. Xue. 2014. "Travel Time Estimation of a Path Using Sparse Trajectories." In *Proceedings of the 20th ACM SIGKDD International Conference on Knowledge Discovery and Data Mining*. New York: ACM, 25–34.

[42] Willams, Louis F., Jr. 1975. "A Modification to the Half-Interval Search (Binary Search) Method." In *Proceedings of the 14th ACM Southeast Conference*. New York: ACM, 95–101. doi:10.1145/503561.503582.

[43] Wolfson, O., P. Sistla, B. Xu, J. Zhou, and S. Chamberlain. 1999. "DOMINO: Databases for Moving Objects Tracking." *ACM SIGMOD Record* 28 (2): 547–549.

[44] Wolfson, O., B. Xu, S. Chamberlain, and L. Jiang. 1998. "Moving Objects Databases: Issues and Solutions." In *Proceedings of the Tenth International Conference on Scientific and Statistical Database Management*. Washington, DC: IEEE Computer Society Press, 111–122.

[45] Xu, X., J. Han, and W. Lu. 1990. "RT-Tree: An Improved R-tree Index Structure for Spatiotemporal Databases." In *Proceedings of the 4th International Symposium on Spatial Data Handling*. Berlin: Heidelberg, 1040–1049.

[46] Yi, B. K., H. Jagadish, and C. Faloutsos. 1998. "Efficient Retrieval of Similar Time Sequences under Time Warping." In *Proceedings of the 14th IEEE International Conference on Data Engineering*. Washington, DC: IEEE Computer Society Press, 201–208.

[47] Yuan, J., Y. Zheng, C. Zhang, X. Xie, and G. Z. Sun. 2010. "An Interactive-Voting Based Map Matching Algorithm." In *Proceedings of the 2010 Eleventh International Conference on Mobile Data Management*. Washington, DC: IEEE Computer Society Press, 43–52.

[48] Zheng, Y. 2015. "Trajectory Data Mining: An Overview." *ACM Transactions on Intelligent Systems and Technology* 6 (3): 29.

[49] Zheng, Y., and X. Zhou. 2011. *Computing with Spatial Trajectories*. Berlin: Springer.

[50] Zhou, Y., X. Xie, C. Wang, Y. Gong, and W. Y. Ma. 2005. "Hybrid Index Structures for Location-Based Web Search." In *Proceedings of the 14th ACM International Conference on Information and Knowledge Management*. New York: ACM, 155–162.

5 Introduction to Cloud Computing

Abstract: Efficiently managing large-scale data from different domains needs the support of infrastructure. Cloud-computing platforms provide urban computing with a good foundation for managing and mining datasets that are continuously generated in cities. This chapter overviews the basic technology of cloud computing from the following three perspectives: storage, computing, and applications. Instead of describing the technical details of each component, we focus on how a technique generally works and how it can be employed to solve a problem. As there are quite a few cloud-computing platforms, such as Microsoft Azure and Amazon Web Services, that have different implementations on similar functions, we choose Microsoft Azure as an example to demonstrate the function of cloud computing.

5.1 Introduction

Cities receive vast and heterogeneous data from different domains every second. However, applications of urban computing usually need to provide instantaneous answers (e.g., predicting traffic conditions and detecting urban anomalies) throughout a very large area, such as an entire city or hundreds of thousands of roads. To bridge the gap between the data received and the applications to enable, it is imperative to have a powerful platform (e.g., cloud-computing platforms) on top of which data can be managed and mined effectively and efficiently. The main contributions that cloud computing can offer urban computing are threefold: storing big and heterogeneous data, offering a powerful computing environment, and providing stable and scalable external services. Hereafter, we use Microsoft Azure as an example to demonstrate these contributions.

Microsoft Azure is a cloud-computing platform and infrastructure created by Microsoft for building, deploying, and managing applications and services through a global network of Microsoft-managed data centers [1, 2]. It provides over six hundred elastic, available, and scalable services. As illustrated in figure 5.1, we group the components in

Microsoft Azure into three categories (storage, computing, and applications), in light of the three contributions mentioned above:

1. *Storing vast and heterogeneous data.* Cloud-computing platforms provide a variety of storage mechanisms for storing heterogeneous datasets from different domains, such as traffic, social media, and meteorology. For example, Microsoft Azure provides complete solutions for both structured and unstructured data storage, as depicted at the bottom level of figure 5.1. SQL (structured query language) Database, a relational database-as-a-service in the cloud built on the Microsoft SQL Server Database Engine, is a good choice for storing structured data, while Azure Storage is fit for unstructured data storage. One can also store data in the Redis Cache for high throughput and low-latency data access. We will elaborate on this further in section 5.2.
2. *Providing a powerful computing environment.* For example, Microsoft Azure supports different components for computing, as illustrated at the middle level of figure 5.1. Virtual Machine (VM) is one of several types of on-demand, scalable computing resources that offer a user more control over the computing environment. Users can do nearly everything with an Azure VM, just as if it were their personal computer or a high-performance server. The cloud service can run highly available, scalable

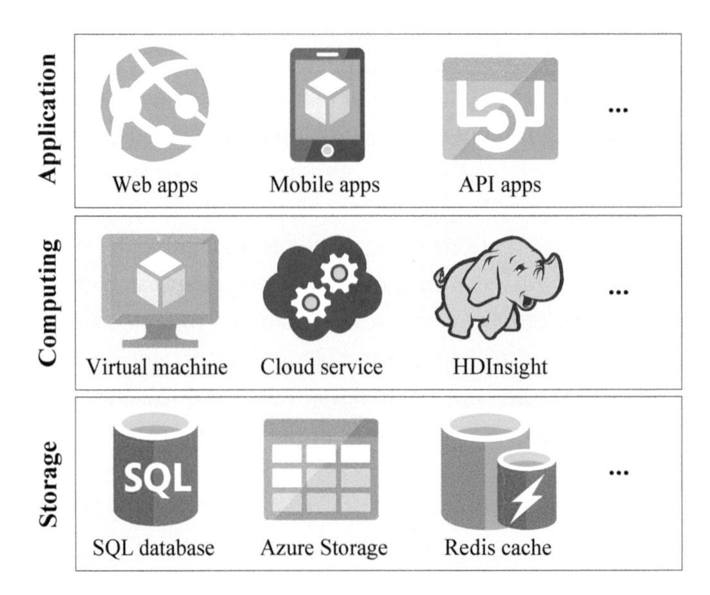

Figure 5.1
The framework of Microsoft Azure.

cloud applications and application programming interfaces (APIs), as a standard web service or a background service, hosted on an Azure VM. HDInsight is a distributed-computing component in Microsoft Azure that performs large-scale data preprocessing, management, and mining. The commonly used Hadoop, Spark, and Storm belong to HDInsight in Azure. We will explain this part further in section 5.3.

3. *Providing stable and scalable external services.* Many urban-computing applications need to deliver reliable services to many users. These services should be able to flexibly scale up and down based on the dynamic flow of user requests. This problem used to frequently block many research projects from being deployed in the real world. In Microsoft Azure, as shown at the top level of figure 5.1, the web apps are used to build and host websites and web applications, while the mobile apps are used to build native mobile applications. API apps make it easier to develop, host, and consume APIs in the cloud. These components have empowered us to walk urban-computing applications through their last miles. Section 5.4 will discuss more about this.

5.2 Storage

5.2.1 SQL Databases

SQL Database [3] is one of the best choices for storing relational and structured data in the cloud. It is a relational database service based on the Microsoft SQL Server Database Engine, capable of handling mission-critical workloads. SQL Database delivers predictable performance at multiple service levels, dynamic scalability with no downtime, built-in business continuity, and data protection, all with near-zero administration. These capabilities allow users to focus on rapid application development and accelerating their time to market, rather than allocating precious time and resources to managing VMs and infrastructure. Because SQL Database is based on the SQL Server Engine, it supports existing SQL Server tools, libraries, and APIs. Thus, it is easy for users to develop new solutions, to move their existing SQL Server solutions, and to extend their existing SQL Server solutions to the Microsoft cloud without having to learn new skills. In short, users can use the SQL Database in the same way they use an ordinary SQL Server.

Figure 5.2 demonstrates the logical architecture through which applications and tools connect to the SQL Database in Azure. First, applications and tools can initiate connection requests through open database connectivity (OBDC) and ADO.NET protocols, both of which support tabular data stream (TDS) protocol. Before an incoming

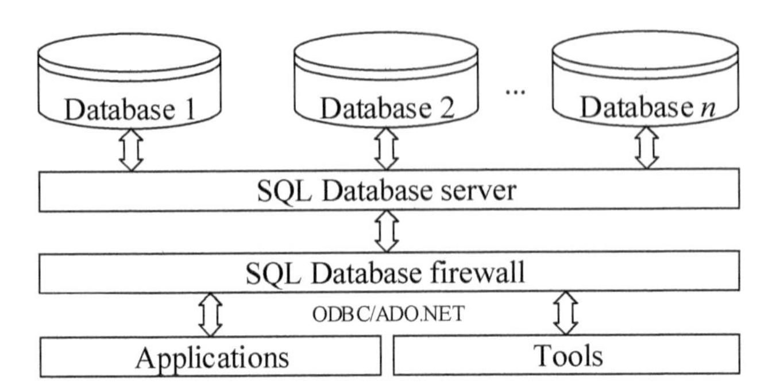

Figure 5.2
The logical structure of SQL Database in Azure.

connection is established, the source IP address is checked against a list of allowed sources by a firewall. If the source address is not in this list, the connection is denied. Otherwise, the commands and requests are intelligently routed to the back-end infrastructure that is really running the SQL Database Server, and finally, the connection is built. An SQL Database logical server acts as a central administrative point for multiple databases.

SQL Database offers three service tiers—Basic, Standard, and Premium—with multiple performance levels to handle different workloads. Higher performance levels provide increasing resources designed to deliver increasingly higher throughput. Users can change service tiers and performance levels dynamically without downtime. Basic, Standard, and Premium service tiers all have an uptime service-level agreement (SLA) of 99.99 percent, flexible business continuity options, security features, and hourly billing.

An SQL Database can be set up with the following procedure:

1. *Create an SQL Database server* in the Azure Portal or through Azure PowerShell, during which we provide the server name, the location of the data center (e.g., West United States or East Asia) that the SQL Database will be running, and a user name and password with which to log in to the SQL Database. To ensure an application's efficient access to an SQL Database, it is recommended to select the same data center where the application is located for the SQL Database.
2. *Create a server-level firewall rule* in the Azure Portal or through Azure PowerShell. By default, only those clients hosted in Azure can connect to the SQL Database. We can open the SQL Database firewall on the server to a single IP address or a range of addresses. Opening the firewall enables SQL administrators and users to log in to any databases on the server to which they have valid credentials.

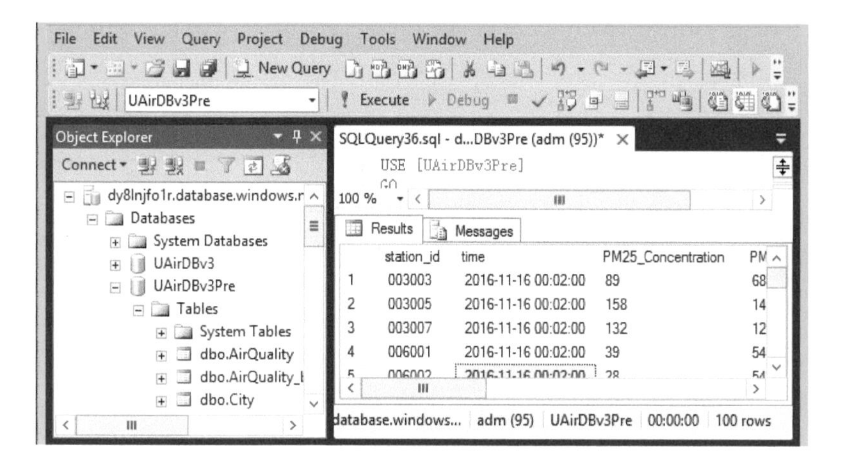

Figure 5.3
An example of SQL Database.

3. *Manage the SQL Database*. We can create, delete, or scale up/down an SQL Database in the Azure Portal, or through Azure PowerShell, or using the latest version of SQL Server Management Studio. When adding a database to a server, we need to create a database name and select a pricing tier. For example, we can create a database named UAirDBv3 with a "standard" pricing tier in the "East Asia" data center.
4. *Access the SQL Database*. An application can access the SQL Database by connecting to a local SQL Server. Please refer to [4] for more details about using an ordinary SQL server in a local computer. Do not forget to add an application's IP address to the firewall if it needs to access an SQL Database on the cloud.

Example. As shown in figure 5.3, dy8Injfo1r is an SQL database server that has two user-created databases, UAirDBv3 and UAirDBv3Pre. AirQuality is one of the tables in the SQL database UAirDBv3Pre, containing some properties like "station_id," "time," "PM25_Concentration," and so on. For more detailed information about SQL Database, please refer to [3].

5.2.2 Azure Storage

Azure Storage [5] is a cloud-storage service suitable for storing unstructured and semi-structured data. Azure Storage provides a storage solution for modern applications that rely on durability, availability, and scalability. It has four main services: Blob (*b*inary *l*arge *ob*ject) storage, Table storage, Queue storage, and File storage.

Blob storage stores unstructured object data. A blob can be any type of text or binary data, such as a document, media file, or application installer. Blob storage is also referred to as *object storage*.

Table storage stores structured datasets, with a NoSQL key–attribute data store format. It allows for the rapid development of and fast access to large quantities of data. Its price is much cheaper than the SQL Server, if users are not concerned with the relationship between data.

Queue storage provides reliable messaging for workflow processing and for communication between components of Cloud Services.

File storage offers shared storage for legacy applications using the standard server message block (SMB) protocol. Azure VMs, Cloud Services, and on-premises applications can share files across application components via mounted shares or representational state transfer (REST) API.

To use Azure Storage, we first create an Azure Storage account (on the Azure Portal or through Azure PowerShell), which gives us access to services in Azure Storage and provides a unique namespace for storage resources. When creating a storage account, we need to provide an account name and a data center location (e.g., West United States or East Asia). Figure 5.4 shows the relationships between Azure Storage resources. We will elaborate on each in the following subsections.

Once a storage account is created, we can use Microsoft Azure Storage Explorer [6], which is a powerful tool from Microsoft, with Windows, MacOS, and Linux versions to easily work with Azure Storage data.

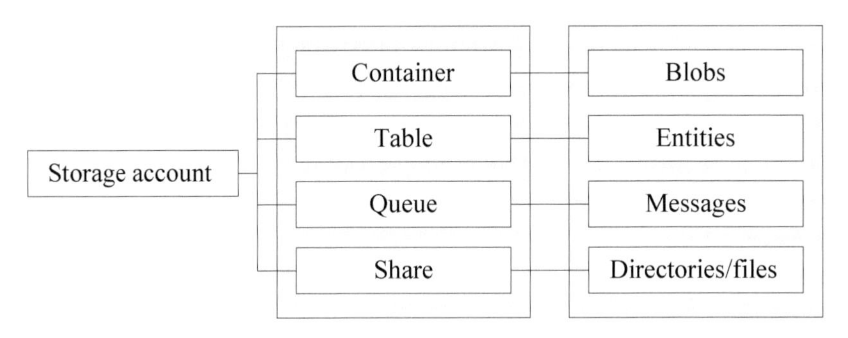

Figure 5.4
The Azure Storage concept.

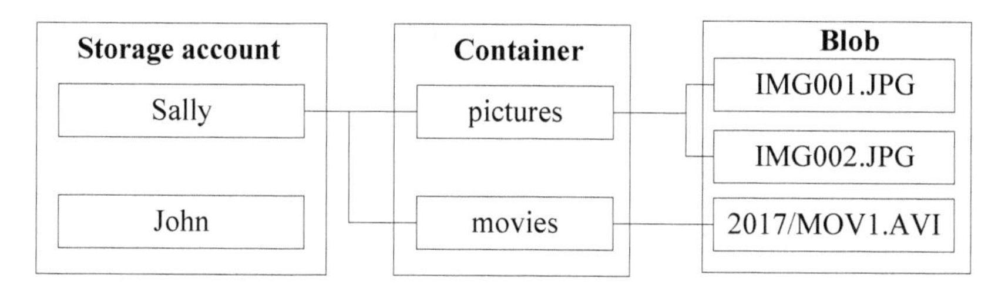

Figure 5.5
The structure of blob files in Azure Storage.

5.2.2.1 Blob storage Blob is a collection of binary data stored as a single entity in the Azure Storage system. It can store any type of text or binary data, such as a document, media file, or even application executable. Figure 5.5 presents an example of the hierarchical structure of blobs. Each storage account can create multiple storage containers, each of which groups a set of blobs. That is, a blob file needs to be stored in a container. In this example, *Sally* is a storage account creating two containers titled *pictures* and *movies*, respectively. The *pictures* container is further composed of two blob files: *IMG001.JPG* and *IMG002.JPG*. Note that the Blob service is based on a flat storage scheme (in a container), which does not have a concept of directory in the underlying structure. But, blobs with the same prefix can be regarded to be in the same directory logically. For instance, the blob name *2017/MOV1.AVI* in the container *movies* can be logically considered a blob named *MOV1.AVI* in a (virtual) directory *2017*.

There are three types of blobs: block blobs, page blobs, and append blobs.

Block blobs are ideal for storing text or binary files, such as documents and media files, enabling us to upload large blobs efficiently. Block blobs comprise blocks, each of which is identified by a block ID. Block IDs are strings of equal length within a blob. Block client code usually uses base-64 encoding to normalize strings into equal lengths. When using base-64 encoding, the preencoded string must be sixty-four bytes or less. Block ID values can be duplicated in different blobs. We create or modify a block blob by writing a set of blocks and committing them by their block IDs. Each block can have a different size, up to a maximum of 100 MB, and a block blob can include up to fifty thousand blocks. Thus, the maximum size of a block is slightly more than 4.75 TB (100 MB × fifty thousand blocks) [7].

We can modify an existing block blob by inserting, replacing, or deleting existing blocks. After uploading the block or blocks that have changed, we can commit a new version of the blob by committing the new blocks with the existing blocks we want to keep using a single commit operation. For any commit operation, if any block is not found, the entire commitment operation fails with an error, and the blob is not modified. A block commitment overwrites the blob's existing properties and metadata and discards all uncommitted blocks.

If we write a block for a blob that does not exist, a new block blob is created, with a length of zero bytes. This blob will appear in blob lists that include uncommitted blobs. If we do not commit any block to this blob, it and its uncommitted blocks will be discarded one week after the last successful block upload. All uncommitted blocks are also discarded when a new blob of the same name is created using a single step (rather than the two-step block upload-then-commit process).

Append blobs are also composed of blocks, but they are optimized for append operations, which make them useful for logging scenarios. For example, they can be employed to store a moving object's trajectory data, in which recently received points are always appended to the end of a file. When we modify an append blob, blocks are added to the end of the blob only. Updating or deleting existing blocks is not supported. Unlike a block blob, an append blob does not expose its block IDs. Each block in an append blob can be a different size, up to a maximum of 4 MB, and an append blob can include up to fifty thousand blocks. The maximum size of an append blob is therefore slightly more than 195 GB (4 MB × fifty thousand blocks) [7].

Page blobs are more efficient for frequent read or write operations. Each page blob is composed of 512-byte pages, used as operating system (OS) and data disks by Azure VMs. When creating a page blob, we initialize the page blob and specify the maximum size the page blob will grow to. To add or update the contents of a page blob, we write a page or multiple pages by specifying an offset and a range that align to 512-byte-page boundaries. A write to a page blob can overwrite just one page, some pages, or up to 4 MB of the page blob. Writes to page blobs happen in place and are immediately committed to the blob. The maximum size for a page blob is 1 TB.

We need to specify the blob type when creating a blob. Once a blob has been created, its type cannot be changed, and it can be updated only by using operations appropriate for that blob type (i.e., writing a block or list of blocks to a block blob, appending blocks to an append blob, and writing pages to a page blob). There is no limitation on how many files or containers can be created for a storage account. However, the total size of a storage account cannot exceed 500 TB [8].

Example. In this example, we use Microsoft Azure Storage Explorer to upload pictures to Azure. As illustrated in figure 5.6, we first create a container named *pictures* under a storage account *stdatamanage*. After clicking the upload button (marked by the red rectangle), we will see a pop-up window where we can choose pictures to upload and the type of blob (e.g., block blob in this example) to store those pictures. The selected pictures will be uploaded to the blob container.

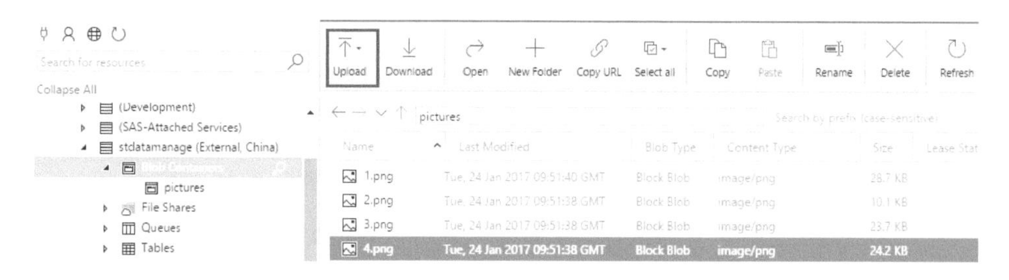

Figure 5.6
An example of using blob files.

Example. We can also upload files to Azure Blob automatically by writing a program. In this example, we use C# language to demonstrate the uploading process consisting of four steps:

1. Connect to the storage account using the storage connection string

   "CloudStorageAccount storageAccount=
   CloudStorageAccount.Parse ([StorageConnectionString]);".

   The [*StorageConnectionString*] is further composed of the following fields:

   "DefaultEndpointsProtocol=[Protocol Type];AccountName=[Account Name];
   AccountKey=[Storage Account Key];EndpointSuffix=[Endpoint Suffix]",

   where [*Protocol Type*] is the name of protocol for communication on the Internet (e.g., http or https); [*Account Name*] is the name of the storage account; [*Endpoint Suffix*] differs by the version of Azure (e.g., *core.windows.net* and *core.chinacloudapi.cn* represent Global Azure and China Azure, respectively); [*Storage Account Key*] is the private key to access the storage account.
2. Create a blob client, with which we can manage blobs in a storage account:

   "CloudBlobClient blobClient=storageAccount.CreateCloudBlobClient();".
3. Retrieve a reference to a container and create the container if it does not exist:

   "CloudBlobContainer container=blobClient.GetContainerReference([Container Name]);
   container.CreateIfNotExists();",

   where [*Container Name*] is the name of the container you want to specify.
4. Retrieve a reference to a blob and upload a file to it.

   "CloudBlockBlob blob=container.GetBlockBlobReference([Blob Name]);
   blob.UploadFromStream([File Stream]);",

where [*Blob Name*] is the name of the blob we want to store a picture in; [*File Stream*] is the stream of the files that we want to upload. Here we choose block blob. In case we want to choose page blob and append blob, we can use "*container.GetPageBlobReference([Blob Name])*" and "*container.GetAppendBlobReference([Blob Name])*", respectively. For more information about Blob storage, please refer to [9].

5.2.2.2 Table storage Table storage is a service that stores structured NoSQL data in the cloud. It is a key-attribute store without a schema design (i.e., schemaless). Because Table storage is schemaless, it is easy to adapt the data of an application as the needs of the application evolve. Moreover, Table storage is significantly lower in cost than traditional SQL for similar volumes of data. Figure 5.7 depicts the hierarchical structure of the Table service, which contains the following components:

- *Storage account.* A storage account is a globally unique entity within the storage system. All access to Azure Storage is done through a storage account. Each storage account can create an unlimited number of tables as long as each table is uniquely named.
- *Table.* A table is a collection of entities. Tables do not enforce a schema on entities (i.e., a single table can contain entities that have different sets of properties).
- *Entity.* An entity is a set of properties similar to a database row. An entity can be up to 1 MB in size. The number of entities that a table can hold is limited only by the storage account capacity limit (i.e., 500 TB) [8]. In each table, a table entity is identified by two properties, PartitionKey and RowKey. Entities with the same partition key are stored in the same partition, which can be queried more quickly, and inserted/updated in atomic operations. An entity's row key is its unique identifier within a

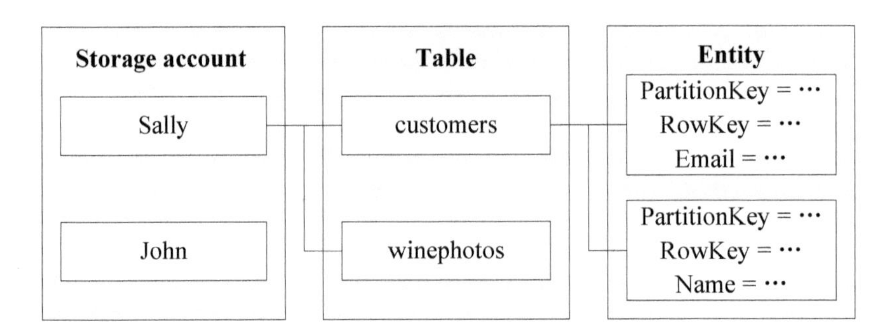

Figure 5.7
The concept of Table service in Azure Storage.

partition. Together, PartitionKey and RowKey uniquely identify every entity within a table.

- *Property*. A property is a name-value pair. Each entity can include up to 255 properties to store data, including three system properties that specify a partition key, a row key, and a time stamp. Thus, a user may include up to 252 custom properties, in addition to the three system properties. We must include the PartitionKey and RowKey properties in every insert, update, and delete operation. The Timestamp property is a date-time value that is maintained on the server side to record the time an entity was last modified.

Azure Table is the best way to store the semistructured datasets that do not require complex joins and foreign keys. The most efficient way to access Azure Table is through point queries (i.e., specifying both the partition key and row key). Azure Table is also very efficient for answering the range queries of the row key within the same partition key.

Example. As shown in figure 5.8, we use Table services to store GPS trajectories generated by taxicabs. In this example, we create a table for each taxi, respectively, based on its identity. The records of a taxi are stored in its own table, with each entity corresponding to a GPS point record. The partition key of an entity is the hour in which the record was generated, and the row key is the exact time stamp when the record was generated. For example, "2017012422" means the record was generated from approximately 22:00 to 23:00 on January 24, 2017, and "20170124221023" means the record was generated at 22:10:23 on January 24, 2017. The properties include speed, heading direction, and the latitude and longitude coordinates of a record.

When inquiring the location of a taxi at a given time stamp, we can first find the table corresponding to the taxi based on its identity and turn the query time into a partition key and a row key. Entities can be retrieved in light of the partition key and row key.

When searching for the location of a taxi at a time interval (e.g., retrieving the trajectory records of "jingxxxx1" during 22:30 to 23:00 on January 24, 2017), we can

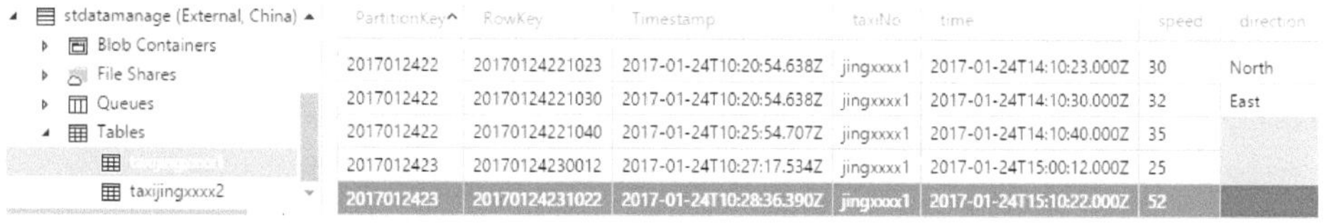

stdatamanage (External, China)
Blob Containers
File Shares
Queues
Tables
taxijingxxxx2

PartitionKey	RowKey	Timestamp	taxiNo	time	speed	direction
2017012422	20170124221023	2017-01-24T10:20:54.638Z	jingxxxx1	2017-01-24T14:10:23.000Z	30	North
2017012422	20170124221030	2017-01-24T10:20:54.638Z	jingxxxx1	2017-01-24T14:10:30.000Z	32	East
2017012422	20170124221040	2017-01-24T10:25:54.707Z	jingxxxx1	2017-01-24T14:10:40.000Z	35	
2017012423	20170124230012	2017-01-24T10:27:17.534Z	jingxxxx1	2017-01-24T15:00:12.000Z	25	
2017012423	20170124231022	2017-01-24T10:28:36.390Z	jingxxxx1	2017-01-24T15:10:22.000Z	52	

Figure 5.8
An example of Table services.

execute the following query in the table "jingxxxx1" (for more information about Table storage, refer to [10]):

> *"PartitionKey ge '2017012422' and PartitionKey lt '2017012423' and RowKey ge '20170124223000' and RowKey lt '20170124230000'"*,

where *ge* means a string greater than or equal to another one, and *lt* denotes a string less than another one, according to the lexicographical order. Including an entity's partition key in its row key can facilitate answering range queries that cross multiple partitions.

5.2.2.3 Queue storage Azure Queue storage provides cloud messaging between application components. In designing applications for scale, application components are often decoupled, so they can scale independently. Queue storage delivers asynchronous messaging for communication between application components, whether they are running in the cloud, on the desktop, on an on-premises server, or on a mobile device [11]. Queue storage also supports managing asynchronous tasks and building process workflows. As demonstrated in figure 5.9, there are three components of the Queue service:

- *Storage account.* Like Table and Blob storage, the access to Queue storage is done through a storage account. Each storage account can create an unlimited number of queues, which are differentiated by queue names.
- *Queue.* A queue contains a set of messages. All messages must be in a queue. Note that the queue name must be all lowercase in Azure.
- *Message.* A message, in any format, can be up to 64 KB. There is no restriction on the number of messages in each queue as long as the total size is no more than the maximum storage account capacity (i.e., 500 TB). Messages are generally added to the end of the queue and retrieved from the front of the queue, although first in, first out (FIFO) behavior is not guaranteed. If there is a need to store messages larger than 64 KB, we can deposit the message data in a blob or table, storing a reference to the data as a message in a queue. The longest time that a message can remain in a queue is seven days.

Example. Suppose there are thousands of pictures to be uploaded. We can use a queue to dispatch uploading tasks across multiple uploaders. As depicted in figure 5.10, Command is a program that receives and forwards users' requests to different uploaders by sending messages to a queue. A message contains the information regarding which picture and where to upload. Uploaders, hosted in different machines, respectively, retrieve messages from Queue storage actively and then upload a corresponding image

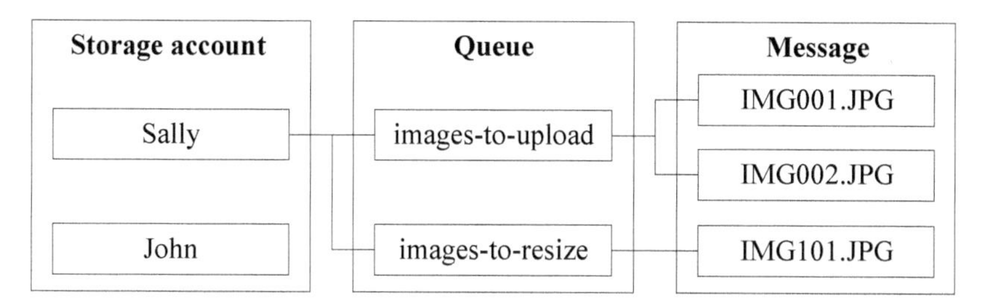

Figure 5.9
The concept of Queue service in Azure Storage.

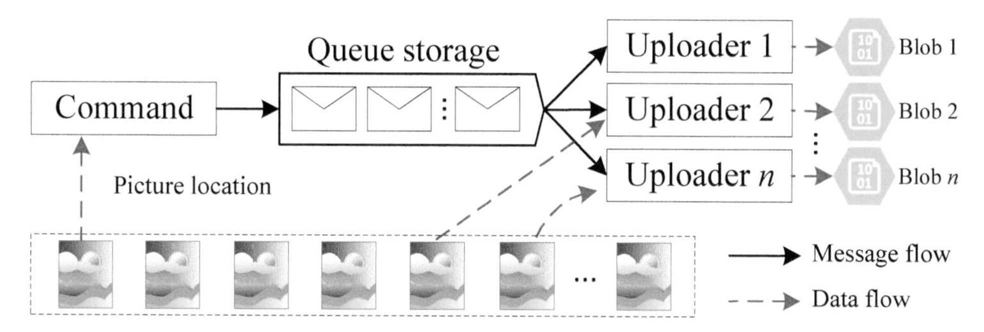

Figure 5.10
An example of using Azure Queue service to upload massive images.

to the cloud according to its location contained in the message. Each image can be stored in a blob.

5.2.2.4 File storage File storage is a service that offers file shares in the cloud using the standard server message block (SMB) protocol [12]. Both SMB 2.1 and SMB 3.0 are supported. With Azure File storage, we can migrate legacy applications that rely on file shares to Azure quickly, without costly rewrites. Applications running in Azure VMs or Cloud Services or from on-premises clients can mount a file share in the cloud, just as a desktop application mounts a typical SMB share. Any number of application components can then mount and access the file storage share simultaneously. Figure 5.11 presents a logical structure of File storage, which has four components:

- *Storage account.* File storage must be accessed through a storage account, which can contain zero or more file shares. The total size of file shares in one storage account cannot be more than 500 TB, which is the capacity limit of a storage account.

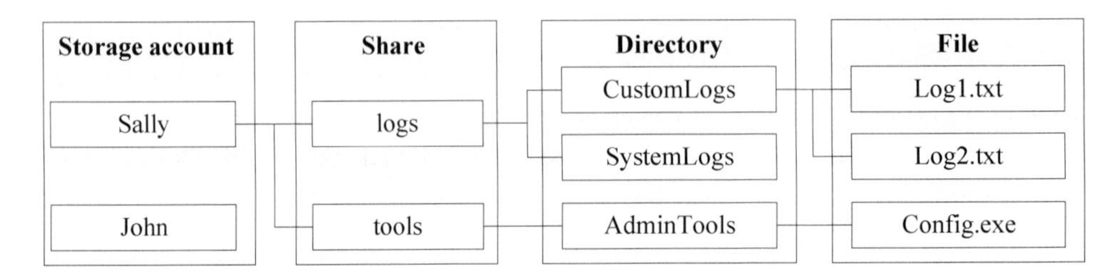

Figure 5.11
The logical structure of files in Azure Storage.

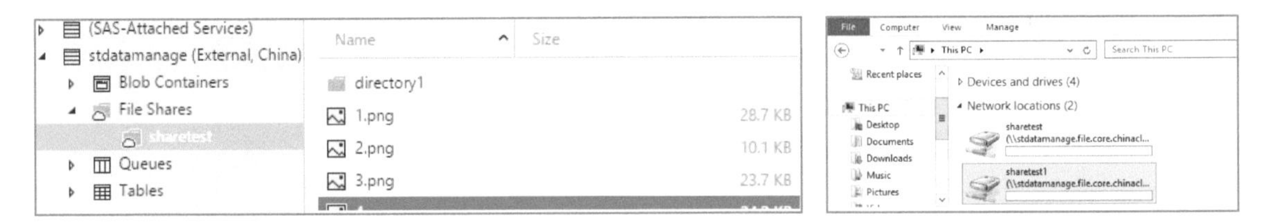

a) Create a share under an Azure Storage account b) Using the file storage in a local machine

Figure 5.12
An example of using File storage in Azure.

- *Share.* A File storage share is an SMB file share in Azure, which is the virtual driver that we can mount. A share can store an unlimited number of files or directories up to the total capacity of the file share (5 TB).
- *Directory.* A directory, which is an optional hierarchy, must be created in a share or another directory. A directory contains zero or more files or directories, as long as the total size does not exceed the restriction of a share (i.e., 5 TB).
- *File.* A file is a single entity comprising binary data, properties, and metadata. Files should be contained in a share or in a directory. A file may be up to 1 TB in size.

Example. To use File storage, we first create shares under a storage account. There are multiple ways to create a share—for example, as shown in figure 5.12a, using Microsoft Azure Storage Explorer, or in the Azure Portal. For example, we created a share, titled *sharetest*, under the Azure Storage account *stdmanage*.

Any machine that supports the standard SMB protocol can connect to a share. For instance, in Windows, we can mount a share by typing the following command in the command line:

"net use [drive letter]: \\[storage account name].file.[endpoint suffix]\[share name]/u: [storage name] [storage account access key]",

where *[drive letter]* is a single alphabetic character (*A* through *Z*) assigned to the mounted share; *[storage account name]* and *[share name]* stand for the names of the storage account and share, respectively; *[endpoint suffix]* differs by the version of Azure (e.g., *core.windows.net* and *core.chinacloudapi.cn* represent Global Azure and China Azure, respectively); and *[storage account access key]* is the private key to access the storage account.

In Linux, we can mount a share using the following command:

"sudo mount -t cifs //[storage account name].file.[endpoint suffix]/[share name] [mount point] -o vers=3.0, username=[storage name], password=[storage account access key], dir_mode=0777, file_mode=0777",

where *[mount point]* is a directory to which the share mounted; *[storage account name]* and *[share name]* stand for the names of the storage account and share, respectively; *[endpoint suffix]* differs based on the version of Azure (e.g., *core.windows.net* and *core.chinacloudapi.cn* represent Global Azure and China Azure, respectively; and *[storage account access key]* is the private key to access the storage account. Note that "0777" represents a directory/file permission code that gives execution/read/write permissions to all users. We can replace it with another file permission code following the Linux file permission document.

After connecting to a share, we can read or write the files in the share, as it is a local disk. As depicted in figure 5.12b, we can see the share *sharetest* has been mounted to a local computer as a disk drive. Remember to open Internet access for port 445 (TCP Outbound) in a local network, which is required by the SMB protocol, as some Internet service providers may block port 445.

5.2.3 Redis Cache

Azure Redis Cache is based on the popular open-source Redis cache, which is widely used as a memory database in the physical computer environment. Azure Redis Cache gives users access to a secure, dedicated Redis cache, managed by Microsoft and accessible from any application within Azure or on-premise [13]. Azure Redis Cache is an advanced in-memory key-value store, which can be used as a database, cache, and message broker. Figure 5.13 describes the logical structure of Redis Cache, which comprises three components:

- *Redis instance.* Users can create multiple Redis instances within an Azure subscription. A Redis instance can be compared to a Redis server, which can be managed, scaled up/down, or monitored independently. There are three tiers for Azure Redis instances: Basic, Standard, and Premium; each tier differs in terms of features and

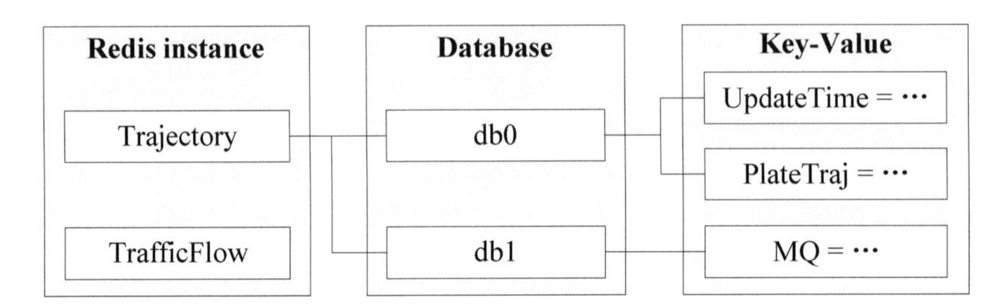

Figure 5.13
The logical structure of Redis Cache.

pricing. Higher performance levels provide more resources designed to deliver increasingly higher throughput. Users can change service tiers and performance levels dynamically without downtime.

- *Database*. By default, there are sixteen databases (indexed from 0 to 15) in each Redis instance. The number of databases can be specified before creating a Redis instance, but the limit is different for each Redis Cache pricing tier. For example, tier Standard, C3 (6 GB) can hold sixteen databases at most, while tier Premium, P2 (13 GB) can contain up to thirty-two databases. Each Redis database has its own key space. By using different databases for "staging" and "production" data, or for different applications, users do not have to worry about key clashes between them [17].
- *Key-value*. In each database, there are multiple key-value pairs, which support data structures such as strings, hashes, lists, sets, sorted sets with range queries, bitmaps, and geospatial indexes with radius queries. Redis supports a set of atomic operations on these data types to ensure data consistency.

Stored in memory, Azure Redis Cache is a good choice to store frequently accessed, high throughput, low-latency, and shared data. It can also be used as a lightweight message queue or message broker. Please refer to [15] for more information.

To use Redis Cache in Microsoft Azure, please follow the instructions shown below:

1. *Create a Redis cache*. On the Microsoft Azure Portal, we click New→Data and Storage→Redis Cache, during which we should provide the name of the cache, choose where in the world to run it, and select a pricing tier to meet the needs of an application. For example, we can create a Redis cache called *Trajectory* in the North Central United States whose pricing tier is Standard, C3 (6 GB).

   Example. Figure 5.14 presents a user interface of Redis Desktop Manager [14], which is a powerful tool to access and revise the content in a Redis. In this example,

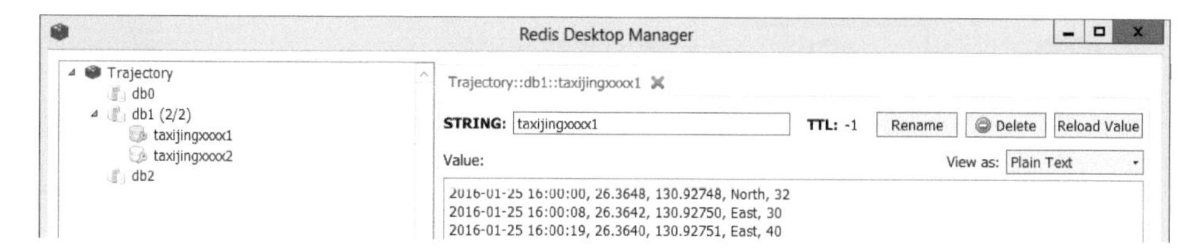

Figure 5.14
An example of using Redis Desktop Manager to browse the content in a Redis cache.

Trajectory is the name of the Redis, which contains three databases, *db0*, *db1*, and *db2*. In *db1*, "taxjingxxxx1" and "taxjingxxxx2" are two key-value pairs denoting two taxis' GPS trajectories. The key "taxjingxxxx1" is a taxi identity, having many lines of values. Each line stands for a GPS point in the taxi's trajectory, consisting of *time*, *latitude*, *longitude*, *direction*, and *speed*.

2. *Manage a Redis cache*. After a Redis cache is created, we can use either the Azure Portal or Azure PowerShell to configure settings and monitor its usage.
3. *Work with a Redis cache*. We can connect to an Azure Redis Cache in the same way as connecting to a local Redis established in our own personal computer. For example, to connect to the Azure Redis Cache using StackExchange.Redis (a high-performance general purpose Redis client for .NET languages), we can use the following connect string:

   "[redis name].redis.cache.[endpoint suffix]: 6380, password=[access key], ssl=True,abortConnect=False",

   where [*redis name*] is the name of the Azure Redis; [*endpoint suffix*] differs by the version of Azure, where *windows.net* and *chinacloudapi.cn* represent Global Azure and China Azure, respectively; and [*access key*] is the private key.

Figure 5.15 presents an example of using Redis to share data between different components. Taxicabs upload their raw trajectories to a Redis cache while other components (i.e., a map match component or index component) read the raw trajectories from the Redis and process them.

5.3 Computing

Microsoft Azure provides different components for computing. In this section, we will introduce some of them, namely Virtual Machine (VM), Cloud Services, and HDInsight.

Figure 5.15
Using Redis to share data between different components.

5.3.1 Virtual Machine

Azure VM is one of several types of on-demand, scalable computing resources that Azure offers, which is an example of infrastructure-as-a-service (IaaS). Typically, we choose a VM when we need more control over the computing environment than other choices offer, especially in the following components:

- *Development and testing.* Azure VMs offer a quick and easy way to create a computer with specific configurations required to code and test an application.
- *Applications in the cloud.* Because demand for an application can fluctuate, it might make economic sense to run it on a VM in Azure. We pay for extra VMs when needed and shut them down when they are not.
- *Extended datacenter.* VMs in an Azure virtual network can be easily connected to an organization's network.

In addition to Windows VMs, Microsoft Azure also supports running a number of popular Linux distributions provided and maintained by a number of partners. We can find distributions, such as Red Hat Enterprise, CentOS, Debian, Ubuntu, CoreOS, RancherOS, FreeBSD, and more in the Azure Marketplace. We can even use our own customized Linux system in an Azure VM.

Azure offers a wide variety of sizes to support many types of uses, from which we can choose a suitable one for our workload. An existing VM can be resized, which determines its processing power, memory, and storage capacity. Azure charges an hourly price based on a VM's size and operating system.

An Azure VM offers the flexibility of virtualization without having to buy and maintain the physical hardware that runs it. However, we still need to maintain a VM by performing tasks, such as configuring, patching, and installing the software that runs on it. To learn more about Microsoft Azure VM, please refer to [18].

To use a VM in Azure, please follow the instructions shown below:

1. *Create a VM.* To create a VM in the Azure Portal or through Azure PowerShell, we need to select an operating system (Windows or Linux) and then set the name of

the VM, a user name and a password with which to log into it, a location for the VM (e.g., West United States or East Asia), and a size that is suitable to the workload of an application.

2. *Connect to a VM*. After a VM is created, we can connect to it by Remote Desktop Connection (for Windows VM) or Putty [16] (for Linux VM). During the log in process, we need to provide the address, user name, and password with which the VM was created. The address is formatted as follows:

   "[*computer name*].[*endpoint suffix*]:[*port*]",

   where [*computer name*] is the name of a VM; according to the version of Azure, [*endpoint suffix*] has two choices, *windows.net* and *chinacloudapi.cn*, representing Global Azure and China Azure, respectively; [*port*] is the number of ports that Azure automatically creates for the VM.
3. *Work with a VM*. We can do nearly everything with an Azure VM, as if it were our personal computer or a high-performance server. For example, we can build a website in a VM to provide a web service and install SQL Server and Redis for data storage. Services in a VM can be accessed from anywhere.

5.3.2 Cloud Services

Microsoft Azure Cloud Services is a container of roles. It aims to host and run highly available, scalable cloud applications and APIs. Cloud Services is an example of platform-as-a-service (PaaS), which is hosted on VMs. There are two types of roles in a cloud service, as shown in figure 5.16:

1. *Web role*. The web role provides a standard web service. Users access this service through hypertext transfer protocol (HTTP). Web role runs a Windows server with a web application deployed to Internet information server (IIS) automatically.
2. *Worker role*. The worker role is another type of role in a cloud service, which runs a Windows server without IIS. In fact, the worker role runs as a background service in a host VM. In addition to HTTP, the worker role provides more services using other protocols, such as user datagram protocol (UDP) and transmission control protocol (TCP).

Each cloud service can have multiple roles (either a web role or a worker role), each of which can have multiple instances of the same implementation. An *instance* is a process playing a type of role in a VM. To handle a heavy access workload to an application, we can create multiple instances for the application. Such a configuration balances users' requests automatically across multiple instances. Microsoft Azure can scale

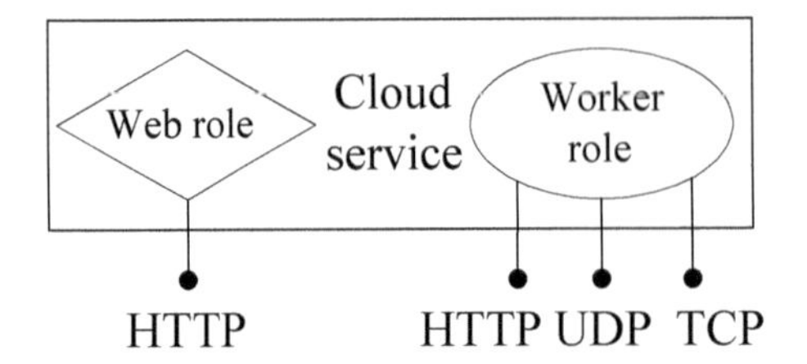

Figure 5.16
Roles in Cloud Services.

and deploy VMs in a cloud service application in a way that avoids a single point of hardware failure.

Even though applications run in VMs, it is important to understand that Cloud Services provides PaaS, not IaaS. With IaaS, such as Azure VMs, we first create and configure the environment our application will run in, then deploy our application into this environment. We are responsible for managing much of this world, such as deploying new patched versions of the operating system in each VM. In PaaS, by contrast, it is as if the environment already exists. All we must do is deploy our applications. The management of the platform these applications run on, including deploying new versions of the operating system, is handled for us. With Cloud Services, we do not create VMs. Instead, we provide a configuration file that tells Azure the number of instances for different roles (e.g., three web role instances and two worker role instances); the platform then creates them for us. Although we still need to choose a size for those backing VMs, we do not explicitly create them ourselves. If our application needs to handle a greater load, we can ask for more VMs by creating more instances. If the load decreases, we can shut those instances down and stop paying for them. More details about Cloud Services can be found in [19].

Example. A simple application might use just a web role, while a more complex application might use a web role to handle incoming requests from users and then pass the work (those requests) to one (or multiple) worker role(s) for processing through a queue, as illustrated in figure 5.17.

To use Cloud Services in Azure, one must follow the instructions below:

1. *Create a cloud service* in the Azure Portal or Azure PowerShell, during which we must provide a service name and select a location where the service will be running.
2. *Create an Azure Cloud Services project* using a developer environment (e.g., Microsoft Visual Studio) in which we need to create some web roles or worker roles. These roles will be associated with the cloud service. As shown in figure 5.18, Web Role 1 and Worker Role 1 are added to the project.

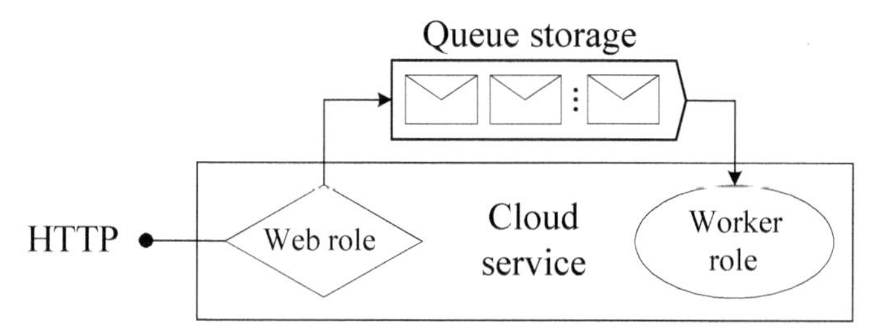

Figure 5.17
An example of using Cloud Services.

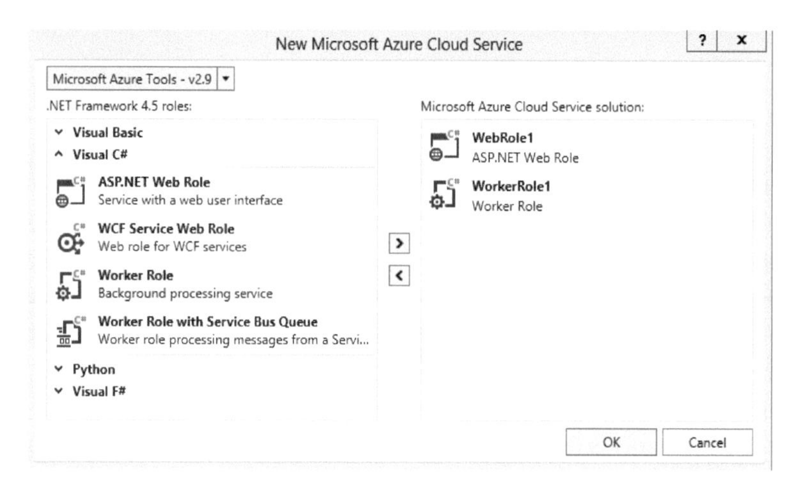

Figure 5.18
Creating a Cloud Services project with a web role and a worker role.

3. *Set each role in its own configuration page*. For example, as depicted in figure 5.19, we can open the configuration page of Web Role 1 by double-clicking it in figure 5.18. We need to specify the number of instances and the size of the VMs the instance will be running on. Here, we use one instance with a small VM (i.e., one-core CPU, 1.75 GB memory). Following the instructions shown in figure 5.19, there are other settings to configure, such as operation logs and ports. We do not enumerate each of them here.
4. *Publish the cloud service project to Azure*. Publishing a cloud service can be simply done by right-clicking the corresponding project name (i.e., AzureCloudService1 shown in figure 5.19) in Visual Studio. After selecting an Azure subscription and the cloud service we built in the first step, we can finally publish the cloud service.

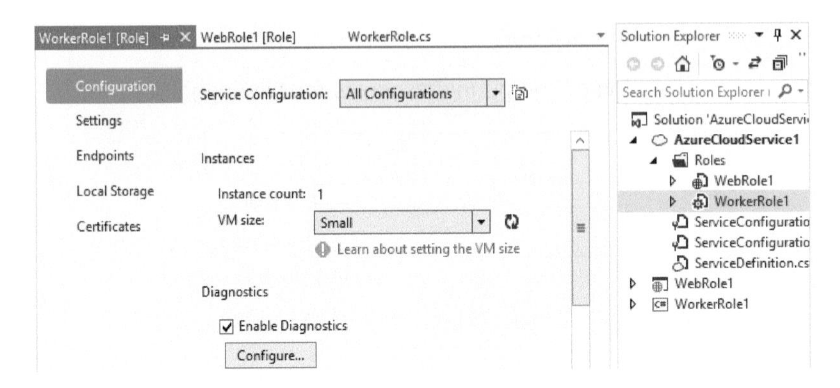

Figure 5.19
Configuring roles in a cloud service.

5.3.3 HDInsight

Besides the conventional VMs and cloud services, Microsoft Azure also supports the following distributed parallel-computing platforms, called HDInsight [20], to perform large-scale data processing. HDInsight integrates Hadoop, Spark, HBase, Storm, and other technologies in the Hadoop ecosystem into Microsoft Azure. We will introduce some of these components, consisting of Hadoop, Spark, and Storm, that can be widely used for spatiotemporal data management.

Figure 5.20 presents the structure of HDInsight, which is composed of two components: HDInsight cluster and Windows Azure Storage Blob (WASB). The HDInsight cluster is a collection of computing resources for running jobs. It abstracts the implementation details of the installation and the configuration of individual nodes. All we need to do is to provide general configuration information. WASB is an implementation of the Hadoop Distributed File System (HDFS) on Azure Blob storage. Storing data in Blob storage enables users to safely delete the HDInsight clusters that are used for computation without losing user data.

To use HDInsight in Azure, we need to follow the procedure below:

1. *To create a cluster in the Azure* Portal or through Azure PowerShell, we first set a cluster's name, the type of the cluster (e.g., Hadoop, Spark, or Storm), the location where the cluster will be running, the user name and password with which to upload jobs, and the number of nodes, as well as their sizes. In addition, we need to select an Azure storage and a container to install programs, which are stored in blob files. Remember that blob files need to be stored in a container (see figure 5.5 for more details).

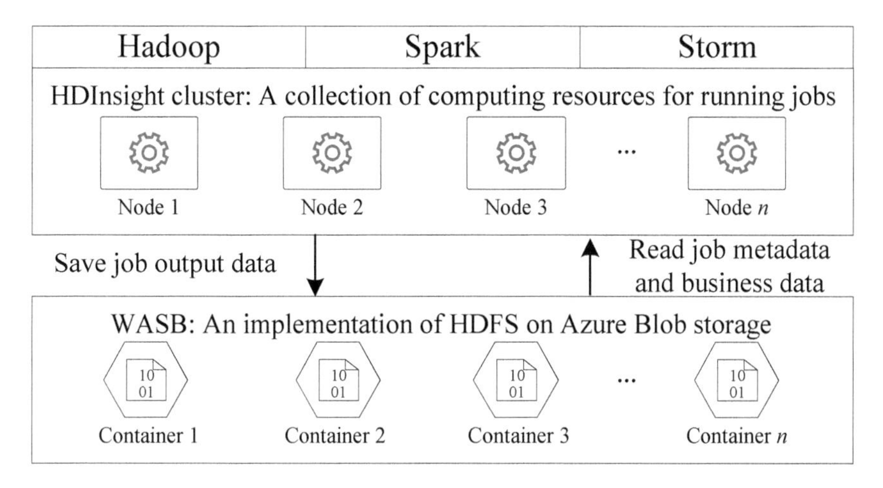

Figure 5.20
The logical structure of HDInsight.

2. *Create a cluster project* using an integrated development environment ([IDE]; e.g., Microsoft Visual Studio).
3. *Upload jobs to a cluster*. There are many ways to upload jobs to a cluster. We can upload a job by remoting to the master node (of the cluster), through the built-in cluster dashboards, or by using an IDE.
4. *Manage the cluster*. We can manage (i.e., monitor, scale, or delete) the cluster by using the Azure Portal. The storage that stores the data of the cluster is not affected unless we delete the storage account.

Hadoop, Spark, and Storm services can be created according to the aforementioned three steps, but their management systems in Azure differ slightly with each other. We will discuss them in the following subsections.

5.3.3.1 Azure Hadoop Hadoop is the most widely used MapReduce [21] framework, which is a good choice for performing offline batch-based big-data processing. Figure 5.21 shows the framework of Hadoop 2.x, which consists of four main components:

1. *Hadoop Common*. Hadoop Common includes libraries and utilities required by other Hadoop modules. These libraries provide file system and OS level abstractions, containing the necessary Java files and scripts required to start Hadoop.
2. *Hadoop YARN*. Hadoop YARN (yet another resource negotiator) is a framework for job scheduling and cluster resource (e.g., CPUs, memory, etc.) management.

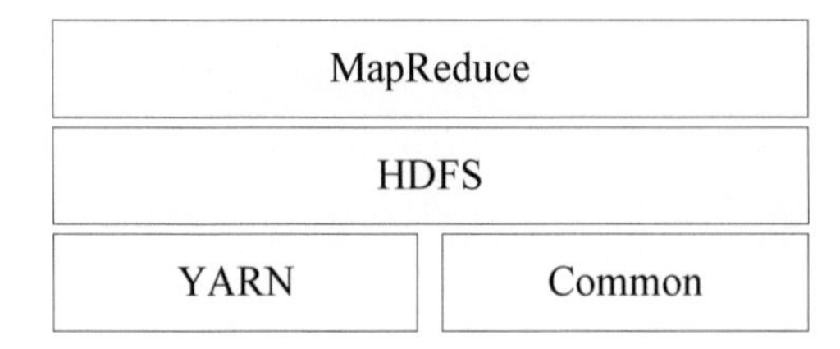

Figure 5.21
Components of Hadoop.

3. *Hadoop Distributed File System* (HDFS). HDFS is a distributed file system for providing permanent, reliable, and distributed storage that supplies high-throughput access to application data. Other alternative storage solutions like WASB can also be used.
4. *MapReduce*. MapReduce is a parallel-processing framework based on YARN.

Azure HDInsight deploys and provisions managed Apache Hadoop [23] clusters in the cloud to process, analyze, and report on big data with high reliability and availability. As part of the Azure cloud ecosystem, Hadoop in HDInsight offers several benefits:

- It provides automatic provisioning of clusters. HDInsight clusters are much easier to create than manually configuring Hadoop clusters. A Hadoop cluster consists of several VMs (nodes) that are used for the distributed processing of tasks on the cluster. HDInsight abstracts the implementation details of the installation and configuration of individual nodes, so we only need to provide general configuration information.
- HDInsight provides many state-of-the-art Hadoop components, from which we can choose the most appropriate for an application.
- Hadoop achieves high availability and reliability by distributing redundant copies of services and data across the nodes in a cluster. However, standard distributions of Hadoop typically have only a single head node. Any outage of the single head node can cause the cluster to stop working. To address this potential problem, HDInsight clusters on Azure provide two head nodes to increase the availability and reliability of Hadoop services and jobs running.
- Cluster scaling enables users to change the number of nodes of a running HDInsight cluster without having to delete or recreate it.
- It is efficient and economical to use Azure Blob storage as a Hadoop storage option. We can also integrate Hadoop in Azure with other Azure services, including SQL Database and Web Apps. (We will talk about Web Apps in the next section.)

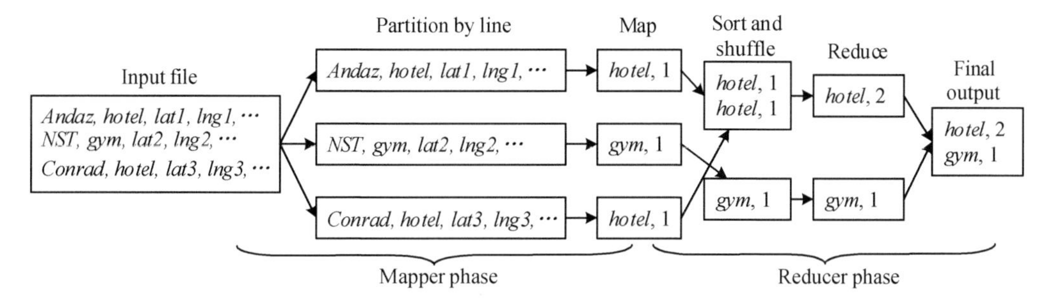

Figure 5.22
Counting POIs of different categories through Hadoop.

Example. Figure 5.22 demonstrates the principal of a general Hadoop. Suppose there is a file in HDFS storing many POIs. Each line of the file stands for a POI record, containing the category and other information about the POI. We want to count the number of POIs in each category. There are two phases in the procedure: a mapper phase and a reducer phase. In the mapper phase, the data file is read, and each line in the file is passed to an individual mapper instance.

The mapper instance forms a key-value pair with each record (e.g., <*hotel*, 1>), where *hotel* is the category, and *1* denotes one hotel. After mappers completely execute on the entire dataset, the reducer starts. In the reducer phase, these key-value pairs are first sorted and shuffled by the keys, with the values for the same key counted. Thus, new key-value pairs are generated (e.g., <*hotel*, 2>), where each key is a category, and the value is a count. Finally, all the category-count pairs are written to files in HDFS.

To turn the aforementioned example into a Hadoop job, we first need to create an HDInsight cluster in the Azure Portal according to the standard procedure for creating an HDInsight, selecting Hadoop as the type of clusters.

We can then submit this Hadoop job using Hive [22], which is a data warehouse tool integrated by the HDInsight Query Console. The HDInsight Query Console can be accessed through the Azure Portal, facilitating users to submit jobs, view job history, and monitor the status of Azure Hadoop. Hive provides an SQL-like language (called HiveQL) and converts queries to MapReduce jobs, enabling users to interact with Hadoop.

Figure 5.23 presents the user interface of Hive Editor, where the *Query Name* text field allows us to give a meaningful name for a job, and corresponding job commands can be placed in the text box below. For example, the aforementioned POI- counting job can be written in HiveQL, consisting of the following two commands:

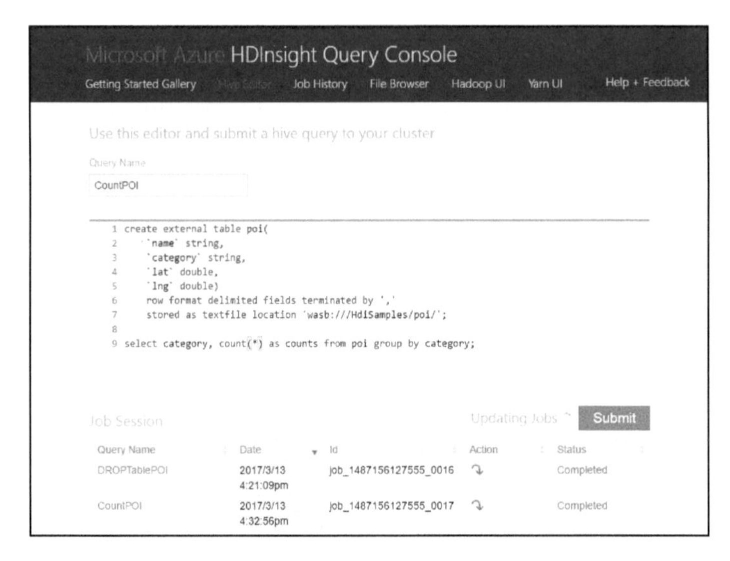

Figure 5.23
Submitting a Hadoop job using Hive in HDInsight Query Console.

1. Create the *poi* table on the comma-separated POI data:

   "create external table poi(`name` string, `category` string, `lat` double, `lng` double) row format delimited fields terminated by ',' stored as textfile location 'wasb://[ContainerName]@[Storage Account Name].blob.[endpoint suffix]/[Blob Virtual Directory Path]/'",

   where [*Container Name*] and [*Storage Account Name*] are the names of the container and the storage account of the cluster, respectively; [*endpoint suffix*] differs according to the version of Azure (e.g., *core.windows.net* and *core.chinacloudapi.cn* represent Global Azure and China Azure, respectively); and [*Blob Virtual Directory Path*] is the virtual directory path of the POI data file(s). The statement above will create a table named *poi*.

2. Count the number of each category:

   "select category, count(*) as counts from poi group by category".

   After clicking the Submit button, the job is uploaded, and its status can be viewed in the bottom *Job Session* table. After the job is completed, we can view its output in the *Job History* page.

For more information about Azure Hadoop, please refer to [20]. To acquire more knowledge about Apache Hadoop, please refer to [23]. Regarding the principle of Hadoop, please read [21].

5.3.3.2 Azure Spark Compared to the conventional Hadoop framework, Spark avoids the disk input/outputs (I/Os) and is able to deliver queries up to one hundred times faster than Hadoop. Spark is widely used in data analytics, machine-learning areas, and graph computations, as it provides a common execution model for tasks like extract-transform-load (ETL), batch queries, interactive queries, real-time streaming, machine learning, and graph processing.

Figure 5.24 depicts the framework of a general Spark, consisting of five components. Spark Core is the underlying general execution engine. Spark SQL allows users to query structured data in Spark programs, using either SQL or a familiar DataFrame API. Spark Streaming enables Spark to perform streaming analytics. MLlib is a machine-learning library, which implements some commonly used machine-learning algorithms, including classification, regression, decision trees, and so on. GraphX supplies a variety of graph algorithms (e.g., PageRank), connected components, and triangle count.

Hadoop stores its intermediate results on disk and thus is not suitable for iterative or interactive scenarios. To address this issue, Spark introduces an abstraction called *resilient distributed datasets* (RDD) [26, 27], which stores data in memory as larger size and longer time as possible. RDD is a read-only collection of objects containing a set of partitions across multiple machines. RDDs can form a directed acyclic graph (DAG) that reflects the dependency between them. If a partition is lost, it can be rebuilt by the information about how it was derived from other RDDs. RDDs support two types of operations: (1) transformations, which create a new dataset from an existing one; and (2) actions, which return a value to the driver program (the main program written by a user) after running a computation on the dataset. Transformations are computed lazily, which means they are not executed until an action operation is triggered.

Figure 5.25 gives an overview of a Spark cluster [25], which is composed of a driver program, a cluster manager, and multiple worker nodes. A driver program is the main program written by a user, in which the SparkContext object coordinates Spark

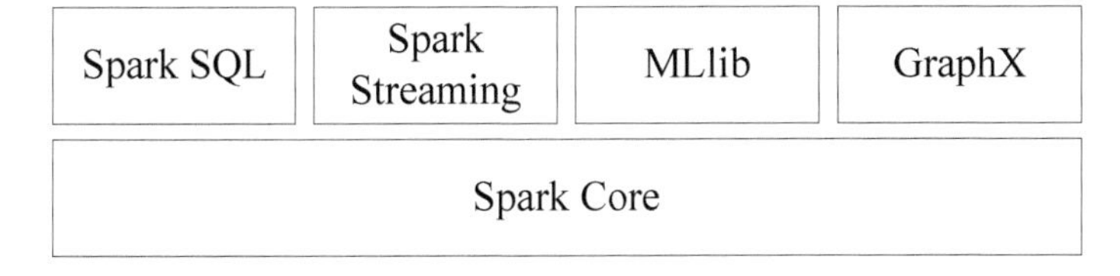

Figure 5.24
Components of Spark.

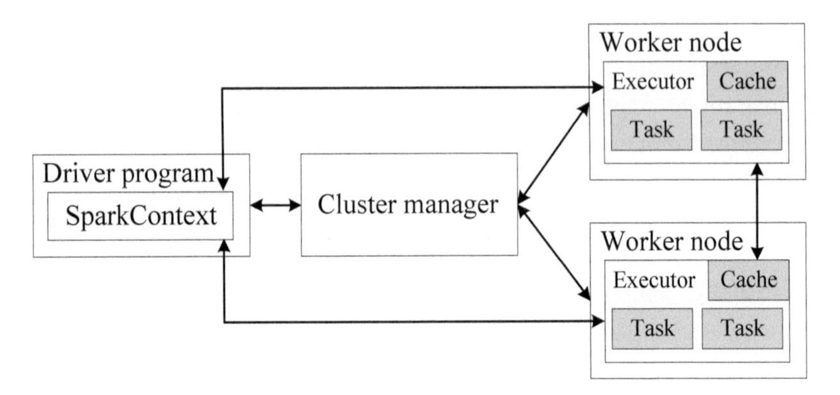

Figure 5.25
An overview of Spark Cluster.

applications running as independent sets of process on the cluster. A cluster manager is an external service for acquiring resources on the cluster (e.g., stand-alone manager, Mesos, or YARN). A worker node can launch multiple processes (a.k.a. executors) to run application code and cache data. A job is a parallel computation consisting of multiple tasks, which will be dispatched to different executors. Specifically, to run a job, first the SparkContext connects to the cluster managers. Once connected, SparkContext acquires executors on worker nodes through a cluster manager. Next, SparkContext sends application code to the executors. Finally, SparkContext sends tasks to the executors to run. Different executors in different worker nodes can communicate with each other. To get more information about Spark, we can look through the Apache Spark website [24].

HDInsight includes Apache Spark [24], an open-source project in the Apache ecosystem that can run large-scale data analytics applications. The Apache Spark provided in HDInsight is also referred to as Azure Spark. The benefits provided by Azure Spark include, but are not limited to, the following:

- *Ease of use*. We can create a new Spark cluster on HDInsight in minutes through the Azure Portal. Azure Spark also integrates many useful tools (i.e., Jupyter, Livy, and Ambari), enabling us to submit jobs, monitor running jobs, and manage the cluster in a convenient way.
- *Concurrent queries*. Azure Spark supports concurrent queries, enabling multiple queries from one user or multiple queries from various users and applications to share the same cluster resources.

- *Caching on solid status disks* (SSDs). We can choose to cache data either in memory or in SSDs attached to the cluster nodes. Caching in memory provides the best query performance but could be expensive; caching in SSDs provides a great option for improving query performance without needing to create a cluster of a size that is required to fit the entire dataset in memory.
- *High scalability and availability*. Although we can specify the number of nodes in our cluster during creation, we may want to grow or shrink the cluster to match the workload. All HDInsight clusters allow us to change the number of nodes in the cluster. Also, Spark clusters can be dropped with no loss of data since all the data is stored in Azure Storage. In addition, Spark on HDInsight comes with enterprise-level 24-7 support and an SLA of 99.9 percent uptime.
- Azure Spark also supports a number of business intelligence tools, such as PowerBI, Tableau, Qlikview, and SAP Lumira, making Azure Spark an ideal platform for data analysts, business experts, and key decision-makers.

To use Azure Spark, we first create an HDInsight cluster in the Azure Portal according to the standard procedure of creating HDInsight, selecting Spark as the type of cluster. We can then use Jupyter, which has been integrated in Azure Spark, to submit jobs in an active fashion, as shown in figure 5.26a. Azure Spark also integrates Apache Ambari by default, enabling users to provision, manage, and monitor its clusters, as illustrated in figure 5.26b.

Example. Following the example shown in figure 5.19, we want to count the POIs of each category using Azure Spark. Figure 5.27 demonstrates the implementation using the Python programming language and corresponding RDDs. First, an RDD (called a

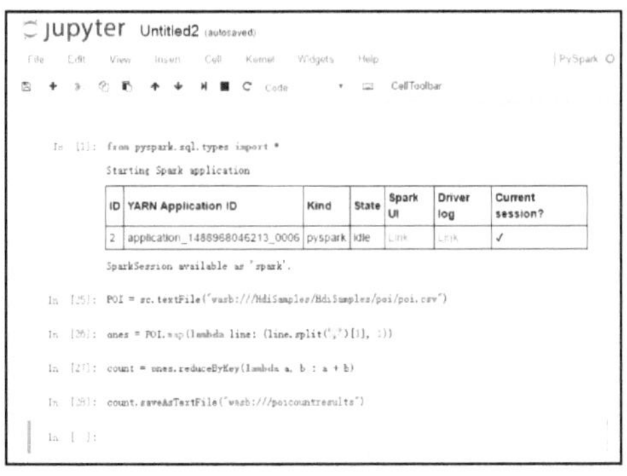

a) Jupyter in Azure Spark

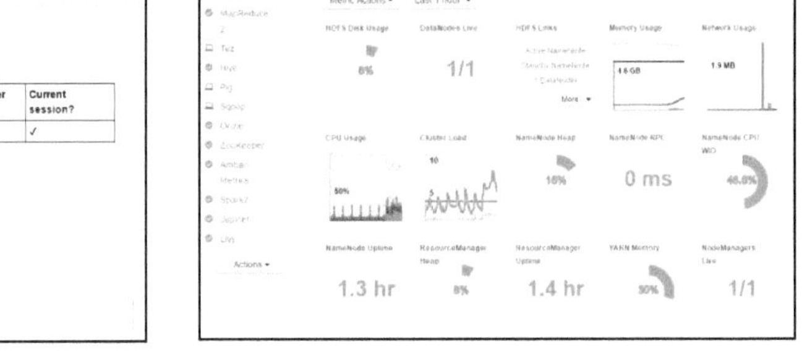

b) Ambari in Azure Spark

Figure 5.26
User interface of Jupyter and Ambari in Azure Spark.

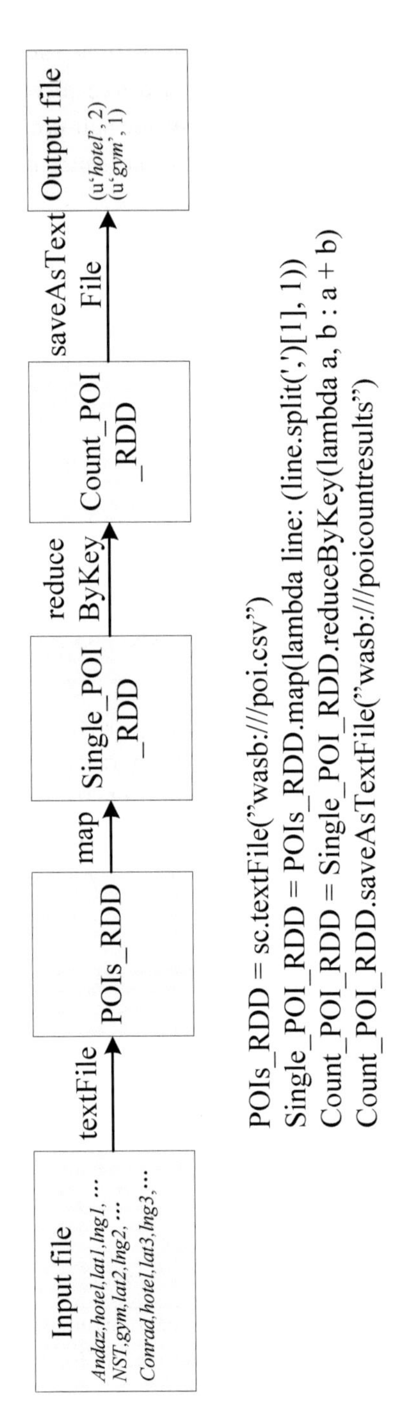

Figure 5.27
Counting the number of POIs using Spark.

POIs_RDD) is created by reading a file containing all POIs from a blob (using the textFile operation). Next, the POIs_RDD is transformed to a new RDD titled Single_POI_RDD, during which each line is mapped to a key-value pair <*category*, 1>. The key-value pairs are then added up based on their keys (i.e., values with the same key are summed up). Finally, we save the results to Azure Blob using saveAsTextFile. The textFile, map, and reduceByKey are transformations, while saveAsTextFile is an action. The former transformations are not executed until the action takes place.

5.3.3.3 Azure Storm Apache Storm [28] is a distributed, real-time event-processing solution for large, fast streams of data. It is the best option for processing real-time data and providing online services. The distributed logic in Storm is called a Storm *topology*, which is analogous to a MapReduce job. The key difference between them is that a MapReduce job eventually finishes, whereas a topology runs forever (or until we kill it).

As illustrated in the gray box in figure 5.28, a storm topology is a graph of spouts and bolts that are connected with streams. A stream is an unbounded sequence of tuples that is processed and created in parallel in a distributed fashion; a spout is a (logical) source of streams in a topology, which can run over multiple machines in a storm cluster. Generally, spouts read tuples from an external source (e.g., a message queue) and emit them into the topology. A bolt is a logical processing unit, the function of which can be defined by users, including filtering, aggregations, joins, talking to databases, and more. Each spout or bolt can execute multiple tasks in different machines of a storm cluster. Each task corresponds to one thread of execution, and the strategy of sending tuples between different tasks is defined by stream groupings. There are several built-in stream groupings in Storm that can be employed to deal with spatiotemporal data:

1. *Shuffle grouping*. Tuples are randomly distributed across the bolt's tasks in a way such that each bolt is guaranteed to get an equal number of tuples.
2. *Field grouping*. The stream is partitioned by the fields specified in the grouping. For example, if the stream is grouped by the "user-id" field, tuples with the same "user-id" will always go to the same task, but tuples with a different "user-id" may go to different tasks.
3. *Global grouping*. The entire stream goes to a single bolt's tasks.
4. *All grouping*. The stream is replicated across all bolts' tasks.

Apache Storm on HDInsight [29] is a managed cluster integrated into the Azure environment, which we call Azure Storm. It provides the following key benefits: First, Storm on HDInsight is more stable than Apache Storm. It performs as a managed service

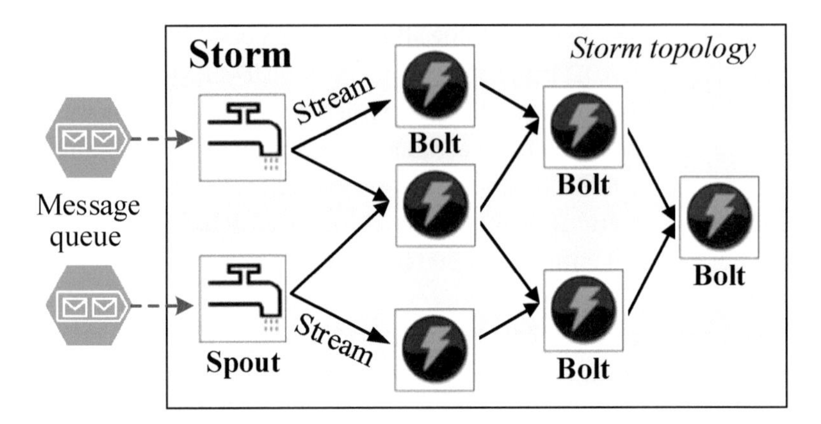

Figure 5.28
The conceptual framework of Storm.

with an SLA of 99.9 percent uptime. Second, we can choose a programming language, such as Java, C#, or Python, or even use mixed languages (e.g., read data using Java and then process the data using C#). Next, scaling an HDInsight cluster has no impact to running Storm topologies, which is important for the product environment. Last, we can integrate Storm with other Azure services, such as SQL Database, Azure Storage, and others.

To use Azure Storm, we first create an HDInsight cluster in the Azure Portal, selecting Storm as the type of cluster. Then we can use the integrated tools to submit jobs and monitor the cluster while running jobs. HDInsight integrates the Storm Dashboard, with which we can submit topologies and monitor the status of the Storm cluster. As shown in figure 5.29, there are two taps in the Storm Dashboard, Submit Topology and Storm UI. In the Submit Topology subpage, we can upload a new jar file (an executable file written in Java) containing a Storm topology and submit the topology after specifying the class name of the topology and additional parameters. In the Storm UI subpage, we can check the status of the cluster and topologies, view the output of a topology, and stop topologies.

Example. Following the example shown in figure 5.19, we use Azure Storm to count the number of POIs of different categories in users' check-in data. Each check-in record contains the category of a POI (as well as other information) and is constantly streamed into a message queue (e.g., Azure Queue storage) once created. Figure 5.30 presents the topology of Storm designed for this task, comprising four components:

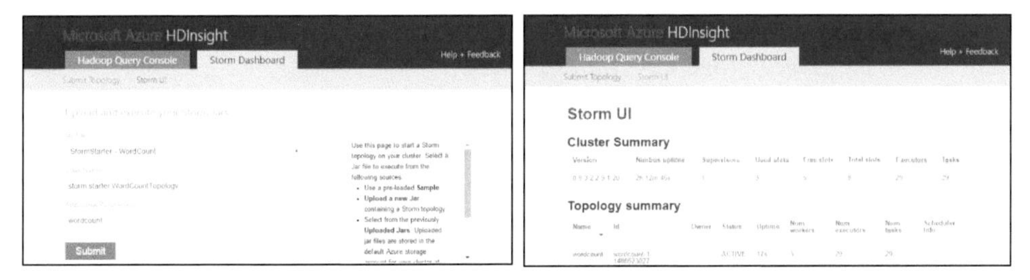

a) Submit topology b) Monitoring Storm status

Figure 5.29
The user interface of Storm Dashboard in Azure HDInsight.

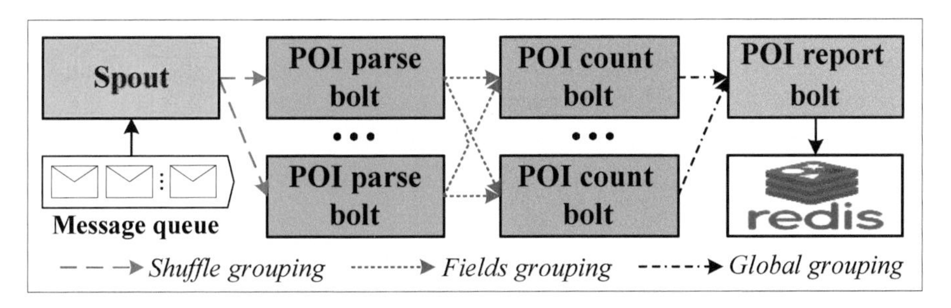

Figure 5.30
Counting POIs of different categories using Storm.

1. *Spout*. The spout reads the check-in records from the message queue and then emits them to POI parse bolts using the *shuffle grouping* mechanism (i.e., random assignment to achieve the workload balance).
2. *POI parse bolts*. The POI parse bolts receives check-in records from the spout, extracting the category of a POI from each record. The POIs are then emitted to POI count bolts using the *fields grouping* mechanism, which assigns POIs of the same category to the same bolt.
3. *POI count bolts*. The POI count bolts counts the POIs of each category it has received. Once it receives a POI, the POI count bolts will increase the count of the corresponding POI category by 1. Later, the counts of different POI categories are emitted to the POI report bolt, based on a *global grouping* mechanism (i.e., all data streams flow into one bolt).
4. *POI report bolt*. The POI report bolt caches the count results and writes them in a batch mode to Redis or Azure Table to reduce the high I/O overhead of frequent small writings.

5.3.3.4 Comparing Hadoop, Spark, and Storm Before comparing Hadoop, Spark, and Storm, let us recall their key features quickly.

Hadoop, an open-source distributed-processing framework, is used for storing huge volumes of data and to run distributed analytics processes on various clusters. Hadoop is efficient because it does not require big-data applications to transmit large volumes of data across a network. Another advantage of Hadoop is that the big-data applications keep running even if the clusters or individual servers fail. As Hadoop MapReduce has a limitation of batch processing one job at a time, Hadoop is mainly used in data warehousing rather than data analytics that need to access data frequently or interact with people instantaneously.

Spark is a data parallel, open-source processing framework that combines batch, streaming, and interactive analytics in one platform via in-memory capabilities. Spark facilitates ease of use, providing the ability to quickly write applications with built-in operators and APIs, along with faster performance and implementation. It also facilitates robust analytics, with out-of-the-box algorithms and functions for complex analytics, machine learning, and interactive queries. Spark also supports a variety of languages, such as Java, Python, and Scala.

Storm is a task parallel, open-source processing framework for streaming data in real time. Storm has its independent workflows in topologies (i.e., DAGs). A topology is composed of spouts, bolts, and streams, where a spout acts as a data receiver from external sources and a creator of streams for bolts to support the actual processing. The topologies in Storm work continuously until there is a flaw, or the system shuts down. Storm does not run on Hadoop clusters but uses Zookeeper and its own minion worker to manage its processes. Storm can read and write files to HDFS.

Similarities among Hadoop, Spark, and Storm include the following:

- All three are open-source processing frameworks.
- All of these frameworks can be used for business intelligence and big-data analytics, though Hadoop can only be used for analyzing historical data with no instantaneously responding requirement.
- Each of these frameworks provides fault tolerance and scalability.
- Hadoop, Spark, and Storm are implemented in Java virtual machine (JVM)-based programming languages: Java, Scala, and Clojure, respectively.

Table 5.1 presents a comparison between Hadoop, Spark, and Storm.

Table 5.1
A comparison of Hadoop, Spark, and Storm.

Features		Hadoop	Spark	Storm
System framework	Implementation language	Java	Scala	Clojure
	Cluster coordinating component	YARN	Stand-alone/ YARN/ Mesos	Zookeeper
	Storage solution	HDFS		
Programming model	Processing models	Batch-only	Batch/microbatch	Stream only
	Primitive	Writable	RDD, DStream	Tuple
	Data source	HDFS	HDFS, network	Spouts
	Computation/ transformation	Map, reduce, shuffle	Action, transformation, window operations	Bolts
	Stateful operations	Yes	Yes	No
	Interactive mode support	No	Yes	No
	Ease of development	Hard	Easy	Easy
	Language options	Any	Java, Scala, Python	Any
Reliability models	At most once	No	No	Yes
	At least once	No	No	Yes
	Exactly once	Yes	Yes	No
Performance	Latency of streaming processing		Seconds	Milliseconds
	Performance of batch processing	Low	High	
	Memory usage	Little	Huge	Little

Hadoop versus Spark When comparing Hadoop and Spark, people actually mean to compare Spark and Hadoop MapReduce (the processing engine for Hadoop), both of which use HDFS to store data in a reliable and secure manner. The differences between them are threefold.

First, Spark processes in-memory data, loading the process into memory and storing it for caching. Thus, Spark requires a lot of memory. Hadoop MapReduce, on the other hand, is limited to writing data to and reading data from disks after a map or a reduce action. Thus, Spark has a better efficiency than Hadoop MapReduce in analyzing data.

Second, Hadoop MapReduce is designed for batch processing one job at a time. Before a job is finished, Hadoop MapReduce cannot run another job. However, Spark, using its own streaming API rather than YARN for functioning, allows independent processes for continuous batch processing at short time intervals. Thus, Spark can be used for batch processing as well as real-time processing.

Third, not limited to data processing, Spark has an array of features for data analytics. For example, it can process graphs by using existing machine-learning libraries. It has a built-in interactive mode that allows people (or algorithms) to interactively modify a model's parameters based on the results generated by the model. Such a feature sufficiently supports the philosophy of machine learning, in which parameters are tuned iteratively based on the errors generated in each iteration. The built-in interactive mode can also enable interactive queries, where people can respecify their query conditions based on the results retrieved by previous queries. Spark also provides high-level APIs in Java, Scala, and Python.

Spark versus Storm When comparing Spark and Storm, we focus on the streaming aspects of each solution. The differences between them are fourfold.

First, Spark Streaming and Storm use different processing models. Spark Streaming uses microbatches to process events while Storm processes events one by one. As a consequence, Spark Streaming has a latency of seconds, while Storm provides a millisecond of latency. Spark's approach lets us write streaming jobs the same way we write batch jobs, facilitating us to reuse most of the code and business logic. Storm focuses on stream processing (some refer to it as complex event processing), the framework of which uses a fault-tolerant approach to complete computations or to pipeline multiple computations of an event as it flows into the system.

Second, Spark Streaming provides a high-level abstraction called a *discretized stream*, or DStream, which represents a continuous sequence of RDDs. Storm works by orchestrating DAGs in a framework, called *topologies*, which describes the various transformations or steps that will be taken on each incoming piece of data as it enters the system. The topologies are composed of streams, spouts, and bolts.

Third, Storm offers at-least-once processing guarantees. That is, each message is processed at least once, but there may be duplicates in some failure scenarios. Storm also offers at-most-once processing guarantees, in which each message is processed no more than once. Storm does not guarantee that messages will be processed in order (to achieve exactly-once, stateful, in-order processing, Storm has introduced an abstraction called Trident [30], but it is beyond the scope of this book). Inversely, Spark Streaming yields perfect, exactly-once, in-order message delivery.

Fourth, different from Spark Streaming, which only supports Java, Scala, and Python, Storm has very wide language support, giving users many options for defining topologies.

Storm versus Hadoop Basically, Hadoop and Storm frameworks are used for analyzing big data. They complement each other and differ in the following aspects:

First, Hadoop is an open-source framework for storing and processing big data in a distributed fashion where data is mostly static and stored in persistent storage on large clusters of commodity hardware. Storm is a free and open-source distributed real-time computation system, which works on a continuous stream of data instead of stored data in persistent storage.

Second, Storm adopts master-slave architecture with Zookeeper-based coordination. The master node is called *nimbus*, and slaves are referred to as *supervisors*. Hadoop takes master-slave architecture with or without Zookeeper-based coordination. The master node is a job tracker, and the slave node is a task tracker.

Third, Storm topology does not guarantee that messages will be processed in order and runs until it is shut down by users or experiences an unexpected and unrecoverable failure. On the contrary, MapReduce jobs in Hadoop are executed in a sequential order, and the process is exterminated as soon as the job is done.

5.4 Applications

In this section, we present some of Azure's platform-as-a-service (PaaS) components, called App Services, that have been widely used for developing web and mobile applications for any platform or device. These services enable developers to focus on their code and reach a stable, highly scalable production state quickly. They include Web Apps, Mobile Apps, and API Apps.

We can utilize these services via the following steps:

1. *Create an app* (e.g., web app, mobile app, or API app) in the Azure Portal or using Azure PowerShell, during which we specify the name of the service and the location at which to run. The app can be accessed using the following URL in any browser:

   "http(s)://[service name].[endpoint suffix]/",

   where [*service name*] is the name of the app, and [*endpoint suffix*] has different values according to the version of Azure (e.g., *azurewebsites.net* and *chinacloudsites.cn* represent Global Azure and China Azure, respectively).
2. *Create a web project* using an editor or integrated development environment (IDE). During the process, we select different templates for different app services. For example, we select an Azure API Apps template in Microsoft Visual Studio for an API app and an Azure Mobile Apps template for a mobile app.
3. *Deploy a project to Azure.* There are many ways to deploy a web project to Azure (i.e., in the Azure Portal, through Azure PowerShell, or using the IDE). Figure 5.31 presents an example of using Microsoft Visual Studio to publish a website project to

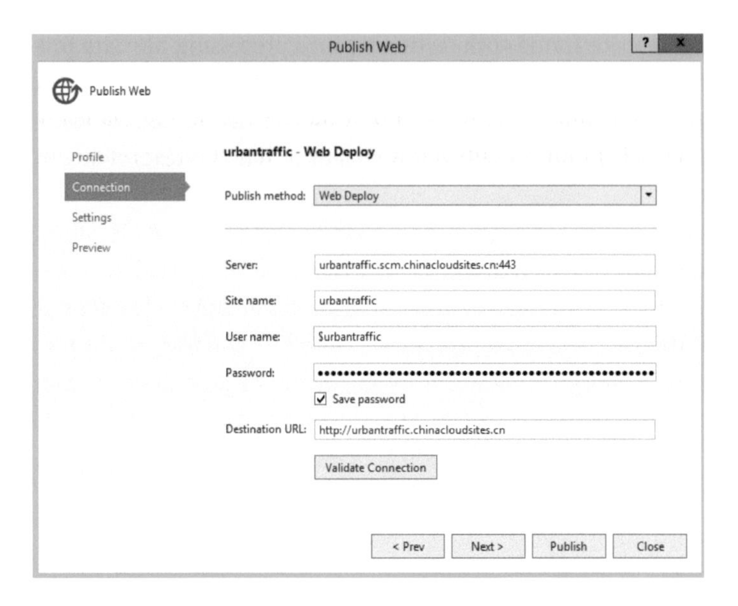

Figure 5.31
Publishing a website to Web Apps using Visual Studio.

the web app *http://urbantraffic.chinacloudsite.cn*. The necessary information, such as server, site name, user name, and password, can be acquired in the Azure Portal.

4. *Manage the app*. We can monitor, scale, and delete the app in the Azure Portal or by using Azure PowerShell. There are few significant differences between Web Apps, Mobile Apps, and API Apps in the underlying implementation, as they are all hosted on a web server (i.e., IIS [Internet Information Services] server). For example, we can even deploy a web project with an Azure API Apps template to a mobile app. The main difference is in the default features they provide. We will elaborate further in the following subsection.

5.4.1 Web Apps

Web Apps is a fully managed computing platform that is optimized for hosting websites and web applications. This PaaS offering from Microsoft Azure enables users to focus on their business logic while Azure takes care of the infrastructure to run and scale apps. The computing resources may be on shared or dedicated VMs, depending on the pricing tier that we choose. The application code runs in a managed VM that is isolated from other customers. A number of languages are supported for web applications, including ASP.NET, Node.js, Java, PHP, and Python. There could be multiple instances in a single web application, and all these instances share a common public IP address. The requests from users automatically balance the loads between them. Besides, this mechanism avoids a single point of hardware failure. The number of instances can be automatically changed according CPU usage. For more details about Web Apps in Azure, you can refer to [31].

5.4.2 Mobile Apps

Mobile Apps offers a highly scalable, globally available mobile application–development platform for enterprise developers and system integrators that brings a rich set of capabilities to mobile developers. As shown in figure 5.32, Mobile Apps provides the following features, which are important to cloud-enabled mobile development:

- *Client SDKs.* Mobile Apps provides a complete set of client software development kits (SDKs) that cover native development (iOS, Android, and Windows), cross-platform development (Xamarin for iOS and Android, Xamarin Forms), and hybrid application development (Apache Cordova). Each client SDK is available with an MIT (Massachusetts Institute of Technology) license and is open-source.
- *Data access.* Azure Mobile Apps provides a mobile-friendly OData v3 data source linked to SQL Azure or on-premises SQL Server. This service can be based on Entity Framework, allowing for easy integration with other NoSQL and SQL data providers,

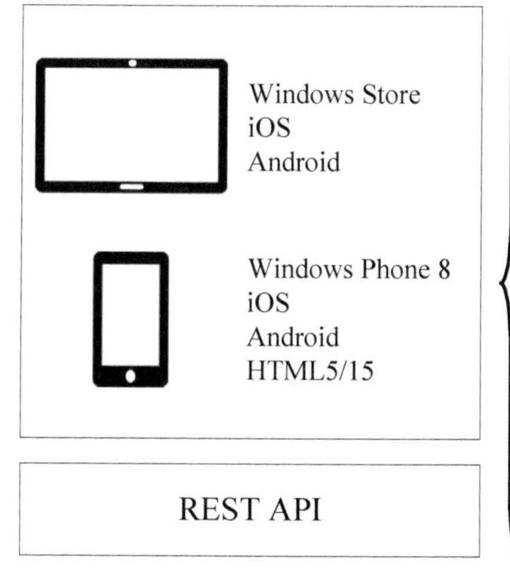

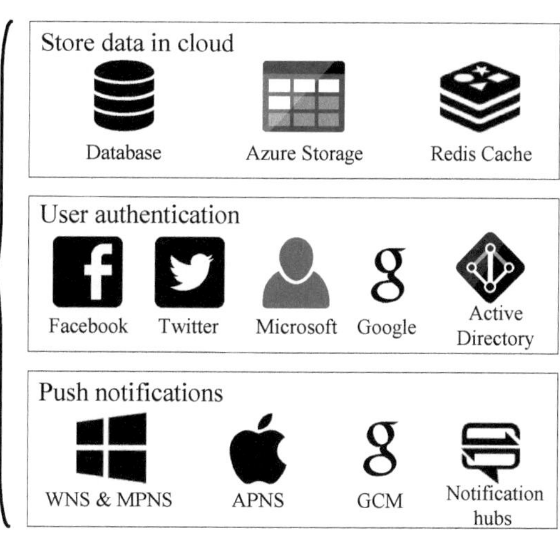

Figure 5.32
Functions of Mobile Apps.

including Azure Table storage, MongoDB, DocumentDB, and SaaS API providers like Office 365 and Salesforce.com.

- *Authentication and authorization.* Select from an ever-growing list of identity providers, including Azure Active Directory for enterprise authentication, plus social providers like Facebook, Google, Twitter, and Microsoft Account. Azure Mobile Apps provides an OAuth 2.0 service for each provider. We can also integrate the SDK for the identity provider for provider-specific functionality.
- *Push notifications.* Our client SDKs seamlessly integrate with the registration capabilities of Azure Notification Hubs, allowing us to send push notifications to millions of users simultaneously.

Like Web Apps, Mobile Apps can also be autoscaled, and it provides dedicated environments. To learn more about Mobile Apps, please refer to [32].

5.4.3 API Apps

API Apps offers features that make it easier to develop, host, and consume APIs in the cloud and on-premises. With API Apps we get enterprise-grade security, simple access control, hybrid connectivity, and automatic SDK generation. Here are some key features:

- *Bring your existing API as is.* We do not have to change any of the code in your existing APIs to take advantage of API Apps—just deploy your code to an API app. Your API can use any language or framework supported by App Service, including ASP .NET and C#, Java, PHP, Node.js, and Python.
- *Easy consumption.* Integrated support for Swagger API metadata makes your APIs easily consumable by a variety of clients. It automatically generates client code for your APIs in a variety of languages, including C#, Java, and JavaScript. It easily configures cross-origin resource sharing (CORS) without changing your code.
- *Simple access control.* Protect an API app from unauthenticated access with no changes to your code. Built-in authentication services secure APIs for access by other services or by clients representing users. Supported identity providers include Azure Active Directory, Facebook, Twitter, Google, and Microsoft Account. Clients can use the Active Directory Authentication Library (ADAL) or the Mobile Apps SDK.
- *Visual Studio integration.* Dedicated tools in Visual Studio streamline the work of creating, deploying, consuming, debugging, and managing API apps.

5.5 Summary

This chapter introduces the main components of cloud-computing platforms from the perspective of storage, computing, and application interfaces. We present the framework of each component and the general procedure for using it. When introducing detailed implementation of these components, we use Microsoft Azure as an example.

There are three main types of storage in Microsoft Azure: SQL Server, Azure Storage, and Redis. The first one is a good choice for storing structured data, while Azure Storage is fit for unstructured data storage. Azure Storage is further composed of four subcategories: Azure Blob, Table, Queue, and File. Redis Cache is designed for high throughput and low-latency data access.

There are three types of computing resources in Microsoft Azure: VMs, Cloud Services, and HDInsight. Users can do nearly everything with an Azure VM, just as if it were their personal computer or a high-performance server. Cloud Services can run highly available, scalable cloud applications and APIs as a standard web service or a background service, hosted on a VM. HDInsight is a distributed-computing component in Microsoft Azure to perform large-scale data preprocessing, management, and mining, containing the widely used Hadoop, Spark, and Storm.

There are three types of application interfaces provided by Azure. Web Apps is used to build and host websites and web applications, while Mobile Apps is used to build native mobile applications. API Apps makes it easier to develop, host, and consume APIs in the cloud. These components have empowered us to ensure the smooth, reliable implementation of urban-computing applications.

References

[1] Global Microsoft Azure. https://azure.microsoft.com/.

[2] *Wikipedia*. 2017. Microsoft Azure. https://en.wikipedia.org/wiki/Microsoft_Azure.

[3] SQL Database. https://docs.microsoft.com/en-us/azure/sql-database/.

[4] Jorgensen, Adam, Bradley Ball, Brian Knight, Ross LoForte, and Steven Wort. 2014. *Professional Microsoft SQL Server 2014 Administration*. Hoboken, NJ: John Wiley and Sons.

[5] Azure Storage. https://docs.microsoft.com/en-us/azure/storage/.

[6] Microsoft Azure Storage Explorer. http://storageexplorer.com/.

[7] *Understanding Block Blobs, Append Blobs, and Page Blobs*. 2017. Microsoft Docs. https://docs.microsoft.com/en-us/rest/api/storageservices/fileservices/understanding-block-blobs—append-blobs—and-page-blobs.

[8] *Azure Storage Scalability and Performance Targets*. 2017. Microsoft Docs. https://docs.microsoft.com/en-us/azure/storage/storage-scalability-targets.

[9] *Blob Service Concepts*. 2017. Microsoft Docs. https://docs.microsoft.com/en-us/rest/api/storage services/fileservices/blob-service-concepts.

[10] *Table Service Concepts*. 2017. Microsoft Docs. https://docs.microsoft.com/en-us/rest/api/storageservices/fileservices/table-service-concepts.

[11] *Queue Service Concepts*. 2017. Microsoft Docs. https://docs.microsoft.com/en-us/rest/api/storageservices/fileservices/queue-service-concepts.

[12] *File Service Concepts*. 2017. Microsoft Docs. https://docs.microsoft.com/en-us/rest/api/storage services/fileservices/file-service-concepts.

[13] Redis Cache. https://redis.io/.

[14] Redis Desktop Manager. https://redisdesktop.com/.

[15] Azure Redis Cache. https://azure.microsoft.com/en-us/services/cache/.

[16] Putty. http://www.putty.org/.

[17] *How to Configure Azure Redis Cache*. 2017. Microsoft Docs. https://docs.microsoft.com/en-us/azure/redis-cache/cache-configure.

[18] *Virtual Machines*. 2017. Microsoft Docs. https://docs.microsoft.com/en-us/azure/virtual-machines/.

[19] *Cloud Services*. 2017. Microsoft Docs. https://docs.microsoft.com/en-us/azure/cloud-services/cloud-services-choose-me.

[20] *An Introduction to the Hadoop Ecosystem on Azure HDInsight*. 2017. Microsoft Docs. https://docs.microsoft.com/en-us/azure/hdinsight/hdinsight-hadoop-introduction.

[21] Dean, J., and S. Ghemawat. 2008. "MapReduce: Simplified Data Processing on Large Clusters." *Communications of the ACM* 51 (1): 107–113.

[22] Apache Hive. https://hive.apache.org/.

[23] Apache Hadoop. http://hadoop.apache.org/.

[24] Apache Spark. http://spark.apache.org/.

[25] *Cluster Mode Overview*. 2017. Spark 2.1.0 document. http://spark.apache.org/docs/latest/cluster-overview.html.

[26] Zaharia, M., M. Chowdhury, M. J. Franklin, S. Shenker, and I. Stoica. 2010. "Spark: Cluster Computing with Working Sets." *HotCloud* 10:10.

[27] Zaharia, Matei, Mosharaf Chowdhury, Tathagata Das, Ankur Dave, Justin Ma, Murphy McCauley, Michael J. Franklin, Scott Shenker, and Ion Stoica. 2012. In *Resilient Distributed*

Datasets: A Fault-Tolerant Abstraction for In-Memory Cluster Computing. Proceedings of the 9th USENIX Conference on Networked Systems Design and Implementation. Berkeley, CA: USENIX, 2.

[28] Apache Storm. http://storm.apache.org/.

[29] *Introduction to Apache Storm on HDInsight: Real-Time Analytics for Hadoop*. 2017. Microsoft Docs. https://docs.microsoft.com/en-us/azure/hdinsight/hdinsight-storm-overview/.

[30] *Trident Tutorial*. 2017. http://storm.apache.org/releases/1.1.2/Trident-tutorial.html.

[31] *Web Apps Overview*. 2017. Microsoft Docs. https://docs.microsoft.com/en-us/azure/app-service-web/app-service-web-overview.

[32] *What is Mobile Apps?*. 2017. Microsoft Docs. https://docs.microsoft.com/en-us/azure/app-service-mobile/app-service-mobile-value-prop.

6 Managing Spatiotemporal Data in the Cloud

Abstract: Massive spatiotemporal data calls for advanced data management techniques on the cloud. However, current commercial cloud-computing platforms lack the capability to handle spatiotemporal data because of the data's unique structures and queries. This chapter introduces data management schemes respectively designed for six types of spatiotemporal data, which enable current cloud-computing platforms to manage massive and dynamic spatiotemporal data with minimal effort. For each type of data, four data management schemes are presented, based on whether a spatial or spatiotemporal index is employed and whether it is deployed on a distributed system. Instead of fundamentally rebuilding a new platform, existing resources and architecture, such as cloud storage and HDInsight, on current clouds are leveraged to create enhanced data management platforms for spatial and spatiotemporal data. The most advanced data management scheme (among the four schemes) is integrating spatial and spatiotemporal indexes (e.g., grid-based indexes, R-trees, and 3DR-trees) into distributed-computing systems, such as Spark and Storm in HDInsight. The advanced scheme combines the strengths of the two sides, empowering us to process a larger scale of spatiotemporal data more efficiently while using fewer computing resources.

6.1 Introduction

6.1.1 Challenges

There are many types of data, such as text, image, traffic, and meteorology, all of which are continuously generated by different sources in urban spaces. For better support of upper-level data mining and services, *large-scale* and *dynamic* datasets need to be managed *across different domains*. Cloud-computing platforms, such as Microsoft Azure and Amazon Web Services, are regarded as ideal infrastructure for large-scale and dynamic data. As most urban data is associated with spatial information and temporal properties (called *spatiotemporal data*), urban computing needs cloud-computing technology for managing spatiotemporal data.

- However, current cloud-computing platforms are not designed specifically for spatiotemporal data. The following are the primary challenges in handling spatiotemporal data for existing cloud-computing platforms: *Unique data structures.* Spatiotemporal data has unique data structures different from texts and images. For example, the size of a photo is fixed once taken, while the length of a trajectory keeps increasing as a taxi travels in a city. The trajectory data of a moving object continues streaming into the cloud with a sequential order that cannot be changed. We do not know when the trajectory terminates and how large it will be in advance. In addition, to support different queries, the storage mechanism for spatiotemporal data should be designed differently.

- *Different queries.* We usually query documents using key words. That is, we match a few key words against a collection of documents precisely or approximately. However, in urban computing, we usually need to handle spatiotemporal range queries or *k*-nearest neighbor queries over spatial temporal data. For example, finding nearby vacant taxis over a sixty-second time span is a spatiotemporal range query over taxi trajectory data. Likewise, searching for the nearest gas station while driving is a continuous 1-nearest neighbor query over points of interest (POIs). Existing components in the cloud cannot handle spatiotemporal queries directly and efficiently, as indexing and retrieval algorithms are not available in most cloud storage systems.
- *Hybrid indexing structures.* Traditional indexing and query algorithms are usually proposed to handle a single type of data. For instance, R-tree is proposed to index spatial point data, and inverted indexing is designed to handle text documents. In urban computing, however, we need to harness multiple datasets with different formats and updating frequencies across different domains. For instance, to study the correlation between traffic conditions, air quality, and POIs, we need to count the co-occurrences of the instances of the three datasets. Without a hybrid indexing structure, we need to use each and every instance from a dataset as a query to search for the instances (in the other two datasets) that are within a spatial and temporal distance to the query. This is a very time-consuming process that prevents interactive visual data analytics from happening. See section 4.5.3 for further details.
- *Integrating spatiotemporal indexing with the cloud.* When processing text and images, a cloud-computing platform mainly relies on a distributed-computing environment, such as Spark, Storm, and Hadoop. As the functions of indexing structures designed for text and images are quite limited, we barely see such indexes involved in a distributed-computing environment. For example, an inverted index is typically employed to maintain the relationship between words and the documents

that contain them. However, for spatiotemporal data, an effective indexing structure can improve the efficiency of a retrieval algorithm by one or several orders of magnitude. Combining the strength of spatiotemporal indexes and the distributed-computing system in a cloud can empower the cloud to process a larger scale of data more efficiently while using fewer computing resources.

Integrating spatiotemporal indexes into the in-parallel computing environment in a cloud is a nontrivial task, particularly when there are massive trajectories streaming into a cloud. For example, to find the vehicles traversing a sequence of road segments over a few minutes requires an organic integration of trajectory indexing structures into a Storm-based computing framework. Managing the trade-off between memory and input/output (I/O) throughput, conditioned on the in-parallel computing framework and indexing structures, is challenging, calling for knowledge from distributed systems, cloud storage, and indexing algorithms. Some indexing structures work very well in a stand-alone machine but may not be a good choice for a distributed environment considering the potential problems with index partitioning and updating.

6.1.2 General Data Management Schemes on the Cloud

Figure 6.1 presents four schemes for managing spatiotemporal data on cloud-computing platforms, based on whether an index is employed and if it is deployed on a distributed system:

1. *Single disk–based data management.* As shown in figure 6.1a, the simplest scheme is to store spatiotemporal data in a single virtual machine (or node) using a type of storage and then query the data directly based on the cloud storage. The cloud storage stores data on disk and may have a very simple index to reach a specific data instance. This scheme is not efficient enough to answer online queries.
2. *Single index–based data management.* As shown in figure 6.1b, to improve the efficiency of query, an index is built in advance over the raw data stored on disk. As the index stores only the identity and summary of data instances, its size is much smaller than the raw data. Thus, it can be held in the memory for efficient access. A query first accesses the index in the memory, retrieving the identities of data instances satisfying the query. Finally, these data instances' complete information can be retrieved from the disk. This scheme is much more efficient than the one shown in figure 6.1a for two reasons. First, the index prunes the search spaces significantly. Second, most computational processes are done in the memory, reducing the number of accesses to the disk.

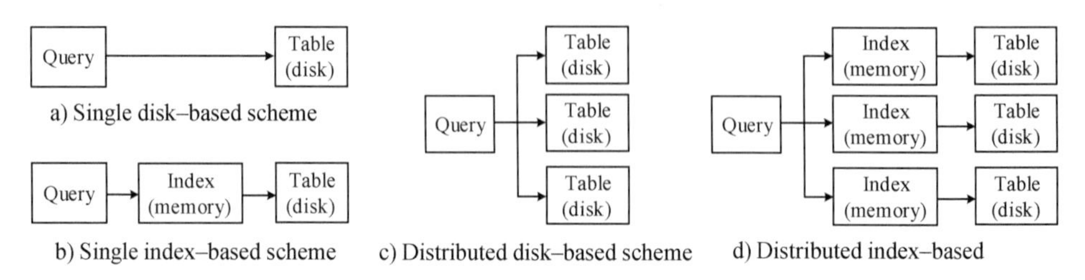

Figure 6.1
General data management schemes for spatiotemporal data on the cloud.

3. *Distributed disk–based data management.* When the scale of spatiotemporal data is large, it cannot be stored in a single machine's disk. Thus, as depicted in figure 6.1c, we partition the data into multiple portions, storing them on distributed cloud storage, such as the Hadoop Distributed File System (HDFS). Each node of distributed storage saves a partition of data on its own disk. When a query comes, it is sent to all nodes, searching for instances satisfying its criteria in each node based on the cloud storage. The search results are then returned to a master node and merged there. Though the data management scheme can handle large-scale data, it is not efficient enough to process queries that need instantaneous answers.
4. *Distributed index–based data management.* To address the issues mentioned above, as illustrated in figure 6.1d, an index is built over each partition of data and is loaded in the memory of each node. When a query comes, it is distributed to every node. In each node, the query first searches the index for a list of instance identities satisfying its criteria and then retrieves the complete information of these instances from the disk. In some cases, these instances' identities can be first sent back to a master node and merged there before we retrieve their complementary information from the disk. By combing indexing structures with distributed-computing systems, such data management schemes can handle instantaneous queries on large-scale spatiotemporal data efficiently.

In the following sections, we focus on introducing the latter three data management schemes, as single disk–based data management is very simple and can be regarded as a special case of the distributed disk–based scheme when only one machine is involved in storing data in a distributed system. Thus, we merge the two data management schemes and call them *disk-based data management* in the rest of the chapter. The implementation examples of these data management schemes are based on Microsoft Azure.

6.2 Managing Point-Based Data

6.2.1 Managing Point-Based Spatiotemporal Static Data

6.2.1.1 Disk-based data management

Storage and indexing Spatiotemporal static data, such as POIs, is associated with a fixed location and static properties. As illustrated in figure 6.2, given a dataset shown in figure 6.2a, we divide the spatial area into uniform grids (e.g., P_1, P_2, and P_3) and store the data by using Azure Table. As presented in figure 6.2b, each grid corresponds to a partition in Azure Table in which data instances share the same partition identity. An instance's identity is used as a row key, as shown in figure 6.2c. Using a partition key and a row key, Azure Table can pinpoint a specific instance, which is recorded by a row. A POI's properties, such as latitude, longitude, and category, are stored in other columns of the same row, right after the partition key and row key columns.

When the scale of the data is small, the single disk–based data management scheme (shown in figure 6.1a) can be employed. To handle large-scale point-based spatiotemporal static data, we generate more partitions by using a fine-grained spatial partition (i.e., grids with a smaller size). Azure Table automatically stores these partitions in different physical machines (users do not need to know where these partitions have been allocated). Thus, the distributed disk–based data management scheme is employed.

Spatial range queries Given a spatial range query (e.g., the red broken rectangle shown in figure 6.2b), we first find the grids (i.e., P_1, P_2, P_4, and P_5) that intersect or are within the range and then retrieve all instances within these partitions. A refinement process is further produced to find the instance (i.e., d_2) truly falling in the query's spatial range. To ensure search efficiency, the number of instances in a partition should not be too large (e.g., less than five hundred in Azure Table). Otherwise, the refinement process will be very time-consuming.

Updating When new data instances (denoted by red points in figure 6.2) arrive, as illustrated in figure 6.3a, we project them onto the grids and insert them into corresponding partitions in the table shown in figure 6.2c. For instance, d_7 belongs to partition P_5. A record with partition key P_5 and row key d_7 is inserted into the table shown in figure 6.2c. Azure Table ensures that the data instances with the same partition key are stored together physically. When the number of instances in a partition becomes very large, we can further split a grid into four subpartitions or even consider rebuilding the partitions.

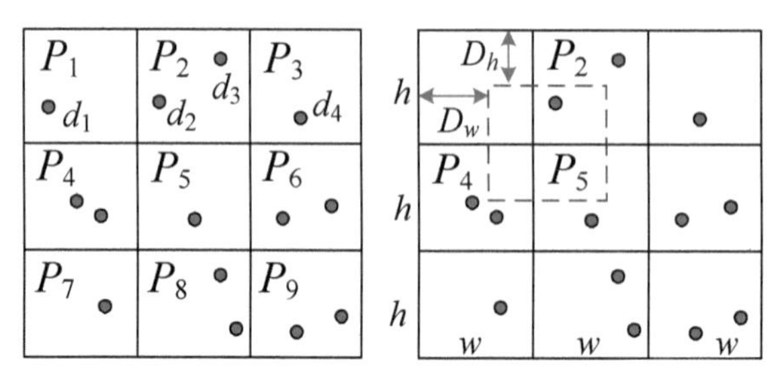

a) Point-based data b) Grid-based index

Partition key	Row key	Lat	Long	A_1	......	A_k
P_1	d_1			*Restaurant*		
P_2	d_2			*Shopping mall*		
P_2	d_3			*Movie theater*		
P_3	d_4			*Museum*		

c) Table-based storage

Figure 6.2
A data management scheme for spatiotemporal static data.

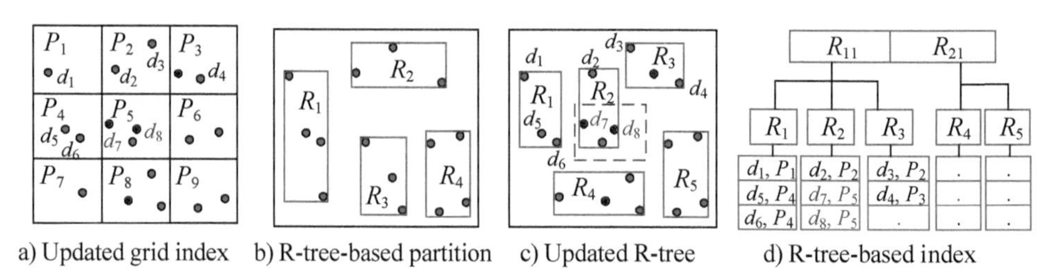

Figure 6.3
An example of an index-based data management scheme for spatiotemporal static data.

When the spatial partition is based on uniform grids, we can easily find the partitions intersecting or within a given spatial range without building any indexes. For example, we know the coordinates of the upper-left and bottom-right points of the entire spatial area. Thus, we can calculate the width W and height H of the spatial area and further compute the uniform width w and height h of a grid after the partitioning process ($w = W/n$, $h = H/n$, where n is the number of partitions on each dimension). Given a spatial range query, we quickly compute the horizontal distance D_w

between its upper-left point and that of the entire space. By calculating D mode w, we know the upper-left point of the range query falls in the first column. Likewise, by calculating D_h mode h, we know the query's upper-left point falls in the first row. Using the same method, we can determine the bottom-right point falls in the second column and second row (i.e., in P_5). As a result, we know P_1, P_2, P_4, and P_5 intersect the range query.

6.2.1.2 Index-based data management scheme When the spatial partition is not based on uniform grids (e.g., after splitting a grid into four parts or using an R-tree), a spatial index must be maintained in the memory to find the grids intersecting or within a spatial range query.

On one hand, spatial indexes improve query efficiency significantly. On the other hand, combining these indexes with cloud storage remains a challenge. For example, R-trees provide better efficiency in querying imbalanced spatial data than uniform grid-based indexing structures. However, its dynamic and data-driven structure poses challenges to integrating it into the cloud. As depicted in figure 6.3b, if employing R-trees to organize point-based spatiotemporal static data, we obtain a collection of minimum bounding rectangles (MBRs), each of which contains a few points. But, we cannot use such MBRs as partitions to store these points in Azure Table because the arrival of new instances may change the partition fundamentally. For instance, after inserting the four new red points, as illustrated in figure 6.3b, five new MBRs are built, and the original four MBRs shown in figure 6.3b disappear. Thus, we cannot build a distributed index–based data management scheme by simply applying indexes like R-trees to a distributed disk–based data management scheme.

Storage and indexing To address this issue, we build an R-tree-based spatial index for the data and maintain the index in the memory, as illustrated in figure 6.3d. Each leaf node of the R-tree stores a collection of points falling in its MBR. For each point, we record its identity and the spatial partition it falls in. For example, (d_7, P_5) under R_2 denotes that data instance d_7 belongs to MBR R_2 and is stored in partition P_5. The complete collection of data is stored in Azure Table, following the approach presented in figure 6.2. The R-tree-based index changes constantly in the memory as new instances arrive, while the grid-based spatial partitions do not change over time.

Spatial range queries When a spatial range query comes, as illustrated in figure 6.3c, we first search the R-tree for a few points—for example, (d_7, P_5) and (d_8, P_5)—that fall in the query's spatial range. More specifically, the search starts from the root node of the

R-tree, checking if a node's MBR overlaps the query's spatial range or not. If yes, the corresponding offspring node has to be further searched. Searching is done this way in a recursive manner until all overlapping nodes have been traversed. When a leaf node is reached, the contained points are tested against the query rectangle. The objects that lie within the query rectangle are returned as results. We then retrieve the full information about these points from Azure Table based on these points' identities and spatial partitions.

Updating When new data instances come in, we update the R-tree index in the memory, whose workload is light, and insert these instances into Azure Table. The structure of partitions in Azure Table does not change as new instances arrive because these partitions are solely for storage rather than search.

6.2.1.3 Distributed index–based data management When the scale of data is very large, the size of a spatial index becomes very large and thus cannot be held in a single machine's memory. To answer instantaneous queries over large-scale data, we need to incorporate indexing structures and distributed-computing systems into the data management scheme (i.e., the distributed index–based data management scheme shown in figure 6.1d).

Storage and indexing We first divide the entire spatial area into uniform grids, as illustrated in figure 6.2a. The data is stored in cloud storage (e.g., Azure Table) using the identity of the grid it belongs to as a partition key and its own identity as the row key, as shown in figure 6.2c. Cloud storage automatically stores different partitions on different physical servers (users do not need to care about it). This is the same as that presented in section 6.2.1.1.

We randomly divide a given dataset into many small portions, each of which has almost the same number of points. A spatial index (e.g., R-tree) is built for each portion of data and loaded in the memory of a machine in a distributed-computing system. Complete information on the data instances is stored in Azure Table, following the method introduced in section 6.2.1.1. The random data partition method ensures a balanced workload for each machine in a distributed-computing system. In addition, if we divide the data by spatial partition (i.e., each machine holds the spatial index for data from a spatial partition), nothing is returned if the machine corresponding to a given query is out of service.

Spatial range queries Figure 6.4 presents an example of answering spatial range queries based on Azure Storm. In this example, there are two types of bolt nodes. One type of bolt node (i.e., Bolt A_1 to Bolt A_n) in the second layer holds the spatial index (e.g., an R-tree) for a portion of data in the memory. The other type of bolt node (i.e., Bolt B) merges the results returned by Bolt A nodes.

Given a spatial range query, the spout node sends it to all Bolt A nodes. Once receiving the query, each Bolt A node searches its own spatial index for the data instances that satisfy the query. If there is no instance from a Bolt A node falling in the spatial range, "null" is returned to the Bolt B node. As the number of points falling in a range query may be very large, retrieving each of them respectively and individually from Azure Table is still a time-consuming process. To this end, the identities of data instances returned by Bolt A nodes are further aggregated in the Bolt B node. Complete information on instances from the same partition is then retrieved collectively by the Bolt B node. For example, as illustrated in figure 6.2a, d_2 and d_3 belong to the same partition P_2, and they form the partition. Thus, we can retrieve information on the entire partition rather than each row individually. Finally, the retrieved results are written into a blob file or in a block of memory (e.g., Redis) for other applications' usage.

The two-layer bolts have two advantages. One is reducing writing conflicts between different bolts when dumping results into a blob or a Redis. This is faster than using multiple bolts to simultaneously write a Redis. The other advantage is aggregating objects falling in the same partition so that these objects' information can be retrieved from Azure Table in a block manner rather than an instance manner. This expedites the efficiency of retrieving data from a disk.

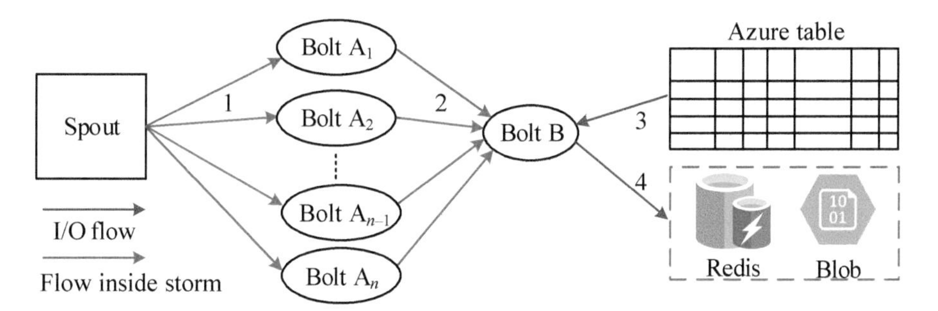

Figure 6.4
Answering spatial range queries based on Storm.

Updating When new data instances arrive, we project them onto grids and insert them into Azure Table using the corresponding grid identity as a partition key and its own identity as a row key. The new instances are assigned to the machines with the least number of points.

6.2.2 Managing Point-Based Spatial Static and Temporal Dynamic Data

This category of data is associated with a static location that does not change over time and dynamic readings that change over time constantly. Most geosensory data belong in this category, whose data structure is presented in figure 4.7.

6.2.2.1 Disk-based data management

Storage and indexing Similar to the disk-based data management scheme for spatiotemporal static data (like POIs), we divide the spatial area into uniform grids (e.g., P_1, P_2, and P_3) as shown in figure 6.5a. Each and every object's spatial information and meta data (such as with d_1, for example) are stored in an Azure table, as illustrated in figure 6.5b. In this table, each row denotes an object. The identity of the grid that an object belongs to is used as the partition key, and the object's identity is employed as the row key. An object's properties, such as latitude, longitude, and category, are stored in other columns of the same row, right after the partition key and row key columns.

Each object's dynamic readings are stored in a separate table, as shown in figure 6.5c, where a row denotes a reading generated at a time stamp. If there are ten objects, ten tables are created for them, respectively. In a table, the partition key is the date (e.g., 2017-08-23) that a reading is generated, and the row key is the time of day (e.g., 8:05:26) the reading is generated. The columns right after the partition and the row keys stand for the specific values of a reading (e.g., temperature and wind speed). Readings of an object are inserted into a table chronologically. When a new reading comes, it is appended to the end of a table. To handle a large-scale dataset, Azure Table

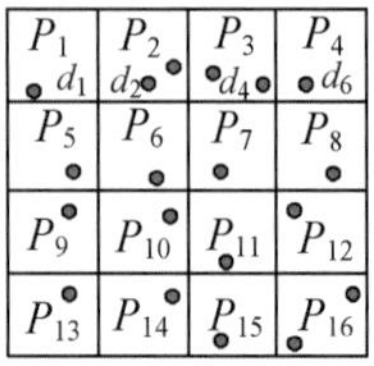

P_1 d_1	P_2 d_2	P_3 d_4	P_4 d_6
P_5	P_6	P_7	P_8
P_9	P_{10}	P_{11}	P_{12}
P_{13}	P_{14}	P_{15}	P_{16}

a) Spatial partitions

Partition key	Row key	Lat	Long	A_1	----	A_k
P_1	d_1					
P_2	d_2					
P_2	d_3					
P_3	d_4					

b) Table-based storage for metadata

Partition key	Row key	V_1	------	V_m
Date	Time			
Date	Time			
Date	Time			
Date	Time			

c) Storage for d_4's dynamic readings

Figure 6.5
A disk-based data management scheme for spatial static and temporal dynamic data.

automatically stores these temporal partitions in different physical machines (users do not need to know where these partitions have been allocated). Thus, the distributed disk–based data management scheme is actually employed.

Spatiotemporal range queries The procedure for answering a spatial range query is the same as that for spatiotemporal static data, which has been described in section 6.2.1.1. When a spatiotemporal range query arrives, we first search the table presented in figure 6.5b for the objects that are within the given spatial range. We then search the table shown in figure 6.5c for the readings within the given temporal range. For instance, we can retrieve d_4's readings from 20:30 on Jan 23, 2017, to 23:00 on Jan 24, 2017, by executing the following query in d_4's table:

"PartitionKey ge '20170123' and PartitionKey lt '20170124' and RowKey ge '20170123203000' and RowKey lt '20170124230000'",

where *ge* means a string is greater than or equal to another one, and *lt* denotes a string is less than another one, according to the lexicographical order. Including an entity's partition key in its row key can facilitate answering range queries that cross multiple partitions. When readings are generated at a high frequency, we can use *Date+Hour* (e.g., "20170102320") as the partition key and *Minute+Second* (e.g., "20170123203000") as the row key. Refer to section 5.2.2.2 for details.

Updating When new objects arrive, we can project them onto the grids and insert them into corresponding partitions in the table shown in figure 6.5b. A new table like that shown in figure 6.5c is created to store a new object's dynamic readings. New readings of an existing object are appended to the end of the object's table, like that shown in figure 6.5c.

6.2.2.2 Index-based data management When the scale of data is large, the disk-based data management scheme cannot answer queries efficiently. In addition, when data is not uniformly distributed in geographical spaces, we need to employ other spatial partition methods (e.g., R-trees). Thus, an index is built and maintained in the memory to expedite query processing.

Storage and indexing Similar to the disk-based data management scheme introduced in figure 6.5, we partition a spatial area into grids, storing all objects' metadata and spatial information into an Azure table. The dynamic readings of an object are stored in a separate table with each row denoting a reading. An R-tree-based index is created to

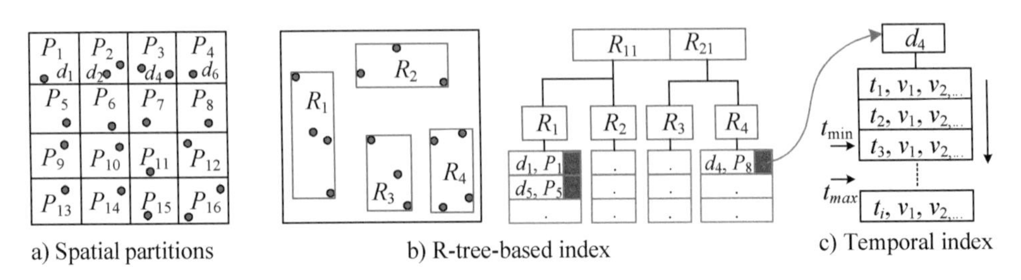

Figure 6.6
Index-based data management for spatial static and temporal dynamic data.

manage objects based on their spatial information, as illustrated in figure 6.6b. Each leaf node of the R-tree contains a list of objects, each of which is associated with an anchor pointing to the object's temporal index, as depicted in figure 6.6c. The temporal index can be a chronologically sorted dynamic array or a B+ tree (refer to section 4.4.1 for details). As the size of the memory is much smaller than the disk while readings are generated constantly, we can only hold the most recent readings in the memory and store historical data in the table.

Spatiotemporal range queries For a spatiotemporal range query, we first search the spatial index for objects within the query's spatial range. The process is the same as searching for POIs, as illustrated in figure 6.3. We then search for these objects' temporal indexes, respectively, for the readings within the given temporal range $[t_{min}, t_{max}]$. If a temporal index is based on a chronologically sorted array, we search for t_{min} and t_{max} in the array, respectively, through two binary search processes. As illustrated in figure 6.6c, the dynamic readings between t_{min} and t_{max} are returned. Because the spatial and temporal indexes are held in the memory, the search processes are very efficient. When objects' metadata are needed, they are retrieved from the table shown in figure 6.5b, based on these objects' partition keys and row keys.

When a given temporal range is concerned with a historical time interval whose data is not stored in the memory, we search the table (as shown in figure 6.5c) that saves an object's dynamic readings for the results. The size of the data held in the memory depends on the applications and varies in computing resources that are affordable. As most queries are concerned with recent time intervals, we can find results efficiently in most cases.

Updating When new objects come, as illustrated in figure 6.6a, we project them onto the grids and insert their metadata into corresponding partitions in the table shown in figure 6.5b. A new table like that shown in figure 6.5c is created to store a new object's dynamic reading. The grid-based spatial partition does not change as new instances arrive. We then update the spatiotemporal index in the memory. In most cases, we insert new objects into corresponding leaf nodes according to their coordinates without changing the structure of the entire index. Sometimes, the arrival of new objects may cause the reconstruction of the entire spatial index (see figure 6.3c for an example). After that, a temporal index is created for each new object, pointed by an entry in the spatial index's leaf node.

New readings of an existing object are chronologically appended to the end of the object's table, like that shown in figure 6.5c. As the size of the memory is limited, we keep the most recent n readings in the memory for each object. The relatively old readings are flashed out of the memory when new readings arrive. But, every reading is recorded in a table on disk, using the time stamp when they are generated as the partition and row keys.

6.2.2.3 Distributed index–based data management When the scale of data is very large, the size of spatial and temporal indexes becomes very large and thus cannot be held in the memory either. To answer instantaneous queries on large-scale data, we need to incorporate indexing structures and distributed-computing systems into the data management scheme (i.e., the distributed index–based data management scheme shown in figure 6.1d).

Storage and indexing We first divide the entire spatial area into uniform grids, as illustrated in figure 6.7a, projecting objects onto these grids. Objects' metadata and dynamic readings are stored in cloud storage (e.g., Azure Table), as shown in figure 6.5b and c. Cloud storage automatically stores different partitions on different physical servers (users do not need to be concerned about it). This is the same as that presented in section 6.2.2.2.

A simple idea is to directly apply the single index–based data management scheme to a distributed-computing setting. More specifically, a node in the distributed system stores the information for a few partitions. For example, as shown in figure 6.8a, P_1, P_2, P_5, and P_6 belong to node S_1, and P_3, P_4, P_7, and P_8 belong to node S_2. In each node, a spatial indexing structure (e.g., R-trees) is built over the spatial information of the objects falling in the partitions the node contains. Each object's dynamic readings are respectively managed by a temporal index and stored in the memory of the node the object belongs to. Figure 6.7b presents the spatiotemporal indexing structure of node S_1, which is the same as that on a single machine.

Using Azure Storm as an example, as illustrated in figure 6.8a, a Bolt A node stores a spatiotemporal index like that shown in figure 6.7b. However, such a simple design

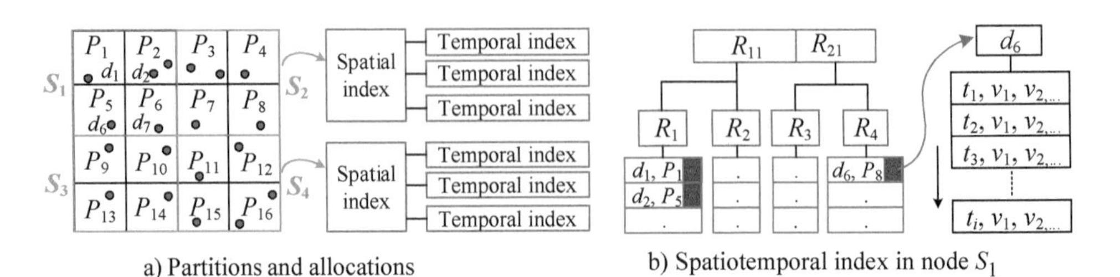

a) Partitions and allocations

b) Spatiotemporal index in node S_1

Figure 6.7
Simple distributed index–based data management.

has two disadvantages if data is highly imbalanced. First, some Bolt *A* nodes with many objects have insufficient memory to store these objects' dynamic readings, while some nodes with few objects waste memory. Second, during the search process, Bolt *A* nodes with many objects will be accessed more often by queries. They also need a longer time to return search results than a Bolt *B* node, thus delaying the entire query-processing procedure. For example, the merging process in a Bolt *B* node can be started only after all results have been received from its predecessor bolts.

To address these issues, an advanced data management method for integrating spatiotemporal indexes into a distributed-computing system is presented in figure 6.8b, using Azure Storm as an example. An R-tree-based spatial index consisting of all objects' spatial information is stored in a spout node, and objects' dynamic readings are evenly stored in different Bolt *A* nodes (i.e., partitioned by objects). Spatial and temporal indexes are separate rather than coupled in figure 6.8a. For example, Bolt A_1 stores the temporal indexes for the dynamic readings of objects d_1, d_2, and d_3, and those of d_4, d_5, and d_6 are stored in Bolt A_2. Different Bolt *A* nodes save the same number of objects' dynamic readings, thus having the same memory consumption and query load (assuming different objects have the same sampling rate). Because objects' spatial information is static, in most cases, we can accommodate them with a single spout node.

In case there are too many objects to manage (e.g., indexing the temperature sensors around the world), we can partition them by spatial area and build a spatial index for each spatial partition. To generate a balanced data distribution, the spatial partition is not necessarily based on uniform grids. We then employ multiple spout nodes in a Storm framework, each of which stores the spatial index of one partition, as illustrated in figure 6.8c. Data is partitioned by objects: each Bolt *A* node saves the temporal index of the same number of objects. This is the same as the method shown in figure 6.8b. Each spout node connects all Bolt *A* nodes in the second layer. These Bolt *A* nodes are finally connected to Bolt *B* for merging results.

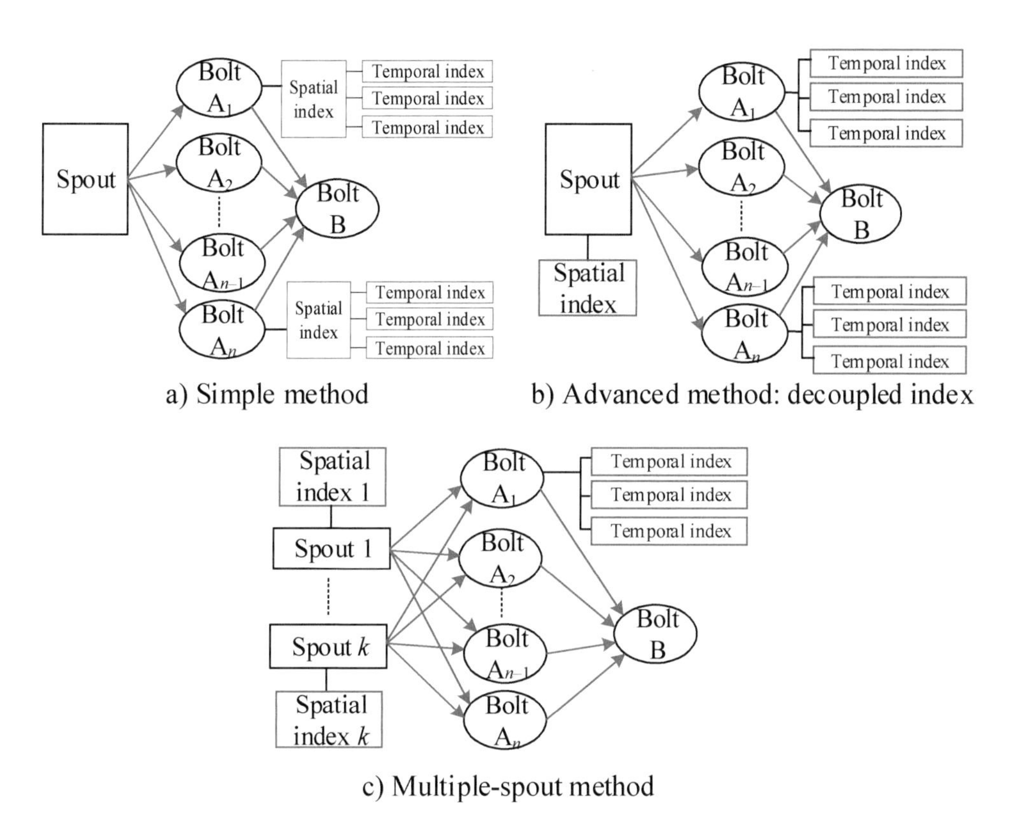

Figure 6.8
Advanced distributed index–based data management.

Spatiotemporal range queries When receiving a query, the distributed-computing system sends the query to all nodes, each of which first searches for objects within the query's spatial range and then seeks these objects' readings within the query's temporal range. The results are then aggregated and dumped into a file or a block of memory.

If the simple data management scheme shown in figure 6.8a is employed, the spout node directly distributes a query to every Bolt *A* node connecting to it. Once receiving the query from the spout node, each and every Bolt *A* node searches its own index for the objects and their readings falling in the query's spatiotemporal range. The identities and dynamic readings satisfying the query are further sent to the Bolt *B* node, which merges objects' identities and retrieves metadata from these objects from corresponding tables. The metadata and dynamic readings of these objects are finally written into a blob file or a Redis by the bolt for other applications' usage.

If the advanced method presented in figure 6.8b is employed, the spout node first searches its own spatial index for objects falling into the query's spatial range. It then distributes these objects' identities and the query's temporal range to all Bolt *A* nodes. If a Bolt *A* node contains the objects whose identities are included in the message from the spout, the Bolt *A* node searches for corresponding temporal indexes for dynamic readings falling in the query's temporal range. The search results and object identities are then passed to the Bolt *B* node, which merges the objects from the same spatial partition to expedite the retrieval process of objects' metadata from the table.

If the multiple-spouts method illustrated in figure 6.8c is employed, we send a query to all spouts. The spouts without overlaps with the query's spatial range return "null" to all Bolt *A* nodes. These Bolt *A* nodes will then do nothing and pass "null" to the Bolt *B* node. For spouts intersecting the query's spatial range, they first find the identities of objects falling in the range and then send these objects' identities to all Bolt *A* nodes. A Bolt *A* node may receive messages from multiple spout nodes, as it may contain objects falling in different spatial partitions. By checking the object identities, a Bolt *A* node can verify if an object's information is stored in it. If yes, the bolt node searches its temporal index for the dynamic readings falling in the query's temporal range. If not, the bolt will do nothing and pass "null" to the Bolt *B* node. The rest is the same as the two methods mentioned above.

Updating When new objects come in, we project them onto the grids and insert their metadata into corresponding partitions in the table shown in figure 6.5b. A new table like that shown in figure 6.5c is created to store a new object's dynamic reading. The spatial partitions do not necessarily change as new instances arrive. However, we update the spatial index in the spout node's memory, as illustrated in figure 6.8b and 6.3c. To balance workloads between different bolt nodes, a new object is assigned to a bolt node with the least number of objects. The bolt node's temporal index is then updated in the memory. The mapping between objects and bolt nodes is also updated.

New readings of an existing object are appended to the end of the object's table, as shown in figure 6.5c. We also update the temporal index in the corresponding bolt node's memory.

6.2.3 Managing Point-Based Spatiotemporal Dynamic Data

This category of data structure is designed for point-based data whose location and reading change over time. For example, in a crowd-sensing program, participants collect data in different locations and at different time intervals. Likewise, in a taxi-dispatching system, passengers submit ride requests from different places and at different time stamps.

Thus, in the aforementioned scenarios, each instance of data is associated with a location and a time stamp. Different instances are independent, having different locations and time stamps.

6.2.3.1 Disk-based data management

Storage and indexing As illustrated in figure 6.9a, we partition a geographical space into grids, which can be uniform grids or quadtree-based partitions. For each spatial partition (i.e., a grid), as illustrated in figure 6.9b, we create a table to store the information of spatiotemporal dynamic data instances falling in the spatial partition. For example, there are sixteen tables created for sixteen grids. As the geographical space is limited, the number of partitions is limited, leading to a limited number of tables. If the space is not uniformly partitioned, we record the spatial coordinates of each partition's upper-left and bottom-right corners.

In such a table, each row corresponds to a data instance. The time stamp when a data instance is generated is used to derive the instance's partition key and row key. More specifically, a coarse representation of a time stamp is used as a partition key so that instances generated at near time stamps are stored in the same partition. The complete representation of a time stamp combined with an object's identity is used as a row key. To avoid two data instances that are generated at the same time stamp, the identity of an object is appended to the end of the time stamp to formulate the row key for the object. For instance, given an instance d_1 generated at time stamp "2017-12-24 20:15:20," we can use "2017122420" as its partition key in a table and "20171224201520_d_1" as its row key. As time goes by, more partitions will be created in a table, which will be automatically stored in different physical machines by Azure Table.

The granularity of the partition key (i.e., the coarse time representation) can be predefined depending on applications (e.g., every hour or day). However, in some cases

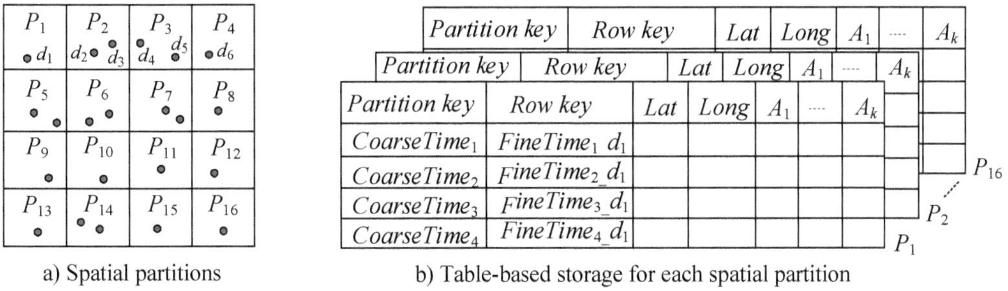

P_1 d_1	P_2 d_2 d_3	P_3 d_4 d_5	P_4 d_6
P_5	P_6	P_7	P_8
P_9	P_{10}	P_{11}	P_{12}
P_{13}	P_{14}	P_{15}	P_{16}

a) Spatial partitions

Partition key	Row key	Lat	Long	A_1	····	A_k
$CoarseTime_1$	$FineTime_1$_d_1					
$CoarseTime_2$	$FineTime_2$_d_1					
$CoarseTime_3$	$FineTime_3$_d_1					
$CoarseTime_4$	$FineTime_4$_d_1					

P_1 P_2 P_{16}

b) Table-based storage for each spatial partition

Figure 6.9
Disk-based data management scheme for spatiotemporal dynamic data.

(such as during major events or at hot spots), there may be many instances generated during a short time interval, leading to many instances stored in the same partition. This compromises the search efficiency of Table when answering range queries.

To address this issue, besides with a default value, the granularity of a partition key can change dynamically based on the number of instances received. For example, once four hundred data instances have been received, a new partition is created even if one hour has not passed. Before a partition is created in Table, these instances are buffered in the memory for a while. The common part of the first and the last data instances' time stamps is used as the partition key of the four hundred instances. For example, the first instance is generated at "2017-12-24 20:10:20," and the four hundredth instance is generated at "2017-12-24 20:19:20." The common part "20171224201" is used as the partition key. However, if the four hundredth instance is generated at "2017-12-24 20:28:20," the common part between it and the first instance is still "2017122420." Thus, the four hundred instances are inserted into the partition created previously (i.e., the strategy is not useful anymore). In this case, we need to divide the four hundred instances into two parts based on time, ensuring that each part of the instances has a consistently finer time representation than "2017122420." For example, we can set "20171224201" as the first part's partition key and "20171224202" for the second part of the instances. The dynamic setting ensures each partition in a table is not too big, resulting in good query performance.

Spatiotemporal range queries Given a spatiotemporal query, we first search for the spatial partitions that partially intersect or fully fall in the query's spatial range. The query's temporal range—(t_{min}, t_{max})—is then sent to the tables corresponding to these spatial partitions. In each of these tables, we find data instances whose time stamps are later than t_{min} but earlier than t_{max}, using the way presented in section 6.2.2.1. These instances are aggregated and returned as a final result.

Updating When new data instances arrive, we project them onto the spatial partitions and then insert them into corresponding tables according to their time stamps. Sometimes, we may receive instances much later than when they are actually generated because of the diversity of communication channels. This results in instances that arrived later having an earlier time stamp than instances received previously. In this case, we can insert such an instance into a previously created partition based on the instance's time stamp, generating a partition key and row key for it according to the partition's temporal granularity.

6.2.3.2 Index-based data management

Storage and indexing To answer instantaneous queries over massive point-based spatiotemporal dynamic data, we need to maintain a spatiotemporal index in the memory in addition to table-based storage. One out of the three indexing structures presented in figure 4.4 can be employed, depending on different applications.

If data is highly imbalanced, three-dimensional (3D) methods, such as 3D R-tree–based indexing structures, are employed. As illustrated in figure 6.10a, each data instance is represented by a point in a three-dimensional space, where time is regarded as the third dimension over the two-dimensional geographical space, based on its spatial coordinates and time stamp. Near instances are first grouped by small cuboids, such as R_{11} and R_{12}, which are further bounded by bigger cuboids like R_1. Figure 6.10b shows the tree structure of the index.

If the data has a balanced distribution in the geographical space, we can employ the indexing structure shown in figure 4.4c (i.e., partitioning the geographical space into disjointed grids and building a temporal index for each spatial partition).

When data has a uniform distribution over the temporal dimension, we can employ the indexing structure depicted in figure 4.4b. That is, we divide the time dimension into (uniform) intervals, building a spatial index for data at each time interval.

As memory size is limited, we keep the most recently received data in the index. Out-of-date data is flushed out of the index. If instances' metadata contains many fields, we only keep these instances' identities, spatial coordinates, and time stamps in the index. All data instances are inserted in tables and stored on disk in the same way presented in figure 6.9. The spatial partition created for storage does not change as new instances arrive.

Spatiotemporal range queries When working with a spatiotemporal range query, which formulates a query cuboid, we search the spatiotemporal index (e.g., the 3D

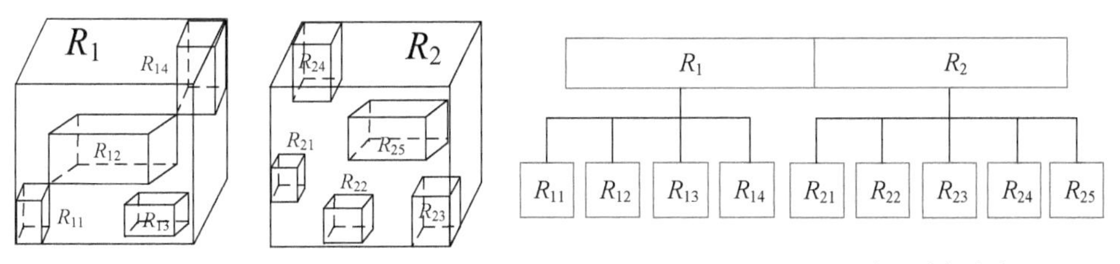

Figure 6.10
3D R-tree-based index for spatiotemporal dynamic data.

R-tree shown in figure 6.10) in the memory for cuboids partially intersecting or completely within the query cuboid. Instances from these cuboids are further checked to see if they really are contained in the query cuboid. After that, the identities of data instances retrieved by the query are aggregated based on spatial partition (i.e., table identity) and time stamps (i.e., partition keys in the table) so that they can be retrieved from the disk collectively and efficiently. Data instances from the same partitions in the same table should be retrieved collectively rather than individually. The latter is a very time-consuming process when the number of instances retrieved by a query is very big.

If a spatiotemporal range query is concerned with a (historical) time period that is not covered by the index in the memory, we need to search tables on the disk for results using the same method introduced in section 6.3.2.1.

Updating When new data instances occur, we project them onto the spatial partitions and then insert them into corresponding tables according to their time stamps, following the approach presented in section 6.2.3.1. In addition, we update the spatiotemporal index in the memory. In most cases, new instances are inserted into an existing spatiotemporal index's leaf nodes without changing the index's structure. This may gradually result in an inefficient index structure (e.g., a highly imbalanced 3D R-tree) that calls for a fundamental reconstruction of the entire index. As the construction of the index takes place in the memory and is independent of storage on disk, it can be done with minimal effort.

6.2.3.3 Distributed index–based data management When the scale of data is very large, the size of a spatiotemporal index becomes very large as well and thus cannot be held in the memory. To answer instantaneous queries on large-scale spatiotemporal dynamic point data, we incorporate indexing structures and distributed-computing systems into the data management scheme.

The general idea is to first divide data into many small partitions, respectively building a proper spatiotemporal index for each partition based on its data distribution (refer to section 6.2.3.2 for details). Each portion's index is maintained in the machine's memory, and the complete collection of data instances is stored on disk, following the approach introduced in section 6.2.3.1.

For example, using the framework shown in figure 6.8a, we randomly divide a given dataset into multiple partitions, each of which has (almost) the same spatiotemporal coverage. We then build a 3D R-tree for each partition and load the index into a Bolt A node's memory. When a spatiotemporal query comes, it is passed to all Bolt A nodes, each of which searches its own 3D R-tree for the data instances falling into the query range. These instances' identities are sent to the Bolt B node for aggregation. Finally, complete information about these data instances is retrieved from tables collectively and efficiently. The random partition makes a distributed-computing system involve all Bolt A nodes to answer a query, and the workload between different nodes is balanced.

We need to be careful if we want to use the framework presented in figure 6.8b and c, which separates the spatial index from the temporal indexes and deploys these indexes in spout and bolt nodes, respectively. When handling point-based spatial static and temporal dynamic data, which has a fixed location, the number of objects returned by the spatial index is limited. Thus, the messages containing these objects' identities sent from a spout node to a bolt node have a limited size. However, if applying these frameworks to spatiotemporal dynamic point data, the number of data instances retrieved from a spatial index can be huge. Thus, the communication cost between the spout and bolt nodes increases. In addition, for a bolt node, checking whether instances belong to it is time-consuming when the list of instance identities is very long.

An adjustment is to divide the geographical space into disjointed grids, based on either uniform grids or other spatial partition methods like quadtrees, and then build a spatiotemporal index (not a temporal index) for instances falling in each grid, respectively. The spatial index (about the grid partition) is held in the spout node, and each Bolt A node maintains the spatiotemporal index of a grid. When a spatiotemporal range query occurs, the spout node first searches its spatial index for grids completely falling in or partially intersecting the query's spatial range. The spout node then sends these grids' identities to all Bolt A nodes. If a Bolt A node finds a grid's identity belonging to it, it searches its own spatiotemporal index for instances completely falling in the query's spatiotemporal range. Otherwise, a Bolt A returns "null" to the second layer without doing anything. Note that each Bolt A node must maintain a spatiotemporal index rather than just a temporal index, even though a spatial index has been held in the spout node. As the grids partially intersecting a query's spatial range may contain instances that do not truly fall in the query's spatial range, bolt nodes maintaining only a temporal index can no longer filter these instances. Thus, it is not a very efficient indexing approach. In addition, the query workload may be carried on by very few bolt nodes. Once these bolt nodes are out of service, nothing will be returned.

6.3 Managing Network-Based Data

6.3.1 Managing Spatiotemporal Static Networks

This category of data structure stores spatial network-based data, such as road networks, which is represented by three types of substructures: nodes, edges, and an adjacent list. A node is a kind of spatiotemporal static point, similar to POIs. An edge consists of an identity, two nodes denoting its starting and ending terminals, a list of spatial points describing its shape, and metadata. The metadata includes the name, the number of lanes, the direction (bidirectional or one-way), the level, and a bounding box of an edge. The adjacent list represents the structure of a (directed) network, showing the neighboring edges of a given edge or the neighboring nodes of a given node. Once such a spatial network is built, the properties of these three substructures do not change over time (i.e., with static properties).

6.3.1.1 Disk-based data management

Storage and indexing Similar to the disk-based data management scheme for point-based spatiotemporal static data (like POIs), we divide the spatial space into disjointed grids (e.g., P_1, P_2, and P_3) as shown in figure 6.11a, projecting a spatial network's edge and nodes onto these grids. Information on nodes and edges is stored in two tables, respectively.

As presented in figure 6.11b, each row stores the information of one node, using the identity of the spatial partition the node falls in as its partition key and its own identity as the row key. The third and fourth columns of the table denote a node's latitude and longitude, respectively. The fifth column stores the identities of the edges that contain the node. For example, n_1 falls in spatial partition P_1 and is contained in edge e_1 and e_5. This is to facilitate the information retrieval of edges for a given node. Other properties of a node (e.g., indegree, outdegree, and level) are stored in the column from A_1 to A_k.

As illustrated in figure 6.11c, each row in this table denotes one edge. The identity of the spatial partition that the edge completely falls in or partially intersects is used as its partition key; the identity of the edge is the row key. The starting and ending nodes of an edge are stored in the third and fourth columns. Each edge is associated with a direction, either one-way (1) or bidirectional (0), as shown in the fifth column. When an edge is not represented by a straight line, a list of intermediate spatial points describing its shape are recorded in the sixth column. Other properties of the edge, such as speed constraints and the number of lanes, are stored in the remaining columns.

As an edge may cross multiple spatial partitions, it may be recorded in multiple partitions in this table. For example, e_1 crosses partition P_1 and P_2. As a consequence,

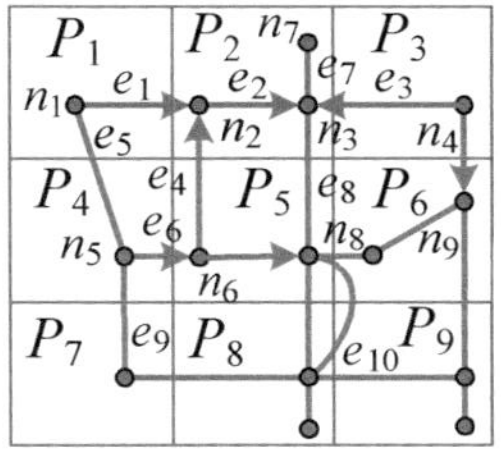

a) Spatial partitions

Partition key	*Row key*	*Lat*	*Long*	*Connected edges*	A_1	......	A_k
P_1	n_1			e_1, e_5			
P_2	n_2			e_1, e_2, e_4			
P_2	n_3			e_2, e_3, e_7, e_8			
P_3	n_4			....			

b) Table-based storage for network nodes

Partition key	*Row key*	*Adjacent nodes*
P_1	n_1	n_2, n_5
P_2	n_2	n_3
P_2	n_3	n_7, n_8
P_3	n_4	n_9

d) Adjacent lists for nodes

Partition key	*Row key*	NID_s	NID_e	*Dir*	*Shape points*	A_2	......	A_k
P_1	e_1	n_1	n_2	1			......	
P_1	e_5	n_1	n_5	0			......	
P_2	e_1	n_1	n_2	1			......	
P_2	e_2	n_2	n_3	1			......	

c) Table-based storage for edges in networks

Partition key	*Row key*	*Adjacent edges*
P_1	e_1	e_2
P_1	e_5	e_1, e_6, e_9
P_2	e_1	e_2
P_2	e_2	e_7, e_8

e) Adjacent lists for edges

Figure 6.11

Disk-based data management for spatiotemporal static network data.

two records are created in the table for e_1, using P_1 and P_2 as partition keys, respectively. This is necessary to ensure a complete search result for spatial range queries. For instance, when a spatial range query partially intersects e_1 in P_2 but does not cover P_1, if e_1 is only stored in P_1's partition, e_1 cannot be found. An extreme case is e_{10}, whose terminal nodes fall in P_5 and P_8, respectively, while its intermediate points belong to four partitions. If a spatial range query covers e_{10}'s part in P_6 and P_9, we cannot correctly return e_{10} as a result. Thus, when storing an edge, we go through its terminal nodes and intermediate spatial points, checking the partitions they fall into and then creating a record for the edge in each of these partitions, respectively.

Besides recording the information of individual nodes and edges, we maintain the structure of a given network from two perspectives. One is the connections between nodes. As illustrated in figure 6.11d, for each node, we record its one-hop neighbors the node can reach. For example, n_1 can reach n_2 and n_5. However, n_2 cannot reach n_1, as e_1 is a one-way edge.

The other perspective to maintain the structure of a given network is the adjacent relationship between edges, as shown in figure 6.11e. This table records each edge's one-hop neighbors that the edge can reach according to the edges' directions. For example, as e_1 and e_4 are one-way edges, we cannot traverse e_4 from e_1. Thus, e_4 is not stored in e_1's adjacent list.

Spatial range queries Given a spatial range query, we can find the nodes falling in the query's range by searching the table shown in figure 6.11b. This is the same as searching for POIs using a table-based data management scheme. We can then retrieve the information on edges containing a node from the table shown in figure 6.11c, based on the partition identity the node falls in and these edges' identities. To retrieve information efficiently from the edge table, an aggregation can be performed to merge edges returned from the node table. Edges from the same partition can be retrieved collectively, and edges with multiple records in different partitions will be retrieved once after the integration.

If the goal is to find edges completely falling in or partially intersecting a given spatial range query, we can search the table shown in figure 6.11c directly. The procedure is almost the same as searching for nodes. We first find spatial partitions completely falling in or partially intersecting the query's spatial range. We then retrieve the detailed information for each edge, particularly an edge's terminal nodes and intermediate nodes, checking whether these spatial points fall in the query's spatial range. If yes, the edge's identity is returned.

Updating When new edges are created, we insert them into the corresponding partitions in the node and edge tables, respectively. For the node table, we insert two new rows in the spatial partition where the new edge's terminals fall in. In most cases, an edge starts from or ends at an existing node. Thus, only one new record is inserted. In this situation, we need to update the record of the existing node, with the new edge's identity added to its adjacent edges field. In some extreme cases, an edge is added to connect two existing nodes. Thus, no new row is created in the node table, but we still need to update the adjacent edges field of the two existing nodes. We then update the adjacent list for nodes shown in figure 6.11d, adding the (new) node identity to the corresponding nodes' adjacent lists.

For the edge table, the new edge's record is inserted into the partitions the edge completely falls into or intersects. We also update the adjacent list for edges shown in figure 6.11e, adding the new edge's identity to the corresponding edges' adjacent lists.

6.3.1.2 Index-based data management Disk-based data management cannot provide an efficient query processing capability because of I/O cost. A more efficient method is to build a spatial index over spatiotemporal network-based data. The spatial index is held in the memory, storing key node and edge information. All information on the nodes and the edges is stored on disk (e.g., using the table-based data management scheme introduced in section 6.3.1.1).

Storage and indexing As shown in figure 6.12a, we divide the spatial space into disjointed grids (e.g., P_1, P_2, and P_3), projecting a spatial network's edge and nodes onto these grids. The spatial partition does not change over time just by storing data on disk. Complete information on the nodes and the edges is stored in two tables, respectively, as presented in figure 6.11b and c, following the method introduced in section 6.3.1.1. We then build a spatial index (e.g., an R-tree) over all spatial points in the data, as illustrated in figure 6.12b. The spatial points comprise nodes and intermediated points from edges. A leaf node stores information on several spatial points, consisting of the spatial point's identity, latitude, longitude, and partition identity it falls in. For example, node n_1 falls in spatial partition P_1. If a spatial point is an intermediate point in an edge, the edge's identity is also included (e.g., s_1 is an intermediate point describing the shape of e_{10}). The index is held in the memory and thus can be updated efficiently.

Besides the spatial index for nodes and edges, we also keep the adjacent list between edges in the memory (e.g., e_1: e_2, e_5 and e_2: e_1, e_7, e_8) as illustrated in figure 6.12c. Depending on applications, some important properties of the edge, such as length and speed

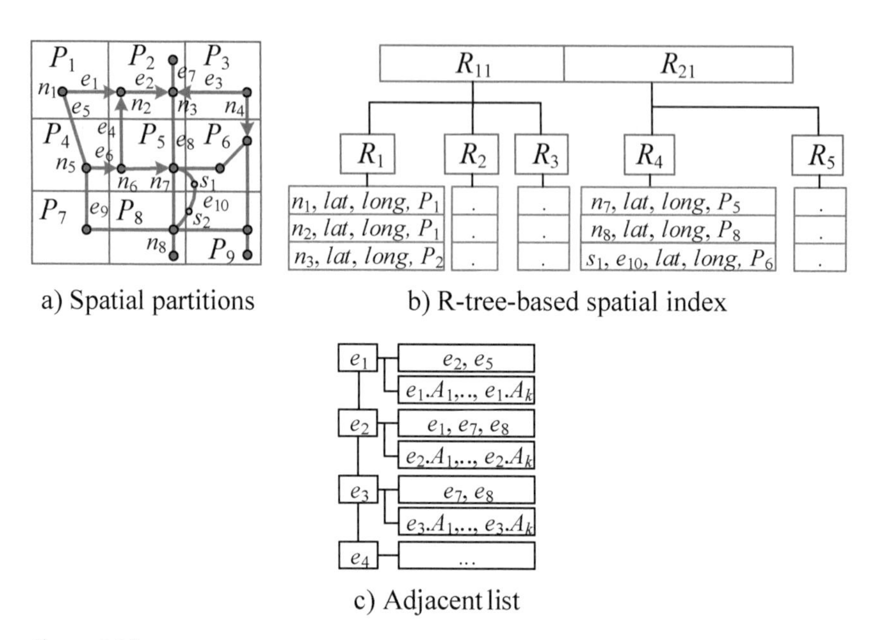

Figure 6.12
Index-based data management scheme for spatiotemporal static network data.

constraints, can be held in the memory too. The information can be attached to an edge's entry in the adjacent list or in a separate array. This is to facilitate applications like routing in a spatiotemporal static network and map-matching trajectories onto such a network.

Query processing When a spatial range query comes in, we search the spatial index for spatial points that fall in the query's range. As there could be multiple spatial points from the same edge returned, we merge the results by edge identity and partition identity. We then retrieve complete information on the nodes from the node table according to the partition identity and node identity. Likewise, we retrieve complete information on an edge from the edge table by its partition identity and edge identity. The *k*-nearest neighbor (KNN) query can also be handled by the spatial index. We do not go into further detail here (refer to section 4.3 for more).

Updating When a new edge is added to a given dataset, we can insert it into the corresponding tables and update the related fields, following the approach introduced in section 6.3.1.1. We then update the spatial index in the memory. In most cases, we can simply insert the spatial points from the edge into the existing leaf nodes of an R-tree based on its spatial coordinates. An entry is created in a leaf node for each newly added spatial point, with its identity, edge identity, partition identity, latitude, and longitude. Sometimes, we need to rebuild the entire tree when the inserted data causes a significantly imbalanced tree. Finally, we update the corresponding entries in the adjacent list and create a new entry for the newly added edge. As the two indexes are held in the memory, updating costs are affordable. In addition, the update occurs very infrequently because of the nature of such data. For example, the structure of a road network does not change constantly.

6.3.1.3 Distributed index–based data management

Storage and indexing When a given spatiotemporal static network dataset is extremely large, we cannot hold its index in a single machine's memory. Thus, we need to partition such network data into small partitions. Considering the load balance and communication cost, the problem of finding an optimal partition for a graph is NP-complete, which has been studied for many years. In addition, different applications require different optimal graph partition approaches. We do not drill down into this problem in this section, only considering a simple partition method for answering spatial range queries and KNN queries.

As illustrated in figure 6.13, we randomly divide a given dataset into (almost) uniform portions, each of which has a similar number of edges. We then build a spatial index for each portion of the data, following the approach shown in figure 6.12b. Each machine in a distributed system loads the index with a portion of the data. An edge can only be stored in one machine's indexes. Note that data is partitioned by edges rather than by spatial grids, as the latter could result in imbalanced workloads and data distribution in different machines. The adjacent list and properties of an edge are also stored in the same machine where the edge's spatial index is held. The storage for nodes and edges on disk is the same as that presented in section 6.3.1.1.

Query processing Using the Storm-based framework presented in figure 6.13 as an example, we demonstrate the procedure of answering a spatial range query or KNN query. When a spatial range query comes in, the spout node distributes the query to all Bolt A nodes, each of which searches its own spatial index for the results falling in the query's range. The Bolt B node then aggregates the results from all Bolt A nodes, merging the spatial points from the same edge. The results are then dumped into a blob file or a block of memory in Redis for other applications' usage. If it is a KNN query, the nearest edge's identity with the distance to the query point is returned by each Bolt A node. The Bolt B node then ranks these edges by their distances to the query point, finally returning the nearest one.

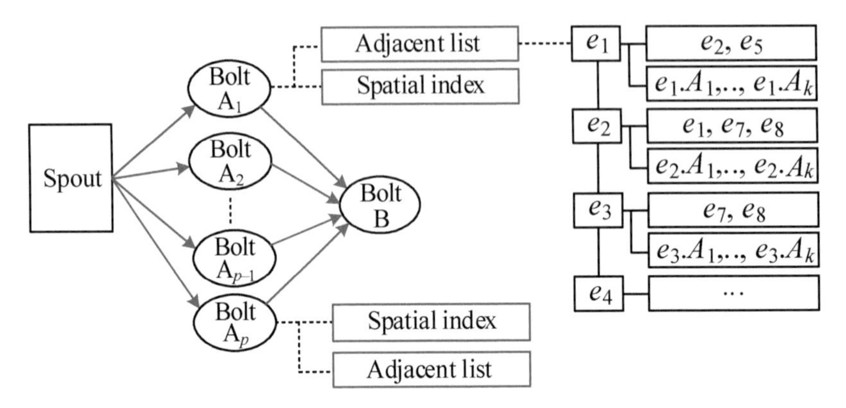

Figure 6.13
Distributed index–based data management scheme for spatiotemporal static network data.

When some aggregated results are needed (e.g., counting the total length of road segments in a given spatial range), each Bolt A node retrieves the corresponding information, such as the length of an edge, from the adjacent list. In this example, we only need the properties of an edge. The adjacent relation is not used. Each Bolt A node sums up the total length of the edges (from its index) that falls in the spatial range. The Bolt B node further aggregates the lengths returned by all Bolt A nodes.

If we want to count the number of road intersections in a given spatial range, each Bolt A machine searches its own spatial index for (network) nodes falling in the query's range. It then sends the number of nodes retrieved from its index to the Bolt B node for a further aggregation.

Updating When a new edge is added to a given dataset, we can insert it into the corresponding tables and update the related fields, following the approach introduced in section 6.3.1.1. The edge is added to the machine with the least number of edges. The spatial index and adjacent list of the machine are then updated, following the method introduced in section 6.3.1.2. The adjacent lists of existing edges that connect to the new edge are updated in corresponding machines.

6.3.2 Managing Network-Based Spatial Static and Temporally Dynamic Data

When dynamic readings are overlapped on a spatial network (e.g., a road network with traffic conditions changing over time), spatial static and temporal dynamic network data are generated. The constantly generated dynamic readings call for a new data management approach beyond that which is used for spatiotemporal static network data.

6.3.2.1 Disk-based data management method

Storage The method for storing the metadata of edges and nodes on a disk is the same as that introduced in section 6.3.1.1, which involves using two tables to store the static information of edges and nodes, respectively. Edges across two spatial partitions have duplicated records in the two partitions. There are another two tables to maintain the structure of a network according to the adjacent relationship of nodes and edges, respectively (see figure 6.11 for details).

Besides that, a table is created to store the dynamic readings of an edge, where each row saves the reading generated at a time stamp. The readings are ranked chronologically by their time stamps. The table is named after an edge's identity. The partition key and the row key are derived from the time stamp. For example, as shown in figure 6.14b, a table is created to store e_1's dynamic reading. If a reading is generated at 20:30:00 on Jan 23, 2017, we can set Jan 23, 2017 (i.e., "20170123"), as the partition key and "20170123203000" as the row key (refer to section 6.2.2.1 for details). Depending on the updating frequency, we can use different temporal granularities for partition keys. Alternatively, a table can be created for storing the dynamic reading of a node, named by the node's identity and using time stamps of readings as partition and row keys.

Spatiotemporal range queries When a spatiotemporal range query arrives, we first search the table presented in figure 6.11c for the edges that completely fall in or partially intersect the given spatial range. The procedure of answering a spatial range query is the same as that described in section 6.3.1.1. We then search these edges' tables (e.g., the one shown in figure 6.14b) for the dynamic readings within the given temporal range. For instance, we can retrieve e_1's readings from 19:00 on Jan 23, 2017, to 23:00 on Jan 24, 2017, by executing the following query in e_1's table:

"PartitionKey ge '20170123' and PartitionKey lt '20170124' and RowKey ge '20170123190000' and RowKey lt '20170124230000'",

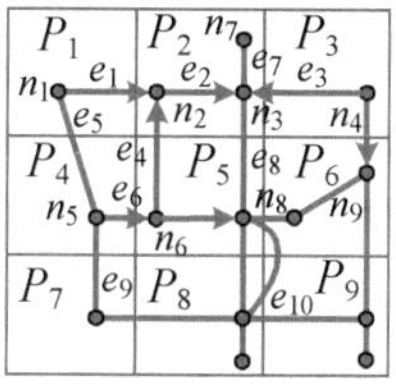

a) Spatial partitions

Partition key	Row key	V_1	......	V_m
Date	Time			
Date	Time			
Date	Time			
Date	Time			

b) Storage for e_1's dynamic readings

Partition key	Row key	V_1	......	V_m
Date	Time			
Date	Time			
Date	Time			
Date	Time			

c) Storage for n_1's dynamic readings

Figure 6.14
On-disk data management scheme for spatial static and temporal dynamic network data.

where *ge* means a string is greater than or equal to another one, and *lt* denotes a string that is less than another one, according to the lexicographical order. Including an entity's partition key in its row key can facilitate answering range queries that cross multiple partitions. Refer to section 5.2.2.2 for details about Azure Table.

Updating When new edges arrive, a new record is created in the edge table shown in figure 6.11c, following the method introduced in section 6.3.1.1. A new table like that shown in figure 6.14b is created to store a new object's dynamic readings. New readings of an existing edge are appended to the end of the edge's dynamic reading table, based on the time stamps when these readings are generated.

6.3.2.2 Index-based data management method

Storage and indexing To answer instantaneous queries over massive spatial static and temporal dynamic data, a spatial index (e.g., an R-tree, illustrated in figure 6.15a) is built over all spatial points from edges. The spatial points are composed of an edge's terminals and intermediate points. Each entry in the index's leaf node corresponds to a spatial point, containing the spatial point's identity, the identity of the edge it belongs to, latitude, longitude, and the spatial partition the point falls in. For example, node n_1 falling in spatial partition P_1 is a terminal of edge n_2.

A temporal index is built over the dynamic reading of an edge, respectively. The temporal index can be a chronologically sorted dynamic array, shown in figure 6.15b, or a B+ tree (refer to section 4.4.1 for details). As the size of the memory is much smaller than the disk while readings are constantly generated, we can only hold the most recent readings in the memory and store historical data in the table.

Spatiotemporal range queries When a spatiotemporal range query comes, we first search the spatial index presented in figure 6.15c for the spatial points that fall completely in the query's spatial range. As there are multiple spatial points at the same edge, the results containing the same edge identity are aggregated. Based on the returned edge identities, we can search these edges' temporal indexes, respectively, for the readings that fall within the query's temporal range.

Updating When new edges come, as illustrated in figure 6.15a, we project them onto the grids and insert their metadata into corresponding partitions in the node and edge tables, respectively, following the approach introduced in section 6.3.1.1.

We then update the spatial index in the memory. In most cases, we can simply insert the spatial points from the new edge into the existing leaf nodes of an R-tree based on

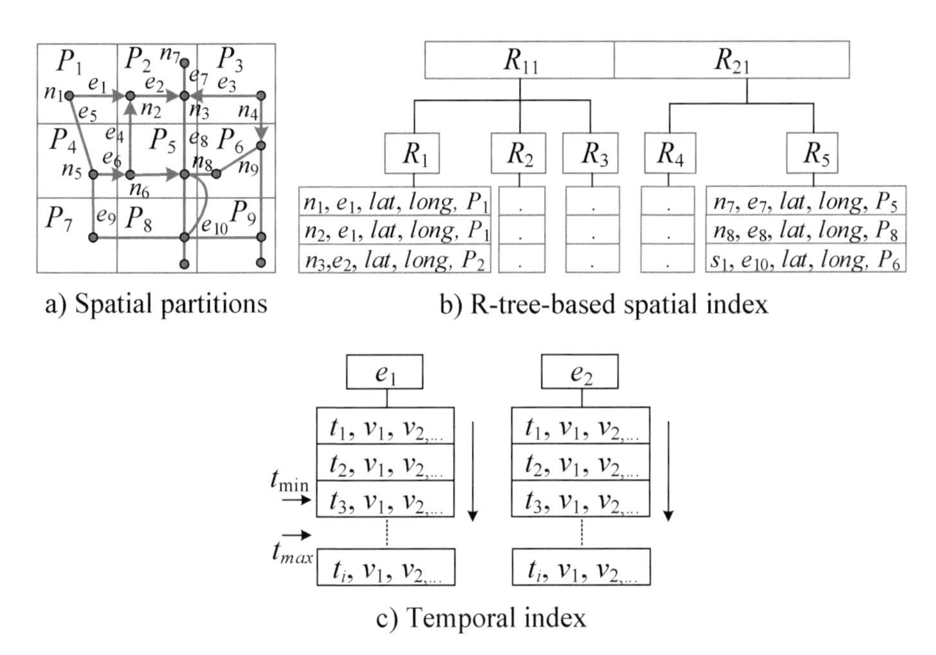

Figure 6.15
Index-based data management scheme for spatial static and temporal dynamic network data.

its spatial coordinates. An entry is created in a leaf node for each newly added spatial point, with its identity, edge identity, partition identity, latitude, and longitude. Sometimes, we need to rebuild the entire tree when the inserted data causes a significantly imbalanced tree. Finally, we update the corresponding entries in the adjacent list and create a new entry for the newly added edge.

New readings of an existing edge are chronologically appended to the end of the object's table, based on the time stamps when these readings are generated. As the memory size is limited, we keep the most recent n readings in the memory for each edge. The relatively old readings are flushed out of the memory when new readings arrive. But, all readings are recorded in the table on disk, based on the time stamp when they were generated as the partition and row keys.

6.3.2.3 Distributed index–based data management method

Storage and indexing When a given spatiotemporal static network dataset is extremely large, we cannot hold its index in a single machine's memory. Thus, we need to partition such network data into small partitions. As the problem of finding an optimal partition for a graph is NP-complete and depends on applications, we consider only a simple partition method for answering spatial range queries and KNN queries.

As illustrated in figure 6.16, we randomly divide a given dataset into (almost) uniform portions, each of which has a similar number of edges. We then build a spatial index for each portion of data, following the approach shown in figure 6.12b, and create a temporal index for each edge according to the method introduced in figure 6.15c. Each machine in a distributed system loads the spatial and temporal indexes of a portion of the data. An edge can only be stored in one machine's indexes. Note that data is partitioned by edges rather than by spatial grids, as the latter could result in imbalanced workloads and data distributions in different machines. The adjacent list and properties of an edge are also stored in the same machine where the edge's spatial index is held. The storage for nodes and edges on disk is the same as that presented in section 6.3.1.1.

Spatiotemporal range queries Using the Storm-based framework presented in figure 6.16 as an example, we demonstrate the procedure of answering a spatiotemporal range query. When such a range query occurs, the spout node distributes the query to all Bolt A nodes, each of which searches its own spatial index for the spatial points that completely fall in the query's spatial range. As there are multiple spatial points from the same edge, the results containing the same edge identity are aggregated in Bolt A nodes, respectively. Each Bolt A node then searches the temporal indexes of the edges returned by the spatial search for readings falling in the query's temporal range. The Bolt B node then aggregates the results from all Bolt A nodes and dumps the results into a blob file or a block of memory in Redis for other applications' usage.

Updating When a new edge is added to a given dataset, we can insert it into the corresponding tables and update the related fields, following the approach introduced in section 6.3.2.1. The edge is added to the machine with the least number of edges. The spatial index, temporal index, and adjacent list of the machine are then updated following the method introduced in section 6.3.2.2. The adjacent lists of existing edges that connect to the new edge are updated in corresponding machines.

6.3.3 Managing Network-Based Spatiotemporal Dynamic Data

This category of data's spatial and temporal information constantly changes over time, with a network structure between different data instances. There are two major subcategories: One is called *trajectory data*, which records traces of a moving object, such as vehicles, people, and animals. The other is called *spatiotemporal graphs*, denoting dynamic connections and interactions between different moving objects. We focus on managing trajectory data in this section because it has been widely used in many systems.

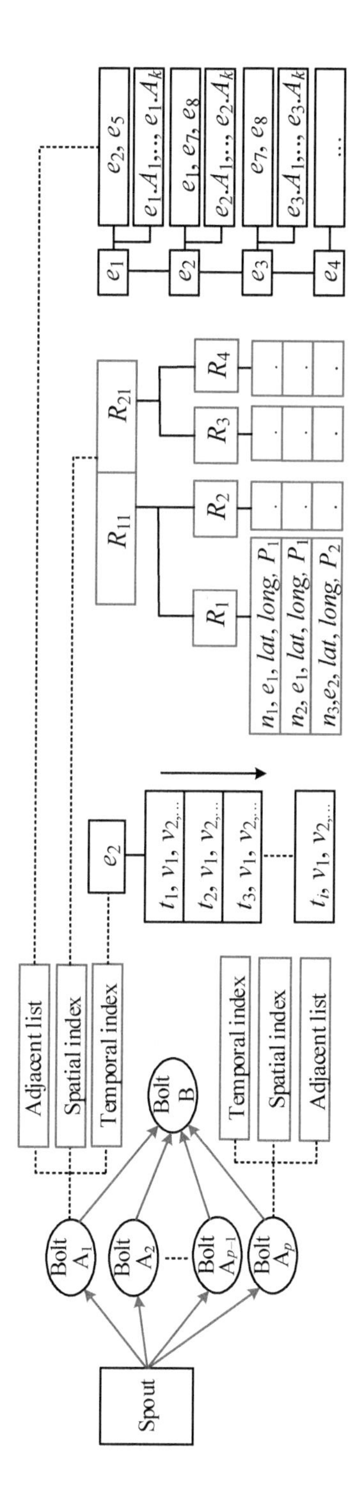

Figure 6.16
Distributed index-based data management for spatial static and temporal dynamic network data.

6.3.3.1 Disk-based data management method

Storage and indexing As illustrated in figure 6.17a, we partition the spatial space into disjointed grids, projecting trajectories generated by different moving objects onto these grids. The grids are not necessarily uniform, depending on the distribution of data. When a nonuniform grid partition is employed to handle imbalanced data, another table is created to store the spatial coordinates of each grid's bounding box.

For each moving object, we create an individual table, named according to the moving object's identity, to store its own data, as shown in figure 6.17b. In such a table, each row denotes the record of a spatiotemporal point from a trajectory, using the point's time stamp to derive its partition key and row key. For example, if a point was generated at 20:30:00 on Jan 23, 2017, we can set Jan 23, 2017 (i.e., "20170123"), as the partition key and "20170123203000" as the row key (refer to section 6.2.2.1 for details). The point's identity (*pid*), latitude, longitude, time stamp, and other properties are stored in the rest of the columns. Points from a trajectory are inserted into the table chronologically based on their time stamps, and the point identity is increased sequentially. Azure Table automatically stores different partitions in different physical machines in the cloud, trying to assign the close partitions to the same machine. Thus, the size of a table can be huge. Users do not need to pay attention to the low-level storage mechanisms on the disk.

For each spatial partition, we create an individual table to store the points (from different moving objects) falling in it, as presented in figure 6.17c. Each row in this table stores the information of a point generated by a moving object. A point's time stamp is employed to derive its partition key and row key, following the approach mentioned above. As there could be multiple moving objects generating a record at the same time stamp, we append an object's identity to the end of a time stamp to formulate a unique row key for a point. The identity of a moving object generating a point and the point's identity, latitude, longitude, time stamp, and other properties are stored in the rest of the columns. A trajectory crossing multiple spatial partitions generates multiple records in these partitions' tables, respectively. For example, moving object O_1 generates six points from p_1 to p_6. p_1 falls in spatial partition P_1, and p_6 falls in spatial partition P_4. Thus, they are recorded in P_1 and P_4's tables, respectively.

Spatiotemporal range queries Given a spatiotemporal range query, we first find the spatial grids that fully fall in or partially intersect the query's spatial range. We then respectively search these grids' tables for points whose time stamps are within the query's temporal range, based on each table's partition key and row key. For instance,

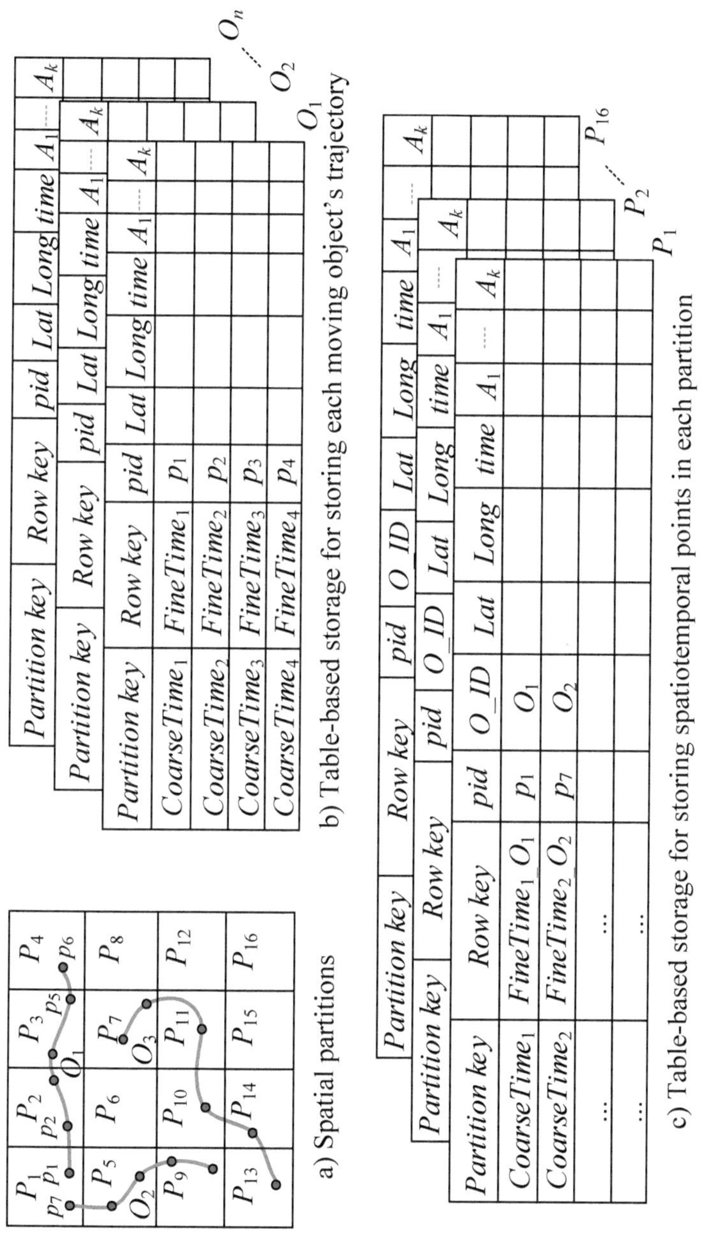

P_1	P_2	P_3	P_4
P_5	P_6	P_7	P_8
P_9	P_{10}	P_{11}	P_{12}
P_{13}	P_{14}	P_{15}	P_{16}

p_7 p_1 p_2 O_1 p_5 p_6 O_2 O_3

a) Spatial partitions

Partition key	Row key	pid	Lat	Long	time	A_1	----	A_k
$CoarseTime_1$	$FineTime_1$	p_1						
$CoarseTime_2$	$FineTime_2$	p_2						
$CoarseTime_3$	$FineTime_3$	p_3						
$CoarseTime_4$	$FineTime_4$	p_4						

O_1 O_2 … O_n

b) Table-based storage for storing each moving object's trajectory

Partition key	Row key	pid	O_ID	Lat	Long	time	A_1	----	A_k
$CoarseTime_1$	$FineTime_1\_O_1$	p_1	O_1						
$CoarseTime_2$	$FineTime_2\_O_2$	p_7	O_2						
...	...								
...	...								

P_1 P_2 … P_{16}

c) Table-based storage for storing spatiotemporal points in each partition

Figure 6.17
Disk-based data management for trajectory data.

we can retrieve points generated in P_1 from 19:00 on Jan 23, 2017, to 23:00 on Jan 24, 2017, by executing the following query in P_1's table:

"PartitionKey ge '20170123' and PartitionKey lt '20170124' and
RowKey ge '20170123190000' and RowKey lt '20170124230000'",

where *ge* means a string is greater than or equal to another one, and *lt* denotes a string is less than another one, according to the lexicographical order. Including an entity's partition key in its row key can facilitate answering range queries that cross multiple partitions in the table. Refer to section 5.2.2.2 for details about Azure Table.

As there are trajectories across multiple grids, we merge the points belonging to the same trajectory according to the moving object identity that a point is associated with. A spatial refinement is further conducted, removing the points in the results that do not truly fall in the query's spatial range.

Updating When a new moving object presents itself, we create a stand-alone table to record its points. When new points generated by a moving object occur, we chronologically insert them into the moving object's table based on the time stamps of these points. These points are projected onto the grids and inserted into the tables of the partitions they fall in. As there can be multiple points generated by different moving objects in the same spatial partition, a moving object's identity is appended to the end of its points' identities.

6.3.3.2 Index-based data management method

Storage and indexing The aforementioned method involves many I/O's on disk and thus is not very efficient. To answer instantaneous queries over a massive amount of trajectory data, a spatiotemporal index is created over the trajectory data. The storage of trajectories on disk is the same as that introduced in section 6.3.3.1.

As illustrated in figure 6.18a, we can create a spatiotemporal index (e.g., 3 DR-tree) over the points from all trajectories (refer to section 4.4.2.4 for details about 3D R-tree), where latitude, longitude, and time stamp formulate the three dimensions. A collection of points is bound by a three-dimensional cuboid. A few small cuboids further formulate a bigger cuboid. Figure 6.18b shows the tree structure representation of the index, where each leaf node consists of a list of entries. Each entry stores the key information of a point, including the point's identity, latitude, longitude, time stamp, and the identity of the moving object generating the point. As memory size is limited, we keep the most recent data in the index while storing the historical data on disk (e.g., by table).

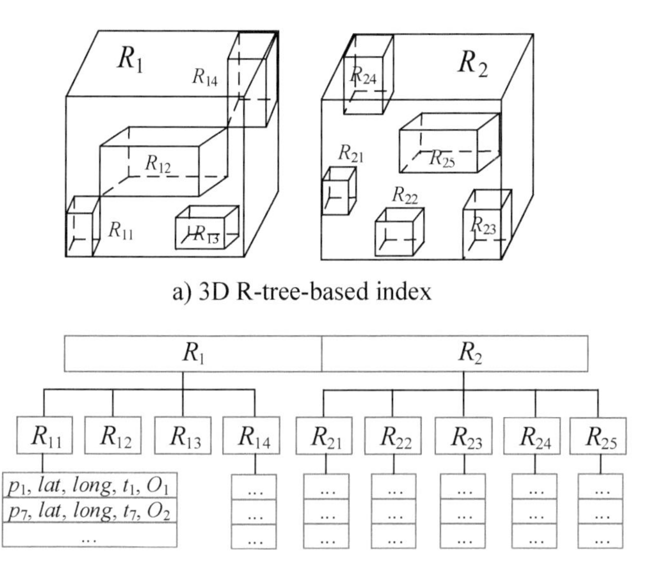

Figure 6.18
Index-based data management for trajectories.

Spatiotemporal range queries Given a spatiotemporal range query, which can be regarded as a three-dimensional cuboid, we first search the spatiotemporal index (e.g., the 3D R-tree in figure 6.18) for points that are completely within the cuboid. In the returned results, points with the same moving object identity are aggregated. Detailed information about a point can be retrieved from the corresponding moving object's table, based on the point's identity and partition key, which can be simply derived from its spatial coordinates.

Other spatiotemporal indexing structures can also be employed to manage trajectory data, depending on different applications, which may have different updating frequencies. Here, we choose only 3D R-tree as an example to illustrate the method that integrates spatiotemporal indexes with cloud storage.

For example, we first divide the spatial space into grids and then build a temporal index, such as sort dynamic arrays and B+ trees, for each grid. Given a spatiotemporal range query, we search for grids that are completely within or partially intersecting a query's spatial range and then search each of these grids' temporal index for points whose time stamp is within the query's temporal range. Finally, a spatial refinement is conducted on returned results, ensuring each and every point is truly within the query's spatial range. The final step is not necessary if a 3D R-tree is employed.

Updating The update consists of two parts. One is updating the records in tables stored on disk. The other is updating indexes in the memory.

When a new moving object comes, we create a stand-alone table to record its points, as shown in figure 6.17b. When new points generated by a moving object arrive, we chronologically insert them into the moving object's table based on the time stamps of these points. These points are projected onto the grids and inserted into the tables of the partitions where they fall, as illustrated in figure 6.17c. As there can be multiple points generated by different moving objects in the same spatial partition, a moving object's identity is appended to the end of its points' identities.

We then update the index in the memory. For example, when a 3D R-tree is employed, we insert newly generated points into corresponding cuboids, based on a point's spatial coordinates and time stamp. To ensure the size of an index is not too large to be held in the memory, points generated a long time ago should be removed from the index. When a 3D R-tree becomes highly imbalanced, we rebuild the tree fundamentally. This inserting-removing-rebuilding process is very complex. Thus, a more practical solution is to create a new 3D R-tree for the data received at a recent time interval (e.g., the last fifteen minutes). Before a new index is created, new points are inserted into an existing 3D R-tree. The tree structure of the existing index does not change, without doing node split and merge operations, even if it becomes imbalanced after new points have been inserted. An existing index is retired when a new index has been built. The frequency of building a new index depends on the applications. For example, we can build a new index every two minutes for the trajectory data received in the last fifteen minutes in a taxi-dispatching system.

If querying data from a long time ago, we search the corresponding tables shown in figure 6.17, following the approach introduced in section 6.3.3.1.

6.3.3.3 Distributed index–based data management method

Storage and indexing When trajectory data is extremely huge, we cannot hold its spatiotemporal index in a single machine's memory. In this situation, we randomly partition trajectory data by moving objects (i.e., by trajectories), assigning different moving objects' trajectories to different machines. Each partition of data has almost the same number of moving objects. The trajectory data of the same object is assigned to the same machine. A spatiotemporal index is built for each portion of the trajectory data and loaded in a machine's memory.

Figure 6.19a presents an example of building and updating an index, using Storm and 3D R-tree as an example. When receiving new points, each of which contains a point identity, moving object identity, latitude, longitude, time stamp, and other

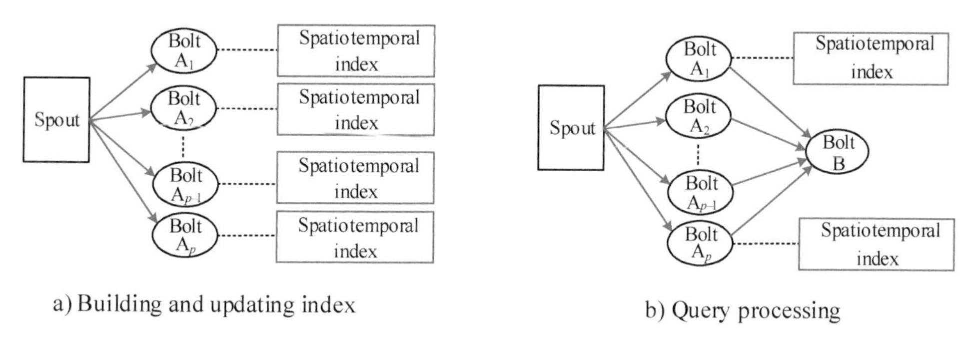

Figure 6.19
Distributed index–based data management for trajectory data.

properties, the spout node distributes these points' information to all Bolt A nodes. Each Bolt A node keeps a list of moving object identities that belongs to its index. If it finds points from these moving objects in the received message, a Bolt A node inserts the points into its spatiotemporal index. To ensure that the size of an index is not too large to be held in the memory, a new index is created at a certain frequency for the recently received trajectory data, following the updating method presented in section 6.3.3.2. The old index is then retired. All points are inserted into tables presented in figure 6.17, which are stored on disk and managed by the cloud.

Spatiotemporal range queries When handling a spatiotemporal range query, the spout node distributes it to all Bolt A nodes. Each Bolt A node searches its own spatiotemporal index for points falling in the query's range, following the method introduced in section 6.3.3.2. These points' identities are first aggregated by moving the object identity (e.g., O_1: $p_1 \sim p_5$) in each Bolt A node and then sending it to the Bolt B node. The Bolt B node collectively retrieves detailed information on points in the aggregated results from the corresponding moving objects' tables—for example, retrieving $p_1 \sim p_5$'s information from O_1's table, as illustrated in figure 6.17b. The Bolt B node then dumps the retrieved results into a file or a block of memory for other applications' usage.

6.4 Urban Big-Data Platform

Figure 6.20 presents a framework of the urban big-data platform, which comprises five layers. In the bottom layer, six types of data models have been defined in terms of data structure and spatiotemporal properties (refer to section 4.1 for details). We use the six data models to accommodate various datasets created in cities, even though

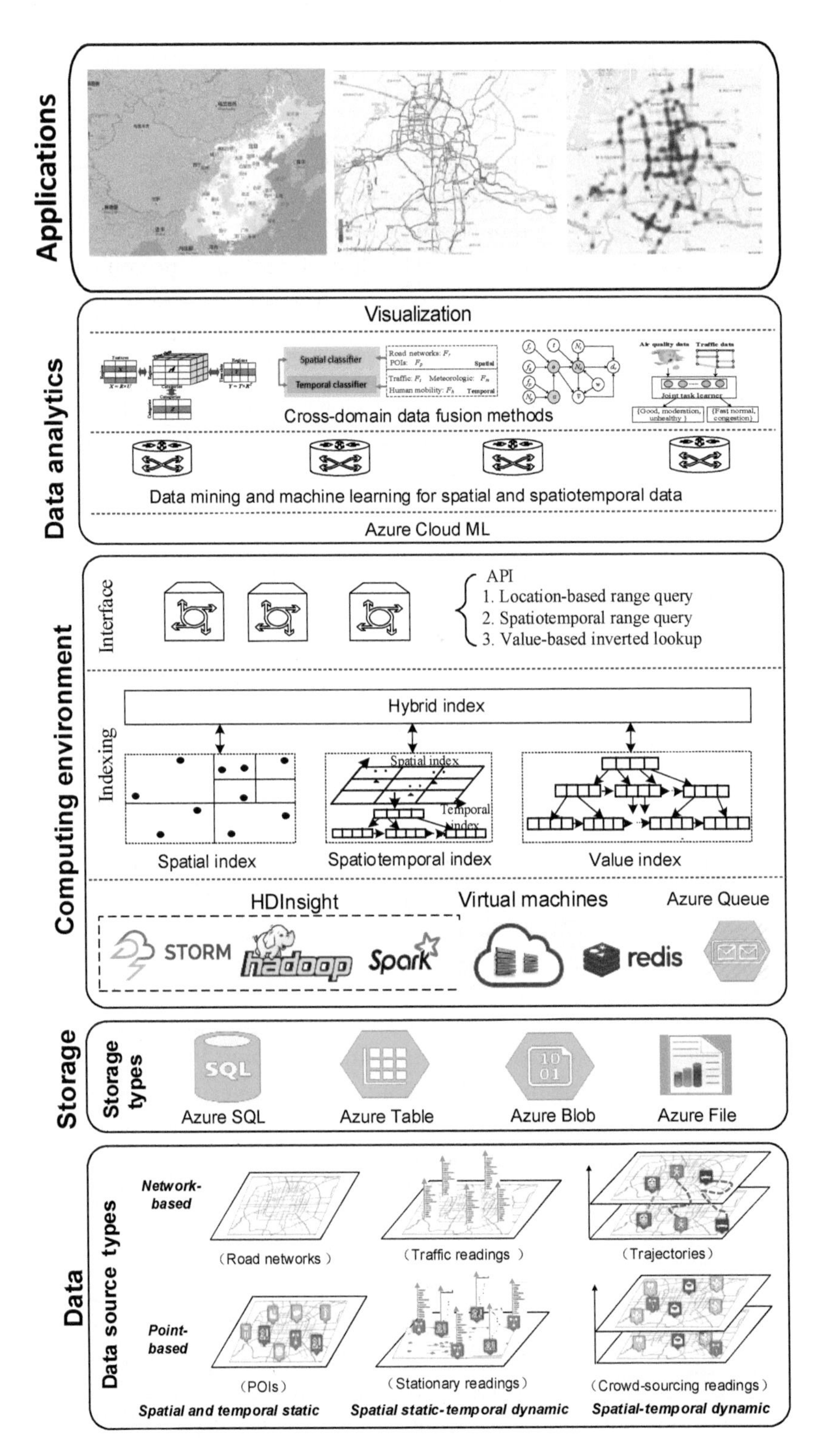

Figure 6.20
The urban big-data platform.

these datasets may look different and are generated in different domains. This enhances the scalability of the urban big-data platform, as we can reuse existing data models rather than create a new one when a new dataset is received.

We leverage the storage, such as Azure Table and Blob, of an existing cloud-computing platform, not necessarily rebuilding everything. Different data models have different storage mechanisms. Refer to previous sections in this chapter for details.

On top of the storage layer, we design spatial and spatiotemporal indexes for different types of data models as well as hybrid indexes for managing multiple datasets across different domains. The indexing structures are then integrated into a distributed-computing environment, such as Hadoop, Spark, or Storm. See previous sections for details. Some application programming interfaces (APIs) are provided to upper-layer machine-learning algorithms, expediting the efficiency of artificial intelligence (AI) techniques.

The urban data analytics is further composed of three sublayers. The bottom sublayer consists of fundamental machine-learning algorithms, including clustering, classification, regression, and anomaly detection models. The middle sublayer contains advanced machine-learning algorithms that are specifically designed for spatiotemporal data (refer to chapter 8 for details). The top sublayer consists of advanced machine-learning models for fusing knowledge from multiple disparate datasets (refer to chapter 9 for details).

On top of the urban big-data platform, there are multiple vertical applications, such as inferring fine-grained air quality, predicting citywide traffic conditions, and deploying charging stations for electrical vehicles. By putting together the components from different layers in the platform, we can quickly enable new applications while keeping the platform scalable and efficient.

6.5 Summary

Current cloud-computing platforms are not designed to handle spatiotemporal data, which has unique data structures and queries. There are quite a few techniques missing in existing clouds (e.g., spatial and spatiotemporal indexing structures). Integrating these indexing structures into cloud-computing platforms calls for knowledge from both sides.

This chapter introduces four schemes for managing spatiotemporal data on the cloud, based on whether an index is employed and deployed on a distributed system. They are single-disk based, single-index based, distributed-disk based, and distributed-index based. The last one is the most advanced and challenging data management scheme, integrating spatial and/or spatiotemporal indexes (e.g., grid-based indexes, R-trees, and 3D R-trees) into distributed-computing systems, such as Spark and Storm in HDInsight. This scheme empowers the cloud to process a larger scale of spatiotemporal data more efficiently while using fewer computing resources.

Implementation processes of the four data management schemes are discussed, respectively, for six types of data models, consisting of point-based spatiotemporal static data, point-based spatial static and temporal dynamic data, point-based spatiotemporal dynamic data, network-based spatiotemporal static data, network-based spatial static and temporal dynamic data, and network-based spatiotemporal dynamic data. For each type of data, the processes for storing, indexing, answering a spatial or spatiotemporal range query, and updating are presented accordingly.

A framework of the urban big-data platform is presented based on an existing cloud-computing platform. Six types of data models are defined to accommodate various types of data. Spatial and spatiotemporal indexes are integrated into distributed-computing environments, such as Spark and Storm, to significantly improve the performance for processing and querying large-scale spatiotemporal data. Advanced machine-learning algorithms are also designed for handling spatiotemporal data and fusing knowledge from multiple datasets across different domains.

IV Urban Data Analytics

7 Fundamental Data-Mining Techniques for Urban Data

Abstract: This chapter introduces the general framework of data mining and presents basic data-mining methods from five perspectives, consisting of association rules and frequent patterns, clustering, regression, classification, and outlier detection. Instead of discussing the technical details of each method, we focus on the general idea of a model and concrete examples demonstrating how it can be employed to mine knowledge from spatial and spatiotemporal data. As data mining is a process of knowledge discovery, which can be conducted using database techniques or machine-learning algorithms, we introduce these data-mining models following the way from databases to machine learning.

7.1 Introduction

7.1.1 General Framework of Data Mining

Data mining, also known as *knowledge discovery from data* (KDD) [99], is the automated or convenient extraction of patterns representing knowledge implicitly stored or captured in data, through a process of collecting, cleaning, and analyzing data [83]. Figure 7.1 presents a general framework of data mining, which is composed of two main parts: data preprocessing and analytical process. The first part includes further components, such as data cleaning, data transformation, and data integration. Detailed techniques of the data preprocessing will be presented in section 7.2. The second part consists of a diversity of data-mining models, results evaluation, and representation methods. We will elaborate on each category of data-mining model in section 7.3.

Data cleaning aims to clean the data by filling missing values, smoothing noisy data, and removing outliers. *Dirty* data can cause confusion during the mining procedure, resulting in unreliable output [99].

Data transformation transfers different forms of data into a format that is user-friendly to data-mining models. Feature construction and (scale) normalization are two main processes of this component.

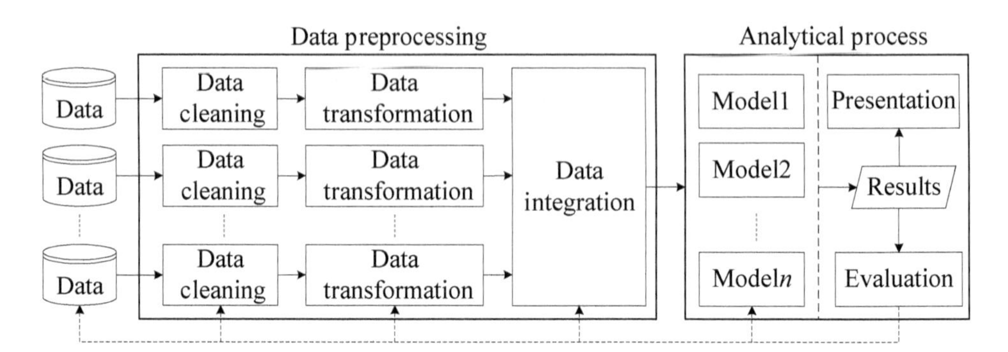

Figure 7.1
A general process of data mining.

Data integration merges data from different sources and reduces the representation of the data while minimizing the loss of information content.

Based on the task that a model is going to complete, *data-mining models* can be divided into five main categories consisting of frequent pattern mining, clustering, classification, regression analysis, and outlier detection:

1. *Frequent patterns* are itemsets, subsequences, or substructures that appear frequently in a dataset. For example, milk and bread that appear frequently together in a transaction dataset is a frequent pattern.
2. *Clustering* is the process of grouping a set of data objects into multiple groups or clusters so that objects within a cluster have high similarity but are very dissimilar to objects in other clusters. For example, we can cluster students into several groups based on their learning interests and behavior at school.
3. *Classification* is a two-step process consisting of a learning step, where a classification model is constructed, and a classification step, where the model is used to predict class labels for given data. A typical example is its use to determine whether a bank should issue a credit card to an applicant in terms of the applicant's characteristics, such as age, gender, income, and job. Classification is usually called *supervised learning*, while clustering is called *unsupervised learning*.
4. *Regression analysis* is a statistical process for estimating the relationships among variables. For instance, we can use a regression model to predict the travel speed on a road segment over certain future hours based on its traffic in the past hours and weather forecasts. Here, the regression model estimates the relationship between the two feature variables and travel speed. While the label of classification models can only be categorical values, the output of regression analysis is a continuous value.

5. *Outlier detection* (also known as *anomaly detection*) is the process of finding data objects with behaviors that are very different from expectations. In addition to fraud detection, outlier detection has a wide range of applications in health care, intrusion detection, and public safety. Outlier detection and clustering analysis are two highly related tasks. Clustering finds the majority patterns in a dataset and organizes the data accordingly, whereas outlier detection tries to capture those exceptional cases that deviate substantially from the majority patterns [99].

Some models can belong to multiple categories simultaneously. For example, artificial neural networks can be used as a regression or a classification model. A data-mining task may employ multiple models with a similar function at the same stage (e.g., ensemble learning) or use models of different functions at different stages.

After data-mining models generate results, we need to *evaluate* their performance. Some tasks, like classification and regression, are easy to evaluate because there are explicit ground truths that we can match the output of a model against. For instance, given the ground truth of a classification task, we can use two metrics, *precision* and *recall*, to measure the performance of a classification model. Likewise, we can use the root mean square error (RSME) to measure the performance of a regression task. Some data-mining tasks, such as clustering and outlier detection, however, may be difficult to evaluate, as there is no ground truth. Note that data mining is an iterative process. In case the performance of a data-mining model is not good enough, we can consider the use of other models, or aggregate data in another way, or extract and select different features, or even add or replace new datasets, denoted by the broken arrows. Generally speaking, data is more important than features, which are more important than models.

After discovering knowledge from data, we need to *present* it in a more intuitive way through visualization. The representation of knowledge may reveal insights that further inspire new designs of features and models. It can also help evaluate some data-mining tasks (e.g., clustering) and enable experts to contribute domain knowledge to a data-mining task.

7.1.2 Relationship between Data Mining and Related Technologies

In recent years, we have been confronted with a series of buzzwords, such as *big data*, *machine learning*, *artificial intelligence*, *deep learning*, *reinforcement learning*, and *cloud computing*. Some of them, like *artificial intelligence* (AI), are not new words but are becoming prevalent again. Some terms, such as *data management* or *databases*, may not be hot terms anymore but are very related to data mining. There is a question in many people's minds: "What is the relationship between data mining and those terms?" After introducing the general framework of data mining, we can further explain its connection to other technologies (or terms).

The term *AI* was coined by Dr. John McCarthy in 1956. In computer science, the field of AI research defines itself as the study of "intelligent agents": any device that perceives its environment and takes actions that maximize its chance of success at some goal [130]. AI has three levels of capability:

1. Interacting with the real world (i.e., to perceive, understand, and act).
2. Reasoning and planning (i.e., modeling the external world, planning, and making decisions).
3. Learning and adaptation (i.e., the ability to solve new problems that have not been experienced).

The first level of capability is relatively easy to achieve, whereas the third level is still difficult. The prevalence of AI in recent years is not its first rise in history. Although many intelligent algorithms were invented before 1990, data and computing infrastructures were still very poor at that moment, thus becoming a bottleneck to AI. Now, they are no longer a problem.

Figure 7.2 presents a taxonomy of AI technology, where machine learning can be regarded as a branch of AI. Of course, a portion of researchers working on machine learning came from the field of mathematical statistics (not the original AI field). In addition to machine learning, there are multiple subfields in AI, including expert systems, planning, evolutionary computing, recommender systems, fuzzy logic, and so on. However, they have not benefited from data and computing infrastructures compared to machine learning. In the meantime, machine learning as a tool has begun contributing to other subfields—for example, recommender systems. That is the reason why people talking AI mainly focus on machine learning. People outside the AI community may even have the misunderstanding that AI is somehow equivalent to machine learning, but this is incorrect. The tasks that machine learning will achieve include regression, classification, clustering, metric learning, outlier detection, causality analysis, and more. Further, machine learning can be divided into six categories in terms of the way to learn from data:

1. *Supervised learning* is the machine-learning task of inferring a function from labeled training data [124]. Each instance in the training data is a pair composed of an input feature vector and a desired output value (sometimes called *class labels*). A supervised learning algorithm analyzes the training data and produces an inferred function,

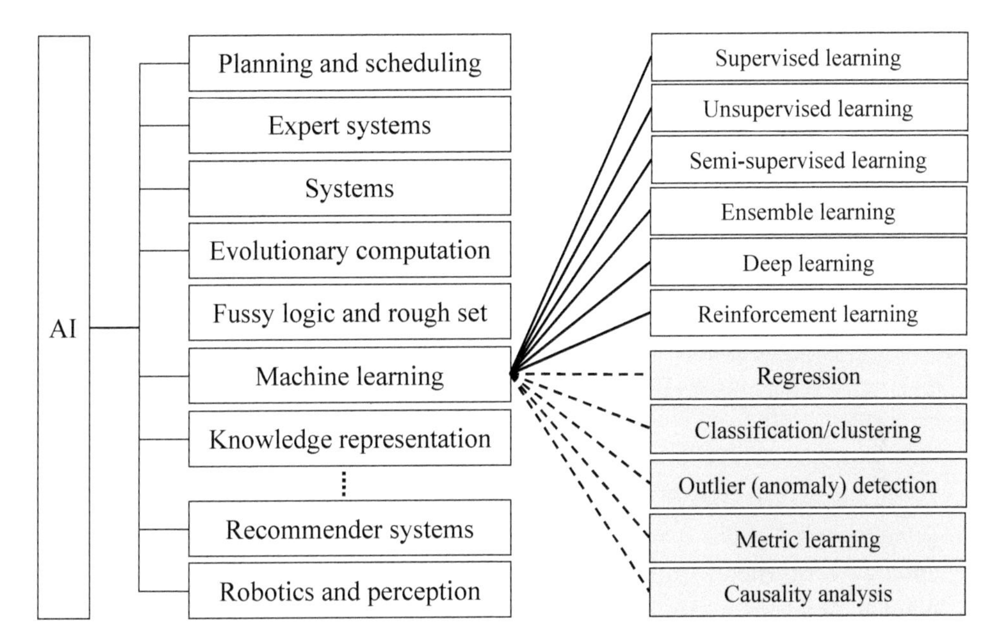

Figure 7.2
Artificial intelligence.

which can be used to determine the class labels for new (unseen) instances. Classification is a typical supervised learning task.

2. *Unsupervised learning* is the machine-learning task of inferring a function to describe hidden structure from unlabeled data (i.e., a classification or categorization is not included in the observations). Since the examples given to the learner are unlabeled, there is no objective evaluation of the accuracy of the structure that is output by the relevant algorithm. This is one way of distinguishing unsupervised learning from supervised learning and reinforcement learning. Clustering is a typical unsupervised learning task.
3. *Semisupervised learning* falls between unsupervised learning (without any labeled training data) and supervised learning (with completely labeled training data). It can also be regarded as a special class of supervised learning techniques that make use of unlabeled data for training—typically, a small amount of labeled data with a large amount of unlabeled data. Many machine-learning researchers have found that unlabeled data, when used in conjunction with a small amount of labeled data, can produce considerable improvement in learning accuracy.

4. *Ensemble learning* uses multiple learning algorithms to obtain better predictive performance than could be obtained from any of the constituent learning algorithms alone [125].
5. *Deep learning* is a branch of machine learning based on a set of algorithms that attempt to model high-level abstractions in data [115]. Deep neural networks (DNNs) and deep learning are similar terms that have been used *almost* equivalently. A DNN is an artificial neural network with multiple (or many) hidden layers of units, composed of linear and nonlinear transformations, between the input and output layers. The term *DNN* emphasizes the complex and deep network structure, while the term *deep learning* focuses on the learning algorithms that train a DNN to model high-level abstraction. Convolutional neural networks (CNNs) [114] and recursive neural networks (RNNs) [103] are the two most popular DNN models applied to computer vision and speech recognition.
6. *Reinforcement learning* is an area of machine learning inspired by behaviorist psychology, concerned with how software agents should take actions in an environment in order to maximize some notion of cumulative reward. Reinforcement learning differs from standard supervised learning, where correct input/output pairs are never presented nor suboptimal actions explicitly corrected. Further, there is a focus on online performance, which involves finding a balance between exploration (of uncharted territory) and exploitation (of current knowledge) [107]. Reinforcement learning is not a new machine technique but has been attracting a lot of attention recently when combined with deep learning. Deep reinforcement learning [123] employs deep-learning techniques to learn an approximate value function (for a reinforcement-learning algorithm), which estimates the reward of a given status and action policy directly without traversing all possible intermediate states.

Figure 7.3 illustrates the connection between data mining and other technical terms, including *machine learning, AI, data management* (databases), *big data*, and *cloud computing*. Generally, data mining is a process of knowledge discovery that uses a variety of tools to mine knowledge from data. The tools can be database techniques (like frequent pattern mining), or machine-learning algorithms (e.g., support vector machines and decision trees), or other AI techniques (such as evolutionary computing and fuzzy logic). Typically, database techniques concentrate on the efficiency of generating results, while machine-learning algorithms (as well as other AI techniques suitable for data mining) focus on the accuracy (or effectiveness) of discovered results. So, we can

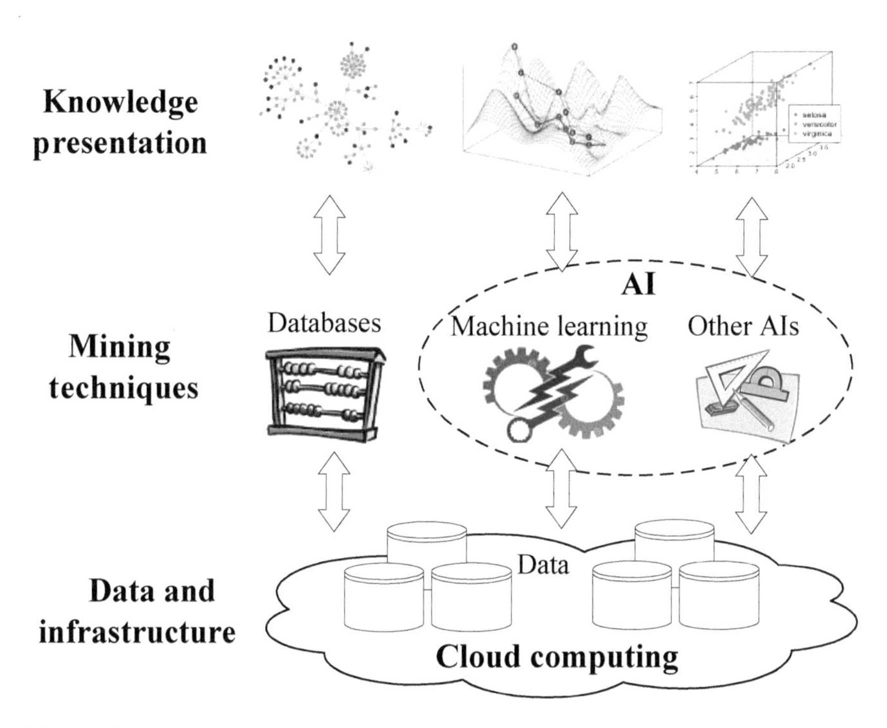

Figure 7.3
The connection between data mining and other terms.

conduct data mining from a database's perspective [90] or from a machine-learning (and related AI's) perspective. Quite a few data-mining methods combine database techniques with machine-learning algorithms to complete a task [138, 147]. This is very important for real-world data-mining applications.

When the scale of data to be mined is big and the application requests online data processing, we need a powerful computing infrastructure like the cloud for efficiently storing, accessing, mining, and representing the data. *Cloud computing* is a more computing environment-oriented term, providing infrastructures and platforms as a service to data-mining tasks. Database techniques, machine-learning algorithms, and other AI methods can be deployed in a cloud and used as a service (i.e., software as a service) by applications and tasks. The representation of knowledge, particularly interactive visual data analytics, can also depend on cloud-computing platforms.

Big data usually includes datasets with sizes and complexity beyond the ability of commonly used software tools to capture, curate, manage, and process data within a tolerable elapsed time [122]. Industries widely describe big data as having the "3*V*s" (i.e., volume [amount of data], velocity [speed of data in and out], and variety [range of data types and sources]) [113]. Some organizations have added additional *V*s, consisting of value and veracity, to describe big data. However, some terms, consisting of *volume*, *velocity*, and *value*, are hard to judge or quantify specifically. For example, the volume of big data changes constantly over time, ranging from a few dozen terabytes to many petabytes of data. It is very difficult to argue if a dozen terabytes can be regarded as big data or not. Compared to these vague terms, variety is very clear and easy to judge. Thus, big data has a new focus on technologies with new forms of integration to reveal insights from datasets that are diverse, complex, and of a massive scale [121], with the main components consisting of techniques for analyzing data (such as machine learning), data management (like cloud computing and databases), and visualization. Thus, big data also emphasizes knowledge discovery from data but describes technology from data's perspective.

Overall, we can say *data mining* is a term describing knowledge discovery from the perspective of the mining *process*, which can use databases, machine learning, and related AI technology as tools and cloud-computing platforms as a computing environment (when dealing with big data). *Big data* is a term describing knowledge discovery from the perspective of *data*, which is an end-to-end capability (from data collection, management, and analysis to visualization) to solve a problem. *Cloud computing* is a term describing a way of computing that can empower big data, data mining, and other large-scale systems.

7.2 Data Preprocessing

7.2.1 Data Cleaning

Real-world data tends to be incomplete and noisy for many reasons. Some data collection technologies are inherently inaccurate because of the hardware limitations associated with collection and transmission or dropping readings because of hardware failure or battery exhaustion [99]. Such *dirty* data causes confusion for the mining procedure, resulting in unreliable output. Data cleaning aims to clean the data by handling missing entries and smoothing noisy data.

7.2.1.1 Handling missing entries Many data entries may remain empty because of imperfect data collection methods. Three classes of techniques are used to handle such missing entries:

- Any data record containing a missing entry may be eliminated entirely. However, this approach may not be practical when most of the records contain missing entries.

In addition, by ignoring a tuple with one or two missing entries, we do not make use of the remaining attributes' values in the tuple.

- The missing values may be imputed based on their neighborhoods or empirical values, such as historical mean. It is a challenging task to impute such missing values in spatiotemporal data [52], as there can be many possible choices for an imputation. For example, we can impute a missing entry with the value of its closet spatial neighborhood or its own value in the past hours (refer to section 3.5 for details). Errors created by the imputation process may affect the results of data-mining algorithms.
- Design analytical models so that they can work with missing values. Quite a few data-mining methods are designed to work robustly with missing values. For example, we can use "*unknown*" to represent a variable's missing value in a Bayesian network. This approach is usually the most desirable, avoiding the error induced by the imputation process.

7.2.1.2 Handling noisy entries Noise is a random error or variance in a measured variable. The key methods to handle noisy entries are in three categories:

- *Rules based on domain knowledge*. Domain knowledge is often available in terms of the ranges of the attributes or rules that specify the relationships across different attributes. For example, it is impossible to see a taxi traveling with a speed faster than 400 km/h in a city. Thus, the GPS point with a travel speed higher than 400 km/h can be considered a noisy point.
- *Outlier detection*. Data points that are inconsistent with the remaining data distribution are often referred to as *outliers*. Outliers may be detected by clustering; for example, values falling outside the set of clusters may be considered outliers. On the other hand, it is risky to assume that all outliers are caused by errors. For example, a record representing credit-card fraud is likely to be inconsistent with respect to the patterns in most of the (normal) data but should not be removed as noisy data [83]. In addition, outlier detection is a research theme in data mining, with many challenges remaining unsolved. Though there are many advanced outlier detection methods (refer to section 7.6 for details), it may not be a good choice to spend too much effort on outlier detection at the stage of data cleaning.
- *Other smoothing techniques*. For spatiotemporal data, we can employ smoothing techniques, such as the Kalman filter, particle filters, and discrete wavelet transformation (DWT), to clean noise. For example, the trajectory estimated from the Kalman filter is a trade-off between the measurements and a motion model. Besides giving estimates that obey the laws of physics, the Kalman filter gives principled estimates of

higher-order motion states such as speed. While the Kalman filter gains efficiency by assuming linear models plus Gaussian noise, the particle filter relaxes these assumptions for a more general, but less efficient, algorithm. A tutorial-like introduction to using the Kalman and particle filters to fix noisy trajectory points can be found in [117]. The DWT is originally a data reduction technique, transforming a data vector X to another vector X' of wavelet coefficients that is larger than a user-specified threshold (see section 7.2.3.2 for details). It can also be employed as a data-cleaning method, removing noise without smoothing out the main features of the data [118]. Given a set of coefficients, an approximation of the original data can be constructed by applying the inverse of DWT.

In the big-data era, with massive data in which the proportion of high-quality data is extremely high compared to noisy data, we may not find it necessary to spend too much effort on removing these noisy values *from the perspective of training an effective machine model*. Note that we are not saying that removing noise is no longer necessary when we have massive high-quality data, as it is also important to ensuring the validity of data normalization (see section 7.2.2.2). As illustrated in figure 7.4, if a small dataset has one noisy point, the linear regression model best fitting the entire dataset (i.e., with the minimum aggregated perpendicular distance to all points) would be the broken red line, although the true model should be the solid blue line (if removing the noisy point). That is, a noisy point in a small dataset will cause a machine-learning model (fitting the entire dataset) to deviate significantly from its true form (fitting the normal points). However, when the volume of high-quality data is huge, even if there are a few noisy points, the linear regression model fitting the entire dataset would still be very

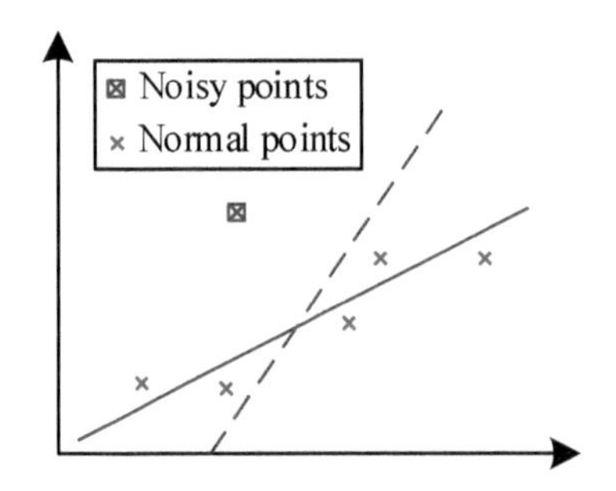

a) A small dataset with noisy points

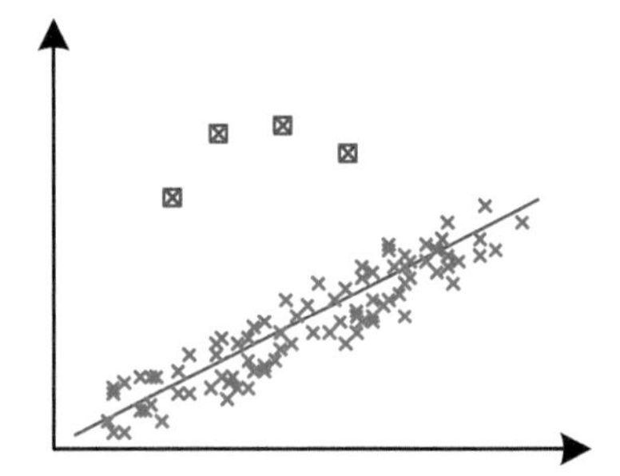

b) A large dataset with noisy points

Figure 7.4
Fitting data with a linear regression model.

close to the true one (i.e., the solid blue line). Under such a circumstance, we may just remove some extremely noisy points based on some (commonsense) rules, thus avoiding having to remove uncertain outliers by clustering (or some advanced method), which is somewhat risky and computationally expensive.

7.2.2 Data Transformation

7.2.2.1 Feature construction When data has been collected, it may be encoded in complex logs or in a free form that is not directly computable for a data-mining model. Thus, it is essential to transform it into a format that is friendly to data-mining algorithms, such as a multidimensional vector or time series. The vector format, in which different entries of the vector correspond to different properties that are referred to as features, attributes, or dimensions, is the most common. For instance, given a POI dataset of a city, we can count POIs of different categories in each neighborhood. Suppose there are five restaurants, three bus stops, one gas station, and one movie theater in a neighborhood. The feature vector we extract from the POI data for the neighborhood is <5, 3, 1, 1 ...>. In this example, a neighborhood is regarded as an instance, which has a diversity of features.

It is crucial to extract features that are relevant to a problem we are going to solve, particularly for classification and regression problems. This calls for knowledge and understanding about the problem itself. For example, to determine if a bank should issue a credit card to an applicant, we need to consider the person's age, job, and income. Those factors can then become features of a binary classification model, which generates a label of yes or no for each applicant based on the same features and labels extracted from historical data. Likewise, to predict traffic conditions in a geographical region, we need to consider the structure of road networks in the region as well as weather conditions. The features, such as the number of road intersections and the total length of highways, extracted from the corresponding road network data influence the region's traffic conditions significantly. If key features of a problem are not extracted from the data, the following data-mining problem may become very difficult to solve.

On the contrary, extracting too many features (particularly irrelevant features) calls for more training data to tune parameters weighting different features. Otherwise, a small dataset with too many features results in overfitting. At this moment, we need to call on subset feature-selection techniques, which will be introduced in section 7.2.3.2.

7.2.2.2 Data normalization Data may be recorded in very different scales. For example, the humidity of a city is generally within [0, 100 percent], while the income of a person may be tens of thousands of dollars per month. The latter feature is typically orders of magnitude larger than the former. Thus, any aggregate function computed on the different features (e.g., Euclidean distances or linear regression models) will be dominated by the feature with a larger magnitude. In other words, the feature with a small value is implicitly ignored. Therefore, it is important to normalize features of different scales before we throw them into a data-mining model. There are two widely used ways to normalize data: *min-max normalization* and *zero-mean normalization*.

Min-max normalization performs a linear transformation on the original data, mapping data to a range of [0, 1]. Suppose that min_A and max_A are the minimum and maximum values of an attribute A. Min-max normalization maps a value v_i of A to v_i' by computing:

$$v_i' = \frac{v_i - min_A}{max_A - min_A}. \tag{7.1}$$

When a dataset contains some outliers, this approach is not effective, mapping most data into a very small range. For example, suppose there is an outlier humidity (10) recorded. Then, most of the normalized humidity attribute will be in the range [0, 0.1]. Consequently, this attribute may be deemphasized. To address this issue, the zero-mean normalization is proposed.

Zero-mean normalization maps an attribute A's value based on its mean $\bar{A}$ and standard deviation σ_A by computing:

$$v_i' = \frac{v_i - \bar{A}}{\sigma_A}, \tag{7.2}$$

where $\bar{A} = \frac{1}{n}(v_1 + v_2 + \cdots + v_n)$ and $\sigma_A = \sqrt[2]{\frac{1}{n}\sum_{i=1}^{n}(v_i - \bar{A})^2}$. This normalization method is useful when the actual minimum and maximum of an attribute are unknown or when there are outliers that dominate the max-min value of an attribute. Most of the normalized values will typically lie in the range [–3, 3] under the normal distribution assumption.

A variation of this zero-mean normalization replaces the standard deviation σ_A by the mean absolute deviation s_A, which is computed by:

$$s_A = \frac{1}{n}(|v_1 - \bar{A}| + |v_2 - \bar{A}| + \cdots + |v_n - \bar{A}|). \tag{7.3}$$

The mean absolute deviation s_A is more robust to outliers than the standard deviation σ_A, as $|v_i - \bar{A}|$ is not squared, hence reducing the effect of outliers [99].

7.2.3 Data Integration

7.2.3.1 Data merging and knowledge integration It is very likely that a data-mining task involves data from multiple sources. As illustrated in figure 7.5a, conventional data integration aims to merge the data (from different sources) that describes the same object with different schemas. For example, there are three POI datasets of Beijing generated by three different data providers. Conventional data integration aims to merge the three datasets into a database with a consistent data schema. Under such a circumstance, the main challenges are schema mapping and object matching. The former transfers different schema into a consistent representation. The latter identifies records of the same object in different sources, merging complementary properties of these records corresponding to the same object and removing duplicate values.

Recent data integration has a new mission, which is a step toward fusing the knowledge of multiple disparate datasets from different domains [145]. As shown in figure 7.5b, there may not be an explicit object based on which we can simply align records from different datasets. For instance, traffic conditions, POIs, and the demography

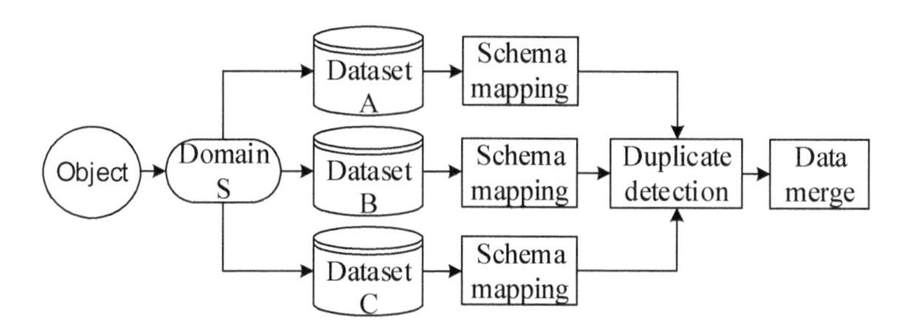

a) Paradigm of the conventional data integration

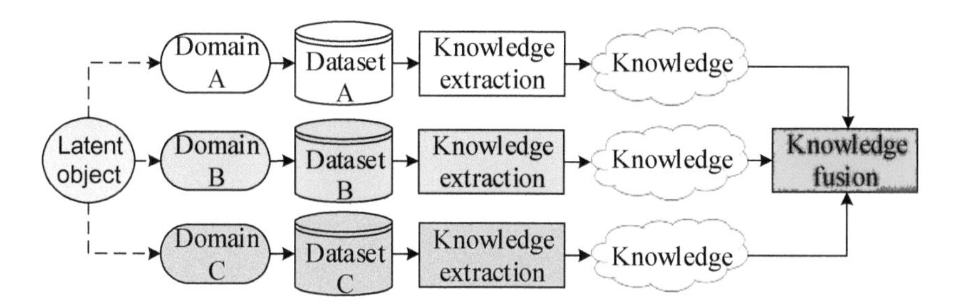

b) Paradigm of data fusion for (cross-domain) big data

Figure 7.5
The concept of data integration.

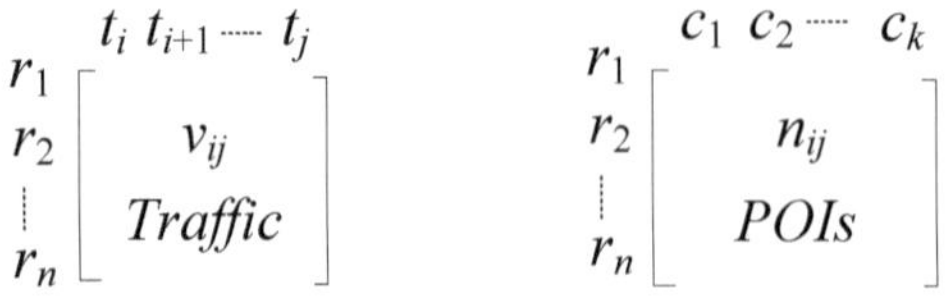

$$\begin{array}{c} \\ r_1 \\ r_2 \\ \vdots \\ r_n \end{array} \overset{t_i \; t_{i+1} \cdots \; t_j}{\begin{bmatrix} \\ v_{ij} \\ Traffic \\ \end{bmatrix}} \qquad \begin{array}{c} \\ r_1 \\ r_2 \\ \vdots \\ r_n \end{array} \overset{c_1 \; c_2 \cdots \; c_k}{\begin{bmatrix} \\ n_{ij} \\ POIs \\ \end{bmatrix}}$$

Figure 7.6
An example of data integration for disparate datasets.

of a region describe the region's latent function collectively, although they are from three different domains. Literally, records from the three datasets describe different objects: a road segment, a POI, and a neighborhood, respectively. Thus, we cannot merge them straightforwardly by schema mapping and object matching. Instead, we need to extract knowledge from each dataset by different methods, fusing the knowledge from them organically to understand a region's function collectively. This is more about knowledge fusion rather than schema mapping.

A possible method to integrate the three data sources in this example is to extract region-level features from each source and place these features in different matrices. As illustrated in figure 7.6, we can accommodate features extracted from the traffic data in the left matrix, where each row denotes a region, and each column stands for a time interval. An entry v_{ij} in this matrix denotes the average travel speed of region r_i at time interval t_j. Likewise, we can place POI-related features in the right matrix, where each column stands for a POI category (e.g., restaurants), and an entry n_{ij} denotes the number of POIs pertaining to category c_j in region r_i. For instance, there are five restaurants, one shopping mall, and one movie theater in a region. The two matrices share the same dimension of region, but the columns have different meanings. As traffic conditions in a region change over time significantly, while POIs in a region may not vary in time, it is not a good choice to integrate the two feature sets into one matrix. With such a data integration approach, we can apply advanced knowledge fusion methods (see chapter 9) [145], such as context-aware matrix factorization, to fuse knowledge from the two disparate datasets [132].

7.2.3.2 Data reduction Data reduction techniques can be applied to obtain a reduced representation of the dataset that is much smaller in volume yet closely maintains the integrity of the original data. That is, mining on the reduced dataset should be more efficient yet produce the same (or almost the same) analytical results [99]. Data reduction strategies include attribute subset selection (a.k.a.

feature selection), axis rotation-based reduction, and summarization-based reduction methods.

Attribute subset selection Datasets for analysis may contain hundreds of attributes, many of which may be irrelevant to the mining task or redundant. Leaving out relevant attributes or keeping irrelevant attributes may cause confusion for mining algorithms. In addition, those irrelevant or redundant attributes would increase the number of parameters in some data-mining models, not only slowing down the mining process but also compromising the accuracy of the model. A small training set with many features may result in an overfitting problem. Attribute subset selection reduces the dataset size by removing irrelevant or redundant attributes while ensuring the data-mining results generated by these selected features are as close as possible to that using all attributes. An exhaustive search for the optimal subset of attributes can be prohibitively expensive, thus heuristic methods that explore a reduced search space are commonly used:

- *Stepwise forward selection.* The procedure starts with an empty set of attributes. The best of the original attributes is determined and added to the reduced set. At each subsequent iteration, the best of the remaining original attributes is added to the set.
- *Stepwise backward elimination.* The procedure starts with the full set of attributes. At each step, it removes the worst attribute remining in the set.
- *Combination of forward selection and backward elimination.* The two aforementioned stepwise methods can be combined. At each iteration, the procedure selects the best attribute and removes the worst from the remaining attributes.
- *Decision tree induction.* Decision tree algorithms were originally designed for classification (see section 7.5.3 for details). When they are used for feature selection, we can assume the features that do not appear in a constructed tree to be irrelevant. The set of attributes appearing in the tree form the reduced subset of attributes.

Recently, advanced machine-learning models have been equipped with the ability (e.g., regularization) to deal with data sparsity and noise. In the big-data era, when we have sufficient training data and advanced machine-learning algorithms, it may not be very necessary to conduct feature selection, or we can spend much less effort than before on feature selection. Given sufficient training data, machine-learning models can determine the weight of each feature automatically, setting a small value for redundant or irrelevant features. An extreme case is deep learning, which uses every pixel of an image as input without doing feature selection.

Axis rotation–based data reduction This category of methods includes wavelet transforms, principal component analysis, and matrix factorization. Unlike attributes subset selection, which reduces the attribute set size by retaining a subset of the initial attributes, this category of methods combines the essence of attributes by creating an alternative, smaller set of variables in another space.

The *discrete wavelet transform* (DWT) is a linear signal-processing technique that transforms a time series or a data vector X to a numerically different vector X' of wavelet coefficients and corresponding basis vectors. X and X' are of the same length, but X' can be retained by storing a small fraction of the strongest wavelet coefficients. The original data vector can be recovered by a production of the wavelet coefficients and basis vectors. The DWT is closely related to the discrete Fourier transformation (DFT), a signal-processing technique involving sines and cosines. In general, the DWT achieves better lossy compression than the DFT. That is, given a data vector, if the same number of coefficients is retained for a DWT and a DFT, the DWT version will provide a more accurate approximation of the original data. Hence, for an equivalent approximation, the DWT requires less space than the DFT. Unlike the DFT, wavelets are quite localized in space, contributing to the conservation of local detail. There are several families of DWTs, including the Haar-2, Daubehies-4, and Daubechies-6.

The Haar wavelet is a popular form of wavelet decomposition because of its intuitive nature and ease of implementation. Suppose a sensor monitors the temperature of a location, generating a reading every minute. While temperatures at adjacent time intervals are very similar, the output of the sensor in a day comprises $60 \times 24 = 1440$ records, most of which are redundant representations. If we only store the average over the entire day, this provides some idea of the temperature but not much else about the variation over the day. Now, if the difference in average temperature between the first half and second half of the day is also stored, we can derive the averages for both the first and second half of the day from the two values. This principle can be applied recursively because the first half of the day can be further divided into the first quarter of the day and the second quarter of the day. Thus, with four stored values, we can perfectly reconstruct the averages in four quarters of the day. This process can be applied recursively right down to the level of granularity of the sensor readings.

Figure 7.7 presents an example of using the Haar wavelet transform to turn a segment of the time series (8, 6, 3, 2, 4, 6, 6, 5) into a few wavelet coefficients. At the bottom layer, the average of the entire time series is 5. If further calculating the average of the first half (4.75) and that of the second half (5.25), we will find the average difference $c_{1,1}$ between the two averages is $(4.75 - 5.25)/2 = -0.25$. Thus, the time series can be roughly represented by the average of the entire time series plus or minus $c_{1,1}$

(i.e., 5 − 0.25 and 5 + 0.25). A basis vector corresponding to this representation is (1, 1, 1, 1, −1, −1, −1, −1). To obtain a more accurate presentation of the time series, we can further compute the average of the first quarter of the time series (i.e., (8 + 6)/2 = 7) and the average of the second quarter (i.e., (3 + 2)/2 = 2.5), respectively. The average difference $c_{2,1}$ is (7 − 2.5)/2 = 2.25. Thus, we can represent the first half of the time series by 4.75 ± 2.25, with a basis vector (1, 1, −1, −1, 0, 0, 0, 0). Likewise, we can calculate $c_{2,2}$ = (5 − 5.5)/2 = −0.25 and represent the second half of the time series with 5.25 − 0.25 and 5.25 + 0.25. We recursively divide each quarter into two halves until the number of the wavelet coefficients equals the length of the original time series. These "difference values," such as $c_{1,1}$, $c_{2,1}$, and $c_{2,2}$, are used to derive wavelet coefficients (5, −0.25, −0.25, 2.25, 0.5, −1, 0.5, 1). The original time series can be recovered based on the wavelet coefficients and corresponding basis vectors. Alternatively, we can say the wavelet representation is a decomposition of the original time series of length q into the weighted sum of a set of q "simpler" time series (or wavelets) that are orthogonal to one another. These "simpler" time series are the basis vectors, and the wavelet coefficients represent the weights of the different basis vectors in the decomposition [83].

We cannot yet achieve any data reduction if the number of wavelet coefficients equals the length of the original time series. Nevertheless, a larger wavelet coefficient corresponds to a more significant variation of values in the time series than the small ones. Thus, for the sake of data reduction, we can maintain the coefficients larger than a user-specified threshold (1 in this example) and set the rest of the coefficients to 0;

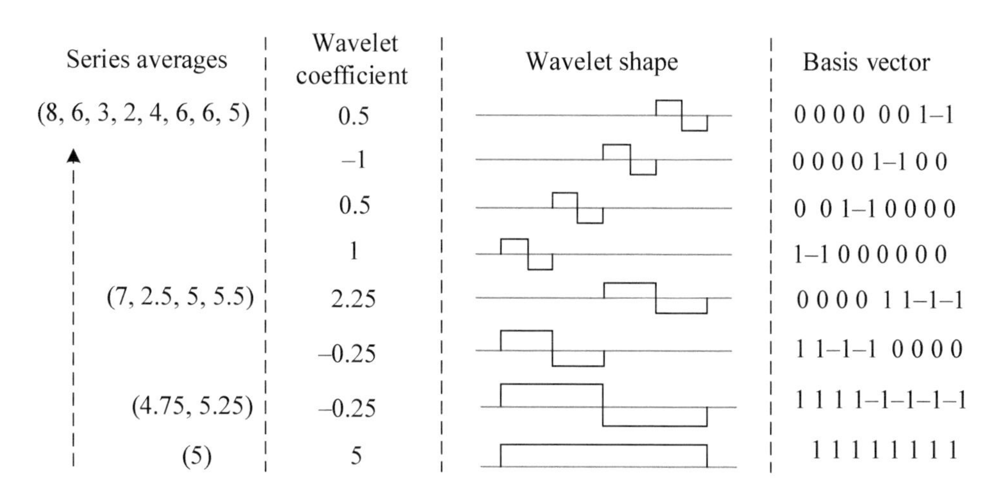

Figure 7.7
An example of using the Haar wavelet transform.

that is, (5, 0, 0, 2.25, 0, −1, 0, 1). For a time series, such as temperature readings, with many similar readings over a short period, there must be many small coefficients, which will be set to 0 after being smoothed by a threshold. Thus, we only need to store the value and indexes (e.g., $c_{2,1}$ standing for the second-level coefficient of the first half) of a few wavelet coefficients. The basis vector can be constructed according to the index of a wavelet coefficient (e.g., the basis vector corresponding to $c_{2,1}$ is [1, 1, −1, −1, 0, 0, 0, 0] in the previous example), thus it is not necessary to store.

Principle component analysis (PCA) computes k p-dimension orthogonal vectors (referred to as *principal components*) that are used as a basis to best represent the original data, which can be represented by an $n \times m$ matrix, $p < m$. The principal components are sorted in order of decreasing significance and essentially serve as a new set of axes for the data. As illustrated in figure 7.8a, the first axis shows the most variance among the data; the second axis shows the next highest variance, and so on. The original data are thus projected onto a much smaller space, resulting in dimensionality reduction. Alternatively, we can say the input data is then represented by a linear combination of the principal components. Note that the data must be mean centered, $x_{ij} - \mu_j$ (i.e., each entry of a row minus the mean value μ_j of the entire column to which the entry belongs). The initial data can then be projected onto this smaller set. Principal components may be used as inputs for multiple regression and cluster analysis.

Matrix factorization decomposes a (sparse) matrix X into the production of two (low-rank) matrices, which denote the latent representation of each row and columns, respectively. The latent representation, whose size is usually much smaller than its original feature set, can be used as features for data-mining algorithms, thereby reducing the size of data. In addition, the production of the two matrices can approximate matrix X, therefore helping fill the missing values in X. There are two widely used matrix factorization methods: singular value decomposition (SVD) [98, 109] and nonnegative matrix factorization (NMF) [105, 116]. SVD factorizes an $m \times n$ matrix X into the production of three matrices $X = U \Sigma V^T$, where U is an $m \times m$ real unitary matrix (a.k.a. left singular vectors), Σ is an $m \times n$ rectangular diagonal matrix with nonnegative real numbers on the diagonal (a.k.a. singular values), and V^T is an $n \times n$ real unitary matrix (a.k.a. left singular vectors). In practice, when trying to approximate matrix X by $U \Sigma V^T$, we only need to keep the top k biggest singular values in Σ and the corresponding singular vectors in U and V, as illustrated in figure 7.8b. That is, $X \approx U_k \Sigma_k V_k^T$. NFM factorizes an $m \times n$ matrix R (with m rows and n columns) into the production of an $m \times K$ matrix P and a $K \times n$ matrix Q, $R = P \times Q$, with the property that all three matrices have no negative elements. SVD has some good properties. First, U and V are orthogonal matrices (i.e., $U \cdot U^T = I$, and $V \cdot V^T = I$). Second, the value of k can be determined by Σ. For

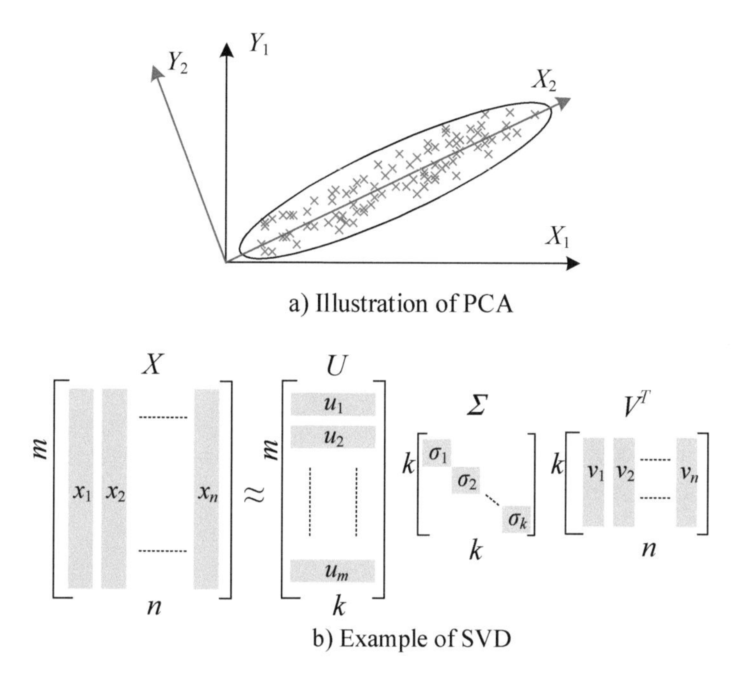

Figure 7.8
Rotation-based data-reduction methods.

example, select the first k diagonal entries (in Σ) whose sum is larger than 90 percent of all diagonal entries' sum. However, SVD is more computationally expensive and harder to parallelize, as compared to NFM.

In comparison with wavelet transforms, PCA tends to be better at handling sparse data, whereas wavelet transforms are more suitable for data of high dimensionality but with similar readings at nearby entries. SVD is closely related to PCA, but SVD is more general than PCA. First, SVD provides a latent representation for both rows and columns of the data matrix, whereas PCA only provides basis vectors of rows of the data matrix. Second, SVD is often applied without mean centering to sparse nonnegative data, such as user-item matrices. When the data is not mean centered, the basis vectors of SVD and PCA will not be the same, and different results may be obtained. Once a matrix is mean centered, PCA can be implemented through SVD.

Summarization-based data reduction This category of data reduction methods tends to use the summary of data to represent the data, including histogram based and clustering based.

- *Histogram-based method.* A histogram partitions the data distribution of an attribute A into disjoint subsets, referred to as *buckets* or *bins*, representing data falling in each bucket with an attribute-value/frequency pair to reduce the size of the data. For example, there are one hundred taxicabs traveling in a region with different speeds. If we are going to extract features related to travel speed, a possible way to reduce the size of the data is to partition the travel speed of taxicabs into several disjoint buckets, such as [0, 20), [20, 40), and [40, 100]. We then count the number of taxicabs falling into each bucket based on their traveling speeds, formulating three attribute-value/frequency pairs—for example, [0, 20)/25, [20, 40)/60, and [40, 100]/15. Later, an attribute value (range) such as [40,100] becomes a feature, and the count corresponding to the range will be used as the value of the feature. Instead of using one hundred taxicabs' travel speeds as features, the histogram-based method reduces the size of the feature set from one hundred to three.
- *Clustering-based methods.* Clustering techniques consider data tuples as objects. They partition the objects into groups, or clusters, so that objects within a cluster are "similar" to one another and "dissimilar" to objects in other clusters. In data reduction, the cluster representations of the data are used to replace the actual data. For instance, there is a dataset about shopping malls, each of which is associated with many features, such as size, the number of floors, and the number of shops and their distribution across different categories, as well as the information about their garages. We can cluster these shopping malls based on their features and then represent a shopping mall by the identity of the cluster it belongs to, thereby reducing the size of features significantly. We can also leverage these clusters as buckets, counting the number of shopping malls falling in each bucket. Then, we can employ the histogram-based methods to formulate attribute-value/frequency pairs, where the identity of a cluster is the attribute value, and the number of shopping malls belonging to the cluster is the frequency.

7.3 Frequent Pattern Mining and Association Rules

Frequent patterns are itemsets, subsquences, or substructures that appear frequently in a dataset. For example, milk and bread that appear frequently together in a transaction dataset is a (*frequent*) *itemset pattern.* If a subsequence, such as first buying a package of bread, then beer, and then diapers, occurs frequently in the transaction records of a shopping mall, it is a (*frequent*) *sequential pattern.* A *substructure* can refer to different structural forms, such as subgraphs, subtresses, or sublattices, which may be combined with itemsets or subsequences. If a substructure occurs frequently, it is called a *(frequent) structured pattern*. Finding frequent patterns plays an essential role in mining associations, correlations, and many other interesting relationships among data [99].

7.3.1 Basic Concepts

The frequency that itemsets, subsequences, or substructures appear in a dataset is called a *support*. If the support of a pattern is not smaller than a threshold, it is called a *frequent pattern*. Figure 7.9 presents three different types of frequent patterns.

As illustrated in figure 7.9a, four transaction records contain five different items: milk, bread, diapers, beer, and yogurt. If we set a threshold of ¾ for the support, (beer, diapers, bread) is a *frequent itemset pattern*, as it appears in three out of the four records. Although the support of (beer, diapers) and (bread, diapers) is also not smaller than ¾, these patterns are subsets of (beer, diapers, bread)—that is, not a maximum frequent pattern. If we further consider the order that these items were purchased in, as presented in figure 7.9b, (beer→bread) is the only sequential pattern satisfying the threshold. Note that the sequence between two items is not required to be consecutive. For example, there is a *yogurt* standing between the *beer* and the *bread* in the fourth record. Figure 7.9c illustrates two categories of frequent subgraph patterns. The top one shows a frequent subgraph pattern (A, B, C) that appears in three independent graphs

Milk, bread, diapers, beer	*Milk→bread→diapers*
Beer, diapers, bread	*Beer→diapers→bread*
Beer, bread, yogurt	*Beer→bread→yogurt*
Diapers, beer, yogurt, bread	*Diapers→beer→yogurt→bread*
(Beer, diaper, bread)	***(Beer→bread)***
a) Frequent itemset patterns	b) Frequent sequential patterns

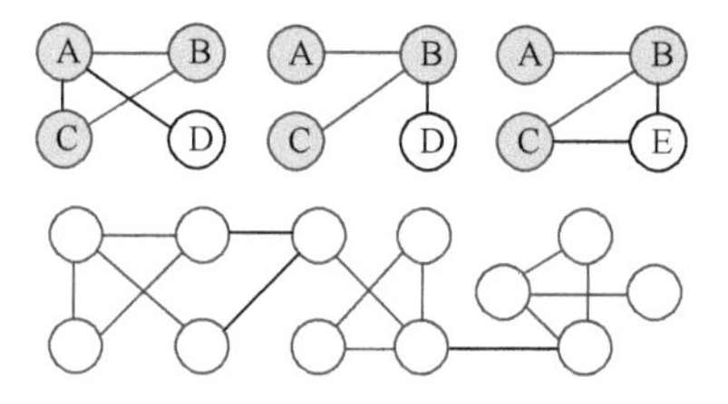

c) Frequent subgraph patterns

Figure 7.9
Examples of frequent patterns.

with the same connecting structure. Thus, its support is 100 percent. This is also called *transactional setting*. The bottom example shown in figure 7.9c finds a frequent substructure in a single graph where nodes in the graph are not differentiated. The blue substructure composed of four nodes occurs three times in the graph.

Frequent patterns can be represented in the form of association rules with two measures (i.e., *support* and *confidence*) denoting the usefulness and certainty of discovered rules, respectively. For example, the information that customers who purchase beer also tend to buy diapers at the same time can be represented in the following association rule:

$$Beer \Rightarrow diaper\ [support = 10\%,\ confidence = 60\%], \tag{7.4}$$

where a support of 10 percent means that 10 percent of all transactions show that beer and diapers are purchased together; a confidence of 60 percent means that 60 percent of the customers who purchased beer also bought diapers. Formally, we can define the rule of support and confidence as follows:

$$support\ (A \Rightarrow B) = P\ (A \cup B), \tag{7.5}$$

$$Confidence(A \Rightarrow B) = P(B|A) = \frac{support(A \cup B)}{support(A)}, \tag{7.6}$$

Where A and B are itemsets like beer and diapers. An itemset can also consist of multiple items, such as

$$(Bread,\ beer) \Rightarrow diaper\ [support = 8\%,\ confidence = 50\%]$$

Association rules are considered interesting if they satisfy both a minimum support threshold and a minimum confidence threshold. In addition, given a support, we want to find the maximum frequent patterns that are not contained in any other patterns. For instance, if (beer, diapers, bread) is a frequent pattern, any subsets of this pattern, such as (beer, diapers) and (diapers, milk), are frequent patterns. Thus, it is not necessary to output its subset anymore.

Figure 7.10 summarizes the advances in algorithms for mining three categories of frequent patterns: frequent itemset patterns, sequential patterns, and subgraph patterns. The figure also presents the relationship between representative frequent pattern-mining algorithms. The solid arrow between algorithm A and algorithm B (i.e., $A \rightarrow B$) denotes algorithm B was derived from A. The dotted lines connecting algorithms at different layers stand for the correspondence of algorithms in different categories. For instance, the equivalent of FreeSpan, which was designed for mining frequent itemsets, in the sequential pattern-mining algorithms is PrefixSpan. They share similar ideas but are designed for mining different frequent patterns. We will go through each category

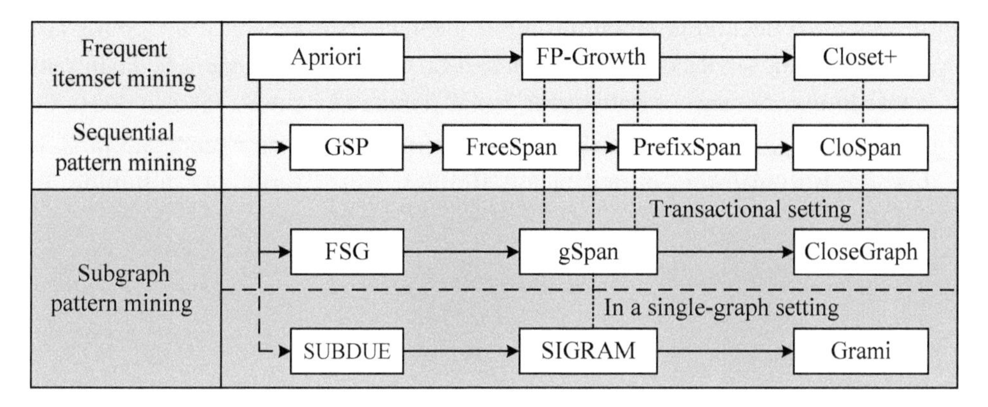

Figure 7.10
Representative algorithms for mining frequent patterns and their relationship.

of algorithms in the following sections, respectively, describing the connections and differences between them.

7.3.2 Frequent Itemset Mining Methods

This section introduces three widely used frequent itemset-mining algorithms: Apriori [83], FP-growth [101], and Closet+ [137], which are depicted in the top layer of figure 7.10. The Apriori algorithm mines frequent itemsets through a generate-and-test approach. However, it may still need to generate a huge number of candidate sets. To improve mining efficiency, the FreeSpan algorithm mines frequent itemset patterns without candidate generation by adopting a divide-and-conquer strategy. Both the Apriori and FreeSpan algorithms detect complete frequent itemsets that are redundant and costly. To address this issue, Closet+ was proposed to mine closed frequent itemsets that are not contained in other superitemsets.

7.3.2.1 Apriori algorithm *Apriori* [83] is a seminal algorithm proposed for mining frequent itemsets for Boolean association rules using prior knowledge of frequent itemsets: *all nonempty subsets of a frequent itemset must also be frequent.* Apriori employs an iterative level-wise search, where k-itemsets are used to explore $(k+1)$-itemsets. First, the set of frequent 1-itemsets is found by scanning the database to accumulate the count for each item and collecting those items that satisfy minimum support. The resulting set is denoted by L_1. Next, L_1 is used to find L_2, the set of frequent 2-itemsets, which is used to find L_3, and so on, until no more frequent k-itemsets can be found. Each level of search is done through the following two steps:

1. *The joint step*. To find L_k, a set of candidate k-itemsets is generated by joining L_{k-1} with itself. This set of candidates is denoted C_k, which is a superset of L_k. That is, its members may or may not be frequent, but all frequent k-itemsets are included in C_k.
2. *The prune step*. A straightforward way to determine whether a candidate in C_k is frequent is to scan the entire database. However, this is very time-consuming. To reduce the size of C_k, the *Apriori property* is used. Any $(k-1)$-itemset that is not frequent cannot be a subset of a frequent k-itemset. In other words, if any $(k-1)$-subset of a candidate k-itemset is not in L_{k-1}, then the candidate cannot be frequent either and thereby can be removed from C_k. This subset testing can be done quickly by maintaining a hash tree of all frequent itemsets [99].

Figure 7.11 illustrates the implementation of the Apriori algorithm, using an example dataset D containing five items $\{I_1, I_2, I_3, I_4, I_5\}$ and nine records $\{T_1, T_2, T_3, T_4, T_5, T_6, T_7, T_8, T_9\}$. In the first iteration, each item is a member of the candidate 1-itemset C_1. The algorithm simply scans all the transactions to count the number of occurrences of each item. Suppose the minimum support count is 2 (i.e., support is $2/9 = 22$ percent), and all candidates in C_1 can be retained as the frequent 1-itemsets L_1. In the second iteration, the algorithm generates a candidate set of 2-itemsets C_2 by joining items in L_1. No candidates can be removed from C_2 during the prune step, as each subset of C_2 is frequent. The transactions in D are scanned again to count the support of each candidate in C_2. Candidates like $\{I_1, I_2\}$ satisfying the minimum support are moved to L_2; others, such as $\{I_2, I_4\}$ and $\{I_4, I_5\}$, are ineligible. In the third iteration, the algorithm generates C_3 by joining the itemsets in L_2:

$$C_3 = \{\{I_1, I_2, I_3\}, \{I_1, I_2, I_5\}, \{I_2, I_3, I_5\}\}.$$

However, $\{I_3, I_5\}$ is not a frequent itemset in L_2. According to the Apriori property, $\{I_2, I_3, I_5\}$ cannot be a frequent itemset and thereby cannot be removed from C_3 without scanning D. Then, the algorithm scans D, counting the number of occurrences for the two candidates. Finally, we find both candidates satisfy the minimum support and thus are frequent itemsets in L_3.

7.3.2.2 FP-growth algorithm The candidate generate-and-test method in the Apriori algorithm significantly reduces the size of candidate sets. However, it may still need to generate a huge number of candidate sets. If there are 10^4 frequent 1-item sets, the algorithm will generate more than 10^7 candidates. Additionally, it may need to repeatedly scan the entire database and check the support of a large number of candidates. To address these issues, Han et al. [101] have proposed the *frequent pattern growth*

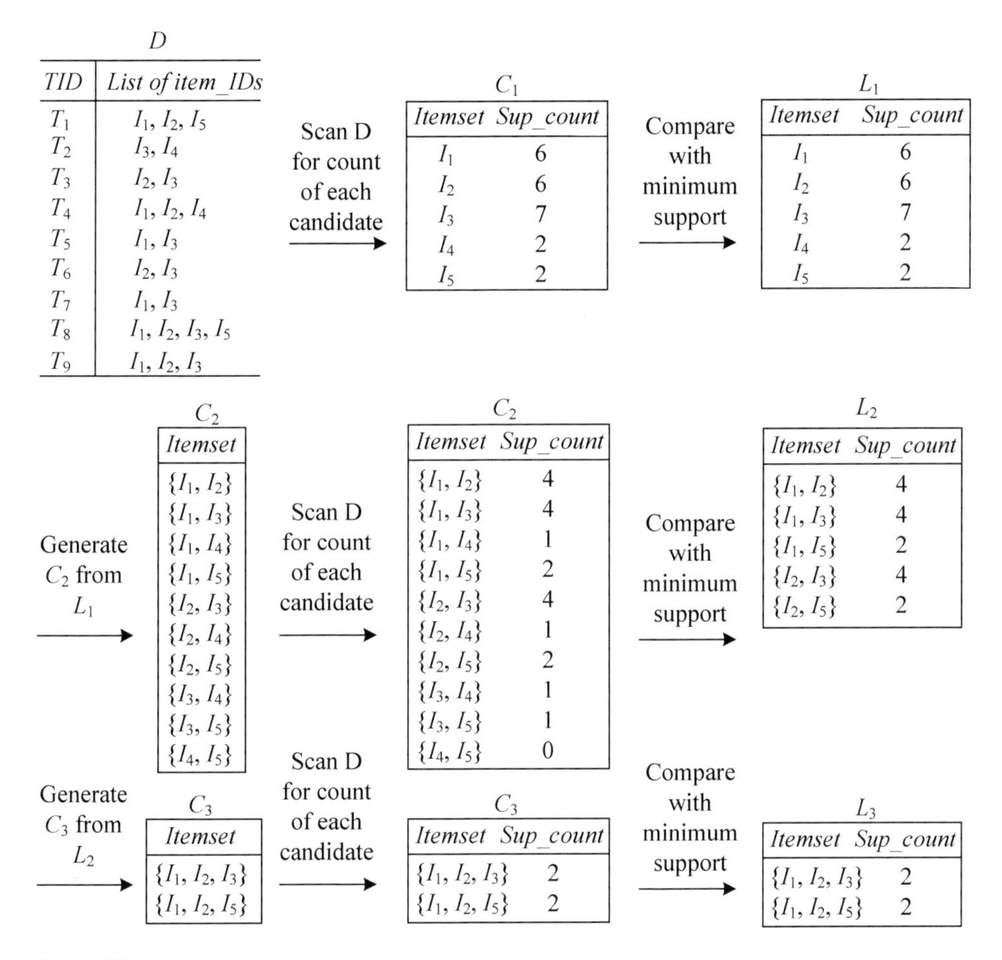

Figure 7.11
An illustration of the Apriori algorithm.

(FP-growth) method, which mines frequent itemset patterns without candidate generation by adopting a divide-and-conquer strategy. First, it compresses the database representing frequent items into a frequent pattern tree, or FP-tree, which retains the itemset association information. It then divides the compressed database into a set of conditional databases (a special kind of projected database), each associated with one of the frequent item or "pattern fragments," and mines each database separately. For each pattern fragment, only its associated datasets need to be examined. Thus, this approach may substantially reduce the size of the data to be searched.

Figure 7.12 presents an example of FP-growth using the same transaction dataset D. The first scan of D is the same as the Apriori algorithm, deriving the set of frequent items (1-itemsets) and their support counts. Let us use the same minimum support count two as the previous example. The frequent items are sorted in a descending order of the support count. The resulting set is denoted by L. An FP-tree is then constructed as follows. First, create the root of the tree, labeled with "null." Scan D a second time, creating a branch in the tree for each transaction. Each transaction is processed in L order. For example, T_9 will be converted to three items $<I_3: 1>$, $<I_1: 1>$, $<I_2: 1>$, which are linked to the PF-tree to formulate a branch. More specifically, $<I_3: 1>$ is linked to the root node, and $<I_1: 1>$ is linked to $<I_3: 1>$ as a children node. Then, $<I_2: 1>$ is linked to $<I_1: 1>$ as a children node. If a transaction to be inserted shares a common prefix with previous branches in the FP-tree, the algorithm increments the count of the nodes on the shared branch. For instance, after inserting T_1, the FP-tree contains a branch $<I_1: 1>$, $<I_2: 1>$, $<I_5: 1>$. When inserting T_4 (i.e., $<I_1: 1>$, $<I_2: 1>$, $<I_4: 1>$), the first two nodes have already existed in the tree. Instead of creating a new branch, the algorithm increments the counts of the common prefix (i.e., $<I_1: 2>$, $<I_2: 2>$), as illustrated in figure 7.12b. To facilitate the tree traversal, an item header table is built so that each item points to its occurrences in the tree via a chain of node links, denoted by the broken arrows in the figure.

The FP-tree is mined as follows. The algorithm starts from each frequent length-1 pattern (as initial suffix pattern), constructing its conditional pattern bases, which consist of the set of prefix paths in the FP-tree co-occurring with the suffix pattern. The algorithm then constructs the conditional FP-tree based on the conditional pattern bases for a length-1 pattern and performs mining recursively on the tree. The pattern growth is achieved by the concatenation of the suffix pattern with the frequent patterns generated from a conditional FP-tree. As illustrated in figure 7.12d, the algorithm first considers I_5, which is the last item in L. I_5 occurs in two branches of the FP-tree: $<I_1, I_2, I_5: 1>$ and $<I_3, I_1, I_2, I_5: 1>$. Considering I_5 as a suffix, its two prefix paths are $<I_1, I_2: 1>$ and $<I_3, I_1, I_2: 1>$, which form the conditional pattern base of I_5. Using this conditional pattern base as a transaction database, we build an I_5 conditional FP-tree, which contains only a single path, $<I_1: 2>$, $<I_2: 2>$; I_3 is not included because its support count of 1 is less than the threshold in the conditional database. By combining the suffix pattern I_5 with frequent patterns generated in the conditional FP-tree (i.e., the single path), the algorithm generates frequent patterns: $\{I_1, I_5: 2\}$, $\{I_2, I_5: 2\}$, $\{I_1, I_2, I_5: 2\}$.

For I_4, its two prefix paths form the conditional pattern base, $\{\{I_1, I_2: 1\}, \{I_3: 1\}\}$, which cannot generate an FP-tree with a branch's support count satisfying the threshold. Consequently, no frequent patterns are generated.

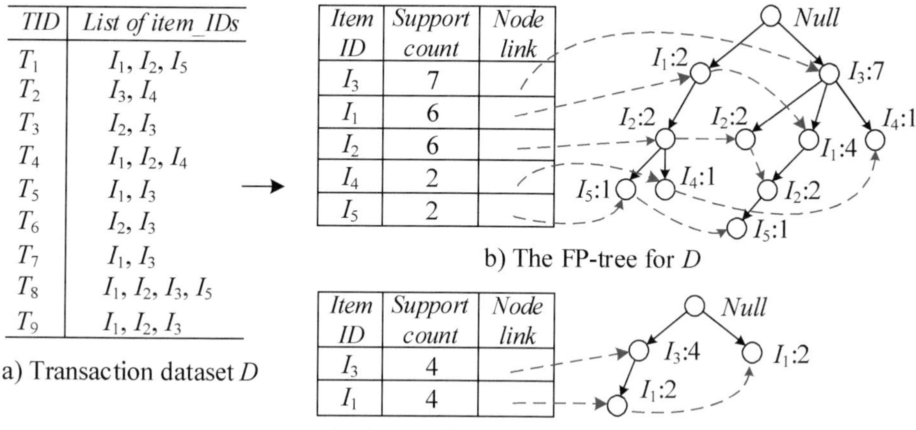

TID	List of item_IDs
T_1	I_1, I_2, I_5
T_2	I_3, I_4
T_3	I_2, I_3
T_4	I_1, I_2, I_4
T_5	I_1, I_3
T_6	I_2, I_3
T_7	I_1, I_3
T_8	I_1, I_2, I_3, I_5
T_9	I_1, I_2, I_3

a) Transaction dataset D

Item ID	Support count	Node link
I_3	7	
I_1	6	
I_2	6	
I_4	2	
I_5	2	

b) The FP-tree for D

Item ID	Support count	Node link
I_3	4	
I_1	4	

c) The conditional FP-tree associated with node I_2

Item	Conditional pattern base	Conditional FP-tree	Frequent patterns generated
I_5	$\{I_1, I_2: 1\}, \{I_3, I_1, I_2: 1\}$	$<I_1: 2, I_2: 2>$	$\{I_1, I_5: 2\}, \{I_2, I_5: 2\}, \{I_1, I_2, I_5: 2\}$
I_4	$\{I_1, I_2: 1\}, \{I_3: 1\}$	null	null
I_2	$\{I_1: 2\}, \{I_3: 2\}, \{I_3, I_1: 2\}$	$<I_3: 4, I_1: 2>, <I_3: 2>$	$\{I_3, I_1: 4\}, \{I_3, I_2: 4\}, \{I_3, I_1, I_2: 2\}$
I_1	$\{I_3: 4\}$	$<I_3: 4>$	$\{I_3, I_1: 4\}$

d) Mining the FP-tree by creating conditional subpattern bases

Figure 7.12
Mining frequent patterns using the FP-growth algorithm.

Similarly, I_2's conditional pattern base is $\{\{I_1: 2\}, \{I_3: 2\}, \{I_3, I_1: 2\}\}$. Its conditional FP-tree has two branches, $<I_3: 4, I_1: 2>$, $<I_3: 2>$, as illustrated in figure 7.12c. Finally, I_3's conditional pattern base is $\{I_3: 4\}$,with an FP-tree that only contains one node $<I_3: 4>$, which generates one frequent pattern, $\{I_3, I_1: 4\}$.

7.3.2.3 Closet+ algorithm The Apriori and the FP-growth algorithms may have good performance when the support threshold is high, and the pattern space is sparse. However, when the support threshold drops low, the number of frequent itemsets increases dramatically. The performance of these algorithms deteriorates quickly because of the generation of a huge number of patterns. Moreover, a complete frequent itemset may not be that useful as there are numerous redundant patterns. Closed frequent itemsets can substantially reduce the number of patterns while preserving complete information regarding the set of frequent items. X is a closed itemset in a dataset $\boldsymbol{S}$ if there exists no proper superitemset Y such that Y has the same support count as X in $\boldsymbol{S}$, and X satisfies the minimum support. A naïve approach would be to first mine the complete set of frequent items and then remove every frequent itemset that is a proper subset of, and carries the same support as, an existing frequent itemset. However, this is quite costly.

To address this issue, Closet+ [137] was proposed to mine closed itemsets using the divide-and-conquer paradigm and the depth-first search strategy. A hybrid tree-projection method is introduced to improve searching efficiency. It employs FP-tree as the compression technique and computes the local frequent items of a certain prefix by building and scanning its projected database (we will provide more details about projected databases in section 7.3.3.2). Unlike frequent itemset mining, the Closet+ algorithm removes the prefix itemsets during the closed itemset mining process that are unpromising for growing closed itemsets. In addition, the algorithm proposes the item-skipping technique to further prune search space.

It is still very costly to maintain all the frequent closed itemsets mined so far in memory in order to check if a newly found closed itemset candidate is really closed. To tackle this challenge, Closet+ designs an efficient subset-checking scheme: a combination of the two-level hash-indexed result tree–based method and the pseudo-projection–based upward-checking method, that can save memory usage and accelerate closure-checking significantly.

7.3.3 Sequential Pattern Mining

This section introduces four widely used algorithms for sequential pattern mining: GSP [133], FreeSpan [126], PrefixSpan [100], and CloSpan [142], as illustrated in the middle layer of figure 7.10. The first algorithm is based on the idea of the Apriori algorithm (i.e., candidate generation and testing). The latter three algorithms are based on the idea of the FP-growth algorithm, generating sequential patterns by pattern growth without candidate generation. Thus, they are more efficient than GSP. Further, PrefixSpan is faster than FreeSpan algorithms in most sequential databases, as it avoids checking every possible combination of a potential candidate sequence. The first three algorithms generate complete frequent sequential patterns, which are redundant and time-consuming. To address this issue, CloSpan mines frequent closed subsequences that are not contained in any supersequences of the same support.

A sequence, denoted by $s=<e_1e_2\dots e_l>$, is an ordered list of itemsets, $I=<I_1, I_2, \dots, I_m>$. Each element e_i of the sequence is an itemset (i.e., $e_i=(x_1x_2\dots x_j)$, where $x_j \in I$ is an item). Figure 7.13 presents a sequence database S, which comprises four sequences $<s_1, s_2, s_3, s_4>$. For example, the first sequence, $s_1=<a(abc)(ac)d(cf)>$, denotes five itemsets that are recorded in an order of $a \to (abc) \to (ac) \to d \to (cf)$. An item can occur at most once in an element of a sequence but can occur multiple times in different elements of a sequence. The number of instances of items in a sequence is called the *length*

SID	*Sequences*
s_1	*<a(abc)(ac)d(cf)>*
s_2	*<(ad)c(bc)(ae)>*
s_3	*<(ef)(ab)(df)cb>*
s_4	*<eg(af)cbc>*

a) A sequence database

4th scan, 4 candidates
4 length-4 sequential patterns

3rd scan, 64 candidates
21 length-3 sequential patterns

2nd scan, 51 candidates
22 length-2 sequential patterns

1st scan, 7 candidates
6 length-1 sequential patterns

☐ Cannot pass the threshold

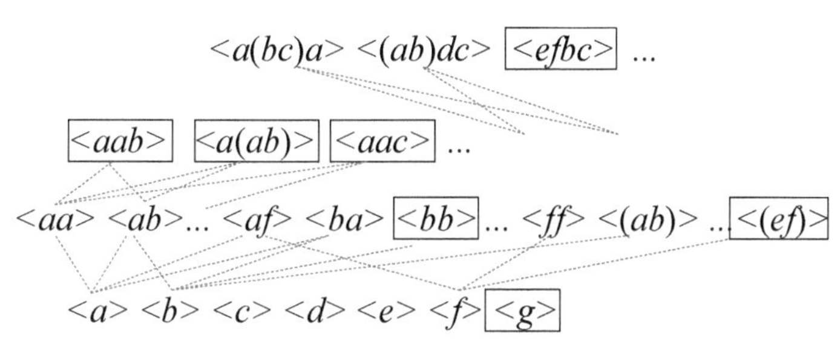

b) An illustration of the GSP algorithm

Figure 7.13
An illustration of the GSP algorithm with a sequence database *S*.

of the sequence. For instance, the first sequence in ***S*** is a 9-length sequence. A sequence $\alpha = <a_1, a_2, \ldots, a_n>$ is called a *subsequence* of another sequence $\beta = <b_1, b_2, \ldots, b_m>$, denoted as $\alpha \sqsubseteq \beta$, if there exists integer $1 \le j_1 < j_2 < \ldots < j_n \le m$ such that $a_1 \subseteq b_{j_1}, a_2 \subseteq b_{j_2}, \ldots, a_n \subseteq b_{j_n}$. Alternatively, we can say β is a supersequence of α. For example, *<a(bc)df>* is a subsequence of s_1. The support of sequence α, denoted as $support(\alpha)$, is the number of sequences (in ***S***) that are supersequences of α. A sequence can only contribute once to its subsequence's support, even if a subsequence may occur multiple times in the sequence. For example, if setting the minimum support to 2, *<(ab)c>* is a frequent sequential pattern, as it is a subsequence of s_1 and s_3. Given a sequence database ***S*** and a minimum support threshold, sequential pattern mining is used to find the complete set of sequential patterns in the database.

7.3.3.1 GSP algorithm A typical general sequential pattern-mining method called *GSP* [133] mines sequential patterns by adopting a candidate subsequence generation-and-test approach, based on the a priori principle. Figure 7.13b illustrates the mining process.

The GSP first scans S, counts the support for each item, and finds the set of frequent items (i.e., frequent length-1 subsequences): $\langle a \rangle$: 4, $\langle b \rangle$: 4, $\langle c \rangle$: 3, $\langle d \rangle$: 3, $\langle e \rangle$: 3, $\langle f \rangle$: 3, $\langle g \rangle$: 1. By filtering the infrequent item g, we obtain the seed set $L_1 = \{\langle a \rangle, \langle b \rangle, \langle c \rangle, \langle d \rangle, \langle e \rangle, \langle f \rangle\}$, each element of which is a one-element sequential pattern. Each subsequent pass starts with the seed set found in the previous pass and uses it to generate new potential sequential patterns, called *candidate sequences*. For example, fifty-one candidate sequences are generated in the second pass based on L_1. In the k-th pass, a sequence is a candidate only if each of its length-$(k-1)$ subsequences is a sequential pattern found at the $(k-1)$th pass. A new scan of the database collects the support for each candidate sequence and finds the new set of sequential patterns. The algorithm repeats this procedure until no sequential pattern is found in a pass or no candidate sequence is generated. The GSP algorithm shares similar strengths and weaknesses as the Apriori algorithm. Though the GSP algorithm can prune many infrequent candidates, the effort for candidate generation is still huge, and the number of scans is at least the maximum length of sequential patterns.

An intuitive idea is to employ the FP-growth algorithm to mine sequential patterns. Because it is not easy to find common data structures among a set of ordered items, however, the FP-growth algorithm cannot be applied to solve this problem directly. To address this issue, two algorithms, called *FreeSpan* [126] and *PrefixSpan* [100], were proposed to mine sequential pattern mining by pattern growth. Instead of repeatedly scanning the entire database and generating and testing large sets of candidate sequences, the two algorithms recursively project a sequence database into a set of smaller databases associated with the set of patterns mined so far and then mine locally frequent patterns in each projected database. The two algorithms differ at the criteria of database projection: FreeSpan creates projected databases based on the current set of frequent patterns without a particular ordering (i.e., growth direction), whereas PrefixSpan projects databases by growing frequent prefixes. As a result, PrefixSpan is substantially faster than FreeSpan in most sequence databases [100].

7.3.3.2 FreeSpan algorithm For a sequence $\alpha = <e_1 e_2 \cdots e_l>$, the itemset $e_1 \cup e_2 \cup \cdots \cup e_l$ is called α's *projected itemset*. FreeSpan is based on the following property: if an itemset X is infrequent, any sequence whose projected itemset is a superset of X cannot be a sequential pattern. Let us illustrate the FreeSpan algorithm using the same example shown in figure 7.13a, setting the support threshold as 2. The first step is similar to GSP—collecting the support for each item through scanning $\boldsymbol{S}$ and finding the frequent items, which are listed in support descending order: *f_list*=$\langle a \rangle$: 4, $\langle b \rangle$: 4, $\langle c \rangle$: 3, $\langle d \rangle$: 3, $\langle e \rangle$: 3, $\langle f \rangle$: 3. According to the *f_list*, the complete set of sequential patterns in $\boldsymbol{S}$ can be divided into six disjoint subsets, those containing only item a (i.e., $\{\langle aaa\rangle, \langle aa\rangle, \langle a\rangle, \langle a\rangle\}$), which is called the *⟨a⟩-projected database*; those containing item b but no item after b in *f_list* (i.e., $\{\langle a(ab)a\rangle, \langle aba\rangle, \langle (ab)b\rangle, \langle ab\rangle\}$), which is called the *⟨b⟩-projected database*; those containing item c but no item after c in *f_list* and so on; and finally, those containing item f. The sequential patterns related to the six partitioned subsets can be mined by constructing six projected databases (obtained by one additional scan of $\boldsymbol{S}$).

By mining the $\langle a\rangle$-projected database $\{\langle aaa\rangle, \langle aa\rangle, \langle a\rangle, \langle a\rangle\}$, only one additional sequential pattern solely containing item a (i.e., $\langle aa\rangle{:}2$) is found. By mining the $\langle b\rangle$-projected database, four additional sequential patterns containing item b but no item after b in *f_list* are found. They are $\{\langle ab\rangle{:}4, \langle ba\rangle{:}2, \langle (ab)\rangle{:}2, \langle aba\rangle{:}2\}$. When mining the $\langle c\rangle$-projected database $\{\langle a(abc)(ac)c\rangle, \langle ac(bc)a\rangle, \langle (ab)cb\rangle, \langle acbc\rangle\}$, the process is as follows.

One scan of the projected database generates the set of length-2 frequent sequences, which are $\{\langle ac\rangle{:}4, \langle (bc)\rangle{:}2, \langle bc\rangle{:}3, \langle cc\rangle{:}3, \langle ca\rangle{:}2, \langle cb\rangle{:}3\}$. One additional scan of the $\langle c\rangle$-projected database generates the projected databases for the six patterns, respectively. For example, the $\langle ac\rangle$-projected database is $\{\langle a(abc)(ac)c\rangle, \langle ac(bc)a\rangle, \langle (ab)cb\rangle, \langle acbc\rangle\}$, and the $\langle bc\rangle$-projected database is $\{\langle a(abc)(ac)c\rangle, \langle ac(bc)a\rangle\}$. The algorithm then mines the $\langle ac\rangle$-projected database to generate the set of length-3 patterns $\{\langle acb\rangle{:}3, \langle acc)\rangle{:}3, \langle (ab)c\rangle{:}2, \langle aca)\rangle{:}2\}$. Four projected databases will be generated for the four length-3 patterns. The mining of the $\langle acb\rangle$-projected database, $\{\langle ac(bc)a\rangle, \langle (ab)cb\rangle, \langle acbc\rangle\}$, generates no length-4 pattern. Thus, this line terminates. Similarly, the mining of the other three projected databases terminates without generating any length-4 patterns for the $\langle ac\rangle$-projected database. Based on the same procedure, the algorithm mines projected databases of other length-2 patterns recursively, deriving the complete set of sequential patterns.

On the one hand, FreeSpan searches a smaller projected database than GSP in each subsequent database projection. On the other hand, FreeSpan may have to generate many nontrivial projected databases. If a pattern appears in each sequence of a database, its projected database does not shrink (except for the removal of some infrequent items). For example, the $\langle f\rangle$-projected database in this example contains three of the same sequences as that in the original sequence database. Moreover, as there are many ways of growing a length-$(k+1)$ candidate sequence from a length-k subsequence, checking possible combinations is costly.

7.3.3.3 PrefixSpan algorithm To avoid checking every possible combination of a potential candidate sequence, PrefixSpan fixes the order of items within each element alphabetically. For example, the sequence s_4 is represented as $\langle a(abc)(ac)d(cf)\rangle$ instead of $\langle a(abc)(ac)d(fc)\rangle$. Then, the algorithm follows the order of the prefix of a sequence

Prefix	Projected (suffix) database	Sequential patterns
<a>	<(abc)(ac)d(cf)>, <(_d)c(bc)(ae)>, <(_b)(df)(cb)>, <(_f)cbc>	<a>,<aa>, <ab>, <a(bc)>, <a(bc)a>, <aba> <abc>, <(ab)>, <(ab)c)>, <(ab)d>, <(ab)f> <(ab)dc>, <ac>, <aca)>, <acb>, <acc>, <ad>, <adc>, <af>
<b>	<(_c)(ac)d(cf)>, <(_c)(ae)> <(df)cb)>, <c>	<b>, <ba>, <bc>, <(bc)>, <(bc)a>, <bd>, <bdc>, <bf>
<c>	<(ac)d(cf)>, <(bc)(ae)>, <b>, <bc>	<c>, <ca>, <cb>, <cc>
<d>	<(cf)>,<c(bc)(ae)>, <(_f)cb>	<d>, <db>, <dc>, <dcb>
<e>	<(_f)(ab)(df)cb>,<(af)cbc>	<e>, <ea>, <eab>, <eac>, <eacb>, <eb>, <ebc>, <ec>, <ecb>, <ef>, <efb>, <efc>, <efcb>
<f>	<(ab)(df)cb>,<cbc>	<f>,<fb>,<fbc>, <fc>, <fcb>

Figure 7.14
An illustration of the PrefixSpan algorithm [127].

and projects only the suffix of a sequence. For instance, ⟨*a*⟩, ⟨*aa*⟩, ⟨*a*(*ab*)⟩, and ⟨*a*(*abc*)⟩ are prefixes of s_4. However, neither ⟨*ab*⟩ nor ⟨*a*(*bc*)⟩ is considered a prefix if every item in the prefix ⟨*a*(*abc*)⟩ is frequent in ***S*** [127]. ⟨(*a*(*bc*) (*ac*)*d*(*cf*)⟩ is the suffix regarding the prefix ⟨*a*⟩, ⟨(_*bc*) (*ac*)*d*(*cf*)⟩ is the suffix with respect to the prefix ⟨*aa*⟩, and ⟨(_*c*)(*ac*)*d*(*cf*)⟩ is the suffix of ⟨*a*(*ab*)⟩.

Figure 7.14 illustrates the PrefixSpan algorithm using the same setting as the previous examples shown in figure 7.12 and figure 7.13. Its first step is the same as GSP and FreeSpan, finding all the length-1 frequent sequences: ⟨*a*⟩: 4, ⟨*b*⟩: 4, ⟨*c*⟩: 3, ⟨*d*⟩: 3, ⟨*e*⟩: 3, ⟨*f* ⟩: 3. The complete set of sequential patterns can be partitioned into the following six subsets according to the six prefixes: 1) the ones with prefix ⟨*a*⟩, 2) the ones with prefix ⟨*b*⟩, and so on, until 3) the ones with prefix ⟨*f* ⟩. The subset of sequential patterns can be mined by constructing the corresponding set of projected databases and mining each recursively. In a sequence containing ⟨*a*⟩, only the subsequence prefixed with the first occurrence of ⟨*a*⟩ should be considered. For example, in sequence ⟨(*ef*)(*ab*)(*df*)*cb*⟩, only the subsequence ⟨(_*b*)(*df*)*cb*⟩ should be considered for mining sequential patterns prefixed with ⟨*a*⟩. (_*b*) means *a* is the last element in the prefix.

The sequences in ***S*** containing ⟨*a*⟩ are projected to form the ⟨*a*⟩-projected database, which consists of four suffix sequences: ⟨(*abc*)(*ac*)*d*(*cf*)⟩, ⟨(_*d*)*c*(*bc*)(*ae*)⟩, ⟨(_*b*)(*df*)*cb*)⟩, and ⟨(_*f*)*cbc*)⟩. By scanning the ⟨*a*⟩-projected database once, its locally frequent items are *a*: 2, *b*: 4, _*b*: 2, *c*: 4, *d*: 2, and *f*: 2. Thus, all the length-2 sequential patterns prefixed with ⟨*a*⟩ are ⟨*aa*⟩: 2, ⟨*ab*⟩: 4, ⟨(*ab*)⟩: 2, ⟨*ac*⟩: 4, ⟨*ad*⟩: 2, and ⟨*af* ⟩: 2. Recursively, all sequential patterns with prefix ⟨*a*⟩ can be partitioned into six subsets: 1) those prefixed with ⟨*aa*⟩, 2) those prefixed with ⟨*ab*⟩, …, and finally, 3) those prefixed with ⟨*af* ⟩. These subsets can be mined by constructing respective projected databases and mining each recursively as follows:

The ⟨aa⟩-projected database is composed of two suffix subsequences prefixed with ⟨aa⟩: ⟨(_bc)(ac)d(cf)⟩ and ⟨(_e)⟩. As a result, no frequent subsequence can be generated from the two sequences. The process for mining the ⟨aa⟩-projected database terminates. The ⟨ab⟩-projected database consists of three suffix sequences: ⟨(_c(ac)d(cf)⟩, ⟨(_c)a)⟩, and ⟨c⟩. Recursively mining the ⟨ab⟩-projected database returns four sequential patterns: ⟨(_c)⟩, ⟨(_c)a⟩, ⟨a⟩, and ⟨c⟩. They form the complete set of sequential patterns prefixed with ⟨ab⟩ (i.e., ⟨a(bc)⟩, ⟨a(bc)a⟩, ⟨aba⟩, and ⟨abc⟩). The ⟨(abc)⟩-projected database consists of two suffix sequences: ⟨(_c)(ac)d(cf)⟩ and ⟨(df)(cb)⟩, which result in the following sequential patterns prefixed with ⟨(ab)⟩: ⟨c⟩, ⟨d⟩, ⟨f⟩, and ⟨dc⟩. Thus, the final sequential patterns are ⟨(ab)c⟩, ⟨(ab)d⟩, ⟨(ab)f⟩, and ⟨(ab)dc⟩, as shown in figure 7.14. The ⟨ac⟩, ⟨ad⟩, and ⟨af⟩-projected databases can be constructed and mined similarly.

Finding sequential patterns with prefix ⟨b⟩, ⟨c⟩, ⟨d⟩, ⟨e⟩, and ⟨f⟩ can be done by constructing the corresponding projected databases and mining them in a similar way as that of ⟨a⟩-projected databases.

7.3.3.4 CloSpan algorithm The aforementioned frequent sequential pattern-mining algorithms confront a problem similar to that faced by frequent itemset mining algorithms—generating the complete sequential pattern set is redundant and time-consuming. To address this issue, the CloSpan algorithm was proposed to mine frequent closed subsequences (i.e., those containing no supersequence with the same support). For example, as shown in figure 7.14, the support of both of the two frequent sequential patterns ⟨fcb⟩ and ⟨efcb⟩ is 2. Thus, ⟨fcb⟩ is considered redundant and not a closed sequential pattern.

To address this issue, the CloSpan algorithm [142] was proposed to mine closed sequential patterns based on the concept of equivalence of projected databases. In addition, a hash-based algorithm was designed to optimize the search space at a negligible cost. Instead of introducing the technical details, which can be found in [142], we present an example of using the CloSpan algorithm to mine sequential patterns from human mobility data (e.g., user-generated GPS trajectories) [143].

As illustrated in figure 7.15, given a user's GPS log, each record of which is composed of a latitude, longitude, and time stamp, as well as other information such as speed, we can formulate a GPS trajectory, which is a sequence of time-ordered spatial points. We detect some stay points, like $S1$ and $S2$, where a user has spent a period of time within a certain spatial distance, using a stay point–detection algorithm [119]. Such stay points carry a semantic meaning, such as visiting a tourist attraction, having

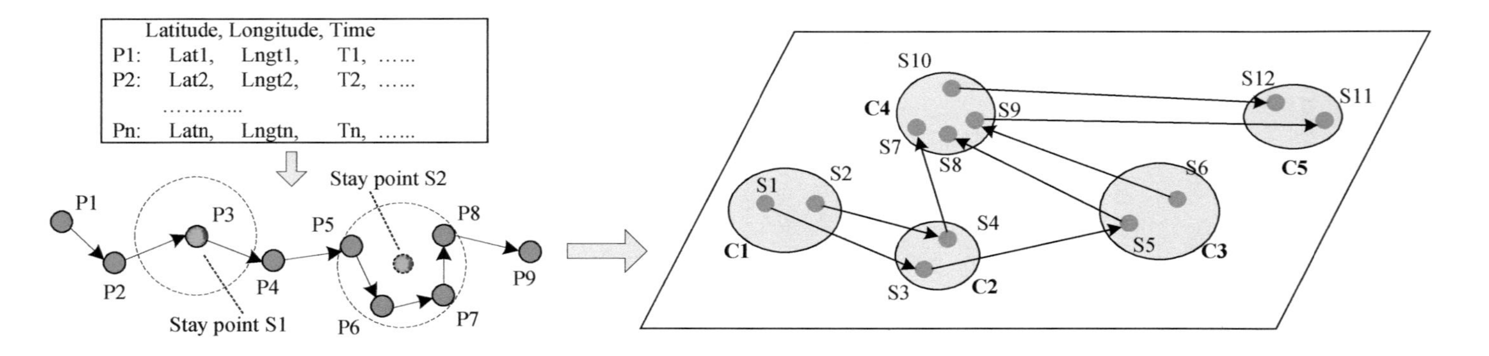

Figure 7.15
Mining sequential patterns from people's GPS trajectories.

dinner in a restaurant, or shopping in a mall; therefore, they are more important than other GPS points in the trajectory. Now, a GPS trajectory is converted to a sequence of stay points, which significantly simplifies the representation of human mobility while capturing the semantic meanings of a trajectory. Having a user's GPS logs of multiple days, we will formulate multiple sequences of stay points, such as $S1 \rightarrow S3 \rightarrow S5 \rightarrow S8$, and $S2 \rightarrow S4 \rightarrow S7$.

Due to the inaccuracy of GPS points, people visiting the same place may generate slightly different stay points. Consequently, sequences are not directly comparable, as they do not exactly share any items. To this end, we further group stay points into clusters (e.g., $S1$ and $S2$ are grouped into cluster $C1$). As a result, the user's mobility illustrated in figure 7.15 is represented by four sequences of cluster IDs: $C1 \rightarrow C2 \rightarrow C4$, $C1 \rightarrow C2 \rightarrow C3 \rightarrow C4$, $C3 \rightarrow C4 \rightarrow C5$, and $C4 \rightarrow C5$. Using the CloSpan algorithm and setting 2 as the support threshold, we find the following closed sequential patterns: $\langle C1 \rightarrow C2 \rightarrow C4 \rangle$, $\langle C3 \rightarrow C4 \rangle$, and $\langle C4 \rightarrow C5 \rangle$.

7.3.4 Frequent Subgraph Pattern Mining

Frequent subgraph mining (FSM) is a fundamental building block of graph mining. The objective of FSM is to extract all the frequent subgraphs that appear more times than a given threshold in a graph dataset. Graph datasets can be divided into two categories: datasets consisting of many stand-alone (small) graphs, called a *transactional setting*, and datasets with a single large graph. In a transactional graph setting, frequent subgraph patterns are subgraphs contained in a proportion of graphs in a graph dataset; the proportion is larger than a given threshold (e.g., 30 percent). In a single graph setting, frequent subgraphs are those occurring in the large graph more than a certain number of times.

7.3.4.1 FSM in a transactional setting As shown on the bottom layer of figure 7.10, representative FSM methods with a transactional setting include the FSG [111], gSpan [140], and CloseGraph [141]. The FSG algorithm is an Apriori-based algorithm [1], which creates candidate frequent subgraphs of size $(k+1)$ by joining frequent subgraphs of size k. gSpan is a pattern growth–based approach, similar to FP-growth, for mining frequent itemsets. It combines the growing and checking of frequent subgraphs into one procedure without candidate generation. CloseGraph find closed frequent subgraph patterns that are not contained in any supergraphs with the same support.

FSG Given a graph database $G = \{G_1 \ldots G_n\}$ and a minimum support θ, an FSM algorithm is used to find frequent subgraphs with a support no less than θ in a fraction of graphs. The basic FSM algorithm, called *FSG*, is an Apriori-based algorithm, which uses level-by-level expansion to generate candidate subgraphs of size $(k+1)$ from the frequent subgraphs of size k. The "size" of a subgraph can be either the number of nodes or the number of edges. We call a graph of size k k-subgraph or k-graph. During the join process, a subgraph isomorphism algorithm [134] is used to determine whether two k-graphs share a common subgraph of size $(k-1)$ so that they can be joined to form $k+1$-graphs. Candidate frequent patterns can be generated by either node joins or edge joins between two graphs.

The node-based join starts with a set of single nodes. The labels of these nodes are present in at least θ graphs in G. When two k-subgraphs with $k-1$ common nodes are joined to create a candidate with $k+1$ nodes, an ambiguity exists (i.e., if an edge exists between the two nonmatching nodes). For example, as illustrated in the top part of figure 7.16, if two 3-subgraphs G_1 and G_2 are merged through a node join, there could be two candidates of 4-subgraphs, C_1 and C_2. The nodes (e.g., $v_1 - v_2$, $v_1 - v_3$, and $v_2 - v_3$) that are connected in a k-subgraph must be connected in a $k+1$-subgraph after the join. However, since the two v_3 link to v_1 (in G_1) and v_2 (in G_2), respectively, they are regarded as different nodes in the node-based join. Thus, the two v_3 could be connected or disconnected in the $k+1$-subgraph.

In the case of edge-based joins, each singleton graph contains a single edge, which appears between specific node labels in at least θ graphs in G. To join two k-graphs (i.e., the number of edges in each graph is k), a matching graph with $k-1$ edges should be found in the two graphs. The resulting candidate will contain exactly $k+1$ edges, while the number of nodes in the candidate may not be greater than that in the original graphs. As shown in the bottom part of figure 7.16, C_4 has the same number of nodes as the original pair of graphs. This is because when finding candidates, subgraph isomorphism is checked. Generally, the edge-based join growth tends to generate fewer candidates in total and is therefore more efficient.

Afterward, the Apriori pruning technique is applied to reduce the number of candidates. That is, if any k-subgraphs contained in a candidate $k+1$-graph are infrequent, the candidate $k+1$-graph can be pruned. For each remaining candidate subgraph, the support is computed with respect to the graph database G. If the support is no less than θ, a frequent $(k+1)$-subgraph pattern is generated. The procedure is repeated iteratively until no frequent subgraph patterns can be found. The drawback of the Apriori-based approach is the costly join operation and the pruning of false positives.

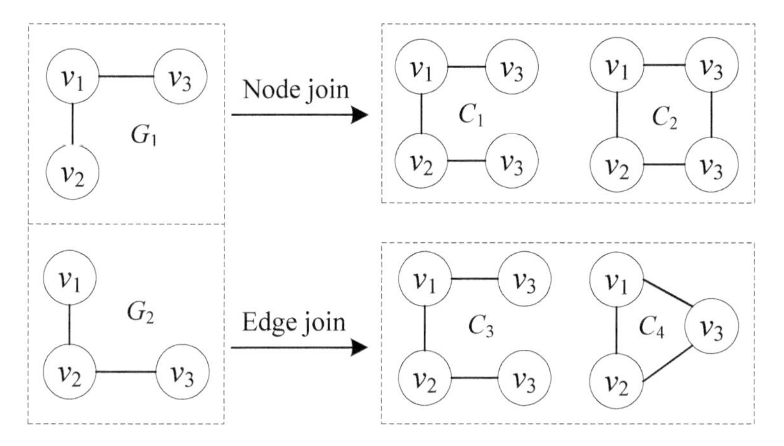

Figure 7.16
Joins between two graphs for subgraph pattern mining.

gSpan To address this issue, gSpan [140] was proposed to extend subgraphs directly from a single subgraph based on a pattern-growth approach, instead of joining two existing frequent subgraphs. The algorithm uses DFS (depth-first search) lexicographic ordering to construct a hierarchical search space, called a *DFS code tree*, over all possible patterns. Each node of this search tree represents a DFS code. The $(k+1)$-th level of the tree has nodes containing DFS codes for k-subgraphs. The nodes on the first level of this tree only contain vertex labels (i.e., 0-edge). Thus, the $(k+1)$-subgraphs are generated by a one-edge expansion from the DFS codes on the $(k+1)$-th level of the tree. This search tree is traversed in a DFS manner, and all subgraphs with nonminimal DFS codes are pruned so that redundant candidate generations are avoided. Since gSpan extends subgraphs directly from a single subgraph instead of joining two previous subgraphs, it has a much lower cost compared to the join-based methods such as FSG. Experiments show that gSpan outperforms FSG by an order of magnitude.

CloseGraph Though gSpan is much faster than FSG, it still generates many redundant subgraph patterns. Just like the closed itemsets and closed sequential patterns, a graph g is closed in a dataset if there exists no proper supergraph of g that has the same support as g. However, gSpan builds a very strict order in graph patterns, preventing itself from generating the closed subgraph patterns. To address this issue, CloseGraph proposes to mine closed subgraph patterns from a given graph database [141] using two novel concepts, *equivalent occurrence* and *early termination,* to prune the search space. Because there could be some cases where the early termination may fail and miss some patterns, a detection of the failure is implemented to remove such cases, guaranteeing the completeness and soundness of the closed graph patterns. Experimental results show that CloseGraph outperforms gSpan and FSG.

7.3.4.2 FSM in a single graph FGM with a single-graph setting finds the subgraphs that frequently occur in a single large graph. The single-graph setting is a generalization of the transactional one, since a set of small graphs can be considered as connected components within a single large graph. However, detecting frequent subgraphs in a single graph is more complicated because multiple instances of identical subgraphs may overlap. The computational load increases exponentially with the size of a graph.

As shown in figure 7.17, a graph comprises seven nodes, and the number associated with each edge is a weight. Subgraph S has three isomorphic subgraphs in G: (u_2, u_4, u_5), (u_2, u_3, u_5), (u_7, u_6, u_3). Two graphs are isomorphic when the vertices of one can be relabeled to match the vertices of the other in a way that preserves adjacency. For instance, if we replace the label of u_2 with v_1, u_4 with v_2, and u_5 with v_3, (u_2, u_4, u_5) share the same adjacency between each pair of nodes. Note that not only the matching nodes but also the matching edges have identical labels. Thus, (u_5, u_3, u_6) is not an isomorphic of S considering the weight of an edge. An isomorphic subgraph is also called an *instance, occurrence, appearance,* or *embedding* in the literature. If we set the support to 3, S is a frequent subgraph pattern in the graph G.

The bottom part of figure 7.10 shows the advances in FSM methods with a single graph setting. SUBDUE [92] is based on the idea of grow-and-store, which is somehow similar to that of Apriori but matches two subgraphs approximately. SIGRAM [111] maintains the antiproperty of the support value to help prune the search space of

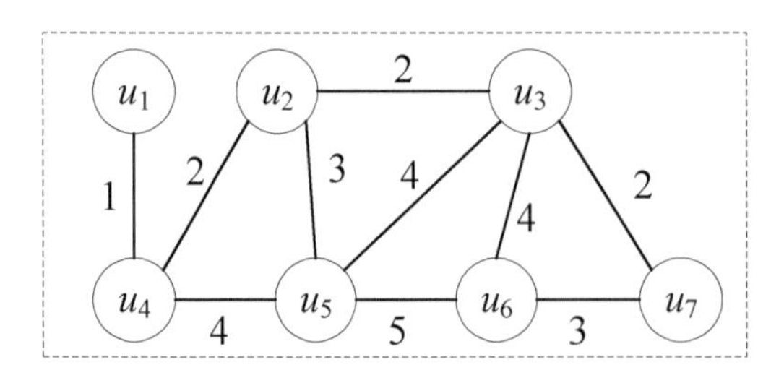

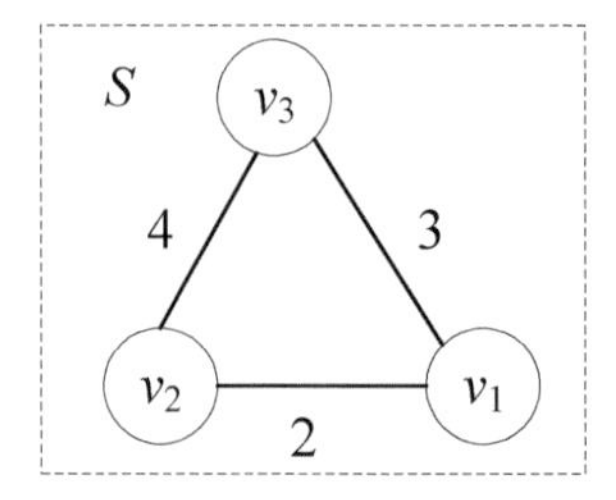

Figure 7.17
The concept of isomorphic and frequent subgraph patterns in a single graph.

candidate subgraphs so that an exhaustive search can be avoided during the grow-and-store. Grami [95] is not a grow-and-store method, having a higher efficiency than SIGRAM.

SUBDUE is an inexact FSM algorithm that uses an approximate measure to compare the similarity of two subgraphs (i.e., two subgraphs are not required to be entirely identical to contribute to the support count). However, the run time of SUBDUE does not increase linearly with the size of the input graph. Furthermore, SUBDUE tends to discover only a small number of patterns. The main bottleneck of typical grow-and-store approaches like SUBDUE is the prohibitively large number of candidate subgraphs, which makes these approaches infeasible in practice.

SIGRAM introduces an antimonotone support value to prune false-positive candidate subgraphs. Given an input graph G and two subgraphs S_1 and S_2, where S_1 is a subgraph of S_2, the antimonotone property states that the support of S_2 in G should never be greater than the support of S_1 in G. In a large graph, the violation of antimonotonicity is usually caused by the overlapping (i.e., the common nodes) among isomorphic subgraphs. Thus, the definition of an appropriate antimonotone support metric is of crucial importance in the FSM over a large graph. To solve the violation problem, only one out of all the overlapping isomorphic subgraphs should be counted. Different types of antimonotone support metrics have been proposed in literature. For instance, the size of the maximum independent set (MIS) of the overlap graph is proposed as an intuitive support measure [6]. The overlap graph is the graph in which each vertex represents the isomorph of the subgraph in the given data graph. An edge between two vertices of an overlap graph exists if the isomorphs of the subgraph overlap (i.e., they share the same vertexes). However, SIGRAM needs to enumerate all isomorphisms and relies on the expensive computation of MIS, which is NP-complete. Thus, the method is very expensive in practice.

Grami does not need to construct all isomorphisms and can scale up to a much larger graph. It stores only the templates of frequent subgraphs, rather than their appearances on the graph. This eliminates the limitations of the grow-and-store methods and allows Grami to mine large graphs and support low-frequency thresholds.

7.4 Clustering

7.4.1 Concepts

Clustering is the process of grouping a set of data objects into multiple groups or clusters so that objects within a cluster have high similarity but are very dissimilar to objects in other clusters. Dissimilarities and similarities are assessed based on the attribute values describing the objects and often involve distance measures [99], such as Euclidean distance, the cosine similarity, and the Pearson correlation. For example, we can cluster restaurants into several groups based on their basic properties and people's visiting patterns, using the Pearson correlation as a distance measure.

Major clustering algorithms can be classified into three categories: partitioning methods, hierarchical methods, and density-based methods, as illustrated in figure 7.18:

1. *Partitioning methods*. As shown in figure 7.18a, given a set of n objects, a partitioning method constructs k partitions of the data ($k \leq n$), where each partition represents a cluster and must contain at least one object. Typically, each object can only belong to one group in most partitioning methods. Some fuzzy algorithms relax the requirement, allowing one object to belong to multiple clusters. Given k, a partitioning method creates an initial partition and then iteratively improves the partition by moving objects from one group to another. The general criterion of a good partition is that objects in the same cluster are "close" to each other, whereas objects in different clusters are "far apart." Achieving a global optimal partition based on such a criterion is computationally impractical. Instead, most applications adopt heuristic methods, such as k-mean and k-medoids algorithms.
2. *Hierarchical methods*. A hierarchical method creates a hierarchical decomposition of a given set of objects, as illustrated in figure 7.18b, through either an agglomerative or divisive approach. The agglomerative approach, also known as a *bottom-up approach*, starts with each object to form a separate group, successively merging the objects of groups close to one another until all the groups are merged into one or a termination condition holds. The divisive approach, also called a *top-down approach*, starts with all the objects in the same cluster, splitting a bigger cluster into multiple smaller clusters in each successive iteration until each object is in one cluster or a termination condition holds. Hierarchical clustering methods can be used as a framework or metamethod in which partitioning methods and density-based methods can be employed in each merging or splitting iteration. Once an object is assigned into a cluster, it will be moved together with the cluster (i.e., we will use the properties of the cluster [like its center] to perform successive clustering). Even if there is another cluster fitting the object better in the successive iteration, it will not be reassigned to other clusters. On the one hand, such a strategy results in a much smaller computational workload. On the other hand, such techniques cannot correct erroneous decisions.
3. *Density-based methods*. Most partitioning methods cluster objects based on the distance between objects, thereby they can only find spherically shaped clusters.

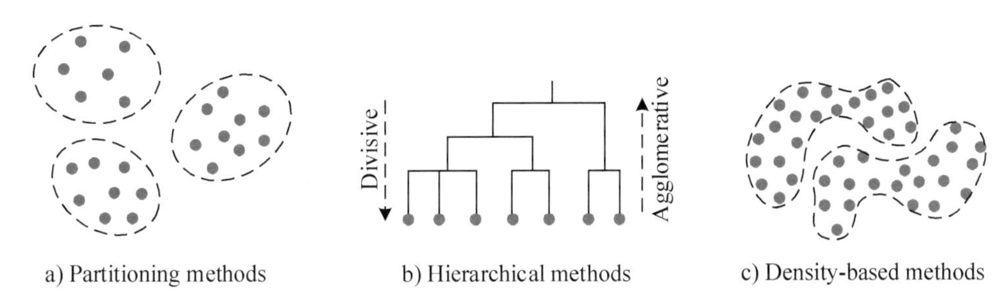

Figure 7.18
Three categories of clustering algorithms.

Density-based methods discover clusters of arbitrary shapes by continuously growing a given cluster as long as the density in the "neighborhood" exceeds a threshold. For example, for each object in a cluster, the number of objects within a neighborhood of a given radius must contain at least a minimum number of points. In addition, such a method may be able to filter noise or outliers, as the density of noisy points is generally much lower than normal points; thus, they cannot form a cluster. Sometimes, to expedite density-based clustering methods, we can quantize the object space into a finite number of grids. All the clustering operations are further performed on the grid structure rather than the original objects. This is an efficient approach to many spatial data-mining problems.

7.4.2 Partitioning Clustering Methods

In this section, we will introduce the most well-known and commonly used portioning methods: k-means and k-medoids algorithms, where k is the number of partitions into which a set of objects will be clustered. The value of k needs to be specified by users based on the requirements of applications. However, this is sometimes not easy to determine.

7.4.2.1 *k*-means algorithm Figure 7.19 illustrates the working procedure of the k-means algorithm. Given a set of objects D and setting $k=3$, the algorithm first randomly selects three objects from D as the initial centroids, based on which D is partitioned into three groups. The algorithm calculates the distance between each object and each centroid, assigning an object to the partition whose centroid has the nearest distance to the object, as shown in figure 7.19a. The k-means algorithm then computes the new centroid based on the objects in each cluster, reassigning each object to the nearest centroid to form three new partitions. As presented in figure 7.19b, the three

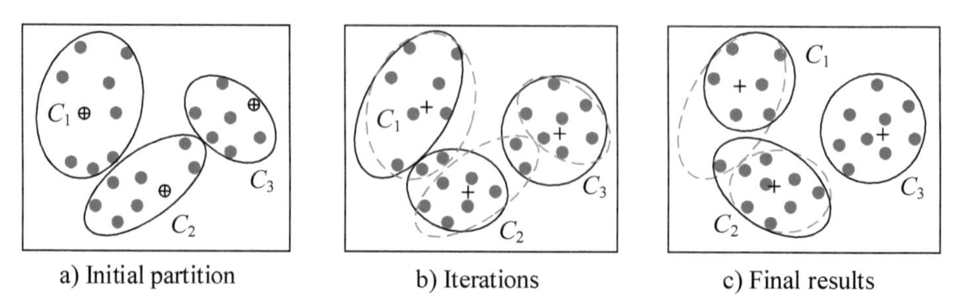

Figure 7.19
An illustration of the *k*-means algorithm.

cross marks are the new centroids, and the solid borders denote the new clusters. There are two points originally belonging to C_1 that have been moved to C_2 because of the reassignment. Likewise, two other points were moved from C_2 to C_3. The *k*-means algorithm repeats this procedure (i.e., computing the new centroids and reassigning objects to the nearest centroid), iteratively improving the quality of clustering until the clustering results do not change anymore (also called *converged*). Finally, we will obtain the results depicted in figure 7.19c.

7.4.2.2 *k*-medoids algorithm The *k*-means algorithm is sensitive to outliers that are far away from the majority of the data, as such objects can dramatically distort the mean of a cluster. To address this issue, the *k*-medoids algorithm proposes to use one representative object (rather than the mean) of a cluster to represent the cluster. It then partitions data into clusters by minimizing the sum of the distance between each object and its representative object. However, this is an NP-hard problem with the complexity $O(k(n-k)^2)$ for each iteration.

The partitioning around medoids (PAM) algorithm [14] is a popular implementation of *k*-medoids clustering, tackling the problem through an iterative and greedy approach. The algorithm considers whether replacing a representative object by a nonrepresentative object would improve the clustering quality. The iterative process of replacing continues until the quality of the clusters cannot be improved anymore. Figure 7.20 illustrates PAM with an example that aims to partition seven objects $\{o_1, o_2, o_3, o_4, o_5, o_6\}$ into two clusters. If we select $\{o_6, o_7\}$ as the initial two representative objects, the overall distance between the objects and the two medoids is 20. If we replace o_6 by o_3, as shown in figure 7.20b, the overall distance is reduced to 15. Likewise, the overall distance is reduced to 14 if we replace o_7 by o_3, as depicted in figure 7.20c. That is,

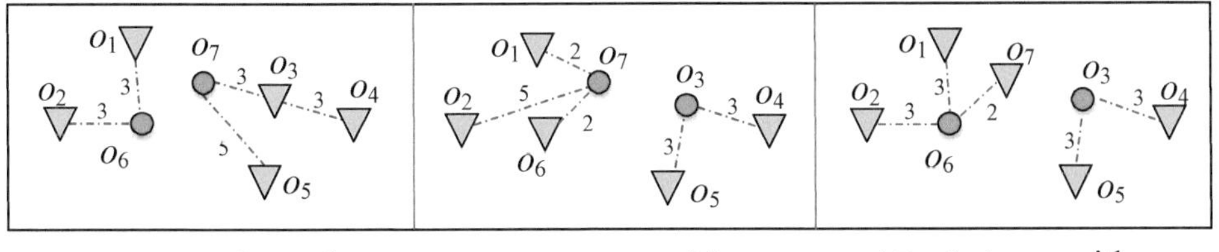

a) Initial R=$\{o_6, o_7\}$ b) Replacing o_6 with o_3 c) Replacing o_7 with o_3

Figure 7.20
An example of using PAM to implement k-medoids clustering.

the latter replacement is more effective. Trying other replacements cannot result in a smaller distance, thus, $\{o_7, o_3\}$ are selected as the medoids of the two clusters.

7.4.3 Density-Based Clustering

Different from the partitioning and hierarchical clustering methods, which are designed to find clusters of a spherical shape, the density-based methods can find clusters of arbitrary shape and can even filter outliers. Hereafter, we introduce three frequently used density-based clustering algorithms: DBSCAN (density-based spatial clustering of applications with noise) [96], OPTICS (ordering points to identify the clustering structure) [84], DENCLUE (density-based clustering) [103], and grid-based clustering methods.

7.4.3.1 DBSCAN The DBSCAN algorithm defines the density of a neighborhood with two user-specified parameters: a distance threshold ϵ and a minimum number of points *MinPts*. An object q is a core object if its ϵ-neighborhood contains at least *MinPts* objects. An object p within the ϵ-neighborhood of q is directly density reachable from q (with respect to ϵ and *MinPts*). Two objects p_1 and p_2 are density connected with respect to ϵ and *MinPts*, if there is an object q such that both p_1 and p_2 are density reachable from q (with respect to ϵ and *MinPts*).

For example, as illustrated in figure 7.21, setting *MinPts* = 3 and ϵ as the radius of the circle, p, q, o, and m are core objects. Object q is directly density reachable from o, and object o is directly density reachable from p. Thus, q is (indirectly) density reachable from p. However, p is not density reachable from q, as q is not a core object. Object s and r are density connected as they are both density reachable from core object m.

The DBSCAN algorithm works as follows: Initially, all objects in a given dataset D are marked as *unvisited*. DBSCAN randomly selects an unvisited object o, marking it *visited* and checking if its ϵ-neighborhood contains at least *MinPts* objects. If not, o is marked as a noise point. Otherwise, a new cluster C is created for o, and all the

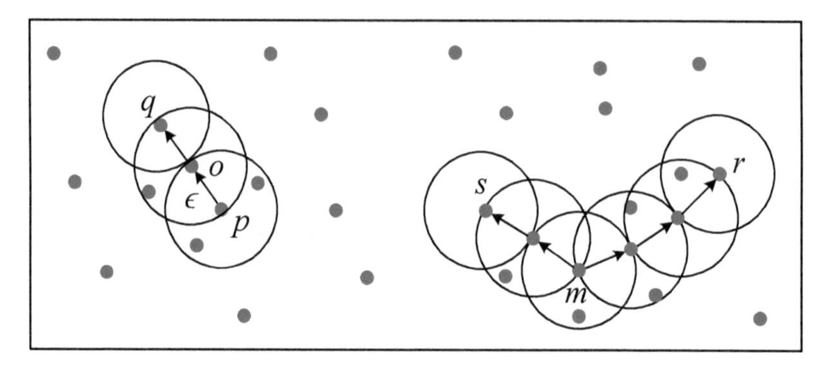

Figure 7.21
Density reachability and density connectivity in density clustering (adapted from [96]).

objects in the ϵ-neighborhood of o are added to a candidate set N. DBSCAN iteratively adds to C those objects (in N) that do not belong to any other cluster. In this process, for an object o' (in N) with the label *unvisited*, DBSCAN marks it as *visited* and checks whether its ϵ-neighborhood of o' contains at least *MinPts* objects; those objects in the ϵ-neighborhood of o' are added to N. The DBSCAN algorithm continues adding objects to C until C can no longer be expanded (i.e., N is empty). At this moment, cluster C is completely formed. To find the next cluster, the algorithm randomly selects an unvisited object from the rest of D, performing the aforementioned process again until a new cluster is formed. After all objects are visited, the DBSCAN algorithm terminates.

The complexity of DBSCAN is $O(n^2)$, where n is the number of objects in a dataset. If a spatial index is employed, the computational complexity can be reduced to $O(nlogn)$. With an appropriate ϵ and *MinPts*, DBSCAN can find clusters of arbitrary shapes. However, setting the two parameters is not an easy task, depending on the empirical understanding of a problem and the distribution of a given dataset. Moreover, real-world, high-dimensional datasets often have very skewed distributions such that their intrinsic clustering structure may not be characterized by a global setting of density parameters.

7.4.3.2 OPTICS To overcome the difficulty in using one set of global parameters in clustering analysis, OPTICS outputs a cluster ordering rather than explicit clusters. This is a linear list of all objects under analysis, representing the density-based clustering structure of the data. Objects in a denser cluster are listed closer to each other in the cluster ordering. This order is equivalent to density-based clustering obtained from a

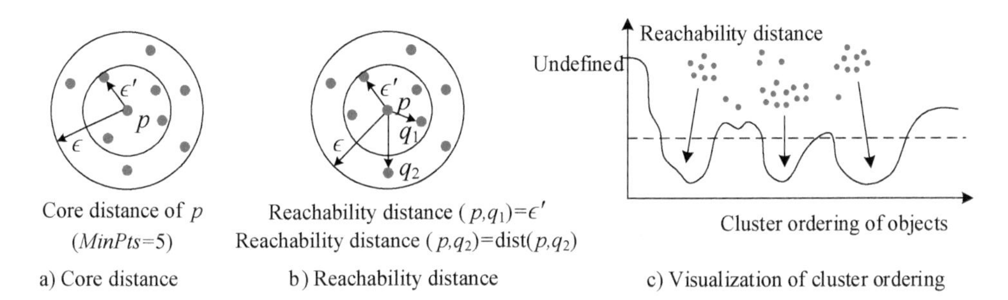

Figure 7.22
An illustration of OPTICS (adapted from [84]).

wide range of parameter settings. Thus, OPTICS does not require users to provide a specific density threshold [84]. OPTICS needs two important properties for an object: core distance and reachability distance.

1. The core distance of an object p is the smallest value ϵ' such that the ϵ'-neighborhood of p has at least *MinPts* objects. In other words, ϵ' is the minimum distance threshold that makes p a core object. If p is not a core object with respect to ϵ and *MinPts*, the core distance of p is undefined. For example, as illustrated in figure 7.22a, the core distance of p is ϵ', which is smaller than ϵ, if we set *MinPts* = 5.
2. The reachability distance to object p from q is the minimum radius value that makes p density reachable from q. According to the definition of density reachability, q has to be a core object, and p must be in the neighborhood of q. Thus, the reachability distance from q to p is max (core distance(q), dista(p, q)). For instance, as demonstrated in figure 7.22b, the reachability distance to p from q_1 is ϵ', while that of q_2 is dist(p, q_2). If q is not a core object with respect to ϵ and *MinPts*, the reachability distance to p from q is undefined.

OPTICS stores the core distance and a suitable reachability distance for each object, maintaining a list (called OrderSeeds) that is sorted by the smallest reachability distance of each object from its respective core objects. OPTICS begins with an arbitrary object from the input database as the current object p, determining its core distance and setting the reachability distance to undefined. Then p is written to output, and the following iteration starts.

If p is not a core object, OPTICS simply moves on to the next object in the databases. Otherwise, OPTICS retrieves the ϵ-neighborhood of p. For each object q in the ϵ-neighborhood of p, OPTICS updates the reachability distance from p to q. If q has not yet been processed, OPTICS inserts q into OrderSeeds based on its reachability distance to p for further expansion and then marks p *processed*. OPTICS retrieves the object k in OrderSeeds with the smallest reachability distance (i.e., object k will be removed from OrderSeeds), determining k's core distance and writing k into the output. If k is a core object, OPTICS will further retrieve objects in k's ϵ-neighborhood and update their reachability-distances from k. If these neighborhood objects have not yet been processed, OPTICS inserts these neighborhood objects into OrderSeeds (based on their reachability distances) for further expansion. Then, these objects will be marked *processed*. The iteration continues until OrderSeeds is empty. OPTICS terminates until all objects in the dataset have been processed. The order of objects in the output file is called *cluster ordering*.

Figure 7.22c presents the visualization (i.e., the reachability plot) of the cluster ordering for a two-dimensional dataset, where the horizontal axis denotes the cluster ordering of an object, and the vertical axis stands for the reachability distance of the object. The visualization can be used to derive the intrinsic clustering structure. For instance, the three hollows represent three clusters. As compared to other clustering methods, the reachability plot of OPTICS is not sensitive to the input parameters ϵ and *MinPts*. The two values just have to be "large" enough to yield a good result. The concrete values are not crucial because there is a broad range of possible values for which we can always see the clustering structure of a dataset when looking at the corresponding reachability plot. The smaller the value of ϵ that we choose, the more objects may have an *undefined* reachability distance. Therefore, we may not see clusters of lower density [84].

7.4.3.3 DENCLUE: Clustering based on density distribution function In DBSCAN and OPTICS, density is estimated by counting the number of objects in a neighborhood defined by a radius parameter ϵ. Such density estimation can be highly sensitive to the radius value used. Sometimes, a slight increase of the radius changes the density tremendously. Kernel density estimation, which is a nonparametric density-estimation approach from statistics, can be used to address this issue. Its general idea is to treat an observed object as an indicator of high-probability density in the surrounding region. The probability density at a point x depends on the distances from this point to the observed objects.

Formally, let $x_1, \ldots, x_n$ be an independent and identically distributed sample of a random variable f. The kernel density approximation of the probability density function at point x is

$$\hat{f}_h(x) = \frac{1}{nh}\sum_{i=1}^{n} K\left(\frac{x - x_i}{h}\right), \tag{7.7}$$

where $K(\ )$ is a kernel, and h is the bandwidth serving as a smoothing parameter. We can choose h as small as data allows. A kernel can be regarded as a function modeling the influence of a sample point within its neighborhood. A frequently used kernel is a standard Gaussian function with a mean of 0 and a variance of 1.

$$K\left(\frac{x-x_i}{h}\right)=\frac{1}{\sqrt{2\pi}}e^{-\frac{(x-x_i)^2}{2h^2}}. \tag{7.8}$$

DENCLUE uses a Gaussian kernel to estimate density based on the given set of objects to be clustered. A point x^* is called a *density attractor* if it is a local maximum of the estimated density function (i.e., a local peak). To avoid trivial points, DENCLUE only considers those density attractors x^* such that $\hat{f}_h(x^*)$ is greater than a noise threshold α

Those nontrivial density attractors are used as the initial centers of clusters. Objects under analysis are assigned to these density attractors using a stepwise hill-climbing procedure [88]. If there is a path between a pair of density attractors where the density is above α, the two density attractors are connected. By merging multiple connected density attractors, DENCLUE can find clusters of arbitrary shape while dealing with noise.

7.4.3.4 Grid-based clustering methods This category of clustering methods partitions the (embedding) space into uniform cells regardless of the distribution of the input objects and then groups a set of connected cells to form clusters. Representing each cell with some statistic properties, such as the density and mean value, such kinds of methods significantly scale down the number of objects that a clustering algorithm needs to process, thereby becoming very efficient and easy to scale up. In addition, such a grid structure facilitates parallel computing and incremental updating. Grid-based clustering algorithms can be regarded as a coarse-grained density-based clustering approach, roughly estimating the density of objects using grids as an approximation.

Figure 7.23a illustrates the principle of grid-based clustering algorithms using a collection of two-dimensional spatial points. The algorithm first partitions the space into nonoverlapping grids, identifying the dense grids with the number of points greater than a given threshold τ. Two grids are *adjacently connected* if they share a common side. Two grids are density connected if a path can be found from one grid to the other containing only a sequence of *adjacently connected* grids. Then, the algorithm uses these dense grids to assemble clusters, each of which is the maximal region formed by these connected grids. It is easy to determine such connected grid regions by using a graph-based model on the grids. Each dense grid is represented by a node in the graph, and each edge of the graph denotes adjacent connectivity between two grids. The connected components in the graph can be determined by using breadth-first or depth-first

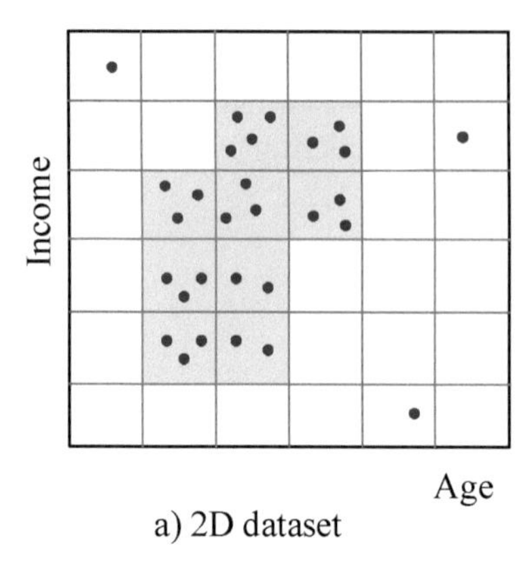

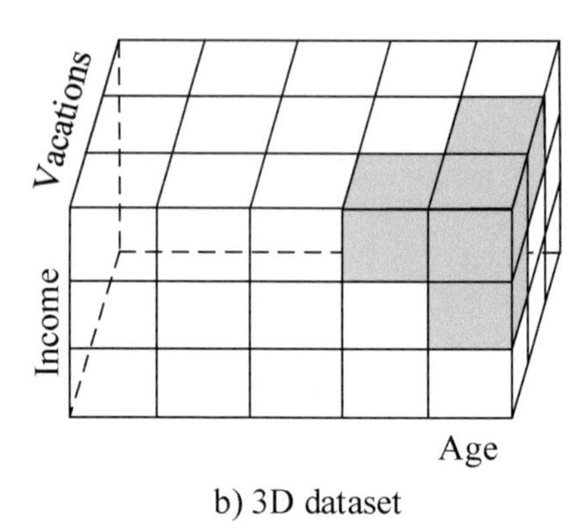

Figure 7.23
Grid-based clustering algorithms.

traversal on the graph. The data points in these connected components are reported as the final clusters.

We can also regard each point in figure 7.23a as an object with two features (e.g., an individual with an income and age). Each feature can be regarded as a dimension, which can be divided by a uniform interval. Then, we can find the maximal connect grids as a cluster following the same procedure mentioned above. Likewise, the density of objects with three features can be represented by the number of points in a three-dimensional cube, as illustrated in figure 7.23b. The clustering results in a collection of connected dense cubes, each of which contains at least τ points. Two cubes are connected if they share one common two-dimensional surface. This kind of algorithm can be generalized to a k-dimensional dataset. Two k-dimensional cubes may be considered adjacent if they share a surface of dimensionality at least r, $r < k$ [88].

The grid-based clustering methods do not need to define the number of clusters and can find clusters of arbitrary shape. However, defining the size of the grid and density threshold τ is not intuitive. Changing the two parameters usually results in very different clustering results.

7.4.4 Hierarchical Clustering Methods

A hierarchical clustering method groups objects into a hierarchy of clusters, through a divisive or agglomerative approach. It can be employed as a clustering framework in which other clustering methods such as partitioning and density-based algorithms can be used.

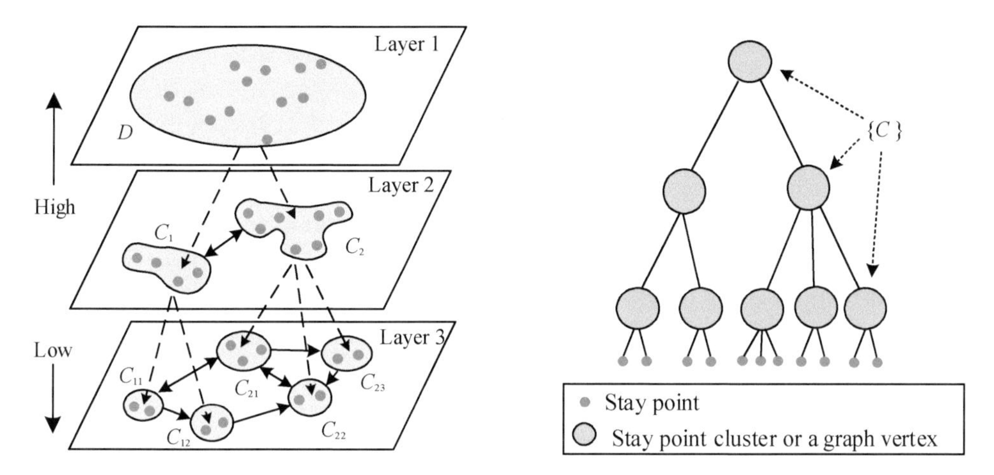

Figure 7.24
Combining hierarchical clustering with density-based clustering methods.

As illustrated in figure 7.24, we aim to identify hot spots (e.g., tourist attractions) based on massive user-generated GPS trajectories. We employ a hierarchical clustering method to discover clusters of different geographical sizes, as hot spots can be represented by the location of different granularities (e.g., a specific restaurant) or a shopping mall, or a park, or even a business district. In addition, we integrate a density-based algorithm into the hierarchical clustering framework to filter those trivial locations that are not dense enough to represent a hot spot.

In figure 7.24, each green point stands for a stay point detected from a GPS trajectory, and a gray circle denotes a cluster. We can put all these stay points together to form a dataset D, which can be regarded as the root node of the hierarchy depicted in the right part of the figure. By applying a density-based clustering method (e.g., OPTICS with $\epsilon=500$ m, and $MinPts=30$) to D, we can obtain two clusters C_1 and C_2, which can be regarded as two children of the root node. We can further apply OPTICS with a finer-grained parameter setting, such as $\epsilon=100$ m and $MinPts=8$, to the stay points in C_1 and C_2, respectively. Finally, two subclusters, C_{11} and C_{12}, are formed based on the stay points in C_1, and three clusters, consisting of C_{21}, C_{22}, and C_{23}, are formed based on C_2. Those clusters construct a hierarchy (or a tree) where high-level nodes stand for hot spots of coarse granularity (i.e., a large geographical size, like a park), and low-level nodes denote hot spots of fine granularity (i.e., a small geographical size, like a restaurant). After that, we can build graphs on different layers of the hierarchy based on users' transitions among these locations.

7.5 Classification

7.5.1 Concepts

Data classification is a two-step process, consisting of a training step, in which a classification is constructed based on historical data, and a classification step, in which the model is used to predict class labels for given data. In the first step, a classification algorithm builds the classifier by analyzing or "learning from" a training set, which is composed of a collection of data tuples X and their corresponding class labels. A tuple, X, is represented by an n-dimensional attribute vector, $X=(x_1, \dots, x_n)$, where each element denotes the value of a type of feature. Each tuple is associated with a predefined class c, which is a discrete valued and unordered attribute. Data tuples can be referred to as *instances*, *samples*, *objects*, or *data points*. The training step can also be regarded as a process to find a mapping function $f(X)$ between a given X and its class label c. In the second step, the classifier is used for classifying tuples in a test set, which is independent of the training set. This can be regarded as a process to predict a tuple's class label c based on its data tuple X and the mapping function $f(X)$. The performance of the classifier is then estimated using some metrics.

The evaluation of a classification model is concerned with two issues. One is the data-partitioning methods for generating training and test sets. The other is metrics for evaluation, such as precision, recall, and F-measure.

7.5.1.1 Data partition methods There are two widely used data partition methods for generating training and testing datasets. One is called the *holdout method*, which typically selects two-thirds of the data for the training set and the remaining one-third as the test set. Training and test sets are not overlapped. The other method is referred to as *k-fold cross-validation*. It randomly partitions a given dataset into k subsets (a.k.a. folds), $D_1, \dots, D_n$, of almost equal size. Training and testing is performed k times. In iteration i, D_i is reserved as the test set, and the remaining folds are collectively used to train the model. The accuracy is the overall number of correct classifications from the k iterations divided by the total number of tuples in the initial data. Unlike the holdout method, each instance is used $k-1$ times for training and once for testing. Thus, the performance of a classifier evaluated by k-fold cross-validation is generally higher than the holdout method. The k-fold cross-validation is typically used when the scale of the data a classification task deals with is small.

7.5.1.2 Evaluation metrics Precision, recall, accuracy, and F-score are commonly used metrics for evaluating the performance of a classification model. Before introducing these metrics, we need to know some other terminology, consisting of true positive

Predicted class

Ground truth	Yes	No	Recall	Total
Yes	TP	FN	$\frac{TP}{TP+FN}$	P
No	FP	TN	$\frac{TN}{FP+TN}$	N
Precision	$\frac{TP}{TP+FP}$	$\frac{TN}{FN+TN}$	$\frac{TP+TN}{P+N}$	P+N

a) Confusion matrix

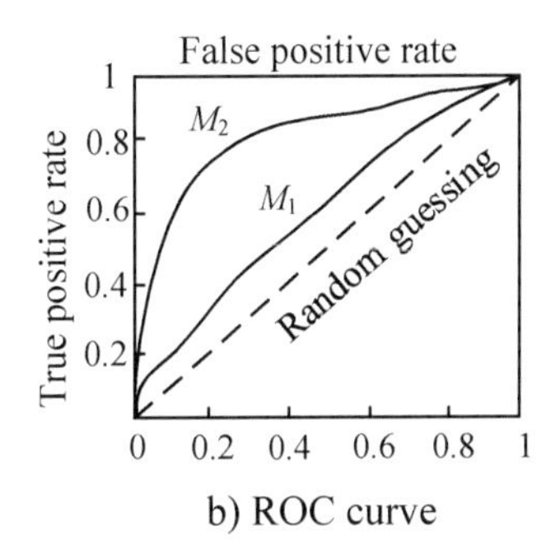

b) ROC curve

Figure 7.25
Evaluation metrics for classification.

(TP), true negative (TN), false positive (FP), and false negative (FN). Figure 7.25a presents these terminologies in a confusion matrix using a binary classification task (i.e., two classes: Yes or No) as an example:

- *TP* refers to the positive instances that were correctly labeled by a classifier.
- *TN* stands for the negative instances that were correctly labeled by a classifier.
- *FP* denotes the negative instances that were incorrectly labeled as positive.
- *FN* represents the positive instances that were incorrectly labeled as negative.

The *accuracy* of a classifier is the percentage of instances (in the test set) that are correctly classified by the classifier:

$$Accuracy = \frac{TP + TN}{TP + TN + FP + FN} = \frac{TP + TN}{P + N}. \tag{7.9}$$

However, when dealing with the class imbalance problem, where the main class of interest is rare, accuracy cannot reveal the true performance of a classifier. For example, in a medical dataset, only 1 percent of the data is derived from people with cancer, while 99 percent of the data is generated from people without cancer. If we want to train a classifier to differentiate between "cancer" and "not cancer" based on the medical data, we will confront a class imbalance problem, which makes a classifier fit the class with the majority of instances (hence, sacrifices the opposite class). More specifically, even if TP for the class of *cancer* is 0, the (overall) accuracy of the classifier we trained may still be higher than 98 percent (as TN is big enough). However, being able to identify these rare classes is of great importance to an application as compared to detecting their opposite classes (i.e., *no cancer*). To address these issues, precision and recall were proposed to see the performance of a classifier on different classes.

Precision is a measure of preciseness (i.e., what is the percentage of instances, labeled as a class, actually belonging to the class?). *Recall* is a measure of completeness (i.e., the percentage of instances [of a class] that have actually been labeled as the class). Each class has its own precision and recall. Formally, the precision of class Yes and No is defined as

$$precision(Yes) = \frac{TP}{TP + FP}; \quad precision(No) = \frac{TN}{TN + FN}. \tag{7.10}$$

The recalls of the two classes are defined as

$$recall(Yes) = \frac{TP}{TP + FN} = \frac{TP}{P}; \quad recall(No) = \frac{TN}{FP + TN} = \frac{TN}{N}. \tag{7.11}$$

Following the aforementioned example about the medical data, we can know the performance of a classifier is not good if the precision or recall of the class *cancer* is very low. Classifiers with a good record on all four metrics are considered a strong classifier. However, different classifiers may have their own strengths at different metrics. For example, some classifiers may have a slightly higher precision but a slightly lower recall than others or even have a slightly good precision and recall on one class but slightly worse precision and recall on the other class. It is difficult to judge the overall performance of a classifier. An alternative way that people try to address this issue is to combine precision and recall to formulate an F-score as follows:

$$F - score = \frac{2 \times precision \times recall}{precison + recall}. \tag{7.12}$$

The F-score cannot measure the performance of classifiers that generate a probability (rather than a class label) of an instance belonging to a class. Given the medical data, the results that classifiers like decision trees generate for each instance are probabilities of the instance belonging to two classes, such as *cancer* (0.75) or *not cancer* (0.25). Instances with a probability of Yes higher than a threshold τ, such as τ=0.7, are finally classified as the class *cancer*. Otherwise, the instance is still labeled as *not cancer*.

To address this issue, the receiver operating characteristic (ROC) curve was proposed to measure the performance of a binary classifier. The curve is created by plotting the TP rate $\left(\text{TPR:}\frac{TP}{P}\right)$ against the FP rate $\left(\text{FPR:}\frac{FP}{N}\right)$ at various threshold τ (e.g., changing τ from 0.9 to 0.1). Figure 7.25b presents the ROC curves of two classifiers M_1 and M_2. The diagonal line represents random guessing. The closer the ROC curve of a classifier is to the diagonal line, the less accurate the model. Thus, M_2 is a better classifier than M_1. Initially, when decreasing τ, we are more likely to encounter TPs. Consequently, the curve moves steeply up from zero. Later, as we start to encounter fewer TPs and more FPs, the curve eases off, becoming more horizontal.

7.5.2 Naïve Bayesian Classification

Let X be a data instance, and C be the class label. Naïve Bayesian classification is to find the class label that maximizes the posterior probability $P(C|X)$, based on Bayes' theorem:

$$P(C|X) = \frac{P(X|C)P(C)}{P(X)}. \tag{7.13}$$

Formally, let D be a training set of instances, each of which is a vector $X=(x_1, x_2, \ldots, x_n)$ comprised of the values of n-dimensional features $(A_1, A_2, \ldots, A_n)$ (e.g., x_1 is the value of feature A_1). Suppose that there are m classes, $C_1, C_2, \ldots, C_m$. The naïve Bayesian classifier predicts that instance X belongs to C_i if and only if $P(C_i|X) > P(C_j|X)$ for $1 \leq j \leq m$, $j \neq i$. In other words, find class C_i with which $P(C_i|X)$ is maximized. This is called the *maximum posterior hypothesis*. Applying the Bayes' theorem, we obtain

$$P(C_i|X) = \frac{P(X|C_i)P(C_i)}{P(X)}. \tag{7.14}$$

As $P(X)$ is constant for all classes, only $P(X|C_i)\ P(C_i)$ needs to be maximized. $P(C_i)$ is usually called *prior probability*, which can be estimated by $P(C_i) = |C_{i,D}| / |D|$, where $|C_{i,D}|$ is the number of training instances of class C_i in D.

Given datasets with many features, it would be extremely computationally expensive to compute $P(X|C_i)$. To reduce the computation workload, the naïve Bayesian classifier assumes that features are conditionally independent of one another, given its class label—that is,

$$P(X|C_i) = P(x_1|C_i) \times P(x_2|C_i) \times \cdots \times P(x_n|C_i). \tag{7.15}$$

We can easily compute $P(x_1|C_i)$, $P(x_2|C_i)$, and $P(x_n|C_i)$ as follows: if feature A_i is categorical, $P(x_i|C_i)$ is the number of instances of class C_i in D having the value x_i for A_i, divided by $|C_{i,D}|$. For example, $P(x_i = male|C_i = buy\ computers)$ can be estimated by counting the number of male consumers who have bought a computer. The count is further divided by the total number of consumers who have bought a computer.

If A_i is continuous valued, we need to calculate $P(x_i|C_i)$ based on a distribution, such as Gaussian distribution with a mean μ and a standard deviation σ.

$$f(x, \mu, \sigma) = \frac{1}{\sqrt{2\pi}\sigma} e^{-\frac{(x-\mu)^2}{2\sigma^2}}. \tag{7.16}$$

$$P(x_i|C_i) = f(x_k, \mu_{C_i}, \sigma_{C_i}), \tag{7.17}$$

where μ_{C_i} and σ_{C_i} are the mean and the standard deviation of the values of feature A_k. They can be estimated based on D. For example, let $X=(36, \$45{,}000)$, where A_1 and A_2 are age and income, respectively. Let the label class be *buy house*=*yes* or *no*. Suppose from the training set, we find that customers in D who buy a house are 37 ± 10 years of age (i.e., $\mu=37$ and $\sigma=10$). Then, we can calculate $P(x_1=36|buy\_house=yes)$ based on equation (7.16).

Though the class-condition independence may not hold in many real datasets, the performance of the naïve Bayesian classifier is acceptable in practice, particularly for classification problems with a small number of features and class labels. Please refer to [88, 99] for more details about the naïve Bayesian classifier.

7.5.3 Decision Trees

Arranged in a treelike structure, decision trees are a classification methodology using a set of hierarchical decisions on the feature variables. The decision at a particular node of the tree, which is referred to as the *split criterion*, is typically a condition on one or more feature variables in the training data. The split criterion divides the training data into two or more parts, trying to minimize the level of "mixing" of the class variables in each branch of the tree. Each node in the decision tree logically represents a subset of the data space defined by the combination of split criteria in the nodes above it. Once a decision tree is built, we can allocate a test instance to a leaf node by traversing the tree based on its feature values. The probabilities of an instance belonging to different classes are represented by the proportion of different classes in the leaf node.

Classical decision tree algorithms include ID3 [135], C4.5 [128], and CART (classification and regression tree) [86]. ID3 is designed for classification with categorical features and class labels. C4.5 handles classification problems with real-valued features and categorical classes. CART can deal with real-valued features and real-valued predictions and can be regarded as a regression model. We will introduce CART in section 7.5.

7.5.3.1 ID3 algorithm As illustrated in figure 7.26, a decision tree shown in figure 7.26b is constructed based on the training data shown in figure 7.26a, using the ID3 algorithm. Each instance stands for a user associated with two features, consisting of *age* and *income*, and a class label *buy computer: yes* or *no*.

ID3 selects the feature with the maximum information gain as the split criterion. The algorithm first computes the information entropy of the given dataset D based on the following equation:

$$Entropy(D) = -\sum_{i=1}^{m} p_i \log_2 p_i, \tag{7.18}$$

UID	*Age*	*Income*	*Class: buy_computer*
u_1	youth	high	no
u_2	youth	high	no
u_3	middle_aged	high	yes
u_4	Senior	medium	yes
u_5	Senior	low	yes
u_6	Senior	low	no
u_7	middle_aged	low	yes
u_8	youth	medium	no
u_9	youth	low	yes
u_{10}	senior	medium	yes

a) Training instances

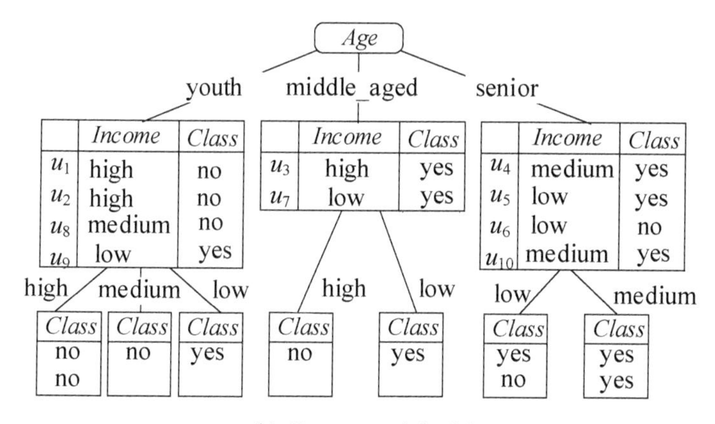

b) Constructed decision tree

Figure 7.26
Constructing a decision tree based on ID3 algorithms.

where p_i is the proportion of class C_i in D, and m is the number of class labels. In this example, there are two class labels: *yes* or *no*. $Entropy(D)$ denotes the impurity of D. If all instances in D belong to the same class (i.e., $p_i = 1$ and $p_j = 0$, $1 \le j \le m$, $j \ne i$), $Entropy(D)$ receives the minimum value 0. If the proportions of different classes in D are equal, the $Entropy(D)$ receives the maximum value (i.e., the most impure). For example, if there are two classes equally distributed in D—that is, $p_1 = p_2 = ½$, then $Entropy(D) = -\frac{1}{2}\log_2\frac{1}{2} - \frac{1}{2}\log_2\frac{1}{2} = \frac{1}{2} + \frac{1}{2} = 1.$

Suppose we partition D by some feature A with k distinct values $(x_1, x_2, \ldots, x_k)$; instances with feature A's value equaling x_i will be assigned into a subset. That is, we can obtain k subsets ($D_1, D_2, \ldots, D_k$). We can then calculate the information entropy of these partitions by

$$Entropy_A(D) = -\sum_{j=1}^{k} \frac{|D_j|}{|D|} Entropy(D_i), \tag{7.19}$$

where $\frac{|D_j|}{|D|}$ is a weight of the j-th partition, and $|D_j|$ denotes the number of instances in the j-th partition. Then we can compute the information gain of the partition criterion by

$$Gain(A) = Entropy(D) - Entropy_A(D). \tag{7.20}$$

In the example shown in figure 7.26a, the proportions of *yes* and *no* are $\frac{3}{5}$ and $\frac{2}{5}$, respectively. Thus, the entropy of D shown is

$$Entropy(D) = -\frac{2}{5}\log_2\frac{2}{5} - \frac{3}{5}\log_2\frac{3}{5} = + = 0.971 \text{ bits.}$$

If we select *age* as the split criterion to divide D into three parts (youth, middle_aged, senior), as illustrated by the first three children nodes in figure 7.26b, the information entropy of D is calculated as follows:

$$Entropy_{age}(D) = \frac{4}{10} \times \left(-\frac{1}{4}\log_2\frac{1}{4} - \frac{3}{4}\log_2\frac{3}{4}\right) + \frac{2}{10} \times (-1\log_2 1 - 0\log_2 0)$$
$$+\frac{4}{10} \times \left(-\frac{3}{4}\log_2\frac{3}{4} - \frac{1}{4}\log_2\frac{1}{4}\right) = 0.649 \text{ bits.}$$

According to equation (7.21), the information gain of $Gain(age) = 0.971 - 0.646 = 0.325$. If we select *income* as the split criterion to divide D into three parts (low, medium, high),

$$Entropy_{income}(D) = \frac{3}{10} \times \left(-\frac{1}{3}\log_2\frac{1}{3} - \frac{2}{3}\log_2\frac{2}{3}\right) + \frac{3}{10} \times \left(-\frac{2}{3}\log_2\frac{2}{3} - \frac{1}{3}\log_2\frac{1}{3}\right)$$
$$+\frac{4}{10} \times \left(-\frac{3}{4}\log_2\frac{3}{4} - \frac{1}{4}\log_2\frac{1}{4}\right) = 0.876 \text{ bits.}$$

$$Gain(income) = 0.971 - 0.876 = 0.095.$$

As $Gain(income) < Gain(age)$, which means *age* is a better split criterion to purify D, we use *age* to construct the nodes of the first layer. Then, we employ *income* to further partition the instances in the node of the first layer into subsets, as illustrated in figure 7.26.

If we thoroughly partition a dataset into leaf nodes with instances belonging to the same class, the structure of a decision tree may become very complex and overfit the training data. In addition, instances in such fine-grained nodes could be some noise and outliers. When applying such a decision tree to a test set, its performance may not be good. A solution is to use statistical measures to remove the least reliable branches in a decision tree, resulting in a smaller and less complex version that is easier to comprehend. There are two common approaches to tree pruning: prepruning and postpruning. In the prepruning approach, a tree is pruned by stopping its construction early. The latter approach removes subtrees from a fully constructed tree.

7.5.3.2 C4.5 algorithm When an attribute of an instance is a real-valued number (e.g., the age of a user is 35), there is no categorical value (of an attribute) that we can use to partition a dataset. Thus, the ID3 algorithm cannot be applied to solving such problems any longer. C4.5 was proposed to handle both continuous and discrete features.

In order to handle continuous attributes, C4.5 creates a threshold and then splits the list into those whose attribute value is above the threshold and those that are less than or equal to it. To find the best threshold, we sort the values of an attribute A and check the midpoint between each pair of consecutive values as a split point (i.e., $(a_i+a_{i+1})/2$). For each split point for A, we calculate $Entropy_A(D)$, where the number of partitions is two. We find the best split point resulting in the maximum information gain $Gain(A)$. Then, the data is divided into two parts; one with attribute A's value $\leq$ the split point and the other with A's value > the split point.

The C4.5 algorithm uses an extension of information gain known as *gain ratio* to handle discrete features. It applies a kind of normalization to information gain using a "split information" value defined analogously with $Entropy_A(D)$ as

$$SplitInfo_A(D) = -\sum_{j=1}^{k} \frac{|D_j|}{|D|}\log_2\left(\frac{|D_j|}{|D|}\right). \tag{7.21}$$

This value represents the potential information generated by splitting the training dataset D into k partitions, corresponding to the k outcomes of a test on attribute A. The gain ration is then defined as

$$GainRatio(A) = \frac{Gain(A)}{SplitInfo_A(D)}. \tag{7.22}$$

The attribute with the maximum gain ratio is selected as the split criterion. This is to avoid the bias that is caused by attributes generating many partitions. For example, if we use product identity (*ID*), which is a unique identifier of an instance, as a split criterion, a large number of partitions will be generated. As each partition contains only one instance, the information entropy of the partitions $Entropy_{ID}(D)=0$. This results in the maximum information gain; therefore, it will be selected as the split criterion. However, such a partition is useless for classification. Using the gain ratio, we will find $SplitInfo_{ID}(D)$ is very big (i.e., the gain ratio is very small). Thus, *ID* will not be selected as a split criterion.

C4.5 can handle training data with missing attribute values, allowing missing attribute values to be marked as *?*. Missing attribute values are simply not used in gain and entropy calculations. C4.5 goes back through the tree once it has been created and attempts to remove branches that do not help by replacing them with leaf nodes.

7.5.4 Support Vector Machines

Support vector machines (SVMs) [93] transform the original training data into a higher dimension using nonlinear mapping and search for the linear optimal separating hyperplane to separate the instances of one class from another. The SVM finds this hyperplane using support vectors (i.e., training instances) and margins (defined by the support vectors).

Figure 7.27 presents an example of finding the best hyperplane for a dataset, where each instance is associated with two features (A_1 and A_2) and a two-class label (yes = 1 and no = −1). If we regard the two features as the two dimensions of a space, each instance can be represented by a plot in the space. More specifically, we use two types of circles to represent instances of the two classes. After the transformation, there are multiple ways to separate the instances of the two classes. Figure 7.27a and b depict two possible separating hyperplanes, respectively, both of which can correctly classify all the given instances. Intuitively, however, the hyperplane with the larger margin could be more accurate at classifying future data instances than the hyperplane with a narrow margin. Formally, we can say the shortest distance from a hyperplane to one side of its margin is equal to the shortest distance from the hyperplane to the other side of its margin, where the “sides” of the margin are parallel to the hyperplane [99].

SVM finds the best separation of the data with the maximum marginal hyperplane (MMH) by solving a constrained (convex) quadratic optimization problem, such as using sequential minimal optimization (SMO) [128]. When there is no linear mapping that can separate the instances of two classes, SVM projects the original data into a higher dimensional space based on nonlinear transformation. For example, in the two-dimensional space shown in figure 7.28a, we cannot find a hyperplane to perfectly separate the instances belonging to different classes. However, if projecting these instances into a three-dimensional space, as illustrated in figure 7.28b, we may find these instances located in different (bottom and top) parts of the space and thus can find a hyperplane separating them.

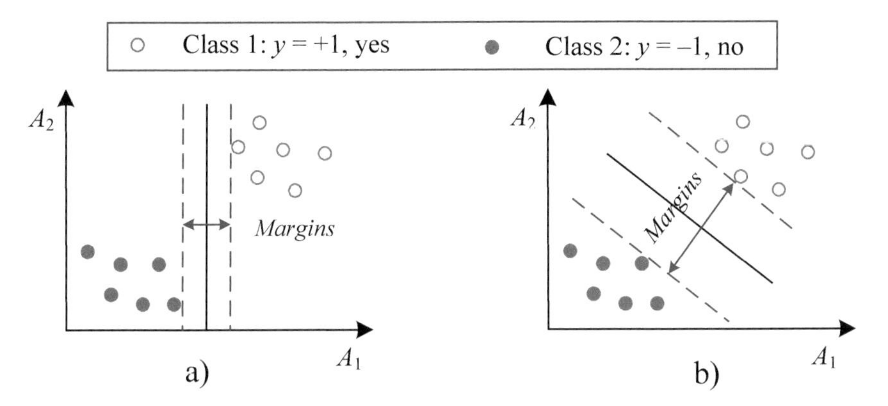

Figure 7.27
Searching for the maximum marginal hyperplane.

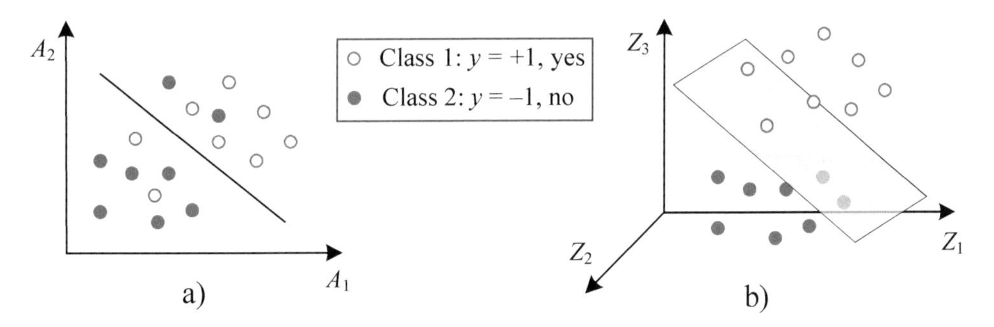

Figure 7.28
Transforming instances to a higher dimensional space for finding the best MMH.

But, there are some problems: First, how do we choose the nonlinear mapping to a higher dimensional space? Second, the computation of finding the MMH in such a high-dimensional space will be costly. Kernel functions are proposed to address the two issues. Suppose $\varphi(X)$ is the nonlinear mapping function applied to transform the original instances. After the transformation, we can find the MMH in the new high-dimensional space by solving the quadratic optimization problem of the linear SVM. The dot products of the instances in the higher dimensional space,

$$\varphi(X_i) \bullet \varphi(X_i) = K(X_i, X_j), \quad (7.23)$$

can be replaced by the calculation of the kernel function in the original space, which is of potentially much lower dimensionality. We can safely avoid the mapping and do not even have to know what the mapping is. After using this trick, we can find an MMH using a way similar to that of finding a linear SVM. Three admissible kernel functions are

$$\begin{aligned}
&\text{polynomial kernel of degree } h\text{: } K(X_i, X_j) = (X_i \bullet X_j + 1)^h, \\
&\text{Gaussian radial basis function kernel: } K(X_i, X_j) = e^{-\|X_i - X_j\|^2/2\sigma^2}, \\
&\text{sigmoid kernel: } K(X_i, X_j) = tanh(kX_i \bullet X_j - \delta)^h.
\end{aligned} \tag{7.24}$$

There is no systematic method to determine which kernel function will result in the most accurate SVM. An SVM with a Gaussian radial basis function kernel generates the same decision hyperplane as a type of neural network known as a *radial basis function network*. An SVM with a sigmoid kernel is equivalent to a simple two-layer neural network known as a *multilayer perceptron* (without a hidden layer) [99].

When using SVMs to deal with a multiclass classification problem, we need to recursively divide a training set into two parts (i.e., instances of a class and the rest of the data), learning a classifier to determine whether an instance belongs to a class or not. For instance, there are three possible class labels (C_1, C_2, and C_3) for an instance in a dataset D. We first partition D into two parts D_1 and D_2: D_1 stores instances belonging to class C_1, and D_2 is a collection of instances of other classes. Using non-C_1 as the class label for instances in D_2, we then train an SVM to determine if an instance belongs to C_1 or not. Further, we partition D_2 into two parts D_{21} and D_{22}, using D_{21} to store instances of C_2 and D_{22} for the rest of D_2 (i.e., C_3). We train another SVM to determine if an instance belongs to C_2 or not. The SVMs can form a tree structure, where each node is an SVM determining if an instance belongs to a class; the height of the tree equals the number of classes. By traversing the tree, the class of an instance in a test set can be determined.

The implementation of SVMs is a nontrivial task, considering the quadratic optimization problem and kernel functions. There are quite a few tools, such as LibSVM [87], that have been developed to help people quickly implement SVMs to solve their problems.

7.5.5 Classification with Imbalanced Data

A dataset is considered imbalanced if there is exhibited significantly unequal distribution between its classes. For example, in health-care data, the number of people with cancer is much smaller than those without cancer. In a stock market or banking system, very few transactions (e.g., less than 0.001 percent) are considered fraud. While the class (such as *cancer* and *fraud*) has a very tiny presence in the data, it is usually more valuable than the majority. If we train a classify straightforwardly based on such an imbalanced dataset, the minority will be ignored by a classification model. By fitting the majority of the data, a classifier can result in a very good performance, even if all the minority is misclassified. This is apparently not what we want.

To address this issue, a series of research, titled *learning from imbalanced data* [102], has been carried out in recent years, proposing four categories of methods: resampling, cost-sensitive learning, ensemble learning, and one-class learning.

7.5.5.1 Resampling This category of methods does not tell the classifier how to learn from imbalanced data. Instead, it provides balanced datasets for standard algorithms through *oversampling* the minority of the data and *undersampling* the majority. The mechanics of random oversampling augment the original set S by replicating the selected examples from the minority set S_{min} and adding them to S. On the contrary, random undersampling randomly selects examples from the majority set S_{maj} and removes these samples from S.

The mechanics of random oversampling and undersampling are under the risks of overfitting and losing information [102], respectively. Thus, resampling should be designed with heuristics based on the imbalanced data provided. A widely used oversampling method is called the *synthetic minority oversampling technique* (SMOTE) [89], which creates artificial data between existing minority examples. More specifically, as shown in figure 7.29, for some minority example X_i and some specified integer K ($K=3$ in this example), the new example X_{new} is created such that

$$X_{new} = X_i + \delta * (X_i - X_i'), \tag{7.25}$$

where X_i' is a minority example that is randomly chosen from the K-nearest minority neighbors of X_i, and δ is a random number between [0, 1]. Eventually, the SMOTE algorithm fulfills the convex hull of the minorities, making them more representative.

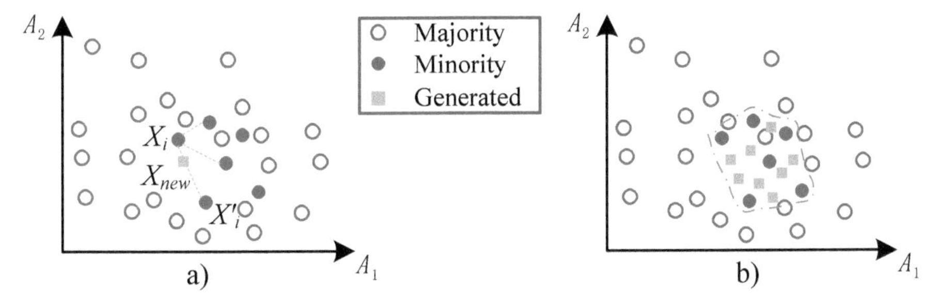

Figure 7.29
An illustration of the SMOTE algorithm.

7.5.5.2 Cost-sensitive learning Cost-sensitive methods consider assigning a larger penalty for misclassifying the minority examples. For example, in two-class imbalanced data, this category of methods defines matrix $C[c_1, c_2]$ as the cost of misclassifying an example of category c_2 as c_1, where $C[Maj, Min] > C[Min, Maj]$ (i.e., the cost of misclassifying a minority class as a majority class is higher than the opposite misclassification). $C[c_1, c_2] = 0$ for a correct classification. By incorporating the cost matrix into a standard classification paradigm, the cost-sensitive learning methods aim to minimize the overall cost $\sum_{X_i} C[f(X_i), y_i]$, where $f(X_i)$ denotes the classifier. For example, in a multilayer perceptron (MLP), the original loss function is

$$E = \frac{1}{2}\sum_{o_i}(\hat{o}_t - o_i)^2, \tag{7.26}$$

where the classes of minority and majority are not differentiated. The cost-sensitivity can be introduced to MLP by adapting the cost matrix to the original loss function:

$$E = C[Min, Maj] * \frac{1}{2}\sum_{o_i \in Maj}(\hat{o}_t - o_i)^2 + C[Maj, Min] * \frac{1}{2}\sum_{o_i \in Min}(\hat{o}_t - o_i)^2, \tag{7.27}$$

where $\hat{o}_t$ is the ground truth, and o_i is the output of the model. Then a cost-sensitive neural network can be learned based on the new loss function [110].

7.5.5.3 Ensemble learning Ensemble learning is the process by which multiple submodels are strategically generated and combined to build the final model. As illustrated in figure 7.30, considering the data imbalance problem, we can generate subdatasets, each of which contains the whole minority data and a part of the majority data, and train submodels using these subdatasets. A straightforward solution is to randomly separate the instances of the majority class into a size equal to the minorities. There are smarter strategies to generate subdatasets for ensemble learning based on some heuristics [108, 120]. The classification results of these individual classifiers are then aggregated.

7.5.5.4 One-class learning One-class learning methods try to identify objects of a specific class among all objects by learning from a training set containing only the objects of that class. One-class learning is useful when we have a lot of "normal" data and few examples of "anomalies." Hence, one-class learning fits the extremely imbalanced data well. A typical one-class classifier is the one-class SVM [131]. Distinguished from the standard SVM, which finds a hyperplane to separate the two classes, as shown in figure 7.31, one-class SVM tries to find a hyperspherical surface that can encircle most of the data belonging to the target class (the majorities in imbalanced learning). When

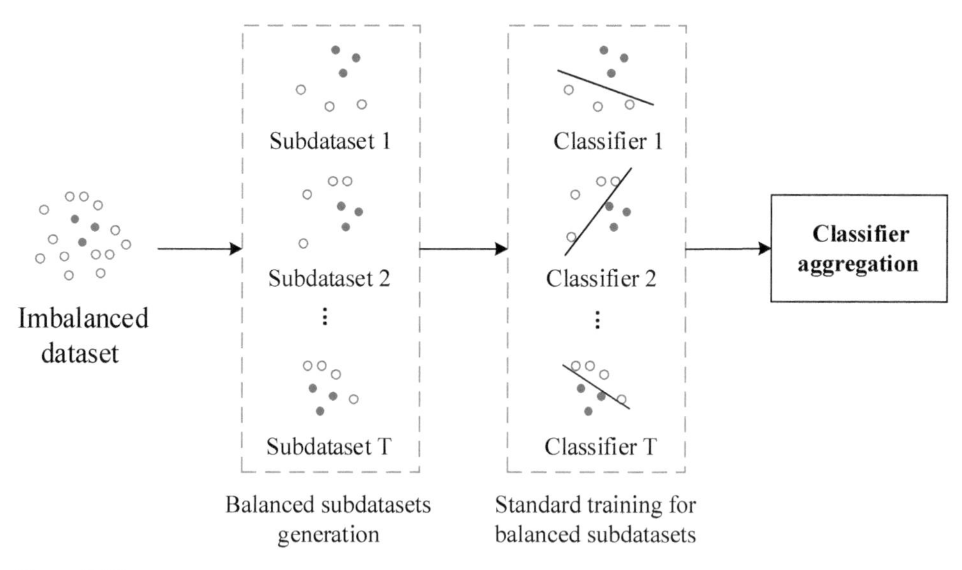

Figure 7.30
Ensemble learning for imbalanced data.

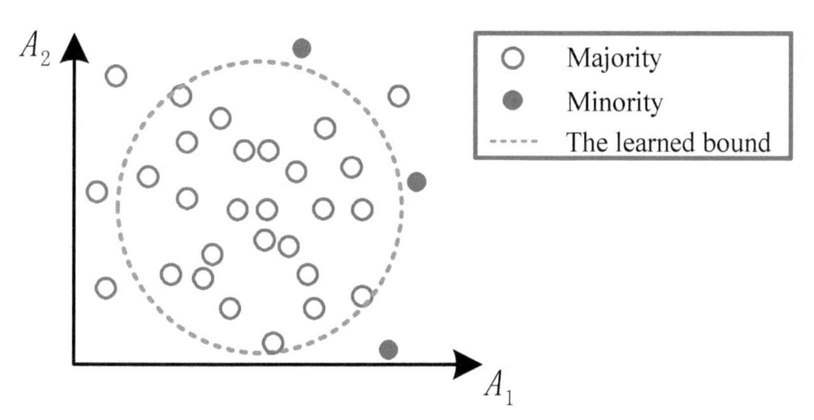

Figure 7.31
One-class SVM for imbalanced data.

the hyperspherical surface is learned, the instances outside are classified as *anomalies*, or minorities. Note that the learning of one-class SVM relies only on the inner product as the standard SVM does, so the kernel trick can be applied to one-class SVM as well when the majorities are inseparable from the minorities in original feature space.

7.6 Regression

Results generated by a classification mode are classes of an instance, which are categorical values, and are sometimes associated with the probabilities of being different classes. There is another category of problems, in which we want to estimate a real-valued number, such as the temperature of the day or the price of a product, based on a set of observations (i.e., features or attributes). Such problems can be handled by a regression model, which is a statistical technique for estimating the relationships among variables. Regression models can be used to interpolate, predict, and forecast unobserved values based on given data.

There are many types of regression models, including linear regression, auto regression, logistical regression, regression tree, and more. Some regression models, such as logistical regression, can be used to solve a classification problem. Some regression models like regression tree are derived from decision trees, which are classical classification models. In this section, we focus on three widely used regression models: linear regression, autoregression and moving average, and regression tree.

7.6.1 Linear Regression

Linear regression is an approach for modeling the relationship between a scalar dependent variable y and one or more explanatory variables (or independent variables) denoted X. The case of one explanatory variable is called *simple linear regression*. For more than one explanatory variable, the process is called *multiple linear regression* [97]. If there are multiple correlated dependent variables (Y) rather than a single scalar variable y to be predicted, the process is called *multivariate linear regression*.

7.6.1.1 Simple linear regression Suppose there is a dataset $D=\{(x_1, y_1), (x_2, y_2), \ldots, (x_m, y_m)\}$; a simple linear regression model aims to find a function

$$f(x_i)=\omega \cdot x_i+\varepsilon, \text{ s.t. } f(x_i)\cong y_i, \tag{7.28}$$

where ω is a parameter denoting the weight of x, and ε is a bias (or error term). We can learn the two parameters by minimizing the square error between estimations and their true values, formally denoted as

$$(\omega, \varepsilon) = argmin_{(\omega,\varepsilon)} \sum_{i=1}^{m} (f(x_i) - y_i)^2 = argmin_{(\omega,\varepsilon)} \sum_{i=1}^{m} (y_i - \omega \cdot x_i - \varepsilon)^2. \quad (7.29)$$

Using the least-square method, we can find the best estimation of the two parameters. Formally, we can calculate the partial derivative of the error $E_{(\omega,\varepsilon)} = \sum_{i=1}^{m} (y_i - \omega \cdot x_i - \varepsilon)^2$ with respect to ω and ε, respectively, as

$$\frac{\partial E_{(\omega,\varepsilon)}}{\partial \omega} = 2\left(\omega \sum_{i=1}^{m} x_i^2 - \sum_{i=1}^{m} (y_i - \varepsilon) x_i\right), \quad (7.30)$$

$$\frac{\partial E_{(\omega,\varepsilon)}}{\partial \varepsilon} = 2\left(m\varepsilon - \sum_{i=1}^{m} (y_i - \omega x_i)\right). \quad (7.31)$$

Letting the above two equations equal 0, we can find a closed-form result as follows:

$$\omega = \frac{\sum_{i=1}^{m} y_i (x_i - \bar{x})}{\sum_{i=1}^{m} x_i^2 - \frac{1}{m}\left(\sum_{i=1}^{m} x_i\right)^2}, \quad (7.32)$$

$$\varepsilon = \frac{1}{m} \sum_{i=1}^{m} (y_i - \omega x_i), \quad (7.33)$$

where $\bar{x} = \sum_{i=1}^{m} x_i$ is the average of x.

Once a linear regression model has been trained (i.e., ω and ε are known), we can infer the value of an instance based on its features x_i and the function $f(x_i)$.

7.6.1.2 Multiple linear regression If there are multiple features affecting a single variable y, this becomes a multiple linear regression. Supposing dataset D has m instances, $X_i = (x_1, x_2, \ldots, x_n)$ is a feature vector of the i-th instance, and y_i is a real-valued label of the instance, a multiple linear regression model aims to find a function $f(X)$:

$$f(X_i) = \omega_1 x_1 + \omega_2 x_2 + \cdots + \omega_n x_n + \varepsilon, \text{ s.t. } f(X_i) \cong y_i, \quad (7.34)$$

where $(\omega_1, \omega_2, \ldots, \omega_n)$ are parameters denoting the weight of each feature, and ε is an error term. Letting $\boldsymbol{\omega} = (\omega_1, \omega_2, \ldots, \omega_n)$, we can write the equation into another form:

$$f(X_i) = \boldsymbol{\omega} \cdot X_i + \varepsilon. \quad (7.35)$$

Letting $\hat{\boldsymbol{\omega}} = (\boldsymbol{\omega}; \varepsilon)$, $\boldsymbol{y} = (y_1; y_2; \ldots, y_m)$, and

$$\boldsymbol{X} = \begin{bmatrix} x_{11} & x_{12} & \ldots & x_{1n} & 1 \\ x_{21} & x_{22} & & x_{2n} & 1 \\ \vdots & & \ddots & \vdots & \\ x_{m1} & x_{m2} & \ldots & x_{mn} & 1 \end{bmatrix}, \quad (7.36)$$

we can write equation (7.25) as

$$\hat{\omega}^* = argmin_{(\hat{\omega})} \sum_{i=1}^{m} (y - X\hat{\omega})^T (y - X\hat{\omega}). \tag{7.37}$$

Letting $E_{\hat{\omega}} = \sum_{i=1}^{m} (y - X\hat{\omega})^T (y - X\hat{\omega})$, we can calculate the partial derivative of $E_{(\hat{\omega})}$ with respect to $\hat{\omega}$:

$$\frac{\partial E_{\hat{\omega}}}{\partial \hat{\omega}} = 2X^T (X\hat{\omega} - y). \tag{7.38}$$

Letting the above equation equal zero, we can find the closed-form result when $X^T X$ is a full-rank matrix or positive definite matrix. When $X^T X$ is not a full-rank matrix (e.g., the number of parameters is more than that of instances), there may be multiple values for $\hat{\omega}$. At this moment, we need to employ a regularization to choose a better result for $\hat{\omega}$.

7.6.1.3 Multivariate linear regression When there are multiple variables we need to infer for an instance (i.e., $Y_i = (y_{i1}, y_{i2}, \ldots, y_{ik})$, based on multiple features $X_i = (x_1, x_2, \ldots, x_n)$ of the instance), this process is called *multivariate linear regression*. This is different from multiple linear regression, in which there is a single dependent variable y to infer. Formally, suppose dataset D has m instances,

$$\mathbf{Y} = \begin{bmatrix} y_{11} & y_{12} & \ldots & y_{1k} \\ y_{21} & y_{22} & & y_{2k} \\ \vdots & & \ddots & \vdots \\ y_{m1} & y_{m2} & \ldots & y_{mk} \end{bmatrix} = \begin{bmatrix} Y_1 \\ Y_2 \\ \vdots \\ Y_m \end{bmatrix},$$
$$\mathbf{X} = \begin{bmatrix} x_{11} & x_{12} & \ldots & x_{1n} \\ x_{21} & x_{22} & & x_{2n} \\ \vdots & & \ddots & \vdots \\ x_{m1} & x_{m2} & \ldots & x_{mn} \end{bmatrix} = \begin{bmatrix} X_1 \\ X_2 \\ \vdots \\ X_m \end{bmatrix}, \tag{7.39}$$
$$\mathbf{B} = \begin{bmatrix} \omega_{11} & \omega_{12} & \ldots & \omega_{1k} \\ \omega_{21} & \omega_{22} & & \omega_{2k} \\ \vdots & & \ddots & \vdots \\ \omega_{n1} & \omega_{n2} & \ldots & \omega_{nk} \end{bmatrix} = [\boldsymbol{\omega}_1^T, \boldsymbol{\omega}_2^T, \ldots, \boldsymbol{\omega}_k^T],$$
$$\boldsymbol{U} = \begin{bmatrix} \varepsilon_{11} & \varepsilon_{12} & \ldots & \varepsilon_{1k} \\ \varepsilon_{21} & \varepsilon_{22} & & \varepsilon_{2k} \\ \vdots & & \ddots & \vdots \\ \varepsilon_{m1} & \varepsilon_{m2} & \ldots & \varepsilon_{mk} \end{bmatrix},$$

then multivariate linear regression tries to find $\boldsymbol{B}$ and $\boldsymbol{U}$ so that $\boldsymbol{Y}=\boldsymbol{X}\boldsymbol{B}+\boldsymbol{U}$. If $\boldsymbol{Y}$, $\boldsymbol{B}$, and $\boldsymbol{U}$ were column vectors, the matrix equation would represent multiple linear regression.

7.6.2 Autoregression

Autoregression aims to predict the value at a time stamp in a time series based on the data points before the time stamp. The autoregressive moving average (ARMA) is a basic model [4], combining an autoregression with a moving-average process to predict a result. Based on ARMA, two advanced models have been proposed. One is called the *autoregressive integrated moving average* (ARIMA) model, which considers the difference between the values of consecutive time stamps. The other is the *seasonal autoregressive integrated moving average* (SARIMA) [27], which further considers the period information of a time series based on ARIMA.

7.6.2.1 Autoregressive moving average Given a time series of data $X=(x_1, x_2, \ldots, x_t)$, the ARMA model is a tool for predicting future values in this series, based on two processes: an autoregressive (AR) process and a moving-average (MA) process. The model is usually referred to as the $ARMA(p, q)$ model, where p is the order of the autoregressive part, and q is the order of the moving-average part. The notation $AR(p)$ refers to the autoregressive model of order p, formally written as

$$x_t = c + \sum_{i=1}^{p} \varphi_i x_{t-i} + \varepsilon_t, \tag{7.40}$$

where $\varphi_1, \varphi_2, \ldots, \varphi_p$ are parameters, c is a constant, and the random variable ε_t is white noise.

The notation $MA(q)$ refers to the moving-average model of order q:

$$x_t = \mu + \varepsilon_t + \sum_{i=1}^{q} \theta_i \varepsilon_{t-i}, \tag{7.41}$$

where $\theta_1, \theta_2, \ldots, \theta_q$ are parameters, μ is the expectation of x_t (often assumed to equal 0), and $\varepsilon_t, \varepsilon_{t-1} \ldots$ are white-noise-error terms.

Thus, $ARMA(p, q)$ is written as

$$x_t = c + \varepsilon_t + \sum_{i=1}^{p} \varphi_i x_{t-i} + \sum_{i=1}^{q} \theta_i \varepsilon_{t-i}. \tag{7.42}$$

When implementing ARMA, we can either use default parameters for φ_i and θ_i or predefine these parameters. ε_t are generally assumed to be independent, identically distributed variables sampled from a normal distribution with a zero mean.

7.6.2.2 ARIMA and SARIMA In order to include more realistic dynamics—in particular, nonstationarity in mean and seasonal behaviors—quite a few variants of the ARMA models, including ARIMA and SARIMA, have been proposed.

The AR part of ARIMA indicates that the evolving variable of interest is regressed on its own prior values. The MA part indicates that the regression error is actually a linear combination of error terms whose values occurred contemporaneously and at various times in the past. The I (for *integrated*) indicates that the data values have been replaced with the difference between their values and the previous values (and this differencing process may have been performed more than once). For example, $x'_t = x_t - x_{t-1}$ is the first-order difference, and $x^*_t = x'_t - x'_{t-1}$ denotes the second-order difference. The purpose of each of these features is to make the model fit the data as well as possible.

Nonseasonal ARIMA models are generally denoted ARIMA(p, d, q), where parameters p, d, and q are nonnegative integers, p is the order (number of time lags) of the autoregressive model, d is the order of differencing (the number of times the data has had past values subtracted), and q is the order of the moving-average model.

SARIMA models are usually denoted by ARIMA$(p, d, q)(P, D, Q)_m$, where m refers to the number of periods in each season, and the uppercase P, D, Q refer to the autoregressive, differencing, and moving-average terms for the seasonal part of the ARIMA model [27].

7.6.3 Regression Tree

Decision trees, such as CART [86], where the target variable can take continuous values (typically real numbers), are called *regression trees*. In general, regression trees hierarchically partition a given dataset into groups based on some discriminative features until the data in a leaf node can be fitted by a simple model.

When the features and target variable are both continuous, regression trees use the decrease in variance (somehow similar to the information gain in a decision tree) in the data to determine partition thresholds. Suppose there is a dataset $D = \{(X_1, y_1), (X_2, y_2), \ldots, (X_n, y_n)\}$, where X_i is an m-dimensional feature vector, and y_i is a real-valued variable. The variance of D is

$$Var(D) = \frac{\sum_{i=1}^{m}(y_i - \bar{y})^2}{m}, \tag{7.43}$$

where $\bar{y} = \frac{\sum_{i=1}^{m} y_i}{m}$ is the mean of the target variable. If we partition D into k subsets $D_1, D_2, \ldots, D_k$ based on a feature A, its variance is

$$Var(D)_A = \sum_{j=1}^{k} \frac{|D_j|}{|D|} Var(D_i), \tag{7.44}$$

where $|D_j|$ denotes the number of instances in D_j, and $Var(D_i)$ is the variance of D_i calculated based on 7.38. The decrease in variance is defined as

$$D\_Var = Var(D) - Var(D)_A. \tag{7.45}$$

The way of finding the splitting point for a real-valued feature is the same as that of C4.5. We sort the values of a feature A and check the midpoint between each pair of consecutive values as a split point (i.e., $(a_i + a_{i+1})/2$). The split point leading to the maximum decrease in variance of the target variable is selected as the threshold for the feature. The feature that results in the most decreases in variance or information gain will be selected as the first node to partition the data into two parts. The process is performed in each part of the data iteratively, until some criteria have been satisfied (e.g., the depth of a tree or the number of instances in a leaf node or the data in leaf nodes is simple to fit).

After partitioning D into leaf nodes of a regression tree, we can fit the data in each node with a simple model (e.g., directly using the mean of the data or using a linear regression model). Figure 7.32 presents an example of a regression tree that predicts the air quality (AQI) of a city based on multiple features [146], such as the spatial factor, temporal factor, wind speed, humidity, and so on. A circle node in the tree denotes a feature selected to partition the data; each square leaf node stands for a linear regression model (LM) that combines different features to calculate AQI (i.e., to fit the data in the node). The number associated with each edge in the tree is the threshold of a selected feature. For instance, when the value of a spatial factor is smaller than 0.003 and the temporal factor is greater than –0.08, we use LM4 to calculate AQI. The weights

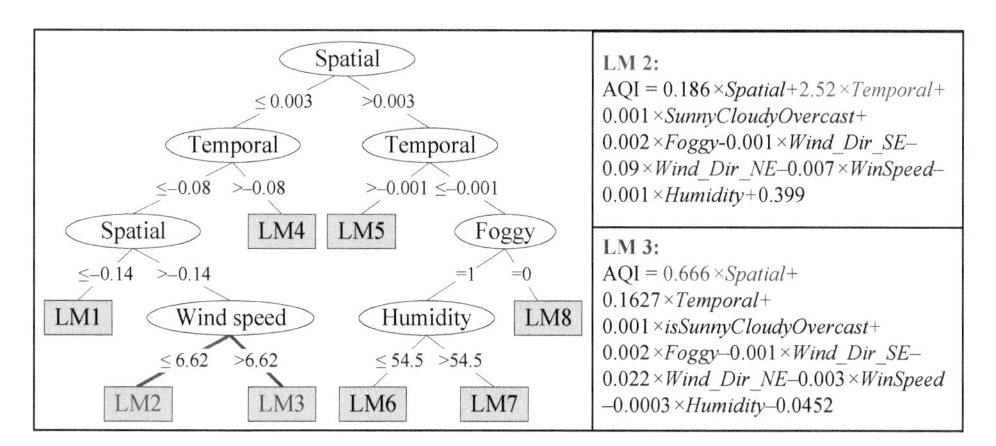

Figure 7.32
An example of using a regression tree to predict air quality.

of a feature in different LMs are different. For example, as presented in the right part of figure 7.32, when wind speed is higher than 6.62, we select LM2, which gives the temporal factor a higher weight, to calculate *AQI*. On the contrary, in LM3, the spatial factor is given a high weight.

7.7 Outlier and Anomaly Detection

Assume that a dataset is generated by a given statistical process. An outlier, also known as an *anomaly*, is a data object that deviates significantly from the rest of the objects, as if it were generated by a different mechanism. Outliers are different from noisy data, which is a random error or variance in a measurable variable.

Han et al. [99] classified outliers into three categories: *global outliers*, *contextual outliers*, and *collective outliers*. Global outliers are the simplest type of outliers, which deviate from the rest of the dataset significantly, as shown in figure 7.33a. A data object is a *contextual outlier* if it deviates significantly with respect to a specific context of the object. This category of outliers is also known as *conditional outliers*. For example, as illustrated in figure 7.33b, suppose the horizontal axis denotes the date, and the vertical axis stands for the daytime temperature. A day with a temperature of 25°C is not an outlier in an entire year but is considered quite anomalous if it is a day from December in Beijing. The temperature records in December become the context of this outlier. A subset of objects forms a *collective outlier* if the objects as a whole deviate significantly from the entire dataset, even though an individual object in the subset may not be an outlier. For instance, as depicted in figure 7.33c, each black object is not an outlier individually. However, these black objects collectively form an outlier, as the density of these objects is much higher than the rest of the dataset.

Methods for detecting outliers can be classified into two main categories:

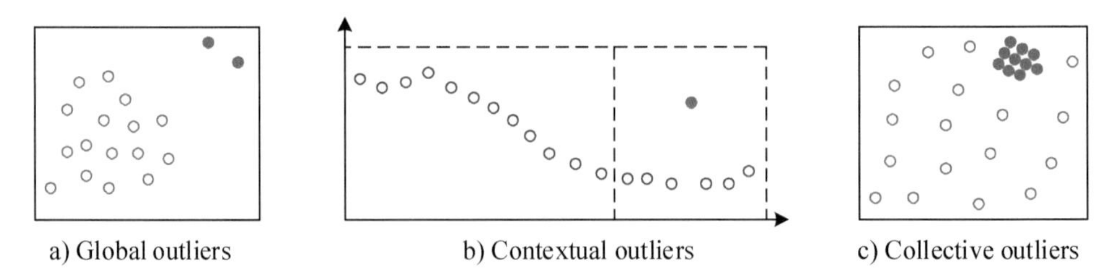

Figure 7.33
Three categories of outliers.

1. *Proximity-based methods*. The first category of methods assumes that a data object is an outlier if the proximity of the object to its nearest neighbors significantly deviates from the proximity of most of the other objects to their neighbors in the same dataset. The proximity-based methods consist of three subcategories: distance-based, clustering-based, and density-based methods.
2. *Statistical methods*. The statistical outlier-detection methods assume that the normal data objects follow a statistical model, where data not following the model is considered an outlier. The statistical methods further comprise two subcategories: statistic assumption–based and predictive model–based methods.

7.7.1 Proximity-Based Outlier Detection

7.7.1.1 Distance-based outlier-detection methods The distance-based outlier-detection methods consult the neighborhood of a data object, defined by a distance function and a radius threshold. Suppose a dataset $D=\{o_1, o_2, \dots o_n\}$ consists of n objects, each of which is associated with m attributes $o_i=(f_1, f_2, \dots f_m)$; $o_i \cdot f_k$ denotes the k-th feature of o_i. $Dist(o_i, o_j)$ stands for the distance between two objects. It can be the Euclidean distance between the attribute vectors of the two objects:

$$Dist(o_i, o_j) = \sqrt[2]{\sum_{k=1}^{m}(o_i \cdot f_k - o_j \cdot f_k)^2} \tag{7.46}$$

or other distances, such as the Pearson correlation and the Mahalanobis distance. The center of D can be obtained by calculating the mean of each attribute across all the objects in D, $\bar{f_k} = \frac{1}{n}\sum_{i=1}^{n} o_i \cdot f_k, 1 \le k \le m, 1 \le i \le n$. The vector of these mean values denotes the center object of D (i.e., $o_c = (\bar{f_1}, \bar{f_2}, \dots, \bar{f_m})$.

A very simple distance-based method is to check if the number of objects falling in an object's r-neighborhood, divided by the size of the dataset $|D||$, is no larger than a ratio threshold p. This is formally written as

$$\frac{||\{o' | Dist(o, o') \le r\}||}{||D||} \le p, \tag{7.47}$$

where r is a distance threshold. We denote this algorithm as $DB(r,p)$. For example, if the proportion of data falling in the 100 m-neighborhood of an object is less than 1 percent, the object is considered an outlier. This distance-based method can detect global outliers but is not very effective for collective outliers. In addition, the distance and ratio threshold is not easy to determine.

Another widely used distance-based outlier-detection method calculates the distance between an object o_i and the center of a given dataset (i.e., $Dist(o_i, o_c)$) and then checks if the distance is three times greater than the standard deviation σ of the dataset, formally written as

$$\sigma = \sqrt[2]{\frac{1}{n}\sum_{i=1}^{n} Dist(o_i, o_c)^2}, \tag{7.48}$$

$$Dist(o_i, o_c) \geq 3\sigma. \tag{7.49}$$

If the condition shown in equation (7.44) holds, o_i is considered an outlier. This distance-based outlier detection is nonparametric and can deal with collective outliers, thus it is widely employed in many applications. However, such a method cannot handle sparse datasets.

7.7.1.2 Clustering-based outlier detection Intuitively, an outlier is an object that belongs to a small and remote cluster or does not belong to any cluster. This leads to three general approaches to clustering-based outlier detection [99]:

1. *Objects that do not belong to any cluster are detected as outliers.* As illustrated in figure 7.34a, using a density-based clustering algorithm, such as OPTICS, we can regard those points like a that do not belong to any cluster as outliers.
2. *Objects in a small and remote cluster are detected as outliers.* As shown in figure 7.34b, using a partition-based clustering method, such as the k-means algorithm, we can divide a given dataset into several clusters. Objects in those clusters like C_3 that are small and far away from other clusters are deemed outliers.
3. *Objects that are far away from the center of the closet cluster are detected as outliers.* As depicted in figure 7.34c, using a partition-based clustering method, we can divide a given dataset into several clusters. After calculating the center of each cluster, we find the closest cluster to object a is C_1 and that of b is C_2. As a is far away from C_1's centers and b is far away from C_2's center, they are detected as outliers.

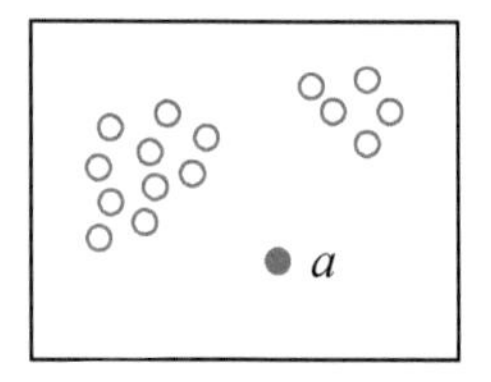

a) Objects not belonging to any cluster

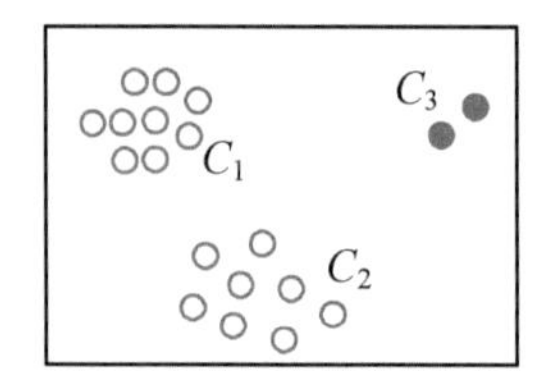

b) Objects in a small and remote cluster

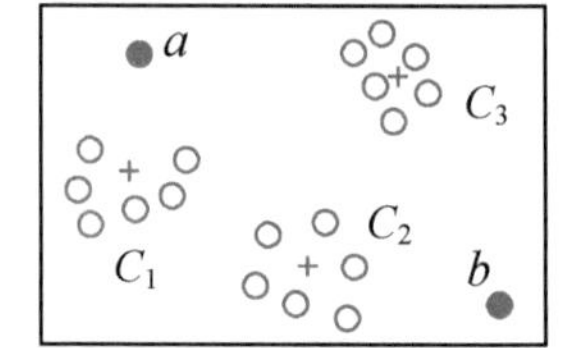

c) Objects far away from the center of the closest cluster

Figure 7.34
Outlier detection based on clustering algorithms.

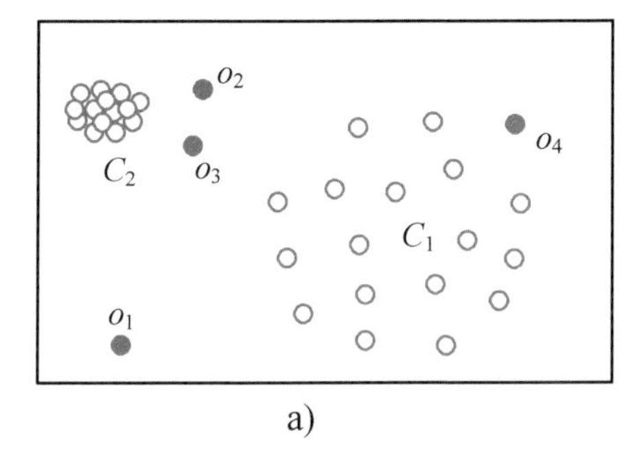

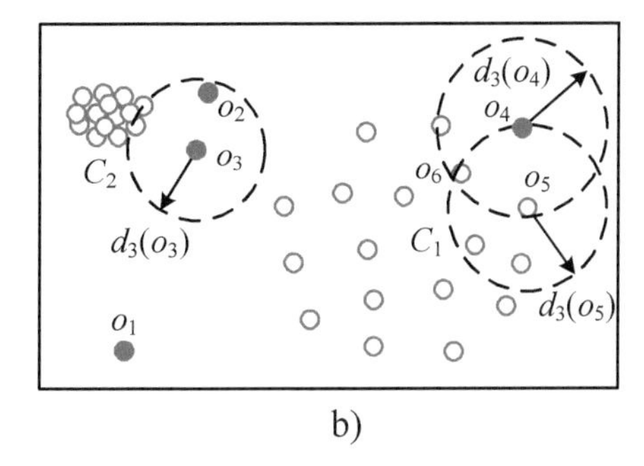

Figure 7.35
An illustration of the density-based outlier-detection methods.

7.7.1.3 Density-based outlier detection Distance-based and clustering-based outlier-detection methods can detect global outliers using a global setting of parameters. Many real-world datasets have a more complex structure, where objects may be considered outliers with respect to their local neighborhoods rather than with respect to the global data distribution.

For instance, as illustrated in figure 7.35, there is a sparse cluster C_1 and a dense cluster C_2. Using a distance-based or clustering-based outlier-detection method, we can detect object o_1 as an outlier, as it is far away from the majority of the data. However, o_2 and o_3 cannot be detected as an outlier by these outlier-detection methods, as the distance between o_2 and o_3 is even smaller than the average distance between an object in C_1 and its nearest neighbor. If expecting to recognize o_2 and o_3 as outliers using the $DB(r, p)$ algorithm introduced in section 7.6.1.1, we need to choose a smaller distance threshold r. Unfortunately, this setting will detect the entire group of objects in C_1 as outliers. To address this issue, we can consider o_2 and o_3 with respect to cluster C_2. As the two objects significantly deviate from other objects in C_2, they can be detected as outliers.

The basic assumption of density-based outlier-detection methods is that the density around a normal object is similar to the density around its neighbors, while the density around an outlier object is significantly different from the density around its neighbors [99]. Based on the assumption, density-based outlier-detection methods use the relative density of an object against its neighbors to indicate the degree to which an object is an outlier. It is formally defined as follows.

Given a dataset D consisting of a collection of objects, the *k-distance* of an object o, denoted by $d_k(o)$, is the distance between o and its k-th nearest neighbor. For example, as shown in figure 7.35b, when setting $k=3$, we can see $d_k(o_3)$, $d_k(o_4)$, and $d_k(o_5)$. The *k-distance neighborhood* of o, denoted by $N_k(o) = \{o' | o' \in D, dist(o, o') \leq d_k(o)\}$, contains all objects with the distance to o no greater than $d_k(o)$. $N_k(o)$ may contain more than k objects because multiple objects may have the same distance to o. For instance, $N_k(o_5)$ has four objects, as o_4 and o_6 have the same distance to o_5.

A straightforward method is to measure the local density of o by using the average distance from the objects in $N_k(o)$ to o. However, the statistical fluctuation of the distance measure can be undesirably high when o has very close neighbors. To overcome this problem, the reachability distance is added as a smoothing factor. For two objects, o and o', the reachability distance from o' to o is denoted by

$$reach\_d_k(o \leftarrow o') = \max\{d_k(o), dist(o, o')\}, \tag{7.50}$$

where k is a user-specified parameter that controls the smoothing effect. Now, the local reachability density of an object o is defined as

$$local\_dense_k(o) = \frac{||N_k(o)||}{\sum_{o' \in N_k(o)} reach\_d_k(o \leftarrow o')}. \tag{7.51}$$

Likewise, we can define the local outlier factor of an object o as

$$LOF_k(o) = \frac{1}{||N_k(o)||} \sum_{o' \in N_k(o)} \frac{local_{dense_k}(o')}{local\_dense_k(o)}. \tag{7.52}$$

If $local\_dense_k(o)$ is low and that of its neighbors is high, $LOF_k(o)$ is high. Thus, o is very likely to be an outlier. As illustrated in figure 7.35b, $local\_dense_k(o_4)$ and $local\_dense_k(o_5)$ are almost similar. Consequently, o_4 is not detected as an outlier. However, $local\_dense_k(o_3)$ is much smaller than that of its neighbors and therefore can be detected as an outlier.

7.7.2 Statistic-Based Outlier Detection

The underlying principle of a statistical-based outlier-detection technique is "an anomaly is an observation which is suspected of being partially or wholly irrelevant because it is not generated by the stochastic model assumed" [10]. It is based on this key assumption: Normal data instances occur in high-probability regions of a stochastic model, while anomalies occur in the low-probability regions of the stochastic model. Statistical techniques fit a statistical model (usually for normal behavior) to the given data and then apply a statistical inference test to determine if an unseen instance belongs to this model or not. Instances that have a low probability from the applied test statistic are declared as outliers.

7.7.2.1 Log-likelihood ratio test In statistics, a likelihood ratio test (LRT) is used to compare the fit of two models, one of which (the null model) is a special case of (or "nested within") the other (the alternative model). This often occurs when testing whether a simplifying assumption for a model is valid, as when two or more model parameters are assumed to be related. Each of the two competing models, the null model and the alternative model, is separately fitted to the data with the log-likelihood recorded. The test statistic (often denoted by Λ) is negative by twice the difference in these log-likelihoods:

$$\Lambda = -2\log \frac{likelihood\ for\ null\ model}{likelihood\ for\ alternative\ model}. \tag{7.53}$$

Whether the alternative model fits the data significantly better than the null model can be determined by deriving the probability or p-value of the obtained difference Λ. In many cases, the probability distribution of the test statistic Λ can be approximated by a chi-square distribution $\chi^2(\Lambda, df)$ with $df = df_2 - df_1$, where df_1 and df_2 represent the number of free parameters of the null model and the alternative model, respectively.

When applying LRT to a dataset s collected in a geographical region r, we assume s follows a certain distribution $\mathcal{P}$ with parameter Θ (e.g., the Poisson distribution with an arrival rate of λ). Supposing the number of occurrences of s observed at time interval t_i is x_i, the likelihood ratio is defined as

$$\Lambda(s) = -2\log\left(\frac{\mathcal{P}(x_i\,|\,\Theta)}{sup\{\mathcal{P}(x_i\,|\,\Theta')\}}\right), \tag{7.54}$$

where Θ' is the new parameter changing over Θ that fits the observed data best; *sup* denotes the supremum function that finds the Θ' maximizing $\mathcal{P}(x_i\,|\,\Theta')$ and returns the latter [139]. The anomalous degree *od* of this test is calculated by

$$od = \chi^2\_\mathrm{cdf}(\Lambda, df), \tag{7.55}$$

where χ^2_cdf denotes the cumulative density function of the chi-square distribution; *df* is the freedom. The time slots, with *od* larger than a given threshold (i.e., Λ's value drops in the tail of χ^2 distribution), are likely to be anomalous.

Figure 7.36 presents two examples using LRT to detect outliers from a spatiotemporal dataset [147]. As shown in figure 7.36a, we first consider a single time slot with an underlying Gaussian distribution whose variance is proportional to the mean ($mean = 200$ and $var = 1300$). Supposing the number of occurrences of s at time slot x_t is 70, then the anomalous degree of s is calculated as follows:

1. Calculate the likelihood of the null model:

   $$L_{null} = Gaussian(70\,|\,mean = 200,\ var = 1300).$$

2. Calculate Θ': in this case, we can achieve the maximum likelihood for the alternative model by setting its mean to 70. Since we assume that the distribution's variance is proportional to its mean, we should multiply the variance by $p = \frac{70}{200} = 0.35$.

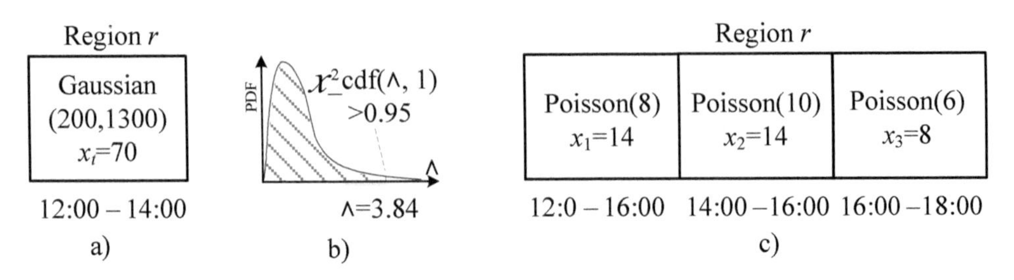

Figure 7.36
An illustration of applying LRT to the data collected in a geographical region.

Thus, the new parameter Θ' for the alternative model is ($mean = 200 \times 0.35 = 70$; $var = 1300 \times 0.35 = 455$).

3. Calculate the likelihood of the alternative model

$$L_{alter} = Gaussian(70 | mean = 70,\ var = 455).$$

4. Calculate $\Lambda(s)$ and od: as we assume the invariant linear relationship between the variance and mean, df is 1. According to equations 7.54 and 7.55, the outlier degree is calculated as follows:

$$\Lambda(s) = -2\log\left(\frac{L_{null}}{L_{alter}}\right) = 14.05,\ od = \chi^2\_\text{cdf}(14.05, fd = 1) = 0.999.$$

As depicted in figure 7.36b, if we set the threshold of od to 0.95, $<r, t>$ is apparently an anomaly. The Λ corresponds to 0.95 in the χ^2 distribution, and 1-freedom is 3.84. So, $\Lambda(s) = 14.05 > 3.84$ is considered the tail of the χ^2 distribution.

In the second example, as illustrated in figure 7.36c, we consider a region r across three consecutive time slots: $\{t_1, t_2, t_3\}$. We suppose the underlying distribution is a Poisson distribution, but different time slots have a different λ: $\lambda_1 = 8$, $\lambda_2 = 10$, and $\lambda_3 = 6$. The number of occurrences of the dataset at the three time slots are 14, 14, and 8.

1. Calculate the likelihood of the null model:

$$L_{null} = Poi(14 | \lambda_1 = 8) \times Poi(14 | \lambda_2 = 10) \times Pois(8 | \lambda_3 = 6).$$

2. Calculate $\Theta' = \{\lambda'_1, \lambda'_2, \lambda'_3\}$: to maximize the likelihood of the alternative model, we multiply λs by (assuming $fd = 1$):

$$p = \frac{14 + 14 + 8}{8 + 10 + 6} = 1.5,$$

$$\lambda'_1 = 8 \times 1.5 = 12,\ \lambda'_2 = 10 \times 1.5 = 15,\ \lambda'_3 = 6 \times 1.5 = 9.$$

3. Calculate the likelihood of the alternative model:

$$L_{alter} = Poi(14|\lambda_1') \times Poi(14|\lambda_2') \times Pois(8|\lambda_3').$$

4. Calculate $\Lambda(s)$ and od:

$$\Lambda(s) = -2\log\left(\frac{L_{null}}{L_{alter}}\right) = 5.19,\ od = \chi^2\_\text{cdf}(5.19, fd = 1) = 0.978.$$

If setting the threshold of od to 0.95, according to figure 7.36b, the three time slots are regarded as an anomaly. An advanced spatiotemporal log-likelihood ratio test (ST_LRT) is proposed in [147] to detect collective anomalies from multiple spatiotemporal datasets.

7.7.2.2 Prediction-based outlier detection Affected by multiple factors, the distribution of a variable could be too complex to be described by an existing distribution function. An approach to addressing this issue is to train a predictive model, which can be regarded as a complex mapping function between the variable and its impacting factors, based on historical data. Then, we can apply the model to predict the value of the variable based on the factors observed. If the prediction significantly deviates from its true value, it can be considered an outlier.

For example, as shown in figure 7.37, a predictive model is trained to estimate the time that an HVAC (heating, ventilation, and air conditioning) system needs to purify a building's indoor air quality to a certain standard [91]. In most cases, the predicted time (PTI) matches the real time very well. However, we see there is a big gap between them at the time interval marked by the broken circle. This is an outlier. An inspection

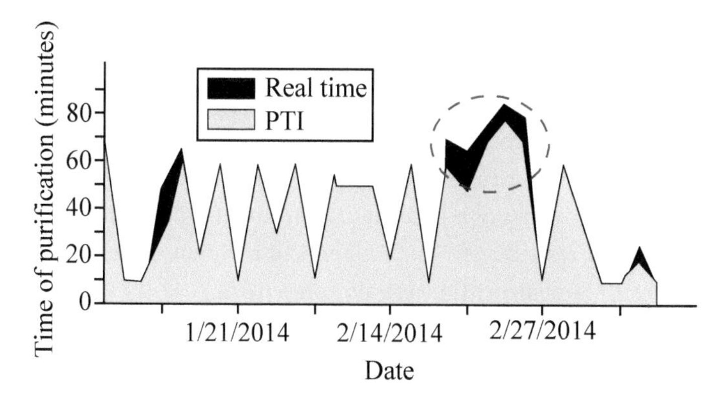

Figure 7.37
An example of outlier detection based on predictive models.

of the floor's HVAC system found the filter sheets to be very dirty and needing to be replaced. After the replacement, the gap disappeared.

Another approach is to train a classification model based on imbalanced data, where normal instances are the majority and outliers are the minority. The classification models, such as the cost-sensitive methods and over/undersampling methods, designed for imbalanced data (see section 7.5.5) can be employed to recognize outliers. This may require a few labeled outliers as training data. In many real-world systems (e.g., *cancer* in health-care data and *fraud* in credit-card data), this may not be a problem, though collecting outliers is not easy.

When outliers have not been collected, we can try one-class SVM. For example, as illustrated in figure 7.31, we can train a one-class SVM model that finds a hyperspherical surface that can encircle most of the data belonging to the normal cases. When the hyperspherical surface is learned, the instances outside are classified as "anomalies."

7.8 Summary

This chapter introduces the general framework of data mining, which comprises two main parts: data preprocessing and data analytics.

The data preprocessing part further consists of data cleaning, data transformation, and data integrating. Data cleaning aims to clean the data by filling missing values, smoothing noisy data, and removing outliers. The data transformation component transfers different forms of data into a format that is user-friendly to data-mining models. The data integration component merges data from different sources and reduces the representation of the data while minimizing the loss of information content.

Data analytics is further composed of a variety of data-mining models, result presentation methods, and evaluation methods. Based on the task that a model is going to complete, data-mining models can be divided into five main categories: frequent pattern mining, clustering, classification, regression analysis, and outlier detection.

Frequent patterns are itemsets, subsquences, or substructures that appear frequently in a dataset. We introduce three categories of algorithms for mining frequent itemsets, sequences, and subgraphs patterns, respectively. The advances in frequent pattern mining and the connection between these algorithms are also presented.

Data clustering is the process of grouping a set of data objects into multiple groups or clusters so that objects within a cluster have high similarity but are very dissimilar to objects in other clusters. Three categories of clustering algorithms are presented: partitioning methods, hierarchical methods, and density-based methods. Clustering is also known as *unsupervised learning*, compared to classification, which is called *supervised learning*.

Data classification is a two-step process consisting of a training step, where a classification is constructed based on historical data, and a classification step, where the model is used to predict class labels for given data. We introduce three well-known classification models consisting of naïve Bayesian, decision trees, and support vector machines. The three algorithms represent three main classification approaches based on the postprobability, the purity of information (i.e., information entropy), and the distance between objects. We also discuss methods for learning from imbalanced data.

Data regression aims to estimate a real-valued number based on a set of observations (i.e., features or attributes). Data regression consists of a training step, where a model is constructed based on historical data, and a prediction step, where the model is used to predict the real-valued number (rather than class labels) for given data. Three well-known regression models, consisting of linear regression, autoregression, and regression tree, are introduced.

Outliers, also known as *anomalies*, are data objects that deviate significantly from the rest of the objects, as if they were generated by a different mechanism. Two categories of outlier-detection algorithms are introduced in this chapter, consisting of proximity-based outlier-detection algorithms and statistic-based outlier detection algorithms. The first category is further composed of distance-based, clustering-based, and density-based outlier-detection algorithms. The second category consists of the likelihood ratio test and prediction-based outlier-detection methods.

References

[1] Agrawal, R., and R. Srikant. 1994. "Fast Algorithms for Mining Association Rules." In *Proceedings of the 20th International Conference on Very Large Data Bases*. San Jose, CA: Very Large Data Bases Endowment (VLDB), 487–499.

[2] Ankerst, M., M. M. Breunig, H. P. Kriegel, and J. Sander. 1999. "OPTICS: Ordering Points to Identify the Clustering Structure." *ACM SIGMOD Record* 28 (2): 49–60.

[3] Box, G., G. M. Jenkins, and G. C. Reinsel. 1994. *Time Series Analysis: Forecasting and Control.* 3rd edition. Englewood Cliffs, NJ: Prentice-Hall.

[4] Breiman, L., J. H. Friedman, R. A. Olshen, and C. J. Stone. 1984. *Classification and Regression Trees*. Monterey, CA: Wadsworth and Brooks.

[5] Chang, C. C., and C. J. Lin. 2011. "LIBSVM: A Library for Support Vector Machines." *ACM Transactions on Intelligent Systems and Technology* 2 (3): 27.

[6] Aggarwal, Charu C. 2015. *Data Mining: The Textbook*. Cham, Switzerland: Springer.

[7] Chawla, Nitesh V., et al. 2011. "SMOTE: Synthetic Minority Over-Sampling Technique." *Journal of Artificial Intelligence Research* 16 (1): 321–357.

[8] Chen, M. S., J. Han, and P. S. Yu. 1996. "Data Mining: An Overview from a Database Perspective." *IEEE Transactions on Knowledge and Data Engineering* 8 (6): 866–883.

[9] Chen, X., Y. Zheng, Y. Chen, Q. Jin, W. Sun, E. Chang, and W. Y. Ma. 2014. "Indoor Air Quality Monitoring System for Smart Buildings." In *Proceedings of the 2014 ACM International Joint Conference on Pervasive and Ubiquitous Computing*. New York: Association for Computing Machinery (ACM), 471–475.

[10] Cook, D. J., and L. B. Holder. 1994. "Substructure Discovery Using Minimum Description Length and Background Knowledge." *Journal of Artificial Intelligence Research* 1:231–255.

[11] Cortes, C., and V. Vapnik. 1995. "Support-Vector Networks." *Machine Learning* 20 (3): 273–297.

[12] Deng, K., S. W. Sadiq, X. Zhou, H. Xu, G. P. C. Fung, and Y. Lu. 2012. "On Group Nearest Group Query Processing." *IEEE Transactions on Knowledge and Data Engineering* 24 (2): 295–308.

[13] Elseidy, M., E. Abdelhamid, S. Skiadopoulos, and P. Kalnis. 2014. "Grami: Frequent Subgraph and Pattern Mining in a Single Large Graph." *Proceedings of the VLDB Endowment* 7 (7): 517–528.

[14] Ester, M., H. P. Kriegel, J. Sander, and X. Xu. 1996. "A Density-Based Algorithm for Discovering Clusters in Large Spatial Databases with Noise." *Proceedings of the Second International Conference on Knowledge Discovery and Data Mining* 96 (34): 226–231.

[15] Freedman, D. A. 2009. *Statistical Models: Theory and Practice*. New York: Cambridge University Press.

[16] Golub, G. H., and C. Reinsch. 1970. "Singular Value Decomposition and Least Squares Solutions." *Numerische mathematik* 14 (5): 403–420.

[17] Han, J., J.. Pei, and M. Kamber. 2011. *Data Mining: Concepts and Techniques*. Waltham, MA: Morgan Kaufmann.

[18] Han, J., J. Pei, B. Mortazavi-Asl, H. Pinto, Q. Chen, U. Dayal, and M. C. Hsu. 2001. "Prefixspan: Mining Sequential Patterns Efficiently by Prefix-Projected Pattern Growth." In *Proceedings of the 17th International Conference on Data Engineering*. Washington, DC: IEEE Computer Society Press, 215–224.

[19] Han, J., J. Pei, and Y. Yin. 2000. "Mining Frequent Patterns without Candidate Generation." *ACM SIGMOD Record* 29 (2): 1–12.

[20] He, H., and E. A. Garcia. 2009. "Learning from Imbalanced Data." *IEEE Transactions on Knowledge and Data Engineering* 21 (9): 1263–1284.

[21] Hinneburg, A., and D. A. Keim. 1998. "An Efficient Approach to Clustering in Large Multimedia Databases with Noise." In *Proceedings of the Fourth International Conference on Knowledge Discovery and Data Mining*. New York: AAAI Press, 58–65.

[22] Hochreiter, S., and J. Schmidhuber. 1997. "Long Short-Term Memory." *Neural Computation* 9 (8): 1735–1780.

[23] Hoyer, P. O. 2004. "Non-Negative Matrix Factorization with Sparseness Constraints." *Journal of Machine Learning Research* 5:1457–1469.

[24] Hyndman, Rob J., and George Athanasopoulos. "8.9 Seasonal ARIMA Models." *Forecasting: Principles and Practice*. oTexts. Retrieved May 19, 2015. www.otexts.org/book/fpp.

[25] Kaelbling, Leslie P., Michael L. Littman, and Andrew W. Moore. 1996. "Reinforcement Learning: A Survey." *Journal of Artificial Intelligence Research* 4:237–285.

[26] Kang, Pilsung, and S. Cho. 2006. "EUS SVMs: Ensemble of Under-Sampled SVMs for Data Imbalance Problems." In *Neural Information Processing. Lecture Notes in Computer Science*. Berlin: Springer, 837–846.

[27] Klema, V., and A. J. Laub. 1980. "The Singular Value Decomposition: Its Computation and Some Applications." *IEEE Transactions on Automatic Control* 25 (2): 164–176.

[28] Kukar, M., and I. Kononenko. 1998. "Cost-Sensitive Learning with Neural Networks." In *Proceedings of the 13th European Conference on Artificial Intelligence* (ECAI). New York: John Wiley and Sons, 445–449.

[29] Kuramochi, M., and G. Karypis. 2001. "Frequent Subgraph Discovery." In *Proceedings of the 2001 IEEE International Conference on Data Mining*. Washington, DC: Institute of Electrical and Electronics Engineers (IEEE) Computer Society Press, 313–320.

[30] Kuramochi, M., and G. Karypis. 2005. "Finding Frequent Patterns in a Large Sparse Graph." *Data Mining and Knowledge Discovery* 11 (3): 243–271.

[31] Laney, D. 2001. "3D Data Management: Controlling Data Volume, Velocity and Variety." *META Group Research Note* 6:70.

[32] LeCun, Y., L. Bottou, Y. Bengio, and P. Haffner. 1998. "Gradient-Based Learning Applied to Document Recognition." *Proceedings of the IEEE* 86 (11): 2278–2324.

[33] LeCun, Y., Y. Bengio, and G. Hinton. 2015. "Deep Learning." *Nature* 521 (7553): 436–444.

[34] Lee, D. D., and H. S. Seung. 2011. "Algorithms for Non-Negative Matrix Factorization." In *Advances in Neural Information Processing Systems*. La Jolla, CA: Neural Information Processing Systems (NIPS), 556–562.

[35] Lee, W.-C., and J. Krumm. 2011. "Trajectory Preprocessing." In *Computing with Spatial Trajectories*, edited by Y. Zheng and X. Zhou, 1–31. Berlin: Springer.

[36] Lang, M., H. Guo, J. E. Odegard, C. S. Burrus, and R. O. Wells. 1996. "Noise Reduction Using an Undecimated Discrete Wavelet Transform." *IEEE Signal Processing Letters* 3 (1): 10–12.

[37] Li, Q., Y. Zheng, X. Xie, Y. Chen, W. Liu, and W. Y. Ma. 2008. "Mining User Similarity Based on Location History." In *Proceedings of the 16th ACM SIGSPATIAL International Conference on Advances in Geographic Information Systems*. New York: ACM, 34.

[38] Liu, Xu Ying, J. Wu, and Z. H. Zhou. 2006. "Exploratory Under-Sampling for Class-Imbalance Learning." In *Proceedings, Sixth International Conference on Data Mining*. Washington, DC: IEEE Computer Society Press, 965–969.

[39] Manyika, J., M. Chui, B. Brown, J. Bughin, R. Dobbs, C. Roxburgh, and A. H. Byers. 2011. *Big Data: The Next Frontier for Innovation, Competition, and Productivity*. New York: McKinsey.

[40] Mashey, J. R. 1997. "Big Data and the Next Wave of InfraStress." Paper presented at Computer Science Division Seminar, October, University of California, Berkeley.

[41] Mnih, V., K. Kavukcuoglu, D. Silver, A. A. Rusu, J. Veness, M. G. Bellemare, A. Graves, M. Riedmiller, A. K. Fidjeland, G. Ostrovski, and S. Petersen. 2015. "Human-Level Control through Deep Reinforcement Learning." *Nature* 518 (7540): 529–533.

[42] Mohri, M., A. Rostamizadeh, and A. Talwalkar. 2012. *Foundations of Machine Learning*. Cambridge, MA: MIT Press.

[43] Opitz, D., and Maclin, R., 1999. "Popular Ensemble Methods: An Empirical Study." *Journal of Artificial Intelligence Research* 11:169–198.

[44] Pei, J., J. Han, and L. V. Lakshmanan. 2001. "Mining Frequent Itemsets with Convertible Constraints." In *Proceedings of the 17th International Conference on Data Engineering*. Washington, DC: IEEE Computer Society Press, 433–442.

[45] Pei, J., J. Han, B. Mortazavi-Asl, J. Wang, H. Pinto, Q. Chen, U. Dayal, and M. C. Hsu. 2004. "Mining Sequential Patterns by Pattern-Growth: The Prefixspan Approach." *IEEE Transactions on Knowledge and Data Engineering* 16 (11): 1424–1440.

[46] Platt, J. 1998. "Sequential Minimal Optimization: A Fast Algorithm for Training Support Vector Machines." PDF. Technical Report MSR-TR-98-14. https://pdfs.semanticscholar.org/59ee/e096b49d66f39891eb88a6c84cc89acba12d.pdf.

[47] Quinlan, J. R. 2014. *C4.5: Programs for Machine Learning*. New York: Elsevier.

[48] Russell, S., and P. Norvig. 1995. *Artificial Intelligence: A Modern Approach*. Englewood Cliffs, NJ: Prentice-Hall, 27.

[49] Schölkopf, B., John C. Platte, John C. Shawe-Taylor, Alex J. Smola, and Robert C. Williamson. 2001. "Estimating the Support of a High-Dimensional Distribution." *Neural Computation* 13 (7): 1443–1471.

[50] Shang, J., Y. Zheng, W. Tong, E. Chang, and Y. Yu. 2014. "Inferring Gas Consumption and Pollution Emission of Vehicles throughout a City." In *Proceedings of the 20th ACM SIGKDD International Conference on Knowledge Discovery and Data Mining*. New York: ACM, 1027–1036.

[51] Srikant, R., and R. Agrawal. 1996. "Mining Sequential Patterns: Generalizations and Performance Improvements." In *Proceedings, 1996 International Conference on Extending Database Technology*. Berlin: Springer, 1–17.

[52] Ullmann, J. R. 1976. "An Algorithm for Subgraph Isomorphism." *Journal of the ACM* 23 (1): 31–42.

[53] Utgoff, P. E. 1989. "Incremental Induction of Decision Trees." *Machine Learning* 4 (2): 161–186.

[54] Vanetik, N., S. E. Shimony, and E. Gudes. 2006. "Support Measures for Graph Data." *Data Mining and Knowledge Discovery* 13 (2): 243–260.

[55] Wang, J., J. Han, and J. Pei. 2003. "Closet+: Searching for the Best Strategies for Mining Frequent Closed Itemsets." In *Proceedings of the Ninth ACM SIGKDD International Conference on Knowledge Discovery and Data Mining*. New York: ACM, 236–245.

[56] Wang, Y., Y. Zheng, and Y. Xue. 2014. "Travel Time Estimation of a Path Using Sparse Trajectories." In *Proceedings of the 20th ACM SIGKDD International Conference on Knowledge Discovery and Data Mining*. New York: ACM, 25–34.

[57] Wu, M., X. Song, C. Jermaine, Sanjay Ranka, and John Gums. 2009. "A LRT Framework for Fast Spatial Anomaly Detection." In *Proceedings of the 15th ACM SIGKDD Conference on Knowledge Discovery and Data Mining*. New York: ACM, 887–896.

[58] Yan, X., and J. Han. 2002. "Gspan: Graph-Based Substructure Pattern Mining." In *Proceedings, 2002 IEEE International Conference on Data Mining*. Washington, DC: IEEE Computer Society Press, 721–724.

[59] Yan, X., and J. Han. 2003. "CloseGraph: Mining Closed Frequent Graph Patterns." In *Proceedings of the Ninth ACM SIGKDD International Conference on Knowledge Discovery and Data Mining*. New York: ACM, 286–295.

[60] Yan, X., J. Han, and R. Afshar. 2003. "CloSpan: Mining: Closed Sequential Patterns in Large Datasets." In *Proceedings of the 2003 SIAM International Conference on Data Mining*. Philadelphia: Society for Industrial and Applied Mathematics (SIAM), 166–177.

[61] Ye, Y., Y. Zheng, Y. Chen, J. Feng, and X. Xie. 2009. "Mining Individual Life Pattern Based on Location History." In *Tenth International Conference on Mobile Data Management: Systems, Services and Middleware*. Washington, DC: IEEE Computer Society Press, 1–10.

[62] Yi, X., Y. Zheng, J. Zhang, and T. Li. 2016. "ST-MVL: Filling Missing Values in Geo-Sensory Time Series Data." In *Proceedings of the 25th International Joint Conference on Artificial Intelligence*. Pasadena, CA: International Joint Conferences on Artificial Intelligence Organization (IJCAI).

[63] Zheng, Y. 2015. "Methodologies for Cross-Domain Data Fusion: An Overview." *IEEE Transactions on Big Data* 1 (1): 16–34.

[64] Zheng, Y., X. Yi, M. Li, R. Li, Z. Shan, E. Chang, and T. Li. 2015. "Forecasting Fine-Grained Air Quality Based on Big Data." In *Proceedings of the 21st ACM SIGKDD International Conference on Knowledge Discovery and Data Mining*. New York: ACM, 2267–2276.

[65] Zheng, Y., H. Zhang, and Y. Yu. 2015. "Detecting Collective Anomalies from Multiple Spatio-temporal Datasets across Different Domains." In *Proceedings of the 23rd SIGSPATIAL International Conference on Advances in Geographic Information Systems*. New York: ACM, 2.

8 Advanced Machine-Learning Techniques for Spatiotemporal Data

Abstract: This chapter presents the unique properties of spatiotemporal data, which call for dedicated designs of data-mining and machine-learning models. The chapter then introduces six categories of advanced machine-learning techniques, consisting of collaborative filtering, matrix factorization, tensor decomposition, probabilistic graphical models, deep learning, and reinforcement learning, within the context of spatiotemporal data.

8.1 Introduction

Real-world problems are often more complex than the examples presented in chapter 7. For instance, there are many factors extracted from different datasets affecting a variable we care about. How do we combine these factors to predict the value of the variable? Data with class labels (i.e., the training data) in a classification task can be very scarce, preventing us from training an effective classifier for the task. How do we handle the label scarcity problem? Sometimes, we only see the results of a variable (e.g., users' ratings of a restaurant), while the factors affecting the variable are invisible. How do we estimate the value of the variable (in another context, such as a user's rating of another restaurant or other users' ratings of this restaurant)? These challenges call for more advanced data-mining and machine-learning models.

However, most machine-learning techniques were originally created for solving computer vision and natural language-processing problems. As spatiotemporal data is very different from text, images, and videos, it is not very effective to simply apply existing machine-learning models to solve the challenges posed by urban computing. Under such a circumstance, we need to adapt advanced machine-learning techniques to spatiotemporal data. This is nontrivial given the unique properties of spatiotemporal data.

8.2 Unique Properties of Spatiotemporal Data

Spatiotemporal data has unique spatial properties and temporal properties as compared to image and text data. The *spatial properties* consist of spatial distance and spatial hierarchy. The *temporal properties* comprise temporal closeness, period, and trend.

8.2.1 Spatial Properties of Spatiotemporal Data

8.2.1.1 Spatial distance

Spatial correlation changing over spatial distances According to the first law of geography, "everything is related to everything else, but near things are more related than distant things." For example, as illustrated in figure 8.1a, the temperature of two locations (n_1 and n_3) with a near distance is generally more similar than those (n_2 and n_3) with a farther distance. Thus, a machine-learning algorithm that aims to infer the temperature of a location based on its neighbors could assign a bigger weight to the readings of near neighbors. The spatial distance is not limited to the Euclidean distance in a geographical space. It can be the network distance in a transportation network, where each node has a spatial coordinate, and each edge is associated with a spatial length. As shown in figure 8.1b, though the Euclidean distance between v_1 and v_4 is shorter than that between v_1 and v_2, traffic flows at the latter two nodes are more similar than the former, which has a longer distance ($v_1 \rightarrow v_2 \rightarrow v_3 \rightarrow v_4$) in the road network. This requires machine-learning algorithms to differentiate between the similarities of objects with different spatial distances.

Though sometimes the distance between two pixels in an image is also used as a feature for object recognition or image segmentation, the depth of the field is squashed in an image; thus, the distance is not very meaningful compared to the distance in spatiotemporal data. For example, as illustrated in figure 8.1c, the two pixels (p_1, p_2) in an image are from the individual's face and the background scene, respectively. Even

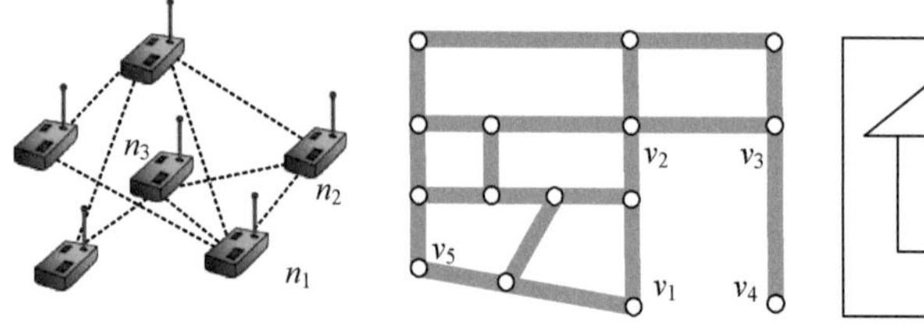

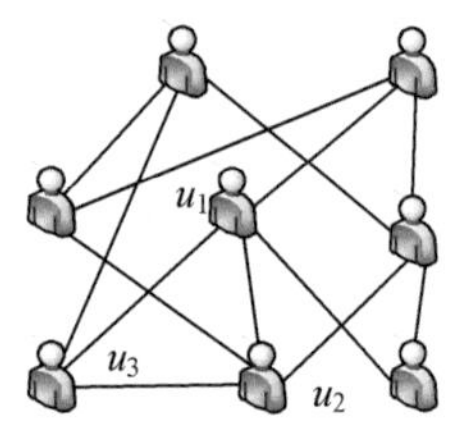

a) Euclidian distance b) Spatial network distance c) Distance in images d) Social networks

Figure 8.1
Spatial distances.

though they are very close in the image, the true locations denoted by the two pixels are actually very far in the real world, thus they are not relevant at all. On the contrary, the two pixels (p_2, p_3) with a farther distance (but both on the individual's body) are more relevant than (p_1, p_2).

The triangle inequality of spatial distances For example, as shown in figure 8.1a, the Euclidean distances among n_1, n_2, and n_3 follow the triangle inequality:

$$[dist(n_1, n_3) - dist(n_2, n_3)] \leq dist(n_1, n_2) \leq [dist(n_1, n_3) + dist(n_2, n_3)]. \quad (8.1)$$

Likewise, as illustrated in figure 8.1b, the length of the shortest path between $v_5 \rightarrow v_3$ is smaller than the shortest path between $v_5 \rightarrow v_2$ plus that between $v_2 \rightarrow v_3$—that is,

$$Shortest(v_5 \rightarrow v_3) \leq Shortest(v_5 \rightarrow v_2) + Shortest(v_2 \rightarrow v_3). \quad (8.2)$$

Thus,

$$Shortest(v_5 \rightarrow v_2) \geq Shortest(v_5 \rightarrow v_3) - Shortest(v_2 \rightarrow v_3). \quad (8.3)$$

The triangle inequality of spatial distance can derive upper and lower bounds that significantly prune the searching spaces for a data-mining task. Suppose v_5 and v_3 are selected as anchor points in their own regions with the shortest path between them precomputed (i.e., $Shortest(v_5 \rightarrow v_3)$ is already known). In addition, $Shortest(v_2 \rightarrow v_3)$ is very easy to compute as they are spatially close. Here, they are on the same road segment. Thus, based on equation (8.3), we can quickly derive the lower bound of the road network distance between v_5 and v_2 without truly computing the shortest path. The lower bound can expedite many computing processes, such as the *k*-nearest neighbor search and the reachability estimation.

This kind of distance does not exist in other types of data. For example, in the social network depicted in figure 8.1d, we cannot differentiate between u_2 and u_3 when considering their distances to u_1, as they are both one-hop connected to u_1. All users that are directly connected to u_1 have the same distance to u_1 (i.e., a one-hop distance).

8.2.1.2 Spatial hierarchy Different granularities of locations result in a spatial hierarchy naturally (e.g., a state comprises many counties, each of which further consists of many cities). The data representation at different levels of the hierarchy implies different levels of knowledge. It is a challenging task for a machine-learning algorithm to capture the information of spatiotemporal data with different granularities.

As illustrated in figure 8.2, different individuals' location histories form a geographical hierarchy, where nodes on an upper level denote location clusters with a coarser granularity. For example, c_{10} consists of three finer-grained location clusters c_{20}, c_{21}, and

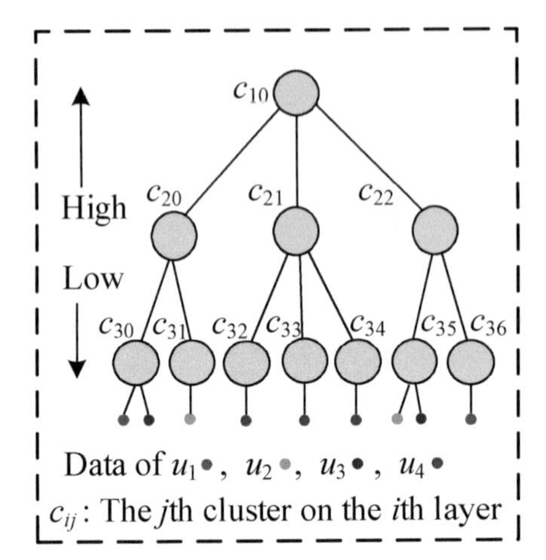

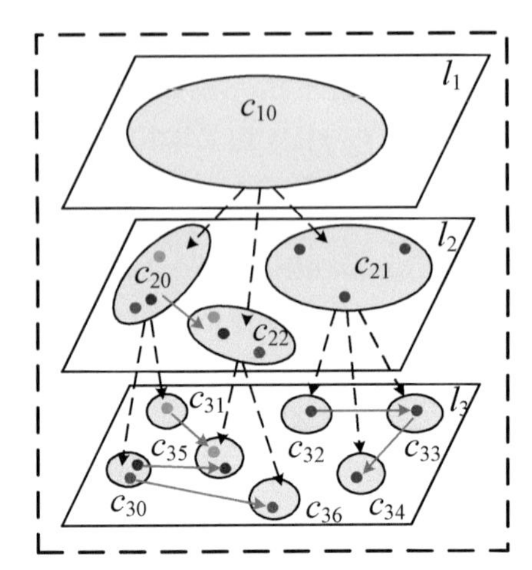

Figure 8.2
Spatial hierarchical properties.

c_{22}. c_{20} is further composed of c_{30} and c_{31}. Dots with different colors stand for the location histories of four different users.

We want to estimate the similarity between four users (u_1, u_2, u_3, u_4) based on their location histories, with a general insight that similar users could share more location histories [38]. If only checking their location histories on the second layer l_2, we cannot differentiate among u_1, u_2 and u_3, as all of them have traveled through $c_{20} \rightarrow c_{22}$. However, if checking the third layer, we find that both u_2 and u_3 have visited c_{35}. Consequently, u_3 is more similar to u_2 as compared to u_1. Likewise, u_3 is more similar to u_1 as compared to u_2, as u_1 and u_3 have both visited c_{30}.

On the contrary, if we only explore these individuals' data on l_3, then u_1, u_2, and u_4 become hard to compare because they do not share any locations on this layer. But by checking their location histories on l_2, we can distinguish u_2 from u_4 because u_1 and u_2 both travel through $c_{20} \rightarrow c_{22}$, whereas u_1 and u_2 do not. Overall, we cannot differentiate the four users without such a spatial hierarchy.

8.2.2 Temporal Properties

1. *Temporal closeness*. The temporal closeness is similar to spatial distances, denoting that data generated at two close time stamps are usually more similar than those from two distant time stamps [67]. For example, as illustrated in figure 8.3a, the horizontal axis denotes the time span between two time stamps, and the vertical axis stands for the similarity between the two time stamps' air quality readings. As

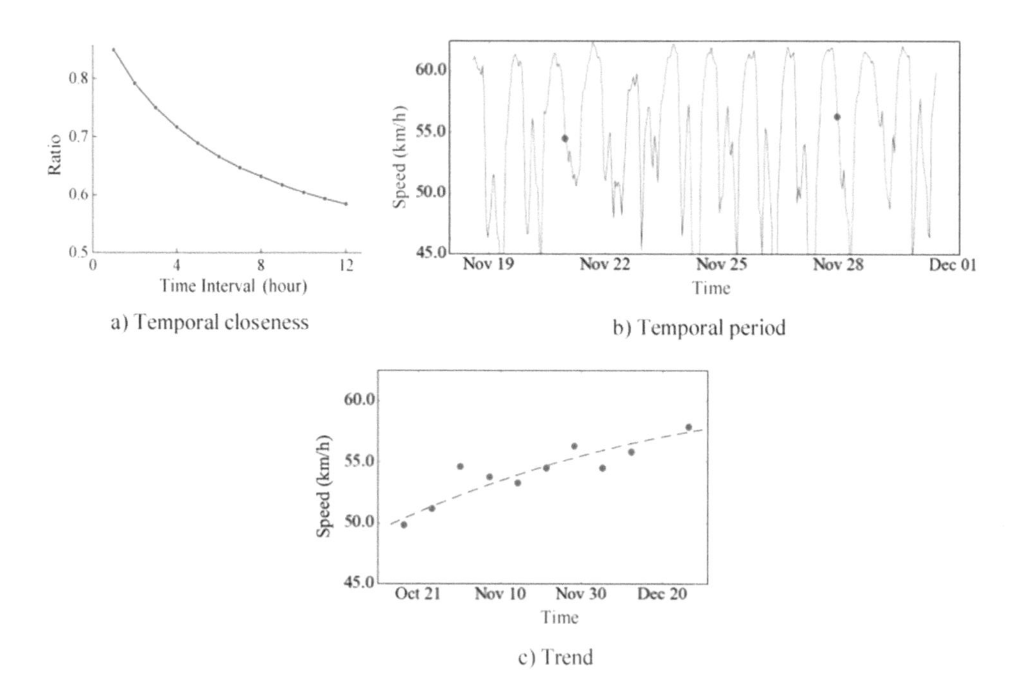

Figure 8.3
Temporal properties.

the time span increases, the similarity decreases, nearly following an exponential distribution.

2. *Temporal period.* The temporal closeness, shown in figure 8.3a, may not always be true because of the periodic patterns of spatiotemporal data. For example, as demonstrated in figure 8.3b, with a daily period, traffic speeds on a road at 8 a.m. are almost identical on consecutive workdays. But, they could be quite different from those of 11 a.m. on the same day, even though 11 a.m. is closer to 8 a.m. than the 8 a.m. of the previous day.
3. *Temporal trend.* The period evolves as time goes by. As illustrated in figure 8.3c, the travel speed 9 a.m.–10 a.m. on weekends keeps increasing as winter approaches. When the temperature drops, people postpone the start times of their weekend activities. Thus, traffic conditions on this road become better and better at 9 a.m.

In literature regarding video analysis, we also observe frequent use of the term *spatiotemporal,* as video is also considered a type of time-series data, though it does not have the spatial and temporal properties mentioned above. For example, a movie would never repeat every few minutes (i.e., no period), and there is no clear trend. The distance between each pair of pixels in an image is not as meaningful as the distance in the geographical spaces. Though we can also formulate a so-called hierarchy in an image by hierarchically splitting it into disjoint regions or scaling it down from its original resolution, the hierarchy does not carry the semantic meanings of the spatial hierarchy introduced in section 8.2.1.2. The splitting of a region in an image and the merging of several pixels do not consider the semantic meaning of the image's content.

Thus, it is still a challenging task for data-mining and machine-learning algorithms to encode the aforementioned spatiotemporal properties.

8.3 Collaborative Filtering

Collaborative filtering (CF) is a well-known model widely used in recommender systems. The general idea behind collaborative filtering is that similar users create ratings in a similar manner for similar items [10]. Thus, if a similarity is determined between users and items, a potential prediction can be made as to the rating of a user with regard to future items. Users and items are generally organized by a matrix, where an entry denotes a user's rating of an item. The rating can be an explicit ranking or an implicit indication, such as the number of visits to a place or the times that a user has browsed an item. In this section, we introduce CF models using location recommendations as an example [44], where a location is regarded as an item.

8.3.1 Basic Models: User Based and Item Based

As depicted in figure 8.4, each row of matrix M denotes a user, and each column stands for a location; the value of each entry is the number of visits a user has paid to a location, which is used as an implicit rating of the user for the location. $\times$ means a location has not been visited by a user, witnessed by the data we can collect. This is also the value a CF model needs to estimate. Once a matrix is formulated, the distance between two rows in the matrix denotes the similarity between two users, as shown in figure 8.4a, while the distance between two columns stands for the similarity between two items (i.e., locations), as illustrated in figure 8.4b. By filling in the missing entries in matrix M, we can predict a user's potential interests in an unvisited location, thus performing a personalized location recommendation.

Memory-based CF is the most widely used algorithm that computes the value of the unknown rating for a user and an item as an aggregate of the ratings of some other (usually, the N most similar) users for the same item. There are two classes of memory-based CF models: user-based and item-based techniques.

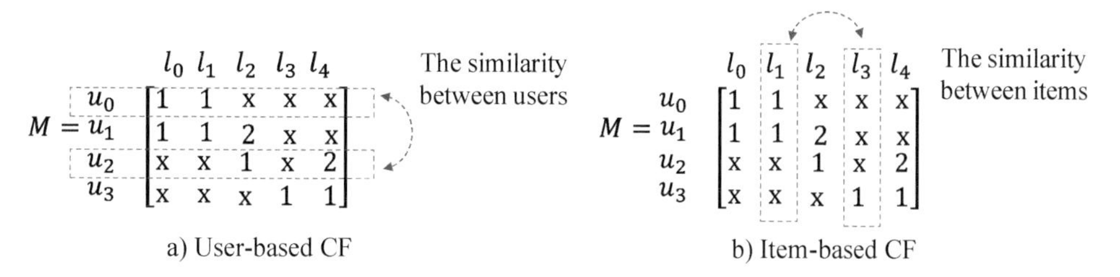

Figure 8.4
Memory-based CF models.

8.3.1.1 User-based CF models User-based CF models estimate the similarity between each pair of users, typically based on the two users' ratings for existing items, as illustrated in figure 8.4a. There are quite a few variants of user-based CF models. We introduce one widely used implementation as follows:

Suppose the ratings from a user u_p, called an *evaluation*, are represented as a vector $R_p = \langle r_{p0}, r_{p1}, \ldots, r_{pn} \rangle$, where r_{pj} is u_p's ratings for location j. User p's interest (r_{pi}) in an unvisited location i can be predicted as

$$r_{pi} = \overline{R_p} + d\sum_{u_q \in U'} sim(u_p, u_q) \times (r_{qi} - \overline{R_q}), \tag{8.4}$$

$$d = \frac{1}{|U'|}\sum_{u_q \in U'} sim(u_p, u_q), \tag{8.5}$$

$$\overline{R_p} = \frac{1}{|S(R_p)|}\sum_{i \in S(R_p)} r_{pi}, \tag{8.6}$$

where $sim(u_p, u_q)$ denotes the similarity between user u_p and u_q; U' is the collection of users most similar to u_q. The number of elements in a set S is $|S|$. $\overline{R_q}$ and $\overline{R_p}$ are the average ratings of u_p and u_q, respectively, denoting their rating scale; $S(R_p)$ is the subset of R_p; $\forall r_{pj} \in S(R_p)$; $r_{pj} \neq 0$ (i.e., the set of locations that has been rated by u_p).

$r_{qi} - \overline{R_q}$ is to avoid the rating biases of different users. For example, supposing 10 is the highest rating score, some conservative users may give a movie they consider good a rating of 7, while others may rate 9 for such a good movie. Likewise, to some users, a rating of 6 for a movie means the movie is very bad, while others may consider scores under 4 as bad.

There are quite a few similarity functions, such as the Pearson correlation or the cosine similarity, that can be employed to compute $sim(u_p, u_q)$, based on R_p and R_q. The following is an implementation of the Pearson correlation:

$$sim(u_p, u_q) = \frac{\sum_{i \in S(R_p) \cap S(R_q)} (r_{pi} - \overline{R_p}) \cdot (r_{qi} - \overline{R_q})}{\sqrt{\sum_{j \in S(R_p) \cap S(R_q)} (r_{pj} - \overline{R_p})^2 \cdot \sum_{j \in S(R_p) \cap S(R_q)} (r_{qj} - \overline{R_q})^2}}. \tag{8.7}$$

The $sim(u_p, u_q)$ can also be derived from advanced models considering more factors, such as the sequence between locations and the popularity of a location [38, 75].

8.3.1.2 Item-based CF models When the number of users becomes big, computing the similarity between each pair of users is impractical for a real system. Given that the number of items (e.g., locations) could be smaller than that of users, item-based CF models (e.g., the Slope One algorithm [37]) have been proposed to address this issue. We still introduce the Slope One algorithm with the context of location recommendations, where an item stands for a location.

Given any two locations i and j with ratings r_{pj} and r_{pi}, respectively, in a user's rating evaluation, we consider the average deviation of location i with regard to location j as

$$dev_{j,i} = \sum_{R_p \in S_{j,i}(\mathcal{X})} \frac{r_{pj} - r_{pi}}{|S_{j,i}(\mathcal{X})|}, \tag{8.8}$$

where χ is the full collection of evaluations; $S_j(\chi)$ means the set of evaluations containing location j, $\forall R_p \in S_j(\chi)$, $i, j \in S(R_p)$; $S_{i,j}(\chi)$ is the set of evaluations simultaneously containing location i and j.

Given that $dev_{j,i} + r_{pi}$ is a prediction for r_{pj} based on r_{pi}, a reasonable predictor might be the average of all the predictions.

$$P(r_{pj}) = \frac{1}{|\mathcal{W}_j|} \sum_{i \in \mathcal{W}_j} (dev_{j,i} + r_{pi}), \tag{8.9}$$

where $\mathcal{W}_j = \{i | i \in S(R_p), i \neq j, |S_{j,i}(\mathcal{X})| > 0\}$ is the set of all relevant locations. Further, the number of evaluations simultaneously containing two locations has been used to weight the prediction regarding different items. Intuitively, to predict u_p's rating of location A given u_p's ratings of item B and C, if two thousand users rated the pair of A and B, whereas only twenty users rated the pair of A and C, then u_p's rating of B is likely to be a far better predictor for A than u_p's rating of C. In short, the bigger the $|S_{j,i}(\mathcal{X})|$ is, the more correlated the two locations i and j. Based on this idea, as shown in equation (8.10), we can further relax the $|S_{j,i}(\mathcal{X})|$ to a type of correlation between two locations cor_{ji}, which can be derived by considering more factors (e.g., people's transition patterns between two locations [75]).

$$P(r_{pj}) = \frac{\sum_{i \in S(R_p) \wedge i \neq j} (dev_{j,i} + r_{pi}) \cdot |S_{j,i}(\mathcal{X})|}{\sum_{i \in S(R_p) \wedge i \neq j} |S_{j,i}(\mathcal{X})|}. \tag{8.10}$$

$$P(r_{pj}) = \frac{\sum_{i \in S(R_p) \wedge i \neq j} (dev_{j,i} + r_{pi}) \cdot cor_{ji}}{\sum_{i \in S(R_p) \wedge i \neq j} cor_{ji}}. \tag{8.11}$$

8.3.2 Collaborative Filtering for Spatiotemporal Data

8.3.2.1 User-based location recommendations Given a user-location matrix M shown in figure 8.4a, in the original user-based CF model, the similarity between users (u_p, u_q) is simply represented by the distance (or the correlation) between the two users' evaluations of locations (i.e., R_p and R_q). However, such a similarity, like the one presented in equation (8.7), does not consider the unique properties of spatiotemporal data, such as the sequence between locations and the spatial hierarchy, and thus cannot sufficiently support location recommendations.

First, a user's location history is represented by a sequence of locations visited. There is a natural connection between two locations in a user's life, indicating the semantic meanings of a user's behavior and interests. For example, two users visiting the same restaurant and shopping mall sequentially would share a similar life pattern. Thus, they should be more similar to each other than those who have visited the two places separately (but never in a trip). Likewise, users who share a longer sequence of locations (e.g., x university$\rightarrow y$ restaurant$\rightarrow z$ movie theater) would be more similar than those sharing a shorter one (e.g., x university$\rightarrow y$ restaurant). The more and the longer the location sequences shared by two users' location histories, the more similar the two users might be.

Second, there is a spatial hierarchy in the geographical space, denoting locations of different granularities, as illustrated in figure 8.2. For example, we can say two users share a location in a classroom, or in a university, or in a city. Thus, a user's location history can also be represented by location sequences on different layers of the spatial hierarchy. Intuitively, users who share similar location histories of a finer granularity might be more correlated than those sharing locations of a coarser granularity.

Third, people co-occurring at locations with different popularities might have different levels of similarity. For instance, lots of people have visited the Great Wall, a famous tourist attraction in Beijing, but this might not mean all these people are similar to one another. However, if two users visited a restaurant that is not well known, they might indeed share some similar preferences or have some (potential) connections. Two users sharing a less-popular location might be more correlated than those who share a location that has been visited by many people.

Considering the aforementioned three factors, we compute the similarity between two users based on their location histories, according to the following procedure.

First, we construct the spatial hierarchy based on the given data, as illustrated in figure 8.5, where each green point stands for a location of the finest granularity, such as a specific point of interest (POI) like a restaurant that a user has visited or a place that a user has stayed. The location can be extracted from a user's check-in data in an online

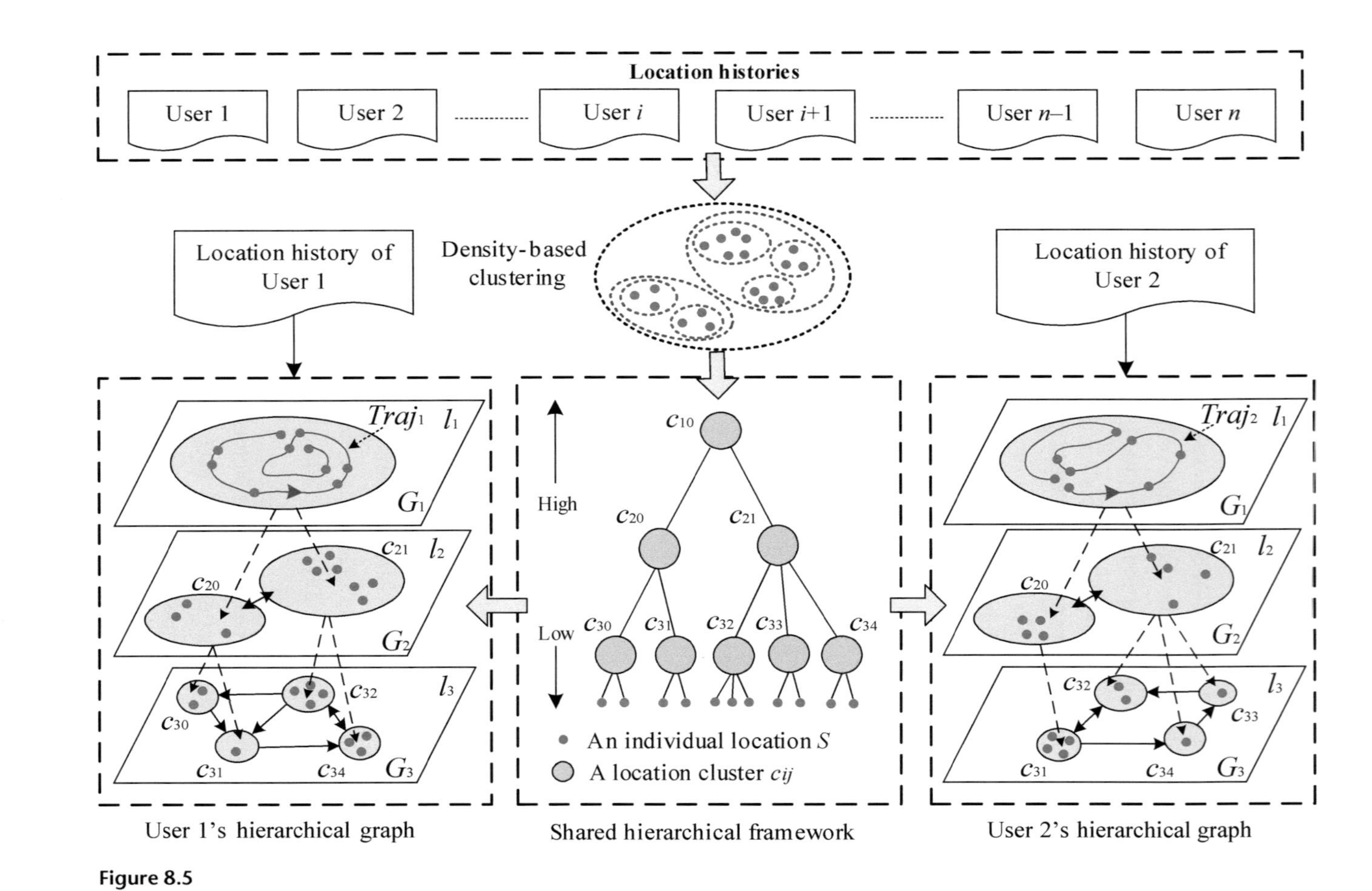

Figure 8.5
Computing the similarity between users based on location histories.

social networking service or detected from a user's GPS trajectories using a stay point–detection algorithm [38]. By putting all users' locations into a set, we can hierarchically cluster the location set into clusters of different sizes. The clusters on the deeper layers of the hierarchy denote locations of a finer granularity and of a smaller size. The hierarchy makes different users' location histories comparable.

We then formulate the location sequences of each user on each layer of the hierarchy, based on his or her own location history and shared spatial hierarchy. For instance, u_1's location history can be represented by a sequence S_1^3 on the third layer of the hierarchy and by S_1^2 on the second layer of the hierarchy.

$$S_1^3 = c_{32} \xrightarrow{\Delta t_1} c_{30} \xrightarrow{\Delta t_2} c_{31} \xrightarrow{\Delta t_3} c_{34} \xrightarrow{\Delta t_4} c_{32} \xrightarrow{\Delta t_5} c_{34}, \tag{8.12}$$

$$S_1^2 = c_{21} \xrightarrow{\Delta t_1'} c_{20} \xrightarrow{\Delta t_1'} c_{21}, \tag{8.13}$$

where Δt_1 is the time interval for traveling from c_{32} to c_{30}.

Second, we find the common subsequences shared by two users, using a sequence-matching algorithm that considers the identity of the locations and the time interval between two consecutive locations [65]. Given two subsequences $S_1[a_1, a_2, \dots, a_k]$ and $S_2[b_1, b_2, \dots, b_k]$ from two sequences S_1 and S_2, respectively, and a temporal constraint factor $\rho \in [0, 1]$, these two subsequences formulate a k-length common subsequence if they hold the following two conditions:

1. $\forall\ i \in [1, k],\ a_i = b_i;$
2. $\forall\ i \in [2, k],\ \dfrac{|\Delta t_i - \Delta t_i'|}{Max(\Delta t_i, \Delta t_i')} \le \rho;$ (8.14)

where Δt_i is the time interval that a user spends on traveling from location a_{i-1} to a_i, and $\Delta t_i'$ is the between location b_i and b_{i-1}. For example, u_1 and u_2's location histories can be represented by two sequences on the third layer of the spatial hierarchy depicted in figure 8.5:

$$S_1^3 = c_{32} \xrightarrow{1} c_{30} \xrightarrow{1} c_{31} \xrightarrow{1.4} c_{34} \xrightarrow{1.8} c_{32} \xrightarrow{0.8} c_{34},$$

$$S_2^3 = c_{31} \xrightarrow{1.5} c_{34} \xrightarrow{0.5} c_{33} \xrightarrow{1} c_{32} \xrightarrow{2.2} c_{31} \xrightarrow{1.6} c_{32} \xrightarrow{0.6} c_{31}.$$

If setting $\rho = 0.2$, we find $c_{31} \to c_{34} \to c_{32}$ is a common sequence shared by the two users. Supposing u_2 spends 0.3 hours at c_{33}, the time interval that u_2 spent traveling from c_{34} to c_{32} is 1.8 (0.5 + 0.3 + 1) hours in total, which is the same as u_1. When doing sequence matching, we allow locations to be nonconsecutive in the original sequence. Though $c_{31} \to c_{34}$ is also a common subsequence shared by the two users, it is not a maximum length subsequence because it is contained in $c_{31} \to c_{34} \to c_{32}$. Likewise, $c_{32} \to c_{31}$ is a maximum-length common subsequence shared by the two users.

Some well-known sequence-matching algorithms, such as longest-common subsequences searching (LCSS) and dynamic time wrapping (DTW), cannot be employed to discover the maximum common subsequence here, as they do not incorporate the travel time between two locations in the matching process. For this reason, a method has been proposed in [65] for detecting the maximum common subsequence between two users' location histories.

Third, we model the popularity of a location c based on the idea of inverse document frequency (IDF) as follows:

$$iuf(c) = \log \frac{N}{n}, \tag{8.15}$$

where N is the total number of users, and n is the number of users who have visited c. The bigger n is, the smaller $iuf(c)$ is. We call $iuf(c)$ the inverse user frequency of location c.

Finally, we calculate a similarity score between user u_1 and u_2 as follows [77]:

$$SimUser(u_1, u_2) = \sum_{l=1}^{L} f_w(l) \times SimSq(S_1^l, S_2^l), \tag{8.16}$$

$$SimSq(S_1, S_2) = \frac{\sum_{j=1}^{m} sg(s_j)}{|S_1| \times |S_2|}, \tag{8.17}$$

$$sg(s) = g_w(k) \times \sum_{i=1}^{k} iuf(c_i), \tag{8.18}$$

where s is a subsequence shared by the two users, $sg(s)$ is the score calculated based on s, $iuf(c_i)$ is the inverse user frequency of each location contained in s, and $g_w(k) = 2^{k-1}$ is a weighting function changing over the length (k) of s. A longer subsequence will be given a bigger weight. $SimSq(S_1, S_2)$ denotes the similarity score calculated based on u_1 and u_2's sequences on a layer of the hierarchy. Supposing there are m common subsequences shared by S_1 and S_2, $SimSq(S_1, S_2)$ is the sum of the similarity scores of the m subsequences. $SimSq(S_1, S_2)$ is further normalized by the production of the two sequences' length $|S_1| \times |S_2|$, as different users may have different scales of data. Otherwise, a user with a longer location history would have more subsequences shared with other users. Thus, the users with more data would be generally more similar to others. S_1^l stands for u_1's location sequence on the l-layer. $f_w(l) = 2^{l-1}$ is a weighting function assigning a bigger weight to the subsequences shared on a deeper layer of the hierarchy. $SimUser(u_1, u_2)$ is a weighted aggregation of the similarity scores of sequences on all layers of the spatial hierarchy.

After obtaining the similarity between users, we apply $SimUser(u_1, u_2)$ to the original user-based CF models (e.g., replacing the $sim(u_p, u_q)$ in equations [8.4] and [8.5]).

To estimate the similarity between users who do not share physical locations (e.g., living in different cities), we first represent a location by its category (e.g., a restaurant). Then, a user's location history is represented by a sequence of POI categories, such as *restaurant → shopping mall → university*. The taxonomy of POI categories can be employed to replace the spatial hierarchy shown in figure 8.5. For instance, the category of restaurant is further composed of Chinese cuisine, Italian food, Japanese food, and so on. Now we are ready to apply the aforementioned sequence-matching method to calculate a similarity score for each pair of users. When the exact POI that a user has visited is unknown (e.g., when trying to detect a sequence of stay points from a user's GPS trajectories), we use the distribution of POI categories around a stay point to denote the location where people have stayed for a while [65]. We then hierarchically cluster these locations based on their category distributions, formulating the hierarchy just like that illustrated in figure 8.5. As there is no physical location existing in this category of methods, the user similarities cannot be applied to location recommendations anymore. Instead, it can be used for friend recommendations and community discovery.

8.3.2.2 Item-based location recommendations In the original item-based CF models, the similarity between two locations is estimated based on users' evaluations of the two locations. Such a similarity does not reveal the correlation between locations as it ignores the sequence between locations and users' travel experiences.

First, the correlation between two locations depends not only on the number of people visiting the two locations but also on these people's knowledge (or travel experience). For instance, overseas tourists might randomly visit some places in Beijing because they are not familiar with the city, while Beijing locals are more capable of arranging a proper and reasonable way to visit certain places in Beijing. Thus, the locations sequentially traveled by people with rich knowledge (about these locations) would be more correlated than locations visited by those having little idea about the region.

Estimating an individual's travel knowledge is a very challenging task, as it depends on the quality of locations the individual has visited. In turn, the quality of a location depends on the knowledge of people who have visited it. Intuitively, a user with rich knowledge would be more likely to find high-quality tourist attractions. In turn, a high-quality tourist attraction would be able to attract more people with rich knowledge. That is, they mutually depend on each other, but we know none of them in advance. In addition, an individual's travel knowledge is different in different places. A HITS (hypertext induced topic search)–based inference model has been proposed to address these issues, computing the quality of each location and the travel knowledge of each individual [75]. We will introduce the model in chapter 9.

Second, the correlation between two locations A and B also depends on the sequences in which the two locations have been visited. This correlation between A and B, $Cor(A,B)$, is asymmetric (i.e., $Cor(A,B) \neq Cor(B,A)$). The semantic meaning of a travel sequence $A \rightarrow B$ might be quite different from $B \rightarrow A$. For example, on a one-way road, people would only go to B from A while never traveling to A from B. Additionally, the two locations continuously visited by a user would be more correlated than those being visited discontinuously. Some users would reach B directly from A ($A \rightarrow B$), while others would visit another location C before arriving at B ($A \rightarrow C \rightarrow B$). Intuitively, the $Cor(A,B)$ indicated by the two sequences might be different. Likewise, in a sequence $A \rightarrow C \rightarrow B$, $Cor(A,C)$ would be greater than $Cor(A,B)$, as the user has continuously accessed $A \rightarrow C$ while traveling to B after visiting C.

In short, the correlation between two locations can be calculated by integrating the travel experience of the users visiting them during a trip in a weighted manner. Formally, the correlation between location A and B can be calculated as

$$Cor(A, B) = \sum\nolimits_{u_k \in U'} \alpha \cdot e_k, \tag{8.19}$$

where U' is the collection of users who have visited A and B on a trip; e_k is u_k's travel experience in the region where A and B belong; and $u_k \in U'$. $0 < \alpha \leq 1$ is a dumping factor, which decreases as the interval between the two locations' indexes on a trip increases. For example, we set $\alpha = 2^{-(|j-i|-1)}$, where i and j are indexes of A and B in the sequence they pertain to. That is, the more discontinuously two locations are visited by an individual ($|i-j|$ would be big, thus α will become small), the smaller the contribution the individual can provide to the correlation between the two locations.

Finally, we can apply the correlation between a pair of locations to equation (8.10), enabling a better item-based CF model for location recommendations.

8.4 Matrix Factorization

8.4.1 Basic Matrix Factorization Methods

When the number of users and the number of items are both huge, matrix factorization–based methods can be employed to implement a CF model. Matrix factorization decomposes a (sparse) matrix X into the production of two (low-rank) matrices, which denote the latent variables of users and items, respectively. The latent representation of users and items can be used as a dimension-reduction technique that benefits clustering and classification problems. In the meantime, the production of the two matrices

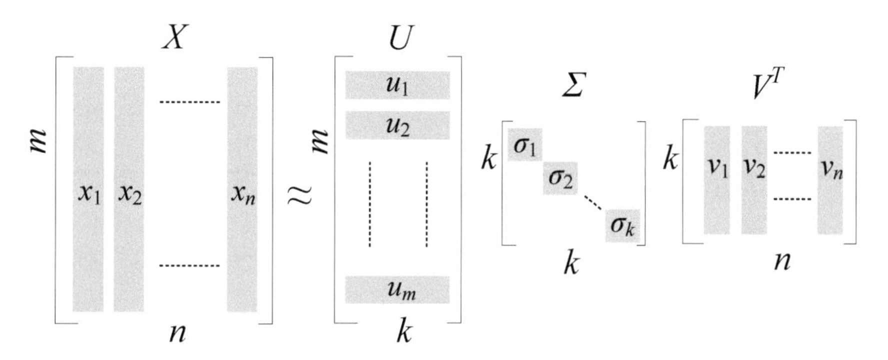

Figure 8.6
The concept of SVD.

can approximate matrix X, thereby helping fill in the missing values in X. Thus, it can be used to estimate a user's rating for an unseen item, simultaneously considering the similarity between users and between items.

There are two widely used matrix factorization methods: *singular value decomposition* (SVD) [1, 30] and *nonnegative matrix factorization* (NMF) [25, 36].

SVD factorizes an $m \times n$ matrix X into the production of three matrices $X = U\Sigma V^T$, where U is an $m \times m$ real unitary matrix (a.k.a. left singular vectors), and Σ is an $m \times n$ rectangular diagonal matrix with nonnegative real numbers on the diagonal (a.k.a. singular values); V^T is an $n \times n$ real unitary matrix (a.k.a. left singular vectors). In practice, as shown in figure 8.6, when trying to approximate matrix X by $U\Sigma V^T$, we only need to keep the top-k biggest singular values in Σ and the corresponding singular vectors in U and V.

SVD has some good properties. First, U and V are orthogonal matrices (i.e., $U \bullet U^T = I$, and $V \bullet V^T = I$). Second, the value of k can be determined by Σ, such as selecting the first k diagonal entries (in Σ) whose sum is larger than 90 percent of the entire diagonal entries' sum. However, SVD is more computationally expensive and harder to parallelize compared to NFM.

NFM *f*actorizes an $m \times n$ matrix R (with m users and n items) into a production of an $m \times K$ matrix P and $K \times n$ matrix Q, $R = P \times Q$, with the property that all three matrices have no negative elements. This nonnegativity makes the resulting matrices easier to inspect [25]. Additionally, nonnegativity is inherent to the data being considered in many applications, such as location recommendations [75], traffic estimation [51], and the processing of audio spectrums. Each row of matrix P denotes the latent feature of a user; each column of matrix Q stands for the latent feature of an item. K can be significantly smaller than m and n, denoting the number of latent features for a user and an item. To predict a rating of an item d_j by u_i, we can calculate the dot product of the two vectors corresponding to u_i and d_j as

$$\hat{r}_{ij} = p_i^T q_j = \sum_{k=1}^{k} p_{ik} q_{kj}. \tag{8.20}$$

To find a proper P and Q, we first initialize the two matrices with some values and calculate the difference between their product and R, as follows:

$$e_{ij}^2 = (r_{ij} - \hat{r}_{ij})^2 = \left(r_{ij} - \sum_{k=1}^{K} p_{ik} q_{kj}\right)^2. \tag{8.21}$$

We then try to minimize e_{ij}^2 iteratively using gradient descent, which finds a local minimum of the difference. To know in which direction we have to modify the values, we differentiate equation (8.21) with respect to p_{ik} and q_{kj} separately:

$$\frac{\partial e_{ij}^2}{\partial p_{ik}} = -2(r_{ij} - \hat{r}_{ij})(q_{kj}) = -2e_{ij} q_{kj}, \tag{8.22}$$

$$\frac{\partial e_{ij}^2}{\partial q_{kj}} = -2(r_{ij} - \hat{r}_{ij})(p_{ik}) = -2e_{ij} p_{ik}. \tag{8.23}$$

Having obtained the gradient, we now formulate the update rules for p_{ik} and q_{kj} as follows:

$$p'_{ik} = p_{ik} + \alpha \frac{\partial e_{ij}^2}{\partial p_{ik}} = p_{ik} + 2e_{ij} q_{kj}, \tag{8.24}$$

$$q'_{kj} = q_{kj} + \alpha \frac{\partial e_{ij}^2}{\partial q_{kj}} = q_{kj} + 2e_{ij} p_{ik}, \tag{8.25}$$

where α is a small value that determines the rate of approaching the minimum. When optimizing p_{ik}, NFM fixes q_{kj} and vice versa; the gradient descent is performed iteratively until the total error $\sum e_{ij}^2$ converges to its minimum. To avoid overfitting, a regularization is introduced to the error function.

$$e_{ij}^2 = \left(r_{ij} - \sum_{k=1}^{K} p_{ik} q_{kj}\right)^2 + \frac{\beta}{2} \sum_{k=1}^{K} \left(\|P\|^2 + \|Q\|^2\right). \tag{8.26}$$

Compared to SVD, NFM is flexible and can be parallelized but is less precise.

8.4.2 Matrix Factorization for Spatiotemporal Data

Within the context of spatiotemporal data, an item in the aforementioned matrix can also be a location, website, or company, while users can be drivers, passengers, or subscribers to a service. We can even generalize a user to an object and an item to a property of the object. When there are multiple datasets concerning an object, we cannot simply deposit different properties from different sources into a single matrix. As different datasets have different distributions and meanings, factorizing them in a single matrix would lead to an inaccurate complementation of missing values in the matrix.

Advanced methods, such as [51, 72], use coupled matrix factorization (also called context-aware matrix factorization) [53] to accommodate different datasets with different matrices, which share a common dimension between one another. By decomposing these matrices collaboratively, we can transfer the similarity between different objects learned from one dataset to another, complementing the missing values in a matrix more accurately. In the following two subsections, we introduce two examples using coupled matrix factorization.

8.4.2.1 Coupled matrix factorization for location recommendations People usually have two types of questions in mind when traveling. They wonder where to go for sightseeing and shopping, and they wonder what they can do at a particular location. The first question corresponds to location recommendations given a particular activity query, which might include restaurants, shopping centers, movies/shows, sports/exercise areas, and sightseeing destinations. The second question corresponds to activity recommendations given a particular location query.

This section introduces a location-activity recommender system [72], which answers the two questions by mining a myriad of social media, such as tips-tagged trajectories or check-in sequences. Regarding the first question, this system provides a user with a list of interesting locations (e.g., the Forbidden City and the Great Wall), that are the top-k candidate locations for conducting a given activity. With respect to the second question, if a user is visiting the Beijing Olympic Park, the recommender suggests that the user try some exercise activities and nice restaurants nearby. This recommender integrates location recommendations and activity recommendations into one knowledge-mining process based on a coupled matrix–factorization method, since locations and activities are closely related in nature.

As illustrated in figure 8.7, a location-activity matrix X is built based on many users' location histories. A row of X denotes a venue, and a column stands for an activity (like shopping and dining). An entry in matrix X denotes the frequency that a particular activity has been performed (by all users) at a particular location. If this location-activity matrix is completely filled, we can recommend a set of locations for a particular activity by retrieving the top-k locations with a relatively high frequency from the column that corresponds to that activity. Likewise, when performing activity

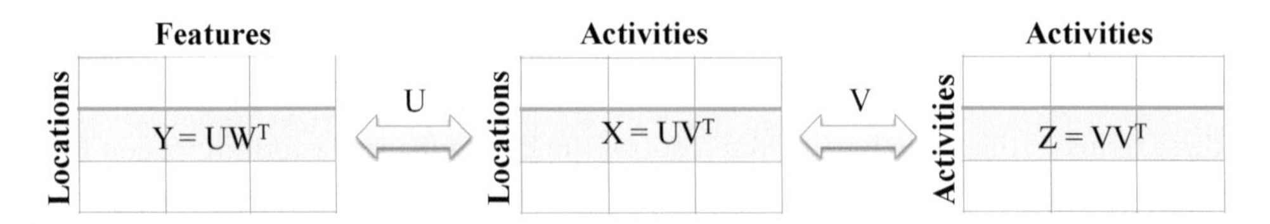

Figure 8.7
Coupled matrix factorization for recommendations.

recommendations for a location, the top-k activities can be retrieved from the row corresponding to the location.

However, the location-activity matrix is incomplete and very sparse, as we only have a portion of users' data (and an individual can visit very few locations). Accordingly, a traditional CF model does not work well in generating quality recommendations. Solely factorizing X does not help much either because the data is overly sparse.

To address this issue, the information from another two matrices (Y and Z), respectively, shown in the left and right part of figure 8.7, are incorporated into the matrix factorization.

One is a location-feature matrix; the other is an activity-activity matrix. Such kinds of additional matrices are usually called *context*, which can be learned from other datasets. In this example, matrix Y, where a row denotes a location and a column stands for a category of POIs (such as restaurants and hotels) that fall in the location, is built based on a POI database. The distance between two rows of matrix Y denotes the similarity between two locations in terms of their geographical properties. The insight is that two locations with a similar geographical property could have similar user behaviors. The other is matrix Z, which models the correlation between two different activities. Matrix Z can be learned from the search results by sending the titles of two activities into a search engine.

The main purpose of coupled matrix factorization is to propagate the information among X, Y, and Z by requiring them to share low-rank matrices U and V in a collective matrix factorization model. As matrix Y and Z are built based on dense data, we obtain an accurate decomposition of them (i.e., matrices U and V). Thus, matrix X can be complemented more accurately by $X = UV^T$. More specifically, an objective function is formulated as

$$L(U,V,W) = \frac{1}{2}\left\|I \circ (X - UV^T)\right\|_F^2 + \frac{\lambda_1}{2}\left\|Y - UW^T\right\|_F^2 + \frac{\lambda_2}{2}\left\|Z - VV^T\right\|_F^2 + \frac{\lambda_3}{2}\left(\|U\|_F^2 + \|V\|_F^2 + \|W\|_F^2\right), \tag{8.27}$$

where $\|\bullet\|_F$ denotes the Frobenius norm. I is an indicator matrix with its entry $I_{ij}=0$ if X_{ij} is missing; $I_{ij}=1$ otherwise. The operator "$\circ$" denotes the entrywise product. The first three terms in the objective function control the loss in matrix factorization, and the last term controls the regularization over the factorized matrices to prevent overfitting. In general, this objective function is not jointly convex to all the variables U, V, and W. Consequently, some numerical methods, such as gradient descent, are used to get local optimal solutions.

8.4.2.2 Matrix factorization for traffic condition estimation The traffic conditions on road networks can be modeled by a matrix, where a row stands for a time interval, and a column stands for a road segment; an entry in the matrix denotes the travel speed (or travel time) on a particular road segment at a particular time interval. We may have observations of traffic conditions on a few road segments (e.g., based on readings of loop detectors or GPS trajectories of vehicles). However, on many road segments (without sensors or GPS-equipped vehicles traversing), we do not know the traffic conditions. Those are what we need to infer. This section introduces a coupled matrix–factorization method [51] that instantly estimates the travel speed on each road segment throughout an entire city, based on the GPS trajectory of a sample of vehicles (such as taxicabs).

As shown in figure 8.8a, after map matching the GPS trajectories onto a road network, a matrix M_r' can be formulated, with a row denoting a time slot (e.g., 2 p.m.–2:10 p.m.) and a column standing for a road segment. Each entry in M_r' contains the travel speed on a particular road segment and in a particular time slot, calculated based on the recently received GPS trajectories. The goal is to fill in the missing values in row t_j, which corresponds to the current time slot. Though we can achieve the goal by solely applying matrix factorization to M_r', the accuracy of the inference is not very high as the majority of road segments are not covered by trajectories.

To address this issue, four context matrices (M_r, M_G, M_G', and Z) are built. Specifically, M_r stands for the historical traffic patterns on road segments. While the rows and columns of M_r have the same meaning as M_r', an entry of M_r denotes the average travel speed derived from historical data over a long period. The difference between the two corresponding entries from M_r' and M_r indicates the deviation of current traffic situations (on a road segment) from its average patterns. As depicted in figure 8.8b, Z contains the physical features of a road segment, such as the shape of a road, the number of lanes, the speed constraint, and the distribution of surrounding POIs. The general assumption is that two road segments with similar geographical properties might have similar traffic conditions at the same time of day.

To capture high-level traffic conditions, as shown in figure 8.8c, a city is divided into uniform grids. By projecting the recently received GPS trajectories into these grids, a

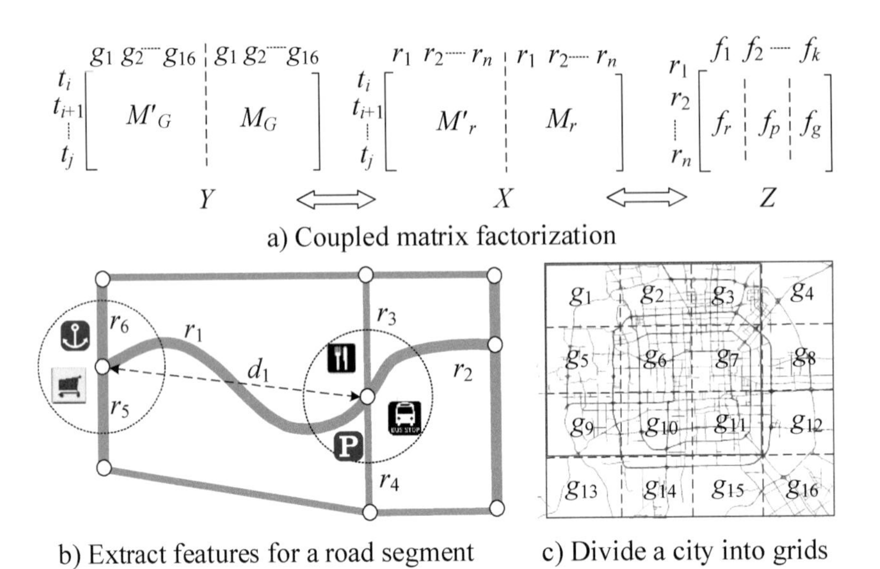

Figure 8.8
Estimated traffic conditions based on matrix factorization and vehicle trajectories.

matrix M_G' is built, with a column standing for a grid and a row denoting a time slot; the entry M_G' indicates the number of vehicles traveling in a particular grid and at a particular time slot. Likewise, by projecting historical trajectories over a long period into the grids, a similar M_G is built, with each entry being the average number of vehicles traveling in a particular grid and at a particular time slot. This means M_G' denotes the real-time high-level traffic conditions in a city, and M_G indicates historical high-level traffic patterns. The difference between the same entries of the two matrices suggests the deviation of current high-level traffic conditions from their historical averages.

By combining these matrices (i.e., $X=M_r'\|M_r$, and $Y=M_G'\|M_G$), a coupled matrix factorization is applied to X, Y, and Z, with the objective function as

$$L(T,R,G,F)=\frac{1}{2}\left\|Y-T(G;G)^T\right\|_F^2+\frac{\lambda_1}{2}\left\|I\circ(X-T(R;R))^T\right\|_F^2 + \frac{\lambda_2}{2}\left\|Z-RF^T\right\|_F^2+\frac{\lambda_3}{2}\left(\|T\|_F^2+\|R\|_F^2+\|G\|_F^2+\|F\|_F^2\right), \tag{8.28}$$

where $\|\cdot\|_F$ denotes the Frobenius norm. I is an indicator matrix with its entry $I_{ij}=0$ if X_{ij} is missing, otherwise $I_{ij}=1$. The operator "$\circ$" denotes the entrywise product. The first three terms in the objective function control the loss in matrix factorization, and the last term is a regularization of penalty to prevent overfitting.

8.5 Tensor Decomposition

8.5.1 Basic Concepts of Tensors

Tensors are multidimensional extensions of matrices describing the relationship between more than two entities. For example, figure 8.9a presents a three-dimensional tensor $\mathcal{X}$, where the first dimension stands for geographical regions, the second dimension denotes categories of POIs, and the third dimension is time of day. An entry $\mathcal{X}(i, j, k)$ in the tensor stores the number of people who have visited POIs of the j-th POI category in the i-th region, at the k-th time interval. For many reasons, the data we can collect is always a sample of the full set. Figure 8.9b shows another tensor $\mathcal{X}'$ modeling different drivers' travel times on different road segments at different time intervals.

By capturing the relationship between multiple entities simultaneously, such a high-order tensor provides us with rich knowledge to better understand each dimension (entity) and a higher possibility of estimating the value of empty entries more accurately than two-dimensional matrices.

We usually refer to the dimensions of a tensor as *modes*. The order of a tensor is its number of modes. For instance, we may refer to a third-order tensor as a three-mode tensor. Figure 8.10a, b, and c present the three modes of a three-dimensional tensor, respectively. The fibers of the three modes are denoted as $\mathcal{X}_{:jk}$, $\mathcal{X}_{i:k}$, and $\mathcal{X}_{ij:}$. Based on different modes, we convert a tensor into different matrices, as illustrated in figure 8.10d. For example, the matrix representation of tensor $\mathcal{X}$ based on mode 1 is $\mathcal{X}_{(1)}$. The matrix representation will be used in decomposing a tensor and computing the production between a tensor and a matrix.

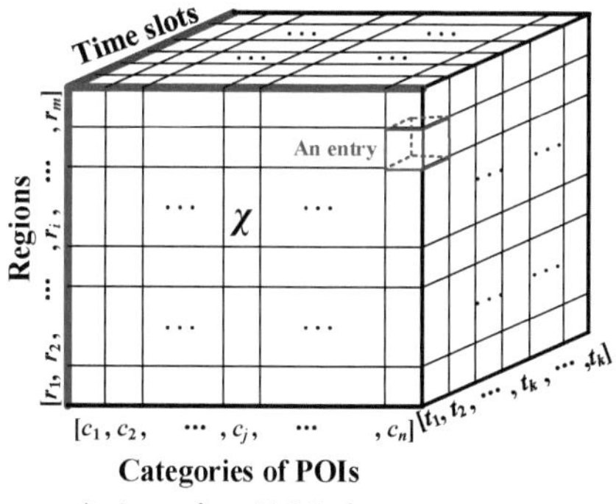

a) A region-POI-time tensor

b) A user-road-time tensor

Figure 8.9
Examples of tensors.

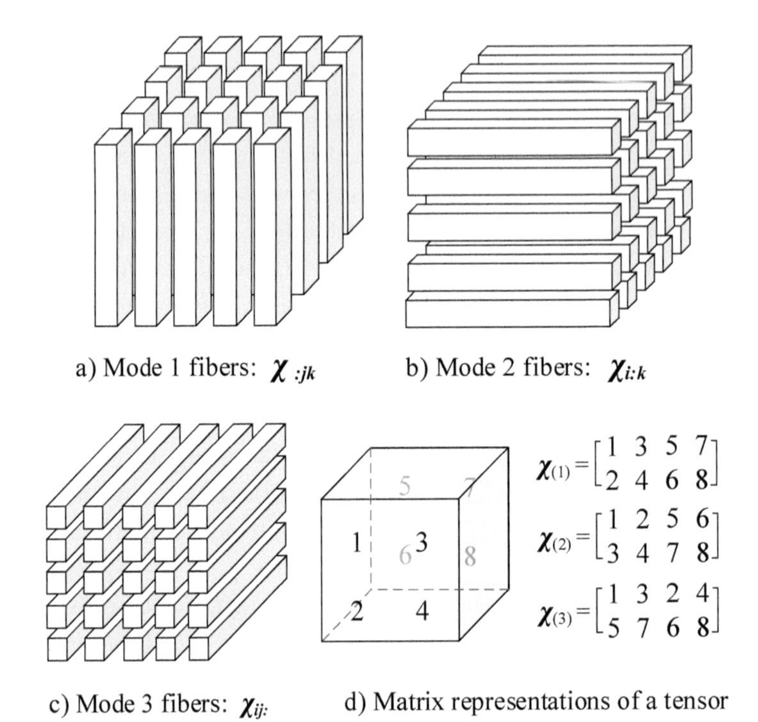

Figure 8.10
Modes and matrix representations of a tensor.

The inner production of two vectors $\boldsymbol{u} \in \mathbb{R}^I$ and $\boldsymbol{v} \in \mathbb{R}^I$ is denoted as $\boldsymbol{u} \cdot \boldsymbol{v} = \sum_i \boldsymbol{u}(i) \times \boldsymbol{v}(i)$

The outer production of the two vectors $\boldsymbol{u} \in \mathbb{R}^I$ and $\boldsymbol{v} \in \mathbb{R}^J$ is denoted as $\boldsymbol{Y} = \boldsymbol{u} \circ \boldsymbol{v}$, $\boldsymbol{Y} \in \mathbb{R}^{I \times J}$,

$$\boldsymbol{Y}(i, j) = \boldsymbol{u}(i) \times \boldsymbol{v}(j). \tag{8.29}$$

The production of an M-mode tensor $\mathcal{X} \in \mathbb{R}^{I_1 \times I_2 \times \cdots \times I_M}$ and a vector $\boldsymbol{V} \in \mathbb{R}^{I_n}$ is denoted as $\boldsymbol{Y} = \mathcal{X}_{\times n} \boldsymbol{V}, \boldsymbol{Y} \in \mathbb{R}^{I_1 \cdots \times I_{n-1} \times I_{n+1} \cdots \times I_M}$.

$$\boldsymbol{Y}(i_1, \ldots, i_{n-1}, i_{n+1}, i_M) = \sum_{i_n=1}^{I_n} \mathcal{X}(i_1, i_2, \ldots, i_M)\, \boldsymbol{V}(i_n). \tag{8.30}$$

For example, for a three-mode tensor $\mathcal{X} \in \mathbb{R}^{I \times J \times K}$ and a vector $\boldsymbol{V} \in \mathbb{R}^I$, $\mathcal{X}_{\times 1} \boldsymbol{V} \in \mathbb{R}^{J \times K}$, where $\boldsymbol{Y}(j, k) = \sum_{i=1}^{I} \mathcal{X}(i, j, k) \boldsymbol{V}(i)$.

Given an M-mode tensor $\mathcal{X} \in \mathbb{R}^{I_1 \times I_2 \times \cdots \times I_M}$ and a matrix $\boldsymbol{A} \in \mathbb{R}^{J \times I_n}$, the n-mode product of $\mathcal{X}$ and A is denoted as $\boldsymbol{Y} = \mathcal{X}_{\times n} \boldsymbol{A}$, where $\boldsymbol{Y} \in \mathbb{R}^{I_1 \times \cdots I_{n-1} \times J \times I_{n+1} \cdots \times I_M}$, and

$$\boldsymbol{Y}(i_1, \ldots, i_{n-1}, j, i_{n+1}, i_M) = \sum_{i_n=1}^{I_n} \mathcal{X}(i_1, i_2, \ldots, i_M)\, \boldsymbol{A}(j, i_n). \tag{8.31}$$

For example, the production between a three-mode tensor $\mathcal{X} \in \mathbb{R}^{I \times J \times K}$ and a matrix $A \in \mathbb{R}^{M \times J}$; the two-mode production $Y = \mathcal{X}_{\times 2} A, Y \in \mathbb{R}^{I \times M \times K}; Y(i, m, k) = \sum_j^J \mathcal{X}(i, j, k)\, A(m, j)$.

The Frobenius norm of a tensor $\mathcal{X} \in \mathbb{R}^{I_1 \times I_2 \times \cdots \times I_M}$ is defined as

$$\|\mathcal{X}\|_F = \sqrt{\sum_{i_1}^{I_1} \sum_{i_2}^{I_2} \cdots \sum_{i_M}^{I_M} \mathcal{X}(i_1, i_2, \ldots, i_M)^2}. \tag{8.32}$$

8.5.2 Methods of Tensor Decomposition

As the dimension of a tensor increases, the tensor becomes sparse more easily than a two-dimensional matrix. For instance, people can only drive on a few road segments in a given time interval. Thus, the more time slots that are included in tensor $\mathcal{X}'$, the more empty entries occur. Similar to matrix factorization, tensor decomposition is proposed to achieve the following two goals:

1. *Latent representation.* One goal is to learn a latent representation of a dimension. For instance, by decomposing tensor $\mathcal{X}$ shown in figure 8.9a, we can obtain a latent representation (usually with a much lower dimension) for different geographical regions, POI categories, and time intervals, respectively. Latent representations can be used for further data-mining and machine-learning tasks, such as clustering similar regions or predicting the traffic of crowds visiting a region during future hours.
2. *Filling missing values.* Another goal of tensor decomposition is to fill in the missing values in a sparse tensor based on those of nonempty entries. For example, we can estimate the travel time that a driver would need to spend to traverse a given road segment by filling in the missing values in tensor $\mathcal{X}'$ shown in figure 8.9b. The general idea is to first decompose a tensor into the production of some low-rank matrices based on nonempty entries and then recover the value of missing entries by the production of these low-rank matrices.

There are two widely used tensor decomposition methods: PARAFAC [1, 19] and Tucker [37, 59] decomposition. As shown in figure 8.11a, PARAFAC, also known as *canonical polyadic decomposition*, decomposes a three-mode tensor $\mathcal{X} \in \mathbb{R}^{I \times J \times K}$ into the sum of three-way outer productions. The Tucker decomposition factorizes $\mathcal{X}$ into the production of three matrices and a core tensor, as illustrated in figure 8.11b.

8.5.2.1 PARAFAC decomposition As shown in figure 8.11a, this method decomposes tensor $\mathcal{X}$ into the three-way outer production as follows:

$$\mathcal{X} \approx \sum_{r=1}^{R} \boldsymbol{a}_r \circ \boldsymbol{b}_r \circ \boldsymbol{c}_r, \tag{8.33}$$

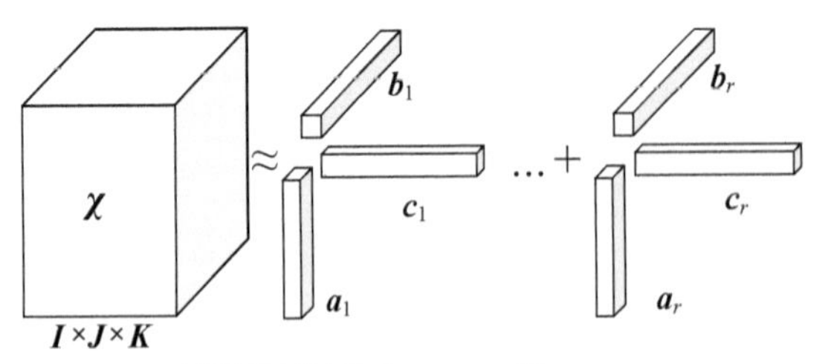

a) PARAFAC decomposition

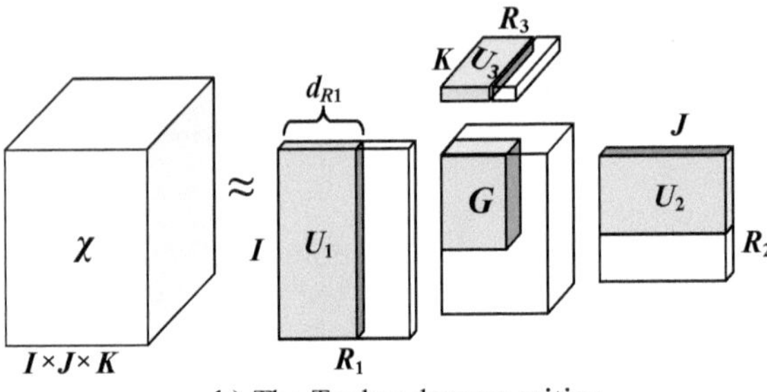

b) The Tucker decomposition

Figure 8.11
Tensor decomposition methods.

where $\boldsymbol{a}_r \in \mathbb{R}^I$, $\boldsymbol{b}_r \in \mathbb{R}^J$, $\boldsymbol{c}_r \in \mathbb{R}^K$ denote the first-, second-, and third-factor vectors, respectively, in the r-th component; R is the number of components shown in figure 8.11a. The three-way outer production of $\boldsymbol{a}_r$, $\boldsymbol{b}_r$, and $\boldsymbol{c}_r$ is given by

$$(\boldsymbol{a}_r \circ \boldsymbol{b}_r \circ \boldsymbol{c}_r)(i, j, k) = \boldsymbol{a}_r(i)\boldsymbol{b}_r(j)\boldsymbol{c}_r(k), \text{ for all } i, j, k. \tag{8.34}$$

With the factor vectors, the factor matrices are defined as

$$\boldsymbol{A} = [\boldsymbol{a}_1, \boldsymbol{a}_2, \ldots, \boldsymbol{a}_R] \in \mathbb{R}^{I \times R},$$
$$\boldsymbol{B} = [\boldsymbol{b}_1, \boldsymbol{b}_2, \ldots, \boldsymbol{b}_R] \in \mathbb{R}^{J \times R},$$
$$\boldsymbol{C} = [\boldsymbol{c}_1, \boldsymbol{c}_2, \ldots, \boldsymbol{c}_R] \in \mathbb{R}^{K \times R}.$$

By minimizing the following objective function, we can solve the PARAFAC decomposition problem. The function is nonconvex; however, if we fix two of the factor matrices, the problem reduces to a linear least-squares problem for the third one.

$$\min_{A,B,C} \left\| \mathcal{X} - \sum_{r=1}^{R} \boldsymbol{a}_r \circ \boldsymbol{b}_r \circ \boldsymbol{c}_r \right\|_F^2. \tag{8.35}$$

To avoid overfitting, some regularization is usually added to the objective function as follows:

$$\min_{A,B,C}\left[\frac{1}{2}\left\|\mathcal{X}-\sum_{r=1}^{R}\boldsymbol{a}_r\circ\boldsymbol{b}_r\circ\boldsymbol{c}_r\right\|_F^2+\frac{\lambda}{2}\left(\|\boldsymbol{A}\|_F^2+\|\boldsymbol{B}\|_F^2+\|\boldsymbol{C}\|_F^2\right)\right], \tag{8.36}$$

where $\frac{\lambda}{2}\left(\|\boldsymbol{A}\|_F^2+\|\boldsymbol{B}\|_F^2+\|\boldsymbol{C}\|_F^2\right)$ is a regularization penalty to avoid overfitting, and λ is a parameter controlling the contribution of the regularization penalty. Apparently, the bigger the three factor matrices, the better $\sum_{r=1}^{R}\boldsymbol{a}_r\circ\boldsymbol{b}_r\circ\boldsymbol{c}_r$ can approximate $\mathcal{X}$. However, the decomposed results become less useful, as tensor decomposition tends to find low-rank latent representations of the original data. The minimal R that yields the quality of proximation is called the *rank* of the tensor. There is no closed-form solution to find the optimal result for equation (8.36). Gradient descent is usually employed to find a local optimal result.

If the data has missing values, we cannot assume the value of these missing entries in a tensor to be zero, which is also a value. One approach to solving this problem is to only consider the error between the approximation and the value of the nonempty entries as follows.

$$\min_{A,B,C}\left\|\boldsymbol{W}\cdot\left(\mathcal{X}-\sum_{r=1}^{R}\boldsymbol{a}_r\circ\boldsymbol{b}_r\circ\boldsymbol{c}_r\right)\right\|_F^2, \tag{8.37}$$

where $\boldsymbol{W}(i,j,k)=\begin{cases}1, & \text{if } (i,j,k) \text{ element is present}\\ 0, & \text{if } (i,j,k) \text{ element is missing}\end{cases}$.

In the case of an N-mode tensor $\mathcal{X}\in\mathbb{R}^{I_1\times I_2\times\cdots\times I_N}$, by adding a new factor matrix $\boldsymbol{A}^{(n)}$ for each additional mode, we can extend the three-way canonical polyadic (CP) decomposition to

$$\mathcal{X}\approx\sum_{r=1}^{R}\boldsymbol{a}_r^{(1)}\circ\boldsymbol{a}_r^{(2)}\circ\cdots\circ\boldsymbol{a}_r^{(N)}, \tag{8.38}$$

where $\boldsymbol{a}_r^{(n)}$ denotes the n-th factor vector in the r-th component. $\boldsymbol{A}^{(n)}=[\boldsymbol{a}_1^{(n)},\boldsymbol{a}_2^{(n)},\ldots,\boldsymbol{a}_r^{(n)}]$. The outer production of the N factors is defined as

$$\mathcal{X}(i_1,i_2,\ldots,i_N)=\boldsymbol{a}_r^{(1)}(i_1)\times\boldsymbol{a}_r^{(2)}(i_2)\times\cdots\boldsymbol{a}_r^{(N)}(i_N). \tag{8.39}$$

8.5.2.2 Tucker decomposition This decomposition method was originally proposed by Tucker [59] and further popularized by De Lathauwer et al. [34], wherein the higher-order singular value decomposition (HOSVD) was proposed for computing the Tucker decomposition. In fact, Tucker proposed three different models. Figure 8.11b presents the third model, also known as *Tucker-3* (henceforth referred to as Tucker). The Tucker decomposition decomposes a three-mode tensor $\mathcal{X} \in \mathbb{R}^{I \times J \times K}$ into three factor matrices $\boldsymbol{A} \in \mathbb{R}^{I \times R_1}$, $\boldsymbol{B} \in \mathbb{R}^{J \times R_2}$, and $\boldsymbol{C} \in \mathbb{R}^{K \times R_3}$, which are also multiplied by a core tensor $\boldsymbol{G}$:

$$\mathcal{X} \approx \boldsymbol{G}_{\times 1} \boldsymbol{A}_{\times 2} \boldsymbol{B}_{\times 3} \boldsymbol{C}. \tag{8.40}$$

Alternatively, the decomposition can be written element-wise as

$$\mathcal{X}(i, j, k) \approx \sum\nolimits_{r_1=1}^{R_1} \sum\nolimits_{r_2=1}^{R_2} \sum\nolimits_{r_3=1}^{R_3} G(r_1, r_2, r_3) A(i, r_1) B(j, r_2) C(k, r_3). \tag{8.41}$$

$\boldsymbol{A}$, $\boldsymbol{B}$, $\boldsymbol{C}$, and $\boldsymbol{G}$ can be computed through HOSVD.

Sometimes, a regularization is added to the objective function to avoid overfitting.

$$L(G, U_1, U_2, U_3) = \frac{1}{2} \left\| \mathcal{X} - \boldsymbol{G}_{\times 1} \boldsymbol{A}_{\times 2} \boldsymbol{B}_{\times 3} \boldsymbol{C} \right\|_F^2 + \frac{\lambda}{2} \left(\|\boldsymbol{G}\|_F^2 + \|\boldsymbol{A}\|_F^2 + \|\boldsymbol{B}\|_F^2 + \|\boldsymbol{C}\|_F^2 \right), \tag{8.42}$$

where the first part of the objective function is to minimize the error between the approximation and the values of the nonempty entries in $\mathcal{X}$; $\frac{\lambda}{2} \left(\|\boldsymbol{G}\|_F^2 + \|\boldsymbol{A}\|_F^2 + \|\boldsymbol{B}\|_F^2 + \|\boldsymbol{C}\|_F^2 \right)$ is a regularization penalty to avoid overfitting, and λ is a parameter controlling the contribution of the regularization penalty.

There is no closed-form solution that can find the optimal $\boldsymbol{A}$, $\boldsymbol{B}$, $\boldsymbol{C}$, and $\boldsymbol{G}$ for the objective function. Alternatively, gradient descent is usually employed to find a local optimal result. Just like equation (8.37), when a tensor is composed of many missing entries, a mask $\boldsymbol{W}$ is added to the objective function so that only nonempty entries are considered when calculating the approximation error.

$$\frac{1}{2} \left\| \boldsymbol{W} \cdot (\mathcal{X} - \boldsymbol{G}_{\times 1} \boldsymbol{A}_{\times 2} \boldsymbol{B}_{\times 3} \boldsymbol{C}) \right\|_F^2. \tag{8.43}$$

After solving for $\boldsymbol{A}$, $\boldsymbol{B}$, $\boldsymbol{C}$, and $\boldsymbol{G}$, we recover the missing entries of $\mathcal{X}$ with the production $\boldsymbol{G}_{\times 1} \boldsymbol{A}_{\times 2} \boldsymbol{B}_{\times 3} \boldsymbol{C}$. More details can be found in the survey papers [31] and [47].

8.5.3 Tensor Decomposition for Spatiotemporal Data

Location and time are two natural dimensions for constructing a matrix. By adding the third dimension (e.g., POI categories or users) to a location-time matrix, we can easily form a tensor that better captures the correlation in spatiotemporal data. On the other hand, the third dimension will create more entries in which we do not have observations, thereby aggravating the data sparsity problem. Simple tensor decomposition techniques do not accurately fill in the missing entries in a very sparse tensor any longer. To address this issue, coupled tensor decomposition (a.k.a. context-aware tensor decomposition) has been proposed, incorporating information from multiple data sources to fill these missing entries more accurately. When coupling multiple data sources, how to preserve the spatiotemporal properties of the original data is also a nontrivial problem.

To demonstrate the technology of coupled tensor decomposition, this section introduces three examples. The first is conducting a personalized location recommendation. The second is estimating the travel time of each driver on a road segment at a particular time interval. The third is diagnosing the noise composition of different locations at different time intervals.

8.5.3.1 Personalized location recommendations As different people may have different preferences for different types of places, we can extend the generic location-activity recommendation [72] introduced in section 8.4.2.1 into a personalized one by adding a user dimension into the original location-activity matrix. As shown in figure 8.12, a user-location-activity tensor A is built, with an entry denoting the times that a particular user has performed a particular activity in a particular location. If we can infer the value of every entry, a personalized recommendation can be enabled. However, tensor A is very sparse as a user usually visits a few places. Thus, a simple tensor completion method cannot fill in its missing entries very well.

To address this issue, four context matrices are built based on additional nonsparse data sources, such as road network and POI datasets. In addition, these matrices share some dimensions with tensor A. For instance, tensor A shares the user dimension with matrix B and the location dimension with matrix E. Consequently, the knowledge from these matrices can be transferred into the tensor to help complete tensor A.

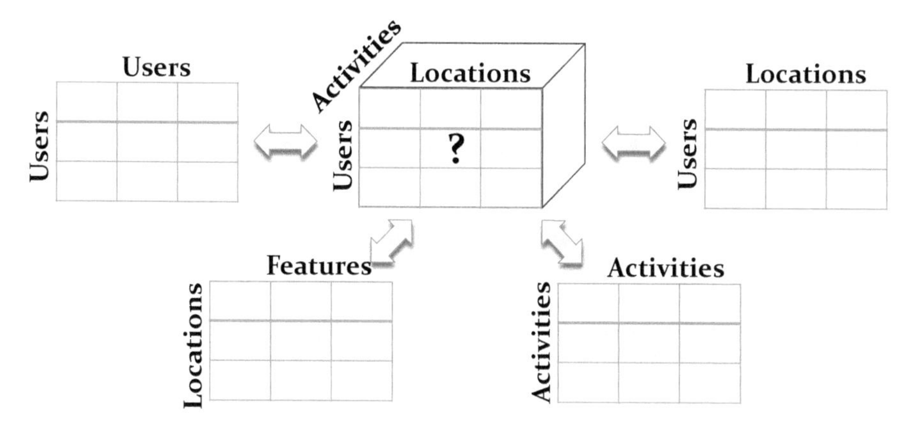

Figure 8.12
Personalized location-activity recommendation based on tensor decomposition.

A PARAFAC-style tensor decomposition framework is employed to incorporate the tensor with these context matrices for a regularized decomposition. More specifically, an objective function is defined as follows:

$$\mathcal{L}(X,Y,Z,U)=\frac{1}{2}\left\|W\cdot(\mathcal{A}-[\![X,Y,Z]\!])\right\|_F^2+\frac{\lambda_1}{2}\operatorname{tr}(X^TL_BX)+\frac{\lambda_2}{2}\left\|C-YU^T\right\|_F^2 \\ +\frac{\lambda_3}{2}\operatorname{tr}(Z^TL_DZ)+\frac{\lambda_4}{2}\left\|E-XY^T\right\|_F^2+\frac{\lambda_5}{2}\left(\|X\|_F^2+\|Y\|_F^2+\|Z\|_F^2+\|U\|_F^2\right), \tag{8.44}$$

where $[\![X, Y, Z]\!]=\sum \boldsymbol{x}_i\circ \boldsymbol{y}_i\circ \boldsymbol{z}_i$; the operator $\circ$ denotes the outer product; W is a mask that chooses nonempty entries. $W(i, j, k)=0$ if $A(i, j, k)$ is empty. L_B is the Laplacian matrix of B, defined as $L_B=Q-B$ with Q being a diagonal matrix whose diagonal entries $Q_{ij}=\sum_j B_{ij}$; $\operatorname{tr}(\bullet)$ denote the trace of a matrix; $\|\bullet\|_F$ denotes the Forbenius norm; λ_i $(i=1,\dots,5)$ are tunable model parameters. Given the objective function, a gradient descent is employed to find a local minimal result for X, Y, and Z.

8.5.3.2 Travel-time estimation of individual drivers using tensor decomposition In section 8.4.2.2, we introduce coupled matrix factorization for estimating the average travel speed on each road segment. As different people may have different driving skills, habits, and knowledge on different roads, their travel times on the same road could be very different. If we want to estimate the time of arrival at a destination for a particular driver, we need to infer different drivers' travel times on different road segments at different time intervals. This is naturally a tensor decomposition problem. To deal with the data sparsity problem, a few context matrices are built to fuse knowledge from other data sources [61].

As illustrated in figure 8.13, a tensor $\mathcal{A}_r\in\mathbb{R}^{N\times M\times L}$ is constructed, with the three dimensions standing for road segments, drivers, and time slots, respectively, based on the GPS trajectories received at the most recent L time intervals and the road network data. An entry $\mathcal{A}_r(i,j,k)=c$ denotes that the ith road segment is traveled by the jth driver with a time cost c in time slot k (e.g., 2–2:30 p.m.). The last time slot denotes the present time slot, combined with the L-1 time slots right before it to formulate the tensor. Clearly, the tensor is very sparse because a driver can only travel a few road segments in a short time period. If the missing entries can be inferred based on the values of nonzero entries, we can obtain the travel time of any driver on any road segment in the present time slot.

To this end, another tensor $\mathcal{A}_h$ is built based on the historical trajectories over a long period of time (e.g., one month). $\mathcal{A}_h$ has the same structure as $\mathcal{A}_r$, while an entry $\mathcal{A}_h(i,j,k)=c'$ denotes the jth driver's average travel time on the ith road segment in time slot k in the recorded past. Intrinsically, $\mathcal{A}_h$ is much denser than $\mathcal{A}_r$, denoting historical traffic patterns and drivers' behavior on an entire road network.

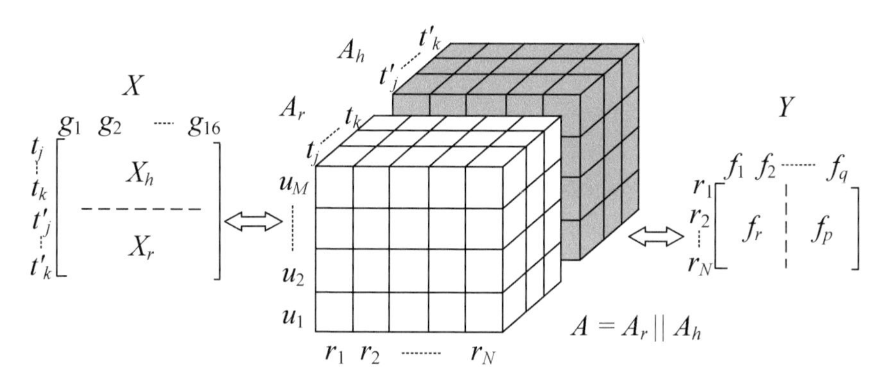

Figure 8.13
Travel-time estimation using tensor decomposition.

Two context matrices (X and Y) are built to help supplement the missing entries of $\mathcal{A}_r$. Matrix X (consisting of X_r and X_h) represents the correlation between different time intervals in terms of coarse-grained traffic conditions. This is similar to the grids shown in figure 8.8c. An entry of X_r denotes the number of vehicles traversing a particular grid at a particular time interval. A row of X_r represents coarse-grained traffic conditions in a city at a particular time slot. Consequently, the similarity of two different rows indicates the correlation of traffic flows between two time intervals. X_h has the same structure as X_r, storing the historical average number of vehicles traversing a grid from t_i to t_j.

Matrix Y stores each road segment's geographical features, which are similar to that of matrix Z shown in figure 8.8a. Later, they decompose $\mathcal{A} = \mathcal{A}_r \,\|\, \mathcal{A}_h$ with matrices X and Y collaboratively, by optimizing the following objective function.

$$\mathcal{L}(S,R,U,T,F,G) = \frac{1}{2}\|W \cdot (\mathcal{A} - S \times_R R \times_U U \times_T T)\|_F^2 + \frac{\lambda_1}{2}\|X - TG\|_F^2 \\ + \frac{\lambda_2}{2}\|Y - RF\|_F^2 + \frac{\lambda_3}{2}\left(\|S\|_F^2 + \|R\|_F^2 + \|U\|_F^2 + \|T\|_F^2 + \|F\|_F^2 + \|G\|_F^2\right), \tag{8.45}$$

where $\|W(\mathcal{A} - S \times_R R \times_U U \times_T T)\|_F^2$ minimizes the error of tensor decomposition, W is a mask, and $W(i, j, k)=0$ if $\mathcal{A}(i,j,k)$ is empty, otherwise, $W(i, j, k)=1$. $\|X - TG\|_F^2$ and $\|Y - RF\|_F^2$ minimize the error for factorizing matrix X and Y, respectively. $\|S\|_F^2 + \|R\|_F^2 + \|U\|_F^2 + \|T\|_F^2 + \|F\|_F^2 + \|G\|_F^2$ is a regularization to avoid overfitting. λ_1, λ_2, and λ_3 are parameters weighting the importance of different parts in the objective function.

8.5.3.3 Diagnosing urban noises using tensor decomposition Zheng et al. [74] infer the fine-grained noise situation by using 311-complaint data together with social media, road network data, and POIs. As shown in figure 8.14, the noise situation of

New York City is modeled with a three-dimensional tensor, where the three dimensions stand for regions, noise categories, and time slots, respectively. An entry $\mathcal{A}(i, j, k)$ stores the total number of 311 complaints of category c_j in region r_i and time slot t_k over the given period of time. This is a very sparse tensor, as there may not be people reporting noise anytime and anywhere. If the tensor can be filled completely, we are able to know the noise situation throughout the city.

To deal with the data sparsity problem, they extract three categories of features—geographical features, human-mobility features, and noise category correlation features (denoted by matrices X, Y, and Z)—from POI/road network data, user check-ins, and 311 data, respectively. For example, a row of matrix X denotes a region, and each column stands for a road network feature, such as the number of intersections and the total length of road segments in the region. Matrix X incorporates the similarity between two regions in terms of their geographic features. Intuitively, regions with similar geographic features could have a similar noise situation. $Z \in \mathbb{R}^{M \times M}$ is the correlation matrix between different categories of noise. $Z(i, j)$ denotes how often a category of noise c_i co-occurs with another category c_j.

These features are used as context in a context-aware tensor decomposition approach to supplement the missing entries of the tensor. More specifically, $\mathcal{A}$ is decomposed into the multiplication of a few (low-rank) matrices and a core tensor (or just a few vectors), based on $\mathcal{A}$'s nonzero entries. Matrix X can be factorized into the multiplication of two matrices, $X = R \times U$, where $R \in \mathbb{R}^{N \times d_R}$ and $U \in \mathbb{R}^{d_R \times P}$ are low-rank latent factors for regions and geographical features, respectively. Likewise, matrix Y can be factorized

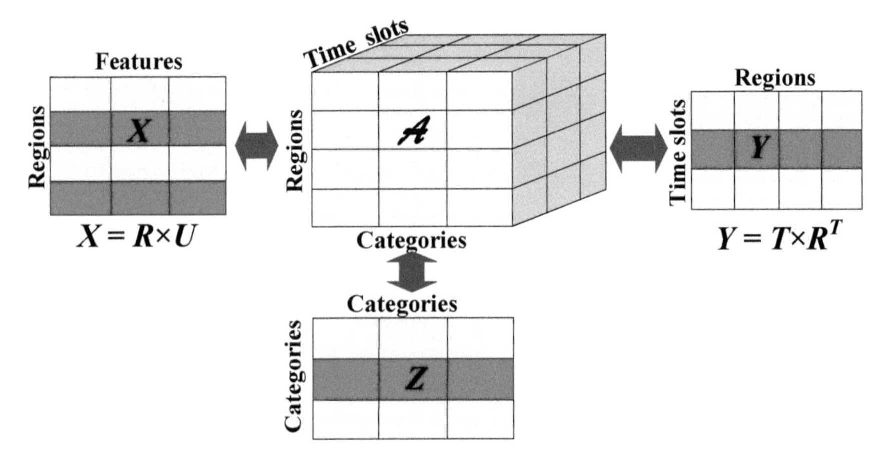

Figure 8.14
Diagnosing urban noise with tensor decomposition.

into the multiplication of two matrices, $Y = T \times R^T$, where $T \in \mathbb{R}^{L \times d_T}$ is a low-rank latent factor matrix for time slots. d_T and d_R are usually very small. The objective function is defined as

$$\begin{aligned}\mathcal{L}(S,R,C,T,U) = &\frac{1}{2}\left\|W \bullet (\mathcal{A} - S \times_R R \times_C C \times_T T)\right\|_F^2 + \frac{\lambda_1}{2}\left\|X - RU\right\|_F^2 + \frac{\lambda_2}{2}\operatorname{tr}(C^T L_Z C) \\ &+ \frac{\lambda_3}{2}\left\|Y - TR^T\right\|_F^2 + \frac{\lambda_4}{2}\left(\left\|S\right\|_F^2 + \left\|R\right\|_F^2 + \left\|C\right\|_F^2 + \left\|T\right\|_F^2 + \left\|U\right\|_F^2\right),\end{aligned} \tag{8.46}$$

where $\left\|W \bullet (\mathcal{A} - S \times_R R \times_C C \times_T T)\right\|_F^2$ controls the error of decomposing $\mathcal{A}$; W is a mask; $W(i, j, k) = 0$ if $\mathcal{A}(i, j, k)$ is empty; otherwise, $W(i, j, k) = 1$. $\left\|X - RU\right\|^2$ controls the error of factorization of X; $\left\|Y - TR^T\right\|^2$ controls the error of factorization of Y; $\left\|S\right\|_F^2 + \left\|R\right\|_F^2 + \left\|C\right\|_F^2 + \left\|T\right\|_F^2 + \left\|U\right\|_F^2$ is a regularization penalty to avoid overfitting; λ_1, λ_2, λ_3, and λ_4 are parameters controlling the contribution of each part during the collaborative decomposition.

Here, matrices X and Y share the same dimensions of region with tensor $\mathcal{A}$. Tensor $\mathcal{A}$ has a common dimension of time with Y and a shared dimension of category with Z. Thus, they share latent spaces for region, time, and category. This idea has been introduced in the coupled matrix factorization. $tr(C^T L_Z C)$ is derived from the following manifold alignment:

$$\begin{aligned}\sum_{i,j}\left\|C(i,.) - C(j,.)\right\|^2 Z_{ij} &= \sum_k \sum_{i,j}\left\|C(i,k) - C(j,k)\right\|^2 Z_{ij} \\ &= tr(C^T(D - Z)C) = tr(C^T L_Z C),\end{aligned} \tag{8.47}$$

where $C \in \mathbb{R}^{M \times d_C}$ is the latent space of category. $D_{ii} = \sum_i Z_{ij}$ is a diagonal matrix, and $L_Z = D - Z$ is the Laplacian matrix of the category correlation graph. $tr(C^T L_Z C)$, which guarantees two (e.g., the ith and jth) noise categories with a higher similarity (i.e., Z_{ij} is bigger) should also have a closer distance in the new latent space C. In this case, only one dataset (i.e., 311 data) is involved in the manifold alignment. So, $\mathbf{D} = D$. As there is no closed-form solution for finding the global optimal result of the objective function (shown in equation [8.46]), a numeric method, gradient descent, is employed to find a local optimization.

8.6 Probabilistic Graphical Models

8.6.1 General Concepts

A probabilistic graphical model is a probabilistic model for which a graph expresses the conditional dependence structure between random variables. Generally, it uses a graph-based representation as the foundation for encoding a complete distribution over a multidimensional space. The graph can be regarded as a compact or factorized representation of a set of independences that hold in the specific distribution.

Two branches of graphical representations of distributions are commonly used: Bayesian networks and Markov networks (also called Markov random fields [29]). Both families encompass the properties of factorization and independences, but they differ in the set of independences they can encode and the factorization of the distribution they induce [7].

A Bayesian network is a directed acyclic graph that factorizes the joint probability of n variables $X_1, X_2, \ldots, X_n$ as $P[X_1, X_2, \ldots, X_n] = \prod_{i=1}^{n} P[X_i \mid PA(X_i)]$. For instance, as shown in figure 8.15a, the joint probability $P(A, B, C, D) = P(A)P(B)P(C|A, B)P(D|B, C)$. A Bayesian network can also be used to represent the causality between different variables.

A Markov network is a set of random variables having a Markov property described by an undirected graph, which may be cyclic. Thus, a Markov network can represent certain dependencies that a Bayesian network cannot (such as cyclic dependencies). On the other hand, it cannot represent certain dependencies that a Bayesian network can (such as induced dependencies).

The structure of a graphical model is usually manually designed and based on human knowledge, although, theoretically speaking, it can be learned automatically from the given data. Automatically learning the structure of a graphical model is still an open challenge, as it is very complicated and computationally expensive. With human knowledge integrated into the model, graphical models can handle machine-learning problems with a small set of training data or even without labeled data. This can help deal with label sparsity issues. It can also handle label imbalance in a classification task, as compared to other classification models like SVM.

On the other hand, if handling a problem with many variables (e.g., inferring traffic conditions for each and every road segment throughout a city), a graphical model will be a very large and complex structure, which drastically increases efforts spent on

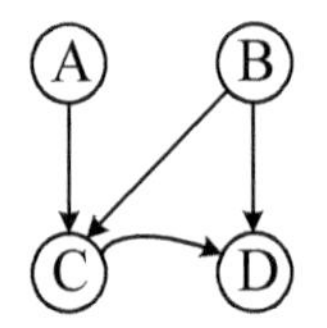

a) Bayesian networks

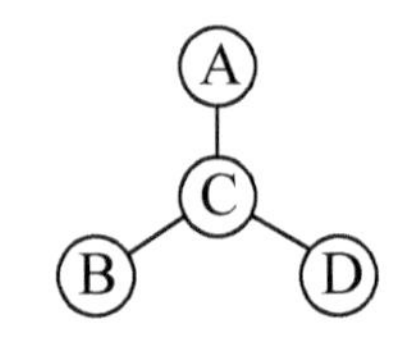

b) Markov networks

Figure 8.15
Two types of graphical models.

learning and inference. Thus, we can say graphical models are not easy to scale up to handling complex problems with massive data.

8.6.2 Bayesian Networks

8.6.2.1 Overview As shown in figure 8.16, there are two main steps when using a Bayesian network:

1. *Learning step*. There are two types of learning in a Bayesian network. One is to learn the structure of a network. The other is to learn the parameters (i.e., the conditional probabilities) given the structure of a Bayesian network. The methods for parameter learning can be divided into three categories: maximum likelihood estimation (MLE) [46], maximum a posterior (MAP), and expectation maximization (EM) [11].
2. *Inference step*. This step aims to infer the marginal probability of a variable with respect to dependent variables, based on some dependent variables' observations and the parameters of distributions obtained from the learning step. There are two categories of inference algorithms: exact inference and approximate inference. Representative exact inference algorithms, including variable elimination, the belief propagation [66], and clique tree, can generate an exact inference result for the variable's marginal probability. Since a Bayesian network is too large to obtain an exact inference result, we need to employ approximate inference algorithms, such as loopy BP [43], variational inference [60], and sampling methods.

To factorize the joint probability of a Bayesian network into the production of conditional probabilities, which will be used in the learning and inference steps, we need to find the conditional independence between variables through D-separation.

A dynamic Bayesian network (DBN) is a Bayesian network that relates variables to each other over adjacent time steps. There are two types of DBNs. The first type shares the same Bayesian network structure at different time intervals, while the other types could have different structures at different time intervals. The second type is a generalization of hidden Markov models (HMMs) and Kalman filters. The inference and learning algorithms of Bayesian networks can be employed in the learning and inference steps of a DBN, though DBNs may have some additional inference algorithms, such as the forward-backward algorithm and the Viterbi algorithm.

8.6.2.2 Independency and D-separation Before digging into the inference and the learning steps in a Bayesian network, we need to introduce the dependencies between variables. Two variables, X and Y, are independent if $P(X, Y) = P(X)P(Y)$, or $P(X|Y) = P(X)$. They are conditionally independent given another variable Z if $P(X, Y|Z) = P(X|Z)P(Y|Z)$.

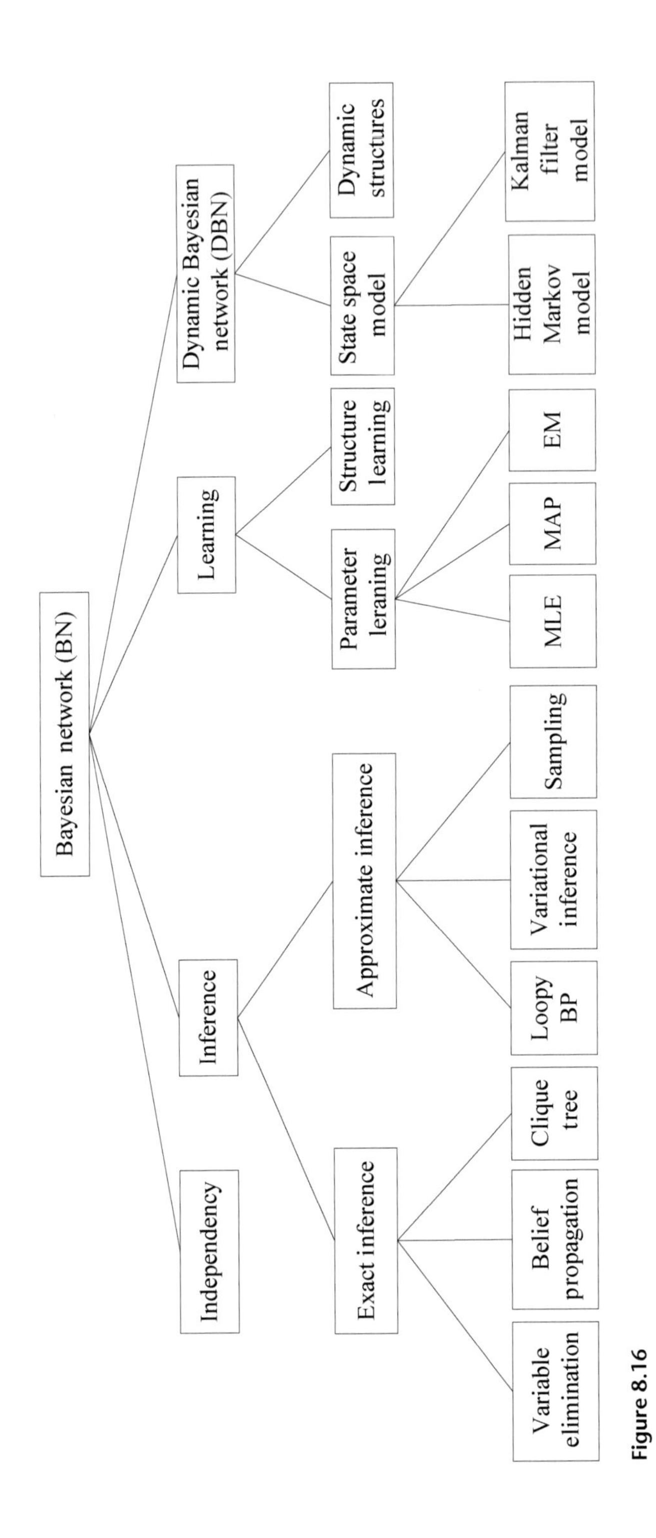

Figure 8.16
An overview of Bayesian networks.

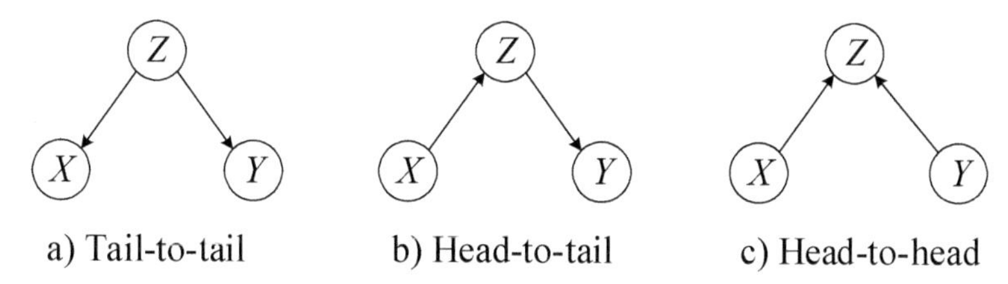

Figure 8.17
Three basic cases of D-separation.

This definition can easily be extended to multiple variables. If we can identify conditional independencies between variables in a Bayesian network, we can factorize the joint probability of the network into the product of conditional probabilities, based on which the inference and learning steps of a Bayesian network can be done much more easily. For example, if X and Y are independent given Z, then $P(X,Y|Z)=P(X|Z)P(Y|Z)$.

A precise way to specify the conditional independency in a Bayesian network is by D-separation [27]. Figure 8.17 presents the three basic cases of D-separation, consisting of tail-to-tail, head-to-tail, and head-to-head, based on which complex cases can be constructed:

1. *Tail-to-tail.* This structure indicates that X and Y are independent and conditionally independent given Z. An observed Z blocks the path from X to Y, denoted by $X-Y$.
2. *Head-to-tail.* This structure indicates that X and Y are dependent and conditionally independent given Z. An observed Z blocks the path $X-Y$.
3. *Head-to-head.* This structure indicates that X and Y are independent. In addition, X and Y are conditionally dependent given Z or any descendants of Z. An observed Z or any descendants of Z unblock the path $X-Y$.

To generalize these D-separation cases, any group of nodes A and B are conditionally independent given another group C if all paths from any node in A to any node in B are blocked.

8.6.2.3 Inference algorithms Given a joint probability $P(X, Y, Z)$, where X has two states (x_1, x_2) and Y has two states (y_1, y_2), if we observe $Z=z_1$, we can infer $P(X|Z=z_1)$ as follows:

$$P(X|Z=z_1)=\frac{P(X,Y,Z=z_1)}{P(Z=z_1)}=\frac{\sum_Y P(X,Y,Z=z_1)}{\sum_{X,Y} P(X,Y,Z=z_1)}. \tag{8.48}$$

Basically, we can calculate the probability by first finding the instances with $Z=z_1$ and then counting the number of different enumerations of X and Y in these instances, respectively (i.e., the number of instances with $(X=x_1, Y=y_1)$, $(X=x_1, Y=y_2)$, $(X=x_2, Y=y_1)$, and $(X=x_2, Y=y_2)$). The computational complexity of such an enumeration algorithm is $O(n^m)$, where n is the number of states, and m is the number of dependent variables except for the given variable $Z (n=m=2$ in this case). When there are multiple variables, the counting processes become too computationally expensive to carry out.

Exact inference To reduce the computational complexity of the enumeration $O(n^m)$, various algorithms have been developed for the inference of a Bayesian network. Hereafter, we introduce three widely used methods: the variable elimination, the belief propagation, and the clique tree algorithms.

Variable elimination If X and Y are conditionally independent given Z, we can factorize the joint probability as $P(X, Y, Z)=P(Z)P(X|Z)P(Y|Z)$. Consequently, we infer the marginal probability of X by

$$P(X) = \sum_Y P(X|Y) \sum_Z P(Z)P(Y|Z),$$

with a time complexity of $O(n \times 2)$, where n is the number of states of Y and Z.

Representing $\sum_Z P(Z)P(Y|Z)$ by $\varphi_1 (Y)$, we can calculate $P(X)$ by

$$P(X) = \sum_Y P(X|Y)\varphi_1(Y).$$

This is equivalent to eliminating variable Z from $P(X, Y, Z)$ during the counting process. We further denote $\sum_Y P(X|Y)\varphi_1(Y)$ by $\varphi_2 (X)$, eliminating Y in the inference.

In general, we eliminate variables (i.e., selecting variables) in an order when utilizing variable elimination in the inference step. Different elimination orders (e.g., first eliminating Z or Y) lead to different time complexities of inferences. Finding the optimal elimination order is NP-hard. Two commonly used methods to find a variable elimination order are maximum cardinality search and minimum deficiency search.

Belief propagation The variable elimination algorithm infers the marginal probability for a specific variable at each time. If there are multiple variables to infer, we need to run the algorithm multiple times, which is not efficient. To address this issue, the belief propagation algorithm calculates the marginal probabilities for multiple variables simultaneously in a Bayesian network with a tree structure. While there are multiple variants of belief propagation algorithms, we focus on the algorithm designed for factor graphs. A factor graph consists of two sets of nodes, variables and factors, representing the joint probability distribution by a product of factors.

For example, as depicted in figure 8.18a, according to the D-separation, X and W are dependent given Z (head-to-head); (W, X) and Y are independent given Z (head-to-tail).

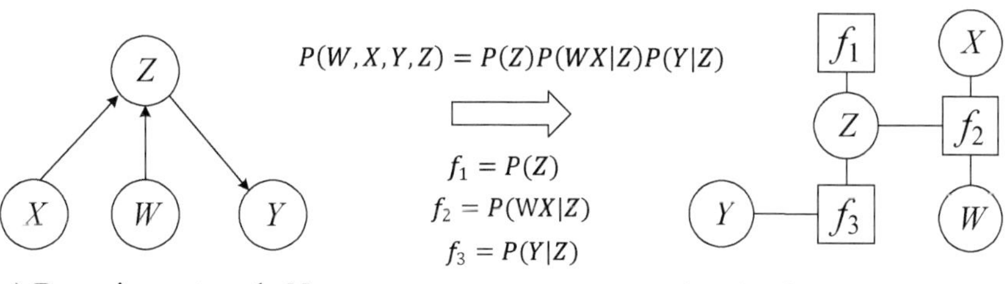

Figure 8.18
Belief propagation for factor graphs.

Thus, $P(W,X,Y,Z)=P(Z)P(WX|Z)P(Y|Z)$. By defining factor node functions: $f_1=P(Z)$, $f_2=P(WX|Z)$, and $f_3=P(Y|Z)$, we can represent $P(X,Y,Z)$ as a factor graph shown in figure 8.18b. The goal of the inference is to compute the marginal probability of $P(X)$, $P(Y)$, and $P(Z)$ based on conditional probabilities (i.e., f_1, f_2, and f_3).

The belief propagation algorithm performs efficient inference based on the conditional independence relationships in the graph. It works by passing a *message* along edges from variables to factors and from factors to variables, based on an iteration where messages from variables to factor nodes are represented by the product of the messages from factor nodes to variables and vice versa. For instance, as illustrated in figure 8.18, a message from a variable node Z to a factor node f_2 (i.e., $message_{Z\to f_2}$) is the product of the messages from Z's neighboring factor nodes (excluding f_2) to Z; that is,

$$message_{Z\to f_2} = message_{f_3\to Z} \times message_{f_1\to Z}. \tag{8.49}$$

In turn, a message from a factor node f_2 to Z (i.e., $message_{f_2\to Z}$) is the product of the messages from the variable nodes it connects to (excluding Z)—that is, $message_{X\to f_2} \times message_{W\to f_2} \times f_2$. $message_{f_2\to Z}$. The message is further marginalized over W and X—that is,

$$message_{f_2\to Z} = \sum_{WX} message_{X\to f_2} \times message_{W\to f_2} \times f_2. \tag{8.50}$$

Randomly selecting a node (e.g., Y) in a factor graph as a root, we turn the factor graph into a tree. The root node can be a variable node or a factor node. Then, starting from the root node, we propagate the messages to the leaf nodes. The messages are constructed in a similar way, shown in equations (8.49) and (8.50). The initial value of the message from the root node is 1 (i.e., $message_{Y\to f_3} = 1$); then $message_{f_3\to Z} = 1 \times f_3 = P(Y|Z)$. After the first round of propagation, we start message propagation from leaf nodes (e.g., W and X) to the root, following the same approach of message construction as the first round. After the iteration, we can find the exact inference.

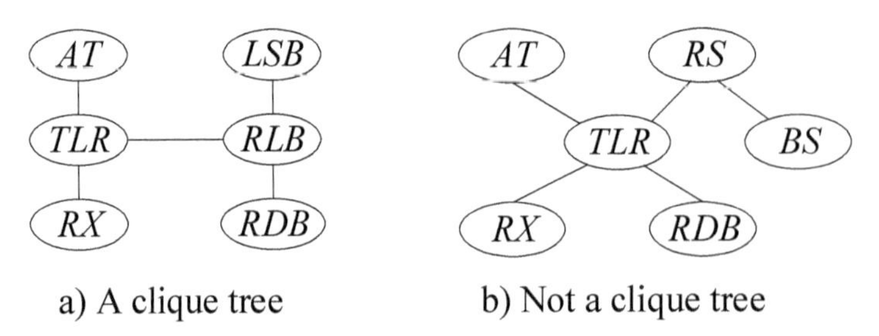

Figure 8.19
Examples of clique trees.

Clique tree The belief propagation algorithm cannot generate exact inference results for a Bayesian network with cycles. In addition, it is not necessary to calculate the marginal probability for all variables in a Bayesian network. To deal with these issues, the clique tree algorithm is proposed to eliminate cycles in a Bayesian network by clustering variables into single nodes. It then infers the joint probability for the variables in each node. After that, the inference for each single variable can be conducted easily, using the enumeration or the variable elimination algorithms.

The clique tree algorithm is a variant of the belief propagation algorithm, which operates on a clique tree. A clique tree is an undirected tree in which a node, called a *clique*, corresponds to a set of variables. If two cliques in a clique tree have common variables, all cliques on the path between these two cliques must contain the common variables. For instance, clique *LSB* has three variables, sharing variable *B* with another clique *RDB*. Thus, the clique *RLB* must contain the common variable *B*; otherwise, it is not a clique tree. For instance, the tree shown in figure 8.19b is not a clique tree because the path between *RDB* and *BS* does not contain the common variable *B*.

After generating a clique tree, we conduct belief propagation on the tree to obtain a joint probability for variables in each node. For example, based on the clique tree shown in figure 8.19a, we obtain $P(T, L, R)$ through the belief propagation algorithm. After that, we can calculate the marginal probabilities of each variable in the clique by $P(T) = \sum_{L,R} P(T, L, R)$, $P(L) = \sum_{T,R} P(T, L, R)$, and $P(R) = \sum_{T,L} P(T, L, R)$.

A clique tree can be constructed according to the following four steps:

1. First, we create a moral graph for a given Bayesian network, connecting the nodes that share at least one child node. For instance, nodes *T* and *L* should be connected with an undirect edge, as illustrated in figure 8.20a, because they share the same offspring node *R*. The same rule applies to nodes *R* and *B*. After that, we turn all the edges into undirected edges.

2. Second, we count the deficiency of each node in the moral graph obtained in the first step, removing the node with the smallest deficiency step by step until all the nodes in a Bayesian network have been removed. The deficiency of a node means the number of edges to be added to maintain the connectivity between the node's neighbors if the node is removed. For example, as depicted in figure 8.20b, node *T*'s deficiency is 2. That is, if we remove node *T* from the moral graph, we need to add an edge between nodes *A* and *R* and another edge between nodes *A* and *L* to maintain the connectivity between node *T*'s neighbors. If we remove node *A*, it is not necessary to add any edges here, as *A* only has one neighbor (i.e., node *T*). After obtaining all nodes' deficiencies, we start with the node with the smallest deficiency. In this example, both nodes *A* and *D* can be selected as a starting node. Different removal orders will result in different cliques. After removing a node from a moral graph, we update the deficiency of each node and then remove the node with the smallest deficiency. The iteration repeats until all nodes have been removed. The example shown in figure 8.20b generates a removing order: *A*, *T*, *D*, *R*, *L*, *S*, *B*.
3. Third, according to the removal order generated in the second step, we construct a clique for each node in the order by merging the node with its one-hop neighbors. The one-hop neighbors of a node, also called *separators*, separate the node from the rest of the moral graph. For example, as shown in figure 8.20c, node *A* is grouped with *T* to form a clique and then removed from the moral graph. After that, node *T* is grouped with *L* and *R* to form a clique—and so on until each node has only one-hop neighborhoods. The remaining nodes form a clique.
4. Fourth, we connect the cliques (generated by the third step) that share common variables step by step to form a clique tree. For instance, as illustrated in figure 8.20d, starting from the last clique LSB, we add other cliques sequentially into a clique tree. A clique *RBL* is connected to *LSB*, as they share common variables *L* and *B*. Clique *DRB* is then added to *RBL*, as they share variables *R* and *B*. Finally, a clique tree without cycles is built. Thus, the belief propagation algorithm can be applied to the clique tree, computing the joint probability of each clique.

Approximate inference In real applications, a large number of random variables with complex dependency structures are involved in a graphical model. As a consequence, inference tasks, such as the calculation of a marginal distribution or joint distribution, become computationally expensive. Thus, the exact inference algorithms introduced above become infeasible for solving the inference problem of such a complex graphical model.

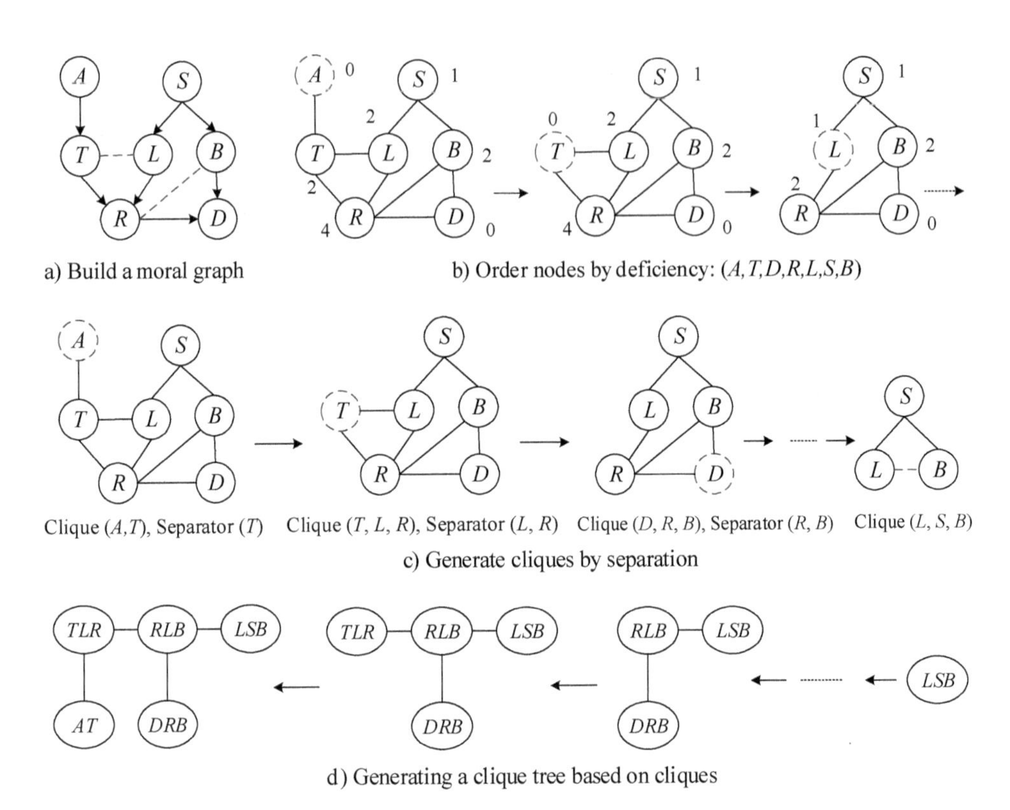

Figure 8.20
Clique tree inference algorithm.

In addition, an exact inference result may not be very necessary for some applications. For instance, to predict if there will be traffic congestion on a road segment at a future time, we only need to know the approximate probability of seeing congestion given some observations, such as the weather and traffic conditions of the past few hours. If the $P(congestion == Yes|observations)$ is significantly higher than $P(congestion == No|observations)$, we know congestion could happen very soon. It is not very important to know the exact value of $P(congestion == Yes|observations) = 0.79$ or 0.81, as long as we estimate it to be around 0.8.

For these two reasons, we introduce three categories of approximate inference methods.

Loopy belief propagation Though the belief propagation algorithm was originally designed for acyclic graphical models, with a slight modification it can be used in general graphs to compute approximate inference results. The algorithm is then sometimes called *loopy* belief propagation because graphs typically contain cycles, or loops. The initialization and scheduling of message updates must be adjusted slightly (compared with the previously described schedule for acyclic graphs) because graphs might not contain any leaves. The loopy belief propagation initializes all variable messages to 1. It uses the same message definitions as the belief propagation algorithm, updating all messages at every iteration. There exist graphs that will fail to converge or that will oscillate between multiple states over repeated iterations. The algorithm converges in most cases on graphs containing a single loop, but the probabilities obtained might be incorrect. The precise conditions under which loopy belief propagation will converge are still not well understood.

Variational inference Variational methods are mainly used for approximating intractable integrals arising in Bayesian inference [60]. The basic idea of variational inference is to simplify the original graphical model and approximate a posterior probability of unobserved variables (i.e., hidden variables and parameters), given the observed data. It provides an analytical approximation to the posterior probability of the unobserved variables through a variational distribution, deriving a lower bound for the marginal likelihood of the observed data.

Variational inference methods can be regarded as an extension of the expectation maximization (EM) algorithm. Variational inference methods consist of a variational expectation-step and a variational maximization-step. However, the original EM algorithm relies on the independence assumption between hidden variables and thus cannot handle cases in which dependencies exist between multiple hidden variables. That is, the EM algorithm cannot generate the derivations of conditional probability $P(Z|D, \theta)$ in the E-step because of the intractability of calculating the marginal probability $P(D, \theta)$ over dependent hidden variables, where Z represents the hidden variables, and D and θ denote observed data and estimation parameters, respectively.

To address this issue, variational inference methods minimize the Kullback–Leibler divergence between the tractable distribution $q(\theta)$ and the intractable distribution $P(Z|D, \theta)$ instead of maximizing the auxiliary function, based on the mean field theory.

Sampling Sampling methods, also known as *Monte Carlo methods*, are based on stochastic numerical sampling from distributions. The general idea behind sampling is to obtain a set of samples independently from the distribution to provide a numerical approximation to the exact posterior. The most commonly used sampling methods are rejection sampling, importance sampling, and Gibbs sampling. For example, the Gibbs sampling algorithm generates a sequence of observations, which are approximated from a specified multivariate probability distribution, when direct sampling is difficult. This sequence can be used to approximate the joint distribution, to approximate the marginal distribution of a variable, or to compute the expected value of one of the variables.

8.6.2.4 Learning algorithms

1. *Structure learning*. The structure of a Bayesian network reflects the conditional dependency between variables. Both inference and learning are conducted based on a Bayesian structure. However, learning the Bayesian network structure is very challenging due to its large search space [13, 58]. Without any prior knowledge about the structure of a graphical model, we need to try all possible combinations between arbitrary numbers of variables. For simplicity, some research assumes that a Bayesian network is a tree when learning its structure, but this does not always hold in practice.
2. *Parameter learning:*
   - *Maximum likelihood estimation (MLE)*. This method estimates the parameters by maximizing the likelihood of the observations given the parameters. For example, assuming the joint probability of a Bayesian network is $P(X, Y, Z|\theta)$, where θ is the estimation parameters and the observations are $\{(x_1, y_1, z_1), (x_2, y_2, z_2), (x_3, y_3, z_3)\}$, then the likelihood of these observations can be defined as $L = \prod_{i=1}^{3} P(x_i, y_i, z_i|\theta)$, from which we can estimate the parameters θ by $max_\theta L$.
   - *Maximum a posteriori (MAP)*. MAP estimates the parameters by maximizing the posterior probability, given its prior knowledge and some observations. Assuming the prior distribution over θ is $P(\theta)$, given the sampling distribution of the observations $f(X, Y, Z|\theta)$, we can estimate the parameters as follows:

     $$\theta = argmax_\theta P(\theta|X, Y, Z) = argmax_\theta \frac{f(X, Y, Z|\theta)P(\theta)}{P(X, Y, Z)}, \tag{8.51}$$

     where $P(X, Y, Z)$ is the probability of the data averaged over all parameters, $P(X, Y, Z) = \int f(X, Y, Z|\theta)P(\theta)d\theta$. MLE can be seen as a special case of the MAP estimation that assumes a uniform prior distribution of the parameters or as a variant of the MAP that ignores the prior and which therefore is unregularized.
   - *Expectation maximization (EM)*. EM is an optimization scheme that is commonly used to solve the MLE and MAP problem where unobserved latent variables exist in the model. In particular, the EM algorithm consists of two important steps: the expectation step (E-step) and the maximization step (M-step). In the E-step, the expectation of the log-likelihood function is calculated by using the current estimates of the model parameters. In the M-step, the new solutions of

model parameters are derived to maximize the expectation of the log-likelihood function (in MLE) or the expected log posterior function (in MAP) in the E-step.

8.6.3 Markov Random Field

The inference and learning algorithms introduced in the previous sections are actually designed for general graphical models, consisting of Bayesian networks and the Markov random field. That is, all these algorithms can be applied to the Markov random field. When a Markov random field is not big and complex, exact inference methods, such as the belief propagation algorithm, can be applied. Otherwise, approximate inference methods (e.g., variational inference and sampling methods) should be employed.

For example, the belief propagation algorithm can be used to generate exact inference results for a Markov random field without cycles. For a Markov random field with cycles, we employ the clique tree algorithm to turn it into a tree-structure graph, based on which the belief propagation algorithm can be used to derive an exact inference. Another method is to employ the loopy belief propagation algorithm to generate approximate inference results for a Markov random field with cycles. When there are many cycles in a Markov random field, we use variational inference and sampling methods for an approximate inference.

We do not introduce further technique details about the learning and inference of the Markov random field in this section. Instead, we present more applications using the Markov random field in section 8.6.5.

8.6.4 Bayesian Networks for Spatiotemporal Data

Bayesian networks can contribute to spatiotemporal data mining from four perspectives. First, based on the probabilistic dependency, Bayesian networks provide a way to fuse knowledge between disparate spatiotemporal datasets. Second, Bayesian networks can handle the inference problem when labeled data cannot be obtained or is very scarce. Third, Bayesian networks can reveal the causal relationship between different spatiotemporal variables. Fourth, dynamic Bayesian networks can model the temporal sequence of spatiotemporal data.

The main challenge of applying Bayesian networks to spatiotemporal data is to model a complex spatiotemporal inference problem with a precise structure so that it can scale up to handling a large-scale inference problem.

8.6.4.1 Basic Bayesian networks for spatiotemporal data This section introduces two examples that use basic Bayesian networks (i.e., not dynamic Bayesian networks) to perform inference with spatiotemporal data. The first example is about traffic volume inference based on a diversity of datasets, such as GPS trajectories of taxicabs and POIs [51]. This example demonstrates the first and second capability of Bayesian networks mentioned above. The second one aims to deduce the causal relationship between air pollutants as well as the meteorological conditions of different places [78], demonstrating the first and third capabilities of Bayesian networks. The first example is concerned with categorical variables, while the second example deals with continuous variables.

Traffic volume estimation In this example, we aim to infer the traffic volume (i.e., the number of vehicles) traversing each road segment in the past few minutes, based on the GPS trajectories of taxicabs, POIs, and road networks. Intrinsically, there exists a certain relationship (e.g., fundamental diagram) among traffic speed, volume, and density. Yet to accurately quantify this relationship requires a large amount of traffic volume data. As many road segments are not equipped with in-road devices, collecting such training data on a city scale is very costly. Thus, it is impractical to use supervised learning algorithms to learn the relationship. In addition, the occurrence of the sampled vehicles (e.g., taxis) on a road segment may be quite different from the entire set of vehicles, though their travel speed could be similar. In other words, observing more taxis on a road segment does not deduce more occurrences of other vehicles. Given this, we cannot estimate the total volume of traffic directly based on the sampled traffic data.

To address this issue, an unsupervised graphical model, called TVI (traffic volume inference), based on a partially observed Bayesian network, is proposed. Figure 8.21 presents the graphical structure of the TVI model, where a white node denotes a hidden variable, and gray nodes are observations. A straightforward idea is to build one Bayesian network to model an entire road network, which might consist of hundreds of thousands of road segments. This leads to a Bayesian network with millions of nodes and edges, which is overly complex to do learning and inference. Instead, we train one TVI model for each level of road segments using the data of all road segments of the same level. The model is then applied to infer the traffic volume for each road segment of the level, respectively.

Specifically, the traffic volume on each road lane N_a (i.e., the number of vehicles per minute per lane) of a road segment is influenced by four major factors: weather conditions w, time of day t, the latent type of road θ, and the volume of observed sample vehicles N_t. Furthermore, a road's θ is codetermined by its road network features f_r (such as the length, indegree, outdegree, and tortuosity of the road segment), global position feature f_g, and surrounding POIs (denoted by latent variable α). α is further

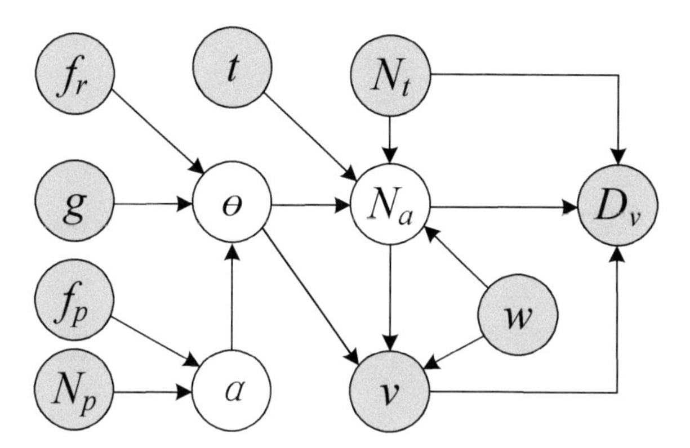

Figure 8.21
Traffic volume inference based on a Bayesian network.

influenced by the distribution of different POI categories f_p and the total number of POIs N_p around a road segment. $\bar{v}$ and d_v are the average travel speed and speed variance, respectively. They can be observed from the taxi trajectory data or inferred from other models (e.g., the TSE proposed in [51]). $\bar{v}$ depends on θ, N_a, and w. dv depends on N_t, N_a, and $\bar{v}$. All the variables in the model are discretized. This reduces inference difficulty while ensuring the inferred results are statistically useful for gas consumption and emission calculation.

Due to hidden nodes, the conditional probability of N_a cannot be drawn simply by counting the occurrence of each condition. Hence, we use the EM algorithm to learn the parameters in an unsupervised manner. In the beginning part, this algorithm sets the parameters—the conditional probabilities (e.g., $P(\alpha|f_p, N_p)$ and $P(\bar{v}|N_a, \theta, w)$—with random values. In the E-step, we use the exact inference method to compute the values of the hidden nodes (α, θ, N_a) for each instance of the observed data. This is an inference process. In the M-step, by scanning the inferred results from the E-step, the algorithm recalculates the conditional probabilities, which will replace the old parameters. We keep iterating until the parameters converge, and we learn a solution for the unknown parameters.

Causality inference of air pollutants Identifying the root cause of air pollution is mission-critical for the sustainability of many cities [78]. In this example, we try to infer the causal relationship between different air pollutants in different cities using Bayesian networks and a variety of datasets, such as air quality data and meteorological data. For instance, as depicted in figure 8.22a, when the wind speed is less than

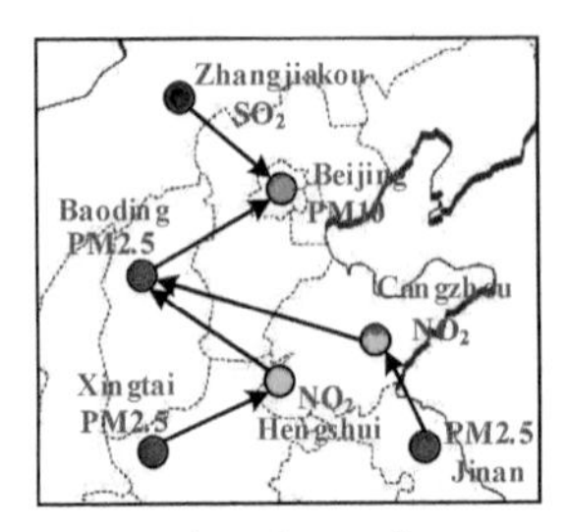

a) Causal pathway between different air pollutants

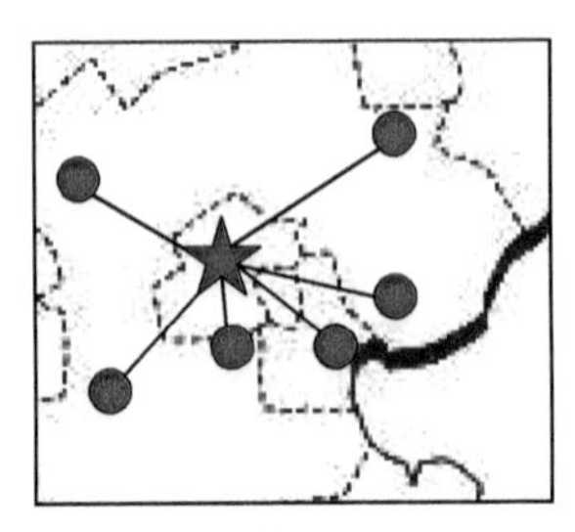

b) Select the top-N most correlated sensors

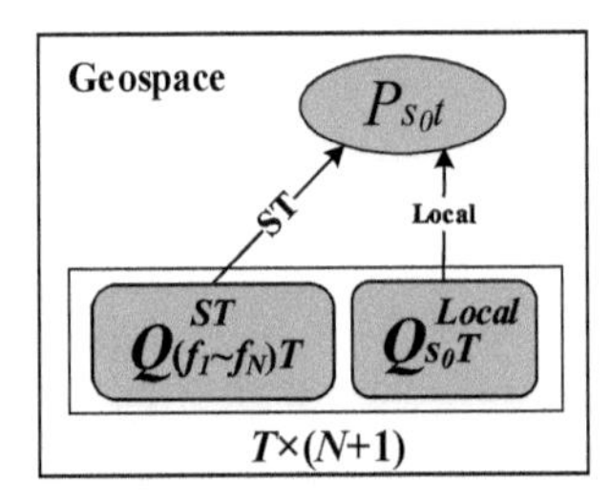

c) Construct a GBN for a sensor based on local and global factors

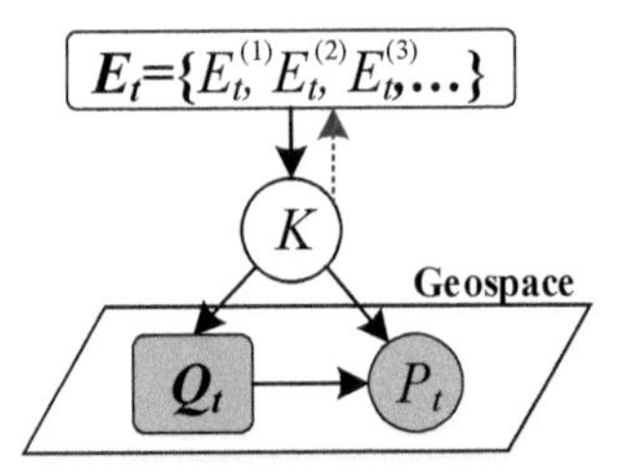

d) Considering the environment factor through a confounder

Figure 8.22
Identifying the root cause of air pollution based on Bayesian networks.

5 m/s, the high concentration of PM10 (particulate matter with a diameter smaller than 10 μm) is mainly caused by the high concentration of SO_2 in Zhangjiakou and that of PM2.5 (particulate matter with a diameter smaller than 2.5 μm) in Baoding. Further, the high concentration of PM2.5 in Baoding is mainly caused by NO_2 from Hengshui and Cangzhou. Knowing such a causal relationship can help tackle air pollution at its sources.

However, generating the causal relationship is a very challenging task for the following reasons. First, the increase and decrease of an air pollutant's concentration in a location depend on many factors, such as the situation over the past few hours in the location and that of its spatial neighborhoods. It is also conditioned on the environment (e.g., the meteorological conditions). If we connect all possible factors to the target variable (e.g., PM2.5 in Beijing) in a Bayesian network, the structure of the model becomes rather complex and thus cannot carry out learning and inference effectively. For instance, there are about one hundred cities within a circle distance of 300 km to Beijing. Each city may have over ten air quality–monitoring stations, each of which generates the concentration of six types of air pollutants. Thus, there are $100\times10\times6$ factors to consider. Each factor may further have over four states. In addition, the target variable would be affected by the values of these factors across multiple time intervals (not only the current time interval). The dependency would also be conditioned on different weather conditions. This increases the complexity of the Bayesian network significantly.

Second, to reduce the complexity of the model, an intuitive idea is to solely keep the top-N most correlated factors in the Bayesian network. However, determining the correlation between a target variable and other factors is also not easy in this application. The readings of the target variable and factors are represented by a geotagged time series, which in most cases changes a bit over time with some trivial fluctuations. If we simply compute the Pearson correlation between such time series, the correlation between them will be dominated by those stable readings and trivial fluctuations, while the readings during tremendous changes truly reveal the dependency between two time series.

For instance, in figure 8.23 the PM2.5 readings of sensor S_2 are generally close to S_3 and relatively far from S_1. Thus, the Pearson correlation between S_1 and S_2 is larger than that between S_2 and S_3. However, when the PM2.5 of S_1 changes, that of S_2 also changes, following almost the same trend. Such tremendous changes may indicate some physical interactions between different air pollutants or chemical processes and thus are more valuable than trivial fluctuations. Besides, we need to consider the spatial location of the two sensors when computing the dependency between them. Air pollutants from two locations far away from each other cannot directly depend on each other even if sometimes their readings look similar, as air pollutants need to travel between locations through propagation and dispersion in the environment.

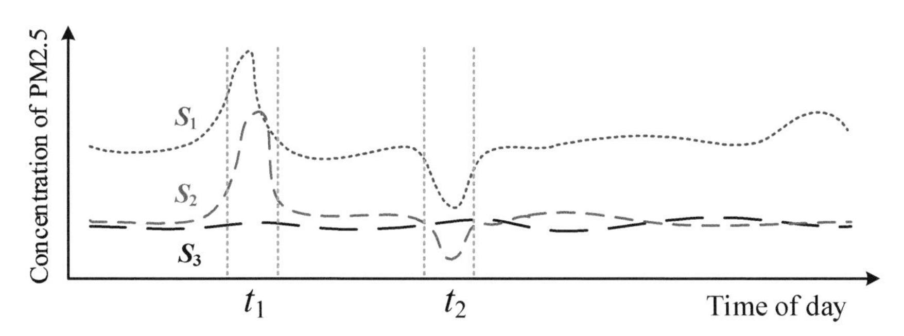

Figure 8.23
Capturing the true dependency between readings of different sensors.

To address the aforementioned two issues, we first mine spatially coevolving patterns between each pair of sensor readings based on an efficient algorithm [70], as shown in figure 8.18b. The patterns can effectively determine the dependency between different air pollutants at different locations, thereby helping us to select the top-N most correlated factors in a Bayesian network. They also find the time intervals that two air pollutants truly depend on each other (e.g., t_1 and t_2 shown in figure 8.19). Thus, we can find readings at more meaningful time intervals when computing the dependency between two time series. This reduces the complexity of a Bayesian network while enhancing the accuracy of the inference result.

We then construct a Bayesian network for a target air pollutant s_0 in a location (e.g., PM2.5 in Beijing) at time interval t, denoted by $P_{s_0,t}$, connecting $P_{s_0,t}$ to the top-N most correlated air pollutants in other locations (such as PM2.5 and NO_2 in Baoding and SO_2 in Zhangjiakou) over the past T hours, denoted by $Q^{ST}_{[s_1,\ldots,s_n]T}$, and its own readings over the past T hours, denoted by $Q^{local}_{s_0,T}$. $Q^{ST}_{[s_1,\ldots,s_n]T}$ and $Q^{local}_{s_0,T}$ formulate the parent node (denoted by $\boldsymbol{Q_t}$) of $P_{s_0,t}$, as illustrated in figure 8.18d. Thus, there are $T\times(N+1)$ factors affecting $P_{s_0,t}$. In addition, we use the one-hour difference (i.e., the value of the current hour minus the value of one hour ago) rather than the original values of a sensor.

Here, we employ Gaussian Bayesian networks (GBNs) to model this problem, as the concentration of air pollutants are continuous values. In addition, the one-hour difference in air quality follows the Gaussian distribution. GBNs have some nice properties, such as the distribution of the target variable $P_{s_0,t}$, which conditioned on its parent node, follows a Gaussian distribution. According to the properties of GBNs,

$$P_{s_0,t} = \mu_{s_0,t} + (Q^{ST}_{[s_1,\ldots,s_n]T} \oplus Q^{local}_{s_0,T})\boldsymbol{A_k} + \varepsilon_{s_0,t},$$

where $\mu_{s_0,t}$ is the mean value of $P_{s_0,t}$; $\oplus$ denotes the concatenation of those features, and $\boldsymbol{A_k}$ is a (regression) parameter vector corresponding to these features; $\varepsilon_{s_0,t}$ is the bias of the regression; and $\mu_{s_0,t}$, A_k, and $\varepsilon_{s_0,t}$ are parameters that we need to learn from the data. It is actually a Gaussian regression model.

To consider the impact of environmental factors, $\boldsymbol{E_t} = \{E_t^{(1)}, E_t^{(2)}, \ldots, E_t^{(n)}\}$, on the causal relationship between air pollutants, we incorporate a (latent) confound variable K into the GBN, as shown in figure 8.22d. As there are multiple environmental factors, each of which has multiple states, combining them will result in a huge space. For example, suppose there are five factors consisting of wind speed, wind direction, humidity, temperature, and weather, each of which has four (discretized) states. Then, there are $4^5=1024$ states. If we directly use $\boldsymbol{E_t}$ as the confounder, the number of samples pertaining to each state is very few. This will result in unreliable inference results. By incorporating a latent variable K with a few discrete values (e.g., $K=1$, 2, 3, each of which indicates a cluster of environmental states), the number of samples falling in each cluster will become large and hence can derive more reliable inference results. We can simply understand the K latent clusters as K types of environmental situations with *good, normal,* and *bad* dispersion conditions, respectively, though the meaning is more sophisticated than those words. Based on the Markov equivalence (directed acyclic graphs [DAGs] that share the same joint probability distribution), we can reverse the arrow $\boldsymbol{E}_t \rightarrow K$ to $K \rightarrow \boldsymbol{E}_t$. K determines the distributions of $\boldsymbol{E}_t$, $\boldsymbol{Q}_t$, and $\boldsymbol{P}_t$, thus enabling us to learn the distribution of the graphical model from a generative process.

The learning process of the Bayesian network is shown in figure 8.24. Based on the spatially coevolving patterns, we select the top-N (out of the M, $M > N$) most correlated air pollutants for a target pollutant, constructing a Gaussian Bayesian network. We cluster the environment conditions into K initial clusters using a K-means algorithm. This expedites the convergence of later iterations. An EM algorithm is then proposed to infer the parameters of this Bayesian network.

After learning the parameters, we assign a K to each time interval according to $P(Q_t, E_t, P_t)$. Based on the readings at the time intervals pertaining to each cluster, we further calculate the Granger causality between the target air pollutant and the top-M candidates. This step refines the causal relationship between the target air pollutant and other candidates. Note that the correlation calculated based on coevolving patterns is not real causality; it is just used to prune the searching space and reduce trivial fluctuations. We then select the top-N candidates with the highest Granger causality score to construct a Bayesian network. Thus, there are K Bayesian networks corresponding to the K clusters. Using the same EM algorithm, we estimate the parameters of each Bayesian network and then recalculate the Granger causality and reconstruct the Bayesian network. The iteration in each cluster stops until the structure of the Bayesian network does not change anymore. The Bayesian network of each cluster denotes the causal relationship between different air pollutants under different environmental situations.

Using the aforementioned method, we can infer the causal relationship for each air pollutant in each location under different environmental situations. By concatenating the one-hop causalities between different locations in the same cluster, we form a causal path. For instance, as illustrated in figure 8.22a, when the wind speed is less than 5 m/s, the high concentration of PM10 in Beijing is mainly caused by the SO_2 in Zhangjiakou and the PM2.5 in Baoding. Further, the PM2.5 in Baoding is mainly caused by the NO_2 from Hengshui and Cangzhou.

8.6.4.2 Hidden Markov models for spatiotemporal data

Basic concepts of hidden Markov models An hidden Markov model (HMM) is a dynamic Bayesian network that repeats the same structure at each time stamp, as shown in figure 8.25. Each white node denotes a hidden state, which is a categorical value

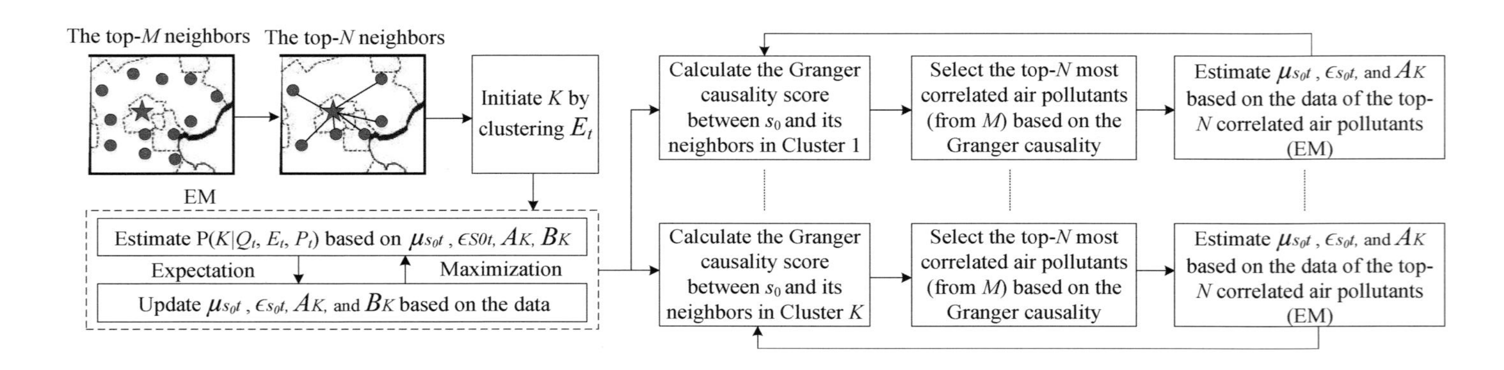

Figure 8.24
The learning process of the Bayesian network for diagnosing air pollution.

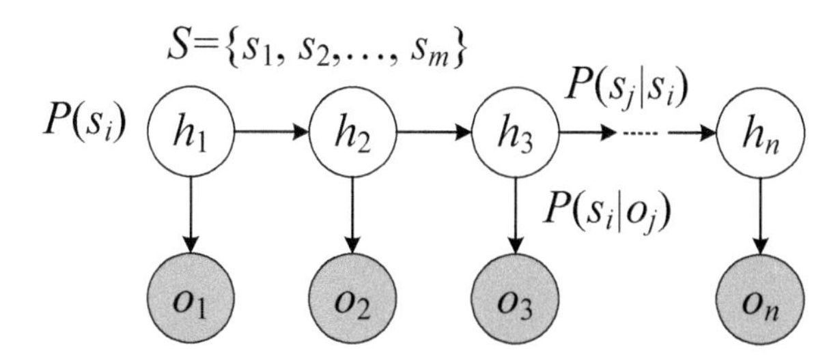

Figure 8.25
The structure of hidden Markov models.

belonging to $\mathbf{S}=\{s_1, s_2, \ldots s_m\}$, and gray nodes are observations (e.g., the features we extracted from the data collected at the corresponding time interval). There are three important conditional probabilities in an HMM. The first one is the transition probability between different states, denoted as $P(s_j|s_i)$, $1 \leq i, j \leq m, i \neq j$. The second is the emission probability, denoted by $P(s|o)$, $s \in \mathbf{S}$, which represents the possibility of seeing a state s given the observation o. The third one is the prior probability of the hidden variable $P(h_1 = s_i)$, $1 \leq i \leq m$.

HMMs can be used to complete three types of tasks. One is to infer the hidden state at the end of a sequence (i.e., $P(h_n|o_1, o_2, \ldots, o_n)$), called *filtering*, given the aforementioned probabilities. This problem can be solved by the forward algorithm. The second one is called *smoothing*, inferring the hidden state of a variable in the middle of a sequence (i.e., $P(h_k|o_1, o_2, \ldots, o_n)$), which can be solved with a forward-backward algorithm. The last one, called the *most likely explanation*, is finding a sequence of hidden states that maximizes the probability of seeing the entire sequence of observations. This problem can be solved by the Viterbi algorithm.

HMM-based map matching Map matching is the process of converting a sequence of raw latitude/longitude coordinates to a sequence of road segments. Knowledge of which road a vehicle was/is on is important for assessing traffic flow, guiding the vehicle's navigation, detecting the most frequent travel path between an origin and a destination, and so on. Map matching is not an easy problem, given parallel roads, overpasses, and spurs.

Figure 8.26 presents a map-matching algorithm [39] for low-resolution GPS trajectories, based on the idea of HMMs. It first finds the candidate road segments that are within a circle distance to each point in a trajectory. For instance, as illustrated in figure 8.26a, road segments e_i^1, e_i^2, and e_i^3 are within the circle distance to p_i, and c_i^1, c_i^2, and c_i^3 are the

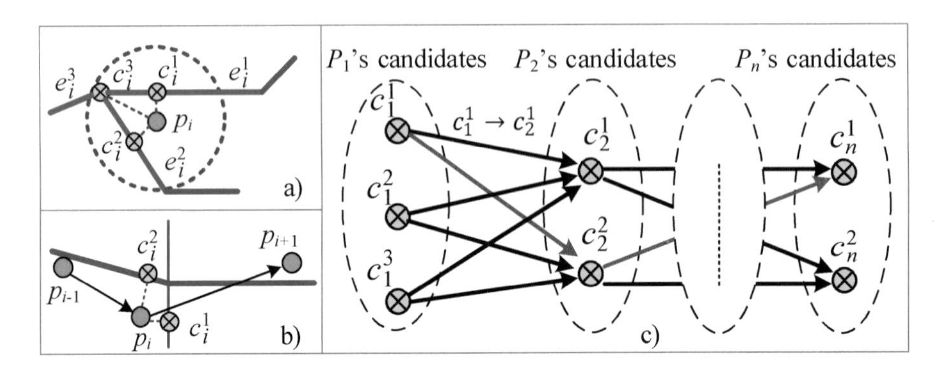

Figure 8.26
Map matching based on HMMs.

candidate points (on these road segments) that p_i should be projected to. The distance between p_i and a candidate point $dist(c_i^j, p_i)$ indicates the probability $N(c_i^j)$ that p_i can be matched to the candidate point. This probability can be regarded as the emission probability of an HMM, which is modeled by a normal distribution:

$$N(c_i^j) = \frac{1}{\sqrt{2\pi}\sigma} e^{-\frac{dist(c_i^j, p_i)^2}{2\sigma^2}}.$$

The algorithm also considers the transition probability between the candidate points of each two consecutive trajectory points. For example, as depicted in figure 8.26b, c_i^2 is more likely to be the true match of p_i, considering p_{i-1} and p_{i+1}. The transition probability between two candidate points is represented by the ratio between their Euclidean distance and the road network distance between them.

Finally, as shown in figure 8.26c, by combining the emission and transition probabilities, the map-matching algorithm finds a sequence of road segments that maximizes the global probability of the match. The idea is similar to the HMM where emission and transition probabilities are considered to find the most possible sequence of state given a sequence of observations. The hidden state in this problem is the road segment on which a GPS point was truly generated, and GPS points are observations.

8.6.4.3 Latent Dirichlet allocation for spatiotemporal data Latent Dirichlet allocation (LDA) is a generative model that includes hidden variables. Its intuition is that a document can be represented by a random mixture over latent topics, each of which is characterized by a distribution over words [8]. With the topic representation of a document, we can reduce the dimension of a document from tens of thousands of words

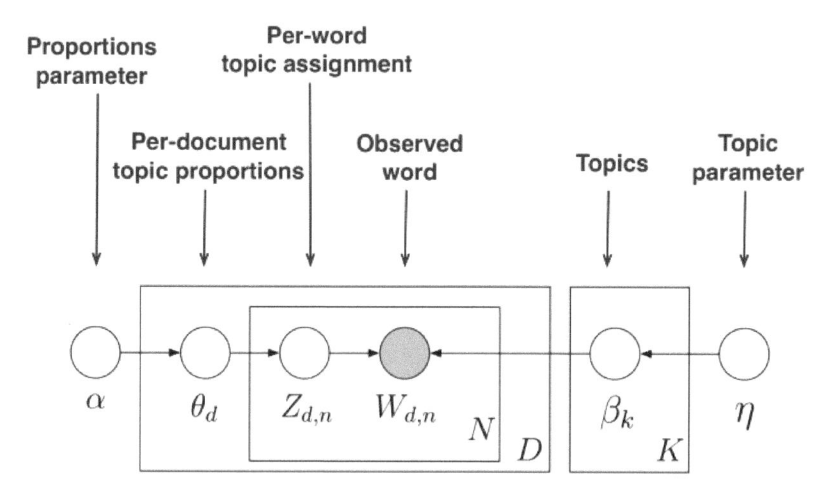

Figure 8.27
Graphical representation of LDA.

to a few topics, thereby enhancing computational efficiency and the accuracy of computing the similarity between two documents. For example, we can cluster documents with similar meanings although they may use different words. The similarity based on the topic distribution can also help recommendation systems and search engines. LDA is an unsupervised learning algorithm. The only thing that a user needs to specify is the number of topics.

Figure 8.27 shows the graphical representation of LDA. α and η are the prior parameters for the Dirichlet document-topic distribution and topic-word distribution, respectively. Assuming there are K topics and M words in the vocabulary, β is a $K \times M$ matrix, where β_k denotes the distribution of the k-th topic over the M words. The topic proportion for the d-th document is θ_d, where $\theta_{d,k}$ stands for the topic proportion for topic k in the d-th document. For instance, a document is concerned with two topics (i.e., $k=2$): dogs and cats. Then $\theta_d=(0.8, 0.2)$ means 80 percent of this document's topic is about dogs, and 20 percent is about cats. The topic assignment for the d-th document is Z_d, where $Z_{d,n}$ denotes the topic assignment for the n-th word in the d-th document. The observed words for document d are W_d, where $W_{d,n}$ is the n-th word in document d.

The generative process of LDA is as follows:

1. For each topic k, draw $\beta_k \sim Dirichlet(\eta)$.
2. For document d, draw $\theta_d \sim Dirichlet(\alpha)$.
3. For the n-th word in document d, $W_{d,n}$,

a. draw $Z_{d,n} \sim Multinomial(\theta_d)$, and

b. draw $W_{d,n} \sim Multinomial(\beta_{Z_{d,n}})$.

The learning process of LDA is to estimate the parameters (θ, z, β) based on observed (W, α, η, K) by maximizing the posterior distribution $P(\theta, z, \beta | W, \alpha, \eta, K)$. This can be achieved by an EM algorithm in which the expectation step (i.e., the inference step) can be handled by Gibbs sampling or variational inference.

Applying LDA to spatiotemporal data faces three challenges. One is to formulate documents and words based on spatiotemporal data. The other is to encode the prior knowledge of spatiotemporal data into an LDA-based model. The third is to fuse the knowledge of multiple spatiotemporal datasets. In the rest of this subsection, we introduce two examples of using LDA's variants to deal with spatiotemporal data.

Inferring functional regions Yuan et al. [68, 69] partition a city into disjointed regions using major roads, such as highways and ring roads, and then infer the functions of each region using POIs and human-mobility patterns in the region. This is a challenging task, as a region's function is compound with a mixture of functions across different categories, such as business, education, and residential areas.

As depicted in figure 8.28a, an LDA variant–based inference model is proposed, regarding a region as a document, a region's functions as topics, categories of POIs (e.g., restaurants and shopping malls) as metadata (like authors, affiliations, and key words), and human-mobility patterns as words. The intuition is that a region with a certain function will generate a certain type of mobility pattern. By feeding POIs and human-mobility patterns into different parts of this model, a region is represented by a distribution of functions, each of which is further denoted by a distribution of mobility patterns.

Figure 8.28b presents the graphical representation of the LDA-variant model, where N stands for the number of mobility patterns; R denotes the number of regions; and K is the number of functions, which should be predefined. In the basic LDA model, all documents share the same Dirichlet prior α, as we do not have prior knowledge of selecting α. When applying the LDA model to this application, we know the POIs in a region, which could describe a region's functions to some extent. Thus, as illustrated in figure 8.28b, regions of different functions can have different α_r, which depends on a region's POI features x_r. A straightforward way to form x_r is to count the number of POIs across different categories in a region. For example, there are five restaurants, one movie theater, two shopping malls, one university, and so on, in a region. Then $x_r = (5, 1, 2, 1, \ldots)$. A more advanced method is to further normalize x_r by the TF-IDF (term frequency-inverse document frequency) of each category of POIs.

Regions r → documents d
Functions k → topics k
Human mobility $M_{r,n}$ → words $W_{d,n}$
POIs x_r → meta data

a) Mapping

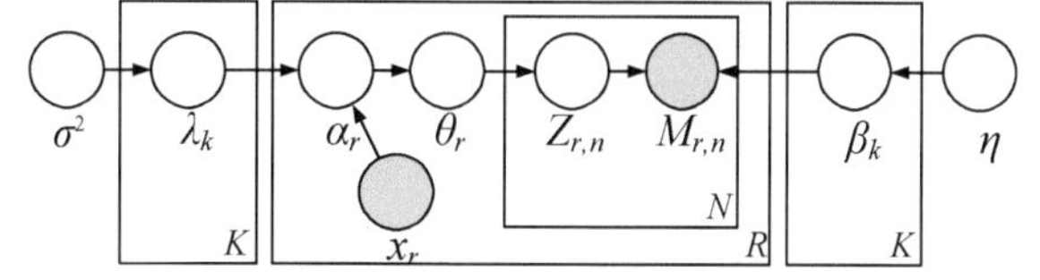

b) Graphical representation of the model

Figure 8.28
Learning functional regions based on an LDA-variant model.

The mobility pattern $M_{r,n}$ is defined as people's commuting patterns between regions. That is, when people depart from region r and where they leave for and when people arrive at region r and where they came from. Each commuting pattern stands for one word describing a region, while the frequency of the pattern denotes the number of occurrences of a word in a document.

The following is the generative process of this model:

1. For each function k,
   a. draw $\lambda_k \sim N(0, \sigma^2 I)$, and
   b. draw $\beta_k \sim Dirichlet\ (\eta)$.
2. Given region r,
   a. for each function k, let $\alpha_{r,k} = \exp(x_r^T \lambda_k)$;
   b. draw $\theta_r \sim Dirichlet\ (\alpha_r)$;
   c. for the n-th mobility pattern in region r, $M_{r,n}$;
      i. draw $Z_{r,n} \sim Multinomial\ (\theta_r)$; and
      ii. draw $M_{r,n} \sim Multinomial(\beta_{Z_{r,n}})$;

where $N(\bullet)$ denotes a Gaussian distribution with σ as a hyperparameter; λ_k is a vector with the same length as x_r; and η is the prior parameter for the topic-word distribution. This model can be estimated using an EM algorithm, in which the inference step is based on Gibbs sampling.

Detecting collective anomalies To determine if an instance is anomalous in a dataset, we usually need to measure how far the instance deviates from its underlying distribution. This calls for an estimation of the underlying distribution of a given dataset, which is very difficult when the dataset is sparse. For example, the occurrence of a specific disease in a region may only occur once per several days. If we concatenate the occurrences into a series with zero values denoting the absences (i.e., <0, 0, 0, 0, 1, 0, 0, 0, 0, 0, 2,...>), the mean and variance of the series are very close to zero. At this moment, if using a distance-based anomaly-detection method, every nonzero entry in the series will be regarded as an anomaly because its distance to the mean value (almost 0) is three times larger than the standard deviation (which is also close to 0).

To address this issue, the multiple-source latent-topic (MSLT) model [76] is proposed to combine multiple datasets to better estimate the distribution of a sparse dataset in a region. After the distribution of each dataset is accurately estimated, we can put it into a statistic model (e.g., log-likelihood ratio test) to detect potential anomalies in a city. For example, MSLT detects the collective anomalies in New York City, based on taxicab data, bike-sharing data, and 311 complaint data.

The MSLT model is motivated by two insights. First, different datasets in a region describe the region from different perspectives, thereby mutually reinforcing each other. For example, POIs and road network data describe the land use of a region, while taxi and bike flows indicate people's mobility patterns in the region. Thus, combining individual datasets results in a better understanding of a region's latent functions. Bridged by a region's latent functions, there is an underlying connection and influence among these datasets. For instance, the land use of a region would somehow determine the traffic flow in the region, while the traffic patterns of a region may indicate the land uses of the region. After working together to better describe a region's latent functions, different datasets in the region can mutually reinforce each other, thereby helping to better estimate their own distributions. Second, a dataset can reference across different regions. For instance, two regions (r_1, r_2) with a similar distribution of POIs and a similar structure of roads could have a similar traffic pattern. So, even if we cannot collect enough traffic data in r_1, we could estimate its distribution based on the traffic data from r_2.

Based on the insights, we can design a latent-topic model to fuse multiple datasets, as shown in figure 8.29a. Similar to the first example, in this model, a geographical region is regarded as a document; the latent functions of a region correspond to the latent topics of a document; and the POIs and road network data in a region are deemed the key words of a document. In addition, the Dirichlet prior α_r of a region also depends on its geographical properties, such as POIs and road networks, rather than an empirical setting. Different from the first example, the words of MSLT come from different datasets. The categories of different datasets are regarded as words. Thus, there are multiple topic-word distributions (β_1, β_2, ... β_s), where s is the number of datasets.

More specifically, the gray nodes in figure 8.29a are observations, and the white nodes are hidden variables. Similar to the first example, K is the number of topics; x_r is a vector storing the features extracted from the road networks and POIs located in a

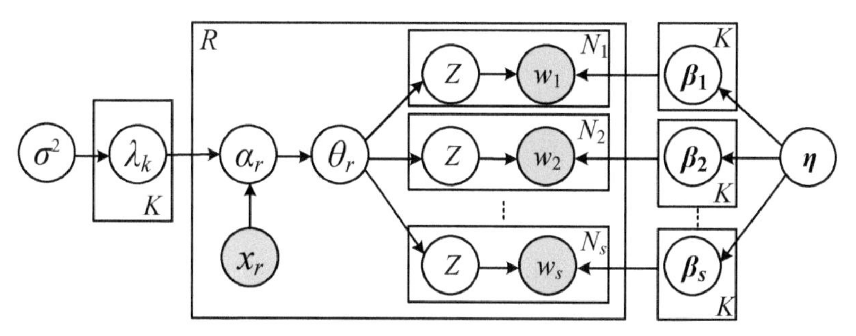

a) Graphic representation of MSLT

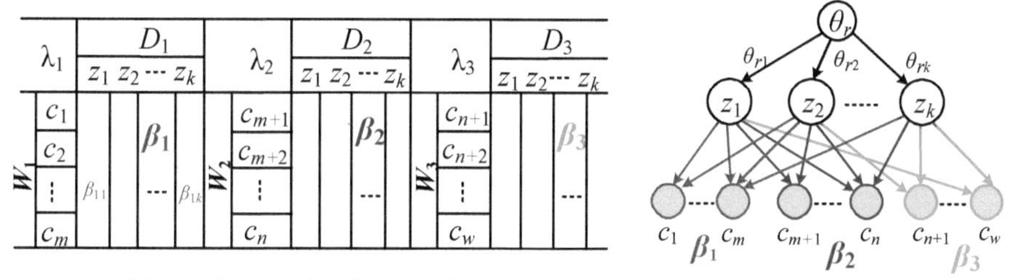

b) Topic-words distribution across different datasets

Figure 8.29
Graphical representation of the MSLT model.

region; λ_k is a vector with the same length as x_r, corresponding to the k-th latent topic; the value of each entry in λ_k follows a Gaussian distribution with a zero mean and a standard deviation σ; and $\alpha_r \in \mathbb{R}^K$ is a parameter of the Dirichlet prior to the per-region topic distributions. $\theta_r \in \mathbb{R}^K$ is the topic distribution for region r.

Different from the first example, $\mathcal{W} = \{W_1, W_2, \ldots, W_s\}$ is a collection of word sets, where W_i is a word set corresponding to the i-th dataset. β_i is a matrix denoting the topic-word distributions of the i-th dataset (i.e., W_i). A word w in W_i is one of the categories the i-th dataset has (e.g., $W_1 = \{c_1, c_2, \ldots, c_m\}$). The category can be the type of noise in a city or a type of mobility pattern defined by time and originate-destination. As illustrated in figure 8.29b, three different datasets share the same distribution of topics controlled by θ_r in region r but have their own topic-word distributions β_i, $1 \le i \le 3$, indicated by arrows with different colors. β_{iz} is a vector denoting the word distribution of topic z in word set W_i. The generative process of the MSLT model is as follows:

1. For each topic z, draw $\lambda_z \sim \mathcal{N}(0, \sigma^2 I)$.
2. For each word set W_i and each topic z, draw $\beta_{iz} \sim Dir(\eta)$.
3. For each region r (i.e., a document r):

a. For each topic k, let $\alpha_{r,k} = \exp(x_r^T \lambda_k)$.

b. Draw $\theta_r \sim Dirichlet(\alpha_r)$.

c. For each word w in document r:

i. Draw $z \sim Multinomial(\theta_r)$.

ii. Choose β_i of the corresponding word set that w belongs to.

iii. Draw $w \sim Multinomial(\beta_{iz})$.

While σ^2, η, and K are fixed parameters, we need to learn λ and β based on observed α and $\mathcal{W}$. We train the model with a stochastic EM algorithm, in which we alternate between the following two steps. One is sampling topic assignments from the current prior distribution conditioned on the observed words and features. The other is numerically optimizing the parameters β given the topic assignments.

We then leverage the latent topic distribution θ_r and the topic-word distribution β in a region to calculate the proportion of each word $prop(w)$ as follows (note that a category is denoted as a word in the MSLT model):

$$prop(w) = \sum_z \theta_{rz} \beta_{iz}(w),$$

where θ_{rz} is the distribution of topic z in region r, and $\beta_{iz}(w)$ denotes the distribution of topic z on word w; β_{iz} is a vector denoting the word distribution of topic z in the word set w belongs to. As the MSLT model learns θ_r and β based on multiple datasets, which mutually reinforce each other, it is more accurate than simply counting the number of occurrences of each category and then calculating its distribution.

8.6.5 Markov Networks for Spatiotemporal Data

8.6.5.1 Linear-chain conditional random field The linear-chain conditional random field (CRF) is a discriminative undirected probabilistic graphical model for parsing sequential data such as natural language text [33]. The advantage of CRFs over HMMs is the relaxation of the independence assumptions between features. Additionally, CRFs avoid the label bias problem exhibited by maximum-entropy Markov models. In the following paragraphs, we introduce an example using CRF to infer the air quality of a location.

Information about urban air quality (e.g., the concentration of PM2.5) is of great importance to protecting human health and controlling air pollution. Many cities monitor PM2.5 by building ground-based air quality measurement stations. However, there are only a limited number of air quality–measurement stations in a city due to the high cost of building and maintaining such a station. Unfortunately, air quality varies by location nonlinearly, depending on multiple complex factors like meteorology,

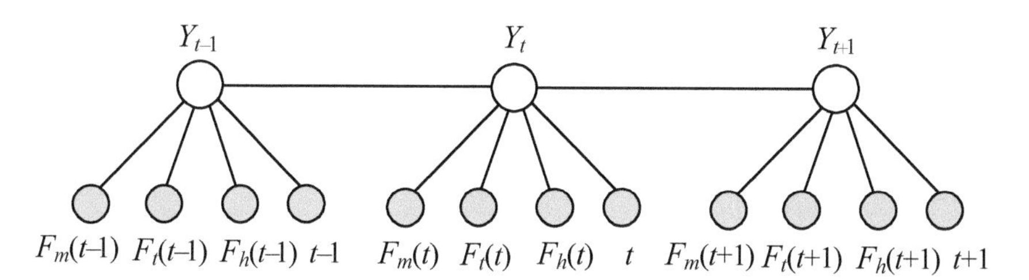

Figure 8.30
Graphical representation of the air quality inference model.

traffic volume, and land use. As a result, we do not really know the air quality of a location without a measurement station.

Figure 8.30 shows the graphical structure $\boldsymbol{G}$ of a model for inferring the air quality of a location based on a diversity of features, such as meteorological conditions F_m, traffic conditions F_t, human-mobility patterns F_h, and time of day and day of the week t, observed in the location [73]. The model consists of two kinds of nodes $\boldsymbol{G}=(\boldsymbol{X}, \boldsymbol{Y})$. The white nodes $\boldsymbol{Y}=\{Y_1, Y_2, \ldots, Y_n\}$ represent hidden state variables to be inferred, given the sequence of observations denoted by gray nodes $\boldsymbol{X}=\{X_1, X_2, \ldots, X_n\}$, $X_i=\{F_m, F_t, F_h, t\}$. The $Y_i \in \boldsymbol{Y}$ is structured to form a chain with an edge between each Y_{i-1} and Y_i, as well as having an air quality index (AQI) "label" belonging to $C=$ {*healthy, moderate, unhealthy for sensitive people, unhealthy, very unhealthy, hazardous*}.

For example, as illustrated in figure 8.30, we can form a sequence with three state nodes to represent the air quality states of the past three hours. The training data is collected from air quality monitor stations. Once the model has been trained, we can infer the states of air quality over the past three hours, in other locations without a station, based on their observations X_i of the past three hours. The output would be like *healthy → moderate → moderate*.

The training process is as follows. When conditioned on $\boldsymbol{X}$, the random variables Y_i obey the Markov property with respect to the graph $\boldsymbol{G}$:

$$P(Y_i|\boldsymbol{X}, Y_j, i \neq j)=P(Y_i|\boldsymbol{X}, Y_j, i \sim j), \tag{8.52}$$

where $i \sim j$ means i and j are neighbors in $\boldsymbol{G}$.

The probability of a particular label sequence $\boldsymbol{y}$ given observation sequence $\boldsymbol{x}$ is defined as a normalized product of potential functions as follows:

$$exp\left(\sum_j \lambda_j t_j(y_{i-1}, y_i, x, i) + \sum_k \mu_k s_k(y_i, x, i)\right), \tag{8.53}$$

where $t_j\,(y_{i-1}, y_i, y_i, x, i)$ is a transition feature function of the entire observation sequence and the label at positions i and $i-1$; $s_k\,(y_i, x, i)$ are a state feature function of the label at position i and the observation sequence; and λ_j and μ_k are parameters to be estimated from training data.

Writing $s_k(y_i, x, i) = s_k(y_{i-1}, y_i, x, i)$, we transfer equation (8.53) to

$$P(y|x, \lambda) = \frac{1}{Z(x)} exp\left(\sum_j \lambda_j f_j(y_{i-1}, y_i, x, i)\right), \tag{8.54}$$

where $Z(x)$ is a normalized factor. This can be informally thought of as measurements on the input sequence that partially determine the likelihood of each possible value for Y_i. The model assigns each feature a numerical weight and combines them to determine the probability of a certain value for Y_i.

Given k sequences of the training data $\{(\mathbf{x}^{(k)}, \mathbf{y}^{(k)})\}$, learning the parameters $\boldsymbol{\lambda}$ is done by maximum likelihood learning $P(\mathbf{y}|\mathbf{x}, \boldsymbol{\lambda})$, which can be solved by gradient descent.

$$L(\boldsymbol{\lambda}) = \sum_k \left[log \frac{1}{Z(x)} + \sum_j \lambda_j f_j(y_{i-1}, y_i, x, i) \right]. \tag{8.55}$$

8.6.5.2 Affinity graph for spatiotemporal data An affinity graph (AG) is a multilayer weighted connected graph $G = \langle \mathcal{G}^1, \mathcal{G}^2, \ldots, \mathcal{G}^n \rangle$, where each layer represents one time interval from $t_1, t_2, \ldots, t_n$; $\mathcal{G}^i = \langle V, E, W^{t_i} \rangle$ is the graph at time interval t_i; V is the set of all grids; E is the set of edges; and W^{t_i} represents edge weights at t_i. As depicted in figure 8.31, we use the AG to infer the AQI of a location. Different from the linear-chain CRF model introduced in section 8.6.5.2, with which the air quality of a location is inferred individually, the AG-based solution infers the air quality of all locations simultaneously.

Before inference, a city is partitioned into uniform and disjoint (e.g., 1 km $\times$ 1 km) grids. A node in the AG is a random variable denoting the state of the air quality in a grid. The node set $V = U \cup \mathcal{V}$ consists of a subset U of grids without air quality–monitoring stations (denoted by the white nodes) and a subset $\mathcal{V}$ of grids with monitoring stations (denoted by the gray nodes). We term nodes in $\mathcal{V}$ as labeled and nodes in U as unlabeled. Each unlabeled node $u \in U$ is associated with an AQI distribution $P(u)$ to be inferred. The AG is constructed as follows [26]:

1. *Connecting to station locations.* Every unobserved node $u \in U$ is connected to all observed nodes $v \in \mathcal{V}$ of the same time stamp, regardless of their geographical distance, denoted by the solid lines connecting a white node and a gray node. As the observed nodes are very sparse, adding those connections does not significantly affect efficiency.
2. *Connecting to nearby locations.* Since the AQI values of nearby locations are naturally highly correlated, within each layer of graph, every node is connected to the

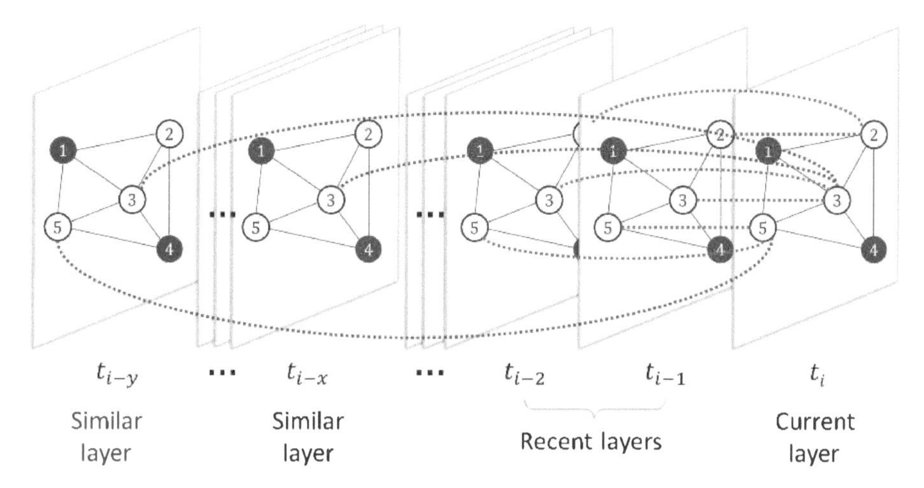

Figure 8.31
Inferring air quality with an affinity graph.

neighboring nodes $w \in U$ within a given geographical radius r, denoted by the solid lines connecting different white nodes.

3. *Connecting to recent layers.* Because the AQI value of a location is highly correlated to its historical AQI values, each node $u \in U$ of time stamp t_i is connected to the previous z time stamps: t_{i-1}, t_{i-2}, ... , and t_{i-z}, denoted by the blue dotted lines depicted in figure 8.31.
4. *Connecting to similar layers.* Since the environmental factors can repeat themselves within a certain period (e.g., every twenty-four hours), a node on the current layer is connected to the corresponding nodes of certain past layers with the most similar environmental features, denoted by the red dotted lines. The similarity between layers is computed based on the features.

Based on the AG and the affinity function learned from data, a graph-based semisupervised-learning approach is proposed, based on the following three insights: First, the observed AQI on labeled nodes $v \in \mathcal{V}$ is utilized to infer the AQI distributions $P(u)$ of unlabeled nodes $u \in U$. Second, nodes with similar features should have similar AQI distributions, hence having a bigger weight of edges. Third, since the AQI values for the unobserved locations are not available and the observed data are sparse, it is less practical to tune our model parameters to minimize the inference error. Instead, we propose tuning the parameters to minimize model uncertainty.

By putting these three ideas together, we seek an optimal set of edge weights W such that a) after inference, the unlabeled nodes shall possess similar AQI distributions with their close neighbors, and b) the learned label distribution should possess small entropy to minimize the uncertainty of inference.

To realize a), we propose to minimize the following loss function:

$$Q(P) = \sum_{(u,v)\in E} w_{u,v} \cdot (P(u) - P(v))^2, \tag{8.56}$$

where $P(u)$ is the probability (or distribution) of air quality across different categories in location u, $u \in U$. $P(v)$ is the distribution of air quality in location v, $v \in \mathcal{V}$. $w_{u,v}$ is the weight of the edge connecting node u and v in the AG. If $w_{u,v}$ is big (i.e., location u and v are very correlated), the difference between the air quality of the two locations should be small, hence $(P(u) - P(v))^2$ should be small. Multiple distance functions, such as Kullback-Leibler (KL) divergence, can be employed to measure $P(u) - P(v)$.

The goal is to find the AQI distributions for unlabeled nodes such that $Q(P)$ is minimized.

The harmonic property of the function P derives the solution to assign the AQI distribution of each unlabeled node using the weighted average of its neighboring nodes:

$$P(u) = \frac{1}{deg_u} \sum_{(u,v)\in E} w_{u,v} \cdot P(v), \tag{8.57}$$

where deg_u is the degree of node u; $w_{u,v}$ is defined as follows:

$$w_{u,v} = \exp\left(-\sum_k \pi_k^2 \times (a \cdot \|u.f_k - v.f_k\| + b)\right), \tag{8.58}$$

where f_k is a type of feature (e.g., the number of traffic lights in a location). $\|u.f_k - v.f_k|$ denotes the difference between u and v's feature f_k, π_k, a, and b are parameters to be learned. This corresponds to the second insight that nodes with similar features should have similar AQI distributions.

An intuitive approach is to adjust $\{\pi_k\}$ to maximize the likelihood of labeled nodes using validation data. Because the observed data is very sparse, however, this approach overfits the model to the validation data. Consequently, we learn π_k by minimizing the entropy of the inferred AQI distribution of the unlabeled nodes. Intuitively, the inference model becomes useless if the inferred distribution has high information entropy. For instance, if $P(u == good) = P(u == moderate) = \cdots = P(u == hazardous) = \frac{1}{6}$, we do not know which label should be chosen as the prediction result. At this moment, the information entropy is the highest (i.e., the most uncertain). On the contrary, if $P(u == good) = 1$, we certainly predict *good* as the final label. The entropy is 0 for this situation.

The average AQI distribution entropy $H(P_U)$ for unlabeled nodes U can be defined as

$$H(P_U) = -\sum_{u\in U} \left[P(u)\log(P(u)) + (1 - P(u))\log(1 - P(u))\right], \tag{8.59}$$

where $|U|$ is the number of unlabeled nodes in the AG.

Then, a mutually reinforced inference flow starts, where the learned feature weights π_k update the AQI distributions $P(u)$ of unlabeled nodes $u \in U$, and $P(u)$ determines the average AQI distribution entropy $H(P_U)$ to be minimized in the next iteration. Each change of feature weights π_k triggers an update of edge weights $w_{u,v}$ that further generates new AQI distribution $P(u)$ based on the AG and proceeds iteratively until convergence.

8.6.5.3 Gaussian Markov random fields A random vector $X=(x_1, x_2, \ldots, x_n)$ is called a *Gaussian Markov random field* (GMRF) with respect to the graph $G=\{V, E\}$ with mean μ and precision matrix Q if and only if its density has the form:

$$\pi(X) = (2\pi)^{-n/2} |Q|^{1/2} \exp\left(-\frac{1}{2}(X-\mu)^T Q(X-\mu)\right), \tag{8.60}$$

and $Q_{ij} \neq 0 \Leftrightarrow \{i, j\} \in E$,

where V and E are the node set and the edge set. The structure of G visually summarizes the conditional dependence among variables in X, and the value of matrix Q decides the special probability density of X.

Based on GMRFs, we can predict the flow of crowds streaming into a region and departing from a region [22]. GMRF is robust to noise and missing data and scalable to big data. For example, we can represent the inflow of a region over n time intervals as a time series $X=(x_1, x_2, \ldots, x_n)$, based on which we can estimate x_{n+1}. As the probability density X may not have the form shown in equation (8.60), before applying GMRF to this problem we need to calculate the first-order forward difference at time t as $\Delta x_t = x_{t+1} - x_t$, $t=1, 2, \ldots, n-1$. According to the empirical study on the real data, Δx_t follows a Gaussian distribution. Thus, the GMRF can be applied to this problem.

To make Q and G sparse, we impose the following assumption on X:

$$\Delta x_t \sim N(0, k^{-1}), \tag{8.61}$$

where $k \in R$ is the precision parameter to be learned from data. Based on the assumption and the structure of the graph shown in figure 8.32, we can construct the precision matrix Q and the density function as:

$$\pi(X \mid k) \propto k^{(n-1)/2} \exp\left(-\frac{1}{2} X^T Q X\right). \tag{8.62}$$

In the learning process, we find the parameter k using a maximum a posterior estimation. As the structure of the GMRF is very simple, it can be learned and inferred very efficiently.

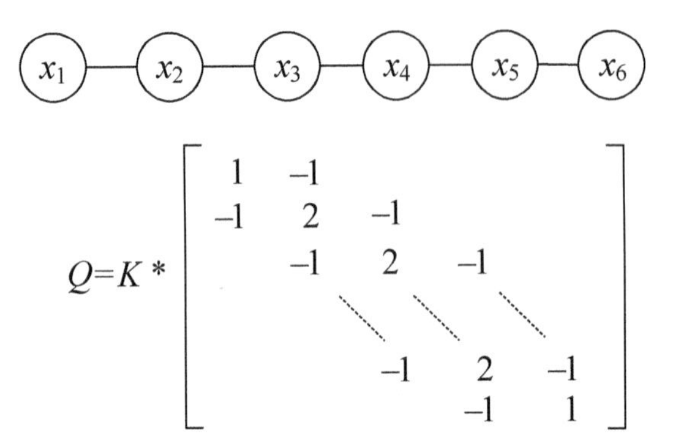

Figure 8.32
Using GMRF to predict a time series.

8.7 Deep Learning

Deep learning is the application to learning tasks of artificial neural networks (ANNs) that contain more than one hidden layer. As illustrated in figure 8.33, an ANN typically has an input layer, a hidden layer, and an output layer. Generally, we call an ANN with multiple hidden layers a deep neural network, as shown in figure 8.33b. Deep learning uses a cascade of many layers of nonlinear processing units for feature extraction and transformation. Each successive layer uses the output from the previous layer as input [12]. Though deep neural networks and deep learning have been used almost equivalently, the former focuses more on the structure of an ANN, while the latter emphasizes the learning algorithms. Deep learning provides two main functions:

1. *Learning representation.* Deep learning is a part of a broader family of machine-learning methods based on learning data representations. Higher-level features are derived from lower-level features to form a hierarchical representation. For example, as shown in figure 8.33b, the output of one of the hidden layers in the deep neural network can be regarded as a latent representation of the input. The latent representation can then be used as input in other classification and clustering models to complete a task. The learning algorithm can be supervised, unsupervised, or partially supervised. Supervised learning is almost the same as learning an ANN and is typically based on a backpropagation algorithm and many input-output training pairs. Unsupervised techniques like autoencoder learn the latent representation of a given input by setting the output of a (three-layer) neural network the same as its input. The value of the middle layer (i.e., the hidden layer) is regarded as the latent representation. The dimension of the latent representation is usually much lower

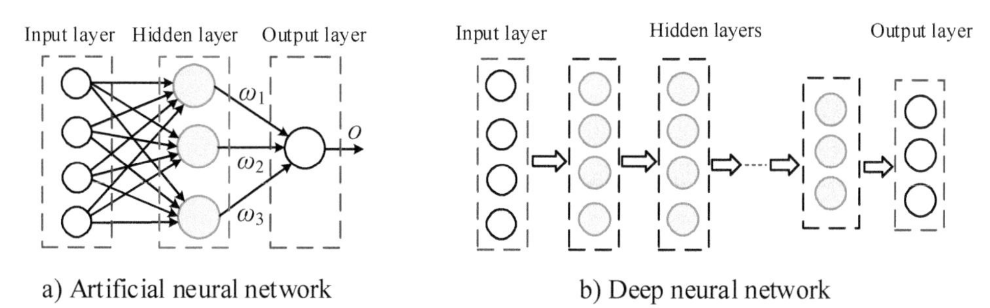

Figure 8.33
ANN and deep neural networks.

than the input, which helps tackle the curse of high dimensionality while improving computational efficiency.

2. *End-to-end prediction.* Besides generating a latent representation of the input, deep learning can also provide end-to-end predictions consisting of regression and classification tasks. For instance, we can use deep learning to predict the flow of crowds in each and every region throughout a city [71]. This is a regression problem, as the output is real valued numbers. We can also employ a deep neural network to classify images of different topics, such as dogs and cats. For such end-to-end applications, we need to provide a large number of training pairs, each of which consists of the input and the corresponding output. In such end-to-end prediction, the value of a hidden layer can also be regarded as a latent representation of the input.

As some of the most successful deep-learning methods involve ANNs, we start this section with an introduction to ANNs. After that, we introduce the two most successful deep-learning methods, consisting of convolutional neural networks (CNNs) and recurrent neural networks (RNNs). The long short-term memory network (LSTM) is also briefly introduced in the RNN subsection.

8.7.1 Artificial Neural Networks

ANNs were inspired by the 1959 biological model proposed by Nobel laureates Hubel and Wiesel, who found two types of cells in the primary visual cortex: simple cells and complex cells. Many ANNs can be viewed as cascading models of cell types inspired by these biological observations.

8.7.1.1 Concepts An ANN typically has an input layer, a hidden layer, and an output layer. The input layer (i.e., the leftmost layer shown in figure 8.33a) receives the input values (e.g., the features extracted from a given dataset). The number of nodes in the input layer usually corresponds to the number of inputs. The rightmost layer shown in figure 8.33a (i.e., the output layer) generates the final results. The number of nodes in the output layer depends on the task an ANN needs to complete. For example, if it is a regression problem aiming to predict a single value, the output layer will have only one node. If it is a binary classification problem, there usually are two nodes, each of which denotes one category. There are no clear rules for setting the number of nodes in the hidden layer.

In an ANN, each node receives a certain number of inputs through the edges that connect the node to that of the predecessor layer. Typically, ANNs employ a fully connected strategy for nodes between different layers. That is, a node connects to all the nodes on the previous layer. Nodes on the same layer are not connected. Each edge is associated with a weight defining the importance of the input from the predecessor node. Each node first calculates a weighted sum of the inputs and then transfers it into a value between (0,1) using an activating function, such as the sigmoid function. For example, the output of the ANN shown in figure 8.32a is

$$o = Sigmoid\ (\omega_1 o_1 + \omega_2 o_2 + \omega_3 o_3 + b), \quad (8.63)$$

where ω_1, ω_2, and ω_3 are weights of corresponding edges; o_1, o_2, and o_3 are the output of the three nodes on the hidden layer, respectively; and b is a bias constant. ω_1, ω_2, ω_3, and b are parameters we need to learn based on data. o_1, o_2, and o_3 can be calculated in the same way shown in equation (8.63), based on the output of the first layer. As the leftmost side is the input we can observe from data, we compute the final output given a neural network.

8.7.1.2 Learning The training process of ANNs is based on the backpropagation algorithm. The algorithm first randomly sets initial values for parameters of an ANN, calculating the output of a training instance based on its input. It then measures the error between the prediction $\hat{p}_i$ and the ground truth p_i of each instance, revising the parameters so as to minimize the sum of the square errors of all instances (a.k.a. MSE):

$$E = \sum_i (\hat{p}_i - p_i)^2, \quad (8.64)$$

where $\hat{p}_i$ is calculated based on the approach shown in equation (8.63). As there is no closed-form solution to find the optimal results, gradient descent is usually employed to find a local optimal result. When adjusting the value of a parameter (e.g., ω_1), the values of other parameters are fixed. The value of ω_1 is adjusted iteratively as follows:

$$\omega_1 \leftarrow \omega_1 - \gamma \cdot \frac{\partial E}{\partial \omega_1}, \quad (8.65)$$

where $\frac{\partial E}{\partial \omega_1}$ is the partial derivative of the error E with respect to ω_1; γ is a learning rate (specified by users) controlling the speed of descent. When the decrease of two consecutive rounds' MSEs is smaller than a predefined threshold (i.e., $|E_n - E_{n-1}| < \delta$), the iteration terminates, and the value of ω_1 is determined for this round. After that, we turn to tuning other parameters, such as ω_2, with ω_1 fixed. A round of tuning terminates after we have adjusted all the parameters. As the values of other parameters change, we need to start a new round of tuning, readjusting ω_1, ω_2, and so on, until the values of all parameters converge or do not change significantly over their values in the previous round.

8.7.1.3 Examples Figure 8.34 presents an example of where we use an ANN to predict the air quality of a location (without a station) based on the air quality readings from existing stations and the geographical data in a city [73]. As depicted in figure 8.34a, each blue point stands for an air quality–monitoring station generating air quality readings every hour.

Figure 8.34b presents the structure of the ANN-based inference model, which consists of two parts: input generation (in the left box) and an artificial neural network. The inference model can be regarded as a nonlinear interpolation algorithm that considers the differences between locations' geographical properties (rather than solely geodistances) when doing interpolation.

a) Air quality monitoring stations

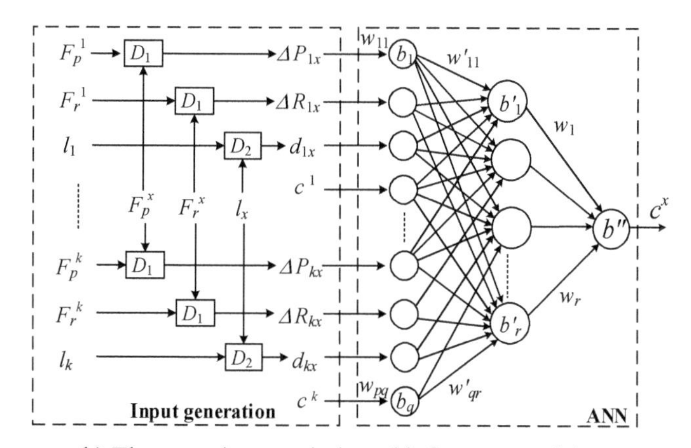

b) The neural network–based inference model

Figure 8.34
Inferring air quality based on ANNs.

Input generation In this phase, we randomly choose n stations to pair with the location to be inferred. For instance, as shown in figure 8.34a, we can infer the air quality of location x based on the readings from station S_1, S_2, and S_3 (we set $n=3$ in this example). Instead of using the original features of these stations as input, we calculate the differences between the features of a station and that of location x by equations (8.66–8.68), where F_p^k, F_r^k, l^k, and c^k denote the POI features, the road network features, the geographical location, and the AQI value of station k, respectively; D_1 is a distance function between features (e.g., the Pearson correlation), and D_2 calculates the geodistance between two locations.

$$\Delta P_{kx} = Pearson\_Cor\,(F_p^k, F_p^x), \tag{8.66}$$

$$\Delta R_{kx} = Pearson\_Cor\,(F_r^k, F_r^x), \tag{8.67}$$

$$d_{kx} = Geo\_Distance\,(l^k, l^x). \tag{8.68}$$

The pairwise differences help the following ANN learn the correlation of air quality between a pair of locations with different geodistances and geographical features.

Artificial neural networks As illustrated in figure 8.34b, the right part of the inference model is a three-layer neural network, formally defined as follows:

$$c^k = \varphi\left(\sum\nolimits_r w_r \varphi\left(\sum\nolimits_q w'_{qr} \cdot \left(\sum\nolimits_p f_p w_{pq} + b_q\right) + b'_n\right) + b''\right), \tag{8.69}$$

where f_p is a feature of input, and p is the number of features; $\varphi(x)$ is a sigmoid function; b_m, b'_n, and b'' are the biases associated with the neuron on different layers; and w_{pq}, w'_{qr}, and w_r denote the weight associated with the input of different layers.

The inference model is trained based on data from existing monitoring stations. For example, we can infer S_1's air quality based on that of S_2, S_3, and S_4. The input of the inference model can be calculated by equations (8.66–8.68). S_1's actual air quality readings are then used as a ground truth to measure the inference error. We can also infer S_1's air quality based on that of (S_4, S_5, S_6) or (S_2, S_5, S_7). Alternatively, we can infer the air quality of S_2 based on (S_4, S_5, S_6) or (S_1, S_3, S_6). By pairing a station with different other stations, we can formulate many input-label pairs as training data. We can find a set of parameters (i.e., b_m, b'_n, b'', w_{pq}, w'_{qr}, and w_r) that minimize the mean square error of all these inferences using the backpropagation algorithm mentioned above. Note that we cannot pair a station only with its three nearest neighbors, as we need to learn the impact (on air quality) caused by the difference between locations' geographical properties (e.g., geodistance and POI distribution). The model trained by such input-label pairs allows us to infer the air quality of an arbitrary location with different kinds of neighboring stations.

In the inference process, we also pair a location to be inferred with multiple sets of stations, each of which has three stations and generates a prediction of air quality. We can then average the predictions of the multiple sets, generating a final real-valued number. The model introduced in figure 8.34 can be easily extended to a classification task, where the prediction is a category of air quality rather than a real-valued number, by setting y nodes on the output layer. Each node corresponds to a category, and the node with the largest real value is selected as the final prediction result. We can then aggregate the predictions of multiple sets of stations, counting the frequency for each category in the prediction. The frequency can be regarded as the probability of a category in the prediction.

8.7.2 Convolutional Neural Networks

8.7.2.1 Basic structure of CNNs Convolutional neural networks (CNNs) originate from a problem that often occurs in the feed-forward neural networks whose layers are fully connected to each other. Such a full connection causes the curse of dimensionality (i.e., too many parameters to learn) when the scale of a neural network is very large. It also removes the spatial information of the inputs that may be useful for many applications.

Inspired by biological processes, in which individual cortical neurons respond to stimuli in a restricted region of space known as the *receptive field*, a CNN is composed of cascades of convolutional layers and sampling (a.k.a. pooling) layers, as illustrated in figure 8.35. The input of a CNN can be a two-dimensional matrix (e.g., an image with each pixel containing a gray level). Here, each entry of the matrix denotes a pixel of an image, and the value of the entry stands for the gray level of the pixel. When there are multiple values contained in an entry, we can turn a matrix into a tensor, where each entry further contains a vector. The number of values in the vector are called *channels*. For example, an image with an RGB representation has three channels at each pixel (denoting the level of red, green, and blue, respectively). Each channel has its own feature map inside each layer of a CNN. Finally, after several convolutional and max pooling layers, the high-level reasoning in the neural network is done via fully connected layers. Neurons in a fully connected layer have full connections to all activations in the previous layer, as seen in regular neural networks. To have a deep understanding of figure 8.35, we need to learn the following three main ideas of CNNs:

1. *Local connectivity*. In the feed-forward neural network, the input is fully connected to the next hidden node for every neuron, as illustrated in figure 8.36a. In contrast, the

input of CNN only makes the connection within a small region. Each neuron in a hidden layer is connected to a small field of the previous layer, which is called a *local receptive field*. Intuitively, objects in an image are usually represented by a group of nearby pixels, while pixels in different parts of an image may not be directly correlated. The idea of local connectivity significantly reduces the complexity of a neural network, helping tackle the curse of the dimensionality and capturing the spatial correlation between inputs.

2. *Share weights*. In the convolutional layer, the neurons are organized into multiple parallel hidden layers, each of which is called a *feature map*. Each neuron in a feature map is connected to a local receptive field on its previous layer. For every feature map, all neurons share the same weight parameter that is known as a filter or kernel in the feature map. Weight sharing further reduces the number of parameters, enabling CNNs to have multiple hidden layers and deal with a large number of inputs.

   Figure 8.36b presents an example of weight sharing, where each grid stands for a pixel in an image or a neuron inside a hidden layer of a CNN. For simplicity, we start with a two-dimensional example, where each grid only has one value. If the local receptive field has 3×3 grids (denoted by the gray area), a neuron of the first convolutional layer connects to nine pixels in an image. Likewise, a neuron on a hidden layer connects to nine neurons inside its previous hidden layer. The small figures on the bottom-right corner of each grid are parameters of the kernel, similar to the weight of an edge in a common neural network. After the convolutional operation, which is actually a dot product between the parameters of the kernel and the input, the values of the first nine grids are aggregated into a single value (i.e., 3). The kernel with the same parameter slides to another nine grids, calculating another value and so on, until all the grids inside the previous layer have been computed. Since the parameters of the kernel do not change during the sliding, we can say all the kernels share the same set of parameters in a feature map.

3. *Pooling*. Besides convolutional layers, a CNN sometimes also contains pooling layers, which are usually used immediately after a convolutional layer. It means the outputs of the convolutional layer are the inputs to the pooling layer of the network. The idea of a pooling layer is to generate translation invariant features by computing statistics of the convolution activations from a small receptive field in a feature map. The size of a small receptive field in here depends on the pooling size or kernel pooling. For example, as illustrated in figure 8.36c, if we select a 2×2 maximum pooling kernel and apply it to the output of the convolutional layer shown in figure 8.36b,

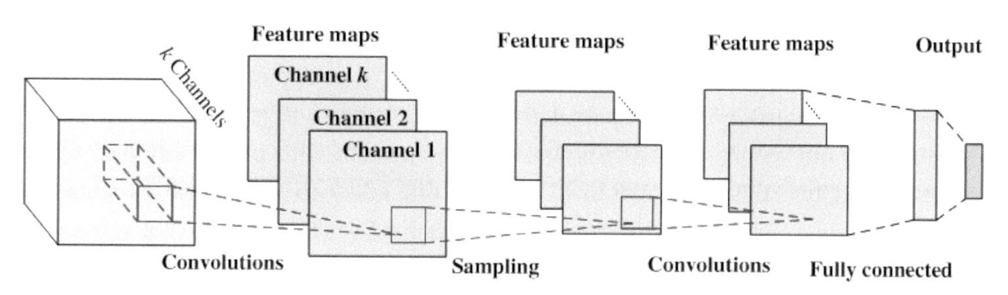

Figure 8.35
General structure of a CNN.

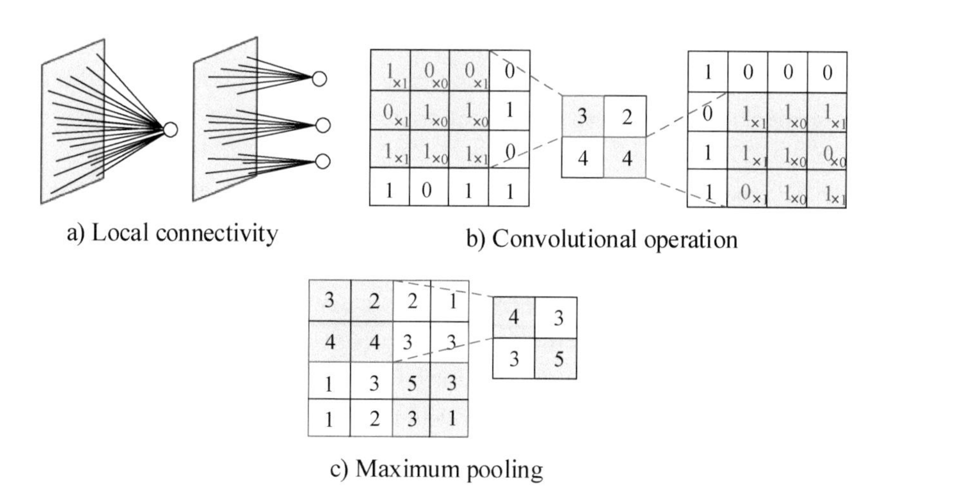

Figure 8.36
Three main ideas of CNNs.

the result of the pooling is 4, which is the maximum value of the four grids. Likewise, the results of other pooling are 3, 3, and 5, respectively. Note that there is no overlap between adjacent receptive fields when a pooling operation is conducted.

8.7.2.2 Empirical settings of CNNs There are some empirical settings for a CNN that do not have theoretical support but have practical contributions:

- *Zero padding.* Sometimes it is convenient to pad the input with zeros on the border of the input volume. Zero padding provides control of the output volume spatial size. In particular, sometimes it is desirable to exactly preserve the spatial size of the input volume—for example, to predict the flow of crowds in each and every grid of a city.

- *Dropout*. Because a fully connected layer occupies most of the parameters, it is prone to overfitting. One method to reduce overfitting is dropping out nodes or connections between nodes [55]. At each training stage, individual nodes are either "dropped out" of the net with probability $1-p$ or kept with probability p so that a reduced network is left; incoming and outgoing edges to a dropped-out node are also removed. Only the reduced network is trained on the data in that stage. The removed nodes are then reinserted into the network with their original weights.
- *ReLU* (rectified linear units). This is a layer of neurons that applies the nonsaturating activation function $f(x)=\max(0, x)$. It increases the nonlinear properties of the decision function and of the overall network without affecting the receptive fields of the convolution layer. Compared to other functions, such as $\tanh(x)$ and $sigmoid(x)$, the usage of ReLU results in the neural network training several times faster [32] without making a significant difference to generalization accuracy.

8.7.2.3 Training process of CNNs The training process of a CNN is almost the same as common ANNs. That is, it involves minimizing the mean square error of all instances through the backpropagation algorithm. However, when the number of hidden layers in a CNN increases, the parameters become hard to tune because of the vanishing gradient problem, which hampers convergence from the beginning. This problem has been largely addressed by normalized initialization and intermediate normalization layers, which enable networks with tens of layers to start converging for stochastic gradient descent with the backpropagation algorithm. As the network depth keeps increasing, however, the accuracy gets saturated and then degrades rapidly. Unexpectedly, such degradation is not caused by overfitting, and adding more layers to a suitably deep model leads to *higher training error* [21].

To address the degradation issue, a deep residual-learning framework is proposed to let layers fit a residual mapping rather than directly fitting a desired underlying mapping. Formally, instead of learning the desired underlying mapping $H(x)$, we let the stacked nonlinear layers fit another mapping of $F(x):=H(x)-x$, where x is the input. The original mapping is recast into $F(x)+x$. Proved by experiments, it is easier to optimize residual mapping $F(x)$ than to optimize the original, unreferenced mapping $H(x)$. Using the deep residual network, a 152-layer deep convolutional network has been trained for several image classification tasks, showing excellent performance beyond baseline methods.

8.7.2.4 Variants of CNNs In 1990, LeCun et al. [35] developed the first CNN model to classify handwritten digits. The increasing availability of labeled data and computing resources, such as graphic processing units (GPUs) and cloud-computing platforms, has fostered a diversity of CNNs. For example, Krizhevsky et al. proposed a classic CNN model (called AlexNet), which won a well-known image competition (ImageNet) in 2012 in an image classification task. Since then, a series of variants, such as ZFNet, VGGNet, GoogleNet, and ResNet, with a deeper and deeper network structure have been created.

The success of CNN in image classification has led to numerous variants of CNN architectures to address other computing vision tasks, such as object detection, object tracking, pose estimation, visual saliency detection, and action recognition. One of the most famous CNN-based object detectors is the region-based CNN (R-CNN) [16]. It first employs the selective search algorithm to extract around two thousand region proposals. These region proposals are then warped into a fixed size and fed into a pretrained CNN model for feature extraction. The feature of a region proposal is used to train multiple binary classifiers, each of which corresponds to a category of objects like cars and people. Once these binary classifiers have been trained, we can feed the feature of each region proposal into all these classifiers, determining if a specific type of object is contained in the region. After determining the category of an object, a linear regression model is employed to output a tighter coordinate for the bounding box of the object.

Although R-CNN shows a significant performance improvement, it is computationally expensive as there are many overlaps between region proposals that have been processed respectively by their own CNN module multiple times. To address this issue, spatial pyramid–pooling networks (SPP-nets) [20] are proposed to expedite R-CNN by sharing computation. That is, the overlapped areas between different region proposals are not necessarily processed multiple times. However, the training procedure of R-CNN and SPP-net is still a multistage pipeline. The CNN-based feature extractor is pretrained in other tasks based on other types of labeled data (e.g., using VGGNet). Thus, the labeled data (dedicatedly for object detection) cannot help tune the parameters of the CNN-based feature extractor.

To solve this problem, Fast R-CNN [15] improves SPP-net by using an end-to-end training procedure that can predict the probability distribution (of an object contained in a region) across different categories and generate the bounding box of the object simultaneously. Though the pretrained CNN module is also employed in the end-to-end procedure, the parameters of the entire network, including a region proposal's bounding box, can be updated during fine-tuning. Fast R-CNN shows compelling accuracy and speed while simplifying the training process. Based on Fast R-CNN, Faster R-CNN [48] introduces a region proposal network to expedite the generation of object proposals.

There is another thread of research that applies sequence-to-sequence learning to CNNs. Designed for text-to-speech (TTS) and speech synthesis tasks, WaveNet [45] is composed of a conditioning network, which upsamples linguistic features to the desired frequency, and an autoregressive network, which generates a probability distribution over discretized audio samples. To handle the long-range temporal dependencies needed for raw audio generation, WaveNet uses a stack of dilated causal convolutions, which exhibit very large receptive fields. Although WaveNet can generate close-to-human-level speech, its inference poses a daunting computational problem due to the high frequency caused by the autoregressive nature of the model. To deal with this problem, Arik et al. [1] propose a variant of WaveNet, called Deep Voice, which varies the number of layers, the number of residual channels, and the number of skip channels for TTS. Deep Voice requires fewer parameters while achieving up to four hundred times the speedups over WaveNet.

Gehring et al. [14] introduce an architecture based entirely on CNNs for sequence-to-sequence learning within the context of machine translation. Compared to recurrent neural networks (e.g., LSTMs), computations over all elements can be fully parallelized during training. The optimization is easier as the number of nonlinearities is fixed and independent of the input length.

8.7.3 Recurrent Neural Networks

8.7.3.1 General framework of recurrent neural networks A recurrent neural network (RNN) is a class of artificial neural network in which connections between units form a directed cycle. This creates an internal state of the network that allows it to exhibit dynamic temporal behavior. Unlike feed-forward neural networks, RNNs can use their internal memory to process arbitrary sequences of inputs. This makes them applicable to tasks such as unsegmented connected handwriting recognition or speech recognition [50]. RNNs are called recurrent because they perform the same task for every element of a sequence, with the output depending on the previous computations. Another way to think about RNNs is that they have a "memory" that captures information about what has been calculated so far. In principle, RNNs can make use of information in arbitrarily long sequences, but in practice they are limited to looking back only a few steps.

Figure 8.37a presents the general framework of an RNN, where x_t and o_t stand for the input and output at time step t, respectively; h_t is the hidden state at time step t; and W, V, and U are parameters of input, output, and transitions. More specifically, s_t is the "memory" of the network, which is calculated based on the previous hidden state and the input at the current step: $h_t = f(Uh_{t-1}, Wx_t)$. The function is usually a nonlinearity

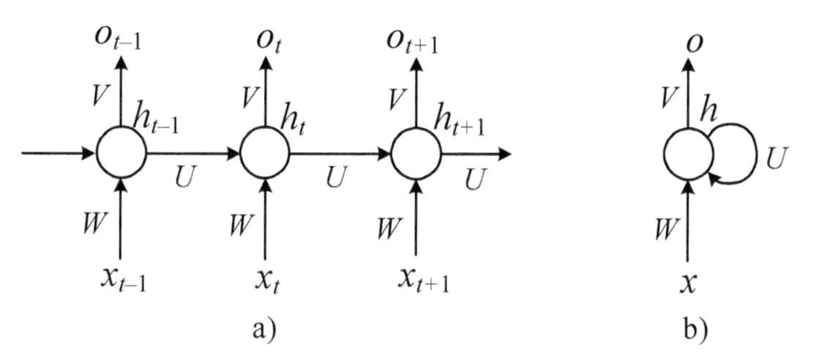

Figure 8.37
The general framework of RNNs.

such as tanh or ReLU. The first hidden state, denoted by h_0, is typically initialized to zeroes. Figure 8.37b shows an aggregated representation of RNNs.

Using natural language processing as an example, if we want to predict the next word of a given sentence, we can regard each word as a time step; x_t is the t-th word of the sentence (it could be represented by a one-hop vector or some word embedding representation); o_t could be a vector of probabilities across our vocabulary. Given a training sentence, where the next word is observed, we can set the probability of the word as 1 and the rest of the words in the vocabulary to 0 for o_t. After the training process, we can apply this RNN model to predict the next word in a partial sentence, selecting the word with the maximum probability as the final prediction.

Unlike a traditional deep neural network, which uses different parameters at each layer, an RNN shares the same parameters (i.e., U, V, and W) across all steps. This reflects the fact that we are performing the same task at each step, just with different inputs. This greatly reduces the total number of parameters we need to learn.

The above diagram has outputs at each time step, but this may not be necessary depending on the task. For example, when predicting the sentiment of a sentence, we may only care about the final output, rather than the sentiment after each word. Similarly, we may not need inputs at each time step. The main feature of an RNN is its hidden state, which captures the information about a sequence.

8.7.3.2 LSTMs The most commonly used type of RNNs are LSTMs [23], which are much better at capturing long-term dependencies than the basic RNNs. LSTMs are essentially the same thing as an RNN but with a different way of computing the hidden state. An LSTM unit is a recurrent network unit that excels at remembering values for either long or short durations of time. The key to this ability is that it uses no activation

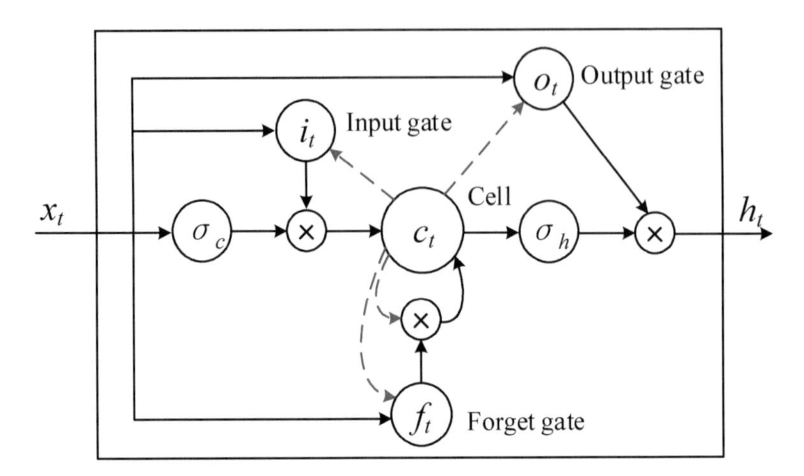

Figure 8.38
Structure of an LSTM block.

function within its recurrent components. Thus, the stored value is not iteratively squashed over time, and the gradient or blame term does not tend to vanish when the backpropagation algorithm (through time) is applied to train it.

There are many types of LSTMs. Figure 8.38 presents the structure of a peephole LSTM block, where x_t is the input vector; h_t is the output vector; c_t is the cell state vector; W, V, and b are parameter matrices and vectors; f_t is a forget gate vector denoting the weight of remembering old information; i_t is an input gate vector denoting the weight of acquiring new information; and o_t is an output gate vector denoting the output candidate. The exit arrows from the c_t node (represented by the blue broken arrows) actually denote the exit arrow from c_{t-1}, except for the single right-to-left arrow.

As compared to traditional LSTMs, c_{t-1}, instead of h_{t-1}, is used in most places. The vectors mentioned above are calculated as follows:

$$f_t = \sigma_g (W_f x_t + U_f c_{t-1} + b_f), \tag{8.70}$$

$$i_t = \sigma_g (W_i x_t + U_i c_{t-1} + b_i), \tag{8.71}$$

$$o_t = \sigma_g (W_o x_t + U_o c_{t-1} + b_o), \tag{8.72}$$

$$c_t = f_t \bullet c_{t-1} + i_t \bullet \sigma_c (W_c x_t + b_c), \tag{8.73}$$

$$h_t = o_t \bullet \sigma_h (c_t), \tag{8.74}$$

where σ_g is a sigmoid function; σ_c is a hyperbolic tangent; σ_h is suggested as $\sigma_h (x) = x$, which is originally a hyperbolic tangent in the traditional LSTM; the symbol $\bullet$ denotes the entrywise product; and the initial state $c_0 = 0$ and $h_0 = 0$.

To minimize LSTM's total error on a set of training sequences, iterative gradient-descent algorithms, such as backpropagation through time [42], can be used to change each weight in proportion to its derivative with respect to the error. A major problem with gradient descent for standard RNNs is that error gradients vanish exponentially quickly with the size of the time lag between important events. With LSTM blocks, however, when error values are backpropagated from the output, the error becomes trapped in the memory portion of the block. This is referred to as an *error carousel*, which continuously feeds errors back to each of the gates until they become trained to cut off the value. Thus, regular backpropagation is effective in training an LSTM block to remember values for very long durations.

8.7.4 Deep Learning for Spatiotemporal Data

8.7.4.1 Challenges Deep learning was originally proposed for processing images, videos, and speeches and later applied to natural-language processing. When using deep learning to deal with spatiotemporal data, we are confronted with the following three challenges:

1. *Data transformation*. The input of a CNN is usually a matrix or tensor. The input of an LSTM could be a sequence of vectors. However, spatiotemporal data has different formats, which could be a collection of points with nonuniform locations, or a trajectory of arbitrary shape and length, or a graph with very sparse connections and dynamic nodes. Transforming spatiotemporal data into a format that can be consumed by a deep-learning model is a nontrivial task.
2. *Encoding spatiotemporal properties*. As introduced in section 8.2, spatiotemporal data has its unique properties, such as the spatial distance, spatial hierarchy, and temporal period and trend, which do not clearly exist in images and text data. Thus, we cannot simply grab existing deep-learning algorithms that do not consider these factors when solving urban-computing problems. How to simultaneously capture these unique spatiotemporal properties in a deep-learning model remains a challenge.
3. *Fusing knowledge across different domains*. In an urban-computing application, we usually need to harness a diversity of datasets. How to fuse the knowledge of multiple datasets across different domains through a deep-learning framework remains a challenge.

8.7.4.2 Predicting the flow of crowds based on deep learning We use an example to demonstrate how the aforementioned three challenges can be tackled by a sophisticated design of deep-learning models for spatiotemporal data [71]. In this example, we aim to predict the inflow and outflow of crowds in each and every region throughout a city. However, the flow of crowds in a region depends on many complex factors, such as the flow of crowds in the region during the past few hours (i.e., the temporal correlation), the flow of crowds in its neighboring regions and that of distant regions (i.e., the spatial correlation), and external factors like weather conditions and events.

Data transformation Figure 8.39 presents the data transformation process, where the input is the GPS trajectories of a large group of people, and the output is a three-dimensional tensor. Each trajectory is a sequence of GPS coordinates with corresponding time stamps, denoting the movement of an individual. We divide a city into uniform grids, projecting people's trajectories onto these grids. By counting the number of trajectories entering and departing from a grid, we can derive the volume of inflow and outflow at a given time interval in each grid. Then, we can turn the trajectory data received at each time interval (e.g., every hour) into a three-dimensional tensor, where the first two dimensions are grids on the geographical space, and the third dimension stores the values of the two flows. For simplicity, figure 8.39 only visualizes the inflow of each grid (i.e., the tensor downgrades to a matrix). The lighter a grid's color, the bigger its volume of traffic. By having the trajectories generated over a period of time, we can generate a series of matrices. In addition, there are other external factors, such as the weather conditions and events, at each time interval. These form the input of the deep-learning model. The prediction is what the next few frames look like given the past frames.

Figure 8.40a presents the framework of the deep-learning model, called ST-ResNet (spatiotemporal residual networks), for solving this problem. It comprises three deep residual CNNs, one early fusion component, and one fully connected neural network. It is an end-to-end prediction framework that predicts each and every region's flows collectively.

Encoding temporal properties The three deep residual CNNs share the same network structure and model temporal closeness, period, and trend properties, respectively. More specifically, the rightmost deep residual CNN takes frames (of flows) at recent hours as an input to model the temporal closeness of crowd flows (i.e., the flows at adjacent time intervals would change smoothly over time). The middle deep residual CNN takes the frames of the same time from yesterday, the day before yesterday, and even a few days before yesterday as an input to model the period information. Intuitively, flows of crowds almost repeat every day. The leftmost deep residual CNN takes the frames from the same time in the last week, the week before last, or even the last month as an input to model trend information. The rest of the frames can be skipped

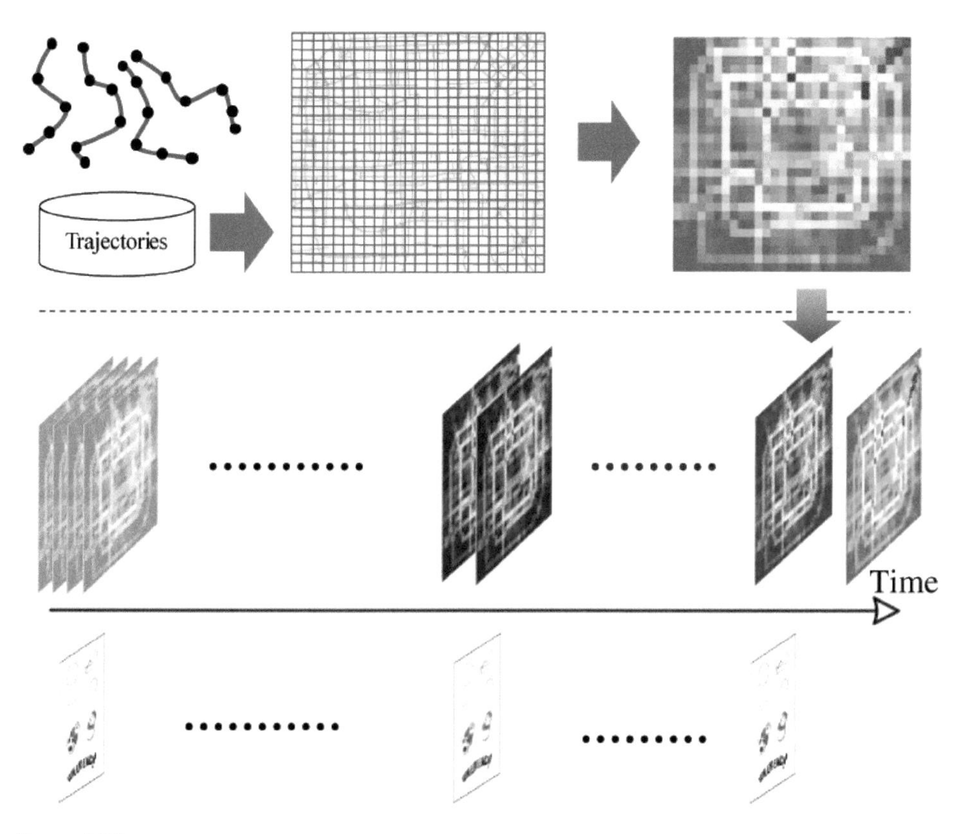

Figure 8.39
Spatiotemporal data transformation for deep learning.

during the learning and prediction processes. This enables a deep-learning model to capture all three temporal properties from the data over a long period of time while not increasing the complexity of network structures and training effort.

Encoding spatial properties In each deep residual CNN, as shown in figure 8.40b, we use a CNN with multiple convolutional layers to capture the spatial correlation of both near and far distances. After one or two rounds of convolutional operation, we aggregate the information of nearby regions into one value. This looks like capturing the spatial correlation between near locations. If we keep on doing the convolutional operation, we can aggregate the information of grids with a far distance, thus capturing the correlation of a farther distance. However, as the number of hidden layers increases, the training of a CNN becomes difficult (see section 8.7.2.3 for reasons). Then, we employ

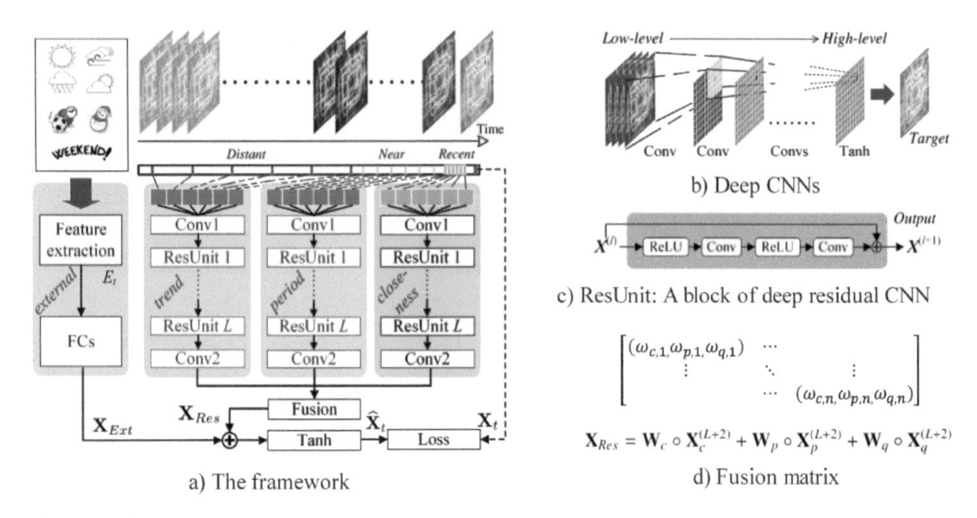

Figure 8.40
Structure of ST-ResNet.

a deep residual structure for the CNN model. A deep residual CNN in the framework is actually a cascade of a few ResUnits as shown in figure 8.40c. To ensure the size of the prediction is the same as the input, zero padding is employed.

Fusion There are two levels of fusion in the framework. First, the model fuses the predictions generated by the three deep residual CNNs using a fusion matrix, as shown in figure 8.40d, where each entry consists of three values corresponding to the weights of the three CNNs in a specific region. Apparently, the three temporal properties have different presences in different regions. Some regions, such as a university, may show a clear period over flows as students need to attend classes regularly regardless of weather conditions, while the trend property may not be obvious in these regions. The value of the matrix is learned with other parameters of neural networks in the end-to-end prediction framework. The second level of fusion aggregates X_{Res} and X_{Ext} through an entrywise summation. X_{Ext} is the output of the fully connected layers, which transforms external factors (e.g., weather conditions) to a vector. The vector is then reshaped to the same matrix representation as X_{Res} before the summation.

ST-ResNet outperforms LSTMs and CNNs on many datasets, such as taxi and bike-sharing data in New York City and Beijing. If we want to capture the trend information of crowd flows using an LSTM, the input of the data must be for a long-enough time (e.g., over three months); otherwise, the trend is not embedded in the data. If setting one hour per frame, the length of an LSTM model is over two thousand. It is impractical to train such a long LSTM model, and the memory cannot keep information from such a long time ago. Moreover, LSTMs do not capture the spatial correlation between locations. Thus, they are less accurate than ST-ResNet in predicting spatiotemporal data.

8.8 Reinforcement Learning

In the real world, people usually need to make a sequence of decisions to receive the maximum reward in a task. For example, when playing a video game like Atari Breakout, a user needs to move a stick (at the bottom of the screen) to the left or right so it can rebound a ball to break the bricks (at the top of the screen) as soon as possible. An optimal sequence of movement decisions leads to the shortest time to break all the bricks. In another example, to reach a destination from an origin, a driver can try different paths by deciding to turn left or right or move forward at different road intersections. There is an optimal decision sequence resulting in the quickest path to the destination. Likewise, to enhance a bike-sharing system's operation efficiency and bike usage, we need to continuously (e.g., every hour) move excessive bikes from overloaded locations to the places that lack bikes. An optimal sequence of reallocation decisions can lead to maximum bike usage.

Reinforcement learning is a branch of machine learning dealing with sequential decision-making problems [29, 57]. It was inspired by behaviorist psychology concerned with how software agents ought to take actions in an environment in order to maximize some notion of cumulative reward. This section introduces the general framework of reinforcement learning and a diversity of methods for finding optimal or approximately optimal actions.

8.8.1 Concepts of Reinforcement Learning

8.8.1.1 The general framework of reinforcement learning Figure 8.41a presents the basic framework of reinforcement learning, in which an agent interacts with the environment to achieve the maximum reward over a period of time. More specifically, the agent takes a decision a_t based on the current state s_t of the environment. After that, the agent receives an instant scalar reward r_t, and the state of the environment evolves to s_{t+1}.

In a reinforcement-learning task, the agent makes a sequence of decisions, receiving rewards and also affecting the next state of the environment, using the following process:

$$s_0 \xrightarrow{a_0,\, r_0} s_1 \xrightarrow{a_1,\, r_1} \cdots \longrightarrow s_t \xrightarrow{a_t,\, r_t} s_{t+1} \longrightarrow \cdots. \tag{8.75}$$

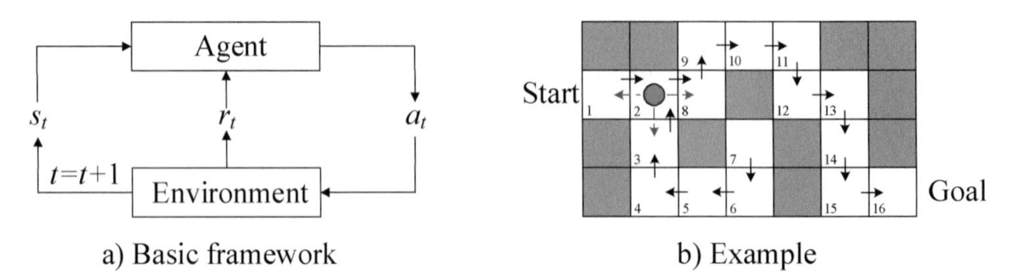

Figure 8.41
Concepts of reinforcement learning.

The goal of reinforcement learning is to learn an optimal policy $\pi^*(a_t|s_t)$ for the agent, such that for any state $s_t \in S$, the agent can receive the maximum long-term reward after state s_t by policy π^*. $\pi^*(a_t \mid s_t) \in [0, 1]$ is the probability of the agent taking action $a_t \in A$ for state s_t, where S and A denote the state space and action space, respectively;

$$\sum_{a_t \in A} \pi^*(a_t \mid s_t) = 1. \tag{8.76}$$

The long-term reward after state s_t can be defined by the discounted reward $\sum_{k=0}^{\infty} \gamma^k r_{t+k}$ or the T-step reward $\sum_{k=0}^{T} \gamma^T r_{t+k}$, where $\gamma \in [0, 1]$ is a discount ratio. The optimal policy π^* is a deterministic policy if there is only one optimal action a_t for a state s_t, and we can denote it as $a_t = \pi^*(s_t)$. Below, we analyze reinforcement-learning tasks with the discounted reward. Similar results can be easily extended for the T-step reward.

We illustrate the idea of reinforcement learning using a maze game shown in figure 8.41b. An agent is denoted by the circle, which starts from the upper-left grid and tries to find the quickest way out of the maze (i.e., reaching the "goal" grid). The environment is the maze, and the state can be defined as the location of the agent (i.e., the grid index). For example, the current state of the agent is $s = 2$. The agent can decide to go up, down, left, or right, which are its possible actions. Currently, the agent is not allowed to go up, as there is no way up. The agent receives a reward of –1 each time it takes an action. With reinforcement-learning methods, given any state s (its location), the agent will find an optimal sequence of actions (i.e., a moving path denoted by the green arrows), such that the agent can leave the maze the most quickly. As each move is associated with a reward of –1, the quickest path achieves the maximum long-term reward.

8.8.1.2 Markov decision process If the environment of a reinforcement-learning task has the Markov property, we call the reinforcement-learning task a Markov decision process (MDP) [5, 6, 64]. In an environment with the Markov property, the next state s_{t+1} of the environment only depends on the current state s_t and the current action a_t. That is, the probability of the next state $s_{t+1} \in S$

$$p(s_{t+1} \mid s_0, a_0, s_1, a_1, \ldots, s_t, a_t) = p(s_{t+1} \mid s_t, a_t). \tag{8.77}$$

The instant reward r_t is the result of s_t, a_t, and s_{t+1} and can be denoted by $r_t = r(s_t, a_t, s_{t+1})$. Many real-world reinforcement-learning tasks can be formulated as MDPs, and most reinforcement-learning methods are proposed for the MDPs. For example, the maze game shown in figure 8.41b is an MDP. Thus, we focus on MDPs in the following sections.

Before presenting the methods for finding the optimal policy $\pi^*(a|s)$, we first need to learn two essential concepts in reinforcement learning: the state value function and the action-value function. Given an arbitrary policy $\pi(a|s)$, we can define its state value function $v_\pi(s)$ and action-value function $q_\pi(s, a)$. The value function $v_\pi(s)$ is the *expected* long-term total reward that the agent can receive after state s by following the policy $\pi(a|s)$. In math, it is

$$v_\pi(s) = \mathbb{E}_\pi\left[\sum_{k=0}^{\infty} \gamma^k r_{t+k} \mid s_t = s\right]. \tag{8.78}$$

Similarly, the action-value function $q_\pi(s, a)$ is defined as the *expected* long-term total reward, starting from state s and taking action a, following the policy $\pi(a \mid s)$. It is

$$q_\pi(s, a) = \mathbb{E}_\pi\left[\sum_{k=0}^{\infty} \gamma^k r_{t+k} \mid s_t = s, a_t = a\right]. \tag{8.79}$$

The optimal policy π^* achieves the maximum state value for each state $s \in S$, beyond other policies:

$$v_{\pi^*}(s) \geq v_\pi(s), \forall s \in S, \forall \pi \neq \pi^*. \tag{8.80}$$

In general, based on the size of the state space S and action space A, methods of learning the optimal policy $\pi^*(a|s)$ can be divided into two classes consisting of the tabular action-value method and the approximate method. When the state space S and the action space A are small, we can calculate each state's value $v_\pi(s)$ and each action's value $q_\pi(s, a)$ and keep them in a table. When the space of S or A is large or even infinite, however, such methods become impractical. In this case, we can use some observed states' values $v_\pi(s)$ to learn an approximation function $v_\pi(s; \theta)$ with parameters θ. We then approximate any unobserved state using the approximation function $v_\pi(s; \theta)$. Likewise, for the action-value function $q_\pi(s, a)$, we can approximate it using a function $q_\pi(s, a; \omega)$ with parameter ω, estimating other (unobserved) actions' values using $q_\pi(s, a; \omega)$. The tabular action-value method is able to obtain the optimal result, while the approximate method achieves approximate results. As the state space and the action space can be extremely large in many real problems, approximate methods are a more practical choice than tabular action-value methods.

8.8.2 Tabular Action-Value Methods

In this category of approaches, there are three fundamental methods: dynamic-programming methods [4, 24, 64], Monte Carlo methods [3, 54], and temporal-difference methods [49, 56, 62]. Dynamic-programming methods require modeling the environment exactly—both the state transition probability $p(s_{t+1}|s_t, a_t)$ and the instant reward $r_t = r(s_t, a_t, s_{t+1})$ are completely given (by equations). In the real world, the environment could be too complex to be modeled exactly. Thus, Monte Carlo methods and temporal-difference methods have been proposed using sampling methods to find the optimal policy iteratively. Table 8.1 presents a comparison among the three methods, which will be detailed in later sections.

8.8.2.1 Dynamic-programming methods We introduce two dynamic-programming methods: the policy iteration method and the value iteration method. Before presenting the two methods, we first introduce the *Bellman equations* [5, 6], which are the foundation of the dynamic-programming methods. The Bellman equations require knowledge of the state transition probability $p(s'|s, a)$ and the reward $r(s, a, s')$.

As shown in figure 8.39a, following the policy π, for the current state s, we have a probability $\pi(a|s)$ to select action a, and then the state of the environment has a probability $p(s'|s, a)$ to become s'. Thus, according to the definition of the state value function in equation (8.78), the relation between the current state's value $v_\pi(s)$ and the next state's value $v_\pi(s')$ is

$$v_\pi(s) = \sum_{a\in A'} \pi(a|s) \sum_{s'\in S'} p(s'|s,a)\left[r(s,a,s') + \gamma v_\pi(s')\right], \tag{8.81}$$

where A' is the collection of possible actions given s; S' is the collection of possible states given A'. For instance, as illustrated in figure 8.42a, there are three possible actions (i.e., $|A'| = 3$) to take given s. After taking these actions, we could reach six different states S'. A' may be a subset of A because of the constraints posed by s. For instance, in the maze game shown in figure 8.42b, we can only move up or down in grid three. Likewise, S' could be a subset of S.

As depicted in figure 8.42b, we can obtain the relation between the current action value $q_\pi(s, a)$ and the next state's value $v_\pi(s')$ as

$$q_\pi(s,a) = \sum_{s'} p(s'|s,a)\left[r(s,a,s') + \gamma v_\pi(s')\right]. \tag{8.82}$$

Table 8.1
Comparison among different tabular action-value methods.

Methods		The policy evaluation step			The policy improvement step
		Initialization	Rules/sampling	Iterations	
Dynamic programing	Policy iteration	Random $v_{\pi_0}^0(s)$	Predefined functions	Multiple steps	Deterministic
	Value iteration	$v_{\pi_0}^1 \rightarrow v_{\pi_1}^0$	Predefined functions	One-step	Deterministic
Monte Carlo methods	Basic methods	Random $q(s, a)$; random a_0	Sampling	Update $q(s, a)$ after one episode	Deterministic
	On-policy methods	Random $q(s, a)$; random a_0	Sampling	Update $q(s, a)$ after one episode	ε-greedy strategy
Temporal-difference methods	Sarsa algorithm	Random $q(s, a)$; random a_0	Sampling	Update $q(s, a)$ after one step in an episode	ε-greedy strategy; $(r_0 + \gamma q\ (s_1, a_1))$
	Q-learning algorithm	Random $q\ (s, a)$; choose a_0 based on the ε-greedy strategy	Sampling	Update $q(s, a)$ after one step in an episode	ε-greedy strategy; $(r_0 + \gamma \max_a q(s_1, a))$

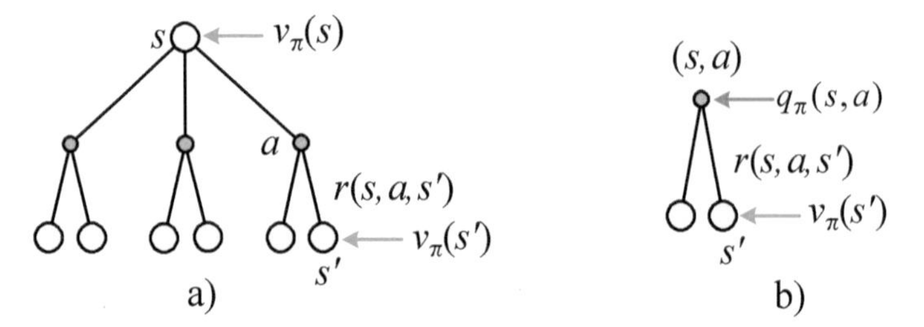

Figure 8.42
The relation between state-action values.

Based on the two Bellman equations, we can derive

$$v_\pi(s) = \sum_a \pi(a \mid s) q_\pi(s, a). \tag{8.83}$$

When $\pi(a|s)$ is a deterministic policy, $v_\pi(s) = q_\pi(s, \pi(s))$.

The policy iteration method This method is composed of two steps: *policy evaluation* (*E*) and *policy improvement* (*I*). Given an arbitrary policy π_0, we obtain its state value function v_{π_0} using the policy evaluation step. Based on v_{π_0}, we improve policy π_0 to policy π_1 through the policy improvement step. We iteratively go through these two steps as follows:

$$\pi_0 \xrightarrow{E} v_{\pi_0} \xrightarrow{I} \pi_1 \xrightarrow{E} v_{\pi_1} \xrightarrow{I} \pi_2 \xrightarrow{E} \cdots \xrightarrow{I} \pi^* \xrightarrow{E} v_{\pi^*}, \tag{8.84}$$

until the policy cannot be improved anymore; we then have the optimal policy π^*.

In the policy evaluation step, for any policy (taking the initial policy π_0, for example), we try to obtain the state value function $v_{\pi_0}(s)$, following policy π_0. This can be done by a secondary iteration inside the evaluation step. Initially, we randomly set the value function as $v_{\pi_0}^0(s)$. Then, according to the Bellman equation (8.81), we update the state value function using

$$v_{\pi_0}^{k+1}(s) = \sum_{a \in A'} \pi_0(a \mid s) \sum_{s' \in S'} p(s' \mid s, a) \left[r(s, a, s') + \gamma v_{\pi_0}^k(s') \right], \tag{8.85}$$

where k is the number of (secondary) iterations, starting from 0. That is, in the $(k+1)$-th iteration, the expected state value starting from s can be computed by the current reward $r(s, a, s')$ plus the (discounted by γ) expected state value starting from the next state s' estimated in the k-th iteration. $v_{\pi_0}^{k+1}(s)$ will gradually converge to the real state value function $v_{\pi_0}(s)$ as k increases.

In the policy improvement step, after obtaining the state value $v_{\pi_0}(s)$ for the initial policy π_0, we try to improve the policy π_0 and obtain a better policy π_1. With $v_{\pi_0}(s)$, we can calculate $q_{\pi_0}(s, a)$ for every action $a \in A'$ according to the Bellman equation (8.82):

$$q_{\pi_0}(s, a) = \sum_{s' \in S'} p(s' \mid s, a)\left[r(s, a, s') + \gamma v_{\pi_0}(s')\right]. \tag{8.86}$$

We then select from A' the action a with the maximum value $q_{\pi_0}(s, a)$; that is,

$$\pi_1(s) = \arg\max_a q_{\pi_0}(s, a), \tag{8.87}$$

to replace the policy π_0. Policy π_1 can be proved to be better than policy π_0—that is, $v_{\pi_1}(s) \geq v_{\pi_0}(s), \forall s \in S$.

Value iteration In the policy evaluation step of the policy iteration method, to obtain the state value function (e.g., $v_{\pi_0}(s)$), we need to conduct many rounds of secondary iterations. If the policy (e.g., π_0) is not a good initial policy, however, many rounds of iterations could be unnecessary and waste time. Thus, the value iteration method is proposed with the following two improvements in the policy evaluation step.

First, given the initial policy π_0 and the initial state value function $v_{\pi_0}^0(s)$, with *only one* round of secondary iteration, denoted by the operation in the bracket shown in figure 8.43, we have

$$v_{\pi_0}^1(s) = \sum_{a \in A'} \pi_0(a \mid s) \sum_{s' \in S'} p(s' \mid s, a)\,[r(s, a, s') + \gamma v_{\pi_0}^0(s')].$$

That is, we stop right after this iteration. However, in the policy iteration method, we need to conduct, as follows, $\pi_0 \to v_{\pi_0}^0 \to v_{\pi_0}^1 \to \cdots \to v_{\pi_0}^k \to \pi_1$.

Second, we use the value of $v_{\pi_0}^1(s)$ to replace $v_{\pi_1}^0(s)$ in the next round of iteration, denoted by the broken arrows shown in figure 8.43. That is, we do not randomly select an initial value of $v_{\pi_1}(s)$ again from the second round of iteration.

The policy improvement step is the same as the previous method. In this way, we obtain the optimal policy π^*.

8.8.2.2 Monte Carlo methods Dynamic-programming methods require knowing the environment of a reinforcement-learning task, with the state transition probability $p(s'|s, a)$ and the reward $r(s, a, s')$ available, which could be too complex to obtain in the real world. Thus, the Monte Carlo methods and temporal-difference methods, which do not require an exact model for the environment, have been proposed.

The basic Monte Carlo method Like dynamic-programming methods, the Monte Carlo methods also use an iteration method to obtain the optimal policy π^*, as follows:

$$\pi_0 \xrightarrow{E} q_{\pi_0} \xrightarrow{I} \pi_1 \xrightarrow{E} q_{\pi_1} \xrightarrow{I} \pi_2 \xrightarrow{E} \cdots \xrightarrow{I} \pi_i \xrightarrow{E} \cdots \xrightarrow{I} \pi^* \xrightarrow{E} q_{\pi^*}, \tag{8.88}$$

where E stands for the policy evaluation step, and I denotes the policy improvement step. However, they differ from the dynamic-programing methods in two ways:

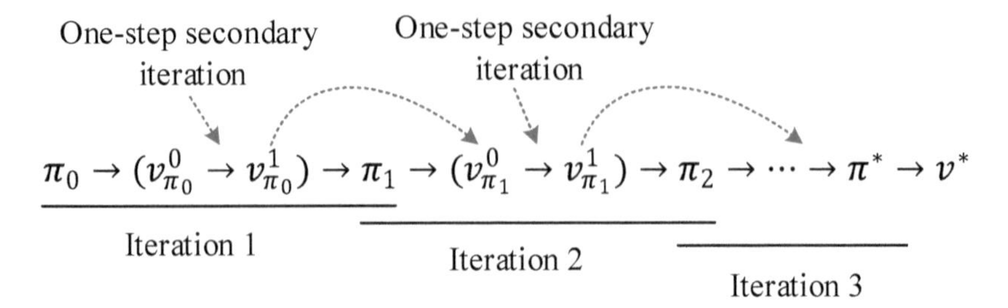

Figure 8.43
An example of the value iteration method.

First, in the policy evaluation step, the Monte Carlo methods evaluate the action-value function $q_{\pi_i}(s,a)$ rather than the state value function $v_{\pi_i}(s)$. As there is no predefined relationship between $v_{\pi_i}(s)$ and $q_{\pi_i}(s,a)$, like equation (8.86) in a complex environment, we have to compute $q_{\pi_i}(s,a)$ directly, without which we cannot derive new deterministic policy by $\pi_{i+1}(s) = \arg\max_a q_{\pi_i}(s,a)$. That is, $v_{\pi_i}(s)$ cannot derive $\pi_{i+1}(s)$, though computing $v_{\pi_i}(s)$ is simpler than $q_{\pi_i}(s,a)$.

Second, in the policy evaluation step, the Monte Carlo methods use a sampling method to obtain the action value $q_{\pi_i}(s,a)$ for each state-action pair (s, a), as the state transition probability $p(s'|s, a)$ and the reward $r(s, a, s')$ are unknown. The reward is obtained through an observation of the result of an action taken in a given environment. The reward is overly complex to be computed based on an existing function such as $r(s, a, s')$.

As illustrated in the top part of figure 8.44, given a random start state s_0 and action a_0, through the interaction with the environment, an agent can receive a reward r_0, and the state of the environment becomes s_1. Following the policy π_0, the agent takes an action a_1, receives a reward r_1, and the state becomes s_2. Based on the interaction with the environment, we observe a sequence of state-action pairs (s_t, a_t), t=0, 1,…, T, where T is the length of time that the agent interacts with the environment. Such a sequence of state-action pairs is called one sampling, also known as one episode, which terminates after the agent stops interacting with the environment (e.g., after the agent works out of the maze depicted in figure 8.38b). For each state-action pair (s_t, a_t) appearing in an episode, we can calculate its return—that is, the long-term reward after (s_t, a_t), denoted by $G(s_t, a_t) = \sum_{l=t}^{T} \gamma^{l-t} r_l$, where $\gamma \in (0, 1)$ is a discount ratio.

In the Monte Carlo methods, we sample one episode at the i-th round of iteration based on π_i. In the meantime, we maintain a global storage for each state-action pair (s,a) occurring in different episodes. If a state-action pair appears in the current episode, we can calculate its return $G_i(s,a)$ based on the observation of the episode;

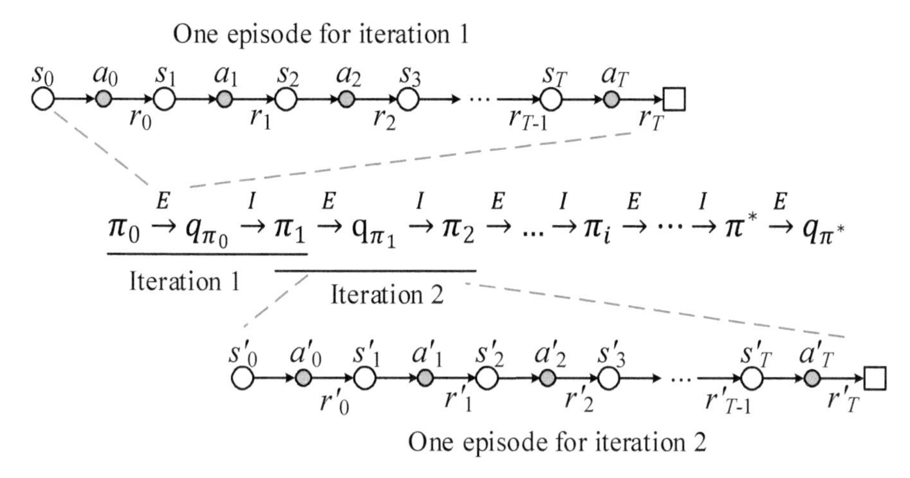

Figure 8.44
One episode in Monte Carlo methods.

otherwise, $G_i(s, a)$ is empty. Then, $q_{\pi_i}(s, a)$ is the average of the nonempty return $G_q(s, a)$, $q = 0, 1, \ldots, i$ over the past i iterations.

In the policy improvement step, based on $q_{\pi_i}(s, a)$, we can obtain a new policy π_{i+1}, which will be used to generate the $(i+1)$-th episode.

The on-policy Monte Carlo method The basic Monte Carlo methods require each state-action pair to appear a sufficient number of times in episodes, such that each state-action pair's value can be evaluated by its historical average. Otherwise, we cannot improve the policy effectively in the policy improvement step. In fact, following a deterministic policy π_k like that shown in equation (8.88), the agent can only explore a very limited part of state-action pairs. This is the *exploring starts* problem. To handle the exploring starts problem, researchers have developed two kinds of methods: on-policy Monte Carlo methods and off-policy Monte Carlo methods. Below, we introduce a classical on-policy Monte Carlo method, while off-policy methods are more complicated and can be found in the literature [28, 56].

The on-policy Monte Carlo method handles the exploring starts problem by incorporating a random probability of selecting other actions, not just the optimal action. Specifically, given the initial (nondeterministic) policy π_0, after updating its action-value function q_{π_0} by sampling one episode, we can obtain a new nondeterministic policy π_1. Given any state s, the policy π_1 has a probability of $1 - \varepsilon$ of selecting the action with the maximum action value and a probability of ε to select an action randomly, where ε is in [0, 1]. We call this the *ε-greedy strategy*. Similarly, we can use the same method to obtain a nondeterministic policy π_2 and so on. ε can be gradually decreased (e.g., we can set $\varepsilon = 1/k$, where k is the number of iterations). In this way, the on-policy Monte Carlo method makes state-action pairs well explored and is thereby more likely to improve the policy to the optimal policy.

8.8.2.3 Temporal-difference methods Similar to the Monte Carlo methods, temporal-difference methods, such as [49, 56, 62], do not require an exact model for the environment. They use the policy evaluation and improvement iterations to obtain the optimal policy and sample real interactions with the environment in each iteration. However, an episode in the Monte Carlo methods may not terminate at all in many reinforcement-learning tasks.

Different from the Monte Carlo methods, which improve a policy only after one episode terminates, the temporal-difference methods can improve the policy continuously in one episode. Thus, they do not necessarily require an episode terminating somewhere. In addition, they may converge faster than the Monte Carlo methods.

Temporal-difference methods also face the exploring starts problem mentioned above. To address this issue, two specific methods, consisting of the on-policy (called the *Sarsa algorithm* [49]) and the off-policy (called *Q-learning*) temporal-difference methods, have been proposed.

The on-policy temporal-difference method The Sarsa algorithm starts with an initial action-value function $q(s, a)$, which is essentially a matrix with each entry corresponding to the value of a specific action at a particular state; a row of the matrix denotes a state, and a column of the matrix stands for an action. The Sarsa algorithm updates one entry of $q(s, a)$ at each step of an episode until $q(s, a)$ converges.

For example, in episode 1 shown in figure 8.44, starting from a random state s_0 and taking a random action a_0, the agent obtains a reward r_0 and reaches s_1 (the two values are observed after taking a_0). According to the randomly initialized $q(s, a)$, the agent follows the ε-greedy strategy to take an action $a_1 = \arg\max_a q(s_i, a)$ with a probability of $1-\varepsilon$ or another random action with a probability of ε. This is to explore more state-action pairs, similar to the Monte Carlo methods. According to $q(s, a)$, we can obtain the value of $q(s_1, a_1)$, which is an entry in $q(s, a)$. Then, we can update $q(s_0, a_0)$ as follows:

$$q(s_0, a_0)' \leftarrow q(s_0, a_0) + \alpha\,((r_0 + \gamma q(s_1, a_1)) - q(s_0, a_0)), \tag{8.89}$$

where $q(s_0, a_0)'$ is the updated version of $q(s_0, a_0)$, and α is a ratio in [0, 1]. As $(r_0 + \gamma q(s_1, a_1))$ is an estimation just derived from one sample (i.e., it may deviate from its expectation that should be computed based on many samples), we cannot replace $q(s_0, a_0)$) by ($r_0 + \gamma q(s_1, a_1)$) aggressively. Instead, we use an α to gradually modify $q(s_0, a_0)$ until it converges. Using this updating strategy, we update the value of one entry in the $q(s, a)$ matrix per step and finally update all entries of the matrix after many steps.

Off-policy time-difference method The Q-learning algorithm [62, 63] is almost the same as the Sarsa algorithm, except for the following two differences:

First, instead of taking a random action a_0, the Q-learning algorithm chooses a_0 based on the ε-greedy strategy.

Second, the Q-learning algorithm updates the action-value function $q(s_t, a_t)$ as follows:

$$q(s_0, a_0)' \leftarrow q(s_0, a_0) + \alpha\left((r_0 + \gamma \max_a q(s_1, a)) - q(s_0, a_0)\right), \tag{8.90}$$

where $\max_a q(s_1, a)$ is the maximum action-value according to current $q(s, a)$ rather than the true value of the action that the agent takes, $q(s_1, a_1)$. The Q-learning algorithm uses the same strategy as the Sarsa algorithm to choose the next action for an agent but chooses different action values to update $q(s, a)$. This updating strategy results in a quicker convergence than the Sarsa algorithm.

8.8.3 Approximate Methods

As mentioned before, tabular action-value methods work only when the state space S and the action space A are small. However, in real-world applications, the state space and action space can be extremely large, even infinite (e.g., the state space could be continuous). Thus, some approximate methods have been proposed to learn an approximation function $v_\pi(s; \theta)$ based on some observed states' values. The function $v_\pi(s; \theta)$ is then used to estimate those unobserved states' values roughly. Similarly, we can also learn $q_\pi(s, a; \omega)$ to approximate each action value $q_\pi(s, a)$, where ω is a set of parameters. The approximation function $v_\pi(s; \theta)$ or $q_\pi(s, a; \omega)$ can be a linear model [9, 40], a decision tree model, an artificial neural network, or a deep neural network (called deep reinforcement learning) [41]. To obtain the approximation function is actually a process of conducting supervised learning.

Approximate methods have been used in many real-world decision-making problems, such as redeploying ambulances [40], playing Atari games [41], playing the game of Go [52], and managing resources in computer systems and networks. [40] applies a linear method to learn the approximation function, while [41, 52] leverage deep neural networks to generate the function.

The approximate function can be applied to all tabular action-value methods, including dynamic-programming methods, the Monte Carlo methods, and the temporal-difference methods. Below, we introduce how to apply approximate methods to the Q-learning method.

8.8.3.1 Approximate Q-learning In the Q-learning method, we start with a random initial action-value function $q(s,a)$. Similarly, for the approximate methods, we start with a random initial weight ω for approximation function $q(s,a;\omega)$. We then update the weight ω at each step in an episode in a similar way as shown in equation (8.90). Assuming the current weight is ω^- and $r_0+\gamma \max_a q(s_1,a;\omega^-)$ is a new estimation for $q(s_t,a_t;\omega)$, a better weight ω can be obtained through minimizing:

$$L(\omega)=[r_0+\gamma \max_a q(s_1,\ a;\ \omega^-)-q(s_0,\ a_0;\ \omega)]^2. \tag{8.91}$$

Based on the gradient descent algorithm, we can update ω using

$$\begin{aligned}\omega &= \omega^- - \frac{\alpha}{2}\frac{\partial L(\omega)}{\partial \omega} \\ &= \omega^- - \alpha\left[(r_0+\gamma \max_a q(s_1,a;\omega^-)-q(s_0,a_0;\omega)\right]\frac{\partial q(s_0,a_0;\omega)}{\partial \omega}.\end{aligned} \tag{8.92}$$

where $\frac{\alpha}{2}$ is the learning rate in the gradient descent algorithm. Note that updating weight ω in $q(s_0,a_0;\omega)$ is equivalent to updating the action value $q(s_0,\ a_0)$ in the Q-learning. Gradually, the weight ω will be updated to the optimal weight ω^*, and the approximation function $q(s,a;\omega^*)$ approximates the optimal action value $q^*(s,\ a)$, based on which we obtain the optimal policy π^*.

8.8.3.2 Deep reinforcement learning As the environment and action space are very complex, it is not easy for traditional approximate methods, which use a linear model or traditional one-hidden-layer neural networks to approximate the value function, to converge to the optimal weight ω^*. With the increasing availability of computing resources and data, deep learning has been employed to learn a more effective value function for a reinforcement-learning task.

As shown in figure 8.45a, we can train a deep neural network, where the input is the current state s_t, to approximate the value of different actions given s_t (i.e., $q(s_t,a_i;\ \omega^*)$, $i=1,\ 2,\ldots,\ m$). When the action space is large, we train an alternative deep neural network, shown in figure 8.45b, to approximate the value of one action at a time. The training process is similar to that shown in equations (8.91) and (8.92), where ω is the parameter of the deep neural network.

For example, the deep reinforcement learning shown in figure 8.45a has been employed to play an Atari boxing game [18]. As illustrated in figure 8.46a, a player controls the white boxer to fight with the black boxer. In this game, a player can take

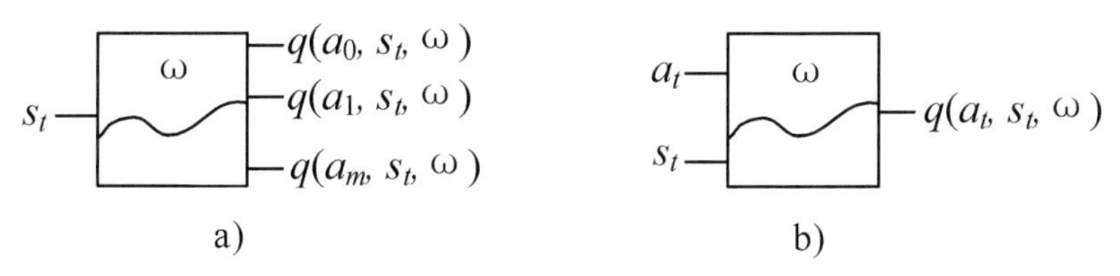

Figure 8.45
Deep neural networks for approximating value functions in reinforcement learning.

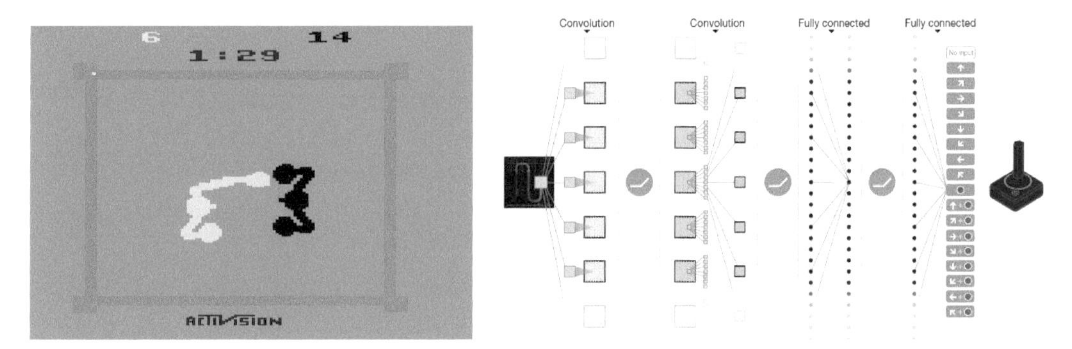

a) Interface of the Atari boxing game b) Deep neural network as the value function

Figure 8.46
Using deep learning to play an Atari boxing game (the figure was derived from [188]).

eighteen actions, including moving left, moving right, moving up, moving down, throwing a punch, and so on. After taking each action, the player will receive a reward (e.g., beating the rival once (1) or not (0)). The state of the game is the image of the current frame plus the previous three frames. Given the current state of the game, deep reinforcement learning finds an optimal sequence of actions for the white boxer, such that the white boxer can receive the maximum score in the future.

Figure 8.46b presents the deep neural network that approximates the value of the eighteen actions. The output layer has eighteen nodes, each of which corresponds to an action and generates a number belonging to [0,1]. The number denotes the value of each action and also stands for the probability that the action should be taken in the next step.

In the examples mentioned above, we have a clear definition or observation of the transition of states after an action has been taken. When the state space is infinite (e.g., continuous numbers), if the next state cannot be predefined or truly observed, we cannot obtain the next state. Thus, we cannot train the deep reinforcement–learning model like equations (8.91) and (8.92). Note that in the real world, not every reinforcement task can be truly running in the environment during the learning process like playing a game, such as when using deep reinforcement learning to control traffic lights or electricity transitions. In these cases, we need to learn a state transition function based on another model (e.g., another deep neural network) using historical data. Then, we can simulate the learning process of reinforcement learning with this transition function and action-value functions.

8.9 Summary

This chapter introduces the unique properties of spatiotemporal data, as compared to images and text data. The *spatial properties* consist of spatial distance and spatial hierarchy. The *temporal properties* are composed of temporal closeness, period, and trend. These unique properties call for advanced machine-learning algorithms that are dedicatedly designed for spatiotemporal data.

This chapter introduces the principle of six categories of machine-learning algorithms and how they should be adapted to handle spatiotemporal data.

It starts with two types of collaborative-filtering algorithms consisting of the user-based CF and the item-based CF models. Two location recommendation systems have been presented, showing how CF models should be used to handle spatiotemporal data.

Following that, two matrix factorization algorithms, comprising SVD and NMF, have been presented. An advanced matrix factorization method, called *coupled matrix factorization*, is introduced using two applications: location recommendation and traffic condition estimation.

This chapter further extends matrices to tensors, introducing two types of widely used tensor decomposition methods consisting of PARAFAC decomposition and the Tucker decomposition. The context-aware tensor decomposition methods have also been introduced through three examples: a personalized location recommendation system, the travel-time estimation of a path, and urban-noise diagnosis. All these scenarios and algorithms are based on spatiotemporal data.

This chapter then introduces the basic concepts of learning and inference algorithms for probabilistic graphical models, which are composed of Bayesian networks and Markov random fields. It shows three examples using Bayesian networks to infer traffic volumes, do map matching, and discover the latent function of a region, respectively. In addition, three examples are presented that use Markov random fields to predict the transportation modes of a user, the air quality of a location, and the flow of crowds in a region, respectively.

The chapter describes the basic idea of neural networks and deep learning, presenting two types of widely used deep neural networks: CNN and LSTM. A deep-learning model that is dedicatedly designed to predict the flow of crowds in each and every region throughout a city is also presented.

Finally, the general framework of reinforcement learning is introduced. Two categories of algorithms, consisting of the tabular action-value methods and approximate methods, for learning the optimal policy have been presented. Deep reinforcement learning, which uses deep neural networks to approximate action-value functions, is also discussed.

References

[1] Arik, S. O., et al. 2017. "Deep Voice: Real-Time Neural Text-to-Speech." arXiv preprint arXiv:1702.07825. Cornell University Library, Ithaca, New York.

[2] Bao, J., Y. Zheng, and M. F. Mokbel. 2012. "Location-Based and Preference-Aware Recommendation Using Sparse Geo-Social Networking Data." In *Proceedings of the 20th International Conference on Advances in Geographic Information Systems*. New York: Association for Computing Machinery (ACM), 199–208.

[3] Barto, A. G., and M. Du. 1994. "Monte Carlo Matrix Inversion and Reinforcement Learning." In *Advances in Neural Information Processing Systems: Proceedings of the 1993 Conference*. San Francisco: Morgan Kaufmann, 687–694.

[4] Bellman, R. E. 1956. "A Problem in the Sequential Design of Experiments." *Sankhya* 16:221–229.

[5] Bellman, R. E. 1957. *Dynamic Programming*. Princeton, NJ: Princeton University Press.

[6] Bellman, R. E. 1957. "A Markov Decision Process." *Journal of Mathematical Mechanics* 6:679–684.

[7] Bishop, Christopher M. 2006. "Graphical Models." In *Pattern Recognition and Machine Learning*. Berlin: Springer, 359–422.

[8] Blei, D., A. Ng, and M. Jordan. 2003. "Latent Dirichlet Allocation." *Journal of Machine Learning Research* 3:993–1022.

[9] Busoniu, L., R. Babuska, B. De Schutter, and D. Ernst. 2010. *Reinforcement Learning and Dynamic Programming Using Function Approximators (Vol. 39)*. Boca Raton: CRC Press.

[10] Carroll, J. D., and J. J. Chang. 1970. "Analysis of Individual Differences in Multidimensional Scaling via an N-way Generalization of 'Eckart-Young' Decomposition." *Psychometrika* 35 (3): 283–319.

[11] Dempster, Arthur P., Nan M. Laird, and Donald B. Rubin. 1977. "Maximum Likelihood from Incomplete Data via the EM Algorithm." *Journal of the Royal Statistical Society. Series B (Methodological)* 39 (1): 1–38.

[12] Deng, L., and D. Yu. 2014. "Deep Learning: Methods and Applications." *Foundations and Trends® in Signal Processing* 7 (3–4): 197–387.

[13] Friedman, N., I. Nachman, and D. Peér. 1999. "Learning Bayesian Network Structure from Massive Datasets: The Sparse Candidate Algorithm." In *Proceedings of the Fifteenth Conference on Uncertainty in Artificial Intelligence*. San Francisco: Morgan Kaufmann, 206–215.

[14] Gehring, J., M. Auli, D. Grangier, Denis Yarats, and Yann N. Dauphin. 2017. "Convolutional Sequence to Sequence Learning." arXiv preprint arXiv:1705.03122. Cornell University Library, Ithaca, New York.

[15] Girshick, R. 2015. "Fast R-cnn." In *Proceedings of the 2015 IEEE International Conference on Computer Vision*. Washington, DC: Institute of Electrical and Electronics Engineers (IEEE) Computer Society Press, 1440–1448.

[16] Girshick, R., J. Donahue, T. Darrell, and J. Malik. 2014. "Rich Feature Hierarchies for Accurate Object Detection and Semantic Segmentation." In *Proceedings of the IEEE Conference on Computer Vision and Pattern Recognition*. Washington, DC: IEEE Computer Society, 580–587.

[17] Goldberg, D., N. David, M. O. Brain, and T. Douglas. 1992. "Using Collaborative Filtering to Weave an Information Tapestry." *Communications of the ACM* 35 (12): 61–70.

[18] Golub, G. H., and C. Reinsch. 1970. "Singular Value Decomposition and Least Squares Solutions." *Numerische mathematik* 14 (5): 403–420.

[19] Harshman, R. A. 1970. "Foundations of the PARAFAC Procedure: Models and Conditions for an 'Explanatory' Multi-Modal Factor Analysis." UCLA Working Papers in Phonetics, 16, 1–84. Ann Arbor, MI: University Microfilms, No. 10,085.

[20] He, K., X. Zhang, S. Ren, and J. Sun. 2015. "Spatial Pyramid Pooling in Deep Convolutional Networks for Visual Recognition." *IEEE Transactions on Pattern Analysis and Machine Intelligence* 37 (9): 1904–1916.

[21] He, K., X. Zhang, S. Ren, and J. Sun. 2016. "Deep Residual Learning for Image Recognition." In *Proceedings of the 2016 IEEE Conference on Computer Vision and Pattern Recognition*. Washington, DC: IEEE Computer Society Press, 770–778.

[22] Hoang, M. X., Y. Zheng, and A. K. Singh. 2016. "FCCF: Forecasting Citywide Crowd Flows Based on Big Data." In *Proceedings of the 24th ACM SIGSPATIAL International Conference on Advances in Geographic Information Systems*. New York: ACM, 6.

[23] Hochreiter, S., and J. Schmidhuber. 1997. "Long Short-Term Memory." *Neural Computation* 9 (8): 1735–1780.

[24] Howard, R. 1960. *Dynamic Programming and Markov Processes*. Cambridge, MA: MIT Press.

[25] Hoyer, P. O. 2004. "Non-Negative Matrix Factorization with Sparseness Constraints." *Journal of Machine Learning Research* 5:1457–1469.

[26] Hsieh, H. P., S. D. Lin, and Y. Zheng. 2015. "Inferring Air Quality for Station Location Recommendation Based on Urban Big Data." In *Proceedings of the 21st ACM SIGKDD International Conference on Knowledge Discovery and Data Mining.* New York: ACM, 437–446.

[27] Jensen, Finn V. 1996. *An Introduction to Bayesian Networks.* Volume 210. London: UCL Press, 6.

[28] Kaelbling, L. P., M. L. Littman, and A. W. Moore. 1996. "Reinforcement Learning: A Survey." *Journal of Artificial Intelligence Research* 4:237–285.

[29] Kindermann, R., and J. L. Snell. 1980. *Markov Random Fields and Their Applications.* Volume 1. Providence, RI: American Mathematical Society.

[30] Klema, V., and A. J. Laub. 1980. "The Singular Value Decomposition: Its Computation and Some Applications." *IEEE Transactions on Automatic Control* 25 (2): 164–176.

[31] Kolda, T. G., and B. W. Bader. 2009. "Tensor Decompositions and Applications." *SIAM Review* 51 (3): 455–500.

[32] Krizhevsky, A., I. Sutskever, and G. E. Hinton. 2012. "Imagenet Classification with Deep Convolutional Neural Networks." In *Proceedings of the 25th International Conference on Neural Information Processing Systems* 1:1097–1105.

[33] Lafferty, J., A. McCallum, and F. Pereira. 2001. "Conditional Random Fields: Probabilistic Models for Segmenting and Labeling Sequence Data." In *Proceedings of the 18th International Conference on Machine Learning.* San Francisco: Morgan Kaufmann.

[34] de Lathauwer, L., B. De Moor, and J. Vandewalle. 2000. "A Multilinear Singular Value Decomposition." *SIAM Journal on Matrix Analysis and Applications* 21 (4): 1253–1278.

[35] LeCun, Y., B. E. Boser, J. S. Denker, D. Henderson, R. E. Howard, W. E. Hubbard, and L. D. Jackel. 1990. "Handwritten Digit Recognition with a Back-Propagation Network." In *Advances in Neural Information Processing Systems 2.* San Francisco: Morgan Kaufmann, 396–404.

[36] Lee, D. D., and H. S. Seung. 2011. "Algorithms for Non-Negative Matrix Factorization." In *Proceedings, Advances in Neural Information Processing Systems.* San Francisco: Morgan Kaufmann, 556–562.

[37] Lemire, D., and A. Maclachlan. 2005. "Slope One: Predictors for Online Rating-Based Collaborative Filtering." *Proceedings of SIAM Data Mining* 5:1–5.

[38] Li, Q., Yu Zheng, Xing Xie, Yukun Chen, Wenyu Liu, and Wei-Ying Ma. 2008. "Mining User Similarity Based on Location History." In *Proceedings of the 17th ACM SIGSPATIAL Conference on Advances in Geographical Information Systems.* New York: ACM Press, 1–10.

[39] Lou, Y., C. Zhang, Y. Zheng, X. Xie, W. Wang, and Y. Huang. 2009. "Map-Matching for Low-Sampling-Rate GPS Trajectories." In *Proceedings of the 17th ACM SIGSPATIAL International Conference on Advances in Geographic Information Systems.* New York: ACM, 352–361.

[40] Maxwell, M. S., M. Restrepo, S. G. Henderson, and H. Topaloglu. 2010. "Approximate Dynamic Programming for Ambulance Redeployment." *INFORMS Journal on Computing* 22 (2): 266–281.

[41] Mnih, V., K. Kavukcuoglu, D. Silver, A. A. Rusu, J. Veness, M. G. Bellemare, A. Graves, M. Riedmiller, A. K. Fidjeland, G. Ostrovski, and S. Petersen. 2015. "Human-Level Control through Deep Reinforcement Learning." *Nature* 518 (7540): 529–533.

[42] Mozer, M. C. 1989. "A Focused Back-Propagation Algorithm for Temporal Pattern Recognition." *Complex Systems* 3 (4): 349–381.

[43] Murphy, Kevin P., Yair Weiss, and Michael I. Jordan. 1999. "Loopy Belief Propagation for Approximate Inference: An Empirical Study." In *Proceedings of the Fifteenth Conference on Uncertainty in Artificial Intelligence*. San Francisco: Morgan Kaufmann.

[44] Nakamura, A., and N. Abe. 1998. "Collaborative Filtering Using Weighted Majority Prediction Algorithms." In *Proceedings of the 15th International Conference on Machine Learning*. San Francisco: Morgan Kaufmann, 395–403.

[45] Oord, A., S. Dieleman, H. Zen, Karen Simonyan, Oriol Vinyals, Alex Graves, Nal Kalchbrenner, Andrew Senior, and Koray Kavukcuoglu. 2016. "Wavenet: A Generative Model for Raw Audio." arXiv preprint arXiv:1609.03499. Cornell University Library, Ithaca, New York.

[46] Pan, Jian-Xin, and Kai-Tai Fang. 2002. "Maximum Likelihood Estimation." In *Growth Curve Models and Statistical Diagnostics*. Berlin: Springer, 77–158.

[47] Papalexakis, E. E., C. Faloutsos, and N. D. Sidiropoulos. 2016. "Tensors for Data Mining and Data Fusion: Models, Applications, and Scalable Algorithms." *ACM Transactions on Intelligent Systems and Technology* 8 (2): 16.

[48] Ren, S., K. He, R. Girshick, and J. Sun. 2015. "Faster R-CNN: Towards Real-Time Object Detection with Region Proposal Networks." *Advances in Neural Information Processing Systems* 28:91–99.

[49] Rummery, G. A., and M. Niranjan. 1994. "On-Line Q-learning Using Connectionist Systems." Technical Report CUED/F-INFENG/TR 166. Engineering Department, Cambridge University, Cambridge.

[50] Sak, H., A. W. Senior, and F. Beaufays. "Long Short-Term Memory Recurrent Neural Network Architectures for Large Scale Acoustic Modeling." In *Proceedings of the Fifteenth Annual Conference of the International Speech Communication Association*. Singapore, 338–342.

[51] Shang, J., Y. Zheng, W. Tong, E. Chang, and Y. Yu. 2014. "Inferring Gas Consumption and Pollution Emission of Vehicles throughout a City." In *Proceedings of the 20th ACM SIGKDD Conference on Knowledge Discovery and Data Mining*. New York: ACM, 1027–1036.

[52] Silver, D., A. Huang, C. J. Maddison, A. Guez, L. Sifre, G. Van Den Driessche, J. Schrittwieser, I. Antonoglou, V. Panneershelvam, M. Lanctot, and S. Dieleman. 2016. "Mastering the Game of Go with Deep Neural Networks and Tree Search." *Nature* 529 (7587): 484–489.

[53] Singh, A. P., and G. J. Gordon. 2008. "Relational Learning via Collective Matrix Factorization." In *Proceedings of the 14th ACM SIGKDD International Conference on Knowledge Discovery and Data Mining*. New York: ACM, 650–658.

[54] Singh, S. P., and R. S. Sutton. 1996. "Reinforcement Learning with Replacing Eligibility Traces." *Machine Learning* 22 (1–3): 123–158.

[55] Srivastava, N., G. Hinton, A. Krizhevsky, I. Sutskever, and R. Salakhutdinov. 2014. "Dropout: A Simple Way to Prevent Neural Networks from Overfitting." *Journal of Machine Learning Research* 15 (1): 1929–1958.

[56] Sutton, R. S. 1988. "Learning to Predict by the Method of Temporal Differences." *Machine Learning* 3 (1): 9–44.

[57] Sutton, R. S., and A. G. Barto. 2017. *Reinforcement Learning: An Introduction*. 2nd edition Cambridge, MA: MIT Press.

[58] Tsamardinos, I., L. E. Brown, and C. F. Aliferis. 2006. "The Max-min Hill-Climbing Bayesian Network Structure Learning Algorithm." *Machine Learning* 65 (1): 31–78.

[59] Tucker, L. R. 1966. "Some Mathematical Notes on Three-Mode Factor Analysis." *Psychometrika* 31 (3): 279–311.

[60] Wainwright, M. J., and M. I. Jordan. 2008. "Graphical Models, Exponential Families, and Variational Inference." *Foundations and Trends® in Machine Learning* 1 (1–2): 1–305.

[61] Wang, Y., Y. Zheng, and Y. Xue. 2014. "Travel Time Estimation of a Path Using Sparse Trajectories." In *Proceedings of the 20th ACM SIGKDD International Conference on Knowledge Discovery and Data Mining*. New York: ACM, 25–34.

[62] Watkins, C. J. C. H. 1989. "Learning from Delayed Rewards." PhD diss., Cambridge University, Cambridge.

[63] Watkins, C. J. C. H., and P. Dayan. 1992. "Q-learning." *Machine Learning* 8 (3–4): 279–292.

[64] White, D. J. 1985. "Real Applications of Markov Decision Processes." *Interfaces* 15:73–83.

[65] Xiao, X., Y. Zheng, Q. Luo, and X. Xie. 2014. "Inferring Social Ties between Users with Human Location History." *Journal of Ambient Intelligence and Humanized Computing* 5 (1): 3–19.

[66] Yedidia, Jonathan S., William T. Freeman, and Yair Weiss. 2000. "Generalized Belief Propagation." *Advances in Neural Information Processing Systems* 13:689–695.

[67] Yi, X., Yu Zheng, Junbo Zhang, and Tianrui Li. 2016. "ST-MVL: Filling Missing Values in Geosensory Time Series Data." In *Proceedings of the 25th International Joint Conference on Artificial Intelligence*. Pasadena, CA: International Joint Conferences on Artificial Intelligence Organization (IJCAI).

[68] Yuan, J., Y. Zheng, and X. Xie. 2012. "Discovering Regions of Different Functions in a City Using Human Mobility and POIs." *Proceedings the 18th ACM SIGKDD International Conference on Knowledge Discovery and Data Mining*. New York: ACM, 186–194.

[69] Yuan, N. J., Y. Zheng, X. Xie, Y. Wang, K. Zheng, and H. Xiong. 2015. "Discovering Urban Functional Zones Using Latent Activity Trajectories." *IEEE Transactions on Knowledge and Data Engineering* 27 (3): 1041–4347.

[70] Zhang, C., Y. Zheng, X. Ma, and J. Han. 2015. "Assembler: Efficient Discovery of Spatial Co-Evolving Patterns in Massive Geo-Sensory Data." In *Proceedings of the 21st ACM SIGKDD International Conference on Knowledge Discovery and Data Mining*. New York: ACM, 1415–1424.

[71] Zhang, Junbo, Yu Zheng, and Dekang Qi. 2017. "Deep Spatio-Temporal Residual Networks for Citywide Crowd Flows Prediction." In *Proceedings of the 31st AAAI Conference on Artificial Intelligence*. New York: AAAI Press.

[72] Zheng, V. W., Y. Zheng, X. Xie, and Q. Yang. 2010. "Collaborative Location and Activity Recommendations with GPS History Data." In *Proceedings of the 19th International Conference on the World Wide Web*. New York: ACM, 1029–1038.

[73] Zheng, Y., F. Liu, and H. P. Hsieh. 2013. "U-air: When Urban Air Quality Inference Meets Big Data." In *Proceedings of the 19th ACM SIGKDD International Conference on Knowledge Discovery and Data Mining*. New York: ACM, 1436–1444.

[74] Zheng, Y., T. Liu, Y. Wang, Y. Zhu, Y. Liu, and E. Chang. 2014. "Diagnosing New York City's Noises with Ubiquitous Data." In *Proceedings of the 2014 ACM International Joint Conference on Pervasive and Ubiquitous Computing*. New York: ACM, 715–725.

[75] Zheng, Y., and X. Xie. 2011. "Learning Travel Recommendations from User-Generated GPS Traces." *ACM Transactions on Intelligent Systems and Technology* 2 (1): 2.

[76] Zheng, Y., H. Zhang, and Y. Yu. 2015. "Detecting Collective Anomalies from Multiple Spatio-temporal Datasets across Different Domains." In *Proceedings of the 23rd SIGSPATIAL International Conference on Advances in Geographic Information Systems*. New York: ACM, 2.

[77] Zheng, Y., L. Zhang, Z. Ma, X. Xie, and W. Y. Ma. 2011. "Recommending Friends and Locations Based on Individual Location History." *ACM Transactions on the Web* 5 (1): 5.

[78] Zhu, Julie Yixuan, Chao Zhang, Huichu Zhang, Shi Zhi, Victor O. K. Li, Jiawei Han, and Yu Zheng. 2017. "pg-Causality: Identifying Spatiotemporal Causal Pathways for Air Pollutants with Urban Big Data." *IEEE Transactions on Big Data*. doi:10.1109/TBDATA.2017.2723899.

9 Cross-Domain Knowledge Fusion

Abstract: Traditional data mining usually deals with data from a single domain. In the big data era, we face a diversity of datasets from different sources in different domains. These datasets consist of multiple modalities, each of which has a different representation, distribution, scale, and density. How to unlock the power of knowledge from multiple disparate (but potentially connected) datasets is paramount in big-data research, essentially distinguishing big data from traditional data-mining tasks. This calls for advanced techniques that can fuse knowledge from various datasets organically in machine learning and data mining. This chapter summarizes knowledge fusion methodologies, classifying them into three categories: stage-based, feature level–based, and semantic meaning–based knowledge-fusion methods. The last category of fusion methods is further divided into four groups: multi-view learning–based, similarity-based, probabilistic dependency–based, and transfer learning–based methods. These methods focus on knowledge fusion rather than schema mapping and data merging, significantly distinguishing between cross-domain data fusion and traditional data fusion studied in the database community. This chapter not only introduces the high-level principles of each category of methods but also gives examples in which these techniques are used to handle real big-data problems. In addition, this chapter positions existing works in a framework, exploring the relationship and difference between different knowledge fusion methods.

9.1 Introduction

In the big-data era, a wide array of data is being generated in different domains, from social media to transportation to health care to wireless communication networks. When addressing a problem, we usually need to harness multiple disparate datasets. Goals of fusing multiple datasets include filling in missing values (of a sparse dataset) [36, 50, 57, 58], predicting the future [60], inferring causalities [63], profiling objects [51, 53], and detecting anomalies [61], among others.

For example, to improve urban planning, we need to consider the structure of a road network, the traffic volume, the points of interest (POIs), and the population of a city. To tackle air pollution, we need to explore air quality data together with meteorological data and emissions from vehicles and factories, as well as the dispersion conditions of a location. To generate more accurate travel recommendations for users, we consider the user's behavior on the Internet and in the physical world. To better understand an image's semantic meanings, we use its surrounding text and features derived from its pixels. Consequently, how to unlock the power of knowledge from multiple datasets across different domains is paramount in big-data research, essentially distinguishing big data from traditional data-mining tasks.

However, data from different domains consists of multiple modalities, each of which has a different representation, distribution, scale, and density. For example, text is usually represented as discrete sparse word-count vectors, whereas an image is represented by pixel intensities or outputs of feature extractors that are real valued and dense. POIs are represented by spatial points associated with a static category, whereas air quality is represented using a geotagged time series. Human mobility data is represented by trajectories [55], whereas a road network is denoted by a spatial graph. Treating different datasets equally or simply concatenating the features from disparate datasets does not result in good performance in data-mining tasks [5, 31, 37]. As a result, fusing data across modalities becomes a new challenge in big-data research, calling for advanced data fusion technology.

This chapter summarizes three categories of methods that can fuse the knowledge of multiple datasets.

The first category of data fusion methods uses different datasets at different stages of a data-mining task. We call them the *stage-based fusion methods*. For example, Zheng et al. [59] first partition a city into disjointed regions by road network data and then detect the pairs of regions that are not well connected based on human mobility data. These region pairs could denote the design that is out of date in a city's transportation network.

The second category, titled feature level–based data fusion, has two subcategories. One subcategory of methods first concatenates the features extracted from different datasets straightforwardly into a feature vector. The feature vector is then used to train a classification or regression model with some regularizations, which avoids overfitting and reduces the redundancy as well as the dependency of features. The other subcategory uses deep neural networks (DNNs) to learn a latent representation of the original features extracted from different datasets. The new feature representation will then be fed into a model for classification or prediction.

The third category of methods is those using semantic meaning–based knowledge fusion. This name is used for two reasons. First, feature level–based knowledge fusion methods do not care about the semantic meanings of each feature and even a dataset. Each feature is just deemed as a real valued or a categorical number by these methods.

However, when using semantic meaning–based methods, we need to understand the meaning of each dataset and feature as well as the relationship between different features. The second reason is that the methods in this category are derived from human thinking (i.e., how people think of multiple datasets to complete a task). This category is further composed of four groups:

1. *Multi-view-based methods.* This group of methods treats different datasets (or features from different datasets) as different views on an object. Different features are fed into different models, describing an object from different perspectives. The results are later merged together or mutually reinforce each other. Co-training [8] is an example of this category.
2. *Similarity-based methods.* This group of methods leverages the underlying correlation (or similarity) between different objects to fuse different datasets. A typical method is coupled collaborative filtering (CF), or context-aware CF, in which different datasets are modeled by different matrices with common dimensions. By decomposing these matrices (or tensors) together, we can achieve a better result than solely factorizing a single matrix (or a tensor). Manifold alignment also belongs to this group.
3. *Probabilistic dependency–based methods.* This group models the probabilistic causality (or dependency) between different datasets using a graphic representation. Bayesian networks and Markov random fields are representative models, representing features extracted from different datasets as graph nodes and the dependency between two features with an edge.
4. *Transfer learning–based methods.* This group of methods transfers the knowledge from a source domain to another target domain, dealing with the data sparsity problems (including the feature structure missing or observation missing) in the target domain. Transfer learning can even transfer knowledge between different learning tasks (e.g., from book recommendations to travel recommendations).

9.1.1 Relationship to Traditional Data Integration

Conventional data fusion [7], which is regarded as a part of data integration, is a process of integrating multiple data representing the same real-world object into a consistent, accurate, and useful representation. Figure 9.1a presents the paradigm of conventional data fusion. For example, there are three POI datasets for Beijing generated by three different data providers. Conventional data fusion aims to merge the three datasets into a database with a consistent data schema through a process of schema mapping and duplicate detection. The records (from different datasets) describing the same POI (e.g., a restaurant) are generated in the same domain (i.e., POI).

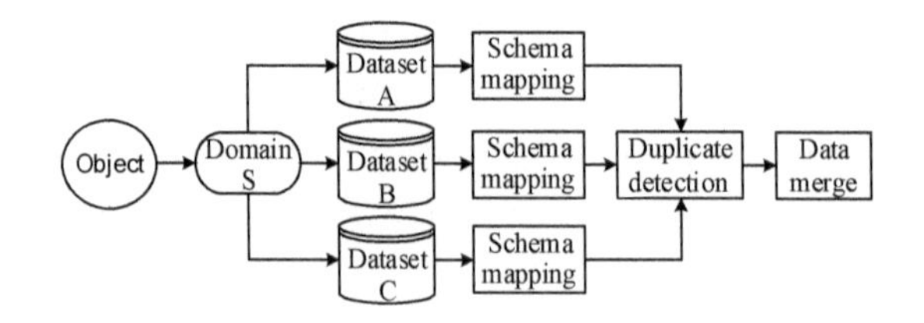

a) Paradigm of the conventional data fusion

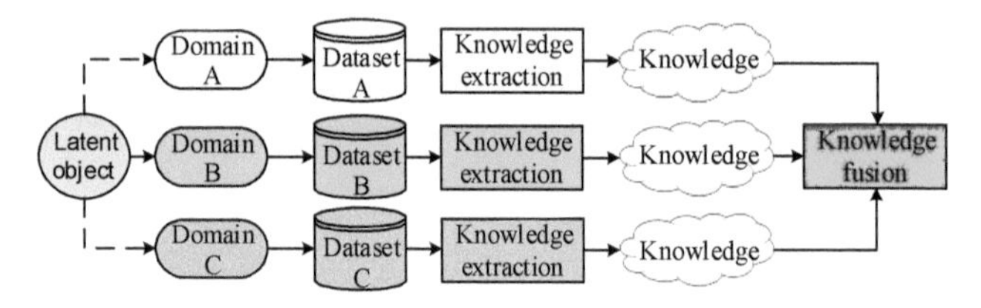

b) Paradigm of the cross-domain knowledge fusion

Figure 9.1
Paradigms of different knowledge fusion methods.

As illustrated in figure 9.1b, however, in the era of big data, there are multiple datasets generated in different domains, which are implicitly connected by a latent object. For instance, the traffic conditions, the POIs, and the demography of a region describe the region's latent function collectively, while they are from three different domains. Literally, records from the three datasets describe different objects (i.e., a road segment), a POI, and a neighborhood, respectively. Thus, we cannot merge them straightforwardly by schema mapping and duplication detection. Instead, we need to extract knowledge from each dataset by different methods, fusing the knowledge from them organically to understand a region's function collectively. This is more about knowledge fusion rather than schema mapping, which significantly differentiates between traditional data fusion (studied in the database community) and cross-domain knowledge fusion.

9.1.2 Relationship to Heterogeneous Information Networks

An information network represents an abstraction of the real world, focusing on objects and interactions between objects. It turns out that this level of abstraction has great power in not only representing and storing essential information about the real world but also providing a useful tool to mine knowledge from it by exploring the power of links [38]. Departing from many existing network models that view interconnected data as homogeneous graphs or networks, a heterogeneous information network consists of nodes and relations of different types. For example, a bibliographic information network consists of authors, conferences, and papers as different types of nodes. Edges between different nodes in this network can denote different semantic meanings, such as an author publishes a paper, a paper is presented at a conference, and an author attends a conference. Quite a few algorithms have been proposed to mine a heterogeneous network (e.g., ranking and clustering [39, 40]).

Heterogeneous information networks can be constructed in almost any domain, such as social networks, e-commerce, and online movie databases. However, they only link the object in a single domain rather than data across different domains. For instance, in a bibliographic information network, people, papers, and conferences are all from a bibliographic domain. In a Flickr information network, users, images, tags, and comments are all from a social media domain. If we want to fuse data across totally different domains (e.g., traffic data, social media, and air quality), such a heterogeneous network may not be able to find explicit links with semantic meanings between objects of different domains. Consequently, algorithms proposed for mining heterogeneous information networks cannot be applied to cross-domain data fusion directly.

9.2 Stage-Based Knowledge Fusion

This category of methods uses different datasets at the different stages of a data-mining task, as shown in figure 9.2a, or throws different data into different models and then aggregates the results of different models, as illustrated in figure 9.2b. Different datasets are loosely coupled, without any requirements on the consistency of their modalities. As this category of methods is somehow straightforward, we briefly introduce them using some examples.

Example 1. As illustrated in figure 9.3a, a city is first partitioned into disjointed regions by major roads using a map-segmentation method [52]. The GPS trajectories of taxicabs are then mapped onto the regions to formulate a region graph, as depicted in figure 9.3b, where a node is a region, and an edge denotes the aggregation of commutes (by taxis in this case) between two regions. The region graph actually blends knowledge from the road network and taxi trajectories. By analyzing the region graph, a body of research has been carried out to identify the improper design of a road network [59] and to detect and diagnose traffic anomalies [12, 28] as well as to find urban functional regions [51, 53].

Example 2. In friend recommendations, as illustrated in figure 9.4, Xiao et al. [45, 46] first detect the stay points from an individual's location history (recorded in a form of spatial trajectories). As different users' location histories may not have any overlaps in the physical world, each stay point is then converted into a feature vector, which describes the distribution of surrounding POIs across different categories. For example,

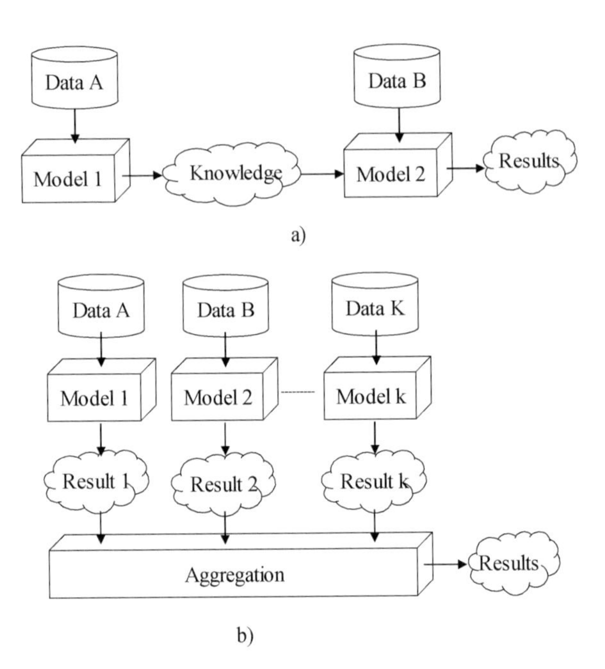

Figure 9.2
The general framework of stage-based knowledge fusion methods.

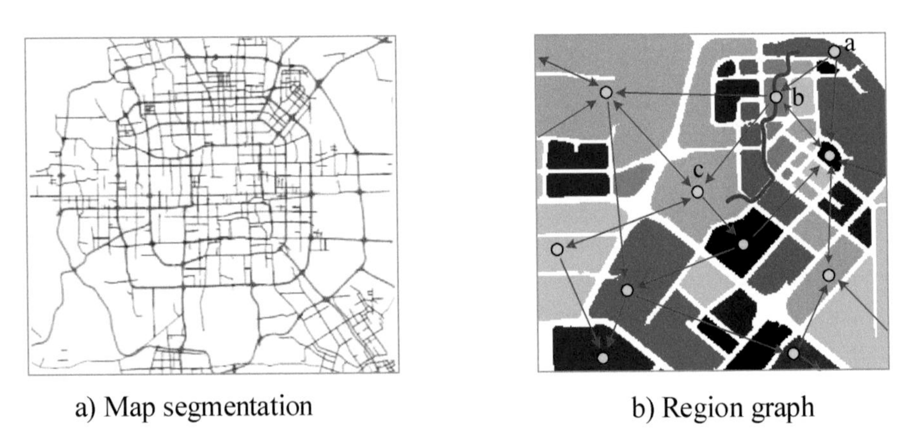

Figure 9.3
Fusing knowledge from road networks and human-mobility data.

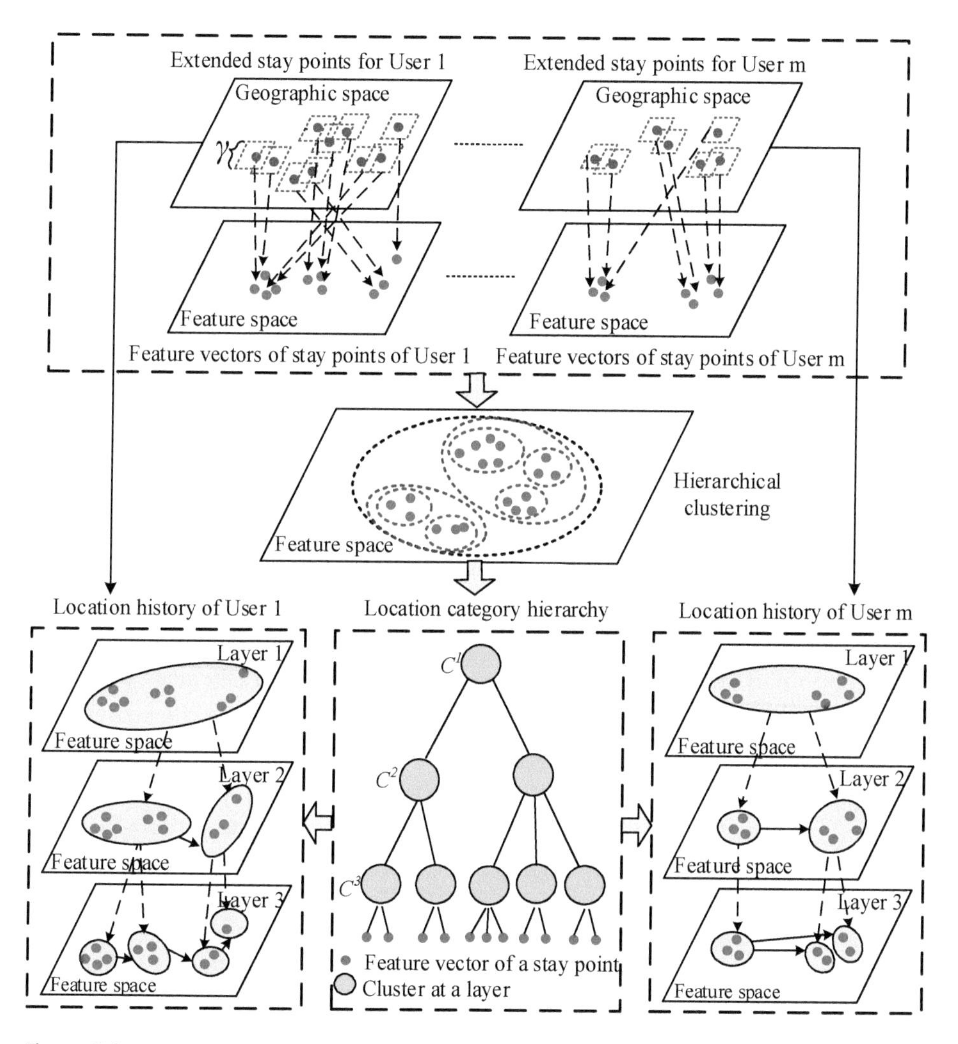

Figure 9.4
Estimating user similarity using trajectories and POIs.

there are five restaurants, one shopping mall, and one gas station around a stay point. The counts are further normalized by term frequency-inverse document frequency (TF-IDF). In other words, the distance between these feature vectors denotes the similarity between the places people have visited.

Later, these stay points are hierarchically clustered into groups based on their feature vectors of POIs, formulating a tree structure, where a node is a cluster of stay points; a parent node is composed of the stay points from its children nodes. By selecting the nodes (from the tree) that a user has at least one stay point in, we can represent the user's location history with a partial tree. A user's partial tree is further converted into a hierarchical graph by connecting two nodes (on the same layer) with an edge if the user has two consecutive stay points occurring in the two nodes within a given time interval. The hierarchical graph contains the information of a user's trajectories and the POIs of the places the user has visited. Because different users' hierarchical graphs are built based on the same tree structure, their location histories become comparable. Finally, the similarity between two users can be measured by the similarity between their hierarchical graphs.

Example 3. In the third example, Pan et al. [33] first detect a traffic anomaly based on the GPS trajectories of vehicles and road network data. An anomaly is represented by a subgraph of a road network where drivers' routing behaviors differ significantly from their original patterns. Using the time span of the detected anomaly and the names of locations that have fallen in the anomaly's geographical scope as queries, they retrieve relevant social media (like tweets) that people have posted at the locations where an anomaly has happened. From the retrieved social media, they then try to describe the detected anomaly by mining representative terms (e.g., *parades* and *disasters*), which rarely occur on normal days but become frequent when the anomaly takes place. The first step scales down the scope of social media to be checked, while the second step enriches the semantic meaning of the results detected by the first step.

This category of methods can be a meta-approach containing other knowledge fusion methods. For example, Yuan et al. [53] first use road network data and taxi trajectories to build a region graph. A graphical model, which is a probabilistic graphical model–based method, is then proposed to fuse the information from POIs and the knowledge of the region graph.

9.3 Feature-Based Knowledge Fusion

9.3.1 Feature Concatenation with Regularization

Straightforward methods in this category treat features extracted from different datasets equally, concatenating them sequentially into a feature vector. The feature vector is then used in clustering and classification tasks. As the representation, distribution, and scale of different datasets may be very different, quite a few studies have suggested limitations to this kind of fusion [2, 31, 37]. First, this concatenation causes overfitting in the case of a small-sized training sample, and the specific statistical property of each view is ignored [47]. Second, it is difficult to discover the highly nonlinear relationships that exist between low-level features across different modalities [37]. Third, there are redundancies and dependencies between features extracted from different datasets that may be correlated.

Advanced learning methods in this subcategory suggest adding a sparsity regularization in an objective function to handle the feature redundancy problem. As a result, a machine-learning model is likely to assign a weight close to zero to redundant features.

Example 4. Combining multiple data sources, we can even predict the ranking of real estate. Fu et al. [16–18] present research on predicting the future ranking of residential properties in a city according to their potential values inferred from a variety of datasets, such as human mobility and urban geography. Here, *value* means the ability to increase faster in a rising market and decrease more slowly than others in a falling market, quantified by discretizing the increasing or decreasing ratio over its previous price into five levels (R1–R5), where R1 stands for the best and R5 denotes the worst. The rank is of great importance to people when settling down or allocating capital investments.

As illustrated in figure 9.5, they consider three categories of factors consisting of geographical utility, neighborhood popularity, and business zone prosperity. These factors actually correspond to the old saying that "real estate is all about three things: location,

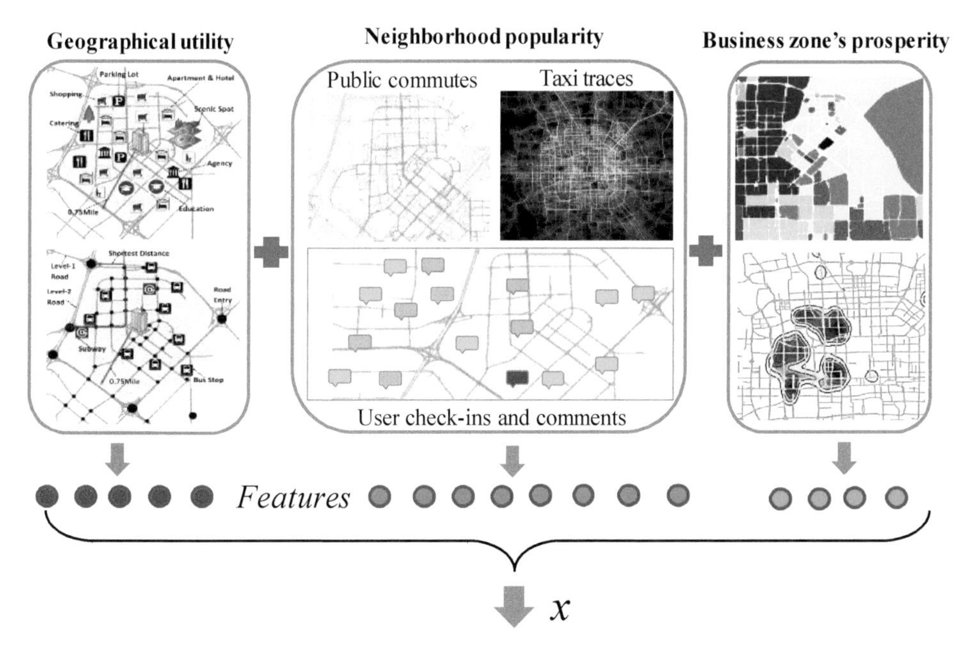

Figure 9.5
Ranking real estate using knowledge from multiple datasets.

location, location." More specifically, they identify a set of discriminative features for each property by mining the surrounding geographic data (e.g., road networks and POIs), traffic data (such as taxi traces and card-swiping records in public transportation systems), and social media.

A straightforward approach is to concatenate these features into a vector x, based on which a linear regression is conducted as follows:

$$y_i = f_i\ (x_i;\ \omega) = \boldsymbol{\omega}^T\ x_i + \epsilon_i, \tag{9.1}$$

where y_i and x_i are the increasing ratio and the feature vector of the real estate property i, respectively; ω is a vector of weights corresponding to different features; ϵ_i is a Gaussian bias with a zero mean and variance σ^2. Writing equation (9.1) into a probabilistic representation,

$$P(y_i|x_i) = N(y_i\,|\,f_i, \sigma^2) = N(y_i\,|\,\boldsymbol{\omega}^T x_i, \sigma^2). \tag{9.2}$$

However, there are redundancies and dependencies between features extracted from different datasets. For example, the traffic patterns in a region depend on the structure of road networks in the region. Thus, the features extracted from human mobility data like taxi trajectories may be correlated to that from road networks. To address this issue, two constraints are added into the objective function.

The first one is a pairwise ranking constraint between real estate property i and h:

$$P(i \rightarrow h) = Sigmoid\,(f_i - f_h) = \frac{1}{1 + \exp\,(-\,(f_i - f_h))}, \tag{9.3}$$

where $i \rightarrow h$ denotes that real estate i is ranked ahead of real estate h in terms of their increasing ratio. If i is ranked ahead of h, the true value of f_i should be larger than f_h. If the predictions follow the ground truth (i.e., $f_i - f_h > 0$), as illustrated in figure 9.6a, $Sigmoid\,(f_i - f_h)$ outputs a result close to 1. This is a reward to the predictions that preserve the correct order between a pair of properties. On the contrary, if i is truly ranked ahead of h while the prediction f_i is smaller than f_h (i.e., $f_i - f_h < 0$), $Sigmoid\,(f_i - f_h)$ outputs a value close to 0. This looks like a punishment to the predictions that incorrectly rank a pair of real estate properties.

As we know the ground truth of each real estate property, we can form many training pairs like $i \rightarrow h$. Then we require the predictions to preserve the order between all these pairs by adding the following constraint into the objective function:

$$\prod\nolimits_{i=1}^{I-1} \prod\nolimits_{h=i+1}^{I} P(i \rightarrow h), \tag{9.4}$$

where I is the total number of properties. By maximizing overall probability, the order between properties can be preserved.

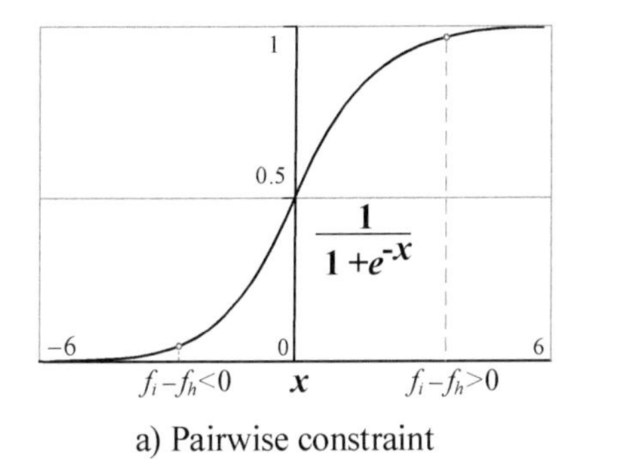

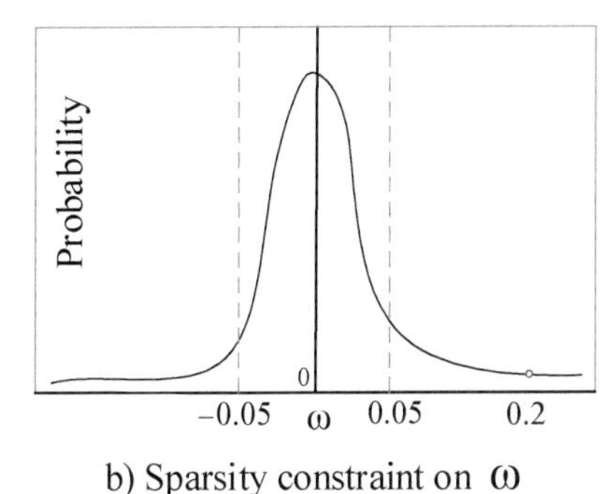

Figure 9.6
Constraints and regularizations on the knowledge fusion problem.

The second constraint is a sparsity regularization on $\boldsymbol{\omega}$. Because of the dependency and correlation between features, there are many redundant features that do not necessarily contribute to the prediction. We hope the weight of these redundant features can be as small as possible. On the other hand, we hope some important features can be highly weighted. As a result, we enforce the distribution of ω to follow a Gaussian distribution with a zero mean and a very small standard deviation β^2. Thus, as illustrated in figure 9.6b, the majority of weights will fall around zero while we can still set a very high weight (with a very small probability) for an important feature. Specifically, we add the following constraint into the objective function:

$$P(\boldsymbol{\omega}|0, \boldsymbol{\beta}^2)P(\boldsymbol{\beta}^2|a, b) = \prod_m N(\omega_m|0, \beta_m^2) \prod_m Inverse - Gamma(\beta_m^2|a, b), \tag{9.5}$$

where ($\boldsymbol{\omega} = \omega_1, \omega_2, \ldots, \omega_m$) is a parameter vector of features; m is the number of features involved in a learning model; $\boldsymbol{\beta}^2 = (\beta_1^2, \beta_2^2, \ldots, \beta_m^2)$ is the variance vector of the corresponding parameters. More specifically, the value of a parameter ω_m is assumed to follow a Gaussian distribution with a zero mean and a variance β_m^2. Setting a zero mean for the distribution reduces the probability of assigning ω_m a big value. A prior distribution (e.g., an inverse gamma) is further placed to regularize the value of β_m^2. To strengthen sparsity, constants a and b are usually set close to zero. Thus, β_m^2 tends to be small. In other words, feature weight ω_m has a very high probability of varying around the Gaussian expectation (i.e., zero).

Through such a dual regularization (i.e., zero-mean Gaussian plus inverse-gamma), we can simultaneously regularize most feature weights to be zero or close to zero via a Bayesian sparse prior while allowing for the possibility of a model learning large weights for significant features. Though the sparsity regularization of a Bayesian sparse prior is not as strong as L_1 and L_2 regularizations, the latter two regularizations may not be a good choice for this application as they aim to minimize every parameter. In addition, the Bayesian sparse prior is a smooth function, and thus its gradient is easy to compute. Given that many objective functions are solved by gradient descent, the sparse regularization can be applied to many data-mining tasks.

Finally, we maximize the following objective function:

$$P(y_i|x_i) = \prod_{i=1}^{I} N(y_i|\omega^T x_i, \sigma^2) \times \prod_{i=1}^{I-1} \prod_{h=i+1}^{I} P(i \to h) \times \prod_m N(\omega_m|0, \beta_m^2). \quad (9.6)$$

9.3.2 Deep Learning–Based Knowledge Fusion

Using supervised, unsupervised, and semisupervised approaches, deep learning learns multiple levels of representation and abstraction that help make sense of data, such as images, sound, and text. Besides being a predictor, deep learning is also used to learn new feature representations [1], which can be fed into other classifiers or predictors. The new feature representations have proven more useful than handcrafted features in image recognition and speech translations. Refer to section 8.7 in chapter 8 for more details about deep learning.

The majority of DNNs are applied to handle data with a single modality. More recently, a series of research has started using DNNs to learn feature presentations from data with different modalities. This representation has proven useful for classification and information-retrieval tasks.

Example 5. Ngiam et al. [31] propose a deep autoencoder architecture to capture the "middle-level" feature representation between two modalities (e.g., audio and video). As shown in table 9.1, three learning settings, consisting of cross-modality learning, shared-representation learning, and multimodal fusion, are studied.

Table 9.1
Multimodal feature-representation learning [31].

	Feature learning	Supervised training	Testing
Classic deep learning	Audio	Audio	Audio
	Video	Video	Video
Cross-modality learning	A+V	A	A
	A+V	V	V
Shared-representation learning	A+V	A	V
	A+V	V	A
Multimodal fusion	A+V	A+V	A+V

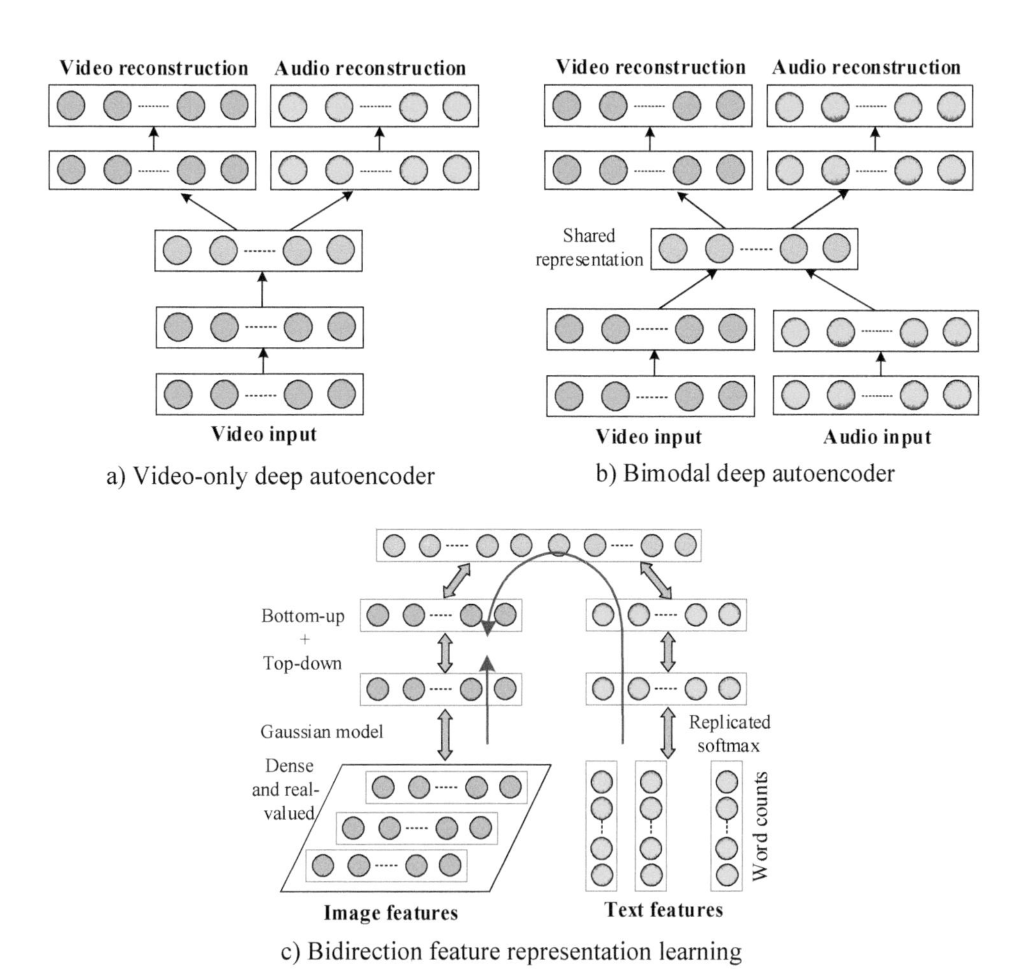

Figure 9.7
Fusing multimodal data with deep learning.

Figure 9.7a presents the structure of the deep autoencoder for cross-modality learning, where a single modality (e.g., video or audio) is used as the input to reconstruct a better feature representation for video and audio, respectively. With respect to shared-representation learning and multimodal fusion, which involve different modalities during training and testing, the study adopts the architecture shown in figure 9.7b. Extensive evaluations on these proposed deep-learning models demonstrate that deep learning effectively learns (1) a better single-modality representation with the help of other modalities and (2) the shared representations capturing correlations across multiple modalities.

Example 6. Deep learning using Boltzmann machines is another piece of work on multimodality data fusion. A deep-learning model called the *multimodal deep Boltzmann machine* (DBM) [37] is proposed to fuse images and text for classification and retrieval problems. The DBM model satisfies the following three criteria: (1) the learned shared-feature representation preserves the similarity of "concepts," (2) the joint-feature representation is easy to obtain in the absence of some modalities and thus fills in missing modalities, and (3) the new-feature representation facilitates the retrieval of one modality when querying from the other.

As shown in figure 9.7c, the multimodal DBM utilizes the Gaussian-Bernoulli RBM (restricted Boltzmann machine) to model dense real-valued image features vectors while employing replicated softmax to model sparse word-count vectors. The multimodal DBM constructs a separate two-layer DBM for each modality and then combines them by adding a layer on top of them. Moreover, the multimodal DBM is a generative and undirected graphic model with bipartite connections between adjacent layers. This graphic model enables a bidirectional (bottom-up and top-down) search (denoted by the two red arrows).

Armed with a well-designed architecture, the key idea of multimodal DBM is to learn a joint-density distribution over text and images (i.e., $P(\mathbf{v}_{img}, \mathbf{v}_{text}; \theta)$, where θ includes the parameters) from a large number of user-tagged images. The study performs extensive experiments on classification as well as retrieval tasks. Both multimodal and unimodal inputs are tested, validating the effectiveness of the model in fusing multimodality data.

In practice, the performance of a DNN-based fusion model usually depends on how well we can tune parameters for the DNN. Finding a set of proper parameters can lead to a much better performance than others. Given a large number of parameters and a nonconvex optimization setting, however, finding optimal parameters is still a labor-intensive and time-consuming process that relies heavily on human experiences. In addition, it is hard to explain what the middle-level feature representation stands for. We do not really understand how a DNN makes raw features a better representation.

9.4 Semantic Meaning–Based Knowledge Fusion

Feature-based knowledge fusion methods do not care about the meaning of each feature, regarding a feature solely as a real-valued number or a categorical value. Unlike feature-based fusion, semantic meaning–based methods understand the insight of each dataset and relations between features across different datasets. We know what each dataset stands for, why different datasets can be fused, and how they reinforce one another. The process of data fusion carries a semantic meaning (and insights) derived from the ways that people think about a problem with the help of multiple datasets. Thus, they are interpretable and meaningful. This section introduces four groups of semantic meaning–based knowledge fusion methods: multi-view based, similarity based, probabilistic-dependency based, and transfer-learning based.

9.4.1 Multi-View-Based Knowledge Fusion

Different datasets or different feature subsets about an object can be regarded as different views on the object. For example, a person can be identified by the information obtained from multiple sources, such as face, fingerprints, or signature. An image can be represented by different feature sets like color or texture. The function of a geographical region can be represented by its POIs, road networks, and human-mobility patterns. As these datasets describe the same object, there is a latent consensus among them. On the other hand, these datasets are complementary to each other, containing knowledge that other views do not have. As a result, combining multiple views can describe an object comprehensively and accurately.

According to [47], the multi-view-learning algorithms can be classified into three groups: (1) co-training, (2) multiple kernel learning, and (3) subspace learning. Notably, co-training-style algorithms [8] train alternately to maximize mutual agreement between two distinct views of the data. Multiple kernel–learning algorithms [19] exploit kernels that naturally correspond to different views and combine kernels either linearly or nonlinearly to improve learning. Subspace-learning algorithms [13] aim to obtain a latent subspace shared by multiple views, assuming that the input views are generated from this latent subspace.

9.4.1.1 Co-training Co-training [8] is one of the earliest schemes for multi-view learning. Co-training considers a setting in which each example can be partitioned into two distinct views, making three main assumptions: (1) sufficiency: each view is sufficient for classification on its own; (2) compatibility: the target functions in both views predict the same labels for co-occurring features with high probability; and (3) conditional independence: the views are conditionally independent given the class label. The conditional independence assumption is usually too strong to be satisfied in practice. Consequently, several weaker alternatives [2] have thus been considered.

In the original co-training algorithm [8] given a set L of labeled examples and a set U of unlabeled examples, the algorithm first creates a smaller pool U' containing u unlabeled examples. It then iterates the following procedures. First, it uses L to train two classifiers f_1 and f_2 on the view v_1 and v_2, respectively. Second, it allows each of these two classifiers to examine the unlabeled set U' and add the p examples it most confidently labels as positive and the n examples it most confidently labels as negative to L, along with the labels assigned by the corresponding classifier. Finally, the pool U' is replenished by drawing $2p+2n$ examples from U at random. The intuition behind the co-training algorithm is that classifier f_1 adds examples to the labeled set that classifier f_2 will then be able to use for learning. If the independence assumption is violated, on average the added examples will be less informative. Thus, co-training may not be that successful. Since then, many variants have been developed.

Instead of assigning labels to the unlabeled examples, Nigam et al. [30] give unlabeled examples probabilistic labels that may change from one iteration to another by running an expectation-and-maximization (EM) algorithm in each view. This algorithm, called co-EM, outperforms co-training for many problems but requires the classifier of each view to generate class probabilities. By reformulating the supported vector machine (SVM) in a probabilistic way, Brefeld et al. [9] develop a co-EM version of SVM to close this gap. Zhou et al. [62] expand the co-training-style algorithms from classification to regression problems. They propose an algorithm, called *CoREG*, that employs two k-nearest neighbor (KNN) regressors. Each regressor labels the unlabeled data for the other during the learning process. For the sake of choosing the appropriate unlabeled examples to label, CoREG estimates the labeling confidence by consulting the influence of the labeling of unlabeled examples on the labeled examples. The final prediction is made by averaging the regression estimates generated by both regressors.

Example 7. Zheng et al. [56, 57] propose a co-training-based model to infer fine-grained air quality throughout a city based on five datasets: air quality, meteorological data, traffic, POIs, and road networks. Figure 9.8a illustrates the philosophy of the model from multi-view learning's perspective. Naturally, air quality has temporal dependency in an individual location (represented by the vertical black broken arrows) and the spatial correlation among different locations (denoted by the horizontal red solid arrows). For example, the current air quality of a location depends on past hours. In addition, the air quality of a place could be bad if the air quality of its surrounding locations is bad. So, the temporal dependency and spatial correlation formulate two distinct views (a temporal view and a spatial view) on the air quality of a location.

As presented in figure 9.8b, a co-training-based framework is proposed, consisting of two classifiers. One is a spatial classifier based on an artificial neural network (ANN), which takes spatially related features (e.g., the density of POIs and the length of highways) as input to model the spatial correlation between air quality levels of different locations. The other is a temporal classifier based on a linear-chain conditional random field (CRF) involving temporally related features (e.g., traffic and meteorology)

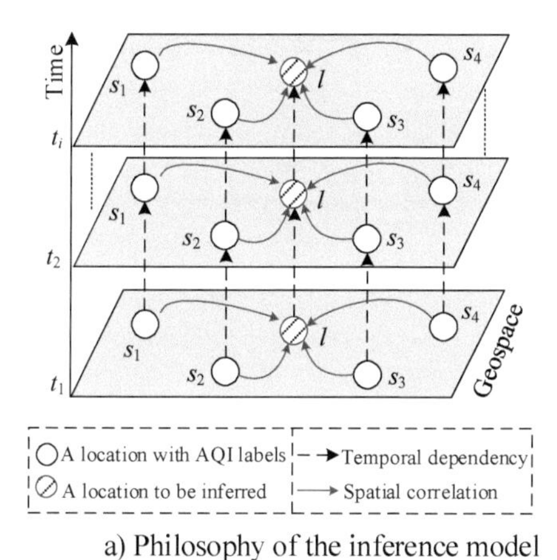

a) Philosophy of the inference model

b) Procedure of co-training

Figure 9.8
Co-training-based air quality inference model.

to model the temporal dependency of air quality in a location. The two classifiers are first trained based on limited labeled data using nonoverlapped features and then infer unlabeled instances, respectively. The instances that are confidently inferred by a classifier in each round are brought to the training set, which will be used to retrain the two classifiers in the next round. The iteration repeats until the unlabeled data has been consumed out, or the inference accuracy does not increase anymore.

When inferring the label of an instance, we send different features to different classifiers, generating two sets of probability distribution across different labels. The label with the maximum product of probabilities from the two classifiers is selected as the result.

9.4.1.2 Multikernel learning *Multiple-kernel learning* (MKL) refers to a set of machine-learning methods that uses a predefined set of kernels and learns an optimal linear or nonlinear combination of kernels as part of the algorithm. A kernel is a hypothesis on the data, which could be a similarity notion, or a classifier, or a regressor. According to [19], there are two uses of MKL (as shown in figure 9.9):

1. Different kernels correspond to different notions of similarity. A learning method picks the best kernel or uses a combination of these kernels. A sample of data is retrieved from the entire set to train a kernel based on all features. As a specific kernel may be a source of bias, allowing a learner to choose among a set of kernels can result in a better solution. For example, there are several kernel functions, such as the linear, polynomial, and Gaussian kernels, successfully used in SVM. This kind of

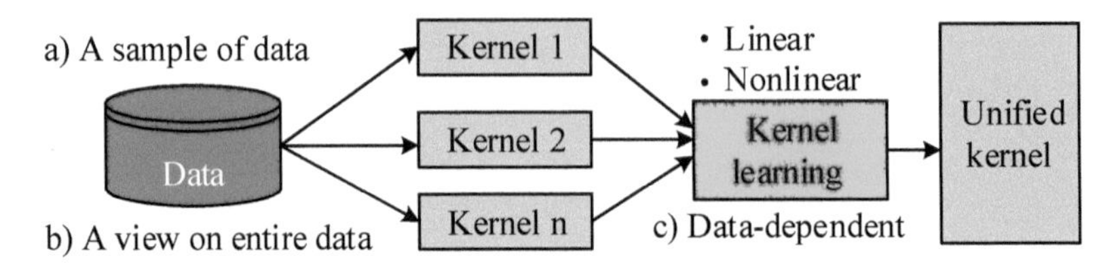

Figure 9.9
The procedure of multikernel learning.

MKL was not originally designed for multi-view learning, as the entire feature set is used for training each kernel.

2. A variation of the first use of MKL is to train different kernels using inputs coming from different representations and possibly from different sources or modalities. Since these are different representations, they have different measures of similarity corresponding to different kernels. In such a case, combining kernels is one possible way to combine multiple information sources. The reasoning is similar to combining different classifiers. Noble [32] calls this method of combining kernels *intermediate combination*, in contrast with *early combination* (where features from different sources are concatenated and fed to a single learner) and *late combination* (where different features are fed to different classifiers whose decisions are then combined by a fixed or trained combiner).

There are three ways to combine the results of kernels: linear, nonlinear, and data-dependent combinations. The linear combination consists of an unweighted (i.e., mean) and a weighted sum. Nonlinear combination methods [42] use nonlinear functions of kernels—namely, multiplication, power, and exponentiation. Data-dependent combination methods assign specific kernel weights for each data instance. By doing this, they can identify local distributions in the data and learn proper kernel-combination rules for each region.

Existing MKL algorithms have two main groups of training methodology:

1. One-step methods calculate the parameters of the combination function and base learners in a single pass, using a sequential approach or a simultaneous approach. In the sequential approach, the combination function parameters are determined first, and then a kernel-based learner is trained using the combined kernel. In the simultaneous approach, both sets of parameters are learned together.
2. Two-step methods use an iterative approach. In each iteration, we first update the parameters of the combination function while fixing that of the base learner. We

then update the parameters of base learners while fixing the parameters of the combination function. These two steps are repeated until convergence.

Example 8. Ensemble and boosting methods [1], such as random forest [10], are inspired by MKL. Random forest combines the idea of bootstrap aggregating (also called *bagging*) and the random selection of features in order to construct a collection of decision trees with a controlled variance. More specifically, it trains multiple decision trees by selecting a portion of training data each time based on bagging and a portion of features according to the principle introduced by [22, 23]. When a test case comes, different selections of the case's features are sent to corresponding decision trees (i.e., kernels) simultaneously. Each kernel generates a prediction, which is then aggregated linearly.

Example 9. Zheng et al. [60] forecast air quality for a location forty-eight hours in advance based on five datasets. Figure 9.10 presents the architecture of the predictive model, which contains two kernels (a spatial predictor and a temporal predictor) and a kernel-learning module (i.e., the prediction aggregator). The temporal predictor predicts the air quality of a station in terms of the data about the station, such as the local meteorology, air quality indexes (AQIs) of the past few hours, and the weather forecast of the place. Instead, the spatial predictor considers spatial neighbor data, such as the AQIs and the wind speed at the other stations, to predict a station's future air quality. The two predictors generate their own predictions independently for a station, which are combined by the prediction aggregator dynamically according to the current weather conditions of the station. Sometimes, local prediction is more important, while spatial prediction should be given a higher weight on other occasions (e.g., when the wind blows strongly). The prediction aggregator is based on a regression tree, learning the dynamic combination between the two kernels from the data.

The MKL-based framework outperforms a single kernel–based model in the air quality forecast example for the following three reasons: (1) from the feature space's perspective: the features used by the spatial and temporal predictors do not have any overlap, providing different views on a station's air quality; (2) from the model's perspective: spatial and temporal predictors model local and global factors, respectively, which have significantly different properties. For example, the local is more about a regression problem, while the global is more about a nonlinear interpolation. Thus, they should be handled by different techniques; and (3) from the parameter-learning perspective: feeding all the features into a single model results in a big model with many parameters to learn. However, the training data is limited. For instance, we only have one-and-a-half years of AQI data for a city. Decomposing a big model into three organically coupled

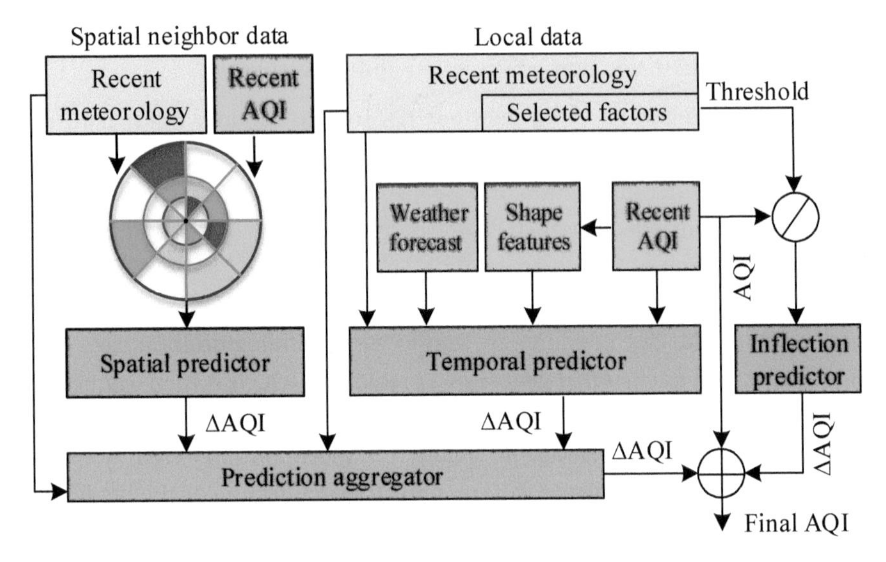

Figure 9.10
MKL-based framework for forecasting air quality.

small models scales down the parameter spaces tremendously, leading to more accurate learning and therefore the prediction.

9.4.1.3 Subspace learning Subspace learning–based approaches aim to obtain a latent subspace shared by multiple views by assuming that input views are generated from this latent subspace, as illustrated in figure 9.11. With the subspace, we can perform subsequent tasks, such as classification and clustering. Additionally, as the constructed subspace usually has a lower dimensionality than that of any input view, the "curse of dimensionality" problem can be solved to some extent.

In the literature on single-view learning, principal component analysis (PCA) is a widely used technique to exploit the subspace for single-view data. Canonical correlation analysis (CCA) [20] can be regarded as the multi-view version of PCA. By maximizing the correlation between two views in the subspace, CCA outputs one optimal projection on each view. The subspace constructed by CCA is linear and thus cannot be straightforwardly applied to nonlinearly embedded datasets. To address this issue, the kernel variant of CCA, namely KCCA [27], is proposed to map each (nonlinear) data point to a higher space in which linear CCA operates. Both CCA and KCCA exploit the subspace in an unsupervised way. Motivated by the generation of CCA from PCA, multi-view Fisher discriminant analysis [25] is developed to find informative projections with

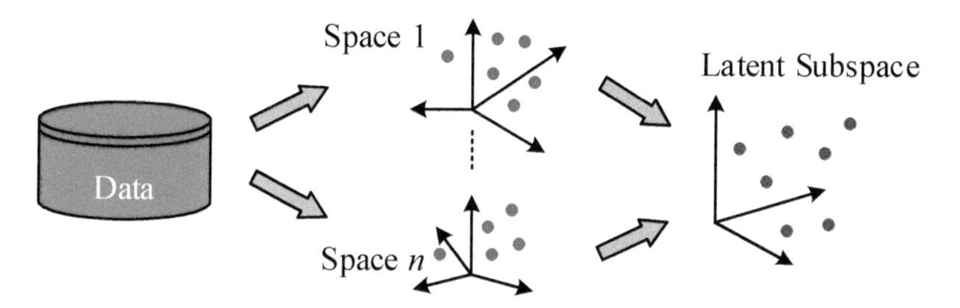

Figure 9.11
The concept of subspace learning.

label information. Lawrence [26] casts the Gaussian process as a tool to construct a latent variable model that could accomplish the task of nonlinear dimensional reduction. Chen et al. [13] develop a statistical framework that learns a predictive subspace shared by multiple views based on a generic multi-view latent-space Markov network.

9.4.2 Similarity-Based Knowledge Fusion

9.4.2.1 General insight Similarity lies between different objects. If we know two objects (X, Y) are similar in terms of some metric, information about X can be leveraged by Y when Y is lacking data, as illustrated in figure 9.12a. This is also the general idea of the collaborative-filtering algorithms that empower many recommendation systems.

When X and Y have multiple datasets (e.g., S_1, $S_2, \ldots, S_n$), respectively, as shown in figure 9.12b, we can learn multiple similarities between the two objects based on each pair of corresponding datasets. For example, we can calculate the similarity between X and Y solely based on S_1, denoted by $Sim_{S_1}(X, Y)$ and $Sim_{S_2}(X, Y)$ based on S_2, and so on. Each similarity describes the correlation between X and Y from one perspective, consolidating the correlation collectively. In turn, the correlation between X and Y can enhance each individual similarity.

For example, X and Y are two geographical regions with three types of datasets, S_1, S_2, and S_3, describing their POIs, road networks, and human-mobility patterns, respectively. The POI dataset S_1 and the road network dataset S_2 are dense enough to represent the similarity between X and Y from their own perspectives. However, the human-mobility patterns S_3 are very sparse, as they are derived from taxicab data, which is only a portion of human traffic. Thus, $Sim_{S_3}(X, Y)$ may not be reliable. If we try to fill the missing values in the two regions' S_3 dataset, the results are not accountable either. By combing S_1, S_2, and S_3, we can better understand the function of region X and Y,

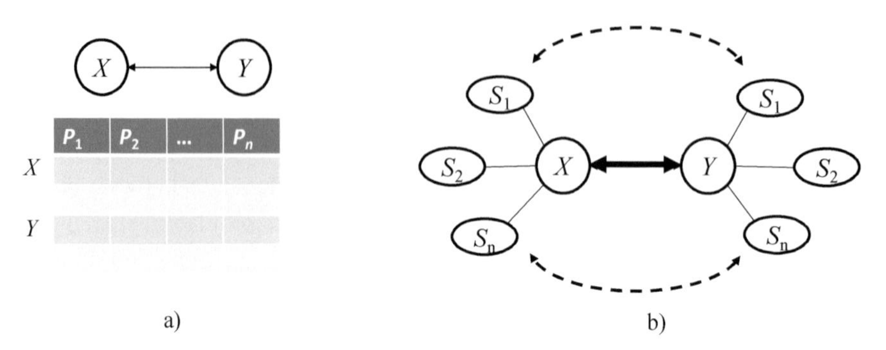

Figure 9.12
The insight of the similarity-based knowledge fusion methods.

thereby being more likely to estimate the true correlation between them. Later, we can leverage the better estimation of the correlation to enhance $Sim_{S3}(X, Y)$ and finally fill the missing values in their own S_3.

The coupled matrix factorization, context-aware tensor decomposition, and manifold alignment are representative methods in this category.

9.4.2.2 Coupled matrix factorization Coupled matrix factorization (or context-aware matrix factorization) accommodates different datasets with different matrices, which share a common dimension between one another. By decomposing these matrices collaboratively, we can transfer the similarity between different objects learned from one dataset to another, therefore complementing the missing values in a matrix more accurately. Chapter 8 introduced the principle of matrix factorization and gave two examples using this technique for location recommendation and traffic condition inference. Thus, we do not present the details of this technique again here.

9.4.2.3 Context-aware tensor decomposition Tensor decomposition algorithms add the third dimension to matrix factorization, modeling the relationship between three entities, such as *location-activity-user* in the personalized location-activity recommendation system and *user-road-time* in a driving direction service. Collectively factorizing a tensor with multiple matrices, which share common dimensions with the tensor and thus are regarded as context, is called *context-aware tensor decomposition*. These context matrices incorporate the knowledge from other datasets, thereby helping supplement the missing entries in the tensor more accurately than decomposing the tensor alone. Section 8.5 introduced the principle of tensor decomposition and gave three examples of using context-aware tensor decomposition to perform personalized location-activity recommendations, estimate the travel time of a path, and diagnose urban noises. Consequently, we do not present further details here.

9.4.2.4 Manifold alignment Manifold alignment utilizes the relationships of instances within each dataset to strengthen the knowledge of the relationships between datasets, thereby ultimately mapping initially disparate datasets to a joint latent space [43]. Manifold alignment is closely related to other manifold-learning techniques for dimensionality reduction, such as Isomap [41], locally linear embedding [35], and Laplacian eigenmaps [4]. Given a dataset, these algorithms attempt to identify the low-dimensional manifold structure of that dataset and preserve that structure in a low-dimensional embedding of the dataset. Manifold alignment follows the same paradigm but embeds multiple datasets. There are two key ideas in manifold alignment.

First, manifold alignment preserves correspondence across datasets; it also preserves the individual structures within each dataset by mapping similar instances in each dataset to similar locations in the Euclidean space. As illustrated in figure 9.13, manifold alignment maps two datasets (X, Y) to a new joint latent space $(f(X), g(Y))$, where locally similar instances within each dataset and corresponding instances across datasets are close or identical in that space. The two similarities are modeled by a lossy function with two parts: one for preserving the local similarity within a dataset and the other for the correspondences across different datasets.

Formally, with c datasets $X^1, X^2, \ldots, X^c$, the local similarity within each data set is modeled by equation (9.7):

$$C_\lambda(F^a) = \sum_{i,j} ||F^a(i,.) - F^a(j,.)||^2 \bullet W^a(i,j), \tag{9.7}$$

where X^a is the a-th dataset, which is an $n_a \times p_a$ data matrix with n_a observations and p_a features; F^a is the embedding of X^a; and W^a is an $n_a \times n_a$ matrix, where $W^a(i,j)$ is the similarity between instance $X^a(i,.)$ and $X^a(j,.)$. The sum is taken over all pairs of instances in that dataset. $C_\lambda(F^a)$ is the cost of preserving the local similarities within X^a. If two data instances from X^a, $X^a(i,.)$ and $X^a(j,.)$, are similar, which happens when $W^a(i,j)$ is larger, their locations in the latent space, $F^a(i,.)$ and $F^a(i,.)$, should be closer (i.e., $||F^a(i,.) - F^a(j,.)||^2$ is small).

To preserve the correspondence information about instances between two datasets, X^a and X^b, the cost of each pair of correspondence is $C_k(F^a, F^b)$:

$$C_k(F^a, F^b) = \sum_{i,j} ||F^a(i,.) - F^b(j,.)||^2 \bullet W^{a,b}(i,j), \tag{9.8}$$

where $W^{a,b}(i,j)$ is the similarity, or the strength of correspondence, of two instances, $X^a(i,.)$ and $X^b(j,.)$. If the two data points are in a stronger correspondence, which happens

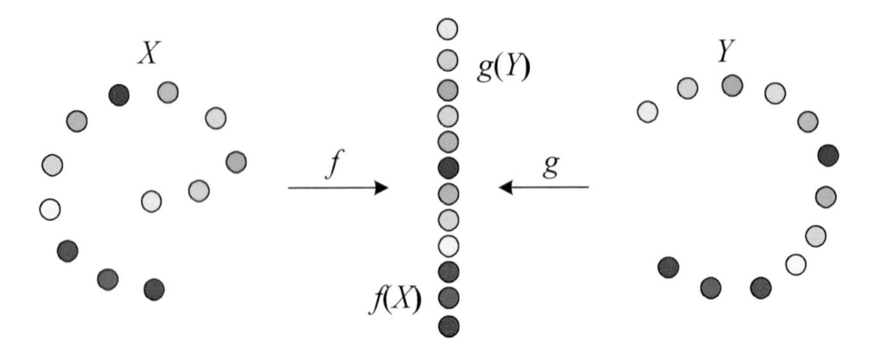

Figure 9.13
Manifold alignment of two datasets (the figure is derived from [43]).

when $W^{a,b}(i,j)$ is larger, their locations in the latent space, $F^a(i,.)$ and $F^b(j,.)$, should be closer together. Typically, $W^{a,b}(i,j)=1$ if $X^a(i,.)$ and $X^b(j,.)$ are in correspondence. So, the complete lossy function is

$$C_1(F^1, F^2, \ldots, F^k) = u \cdot \sum_a C_\lambda(F^a) + v \cdot \sum_{a \neq b} C_\kappa(F^a, F^b); \tag{9.9}$$

typically, $u=v=1$.

Second, at the algorithmic level, manifold alignment assumes the aligned disparate datasets have the same underlying manifold structure. The second loss function is simply the loss function for Laplacian eigenmaps using the joint adjacency matrix:

$$C_2(\mathbf{F}) = \sum_{i,j} \|\mathbf{F}(i,.) - \mathbf{F}(j,.)\|^2 \cdot \mathbf{W}^{a,b}(i,j), \tag{9.10}$$

where the sum is taken over all pairs of instances from all datasets, $\mathbf{F}$ is the unified representation of all the datasets, and $\mathbf{W}$ is the $(\sum_a n_a \times \sum_a n_a)$ joint adjacency matrix of all the datasets.

$$\mathbf{W} = \begin{pmatrix} vW^1 & uW^{1,2} & \cdots & uW^{1,c} \\ & \vdots & \ddots & \vdots \\ uW^{c,1} & uW^{c,2} & \cdots & vW^c \end{pmatrix}. \tag{9.11}$$

Equation (9.10) denotes that if two data instances, $X^a(i,.)$ and $X^b(j,.)$, are similar, regardless of whether they are in the same dataset $(a=b)$ or from different datasets $(a \neq b)$, which happens when $\mathbf{W}(i,j)$ is larger in either case, their locations in the latent space, $\mathbf{F}(i,.)$ and $\mathbf{F}(i,.)$, should be closer together. Making use of the fact that $\|M(i,.)\|^2 = \sum_k M(i,k)^2$ and that the Laplacian is a quadratic difference operator,

$$
\begin{aligned}
C_2(\mathbf{F}) &= \sum_{i,j}\sum_k \|\mathbf{F}(i,k)-\mathbf{F}(j,k)\|^2 \cdot \mathbf{W}^{a,b}(i,j) \\
&= \sum_k\sum_{i,j} \|\mathbf{F}(i,k)-\mathbf{F}(j,k)\|^2 \cdot \mathbf{W}^{a,b}(i,j) \\
&= \sum_k tr(\mathbf{F}(.,k)'\mathbf{L}\mathbf{F}(.,k)) = tr(\mathbf{F}'\mathbf{L}\mathbf{F}),
\end{aligned} \tag{9.12}
$$

where $tr(\cdot)$ denotes the matrix trace, $\mathbf{L}=\mathbf{D}-\mathbf{W}$ is the joint Laplacian matrix of all the datasets, and $\mathbf{D}$ is an $(\Sigma_a n_a \times \Sigma_a n_a)$ diagonal matrix with $\mathbf{D}^a(i,j)=\sum_j \mathbf{W}(i,j)$. Standard manifold-learning algorithms are then invoked on $\mathbf{L}$ to obtain a joint latent representation of the original datasets. Manifold alignment can therefore be viewed as a form of constrained joint dimensionality reduction that finds a low-dimensional embedding of multiple datasets that preserves any known correspondence across them.

Example 10. Zheng et al. [58] infer the fine-grained noise situation by using 311 complaint data together with social media, road network data, and POIs. As shown in figure 9.14, they model the noise situation of New York City with a three-dimensional tensor, where the three dimensions stand for regions, noise categories, and time slots, respectively. An entry $\mathcal{A}(i,j,k)$ stores the total number of 311 complaints of category c_j in region r_i and time slot t_k over the given period of time. This is a very sparse tensor, as there may not be people reporting a noise situation anytime and anywhere. If the tensor can be filled completely, we are able to know the noise situation throughout the city.

To deal with the data sparsity problem, they extract three categories of features—geographical features, human-mobility features, and noise category correlation features (denoted by matrices X, Y, and Z)—from POI/road network data, user check-ins, and 311 data, respectively. For example, a row of matrix X denotes a region, and each column stands for a road network feature, such as the number of intersections and the total length of road segments in the region. Matrix X incorporates the similarity between two regions in terms of their geographic features. Intuitively, regions with similar geographic features could have a similar noise situation. $Z \in \mathbb{R}^{M \times M}$ is the correlation matrix between different categories of noise. $Z(i,j)$ denotes how often a category of noise c_i co-occurs with another category c_j.

These features are used as contexts in a context-aware tensor decomposition approach to supplement the missing entries of the tensor. More specifically, $\mathcal{A}$ is decomposed into the multiplication of a few (low-rank) matrices and a core tensor (or just a few vectors), based on $\mathcal{A}$'s nonzero entries. Matrix X can be factorized into the multiplication of two matrices, $X=R \times U$, where $R \in \mathbb{R}^{N \times d_R}$ and $U \in \mathbb{R}^{d_R \times P}$ are low-rank latent factors for regions and geographical features, respectively. Likewise, matrix Y can be factorized

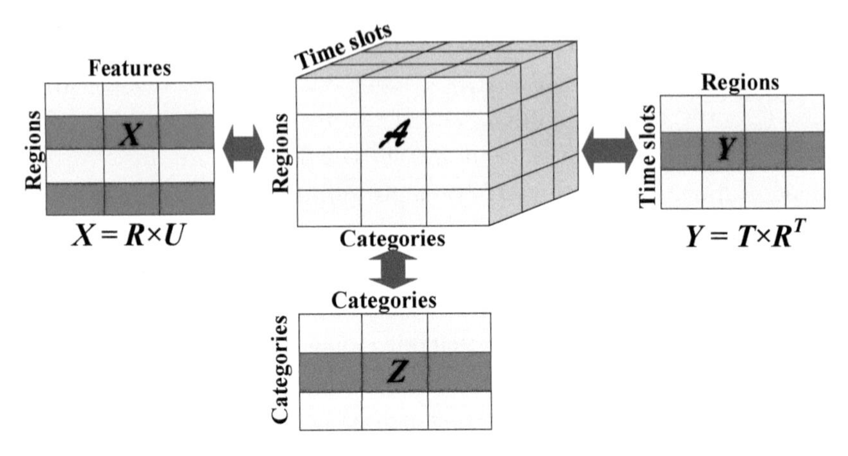

Figure 9.14
Diagnosing urban noise with tensor decomposition.

into the multiplication of two matrices, $Y = T \times R^T$, where $T \in \mathbb{R}^{L \times d_T}$ is a low-rank latent factor matrix for time slots. d_T and d_R are usually very small. The objective function is defined as

$$\begin{aligned}\mathcal{L}(S,R,C,T,U) = &\frac{1}{2}\|\mathcal{A} - S \times_R R \times_C C \times_T T\|^2 \\ &+ \frac{\lambda_1}{2}\|X - RU\|^2 + \frac{\lambda_2}{2} tr(C^T L_Z C) + \frac{\lambda_3}{2}\|Y - TR^T\|^2 \\ &+ \frac{\lambda_4}{2}(\|S\|^2 + \|R\|^2 + \|C\|^2 + \|T\|^2 + \|U\|^2),\end{aligned} \tag{9.13}$$

where $\|\mathcal{A} - S \times_R R \times_C C \times_T T\|^2$ controls the error of decomposing $\mathcal{A}$, $\|X - RU\|^2$ is to control the error of factorization of X, $\|Y - TR^T\|^2$ is to control the error of factorization of Y, $\|S\|^2 + \|R\|^2 + \|C\|^2 + \|T\|^2 + \|U\|^2$ is a regularization penalty to avoid overfitting; and λ_1, λ_2, λ_3, and λ_4 are parameters controlling the contribution of each part during the collaborative decomposition. Here, matrixes X and Y share the same dimension of region with tensor $\mathcal{A}$. Tensor $\mathcal{A}$ has a common dimension of time with Y and a shared dimension of category with Z. Thus, they share latent spaces for region, time, and category. This idea has been introduced in the coupled matrix factorization. $tr(C^T L_Z C)$ is derived from the manifold alignment (i.e., according to equation [9.12]):

$$\begin{aligned}\sum_{i,j}\|C(i,.) - C(j,.)\|^2 Z_{ij} &= \sum_k \sum_{i,j}\|C(i,k) - C(j,k)\|^2 Z_{ij} \\ &= tr(C^T(D - Z)C) = tr(C^T L_Z C),\end{aligned} \tag{9.14}$$

where $C \in \mathbb{R}^{M \times dc}$ is the latent space of category, $D_{ii} = \sum_i Z_{ij}$ is a diagonal matrix, and $L_Z = D - Z$ is the Laplacian matrix of the category correlation graph. $tr(C^T L_Z C)$, which guarantees two (e.g., the ith and jth) noise categories with a higher similarity (i.e., Z_{ij} is bigger) should also have a closer distance in the new latent space C. In this case, only one dataset (i.e., 311 data) is involved in the manifold alignment, so $\mathbf{D} = D$. As there is no closed-form solution for finding the global optimal result of the objective function, a numeric method, gradient descent, is employed to find a local optimization.

9.4.3 Probabilistic Dependency–Based Knowledge Fusion

This category of methods fuses knowledge from different datasets based on the probabilistic dependency between them, which could be correlation or causality. More specifically, it connects attributes of different datasets with directed or undirected edges, formulating a graphical presentation of the dependency, based on which an inference is carried out to predict the value of some variables.

For example, as illustrated in figure 9.15, a factor C from dataset S_1 is caused by another factor B from another dataset S_2, which further depends on factor C from dataset S_3. Alternatively, we may know factor A and B jointly result in C, which mutually depends on D. Probabilistic graphical models, consisting of Bayesian networks and Markov random fields, are representative methods in this category of knowledge fusion approach.

The structure of a graphical model is usually manually designed based on human knowledge, although, theoretically speaking, it can be learned automatically from the given data. Automatically learning the structure of a graphical model is still an open challenge, which is very complicated and computationally expensive. With human knowledge integrated into the model, graphical models can handle machine-learning problems with a small set of training data or even without labeled data. This can help

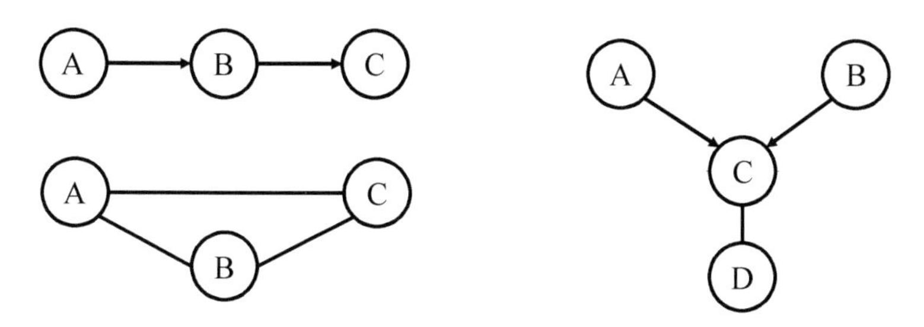

Figure 9.15
Examples of probabilistic dependency–based knowledge fusion methods.

deal with the label sparsity problem. It can also handle the label imbalance problem in a classification task, compared to other classification models, such as SVM.

9.4.3.1 Bayesian networks A Bayesian network is a directed acyclic graph that factorizes the joint probability of n variables. There are two main steps when we use a Bayesian network, composed of learning parameters and inferring the marginal probability of a variable with respect to dependent variables. Refer to section 8.6.2 for more details about Bayesian networks.

9.4.3.2 Markov random networks A Markov network, also known as a *Markov random field*, is a set of random variables having a Markov property described by an undirected graph, which may be cyclic. Thus, a Markov network can represent certain dependencies that a Bayesian network cannot (such as cyclic dependencies). On the other hand, it cannot represent certain dependencies that a Bayesian network can (such as induced dependencies). The learning and inference algorithm of Markov random networks is similar to that of Bayesian networks. Refer to section 8.6.3 for more details about Markov random networks.

9.4.4 Transfer Learning–Based Knowledge Fusion

A major assumption in many machine-learning and data-mining algorithms is that the training and future data must be in the same feature space and have the same distribution. However, in many real-world applications, this assumption may not hold. For example, we sometimes have a classification task in one domain of interest, but we only have sufficient training data in another domain of interest, where the latter data may be in a different feature space or follow a different data distribution. Different from semisupervised learning, which assumes that the distributions of the labeled and unlabeled data are the same, transfer learning, in contrast, allows the domains, tasks, and distributions used in training and testing to be different.

In the real world, we observe many examples of transfer learning. For instance, learning to recognize tables may help to recognize chairs. Learning to ride a bike may help with riding a motorcycle. Such examples are also widely witnessed in the digital world. For instance, by analyzing a user's transaction records on Amazon, we can diagnose that person's interests, which may be transferred into another application or travel recommendation. The knowledge learned from one city's traffic data may be transferred to another city.

Table 9.2
Taxonomy of transfer learning (TL) [34].

Learning settings		Source and target domains	Source and target tasks
Traditional ML		Same	Same
Transfer learning (TL)	Inductive learning/ unsupervised TL	Same	Different but related
		Different but related	Different but related
	Transductive learning	Different but related	Same

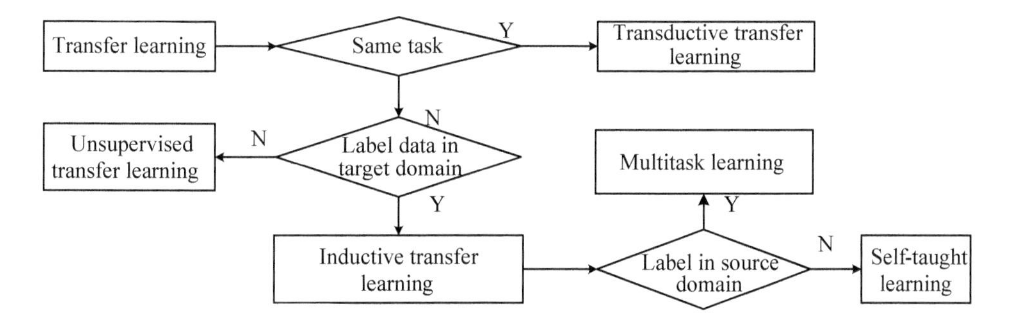

Figure 9.16
Another taxonomy of transfer learning.

9.4.4.1 Transfer between the same types of datasets Pan and Yang et al. [34] present a good survey that classifies transfer learning into three categories based on different tasks and situations between the source and target domains, as shown in table 9.2. Figure 9.16 presents another taxonomy of transfer learning according to whether label data are available in source and target domains.

Transductive learning is proposed to handle the cases in which the task is the same, but the source and target domain are different. Furthermore, there are two subcategories of the differences between source and target domains.

In the first category, the feature spaces between domains are the same, but the marginal probability distributions are different. Most existing works on transfer learning fall into this category. For example, in a traffic prediction task, we can transfer traffic data from one city to another where training data is limited.

In the second category, the feature spaces between domains are different. For example, one domain has Chinese web pages; the other has English web pages. But, the task is the same (i.e., clustering web pages according to the similarity of their semantic meanings).

Yang et al. [49] initiate the setting called *heterogeneous transfer learning* to handle this category of situation. There are two directions in this stream of work: (1) translation from the source to the target [14] or (2) projection of both domains into a common latent space [64]. Although the source and the target domains are from different feature spaces in heterogeneous transfer learning, each domain itself is homogeneous with a single data source.

Different from transductive learning, inductive learning handles learning cases in which the tasks are different in the source and target domains. It focuses on storing knowledge gained while solving one problem and applying it to a different but related problem. Multitask learning (MTL) [11] is a representative approach to inductive transfer learning. MTL learns a problem together with other related problems at the same time, using a shared representation. This often leads to a better model for the main task because it allows the learner to use the commonality among tasks [11]. MTL works well if these tasks have some commonality and are slightly undersampled.

Figure 9.17 presents two examples of MTL. Figure 9.17a illustrates the transfer of learning between two classification tasks. One task is to infer an individual's interests in different travel packages in terms of her location history in the physical world (e.g., check-ins from a social networking service). The other task is to estimate a user's interests in different book styles based on the books the user has browsed on the Internet. If we happen to have two datasets from the same user, we can associate the two tasks in an MTL framework, which learns a shared representation of a user's general interests. The books a user has browsed may imply her general interests and characteristics, which can be transferred into a travel package recommendation. Likewise, the knowledge from a user's physical location can also help estimate a user's interests in different book styles. MTL is particularly helpful when the dataset we have is sparse (e.g., we only have a small amount of check-in data from a user).

Figure 9.17b presents another example of MTL, which copredicts the air quality and traffic conditions in the near future simultaneously. The general insight is that different traffic conditions will generate different volumes of air pollutants, therefore having different impacts on air quality. Likewise, people tend to go hiking or picnic on a day with good air quality while preferring to minimize travel on a day with hazardous air quality. As a result, the traffic conditions are also affected by air quality. The shared feature representation of the two datasets can be regarded as the latent space of a location in a time slot.

Example 11. Liu et al. [29] propose a novel multitask, multi-view learning framework to forecast water quality (e.g., residual chlorine) at fifteen monitoring points in a pipe network over a two-hour time period. The prediction can inform a waterworks'

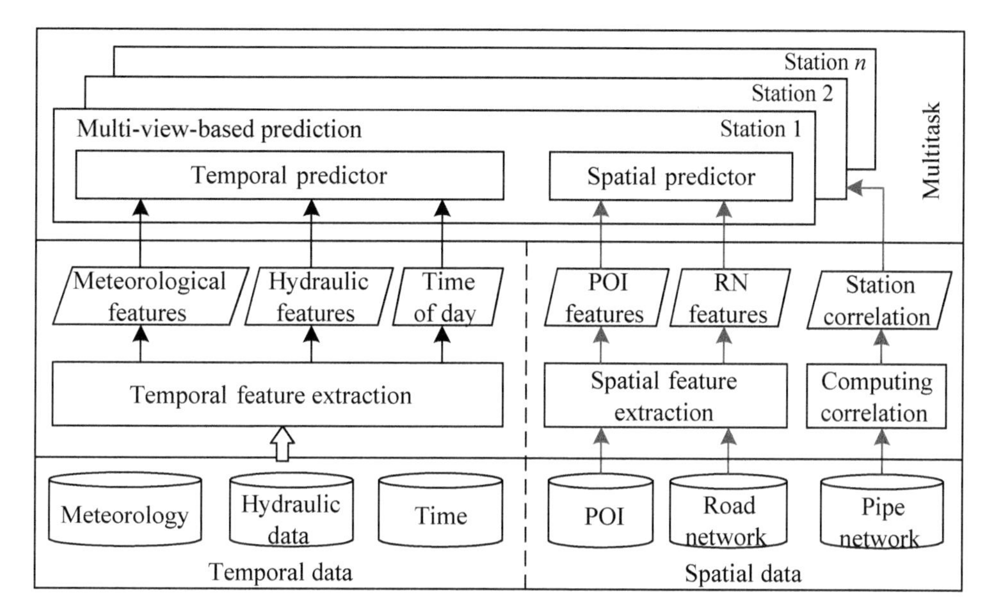

Figure 9.17
The framework of multitask, multi-view learning for forecasting urban water quality.

decision-making on adjusting water quality at its source, issuing pollution alerts, and suggesting maintenance (e.g., the replacement of certain pipelines).

In this framework, different sources of urban data are regarded as spatial or temporal views on water quality, respectively. For example, as shown in figure 9.17, features such as road networks, POIs, and the water quality of neighboring points form the spatial view on water quality, while properties like meteorology, hydraulic features, time of day, and the water quality at a point over the past few minutes are regarded as a temporal view. Two predictors, consisting of a temporal predictor and a spatial predictor, take the features of the corresponding view as an input, respectively, and generate a prediction result individually.

In addition, the prediction at each monitoring point is deemed a task. The water quality at the fifteen monitoring points is predicted collectively. This is MTL. Formally, the prediction results of different views and tasks are aggregated according to the following objective function:

$$\min_W \frac{1}{2}\sum_{l=1}^{M}\left\|y_l - \frac{1}{2}X_l w_l\right\|_2^2 + \lambda\sum_{l=1}^{M}\left\|X_l^s w_l^s - X_l^t w_l^t\right\|_2^2 + \gamma\sum_{l,m,\, l\neq m}^{M} c_{lm}\left\|w_l - w_m\right\|_2^2 + \theta\left\|W\right\|_{2,1}, \tag{9.15}$$

where $X_l w_l = X_l^s w_l^s + X_l^t w_l^t$ is the aggregation of the two predictors in a task. X_l^s is the feature set of the spatial predictor at task l, and w_l^s is the corresponding parameter. For simplicity, a linear regression model is employed in the two predictors. Likewise, X_l^t is the feature set of the temporal predictor at task l, and w_l^t is the corresponding parameter. w_l denotes the parameter set of task l, consisting of w_l^s and w_l^t•$||x||_2$ stands for the l_2-norm of x.

$\left\| X_l^s w_l^s - X_l^t w_l^t \right\|_2^2$ denotes the view alignment. That is, the prediction result of the spatial and temporal views in the same task should be close to each other.

$\sum_{l,m,l \neq m}^{M} c_{lm} \left\| w_l - w_m \right\|_2^2$ stands for the task alignment. c_{lm} is the correlation between two tasks, which can be measured by the connectivity between the two monitoring points in a pipe network. If two monitoring points are well connected with short and bold pipes, the water can easily flow from one to another point. Consequently, the correlation between their water quality should be high. If c_{lm} is big (i.e., the correlation between task l and m is strong), then the parameters of the two tasks should be similar too, thus $\left\| w_l - w_m \right\|_2^2$ should be small.

λ, γ, and θ are regularization parameters. W is an $M \times N$ parameter matrix, where each row denotes one task (i.e., one monitoring point), and each column stands for a feature. M is the number of tasks, and $N = \left\| X_l^s \right\| + \left\| X_l^t \right\| \cdot \left\| W \right\|_{2,1} = \sum_{i=1}^{M} \sqrt[2]{\sum_{j=1}^{N} w_{ij}}$ is a regularization to avoid overfitting.

Experiments show that the multitask, multi-view framework outperforms classical prediction models and other machine-learning algorithms using a single view or based on a single task.

9.4.4.2 Transfer learning among multiple datasets

In the big-data era, many machine-learning tasks have to harness a diversity of data in a domain in order to achieve better performance. This calls for new techniques that can transfer the knowledge of multiple datasets from a source to a target domain. For example, a major city like Beijing may have sufficient datasets (such as traffic, meteorological, and human mobility) to infer its fine-grained air quality. But, when applying the model to another city, we may not have some kinds of dataset (e.g., traffic) at all or not enough observations in some datasets (e.g., human mobility). Can we transfer the knowledge learned from multiple datasets of Beijing to another city?

Figure 9.18 presents the four situations of transfer learning when dealing with multiple datasets, where different shapes denote different datasets (a.k.a. views). As depicted in figure 9.18a, a target domain has all kinds of datasets (that the source domain has), each of which has sufficient observations (as the source domain). That is,

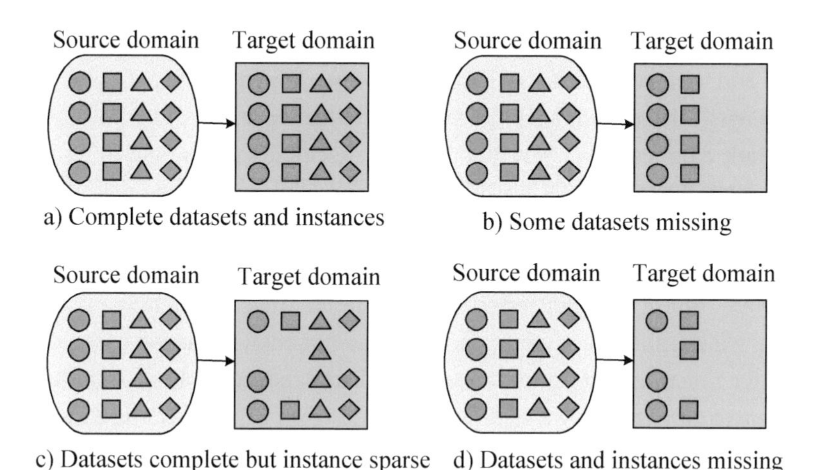

Figure 9.18
Different paradigms of transfer learning among multiple datasets.

the target domain has the same (and sufficient) feature spaces as the source domain. This kind of situation can be handled by multi-view transfer learning [15, 48, 54]. For example, Zhang et al. [54] propose multi-view transfer learning with a large margin approach (MVTL-LM), which leverages both labeled data from the source domain and features from different views. DISMUTE [15] performs feature selection for multi-view cross-domain learning. Multi-view discriminant transfer (MDT) [48] learns discriminant weight vectors for each view to minimize the domain discrepancy and the view disagreement simultaneously.

As shown in figure 9.18b, some datasets do not exist in the target domain, while other datasets are as sufficient as the source domain. To deal with such a dataset- (a.k.a. view structural) missing problem, a line of research on multi-view, multitask learning [21, 24] has been proposed. However, these algorithms cannot handle the situations shown in figure 9.18c, where a target domain has all kinds of datasets but some datasets may have very few (or very sparse) observations, or the situation presented in figure 9.18d, where some datasets do not exist (view missing) and some datasets are very sparse (observation missing). To address these issues, Wei et al. [44] propose a transfer-learning approach that can handle the view structural missing and observation missing problems.

Example 12. In urban-computing scenarios, it is very common to have sufficient data in some major cities like Beijing and New York. In small or new cities, where infrastructure for data collection is not available or just built, however, we may lack a type of data or have very sparse observations on a type of data. This leads to the structural-missing and observation-missing problems mentioned above. If we can transfer the knowledge learned from a city with rich data (called a *source city*) to other cities with insufficient data (called *target cities*), we can quickly deploy a new technology in many cities even if their data is not very ready. This helps solve the cold start problem in urban computing.

When transferring knowledge between cities, we need to understand what can be transferred and what cannot be transferred. Using the air quality–inference problem as an example, we cannot directly transfer the inference model learned based on Beijing's data to infer the air quality of another city called Baoding, as the air quality distributions in the two cities are very different. We cannot grab Beijing's traffic data to train the inference model of Baoding either, as the road networks of the two cities are very different. Thus, the two kinds of transfer do not work here.

Instead, the knowledge about the relationship between different types of datasets may be universal and thus can be transferred between different cities. For instance, air pollution might be correlated to traffic congestion. This could hold in any city. In addition, the same type of dataset in different cities may have a similar latent representation. If we can project the data of different cities into the same latent space, we might be able to use it to tackle the data scarcity problem. For example, although the traffic patterns of different cities have different geographical distributions shaped by different road networks, the latent representation may be the same (i.e., the travel speed is slow in the morning and evening rush hour and relatively fast the rest of the time). Thus, if we can project the traffic data of the two cities into the latent space, we can use it to tackle data scarcity.

As shown in figure 9.19, the transfer-learning framework first builds a dictionary based on the sufficient data from the source city through a graph clustering–based dictionary learning algorithm. The dictionary encodes the relationship between different types of datasets and is used as a base (for a sparse coding model) to project the data from both source and target cities to a latent space. Then the data from the two cities can be used together in the latent space to train a multimodal transfer AdaBoost model. As the data in the target city may have a type of data missing, a maximum pooling is employed to aggregate its data. In short, the transfer-learning framework transfers the dictionary and instances from a source city to a target city.

Regarding the dictionary-learning algorithm, it creates a graph by connecting features of different instances with two types of edges. As illustrated in figure 9.19b, each circle denotes a location (i.e., an instance), and icons of different shapes stand for

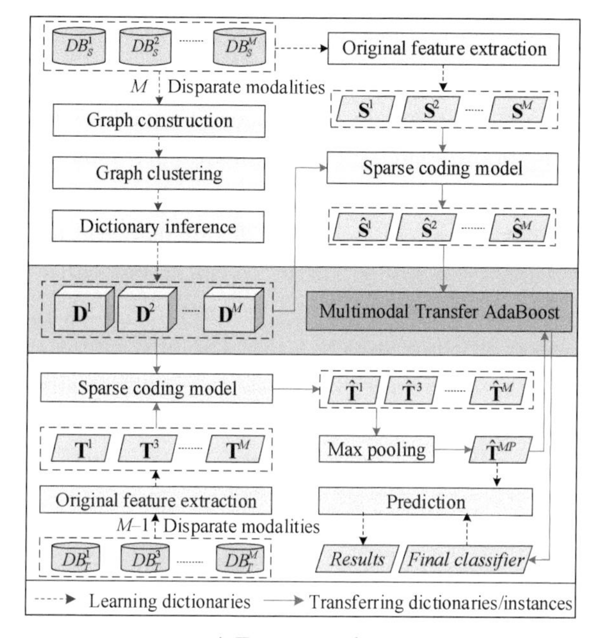

a) Framework

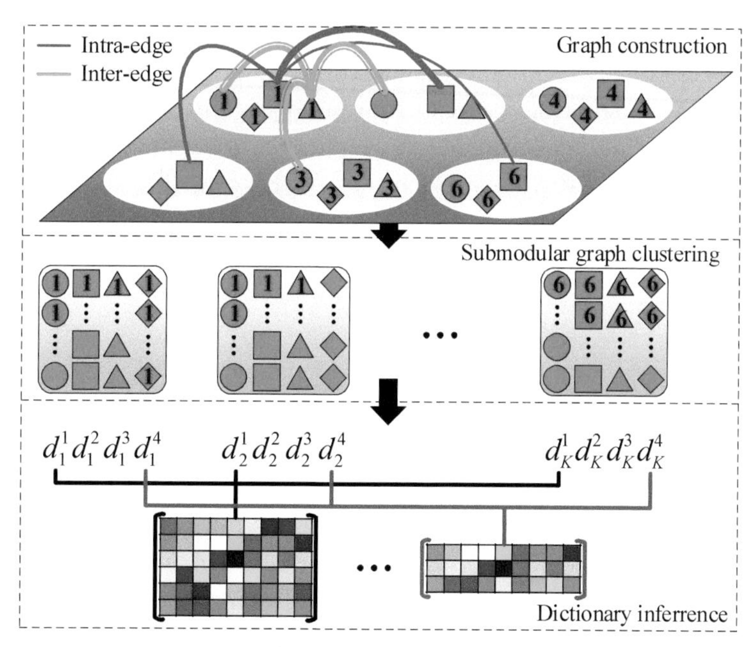

b) Dictionary learning

Figure 9.19
Transfer knowledge between cities.

different types of features. For example, the square icons may denote POI features, and the triangle icons may represent meteorological features. Each type of feature is essentially a feature vector. For instance, a POI feature vector may store the distribution of POIs across different categories. More specifically, the dictionary-learning algorithm connects the same type of feature of two different locations (i.e., two instances) with an intra-edge, if the similarity between the two instances' features is over a given threshold. The algorithm also connects different types of features of different instances with an inter-edge if their geographical distances are smaller than a given threshold.

The dictionary-learning algorithm then clusters these features in the graph into k groups with the following three criteria. First, each cluster should contain all types of features. Second, the label data (i.e., the air quality label) should be equally distributed in the k groups. Third, the label data in each group should be as consistent as possible. By averaging the feature values of the instances falling in each cluster, a dictionary can be created for each type of dataset. This algorithm encodes the relationship between different types of features in each dictionary, which will be used as a base of sparse coding to project a type of data into the latent space.

The multimodal transfer AdaBoost model trains a predictor based on features extracted from one type of dataset, aggregating the results from different predictors through a boosting approach. More specifically, the algorithm assigns initial weights to all instances in the source and target cities. It then iteratively updates the weights of these instances by assigning a smaller weight to the misclassified instances from the source city and a bigger weight to the misclassified instances from the target city, until the performance converges.

9.5 Comparison between Different Fusion Methods

Table 9.3 presents a comparison among these knowledge fusion methods (listed in the first column), where the second column (Meta) indicates if a method can incorporate other approaches as a metamethod. For instance, a semantic meaning–based data fusion method can be employed in a stage-based fusion method.

It is hard to judge which knowledge fusion method is the best, as different methods behave differently in different applications. Though a comparison is given, selecting a proper method for a given problem is still a challenging problem depending on quite a few factors.

Table 9.3
Comparison of different data fusion methods.

Methods		Meta	Labels		Goals	Training	Scalability
			Volume	Position			
Stage-based		Y	NA	NA	NA	NA	NA
Feature	Concatenation	N	Large	Flexible	F, P, C, O	S	Y
	DNN	N	Large	Flexible	F, P, A, O	U/S	Y
Semantic	Multi-view	Y	Small	Fixed	F, P, O	S, SS	Y
	Probabilistic	N	Small	Fixed	F, P, C, O, A	S/U	N
	Similarity	N	Small	Flexible	F, A, O	U	Y
	Transfer	Y	Small	Fixed	F, P, A, O	S/U	Y

9.5.1 Volume, Properties, and Insight of Datasets

First, as shown in the third column (Volume) of table 9.3, feature-based data fusion methods need a large number of labeled instances as training data, while the semantic meaning–based methods can be applied to a dataset with a small number of labeled instances. Generally, given the same amount of training data, feature concatenation–based methods are not as good as semantic meaning–based approaches, as there are dependencies and correlations between features. Adding a sparsity regularization can alleviate the problem to some extent but cannot solve it fundamentally. In some cases with a large amount of labeled data, particularly for image and speech data, the feature-based fusion using DNNs can perform well. However, the performance of the model relies heavily on tuning parameters. Given a huge model with many parameters to learn, this is usually a time-consuming process that needs the involvement of human experiences.

Second, when studying a type of object (e.g., geographical regions), we need to consider whether there are some object instances that can constantly generate labeled data (titled Fixed or Flexible in the fourth column, Position). For example, we can have fixed monitoring stations constantly generating air quality data in some regions in example 7. On the contrary, we cannot ensure that there are 311 complaints (mentioned in example 10) constantly reported by people in a region. Sometimes regions A and B have 311 complaints, while at other time intervals region C, D, and E receive these complaints. In some extreme cases, there are no regions with 311 data. That is, regions with 311 data occur flexibly and cannot formulate stable view-class label pairs for a region. As a consequence, it is not appropriate to use a multi-view-based fusion method or the feature-based fusion methods to handle the 311 example. On the other hand, the former problem cannot be solved by a similarity-based fusion method either.

As the location of a station is fixed, regions with and without labels are both fixed. We cannot calculate the similarity between a region with labeled data and another always without data.

9.5.2 The Goal of a Machine-Learning Task

The goals of fusing multiple datasets include *f*illing missing values (of a sparse dataset), *p*redicting the future, *c*ausality inference, *o*bject profiling, and *a*nomaly detection. As presented in the fifth column, probabilistic dependency–based data fusion methods can achieve all these goals (F, P, C, O, A). Particularly, Bayesian networks and feature concatenation–based fusion methods (e.g., when using a linear regression model) are usually good at dealing with causality inference problems (C). The directed edges of a Bayesian network reveal the causality between different factors (i.e., nodes), and the weight of a feature in a linear regression model denotes the importance of a factor to a problem. On the contrary, the DNN-based knowledge fusion method may not work very well for the causality inference. As raw features have been converted into a middle-level feature representation by DNNs, the semantic meaning of each feature is not clear anymore.

9.5.3 Learning Approach of Machine-Learning Algorithms

The learning methods consist of supervised (S), unsupervised (U), and semisupervised (SS) learning, as denoted in the sixth column. We can choose a corresponding method based on whether labeled ground truth is available or the volume of the labeled data. For example, supervised- and semisupervised-learning approaches can be applied to multi-view-based data fusion methods. The DNN-based fusion methods can take a supervised-learning approach to train an end-to-end classification model or use an unsupervised-learning approach to learn a latent representation of raw features. Feature concatenation–based methods cannot be applied to machine-learning tasks without labels, as they can only be trained through a supervised-learning approach.

9.5.4 Efficiency and Scalability

There are some requirements for a data-mining task, such as efficiency and scalability (shown in the right column). Generally speaking, it is not easy for probabilistic dependency–based approaches to scale up (N). A graphical model with a complex structure (e.g., many [hidden] nodes and layers) may become intractable. With respect to the similarity-based data fusion methods, when a matrix becomes very large, nonnegative matrix factorization (NMF), which can be operated in parallel, can be employed to expedite decomposition (Y).

9.6 Summary

Knowledge fusion across multiple disparate datasets is a fundamental problem in big-data research and urban computing, significantly distinguishing between cross-domain data fusion and traditional data fusion studied in the database community. This chapter introduces three categories of knowledge fusion methods, consisting of the stage-based, feature level–based, and semantic meaning–based methods.

The second category of methods is further composed of two components: the feature concatenation–based and the DNN-based methods. The feature concatenation–based methods usually come up with a regularization to avoid overfitting and reduce redundancy between features. The middle-level feature representation in the DNN-based knowledge fusion methods can be regarded as the fusion of knowledge from multiple datasets. The latent representation can be used in another classification model like SVM or a clustering algorithm (e.g., the K-means algorithm to complete a data-mining task).

The third category consists of four components: multi-view learning–based, similarity-based, probabilistic dependency–based, and transfer learning–based methods.

The multi-view learning–based method is motivated by the following insight. Different datasets or different feature subsets about an object can be regarded as different views on the object. These views are complementary to each other, containing knowledge that other views do not have. As a result, combining multiple views can describe an object comprehensively and accurately. Representative methods in the multi-view learning–based approach include co-training, multikernel learning, and subspace learning.

The similarity-based approach is motivated by the following insight. Similarity lies between different objects. If we know two objects (X, Y) are similar in terms of some metric, the information of X can be leveraged by Y when Y is lacking in data. When X and Y have multiple datasets, respectively, we can learn multiple similarities between the two objects based on each pair of corresponding datasets. Each similarity describes the correlation between X and Y from one perspective, consolidating the correlation collectively. In turn, the correlation between X and Y can enhance each individual similarity. Representative methods in the similarity-based approach are coupled matrix factorization, context-aware tensor decomposition, and manifold alignment.

Probabilistic dependency–based methods connect attributes of different datasets with directed or undirected edges, formulating a graphical presentation of the dependency, based on which an inference is carried out to predict the value of some variables. Representative methods in the probabilistic dependency–based methods are Bayesian networks, such as hidden Markov models, latent Dirichlet allocation, and Markov random fields (e.g., conditional random fields and Gaussian Markov random fields).

Motivated by the phenomenon that people can transfer the knowledge learned in a domain to solve a problem in another domain, the last category of approaches can transfer knowledge between a single type of dataset or multiple types of datasets. Multitask learning is a representative method in this category, which can be combined with multi-view learning to form a multitask, multi-view learning approach.

A comparison of these knowledge fusion methods is given at the end of this chapter, helping to choose a proper method for a given problem, which is a challenge requiring expert knowledge.

References

[1] Abney, S. 2002. "Bootstrapping." In *Proceedings of the 40th Annual Meeting of the Association for Computational Linguistics*. Stroudsburg, PA: Association for Computational Linguistics, 360–367.

[2] Balcan, M. F., A. Blum, and Y. Ke. 2004. "Co-Training and Expansion: Towards Bridging Theory and Practice." *Advances in Neural Information Processing Systems* 14:89–96.

[3] Baxter, J. 2000. "A Model of Inductive Bias Learning." *Journal of Artificial Intelligence Research* 12:149–198.

[4] Belkin, M., and P. Niyogi. 2003. "Laplacian Eigenmaps for Dimensionality Reduction and Data Representation." *Neural Computation* 15 (6): 1373–1396.

[5] Bengio, Y., A. Courville, and P. Vincent. 2013. "Representation Learning: A Review and New Perspectives." *IEEE Transactions on Pattern Analysis and Machine Intelligence* 35 (8): 1798–1828.

[6] Blei, D., A. Ng, and M. Jordan. 2003. "Latent Dirichlet Allocation." *Journal of Machine Learning Research* 3:993–1022.

[7] Bleiholder, J., and F. Naumann. 2008. "Data Fusion." *ACM Computing Surveys* 41 (1): 1–41.

[8] Blum, A., and T. Mitchell. 1998. "Combining Labeled and Unlabeled Data with Co-Training." In *Proceedings of the Eleventh Annual Conference on Computational Learning Theory*. New York: ACM, 92–100.

[9] Brefeld, U., and T. Scheffer. 2004. "Co-em Support Vector Learning." In *Proceedings of the Twenty-first International Conference on Machine Learning*. New York: Association for Computing Machinery (ACM), 16.

[10] Breiman, L. 2001. "Random Forests." *Machine Learning* 45 (1): 5–32.

[11] Caruana, R. 1997. "Multitask Learning: A Knowledge-Based Source of Inductive Bias." *Machine Learning* 28:41–75.

[12] Chawla, S., Y. Zheng, and J. Hu. 2012. "Inferring the Root Cause in Road Traffic Anomalies." In *Proceedings of the 2012 IEEE 12th International Conference on Data Mining*. Washington, DC: Institute of Electrical and Electronics Engineers (IEEE) Computer Society Press, 141–150.

[13] Chen, N., J. Zhu, and E. P. Xing. 2010. "Predictive Subspace Learning for Multi-View Data: A Large Margin Approach." *Advances in Neural Information Processing Systems* 23:361–369.

[14] Dai, W., Y. Chen, G.-R. Xue, Q. Yang, and Y. Yu. 2008. "Translated Learning: Transfer Learning across Different Feature Spaces." In *Proceedings of the 21st International Conference on Neural Information Processing Systems*. La Jolla, CA: Neural Information Processing Systems (NIPS), 353–360.

[15] Fang, Z., and Z. M. Zhang. 2013. "Discriminative Feature Selection for Multi-View Cross-Domain Learning." In *Proceedings of the 22nd International Conference on Information and Knowledge Management*. New York: ACM, 1321–1330.

[16] Fu, Y., Y. Ge, Y. Zheng, Z. Yao, Y. Liu, H. Xiong, and N. Jing Yuan. 2014. "Sparse Real Estate Ranking with Online User Reviews and Offline Moving Behaviors." In *Proceedings of the 2014 IEEE International Conference on Data Mining*. Washington, DC: IEEE Computer Society Press, 120–129.

[17] Fu, Y., H. Xiong, Y. Ge. Z. Yao, and Y. Zheng. 2014. "Exploiting Geographic Dependencies for Real Estate Appraisal: A Mutual Perspective of Ranking and Clustering." In *Proceedings of the 20th SIGKDD Conference on Knowledge Discovery and Data Mining*. New York: ACM.

[18] Fu, Y., H. Xiong, Y. Ge, Y. Zheng, Z. Yao, and Z. H. Zhou. 2016. "Modeling of Geographic Dependencies for Real Estate Ranking." *ACM Transactions on Knowledge Discovery from Data* 11 (1): 11.

[19] Gonen, M., and E. Alpaydn. 2011. "Multiple Kernel Learning Algorithms." *Journal of Machine Learning Research* 12:2211–2268.

[20] Hardoon, D., S. Szedmak, and J. Shawe-Taylor. 2004. "Canonical Correlation Analysis: An Overview with Application to Learning Methods." *Neural Computation* 16 (12): 2639–2664.

[21] He, J., and R. Lawrence. 2011. "A Graph-Based Framework for Multi-Task Multi-View Learning." In *Proceedings of the 28th International Conference on Machine Learning*. Madison, WI: Omnipress, 25–32.

[22] Ho, T. K. 1995. "Random Decision Forest." In *Proceedings of the 3rd International Conference on Document Analysis and Recognition*. Washington, DC: IEEE Computer Society Press, 278–282.

[23] Ho, T. K. 1998. "The Random Subspace Method for Constructing Decision Forests." *IEEE Transactions on Pattern Analysis and Machine Intelligence* 20 (8): 832–844.

[24] Jin, X., F. Zhuang, H. Xiong, C. Du, P. Luo, and Q. He. 2014. "Multi-Task Multi-View Learning for Heterogeneous Tasks." In *Proceedings of the 23rd ACM International Conference on Information and Knowledge Management*. New York: ACM, 441–450.

[25] Kan, M., S. Shan, H. Zhang, S. Lao, and X. Chen. 2012. "Multi-View Discriminant Analysis." In *Proceedings of the 12th European Conference on Computer Vision*. Berlin: Springer, 808–821.

[26] Lawrence, N. D. 2004. "Gaussian Process Latent Variable Models for Visualisation of High Dimensional Data." *Advances in Neural Information Processing Systems* 16:329–336.

[27] Lai, P. L., and C. Fyfe. 2000. "Kernel and Nonlinear Canonical Correlation Analysis." *International Journal of Neural Systems* 10 (5): 365–377.

[28] Liu, W., Y. Zheng, S. Chawla, J. Yuan, and X. Xie. 2011. "Discovering Spatio-Temporal Causal Interactions in Traffic Data Streams." In *Proceedings of the 17th ACM SIGKDD Conference on Knowledge Discovery and Data Mining*. New York: ACM, 1010–1018.

[29] Liu, Y., Y. Zheng, Y. Liang, S. Liu, and D. S. Rosenblum. 2016. "Urban Water Quality Prediction Based on Multi-Task Multi-View Learning." In *Proceedings of the Twenty-Fifth International Joint Conference on Artificial Intelligence*. New York: AAAI Press.

[30] Nigam, K., and R. Ghani. 2000. "Analyzing the Effectiveness and Applicability of Co-training." In *Proceedings of the Ninth International Conference on Information and Knowledge Management*. New York: ACM, 86–93.

[31] Ngiam, J., A. Khosla, M. Kim, J. Nam, H. Lee, and A. Y. Ng. 2011. "Multimodal Deep Learning." In *Proceedings of the 28th International Conference on Machine Learning*. Madison, WI: Omnipress, 689–696.

[32] Noble, W. S. 2004. "Support Vector Machine Applications in Computational Biology." In *Kernel Methods in Computational Biology*, edited by Bernhard Schölkopf, Koji Tsuda, and Jean-Philippe Vert. Cambridge, MA: MIT Press.

[33] Pan, B., Y. Zheng, D. Wilkie, and C. Shahabi. 2013. "Crowd Sensing of Traffic Anomalies Based on Human Mobility and Social Media." In *Proceedings of the 21st ACM SIGSPATIAL International Conference on Advances in Geographic Information Systems*. New York: ACM, 334–343.

[34] Pan, S. J., and Q. Yang. 2010. "A Survey on Transfer Learning." *IEEE Transactions on Knowledge Discovery and Data Engineering* 22 (10): 1345–1359.

[35] Roweis, S., and L. Saul. 2000. "Nonlinear Dimensionality Reduction by Locally Linear Embedding." *Science* 290 (5500): 2323–2326.

[36] Shang, J., Y. Zheng, W. Tong, E. Chang, and Y. Yu. 2014. "Inferring Gas Consumption and Pollution Emission of Vehicles throughout a City." In *Proceedings of the 20th ACM SIGKDD International Conference on Knowledge Discovery and Data Mining*. New York: ACM, 1027–1036.

[37] Srivastava, N., and R. Salakhutdinov. 2012. "Multimodal Learning with Deep Boltzmann Machines." In *Proceedings of the Neural Information and Processing Systems*.

[38] Sun, Y., and J. Han. 2012. "Mining Heterogeneous Information Networks: Principles and Methodologies." *Synthesis Lectures on Data Mining and Knowledge Discovery* 3 (2): 1–159.

[39] Sun, Y., J. Han, P. Zhao, Z. Yin, H. Cheng, and T. Wu. 2009. "Rankclus: Integrating Clustering with Ranking for Heterogeneous Information Network Analysis." In *Proceedings of the*

12th International Conference on Extending Database Technology: Advances in Database Technology. New York: ACM, 565–576.

[40] Sun, Y., Y. Yu, and J. Han. 2009. "Ranking-Based Clustering of Heterogeneous Information Networks with Star Network Schema." In *Proceedings of the 15th ACM SIGKDD International Conference on Knowledge Discovery and Data Mining*. New York: ACM, 797–806.

[41] Tenenbaum, J., Vin de Silva, and J. Langford. 2000. "A Global Geometric Frame-Work for Non-Linear Dimensionality Reduction." *Science* 290 (5500): 2319–2323.

[42] Varma, M., and B. R. Babu. 2009. "More Generality in Efficient Multiple Kernel Learning." In *Proceedings of the 26th Annual International Conference on Machine Learning*. New York: ACM, 1065–1072.

[43] Wang, C., P. Krafft, and S. Mahadevan. 2011. "Manifold Alignment." In *Manifold Learning: Theory and Applications*, edited by Yunqian Ma and Yun Fu. Boca Raton, FL: CRC Press.

[44] Y. Wei, Y. Zheng, and Q. Yang. 2016. "Transfer Knowledge between Cities." In *Proceedings of the 22nd ACM SIGKDD International Conference on Knowledge Discovery and Data Mining*. New York: ACM, 1905–1914.

[45] Xiao, X., Y. Zheng, Q. Luo, and X. Xie. 2010. "Finding Similar Users Using Category-Based Location History." In *Proceedings of the 18th ACM SIGSPATIAL Conference in Advances in Geographic Information Systems*. New York: ACM, 442–445.

[46] Xiao, X., Y. Zheng, Q. Luo, and X. Xie. 2014. "Inferring Social Ties between Users with Human Location History." *Journal of Ambient Intelligence and Humanized Computing* 5 (1): 3–19.

[47] Xu, C., T. Dacheng, and X. Chao. 2013. "A Survey on Multi-View Learning." arXiv:1304.5634.

[48] Yang, P., and W. Gao. 2013. "Multi-view Discriminant Transfer Learning." In *Proceedings of the Twenty-Third International Joint Conference on Artificial Intelligence*. New York: AAAI Press, 1848–1854.

[49] Yang, Q., Y. Chen, G.-R. Xue, W. Dai, and Y. Yu. 2009. "Heterogeneous Transfer Learning for Image Clustering via the Social Web." In *Proceedings of the Joint Conference of the 47th Annual Meeting of the ACL and the 4th International Joint Conference on Natural Language Processing of the AFNLP: Volume 1*. Stroudsburg, PA: Association for Computational Linguistics, 1–9.

[50] Yi, X., Y. Zheng, J. Zhang, and T. Li. 2016. "ST-MVL: Filling Missing Values in Geo-Sensory Time Series Data." In *Proceedings of the Twenty-Fifth International Joint Conference on Artificial Intelligence*. New York: AAAI Press.

[51] Yuan, J., Y. Zheng, and X. Xie. 2012. "Discovering Regions of Different Functions in a City Using Human Mobility and POIs." In *Proceedings of the 18th ACM SIGKDD International Conference on Knowledge Discovery and Data Mining*. New York: ACM, 186–194.

[52] Yuan, N. J., Y. Zheng, and X. Xie. 2012. "Segmentation of Urban Areas Using Road Networks." *Microsoft Technical Report*, MSR-TR-2012-65.

[53] Yuan, N. J., Y. Zheng, X. Xie, Y. Wang, K. Zheng, and H. Xiong. 2015. "Discovering Urban Functional Zones Using Latent Activity Trajectories." *IEEE Transactions on Knowledge and Data Engineering* 27 (3): 1041–4347.

[54] Zhang, D., J. He, Y. Liu, L. Si, and R. Lawrence. 2011. "Multi-View Transfer Learning with a Large Margin Approach." In *Proceedings of the 17th SIGKDD Conference on Knowledge Discovery and Data Mining*. New York: ACM, 1208–1216.

[55] Zheng, Y. 2015. "Trajectory Data Mining: An Overview." *ACM Transactions on Intelligent Systems and Technology* 6 (3): 1–29.

[56] Zheng, Y., X. Chen, Q. Jin, Y. Chen, X. Qu, X. Liu, E. Chang, W.-Y. Ma, Y. Rui, and W. Sun. 2013. "A Cloud-Based Knowledge Discovery System for Monitoring Fine-Grained Air Quality." *Microsoft Technical Report*, MSR-TR-2014-40.

[57] Zheng, Y., F. Liu, and H. P. Hsieh. 2013. "U-Air: When Urban Air Quality Inference Meets Big Data." In *Proceedings of the 19th SIGKDD Conference on Knowledge Discovery and Data Mining*. New York: ACM, 1436–1444.

[58] Zheng, Y., T. Liu, Y. Wang, Y. Zhu, Y. Liu, and E. Chang. 2014. "Diagnosing New York City's Noises with Ubiquitous Data." In *Proceedings of the 2014 ACM International Joint Conference on Pervasive and Ubiquitous Computing*. New York: ACM, 715–725.

[59] Zheng, Y., Y. Liu, J. Yuan, and X. Xie. 2011. "Urban Computing with Taxicabs." In *Proceedings of the 13th International Conference on Ubiquitous Computing*. New York: ACM, 89–98.

[60] Zheng, Y., X. Yi, M. Li, R. Li, Z. Shan, E. Chang, and T. Li. 2015. "Forecasting Fine-Grained Air Quality Based on Big Data." In *Proceedings of the 21st ACM SIGKDD International Conference on Knowledge Discovery and Data Mining*. New York: ACM, 2267–2276.

[61] Zheng, Y., H. Zhang, and Y. Yu. 2015. "Detecting Collective Anomalies from Multiple Spatio-Temporal Datasets across Different Domains." In *Proceedings of the 23rd SIGSPATIAL International Conference on Advances in Geographic Information Systems*. New York: ACM, 2.

[62] Zhou, Z. H., and M. Li. 2005. "Semi-Supervised Regression with Co-Training." In *Proceedings of the 19th International Joint Conference on Artificial Intelligence*. San Francisco, CA: Morgan Kaufmann.

[63] Zhu, Julie Yixuan, Chao Zhang, Huichu Zhang, Shi Zhi, Victor O. K. Li, Jiawei Han, and Yu Zheng. 2017. "pg-Causality: Identifying Spatiotemporal Causal Pathways for Air Pollutants with Urban Big Data." *IEEE Transactions on Big Data*. doi:10.1109/TBDATA.2017.2723899.

[64] Zhu, Y., Y. Chen, Z. Lu, S. J. Pan, G.-R. Xue, Y. Yu, and Q. Yang. 2011. "Heterogeneous Transfer Learning for Image Classification." In *Proceedings of the Twenty-Fifth AAAI Conference on Artificial Intelligence*. New York: AAAI Press.

10 Advanced Topics in Urban Data Analytics

Abstract: This chapter discusses a few advanced topics in urban data analytics, based on the fundamental techniques introduced in previous chapters. First, given an urban-computing problem, we usually need to answer the following question: Which datasets should be selected to solve the problem? By choosing the right datasets, we are more likely to solve a problem efficiently and effectively. Second, trajectory data has a complex data model and contains rich knowledge about moving objects, calling for unique data-mining techniques. Third, to extract (profound) knowledge from large-scale datasets requires both efficient data management techniques and machine-learning models. An organic integration of the two techniques is imperative to complete an urban-computing task. Finally, solving an urban-computing problem requires both data science and domain knowledge. How to involve human intelligence with machine intelligence is also an advanced topic that deserves discussion. Interactive visual data analytics may be an approach to addressing this issue.

10.1 How to Select Useful Datasets

Given an urban-computing problem (e.g., forecasting the air quality of a city over the next forty-eight hours), the first step is to select a few datasets, such as traffic, points of interest (POIs), and meteorological data, that can help solve this problem. As there are tens of thousands of datasets in cities, how these datasets come to mind and why these datasets could help are indeed important questions to answer. Selecting the right datasets may help solve a problem quickly, even if we do not employ the most advanced data analytics model. On the contrary, if we select a low-quality dataset that does not contain insights about the problem to be solved, we could waste a lot of effort on feature engineering and model learning. In the big-data era, data is more important than features, which are more important than models.

Choosing an effective and valuable dataset depends on two aspects. One is an understanding of the problem we are going to solve. The other is the insight behind a dataset. With the understanding of a target problem, we know the factors relating to the problem. Knowing the insights a dataset implies, we can determine which type of data can represent the factor and to what extent it can represent the factor. With the knowledge about the aforementioned two aspects, we can further validate our assumption on the correlation between the datasets we want to use and the problem we are going to solve by using correlation analysis tools, visualizations, and experiments.

10.1.1 Understanding Target Problems

With the understanding of a problem, we know the factors that could cause or correlate with the problem. Thus, we can select the data and extract corresponding features that can represent these factors. If one dataset does not contain all these factors, we will then complement it with other datasets. For example, if knowing that the air quality is related to the flow of vehicles that might issue air pollutants, we can consider the traffic flow derived from the loop detector data when predicting future air quality. However, as loop detectors are usually installed on major roads, which are a small subset of an entire road network, the data only represents a portion of traffic in the city. Then, we need to consider more datasets (e.g., using GPS trajectories from taxis, points of interest [POIs], and road network structures) as a complement. Those datasets are highly related to the traffic flow of a city, thereby helping to predict the air quality in a city.

The understanding of a problem comes from three parts. The first part is the commonsense knowledge that we have accumulated in our lives. For instance, when the weather is foggy, the air quality tends to be bad. Or, traffic repeats almost every workday in a city, showing congestion during the morning rush hour.

The second part is the knowledge learned from existing literature published by other researchers. For example, we can learn from publications in the environmental protection domain that vehicles can generate more air pollutants during traffic congestion than driven at a speed of 80 km/h. An area with a high density of buildings would have a poor dispersion condition for air pollutants and thus is more likely to experience a bad situation of air quality than open fields.

The third part comes from a simple analysis on the data we have. For example, by plotting the air quality data of a city over a period of time, we will observe a certain periodic pattern in which air quality tends to be bad during the nighttime and then goes back to normal during the daytime. This may conflict with people's ordinary understanding. Likewise, by visualizing POIs and the water quality in a pipe network together, we observe that a region with a high density of POIs tends to have good water quality. The reason could be the water-usage pattern. A region with many POIs tends to use more water and use it more often during the day, circulating fresh water more quickly in its pipe network than regions with fewer POIs. Such understanding can be obtained by simply correlating disparate datasets but is missing in the research reported in the traditional domains.

10.1.2 Insights behind Data

We usually deal with many datasets in urban computing. Besides knowing its source, format, and original meanings, we need to have a deep understanding of the insights the data implies. The insights include what factors can be learned from a dataset and to what extent the dataset can represent the factor.

For instance, the GPS trajectory of a taxicab denotes the movement of a taxi and whether there are passengers on board in different locations. This is the origin meaning of the data. However, there are many more insights, such as traffic conditions, people's commuting patterns, and a region's function, that can be derived from the data. More specifically, as taxis usually travel with other vehicles at a similar speed on the same road, the speed information derived from taxis' trajectory data indicates traffic conditions on a road. Further, as we know the pickup and drop-off point of each taxi trip, the trajectory data denotes people's commute patterns in cities. That is, massive taxi trajectories can represent people's mobility patterns in a region. The traffic conditions and people's commuting patterns can further derive the function, environment, and economy of the region. Consequently, we might use taxi trajectories as one type of source to infer a region's function.

Knowing the insight of a dataset enables us to use data from a domain to solve the problems in another domain. When one type of data is not available, we can solicit other datasets that look disparate but have an underlying correlation to the target problem. Then, we may not feel that data is enough anymore.

On the other hand, we need to know to what extent a type of data can represent a factor. For example, we know the taxi flow is a partition of traffic flow on roads. If we want to infer the traffic volume on each road segment, this type of data is not enough. Likewise, taxi is only one of many transportation modes that people can choose (there are subways, bikes, and buses, etc.). The mobility patterns derived from taxis may not be sufficient to infer the function of a region alone. To this end, we need to incorporate more datasets to supplement the missing information. For example, POIs, such as shopping malls, movie theaters, subway stations, universities, and residential real estate, would determine people's commuting patterns in the region. Likewise, the road network data significantly affects the traffic conditions on roads, as well as people's travel patterns. Thus, the two datasets can be used together with taxi trajectories to infer a region's functions, complementing the information that is missing in taxi trajectories.

10.1.3 Validating Assumptions

With the understanding of the target problem and the insights that a dataset implies, we can derive some assumptions on the correlation between the factors affecting the problem and the features extracted from the dataset. These assumptions can be further verified through the following three approaches: using correlation analysis tools, visualizations, and experimental tests.

10.1.3.1 Using the correlation analysis tools There are quite a few tools, such as the Pearson correlation and the KL-divergence, that can be used to measure the correlation between two vectors. By transforming the observations of a target problem and the features extracted from the potentially correlated dataset into two vectors of the same length, we can apply such correlation tools to validate the assumption. If the correlation between the two vectors is high or the distance between them is small, the assumption is correct. Then, we can employ this dataset to solve the problem. We demonstrate this category of approach using an example.

Example 1: Diagnose urban noise. Since 2001, New York City has operated the 311 platform to allow people to register nonemergency city disturbances by using a mobile app or making a phone call; noise is the system's third-largest category of complaints. Each complaint about noise is associated with a location, a time stamp, and a fine-grained noise category, such as loud music or construction noises. Thus, the 311 data is actually the result of "human as a sensor" and "crowd sensing," where each individual contributes her or his own information about the ambient noises, helping to diagnose the noise pollution throughout a city collectively. However, the 311 data is rather sparse, as people do not report on the noise situation around them at anytime and anywhere (refer to section 2.4.4.2 for details).

To address this issue, Zheng et al. [114] employ three additional datasets consisting of check-in data from social media, POIs, and road network data to complement the sparse 311 data, as the three datasets are correlated to urban noise according to commonsense knowledge and the following analysis. Figure 10.1 presents the number of check-ins and 311 complaints received in different categories and at different time intervals. The numbers are normalized into a value falling in [0, 1] and stored in vectors of the same length, respectively. We can see the two datasets sharing a very similar trend changing over time. For example, as shown in figure 10.1a, there is a strong correlation (Pearson correlation 0.873; P-value of T-test $\ll 0.001$) between the number of check-ins in the *Art & Entertainment* category and the number of noise complaints about vehicles in each hour of a day. Likewise, the number of user check-ins in the *Nightlife Spot* category also has a positive correlation with the number of complaints

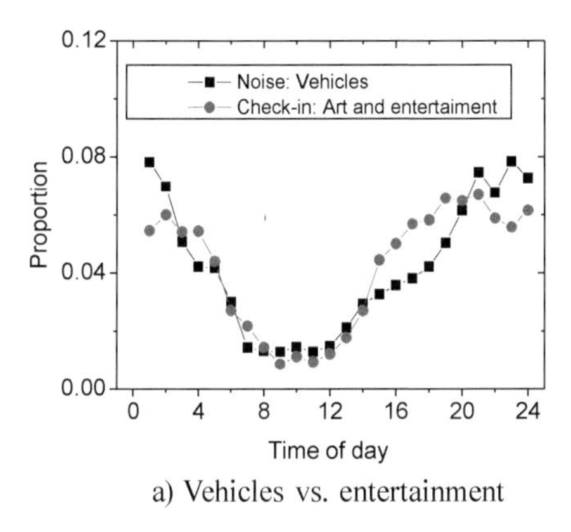

a) Vehicles vs. entertainment

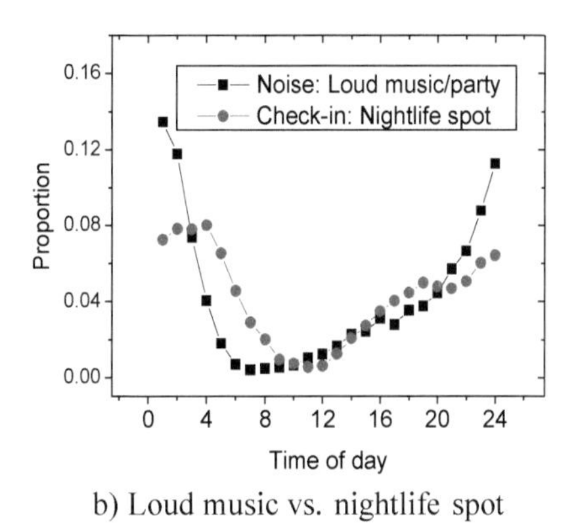

b) Loud music vs. nightlife spot

Figure 10.1
The distribution of check-ins vs. noise across different times of day.

in the category *Loud Music/Party* (Pearson correlation 0.745; *P*-value of T-test << 0.001). Figure 10.1b respectively presents the geospatial distributions of user check-ins (in the *Art & Entertainment* and *Nightlife Spot* categories) and noise complaints (in the *Loud Music/Party* category).

10.1.3.2 Through using visualizations As a target problem may be affected by multiple factors (represented by multiple datasets), the correlation between the observations of a target problem and the feature vector extracted from a single dataset may not be that obvious. Thus, the correlation analysis tools may not be able to reveal the relevance. At this moment, we can use some visualization approaches that can visually disclose the latent correlation between a dataset and a target problem in other spaces.

Following example 1, we study the correlation between POIs and 311 data using the visualizations on a geographical space. As illustrated in figure 10.2a and b, the geospatial distribution of *Loud Talking* noise complaints shares some similar regions (marked by the dotted circles) with the distribution of POIs in the *Food* category. The more POIs in the *Food* category located in a region, the more *Loud Talking* noise complaints received in the region. We also find a similarity between the distributions of noises in *Loud Music* and the POIs in *Art & Entertainment*. So, POIs and road network data can be treated as complementary information, helping supplement the noises of regions without sufficient 311 data. There are still some differences between these distributions, as each piece of data may only tell us a part of the panoramic view of urban noises. That is the reason why we need to embrace multiple data sources.

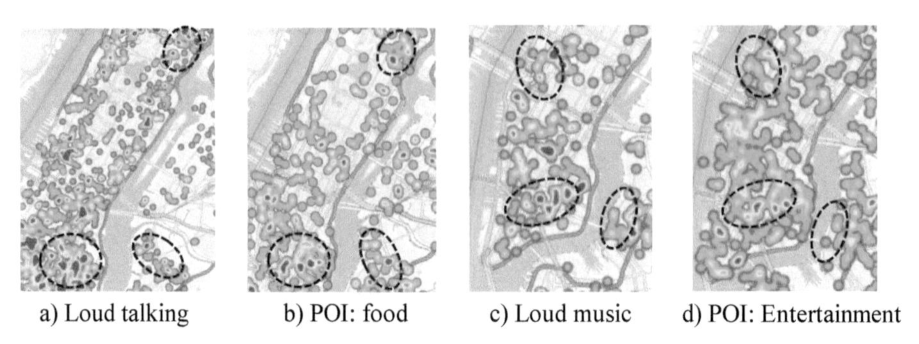

Figure 10.2
Geospatial distributions of POIs and noise complaints.

Example 2. Zheng et al. [113] infer the air quality of locations without a monitoring station using meteorological data, traffic data, POIs, and road networks. Figure 10.3 shows the correlation matrix between the air quality index (AQI) of PM10 (particulate matter with a diameter smaller than 10 μm) and four meteorological features using the data collected from August to December 2012 in Beijing.

Each row/column in this figure denotes one feature. For instance, the vertical axis of the first row and the horizontal axis of the first column denote temperature. Two features formulate a two-dimensional space, where a plot means the AQI label of a location and the coordinate of the plot is set by the two features' values in the location. Different AQI labels are represented by different shapes with different colors. For example, a green square denotes a location with good air quality, and a purple star denotes a location with hazardous air quality. (*humidity*=80, *temperature*=17, AQI=*hazardous*) is a plot in the humidity-temperature space (i.e., the second block in the first row). Supposing there are five hundred locations, then there are five hundred plots of different shapes.

In the visualization, more green squares appear on the right side of the last column where wind speeds are high, indicating that a high wind speed disperses the concentration of PM10. In addition, more purple stars occur on the top of the second row, denoting that high humidity usually causes a high concentration of PM10. The visualizations suggest the correlation between these meteorological features and PM10. Thus, these features are considered in the machine-learning model to infer the air quality of a location.

10.1.3.3 Quantitative experimental results After the two validation approaches mentioned above, we need to check whether the performance of a data analytics model increases after adding a dataset. Sometimes, even if a dataset is relevant to a problem

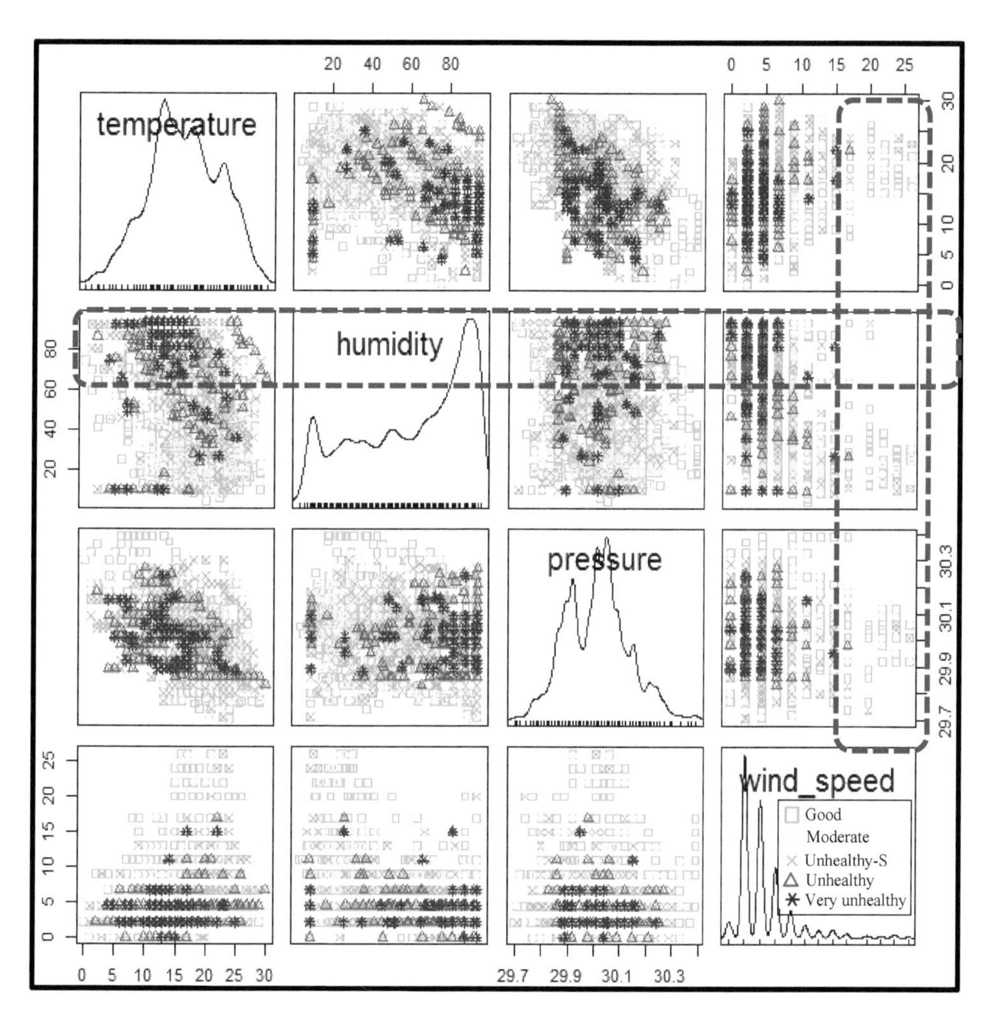

Figure 10.3
Correlation matrix between meteorological features and PM10.

according to the correlation analysis and visualization, the dataset may not enhance the performance of a machine-learning model, as its value may have already been fulfilled by the combination of other datasets.

Table 10.1 presents the evaluation of example 1, where X denotes the feature matrix extracted from POIs and road network data; Y is the feature matrix derived from social media. After additional data sources X and Y have been added one by one, the performance of the context-aware tensor decomposition model increases (i.e., the inference

Table 10.1
The performance of example 1.

	Weekdays		Weekends	
Methods	RMSE	MAE	RMSE	MAE
TD	4.391	2.381	4.141	2.393
TD+X	4.285	2.279	4.155	2.326
TD+X+Y	4.160	2.110	4.003	2.198
TD+X+Y+Z	4.010	2.013	3.930	2.072

errors decrease) gradually. This indicates that the features extracted from the two datasets are useful to solving the problem.

10.2 Trajectory Data Mining

A spatial trajectory is a trace generated by a moving object in geographical spaces, usually represented by a series of chronologically ordered points (e.g., $p_1 \rightarrow p_2 \rightarrow \cdots \rightarrow p_n$), where each point consists of a geospatial coordinate set and a time stamp such as $p=(x, y, t)$.

The advance in location acquisition technologies has generated a myriad of spatial trajectories representing the mobility of various moving objects, such as people, vehicles, and animals. Such trajectories offer us unprecedented information to understand moving objects and locations, fostering a broad range of applications in location-based social networks, intelligent transportation systems, and urban computing. The prevalence of these applications in turn calls for systematic research on new computing technologies for discovering knowledge from trajectory data. Under the circumstance, trajectory data mining [112] has become an increasingly important research theme, attracting attention from numerous areas, including computer science, sociology, and geography.

Intensive and extensive individual research has been done in the field of trajectory data mining. However, we lack a systematic review that can shape the field and position existing research. Facing a huge volume of publications, the community is still not very clear about the connections, correlations, and differences among these existing techniques. To this end, in this section, we conduct a comprehensive survey that thoroughly explores the field of trajectory data mining according to the paradigm shown in figure 10.4.

First, we classify the sources generating trajectory data into four groups, listing a few key applications that trajectory data can enable in each group.

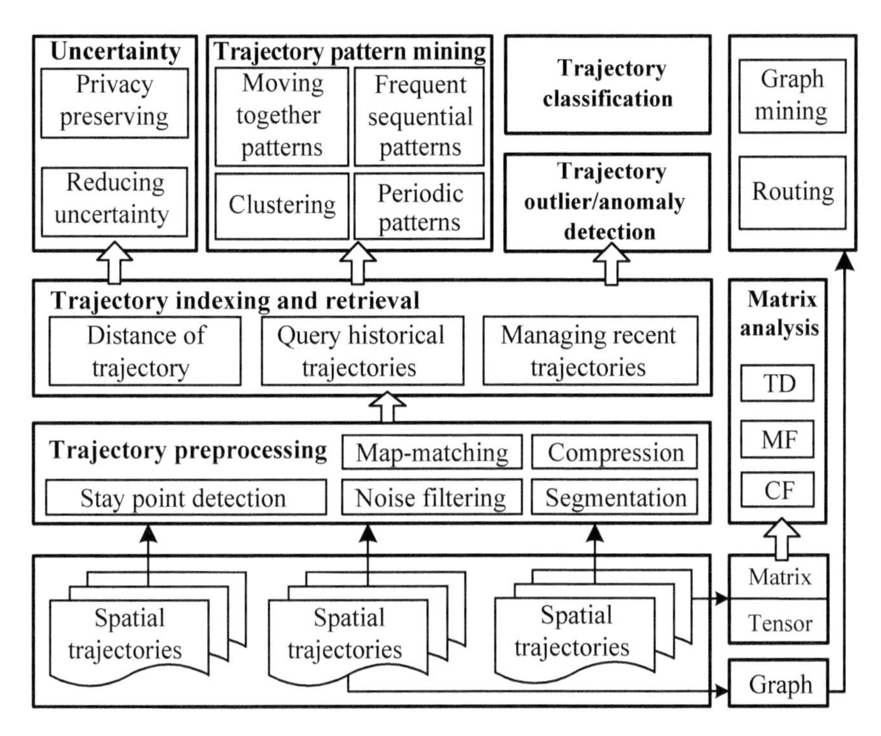

Figure 10.4
The paradigm of trajectory data mining.

Second, before using trajectory data, we need to deal with a number of issues, such as noise filtering, segmentation, and map matching. This stage is called *trajectory preprocessing*, which is a fundamental step of many trajectory data-mining tasks. The goal of noise filtering is to remove from a trajectory some noise points that may be caused by the poor signal of location-positioning systems (e.g., when traveling in a city canyon). Trajectory compression is to compress the size of a trajectory (for the purpose of reducing overhead in communication, processing, and data storage) while maintaining the utility of the trajectory. A stay point–detection algorithm identifies the location where a moving object has stayed for a while within a certain distance threshold. A stay point could stand for a restaurant or a shopping mall that a user has been to, carrying more semantic meanings than other points in a trajectory. Trajectory segmentation divides a trajectory into fragments by time interval, or spatial shape, or semantic meanings, for a further process like clustering and classification. Map matching aims to project each point of a trajectory onto a corresponding road segment where the point was truly generated.

Third, many online applications require the instant mining of trajectory data (e.g., detecting traffic anomalies), calling for effective data management algorithms that can quickly retrieve particular trajectories satisfying certain criteria (such as spatiotemporal constraints) from a big trajectory corpus. There are usually two major types of queries: the nearest neighbors and the range queries. The former is also associated with a distance metric (e.g., the distance between two trajectories). Additionally, there are two types (historical and recent) of trajectories, which need different managing methods. This part has been discussed in chapter 4.

Fourth, based on the first two steps, we can then conduct mining tasks, like trajectory pattern mining, trajectory uncertainty, trajectory outlier detection, and trajectory classification.

- *Trajectory pattern mining*. The huge volume of spatial trajectories enables opportunities for analyzing the mobility patterns of moving objects, which can be represented by an individual trajectory containing a certain pattern or a group of trajectories sharing similar patterns. In section 9.2.6, we survey the literature that is concerned with four categories of patterns: moving-together patterns, trajectory clustering, periodic patterns, and frequent sequential patterns.
- *Trajectory uncertainty*. Objects move continuously while their locations can only be updated at discrete times, leaving the location of a moving object between two updates uncertain. To enhance the utility of trajectories, a series of research has tried to model and reduce the uncertainty of trajectories. On the contrary, a branch of research aims to protect a user's privacy when the user discloses her trajectories. We review the uncertainty of trajectory in section 9.2.5.
- *Trajectory outlier detection*. Different from trajectory patterns that frequently occur in trajectory data, trajectory outliers (a.k.a. anomalies) can be items (a trajectory or a segment of a trajectory) that are significantly different from other items in terms of some similarity metric. They can also be events or observations (represented by a collection of trajectories) that do not conform to an expected pattern (e.g., traffic congestion caused by a car accident). Section 8 introduces outlier/anomaly detection from trajectory data.
- *Trajectory classification*. Using supervised-learning approaches, we can classify trajectories or segments of a trajectory into certain categories, which can be activities (like hiking and dining) or different transportation modes, such as walking and driving. We show examples of trajectory classification in section 9.2.7.

Finally, besides studying trajectories in their original form, we can transform trajectories into other formats, such as graph, matrix, and tensor (see the right part of figure 10.4). The new representations of trajectories expand and diversify the approaches for trajectory data mining, leveraging existing mining techniques, such as graph mining, collaborative filtering (CF), matrix factorization (MF), and tensor decomposition (TD). In section 9.2.9, we present representative examples of the transformation.

The framework defines the scope and road map for trajectory data mining, providing a panorama for people who want to step into this field. Individual research works are well positioned, categorized, and connected in each layer of this framework. Professionals can easily locate the methods they need to solve a problem or find the unsolved problems.

10.2.1 Trajectory Data

We classify the derivation of trajectories into four major categories.

1. *Mobility of people.* People have been recording their real-world movements in the form of spatial trajectories, actively and passively, for a long time.
   - *Active recording.* Travelers log their travel routes with GPS trajectories for the purpose of memorizing a journey and sharing experiences with friends. Bicyclers and joggers record their trails for sports analysis. In Flickr, a series of geotagged photos can formulate a spatial trajectory as each photo has a location tag and a time stamp corresponding to where and when the photo was taken. Likewise, the "check-ins" of a user in a location-based social network can be regarded as a trajectory when sorted chronologically.
   - *Passive recording.* A user carrying a mobile phone unintentionally generates many spatial trajectories represented by a sequence of cell tower IDs with corresponding transition times. Additionally, transaction records of a credit card also indicate the spatial trajectory of the cardholder, as each transaction contains a time stamp and a merchant ID denoting the location where the transaction occurred.
2. *Mobility of transportation vehicles.* A large number of GPS-equipped vehicles (such as taxis, buses, vessels, and aircraft) have appeared in our daily life. For instance, many taxis in major cities have been equipped with GPS sensors, which enables them to report a time-stamped location with a certain frequency. Such reports formulate a large number of spatial trajectories that can be used for resource allocation, traffic analysis, and improving transportation networks.
3. *Mobility of animals.* Biologists have been collecting the moving trajectories of animals like tigers and birds for the purposes of studying animals' migratory traces, behaviors, and living situations.

4. *Mobility of natural phenomena.* Meteorologists, environmentalists, climatologists, and oceanographers are busy collecting the trajectories of certain natural phenomena such as hurricanes, tornados, and ocean currents. These trajectories capture the change of the environment and climate, helping scientists deal with natural disasters and protect the natural environment we live in.

10.2.2 Trajectory Preprocessing

This section introduces four types of basic techniques that we need to process a trajectory before starting a mining task, consisting of noise filtering, stay point detection, trajectory compression, and trajectory segmentation.

10.2.2.1 Noise filtering Spatial trajectories are never perfectly accurate due to sensor noise and other factors, such as receiving poor positioning signals in urban canyons. Sometimes, the error is acceptable (e.g., a few GPS points of a vehicle fall out of the road the vehicle was actually driven on) and can be fixed by map-matching algorithms. In other situations, as shown in figure 10.5, the error of a noise point like p_5 is too big (e.g., several hundred meters away from its true location) to derive useful information, such as travel speed. So, we need to filter such noise points from trajectories before starting a mining task. Although this problem has not been completely solved, existing methods fall into three major categories.

Mean (or median) filter For a measured point z_i, the estimate of the (unknown) true value is the mean (or median) of z_i and its $n-1$ predecessors in time. The mean (median) filter can be thought of as a sliding window covering n temporally adjacent values of z_i. In the example shown in figure 10.5, $p_5 \cdot z = \sum_{i=1}^{5} p_i \cdot z/5$ if we use a mean filter with a sliding window size of 5. The median filter is more robust than the mean filter when

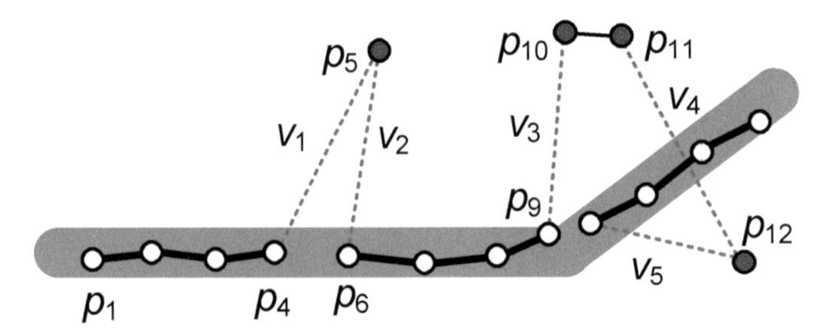

Figure 10.5
Noise points in a trajectory.

handling extreme errors and using the median rather than the mean value of the window to smooth a point.

The mean (median) filters are practical for handling individual noise points like p_5 in a trajectory with a dense representation. However, when dealing with multiple consecutive noise points (e.g., p_{10}, p_{11}, and p_{12}), a larger size of sliding window is needed. This results in a bigger error between the calculated mean (or median) value and a point's true position. When the sampling rate of a trajectory is very low (i.e., the distance between two consecutive points could be longer than several hundred meters), the mean and median filters are not good choices anymore.

Kalman and particle filters The trajectory estimated from the Kalman filter is a trade-off between the measurements and a motion model. Besides giving estimates that obey the laws of physics, the Kalman filter gives principled estimates of higher-order motion states like speed. While the Kalman filter gains efficiency by assuming linear models plus Gaussian noise, the particle filter relaxes these assumptions for a more general, but less efficient, algorithm. A tutorial-like introduction to using the Kalman and particle filters to fix noisy trajectory points can be found in [41].

The initialization step of the particle filtering is to generate P particles $\mathbf{x}_i^{(j)}, j = 1, 2, \ldots, P$ from the initial distribution. For example, these particles would have zero velocity and be clustered around the initial location measurement with a Gaussian distribution. The second step is *importance sampling*, which uses the dynamic model $P(\mathbf{x}_i|\mathbf{x}_{i-1})$ to probabilistically simulate how the particles change over one time step. The third step computes *importance weights* for all the particles using the measurement model $\omega_i^{(j)} = P(z_i \mid \hat{\mathbf{x}}_i^{(j)})$. Larger importance weights correspond to particles that are better supported by the measurement. The importance weights are then normalized so they sum to one. The last step in the loop is the *selection step*, when a new set of P particles $\mathbf{x}_i^{(j)}$ is selected from $\hat{x}_i^{(j)}$ proportional to the normalized importance weights $\omega_i^{(j)}$. Finally, we can compute a weight sum by $\hat{\mathbf{x}}_i = \sum_{i=1}^{P} \omega_i^{(j)} \hat{\mathbf{x}}_i^{(j)}$.

The Kalman and particle filters model both the measurement noise and the dynamics of the trajectory. However, they depend on the measurement of an initial location. If the first point in a trajectory is noisy, the effectiveness of the two filters drops significantly.

Heuristics-based outlier detection While the above-mentioned filters replace a noise measurement in a trajectory with an estimated value, the third category of methods removes noise points directly from a trajectory by using outlier detection algorithms. The noise-filtering method, which has been used in T-Drive [97] and GeoLife [111] projects, first calculates the travel speed of each point in a trajectory based on the time interval and distance between a point and its successor (we call this a *segment*). The segments, such as $p_4 \rightarrow p_5$, $p_5 \rightarrow p_6$, and $p_9 \rightarrow p_{10}$ (illustrated by the dotted lines in figure 10.5), with a speed larger than a threshold (e.g., 300 km/h) are cut off. Given the fact that the number of noise points is much smaller than common points, the separated points like p_5 and p_{10} can be regarded as outliers. Some distance-based outlier detection can easily find whether the number of p_5's neighbors within a distance d is smaller than the p proportion of the points in the entire trajectory. Likewise, p_{10}, p_{11}, and p_{12} can be filtered. While such algorithms can handle the initial error in a trajectory and data sparsity problems, setting the threshold d and p is still based on heuristics.

10.2.2.2 Stay point detection Spatial points are not equally important in a trajectory. Some points denote locations where people have stayed for a while, such as shopping malls and tourist attractions or gas stations where a vehicle was refueled. We call these kinds of points *stay points*. As shown in figure 10.6a, there are two types of stay points occurring in a trajectory. One is a single-point location (e.g., Stay Point 1), where a user remains stationary for a while. This situation is very rare, because a user's positioning device usually generates different readings even in the same location. The second type, like Stay Point 2 shown in figure 10.6a, is more generally observed in trajectories, representing the places where people move around (e.g., as depicted in figure 10.6b and c) or remain stationary but with shifting positioning readings.

With such stay points, we can turn a trajectory from a series of time-stamped spatial points $\boldsymbol{P}$ into a sequence of meaningful places $\boldsymbol{S}$:

$$\boldsymbol{P} = p_1 \rightarrow p_2 \rightarrow \cdots \rightarrow p_n, \Rightarrow \boldsymbol{S} = s_1 \xrightarrow{\Delta t_1} s_2 \xrightarrow{\Delta t_2}, \cdots, \xrightarrow{\Delta t_{n-1}} s_n,$$

therefore facilitating a diversity of applications, such as travel recommendations, destination predictions, taxi recommendations, and gas consumption estimations. On the other hand, in some applications (e.g., estimating the travel times of a path and

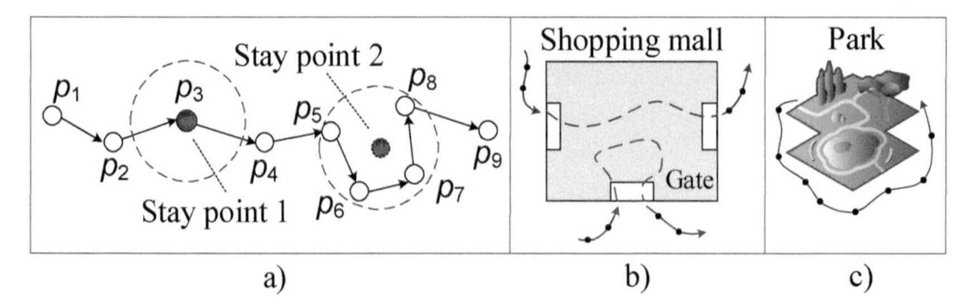

Figure 10.6
Stay points in a trajectory.

driving-direction suggestions) such stay points should be removed from a trajectory during preprocessing.

Li et al. [42] first proposed the stay point–detection algorithm. This algorithm first checks if the distance between an anchor point (e.g., p_5 depicted in figure 10.6a) and its successors in a trajectory are larger than a given threshold (e.g., 100 m). It then measures the time span between the anchor point and the last successor (i.e., p_8) that is within the distance threshold. If the time span is larger than a given threshold, a stay point (characterized by p_5, p_6, p_7, and p_8) is detected. After that, the algorithm starts detection on the next stay point from p_9.

Yuan et al. [100, 101] improved this stay point–detection algorithm based on the idea of density clustering. After finding p_5 to p_8 is a candidate stay point (using p_5 as an anchor point), their algorithm further checks the successor points from p_6. For instance, if the distance from p_6 to p_9 is smaller than the threshold, p_9 will be added into the stay point.

10.2.2.3 Trajectory compression Basically, we can record a time-stamped geographical coordinate every second for a moving object. But, this costs a lot of battery power and overhead for communication, computing, and data storage. In addition, many applications do not really need such a precision of location. To address this issue, two categories of trajectory compression strategies have been proposed, aiming to reduce the size of a trajectory while not compromising much precision in its new data representation [41]:

1. One is based on the shape of a trajectory, which is further composed of the off-line compression (a.k.a. batch mode) and online compression modes. The off-line compression mode reduces the size of a trajectory after the trajectory has been fully generated. The online mode compresses a trajectory instantly as an object travels.
2. The other category of methods compresses a trajectory based on the semantic meaning (e.g., speed and the place that a photo was taken) of each point in a trajectory, retaining not only the shape but also the semantic meaning of the trajectory.

Besides the two strategies, there are two distance metrics to measure the error of a compression: the *perpendicular Euclidean distance* and the *time-synchronized Euclidean distance*. As illustrated in figure 10.7, supposing we compress a trajectory with twelve points into a representation of three points (i.e., p_1, p_7, and p_{12}), the two distance metrics are the summation of the lengths of the segments connecting p_i and p_i', in figure 10.7a and b, respectively. The latter distance assumes a constant speed traveling between p_1 and p_7, calculating the projection of each original point on $\overline{p_1p_7}$ by time intervals.

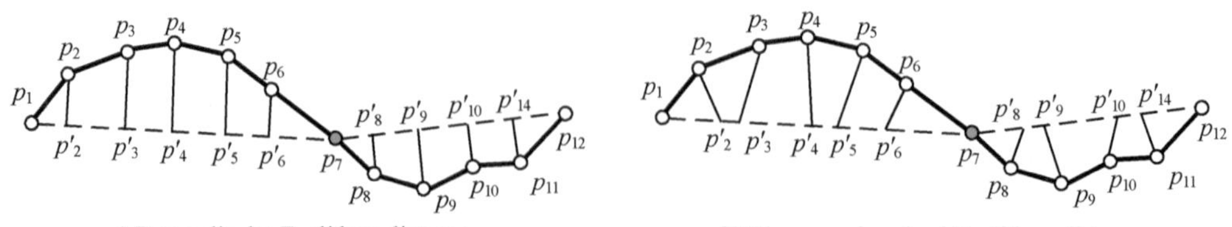

Figure 10.7
Distance metric measuring the compression error.

Shape-based trajectory compression

Off-line compression Given a trajectory that consists of a full series of time-stamped points, a batched compression algorithm aims to generate an approximated trajectory by discarding some points with a negligible error from the original trajectory. This is similar to the line simplification problem, which has been studied in the computer graphics and cartography research communities [53].

A well-known algorithm called Douglas-Peucker [20] is used to approximate the original trajectory. As demonstrated in figure 10.8a, the idea of the Douglas-Peucker algorithm is to replace the original trajectory by an approximate line segment (e.g., $\overline{p_1 p_{12}}$). If the replacement does not meet the specified error requirement (the perpendicular Euclidean distance is used in this example), it recursively partitions the original problem into two subproblems by selecting the point contributing the biggest error as the splitting point (i.e., p_4). This process continues until the error between the approximation and the original trajectory is below a specified error. The complexity of the original Douglas-Peucker algorithm is $O(N^2)$, where N is the number of points in a trajectory. Its improvement achieves $O(N \log N)$ [28]. To ensure that the approximated trajectory is optimal, Bellman's algorithm [5] employs a dynamic-programming technique with a complexity of $O(N^3)$.

Online data reduction As many applications are required to transmit trajectory data in a timely fashion, a series of online trajectory compression techniques has been proposed to determine whether a newly acquired spatial point should be retained in a trajectory. There are two major categories of online compression methods. One is the window-based algorithms, such as the *sliding window algorithm* [34] and the *open window algorithm* [52]. The other is based on the speed and direction of a moving object.

The idea of the sliding window algorithm is to fit the spatial points in a growing sliding window with a valid line segment and continue to grow the sliding window

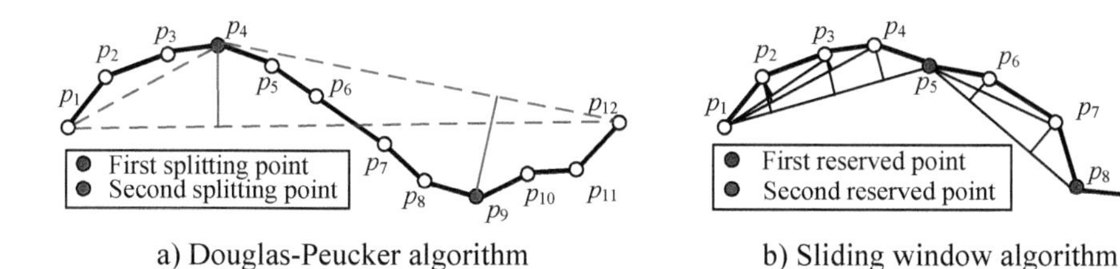

Figure 10.8
An illustration of the Douglas-Peucker algorithm.

until the approximation error exceeds some error bound. As illustrated in figure 10.8b, p_5 will be first reserved as the error for p_3 exceeds the threshold. Then, the algorithm starts from p_5 and reserves p_8. Other points are negligible. Different from the sliding window algorithm, the open window algorithm applies the heuristic of the Douglas-Peucker algorithm to choose the point with the maximum error in the window (e.g., p_3 in figure 10.8b) to approximate the trajectory segment. This point is then used as a new anchor point to approximate its successors.

Another category of algorithms considers speed and direction as key factors when doing online trajectory compression. For instance, Potamias et al. [67] use a safe area, derived from the last two locations and a given threshold, to determine whether a newly acquired point contains important information. If the new data point is located within the safe area, then this location point is considered redundant and thus can be discarded; otherwise, it is included in the approximated trajectory.

Trajectory compression with semantic meanings A series of research [15, 70] aims to keep the semantic meanings of a trajectory when compressing the trajectory. For instance, in a location-based social network, some special points where a user stayed, took photos, or changed direction greatly would be more significant than other points in representing the semantic meanings of a trajectory. Chen et al. [15] have proposed a trajectory simplification algorithm (TS) that considers both the shape skeleton and the aforementioned special points. TS first divides a trajectory into walking and nonwalking segments using a trajectory segmentation algorithm [115] (see section 9.2.2.4). A point is weighted by its heading change degree and the distance to its neighbors.

Another branch of research [33, 72] considers trajectory compression with the constraints of transportation networks. For example, we can reduce the redundant points on the same road segment. We can even discard all the newly acquired points after an anchor point, as long as the moving object is traveling on the shortest path from the anchor point to its current location. This branch of work usually needs the support of map-matching algorithms (refer to section 9.2.2.5). In 2014, PRESS [72] was proposed to separate the spatial representation of a trajectory from its temporal representation. PRESS consists of a hybrid spatial compression algorithm and an error-bounded temporal compression algorithm, compressing the spatial and temporal information of trajectories, respectively. The spatial compression combines frequent sequential pattern-mining techniques with Huffman coding to reduce the size of a trajectory (i.e., a frequently traveled path can be represented by a shorter code), therefore saving storage.

10.2.2.4 Trajectory segmentation In many scenarios, such as trajectory clustering and classification, we need to divide a trajectory into segments for a further process. The segmentation not only reduces the computational complexity but also enables us to mine richer knowledge, such as subtrajectory patterns, beyond what we can learn from an entire trajectory. In general, there are three types of segmentation methods.

The first category of methods is based on *time interval*. For example, as illustrated in figure 10.9a, if the time interval between two consecutive sampling points is larger than a given threshold, a trajectory is divided into two parts at the two points (i.e., $p_1 \rightarrow p_2$ and $p_3 \rightarrow \cdots \rightarrow p_9$). Sometimes, we can divide a trajectory into segments of the same time length.

The second category of methods is based on the *shape of a trajectory*. For example, as demonstrated in figure 10.9b, we can partition a trajectory by the turning points with the heading direction changing over a threshold. Alternatively, we can employ the line simplification algorithms, such as the Douglas-Peucker algorithm, to identify the key points retraining a trajectory's shape, as depicted in figure 10.9c. The trajectory is then partitioned into segments by these key points.

Similarly, Lee et al. [40] have proposed to partition a trajectory by using the concept of minimal description language (MDL), which comprises two components: $L(H)$ and $L(D|H)$. $L(H)$ is the length, in bits, of the description of the hypothesis H, and $L(D|H)$ is the length, in bits, of the description of the data when encoded with the help of the hypothesis. The best hypothesis H to explain D is the one that minimizes the sum of $L(H)$ and $L(D|H)$. More specifically, they use $L(H)$ to denote the total length of partitioned segments (like $\overline{p_1p_7}$ and $\overline{p_1p_9}$), while letting $L(D|H)$ represent the total (perpendicular and angle) distance between the original trajectory and the new partitioned segments. Using an approximation algorithm, they find a list of characteristic points that minimize $L(H)+L(D|H)$ from a trajectory. The trajectory is partitioned into segments by these characteristic points.

The third category of methods is based on the *semantic meanings* of points in a trajectory. As illustrated in figure 10.9d, a trajectory can be divided into segments (i.e.,

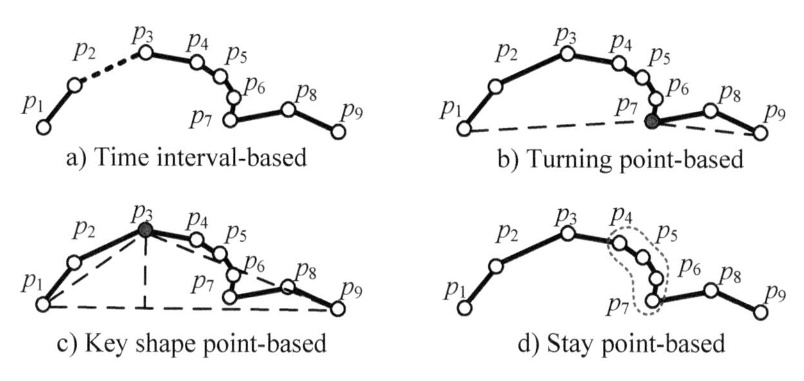

Figure 10.9
Methods of trajectory segmentation.

$p_1 \rightarrow p_2 \rightarrow p_3$ and $p_8 \rightarrow p_9$) based on the stay points it contains. Whether we should keep the stay points in the divided results depends on the applications. For example, in a task of travel speed estimation, we should remove the stay points (from a taxi's trajectory) where a taxi was parked to wait for passengers [103]. On the contrary, to estimate the similarity between two users [42], we can only focus on the sequences of stay points while skipping other raw trajectory points between two consecutive stay points.

Another semantic meaning–based trajectory segmentation is to divide a trajectory into segments of different transportation modes, such as driving, taking a bus, and walking. For example, Zheng et al. [109, 115, 119] have proposed a walk-based segmentation method. The key insight is that people have to walk through the transition between two different transportation modes. Consequently, we can first distinguish *Walk Points* from *Nonwalk Points* in a trajectory based on a point's speed ($p.v$) and acceleration ($p.a$). The trajectory can then be divided into alternate *Walk Segments* and *Nonwalk Segments*, as illustrated in figure 10.10a.

In reality, however, as shown in figure 10.10b, a few points from *Nonwalk Segments* may be detected as possible *Walk Points*, such as when a bus moves slowly in traffic congestion. On the other hand, due to the locative error, a few points from *Walk Segments* might exceed the upper bound of travel speed (v_t) and therefore be recognized as *Nonwalk Points*. To address this issue, a segment is merged into its backward segment if the distance or time span of the segment is less than a threshold. After that, a segment is regarded as a *Certain Segment* if its length exceeds a threshold, as presented in figure 10.10c. Otherwise, it is deemed to be an *Uncertain Segment*. As common users do not frequently change their transportation modes within a short distance, *Uncertain Segments* are merged into one *Nonwalk Segment* if the number of consecutive

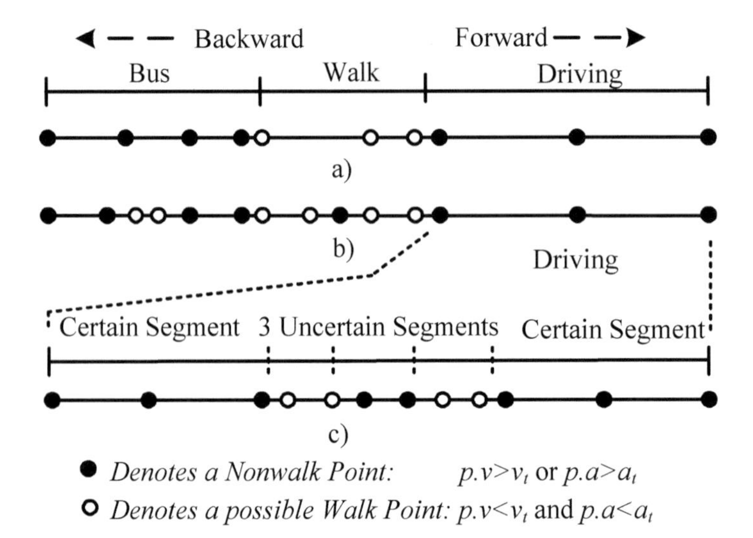

Figure 10.10
The change point–based segmentation method.

Uncertain Segments exceeds a certain threshold (three in this example). Later, features are extracted from each segment to determine its exact mode.

10.2.2.5 Map matching Map matching is a process to convert a sequence of raw latitude/longitude coordinates to a sequence of road segments. Knowledge of which road a vehicle was/is on is important for assessing traffic flow, guiding the vehicle's navigation, predicting where the vehicle is going, and detecting the most frequent travel path between an origin and a destination. Map matching is not an easy problem, given parallel roads, overpasses, and spurs [37]. There are two approaches to classifying map-matching methods based on the additional information used or the range of sampling points considered in a trajectory.

According to the additional information used, map-matching algorithms can be categorized into four groups: *geometric* [25], *topological* [14, 94], *probabilistic* [59, 66, 69] and other *advanced* techniques [51, 56, 99]. Geometric map-matching algorithms consider the shape of individual links in a road network—for example, matching a GPS point to the nearest road. Topological algorithms pay attention to the connectivity of a road network. Representative algorithms are those that use the Fréchet distance to measure the fit between a GPS sequence and a candidate road sequence [7]. To deal with noisy and low-sampling-rate trajectories, probabilistic algorithms [59, 66, 69] make explicit provisions for GPS noise and consider multiple possible paths through the road network to find the best one. More advanced map-matching algorithms have emerged recently that embrace both the topology of the road network and the noise in trajectory data, exemplified by [51, 56, 99]. These algorithms find a sequence of road segments that simultaneously come close to the noisy trajectory data and form a reasonable route through the road network.

According to the range of sampling points considered, map-matching algorithms can be classified into two categories: *local/incremental* and *global* methods. The local/incremental algorithms [11, 19] follow a greedy strategy of sequentially extending the solution from an already matched portion. These methods try to find a local optimal point based on the distance and orientation similarity. Local/incremental methods run very efficiently and are often adopted in online applications. However, when the sampling rate of a trajectory is low, the matching accuracy degrades. Instead, global algorithms [1, 7] aim to match an entire trajectory with a road network (e.g., considering the predecessors and successors of a point). Global algorithms are more accurate, but less efficient, than local methods and are usually applied to off-line tasks (e.g., mining frequent trajectory patterns) where entire trajectories have already been generated.

Advanced algorithms [51, 56, 99] embrace local and global information (or geometric, topological, and probability) to deal with the mapping of a low-sampling-rate trajectory. As shown in figure 10.11a, the algorithm proposed in [51] first finds the local candidate road segments that are within a circle distance to each point in a trajectory. For instance, road segments e_i^1, e_i^2, and e_i^3 are within the circle distance to p_i, and c_i^1, c_i^2, and c_i^3 are the candidate points on these road segments. The distance between p_i and a candidate point $dist(c_i^j, p_i)$ indicates the probability $N(c_i^j)$ that p_i can be matched to the candidate point. This probability can be regarded as the local and geometric information, which is modeled by a normal distribution:

$$N(c_i^j) = \frac{1}{\sqrt{2\pi}\sigma} e^{-\frac{dist(c_i^j, p_i)^2}{2\sigma^2}}. \tag{10.1}$$

The algorithm also considers the transition probability between the candidate points of each two consecutive trajectory points. For example, as depicted in figure 10.11b, c_i^2 is more likely to be the true match of p_i, considering p_{i-1} and p_{i+1}. The transition probability between two candidate points is denoted by the ratio between their Euclidean distance and the road network distance. The transition is actually based on the topologic information of a road network. Finally, as shown in figure 10.11c, combining the local and transition probabilities, the map-matching algorithm finds a path (on a candidate graph) that maximizes the global probability of matching. The idea is similar

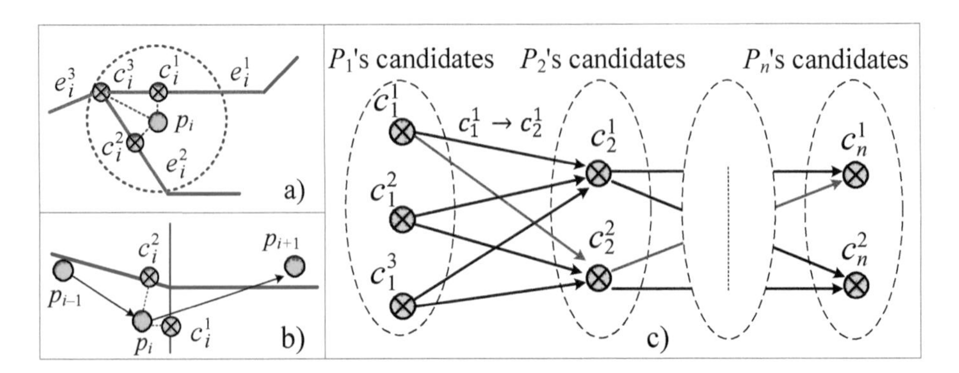

Figure 10.11
An advanced map-matching algorithm.

to the hidden Markov model (HMM) where emission and transition probabilities are considered to find the most possible sequence of status given a sequence of observations [56].

10.2.3 Trajectory Data Management

Mining massive trajectories is very time-consuming, as we need to access different samples of the trajectories or different parts of a trajectory many times. This calls for effective data management techniques that can quickly retrieve the trajectories (or parts of a trajectory) needed. Different from moving-object databases that are concerned with the current location of a moving object, trajectory data management deals with the traveling history of a moving object. As a result, the techniques include indexing structures and retrieval algorithms as well as cloud-computing platforms dedicatedly designed for trajectory data. The former have been introduced in section 4.4, and the latter have been discussed in chapter 5. Thus, we do not present them again in this section.

10.2.4 Uncertainty in a Trajectory

As the location of a moving object is recorded at a certain time interval, the trajectory data we obtain is usually a sample of the object's true movement. On the one hand, the movement of an object between two consecutive sampling points becomes unknown (or called *uncertain*). To this end, we expect to reduce the uncertainty of a trajectory. On the other hand, in some applications, to protect a user's privacy that could be leaked from her trajectories, we need to make a trajectory even more uncertain.

10.2.4.1 Reducing uncertainty from trajectory data Many trajectories have been recorded with a very low sampling rate, leading to an object's movement between sampling points being uncertain; we call them *uncertain trajectories*. For instance, as shown in figure 10.12a, the GPS coordinates of a taxi (p_1, p_2, p_3) were recorded every few minutes to reduce communication loads, resulting in multiple possible paths between two consecutive sampling points. As illustrated in figure 10.12b, people's check-in records in a location-based social-networking service like Foursquare can be regarded as trajectories if we connect them chronologically. As people do not check in very often, the time interval (and distance) between two consecutive check-ins may be hours (and several kilometers). Consequently, we have no idea how a user traveled between two check-ins. As demonstrated in figure 10.12c, to save energy, the GPS logger installed on a migratory bird can only send a location record every half day. Consequently, the path that a bird flew over two particular locations is quite uncertain.

Modeling uncertainty in a trajectory for queries Several models of uncertainty paired with appropriate query evaluation techniques [16, 65] have been proposed for moving-object databases to answer queries (e.g., "Is it possible for an object to intersect a query window?"). As illustrated in figure 10.12b, we do not know whether the trajectory formulated by the three blue check-ins should be retrieved or not by the range query R without modeling the uncertainty of the trajectory. Many of these techniques aim to provide conservative bounds for the positions of uncertain objects between two sampling points. This is usually achieved by employing geometric objects, such as cylinders [80, 81] or beads [79], as trajectory approximations. These models have little concern with regard to data mining and therefore are not the focus of this section. Recent approaches use independent probability density functions at each point in time [17], or

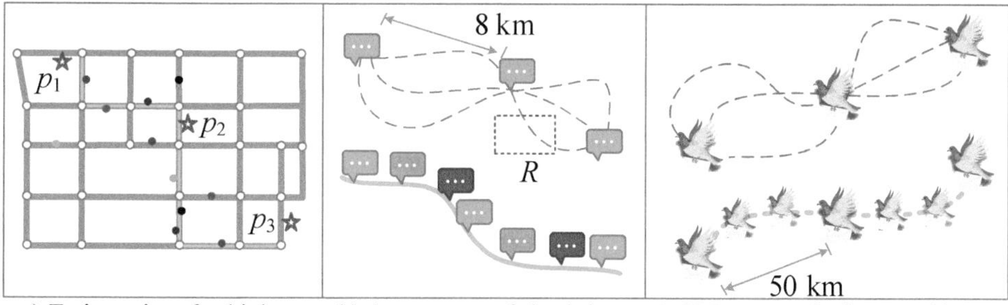

a) Trajectories of vehicles b) A sequence of check-ins c) GPS traces of migratory birds

Figure 10.12
Examples of uncertain trajectories.

stochastic processes [21, 58, 68, 87] (e.g., Markov chains), to better model the uncertain positions of an object and answer different queries.

Path inference from uncertain trajectories Different from the aforementioned models aiming at the retrieval of existing trajectories by different queries, a new series of techniques infers (or say "constructs") the most likely *k*-route(s) that a moving object could travel (i.e., the missing subtrajectory) between a few sample points based on a bunch of uncertain trajectories. The major insight is that trajectories sharing (or partially sharing) the same/similar routes can often supplement each other to make themselves more complete. In other words, it's possible to interpolate an uncertain trajectory by cross-referring other trajectories on (or partially on) the same/similar route (i.e., "uncertain + uncertain → certain").

For example, given the uncertain trajectories of many taxicabs (marked by different-colored points in figure 10.12a), we could infer that the blue path is the most likely route traversing (p_1, p_2, p_3). Likewise, based on the check-in data of many users, as depicted in figure 10.12b, we could find the blue curve the most likely travel path between the three blue check-ins. Similarly, given the uncertain GPS traces of many birds, we can identify the path that birds have flown over a few locations. Reducing the uncertainty of trajectories can support scientific studies and enable many applications, such as travel recommendations and traffic management. There are two categories of methods to complement an uncertain trajectory:

1. One is designed for the trajectories generated in a road network setting [108]. What sets this category of methods apart from map-matching algorithms lies in two aspects. First, the methods for reducing the uncertainty of trajectories leverage the data from many other trajectories, while map-matching algorithms only use the geometric information from a single trajectory and the topological information of road networks. Second, the sampling rate of trajectories handled by the uncertainty methods can be very low (e.g., more than ten minutes). This seems nearly impossible for a map-matching algorithm.
2. The other is for a free space, where moving objects (like flying birds or people hiking a mountain) do not follow paths in road networks [83], as illustrated in figure 10.12b and c. The main challenges are twofold. One is to determine those trajectories that may be relevant to a series of query points. The other is to construct a route that can approximate a bunch of relevant trajectories. As shown in figure 10.13a, the method proposed in [83] first partitions a geospace into uniform grids (the size of a grid depends on the required inference accuracy) and then maps trajectories onto

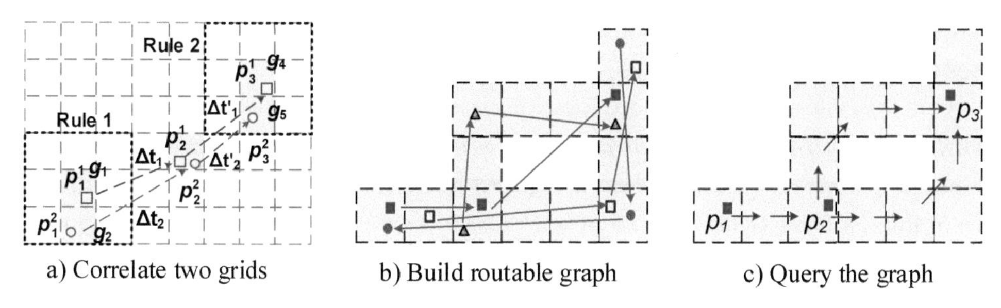

Figure 10.13
The most likely route based on uncertain trajectories.

these grids. Some grids can be connected to form a region if the trajectories passing them satisfy one of the following two rules:

i. If the starting points (p_1^1, p_1^2) of two trajectory segments are located in two grids (g_1, g_2) that are geospatial neighbors and the ending points (p_2^1, p_2^2) of the two segments are located in the same grid and the travel times $(\Delta t_1, \Delta t_2)$ of the two segments are similar, then the two grids (g_1, g_2) can be connected.

ii. If the starting points (p_2^1, p_2^2) are located in the same grid and the ending points (p_3^1, p_3^2) fall in the grids (g_4, g_5) that are neighbors and the travel times $(\Delta t_1', \Delta t_2')$ of the two segments are similar, then grids (g_4, g_5) can be connected.

After turning disjoint grids into connected region(s), as demonstrated in figure 10.13b, we can build a routable graph where a node is a grid. The direction and travel time between two adjacent grids in the graph is inferred based on the trajectories passing the two grids. Finally, as depicted in figure 10.13c, given three query points, we can find the most likely route on the graph based on a routing algorithm. To find a more detailed path, a regression can be performed over the trajectories passing the identified route.

Su et al. [74] have proposed an anchor-based calibration system that aligns trajectories to a set of fixed anchor points. The approach considers the spatial relationship between anchor points and trajectories. It also trains inference models from historical trajectories to improve the calibration.

10.2.4.2 Privacy of trajectory data Instead of reducing the uncertainty of a trajectory, a series of techniques aim to protect a user from the privacy leak caused by the disclosure of the user's trajectories [1, 18, 88]. This kind of technology tries to blur a user's location while ensuring the quality of a service or the utility of the trajectory data. There are two major scenarios that we need to protect a user's trajectory data from a privacy leak.

One is in real-time, continuous location-based services (e.g., tell me the traffic conditions that are 1 km around me). In this scenario, a user may not want to exactly disclose her current location when using a service. Different from simple location privacy, the spatiotemporal correlation between consecutive samples in a trajectory may help infer the exact location of a user. Techniques trying to protect privacy in this scenario include spatial cloaking [54], mix zones [6], path confusion [29], Euler histograms based on short IDs [86], dummy trajectories [36], and so on.

The second is the publication of historical trajectories. Collecting many trajectories of an individual may allow attackers to infer her home and workplace, therefore identifying the individual. Techniques for protecting users' privacy in such scenarios include clustering-based [1], generalization-based [55], suppression-based [77], and grid-based approaches [24]. A comprehensive survey on trajectory privacy can be found in [18].

10.2.5 Trajectory Pattern Mining

In this section, we study four major categories of patterns that can be discovered from a single trajectory or a group of trajectories. They are moving-together patterns, trajectory clustering, sequential patterns, and periodic patterns.

10.2.5.1 Moving-together patterns A branch of research focuses on discovering a group of objects that move together for a certain time period, such as a *flock* [26, 27], *convoy* [31, 32], *swarm* [43], *traveling companion* [75, 76], and *gathering* [107, 121]. These patterns can help the study of species' migration, military surveillance, traffic event detection, and so on. These patterns can be differentiated between each other based on the following factors: the shape or density of a group, the number of objects in a group, and the duration of a pattern.

Specifically, a *flock* is a group of objects that travel together within a disc of some user-specified size for at least k consecutive time stamps. A major concern with a *flock* is the predefined circular shape, which may not well describe the shape of a group in reality, therefore resulting in the so-called lossy-flock problem. To avoid rigid restrictions on the size and shape of a moving group, the *convoy* is proposed to capture the generic trajectory pattern of any shape by employing density-based clustering. Instead of using a disc, a *convoy* requires a group of objects to be density-connected during k consecutive time points. While both the *flock* and the *convoy* have a strict requirement on consecutive time periods, Li et al. [43] have proposed a more general type of trajectory pattern, called a *swarm*, which is a cluster of objects lasting for at least k (possibly nonconsecutive) time stamps. While the *convoy* and the *swarm* need to load entire trajectories into memory for a pattern mining, the *traveling companion* [75, 76] uses a data structure (called a *traveling buddy*) to continuously find *convoy/swarm*-like patterns from trajectories that are being streamed into a system. So, the *traveling companion* patterns can be regarded as an online (and incremental) detection fashion of the *convoy* and the *swarm*.

To detect some incidents, such as celebrations and parades, in which objects join and leave an event frequently, the *gathering* pattern [107, 121] further loosens the constraints of the aforementioned patterns by allowing the membership of a group to evolve gradually. Each cluster of a *gathering* should contain at least m_p participants, which are the objects appearing in at least k_p clusters of this gathering. As the *gathering* pattern is used to detect events, it also requires the geometric property (like location and shape) of a detected pattern to be relative stable.

Figure 10.14a illustrates these patterns. If setting the requirement of time stamp $k=2$, a group $<o_2, o_3, o_4>$ is a *flock* from t_1 to t_3. Though o_5 is a companion of the group, it cannot be included due to the fixed size of the disc employed by the *flock* definition. On the other hand, a *convoy* can include o_5 into the group, since $<o_2, o_3, o_4, o_5,>$ is density-based connected from t_1 to t_3. The five objects also form a *swarm* during the nonconsecutive time period t_1 and t_3. As demonstrated in figure 10.14b, if we set $k_p=2$ and $m_p=3$, then $<C_1, C_2, C_4>$ is a gathering. $<C_1, C_3, C_5>$ is not a gathering because C_5 is too far away from C_2 and C_3.

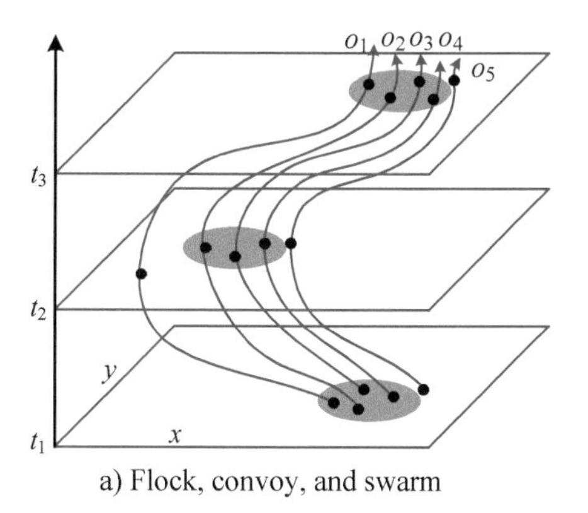

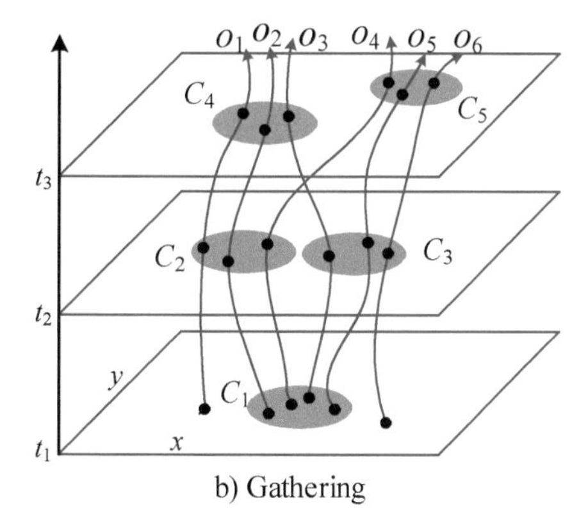

Figure 10.14
Examples of moving-together patterns.

The aforementioned pattern-mining algorithms usually use a density-based distance metric (in a Euclidean space) to find a cluster of moving objects. Christian et al. [30] extend the distance metric by considering semantic factors, such as heading direction and speed, of a moving object.

10.2.5.2 Trajectory clustering To find representative paths or common trends shared by different moving objects, we usually need to group similar trajectories into clusters. A general clustering approach is to represent a trajectory with a feature vector, denoting the similarity between two trajectories by the distance between their feature vectors. However, it is not easy to generate a feature vector with a uniform length for different trajectories, as different trajectories contain different and complex properties, such as length, shape, sampling rate, and the number of points and their orders. In addition, it is difficult to encode the sequential and spatial properties of points in a trajectory into its feature vector.

Given the challenges mentioned above, a series of techniques have been refined. Since the distance metrics between trajectories have been introduced in section 4.4, we hereafter focus on the clustering methods proposed for trajectories. Note that the clustering methods discussed in this section are dedicated to trajectories in free spaces (i.e., without a road network constraint). Though there are a few publications (e.g., [35]) discussing trajectory clustering with a road network setting, this problem can actually be solved by a combination of map-matching and graph-clustering algorithms. That is, we can first use map-matching algorithms to project trajectories onto a road network and then employ graph-clustering algorithms to find a subgraph (i.e., a collection of roads) on the road network.

Gaffney et al. [8, 22] proposed grouping similar trajectories into clusters by using a regression mixture model and the EM algorithm. This algorithm clusters trajectories with respect to the overall distance between two entire trajectories. However, moving objects rarely travel together for an entire path in the real world. To this end, Lee et al. [40] proposed to partition trajectories into line segments and to build groups of close trajectory segments using the trajectory Hausdorff distance, as illustrated in figure 10.15a. A representative path is later found for each cluster of segments.

Since trajectory data is often received incrementally, Li et al. [45] further proposed an incremental clustering algorithm, aiming to reduce the computational cost and storage of received trajectories. Both Lee [40] and Li [45] adopted a *micro-and-macro-clustering framework*, which was proposed by Aggarwal et al. [2] to cluster data streams. That is, their methods first find microclusters of trajectory segments (as demonstrated

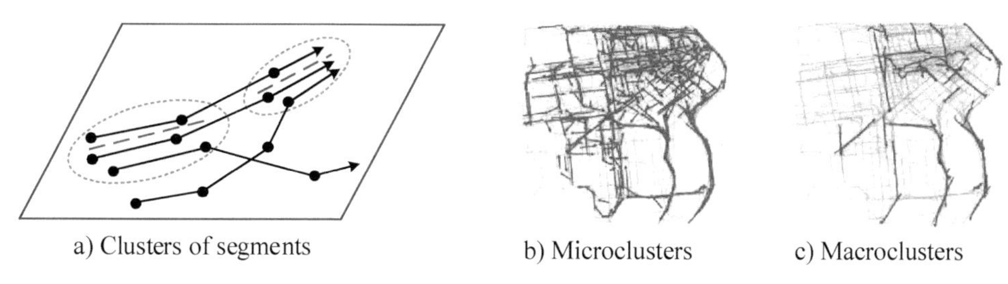

Figure 10.15
Trajectory clustering based on partial segments [108].

in figure 10.15b) and then group microclusters into macroclusters (as shown in figure 10.15c). A main insight of Li's work [45] is that new data will only affect the local area where the new data was received rather than faraway areas.

10.2.5.3 Mining sequential patterns from trajectories A branch of research aims to find the *sequential patterns* from a single trajectory or multiple trajectories. Here, a sequential pattern means a certain number of moving objects traveling a common sequence of locations over a similar time interval. The locations in a travel sequence do not have to be consecutive. For instance, two trajectories A and B,

$$A: l_1 \xrightarrow{1.5h} l_2 \xrightarrow{1h} l_7 \xrightarrow{1.2h} l_4 \quad B: l_1 \xrightarrow{1.2h} l_2 \xrightarrow{2h} l_4,$$

share a common sequence $l_1 \rightarrow l_2 \rightarrow l_4$, as the visiting orders and travel times are similar (though l_2 and l_4 are not consecutive in trajectory A). When the occurrence of such a common sequence in a corpus, usually called a *support*, exceeds a threshold, a *sequential trajectory pattern* is detected. Finding such kinds of patterns can benefit travel recommendations, life-pattern understanding, next-location prediction, estimating user similarity, and trajectory compression.

To detect the sequential patterns from trajectories, we first need to define a (common) location in a sequence. Ideally, in trajectory data like user check-in sequences from a social-networking service, each location is tagged with a unique identity (such as the name of a restaurant). If two locations share the same identity, they are common. In many GPS trajectories, however, each point is characterized by a pair of GPS coordinates, which do not repeat themselves exactly in every pattern instance. This makes the points from two different trajectories not directly comparable. In addition, a GPS trajectory may consist of thousands of points. Without proper handling, these points will result in a huge computational cost.

Sequential pattern mining in a free space

Line simplification–based methods An early solution aiming to deal with the aforementioned issues was proposed in 2005 [9]. The solution first identifies key points shaping a trajectory using a line-simplification algorithm like Douglas-Peucker [20]. It then groups the fragments of a trajectory that are close to each simplified line segment in order to count the support of each line segment. The travel time between two points in a trajectory is not considered.

Clustering-based methods Recently, a more general way to solve the above-mentioned problems is to cluster points from different trajectories into regions of interest. A point from a trajectory is then represented by the cluster ID the point belongs to. As a consequence, a trajectory is reformed as a sequence of cluster IDs, which are comparable among different trajectories. For example, as shown in figure 10.16a, the three trajectories can be represented as

$$Tr_1 : l_1 \xrightarrow{\Delta t_3} l_3,\ Tr_2 : l_1 \xrightarrow{\Delta t_1} l_2 \xrightarrow{\Delta t_2} l_3,\ Tr_3 : l_1 \xrightarrow{\Delta t_1'} l_2 \xrightarrow{\Delta t_2'} l_3,$$

where l_1, l_3, and l_3 are clusters of points. After the transformation, we can mine the sequential patterns from these sequences by using existing sequential pattern-mining algorithms, such as PrefixSpan [64] and CloseSpan [91], with time constraints. In this example, setting the support threshold to 3, we can find $l_1 \rightarrow l_3$ is a sequential pattern if

$$\frac{|\Delta t_3 - (\Delta t_1 + \Delta t_2)|}{\max(\Delta t_3, \Delta t_1 + \Delta t_2)} < \rho, \quad \text{and} \quad \frac{|\Delta t_3 - (\Delta t_1' + \Delta t_2')|}{\max(\Delta t_3, \Delta t_1' + \Delta t_2')} < \rho,$$

where ρ is a ratio threshold guaranteeing that two travel times are similar. Likewise, setting the threshold of support to 2, $l_1 \rightarrow l_2 \rightarrow l_3$ is a sequential pattern if Δt_1 is similar to $\Delta t_1'$ and Δt_2 is similar to $\Delta t_2'$. Toward this direction, Giannotti et al. [23] divide a city into

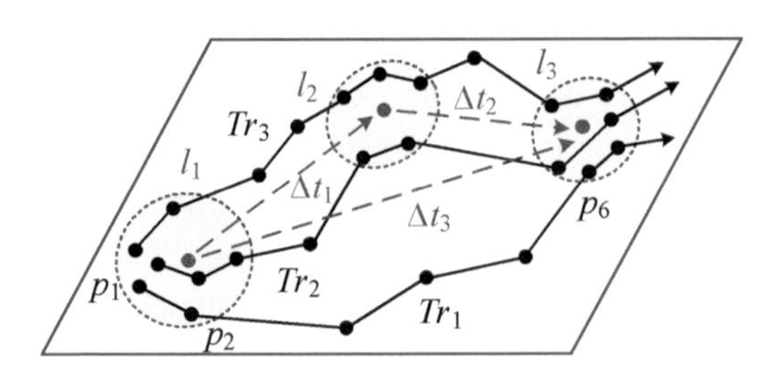

a) Sequential trajectory patterns

b) Suffix tree–based sequential pattern mining

Figure 10.16
Sequential pattern mining in trajectory data.

uniform grids, grouping these grids into regions of interest based on the density of the GPS points that have fallen into each grid. An Apriori-like algorithm is then proposed to detect the sequential patterns of the regions of interest.

With respect to the applications caring more about the semantic meaning of a location, we can first detect stay points from each trajectory, turning a trajectory into a sequence of stay points. Later, we can cluster these stay points to formulate regions of interest and use the cluster ID to which a stay point belongs to represent a trajectory. Following this strategy, Ye et al. [93] proposed to mine life patterns from an individual's GPS trajectories. Xiao et al. [84, 85] proposed a graph-based sequence-matching algorithm to find the sequential pattern shared by two users' trajectories. These patterns are then used to estimate the similarity between two users.

Sequential pattern mining in a road network When the sequential pattern-mining problem is applied to a road network setting, we can first map each trajectory onto a road network by using map-matching algorithms. A trajectory is then represented by a sequence of road segment IDs, which can be regarded as strings. As a result, some sequential pattern-mining algorithms designed for strings, such as PrefixSpan, can be adapted to finding sequential trajectory patterns.

Figure 10.16b presents a suffix tree that represents the four trajectories depicted in figure 4.27. Here, a node is a road segment, and the path from the root to a node corresponds to a suffix of the string representing a trajectory. For example, Tr_1 is represented by a string $r_1 \rightarrow r_2 \rightarrow r_6$, where $r_2 \rightarrow r_6$ and r_6 are suffixes of the string. The number associated with each link denotes the number of trajectories traversing the path (i.e., the support of the string pattern). For instance, there are two trajectories (Tr_1 and Tr_2) traversing $r_1 \rightarrow r_2$ and one trajectory traversing $r_1 \rightarrow r_2 \rightarrow r_6$. After building such a suffix tree, we can find the frequent patterns (i.e., the paths on the tree) with a support greater than a given threshold with a complexity of $O(n)$. Note that the size of a suffix tree can be much bigger than the original trajectories. So, when the size of a trajectory dataset is very large, we need to set a constraint on the depth of its suffix tree. Additionally, the sequential patterns derived from the suffix tree have to be consecutive. Though the temporal constraint is not explicitly considered, two objects' travel times on the same path should be similar, given the speed constraint of a path.

Toward this direction, Song et al. [72] use a suffix tree to detect frequent trajectory patterns, which are then leveraged to compress trajectories in conjunction with Huffman encoding. Wang et al. [82] employ suffix trees to find frequent trajectory patterns, which are used to reduce the candidates of a combination of subtrajectories when estimating the travel time of a query path.

10.2.5.4 Mining periodical patterns from trajectories Moving objects usually have periodic activity patterns. For example, people go shopping every month, and animals migrate yearly from one place to another. Such periodic behaviors provide an insightful and concise explanation over a long moving history, helping to compress trajectory data and predict the future movement of a moving object.

Periodic pattern mining has been studied extensively for time series data. For example, Yang et al. tried to discover asynchronous patterns [89], surprising periodic patterns [90], and patterns with gap penalties [92] from (categorical) time series. Due to the fuzziness of spatial locations, existing methods designed for time series data are not directly applicable to trajectories. To this end, Cao et al. [9] proposed an efficient algorithm for retrieving maximal periodic patterns from trajectories. This algorithm follows a paradigm that is similar to *frequent* pattern mining, where a (global) minimum support threshold is needed. In the real world, however, periodic behaviors could be more complicated, involving multiple interleaving periods, partial time spans, and spatiotemporal noises and outliers.

To deal with these issues, Li et al. [44] have proposed a two-stage detection method for trajectory data. In the first stage, the method detects a few reference spots where a moving object has visited frequently by using a density-based clustering algorithm, such as kernel density estimation (KDE). The trajectory of a moving object is then transformed into several binary time series, each of which indicates the *in* (1) and *out* (0) status of the moving object at a reference spot. Through applying Fourier transform and autocorrelation methods to each time series, the values of periods at each reference spot can be calculated. The second stage summarizes the periodic behaviors from partial-movement sequences by using a hierarchical-clustering algorithm. In 2012, Li et al. [46] further extend the research [44] to mining periodic patterns from incomplete and sparse data sources.

10.2.6 Trajectory Classification

Trajectory classification aims to differentiate between trajectories (or their segments) of different status, such as motions, transportation modes, and human activities. Tagging a raw trajectory (or its segment) with a semantic label raises the value of trajectories to the next level, which can facilitate many applications, such as trip recommendations, life experiences sharing, and context-aware computing.

In general, trajectory classification comprises three major steps:

1. Divide a trajectory into segments using segmentation methods. Sometimes, each single point is regarded as a minimum inference unit.

2. Extract features from each segment (or point).
3. Build a model to classify each segment (or point).

As a trajectory is essentially a sequence, we can leverage existing sequence inference models, such as a dynamic Bayesian network (DBN), HMM, and conditional random field (CRF), which incorporate the information from local points (or segments) and the sequential patterns between adjacent points (or segments).

Using a sequence of 802.11 radio signals, LOCADIO [38] employs an HMM to classify the motion of a device into two statuses: Still and Moving. Based on a trajectory of Global System for Mobile communication (GSM) signals, Timothy et al. [78] attempted to classify the mobility of a user into three statuses consisting of stationary, walking, and driving. Zhu et al. [123] aim to infer the status of a taxi, consisting of *Occupied, Nonoccupied,* and *Parked,* according to its GPS trajectories. They first seek the possible *Parked* places in a trajectory using a stay point–based detection method. A taxi trajectory is then partitioned into segments by these *Parked* places (refer to figure 10.9d for an example). For each segment, they extract a set of features incorporating the knowledge of a single trajectory, historical trajectories of multiple taxis, and geographic data such as road networks and POIs. After that, a two-phase inference method is proposed to classify the status of a segment into either *Occupied* or *Nonoccupied.* The method first uses the identified features to train a local probabilistic classifier and then globally considers travel patterns via a hidden semi-Markov model.

Zheng et al. [115, 119] classify a user's trajectory by transportation modes, which are composed of *Driving, Biking, Bus,* and *Walking.* As people usually change transportation modes in a single trip, a trajectory is first partitioned into segments based on the *Walk*-based segmentation method (refer to figure 10.10 for details). A set of features such as the heading change rate, stop rate, and velocity change rate are extracted and fed into a decision tree classifier. Based on the inference results, a graph-based postprocessing step is conducted to fix the possibly wrong inference, considering the transition probability between different transportation modes at different places.

Lin et al. [47, 63] proposed a hierarchical inference model for location-based activity recognition and significant place discovery, as shown in figure 10.17a. A GPS trajectory is first divided into 10 m segments, each of which is then projected onto corresponding street patches by using a CRF-based map-matching algorithm. Based on the features extracted from these street patches, the model classifies a sequence of GPS points into a sequence of activities like $a_1, a_2, \dots, a_n$ (such as *Walk, Driving,* and *Sleep*) and identifies a person's significant places like P_1 and P_2 (e.g., home, work, and bus stops) simultaneously.

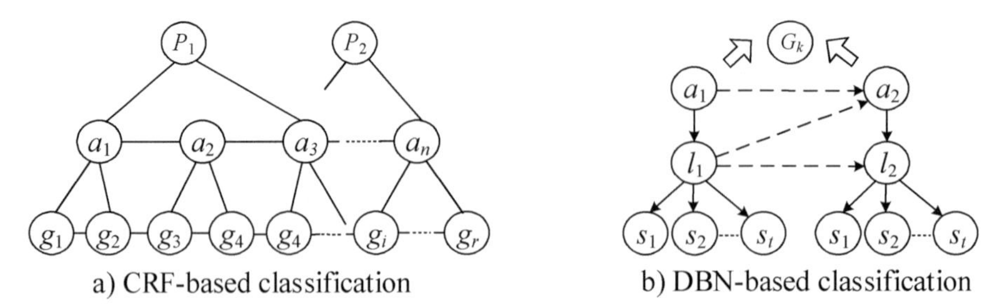

Figure 10.17
Trajectory classification for activity recognition.

Yin et al. [94] proposed a DBN-based inference model to infer a user's activities as well as high-level goals, according to a sequence of Wi-Fi signals. Figure 10.17b presents the structure of the DBN, where the bottom layer contains the input of raw Wi-Fi signals; the second layer is a list of locations where these signals are received, and the top level corresponds to user activities. Finally, the high-level goal is inferred based on the sequence of inferred activities.

10.2.7 Anomalies Detection from Trajectories

Trajectory outliers (a.k.a. anomalies) can be items (e.g., a trajectory or a segment of trajectory) that are significantly different from other items in terms of some similarity metric. They can also be events or observations (represented by a collection of trajectories) that do not conform to an expected pattern (e.g., traffic congestion caused by a car accident). A survey on general anomaly detection methods can be found in [10].

10.2.7.1 Detecting outlier trajectories An outlier trajectory is a trajectory or a part of a trajectory that is significantly different from others in a corpus in terms of a distance metric, such as shape and travel time. The outlier trajectories could be a taxi driver's malicious detour [49, 105] or unexpected road changes (due to traffic accidents or construction). It can also remind people when traveling on a wrong path.

A general idea is to leverage existing trajectory clustering or frequent pattern-mining methods. If a trajectory (or a segment) cannot be accommodated in any (density-based) clusters or is not frequent, it may be an outlier. Lee et al. [39] have proposed a partition-and-detection framework to find anomalous segments of trajectories from a trajectory dataset. This method can be an extension of the trajectory clustering proposed in [40].

10.2.7.2 Identifying anomalous events by trajectories Another direction is to detect traffic anomalies (rather than a trajectory itself) by using many trajectories. The traffic anomalies could be caused by accidents, controls, protests, sports, celebrations, disasters, and other events.

Liu et al. [50] partition a city into disjointed regions with major roads and glean the anomalous links between two regions according to the trajectories of vehicles traveling between the two regions. They divide a day into time bins and identify for each link three features: the number of vehicles traveling a link in a time bin, the proportion of these vehicles among all vehicles entering the destination region, and those vehicles departing from the origin region. The three features of a time bin are respectively compared with those in the equivalent time bins of previous days to calculate the minimum distort of each feature. Then, the link of the time bin can be represented in a three-dimensional space, with each dimension denoting the minimum distort of a feature. Later, the Mahalanobis distance is used to measure the extreme points (in the three-dimensional space), which are regarded as outliers. Following the aforementioned research, Sanjay et al. [12] proposed a two-step mining and optimization framework to detect traffic anomalies between two regions and explain an anomaly with the traffic flows passing the two regions (see section 10 for details).

Pan et al. [60] identify traffic anomalies according to drivers' routing behavior on an urban road network. Here, a detected anomaly is represented by a subgraph of a road network where drivers' routing behaviors differ significantly from their original patterns. They then try to describe the detected anomaly by mining representative terms from the social media that people posted when the anomaly was happening.

Pang et al. [61, 62] adapt the likelihood ratio test, which has previously been used in epidemiological studies, to describe traffic patterns. They partition a city into uniform grids and count the number of vehicles arriving in a grid over a time period. The objective is to identify a contiguous set of cells and time intervals that have the largest statistically significant departure from expected behavior (i.e., the number of vehicles). The regions whose log-likelihood ratio statistic value drops in the tail of χ^2 distribution are likely to be anomalous.

Zheng et al. [106] detect a collective anomaly, which denotes a collection of nearby locations that are anomalous during a few consecutive time intervals, in terms of phenomena collectively witnessed by multiple datasets consisting of taxi trips, bike-sharing data, 311 complaints, POIs, and road networks. The proposed method first estimates the distribution of a sparse dataset, such as the 311 complaint data, using a multisource latent Dirichlet allocation (LDA) algorithm. It then measures the anomalous degree of different combinations of regions and time intervals based on a spatiotemporal likelihood ratio test model.

10.2.8 Transferring Trajectories to Other Representations

10.2.8.1 From trajectory to graph Trajectories can be transformed into other data structures, besides being processed in their original form. This enriches the methodologies that can be used to discover knowledge from trajectories. Turning trajectories into graphs is one of the representative types of transformation. When conducting such a transformation, the main effort is to define what a node and an edge are in the transformed graph. The methods for transforming trajectories into a graph differentiate between one another depending on whether a road network is involved in the transformation.

In a road network setting A road network is essentially a directed graph, where a node is an intersection, and an edge denotes a road segment. Consequently, the most intuitive approach to turning trajectories into a graph is to project trajectories onto a road network. We can then calculate some weights, such as speed and traffic volume, for the edges based on the projected trajectories. Later, given the weighted graph, we can find the most likely route (traveled by people) between a few query points, identify the most popular route between a source and a destination, detect traffic anomalies, and update maps automatically.

The second approach is to build a *landmark* graph. For example, Yuan et al. [97, 103] proposed an intelligent driving-direction system, titled T-Drive, based on the GPS trajectories generated by a large number of taxicabs. After the map-matching process, T-Drive regards the top-*k* road segments frequently traversed by taxicabs as *landmark* nodes (i.e., the red points shown in figure 10.18a). The trajectories traversing two landmarks consecutively are aggregated into a landmark edge (denoted by a blue line) used to estimate the travel time between two landmarks. A two-stage routing algorithm is proposed to find the fastest driving path. The algorithm first searches the landmark graph for a rough route (represented by a sequence of landmarks) and then finds a detailed route connecting consecutive landmarks on the original road network.

The third approach is to build a *region graph*, where a node denotes a region, and an edge stands for the aggregation of commutes between the two regions. For instance, as illustrated in figure 10.18b, using an image segmentation–based algorithm [101], Zheng et al. [120] partition a city into regions by major roads to detect the underlying problems in a city's road network. A region bounded by major roads is then represented by a node, and two regions are connected with an edge if there are a certain

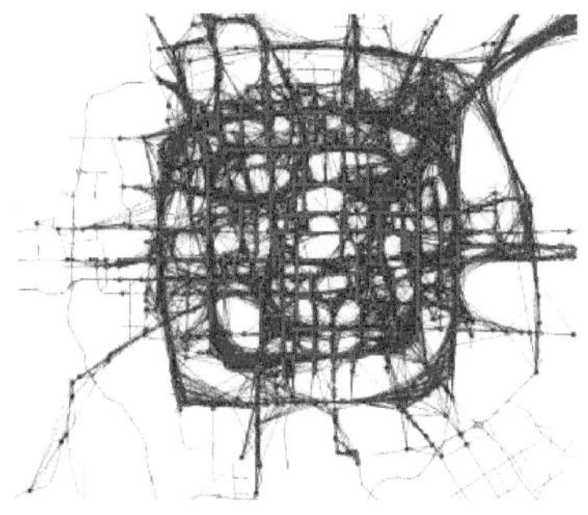

a) A landmark graph

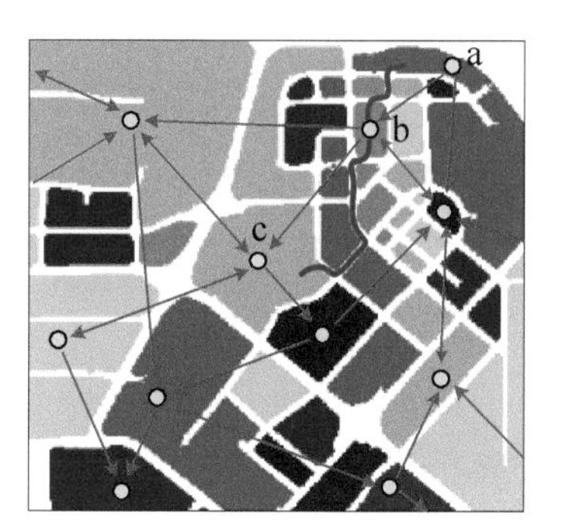

b) A region graph

Figure 10.18
Transforming trajectories into graphs.

number of commutes between them. After the transformation, they glean the region pairs (i.e., edges) that are not well connected (i.e., with a huge traffic volume, a slow travel speed, and a long detour between them) using a skyline algorithm. The region graphs are also employed to detect traffic anomalies [50, 12] and urban functional regions [102, 104].

In free spaces Another branch of research transfers trajectories into a graph without using a road network, according to two major steps:

1. Identify key locations as vertexes from raw trajectories by using clustering methods.
2. Connect the vertexes to formulate a routable graph based on trajectories passing two locations.

Travel recommendations Zheng et al. [109, 118] proposed to find the interesting locations and travel sequences from trajectories generated by many people. In this method, they first detect stay points from each trajectory and then cluster the stay points from different people into locations, as shown in figure 10.19a. Based on these locations and raw trajectories, they build a user-location bipartite graph as illustrated in figure 10.19b, as well as a routable graph between locations, as depicted in figure 10.19c.

In the bipartite graph, a user and a location are regarded as two different types of nodes. An edge is built between a user node and a location node if the user has visited the location. A HITS (hypertext-induced topic search)-based model is then employed to infer the interest level of a location (i.e., the authority score) and the travel knowledge of a user (i.e., the hub score). According to the inferred scores, we can identify the top-k

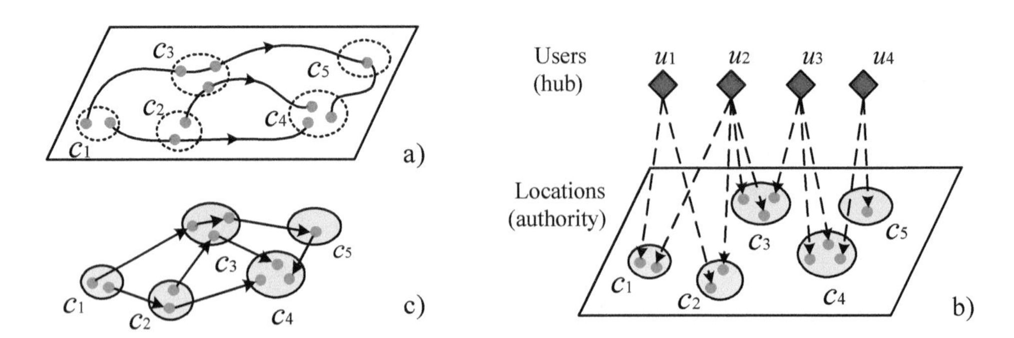

Figure 10.19
Mining interesting locations and travel sequences.

most interesting locations and travel experts in a city. Jie et al. [4] apply a similar idea in a collaborative filtering framework to conduct a travel recommendation concerned with a user's preferences, social environment, and current location.

In the location graph, as shown in figure 10.19c, an edge denotes the aggregation of raw trajectories traveling through two locations. To calculate the importance (or the representativeness) of an edge in this graph, three factors are considered: (1) the authority score of the source location (of the edge) weighted by the probability of people moving out by this edge, (2) the authority score of a destination location (in the edge) weighted by the probability of people moving in by this edge, and (3) the hub scores of the users who have traveled this edge. The score of a path is calculated by summing up the score of the edges the path contains.

Inspired by [118], a series of research was conducted to identify the popular routes from massive trajectories since 2010. Specifically, Yoon et al. [95, 96] suggest the best travel route, consisting of a sequence of locations with a typical stay time interval at each location, to a user, given the user's source and destination as well as time period. Chen et al. [13] identify turning points from each raw trajectory, clustering these turning points into groups. These clusters are then used as vertexes to build a transfer network. Afterward, the probability that people would travel from one vertex to another is calculated based on the counts of the trajectories passing the two vertexes. Finally, given a source and a destination, the path with the maximum production of probabilities is found in the transfer network as the most popular route. However, the proposed method is not applicable to low-sampling-rate trajectories. To this end, Wei et al. [83] divide a geographical space into uniform grids and then construct routable graphs based on the grids and raw trajectories.

Another branch of research is to detect the community of places based on the graph that is learned from trajectories, using certain community discovery methods.

A community of places is a cluster of locations with denser connections between locations in the cluster than between clusters. For example, Rinzivillo et al. [71] attempt to find the borders of human mobility at the lower spatial resolution of municipalities or counties. They map vehicle GPS tracks onto regions to formulate a complex network in Pisa. A community discovery algorithm, namely Infomap, is then used to partition the network into nonoverlapped subgraphs. More semantic meanings of a trajectory, such as a user's travel speed and experiences, have been considered in [48, 110] to estimate the strength of interaction between two locations.

Estimating user similarity Another series of research transfers users' trajectories into hierarchical graphs to compute the similarity between different users. This is a foundation of many social applications, such as friend recommendation and community discovery.

As illustrated in figure 10.20, Zheng et al. [117] deposit the stay points detected from different users' trajectories, clustering them divisively by using a density-based clustering algorithm iteratively. As a result, a tree-based hierarchy is built, where a node on a higher level is a coarse-grained cluster (of stay points), and the nodes on a lower level are fine-grained clusters. The hierarchy is shared by different users as it is derived from all users' stay points. By projecting a user's trajectories onto this shared hierarchy, an individual hierarchical graph can be constructed for a user.

As demonstrated at the bottom-left and bottom-right part of figure 10.20, two users' location histories are transformed from a collection of trajectories (which are not comparable between one another) to two individual graphs with common nodes. By matching the two graphs, common sequences of clusters are found on each level of the graphs. For example, $c_{32} \rightarrow c_{31} \rightarrow c_{34}$ is a common sequence shared by the two users on the third level. Considering the popularity of a cluster in a common sequence and the length and the level (on the hierarchy) of the common sequences, a similarity score is calculated for a pair of users.

Xiao et al. [85] extend similarity computing from physical locations to a semantic space, aiming to facilitate the similarity estimation between users living in different cities or countries. A stay point detected from a trajectory is represented by the distribution of POIs (across different categories) within the scope of the stay point. The stay points from different users are then clustered into a hierarchy according to their distributions on different POI categories, in a way similar to that of figure 10.20.

10.2.8.2 From trajectory to matrix Another form that we can transform trajectories into is a matrix. Using existing techniques, such as collaborative filtering and matrix factorization, a matrix can help complement missing observations. The key of the

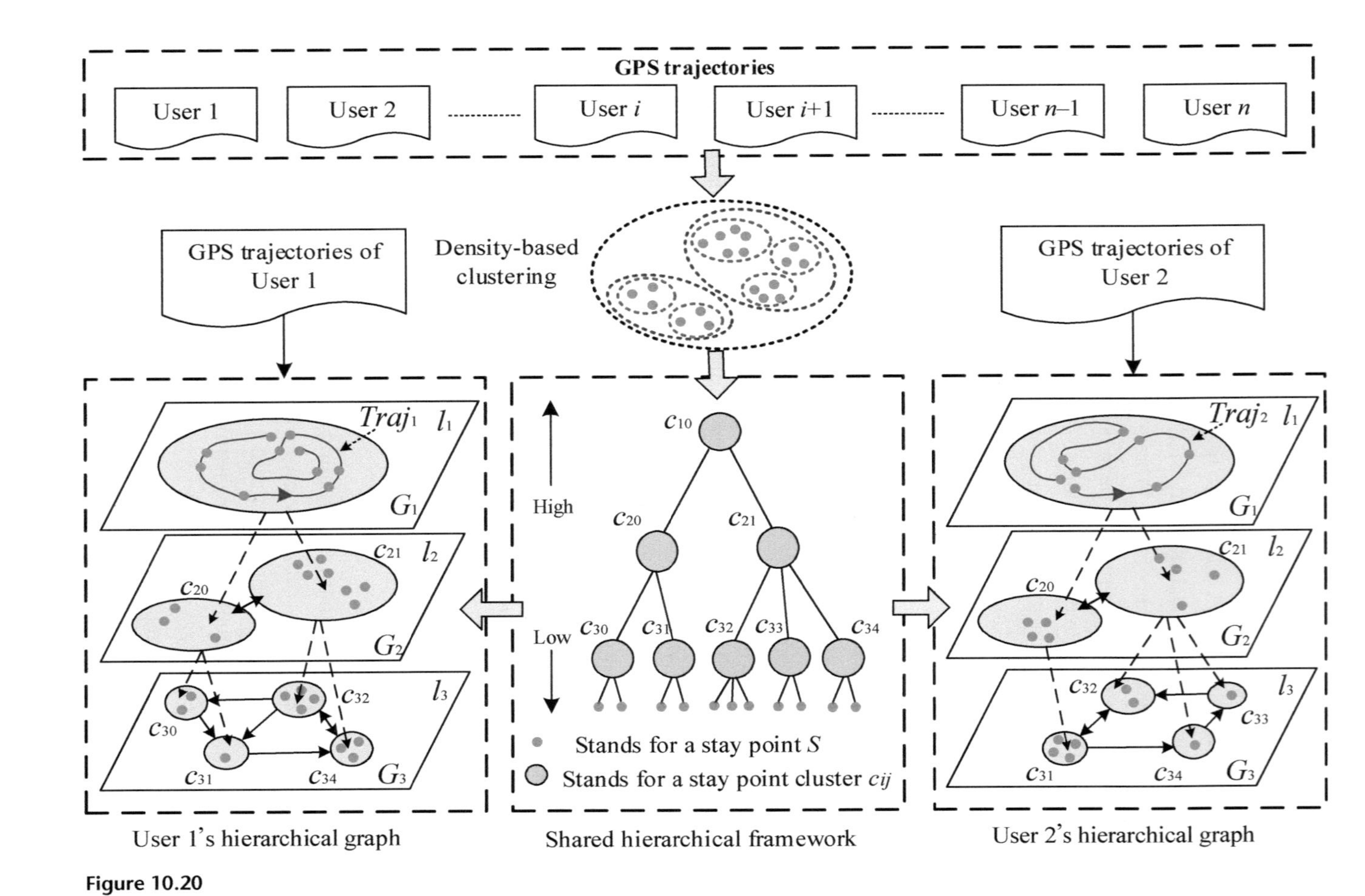

Figure 10.20
Hierarchical graph-based user similarity estimation.

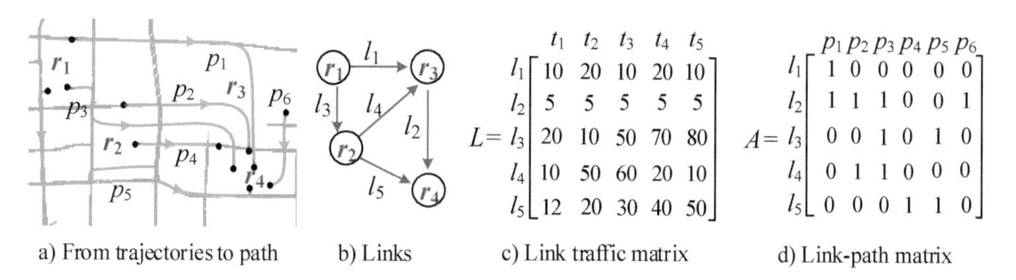

Figure 10.21
Detecting traffic anomalies using the PCA-based method.

transformation lies in three aspects: (1) What does a row mean, (2) what is a column, and (3) what does an entry denote? As techniques and examples about matrix factorization have been introduced in section 8.4, we do not detail them here.

Additionally, a matrix can also be used as an input to identify anomalies. Chawla et al. [12] aim to identify the traffic flows that cause an anomaly between two regions. In the methodology, they first partition a city into regions by major roads, building a region graph based on trajectories of taxicabs, as illustrated in figure 10.21a. A trajectory is then represented by a path on the graph (i.e., a sequence of links between regions), as shown in figure 10.21b.

Two matrices are built based on the trajectories and the graph. One is a link-traffic matrix L, as shown in figure 10.21c, where a row is a link, and a column corresponds to a time interval. An entry of L denotes the number of vehicles traversing a particular link at a specific time interval. For instance, there are ten vehicles traversing link l_1 at time interval t_1. The other is a link-path matrix A, with a row standing for a link and a column denoting a path. An entry of A is set to 1 if a particular link is contained in a particular path.

Given matrix L, a principal component analysis (PCA) algorithm is employed to detect some anomalous links, which are represented by a column vector b with 1 denoting an anomaly detected on the link. More specifically, let $\tilde{L} = L - \mu$, where μ is the column sample mean. Then, a $t \times t$ matrix $C = \tilde{L}^T \tilde{L}$ is formed, where t is the number of time intervals. The eigenvalue-eigenvector pairs (λ_i, v_i) can be obtained by the eigen-decomposition of matrix C (i.e., $Cv_i = \lambda_i v_i$). The eigenvalues λ_i $(i = 1, 2, \ldots, 5)$ of this example are

$$(1.9 \times 10^3, 0.67 \times 10^3, 0.02 \times 10^3, 0.01 \times 10^3, 0).$$

The pairs (λ_i, v_i) are sorted by a decreasing order of eigenvalues λ_i. The first r eigenvectors $[v_1, \ldots, v_r]$ form the normal subspace P_n, while the remaining eigenvectors $[v_{r+1}, \ldots, v_t]$

form the abnormal subspace P_a. In this example, the first eigenvector is selected. Later, all data points are projected onto P_a: $x \rightarrow x_a$, where x is the origin point, and x_a is the projected value. In space P_a, for all points, we compute the square of the deviation from the mean μ_a as follows:

$$(0.4\times10^3,\ 0.06\times10^3,\ 0.5\times10^3,\ 1.47\times10^3,\ 0.49\times10^3).$$

Select all points, which $\|x_a - \mu_a\| > \theta$, as anomalous links, where θ is a threshold. In this example, 1.47×10^3 is much larger than the others. Thus, l_4 is detected as the anomalous link (i.e., $b=(0, 0, 0, 1, 0)$).

Then, the relationship between anomalous links and paths is captured by solving the equation $Ax=b$, where x is a column vector denoting which paths contribute to the emergency of these anomalies shown in b. Using L_1 optimization techniques, x can be inferred.

10.2.8.3 From trajectory to tensor A natural extension of the matrix-based transformation is turning trajectories into a (three-dimensional) tensor, where the third dimension is added to a matrix in order to accommodate additional information. The goal of the transformation is usually to fill the missing entries (in a tensor) or to find the correlation between two objects, like two road segments or gas stations. A common approach to solving this problem is to decompose a tensor into the multiplication of a few (low-rank) matrices and a core tensor (or just a few vectors), based on the tensor's nonzero entries. When a tensor is very sparse, in order to achieve a better performance, the tensor is usually decomposed with other (context) matrices in a framework of collaborative filtering. Detailed techniques and examples can be found in section 8.5.

10.3 Combining Machine Learning with Data Management

10.3.1 Motivation

Data management (a.k.a. databases) and machine learning seem to be two separate fields with very tiny overlaps between their theories, conferences, and communities. The former focuses on the efficiency of storing, updating, and retrieving data in an information system (i.e., go fast), while the latter cares more about the effectiveness of an algorithm in learning knowledge from data (i.e., go deep). In a real big-data project, particularly in urban computing, however, we need to organically integrate the knowledge from both sides to tackle an urban challenge.

For example, a machine-learning model needs to use the travel speed of vehicles in a region as an input to predict the region's traffic conditions over the next hour. It is a very time-consuming process to find vehicles traveling in the region over the past hour and then retrieve the corresponding GPS trajectories to calculate travel speeds. Without data management techniques (e.g., a spatiotemporal indexing structure), we need to scan the entire dataset in a brute-force manner, checking if each GPS point falls in a given region or a given time interval many times. It would cost more than two hours to extract the travel speed feature for every region throughout a city. This prevents the urban-computing application from generating traffic predictions for the entire city over the next hour in a few seconds.

Different from other artificial intelligence systems, many machine-learning algorithms cannot handle urban-computing problems alone for the following three reasons:

1. *Dynamic environment.* The environment of an urban-computing application is highly dynamic, changing over time almost every second. This does not allow us to build an off-line index to serve online queries. For instance, as illustrated in figure 10.22a, in a search engine, we can build a word-document index for the web pages crawled by the entire Internet every few days in an off-line process. The index is then used to serve online queries and does not change once it is built. However, in an urban-computing application (e.g., a taxi-dispatching system), the location of taxis continuously changes, forcing us to update the index every few seconds. Otherwise, the search results of a query are not very useful.
2. *Complex data transformation.* The feature-extraction process is much more complicated in urban-computing applications, as compared to other machine-learning systems. In urban computing, we usually cannot observe the features that a machine-learning model needs. In many cases, we may not be able to derive them from a single dataset. Instead, features must be derived from (different types of) raw data through nontrivial computing processes, as illustrated in figure 10.22b. For example, vehicles' travel speeds are computed based on their GPS trajectories and road networks. The feature-extraction process calls for a map-matching algorithm to project a GPS trajectory onto the corresponding road segments and spatiotemporal range queries to retrieve GPS trajectories within a given region and time span. There are two types of data, consisting of trajectories (D1) and road networks (D2), involved in the feature-extraction process. None of the two datasets can derive the speed features alone. Without efficient data management techniques, the feature-extraction process may last for a couple of hours.

   This is very different from other machine-learning systems, like a search engine, where the input is just a few key words or a short phrase, or a computer vision system, where the input is just one image or a few video frames. Features are generated within a single domain through simple transformations, such as calculating mean, standard deviation, term-frequency, inverse document frequency, or scale-invariant feature transform (SIFT) features.

3. *Citywide and multisource input.* As there are strong dependencies between different locations as well as different time intervals, the input of a machine-learning model in an urban-computing system is usually the information of an entire city rather than an individual region or road, even if the target is a prediction in an individual region. For example, the flow of crowds in a region depends on that of its neighbors as well as the locations that are far from the region. Thus, to predict the flow of crowds in an individual region, we need to take into account the traffic conditions of all regions in a city. As depicted in figure 10.22c, traditional machine-learning systems are much more simple than urban-computing systems. For example, in an image classification problem, the input of a machine-learning model is just one image, as the results of the recognition are independent of other images. Thus, to classify one image, it is not necessary to use tens of thousands of images as the model's input.

   Additionally, in an urban-computing application, we usually need to use multiple datasets from different domains to complete a task—for instance, predicting air quality based on traffic, meteorology, and POIs. Consequently, we can see multiple circles of different colors occurring in an urban-computing system's inputs, where different colors denote data from different domains.

   The two issues increase the scale and complexity of a machine-learning model's inputs, calling for advanced data management techniques.

A straightforward way of trying to address these issues is to use more servers to process different datasets or different partitions of data in parallel. However, it wastes a lot of computing resources, hence limiting us to try a more advanced machine-learning model. Moreover, there is a bottleneck of performance improvement if solely increasing the number of machines, as the more machines used, the more input/output costs involved. Thus, the integration of data management into machine-learning enables us to learn deeper knowledge from data more efficiently with fewer computational resources.

Currently, data management and machine learning are two consecutive steps in many big-data systems. The combination between them is very loose, or at a system level, or almost zero. For example, some data management techniques are used to store the data, which will be used later by machine-learning algorithms. Or, when handling big data, machine-learning algorithms are usually run in a distributed system. There is no integration of the two parts at an algorithmic level.

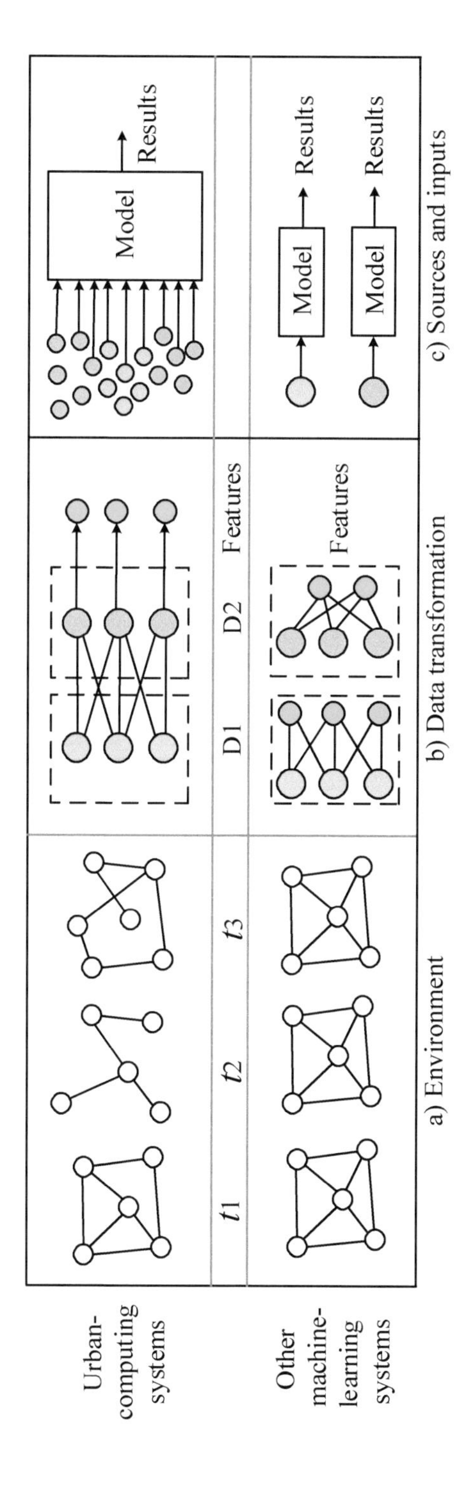

Figure 10.22
Differences between urban-computing and traditional machine-learning systems.

To this end, this section introduces three categories of approaches to organically integrating data management techniques into machine-learning algorithms at an algorithmic level:

1. The first category of approaches boosts the efficiency of the feature extraction process in a machine-learning model, using spatial and spatiotemporal indexing structures.
2. The second category of combinations employs data management techniques, such as pattern-mining algorithms, to find frequent patterns from raw data, using these patterns as the machine-learning algorithms' input. As the scale of frequent patterns is significantly smaller than the original data, the combination refines the input of a machine-learning algorithm, thus reducing its computing complexity while boosting its efficiency.
3. The third category of combinations uses data management techniques to derive upper or lower bounds for a machine-learning algorithm, pruning the computational space of the algorithm.

10.3.2 Boosting Machine Learning with Indexing Structures

This category of methods builds spatial or spatiotemporal indexing structures to expedite the data transformation (e.g., feature extraction) of urban-computing systems. In this section, we use three examples to demonstrate the integration.

Example 1: Urban air. Zheng et al. [113] aim to infer the air quality at arbitrary locations throughout a city, based on a machine-learning model and a variety of datasets consisting of POIs, road networks, traffic conditions, and air quality data from existing monitoring stations, such as s_1 and s_2. As depicted in figure 10.23a, the air quality in a location l is assumed to be influenced by the data within a circle distance of the location. The region within the circle is called l's *affecting region.*

To infer the air quality of l, the inputs of the machine-learning model include (1) the features extracted from POIs and networks located in l's affecting region and (2) the features (such as travel speed and the number of pickups and drop-offs) extracted from GPS trajectories that are generated by taxicabs traveling in (or through) the affecting region over the past hour. To extract the speed feature, a straightforward method is first scanning each and every GPS point in the taxi trajectories received recently to check if they fall in the spatial range of l's affecting region. The travel speed of each GPS point falling in the affecting region is then calculated based on the distance and time interval between a GPS point and its predecessor.

However, this process is very time-consuming. Supposing there are n GPS points received over the past hour and m locations to infer, the computational complexity is

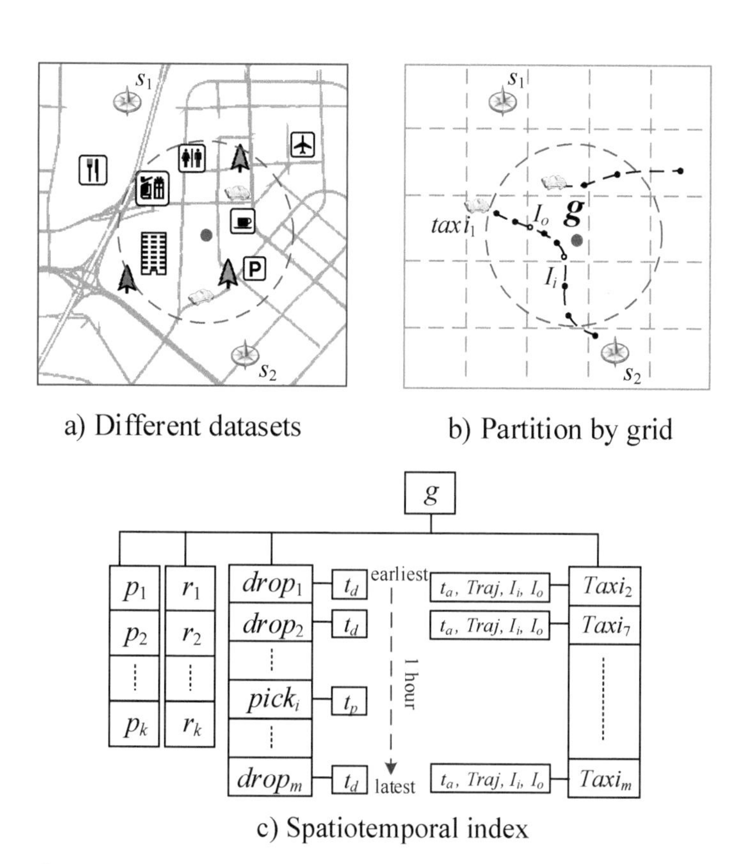

Figure 10.23
Indexing structures for expediting the feature extraction process in urban air.

$O(n \times m)$. Typically, n is huge. For example, there are one hundred thousand vehicles generating a GPS point every five seconds. The number of GPS points received over an hour is $100{,}000 \times 3{,}600/5 = 72{,}000{,}000$. If trying to infer ten thousand locations' air quality, we need to check seventy-two million points against ten thousand affecting regions, respectively (i.e., 720 billion times of matching). Without a spatial indexing structure, the scan may cost a few hours even if all these points have been loaded in memory. In addition, there other features, such as the number of pickups and the distribution of POIs across different categories in l's affecting region, to extract. This prevents a machine-learning algorithm from predicting the air quality of the entire city in a few minutes.

As shown in figure 10.23b, a city is portioned into uniform grids, where a spatiotemporal indexing structure is built to maintain the relationship between a grid g and the POIs, road networks, and trajectories relating to the grid, expediting the feature extraction significantly.

More specifically, as illustrated in figure 10.23c, the identities of POIs falling in grid g are listed in an array. Likewise, the identities of the road segments falling in or intersecting g are stored in another array. The pickups and drop-offs extracted from taxi trajectories are sorted by the time (i.e., t_p and t_d) they were generated and are stored in another array. Finally, the identities of taxis entering g are sorted by their arrival times t_a in an array. For each record in the array, the trajectory identity corresponding to a taxi is also stored with the two indexes of points in the trajectory, I_i and I_o. As illustrated in figure 10.23b, the first point right after a taxi entering g is the fourth point in the trajectory (i.e., $I_i = 4$), while the last point right before a taxi departs from the region is the seventh point in the trajectory (i.e., $I_o = 7$). The taxi identity array and pickup/drop-off array only store the data from the most recent hour, flushing out the out-of-date data.

When extracting features for a given location, we first find the grids $\boldsymbol{G}$ that are intersecting or within the location's affecting region. Based on the index in each grid of $\boldsymbol{G}$, we retrieve the POIs within the affecting region, counting the number of POIs in different categories. Other grids that are completely out of the circle can be neglected. This saves a lot of computational load. Likewise, the pickups, drop-offs, and taxi identities generated within an hour can be retrieved from these indexes efficiently. As a taxi may traverse several grids in $\boldsymbol{G}$, we need to merge the records generated by the same taxi. For example, as illustrated in figure 10.23b, $Taxi_1$ has traversed three grids in l's affecting region. Thus, there are three records retrieved from the three grid indexes. After merging the three records, the result is ($Taxi_1$, $I_i = 2$, $I_o = 9$). As GPS trajectories are too big to be held in memory, they are usually stored on disk. Merging the records of the same taxi can reduce the number of accesses to the disk, thereby enhancing the efficiency of the machine-learning algorithm.

Example 2: Suffix tree–based index for travel time estimation. Wang et al. [82] estimate the travel time of an arbitrary path using sparse vehicular trajectories combined with other datasets, such as POIs, road networks, and weather. For example, as depicted in figure 10.24, the travel time of path P: $r_1 \rightarrow r_2 \rightarrow r_3$ at the current time interval can be estimated based on the four trajectories, consisting of Tr_1, Tr_2, Tr_3, and Tr_4, received recently. After a map-matching process, each trajectory is converted to a sequence of road segments (e.g., $Tr_1 : r_1 \rightarrow r_2 \rightarrow r_6$).

As there may not be enough trajectories traversing an entire path, we need to combine fragments of different trajectories to estimate the travel time of the path. For instance, we can calculate the travel time of $r_1 \rightarrow r_2 \rightarrow r_3$ solely based on Tr_2. Or, we can

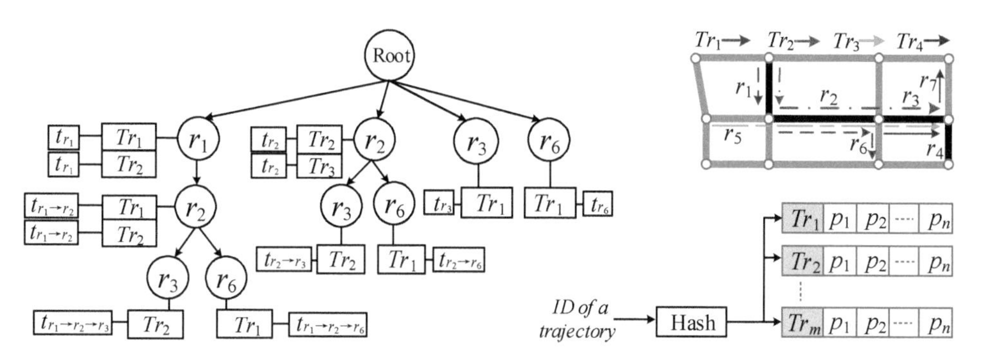

Figure 10.24
A suffix tree–based index for answering path queries.

compute the travel time for r_1 (based on Tr_1 and Tr_2), r_2 (based on Tr_1, Tr_2, and Tr_3), and r_3 (using Tr_2, Tr_3, and Tr_4) separately. Later, the travel time of $r_1 \rightarrow r_2 \rightarrow r_3$ can be obtained by summing the travel times of each road segment. We can also use Tr_2 and Tr_3 to estimate the travel time of $r_2 \rightarrow r_3$ and then concatenate it with that of r_1. Alternatively, we can do $r_1 \rightarrow r_2$ first based on Tr_1 and Tr_2 and then concatenate it with r_3.

Different concatenations have their own advantages and disadvantages subject to a trade-off between their support and length. As the length of a path increases, the number of trajectories (i.e., the support) traveling on the path decreases. Consequently, the confidence of the travel time (derived from few drivers) decreases. On the other hand, using the concatenation of shorter subpaths can have more occurrences of trajectories on each subpath but result in more fragments, across which more uncertainties are involved. Thus, a learning algorithm is proposed to find an optimal concatenation method, which results in an accurate estimate of a path's travel time.

The learning algorithm invokes many times of queries that search for trajectories traversing a given path. Without building an effective indexing structure, we need to scan every trajectory recently received when answering such a path query. This becomes very time-consuming, as there could be hundreds of thousands of trajectories generated during a short time interval; each trajectory can traverse hundreds of road segments. Thus, the learning algorithm cannot estimate the travel time of a path efficiently.

To address this issue, a suffix tree–based indexing structure is proposed to maintain the trajectories received in the current time slot, as shown in the left part of figure 10.24. Each node in the indexing tree stands for a road segment; each path on the tree corresponds to a route in the road network. Each node stores the IDs and travel times of the trajectories that traverse the path from the root to the node. For example, $t_{r_1 \rightarrow r_2 \rightarrow r_3}$ stands for the time for traveling path $r_1 \rightarrow r_2 \rightarrow r_3$. In order to find any existing paths, a suffix tree finds all suffixes of a string (i.e., a map-matched trajectory) and inserts them into the tree. For instance, the suffix of $r_1 \to r_2 \to r_6$ consists of r_6 and $r_2 \to r_6$. To reduce the size of the index, points (like p_1 and p_2) of a trajectory are not stored in the tree.

During a search, we can easily find a query path in such a suffix tree, starting from the root and ending at the node denoting the last road segment of the query path. We can then retrieve the trajectory IDs and their corresponding travel times from the ending node. Based on a trajectory's ID, we can retrieve its points through a hash table (as shown in the bottom-right part of figure 4.27). If there is no trajectory passing a given path query, the path does not exist in the suffix tree.

10.3.3 Scale Down Candidates for Machine Learning

This category of methods scales down the input of a machine-learning algorithm by pattern mining, enabling a machine-learning algorithm to solve more complex problems more efficiently. We present three examples of this category of methods.

Example 1: Travel time estimation of a path. Following the second example introduced in section 10.3.2, we further introduce how database techniques can help machine-learning algorithms. When a path is very long (i.e., containing many road segments), there are many ways of concatenating the road segments in the path to estimate the travel time. This prevents the later machine-learning algorithm from estimating the travel time of a path efficiently.

In practice, it is not necessary to check every concatenation of a path, as many subpaths may not be traversed by any trajectory in the current time slot. To further improve the efficiency of the solution, we can mine frequent trajectory patterns from historical trajectories in advance by using a suffix tree-based pattern-mining algorithm (refer to figure 10.16b for details). Then, we just need to check the concatenation of the trajectories' patterns, the scale of which is much smaller than the original concatenations. For instance, when estimating the travel time of path $r_1 \to r_2 \to r_3 \to r_4$, we do not need to check the combination of $r_3 \to r_4$, as there is no trajectory traversing it.

Example 2: Causality of air pollution. Zhu et al. [123] aim to identify the causal relationship between different air pollutants in different cities, using Bayesian networks and a variety of datasets, such as air quality data and meteorological data. However, this is a very challenging task, as an air pollutant's concentration in a location depends on many factors, such as the situation of the past few hours in the location and that of its spatial neighborhoods as well as the environment. This increases the complexity of the Bayesian network significantly. To reduce the complexity of the Bayesian model, an intuitive idea is to keep solely the top-N most correlated factors in the Bayesian network. However, without a special treatment, the correlation between two air pollutants' readings are dominated by those stable readings and trivial fluctuations, and thus they cannot truly reveal the dependency between two air pollutants (see figure 8.23).

To address this issue, the spatially coevolving patterns between each pair of sensor readings are mined from the original input using an efficient pattern-mining algorithm. The patterns can effectively determine the dependency between different air pollutants in different locations, thereby helping to select the top-N most correlated factors in a Bayesian network. They also find the time intervals that two air pollutants truly depend on each other, thus computing the dependency between two air pollutants more accurately. This reduces the complexity of a Bayesian network while enhancing the accuracy of the inference result. Without the pattern-mining algorithm, a Bayesian network cannot handle the input at such a scale.

Example 3: Discovering congestion propagation patterns. A similar idea has been proposed to discover congestion propagation patterns from traffic data. Nguyen et al. [57] construct congestion trees by connecting spatially adjacent road segments with congestion occurring at consecutive time intervals. Frequent subtree patterns are mined from those congestion trees. A dynamic Bayesian network is then built based on these frequent subtree patterns, modeling the congestion propagation and estimating the probability with which a congestion would occur. The frequent subtree patterns help design the structure of a dynamic Bayesian network, significantly scaling down the number of candidate factors that could cause congestion on a road segment.

10.3.4 Derive Bounds to Prune Computing Spaces for Machine Learning

This category of approach derives upper or lower bounds for a machine-learning algorithm, avoiding unnecessary computational load significantly. For example, if the upper bound of candidate group A is smaller than the lower bound of another candidate group B, we can ignore all the candidates in group A when seeking the candidate with the biggest value. Alternatively, we can say group A is pruned before we compute the exact value for each candidate in it. As the effort to estimate the upper or lower bounds of a group of candidates is much lighter than calculating each candidate's exact value, the computational load is reduced significantly.

For instance, Zheng et al. [106] propose using the log-likelihood ratio test (LRT) to detect the *collective* anomalies in a city instantaneously based on multiple datasets across different domains. Here, *collective* has two types of meaning. One denotes the spatiotemporal collectiveness. That is, a collection of nearby locations is anomalous during a few consecutive time intervals, while a single location in the collection may not be anomalous at a single time interval if being checked individually. The other is that an anomaly might not be that anomalous in terms of a single dataset but is considered an anomaly when checking multiple datasets simultaneously. Such collective anomalies could denote an early stage of an epidemic disease, the beginning of a natural disaster, an underlying problem, or a potentially catastrophic accident. For instance, a_1, a_2, and a_3 are three collective anomalies, as shown in figure 10.25a.

As where and when an anomaly will occur is unknown in advance, we need to check the LRT result of different combinations of time intervals and locations. As there are many combinations to check, this is very time-consuming, preventing the LRT algorithm from detecting anomalies instantaneously.

To address this issue, an anomalous degree upper bound is derived for a set of <location, time> entry combinations. In addition, the skyline of anomalous degrees of the entry combinations that have been computed is maintained. For example, as illustrated in figure 10.25b, each green circle denotes an entry combination, and each dimension stands for the anomalous degree derived from a dataset. In this example, three datasets, consisting of taxi trip data, bike-sharing data, and urban noise data, are used to detect collective anomalies. Those solid red points are skyline combinations that are not dominated by other green circles.

When a new set of entries (e.g., $<l_1, t_1>$, $<l_2, t_2>$, $\cdots <l_k, t_k>$) comes, we first compute the upper bound of the set's anomalous degree based on each type of dataset. If the set's upper bound is dominated by existing skyline combinations, all the combinations of its subsets will be dominated by the skyline too. Thus, we do not need to further check all possible combinations between its subsets, such as ($<l_1, t_1>$, $<l_2, t_2>$), and ($<l_1, t_1>$, $<l_3, t_3>$). Otherwise, the set's anomalous degree is inserted into the skyline to filter coming entry combinations. If an original entry combination in the skyline is dominated by the newly inserted one, the original one is removed from the skyline. Using the skyline and upper bound, we can avoid checking many entry combinations.

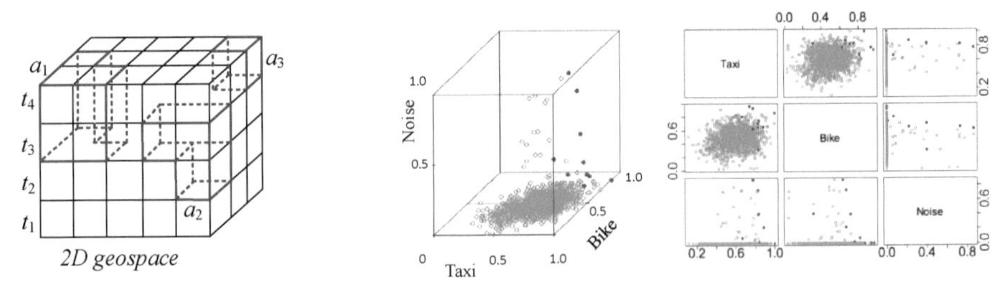

a) Examples of collective anomaly

b) Using skyline to prune candidates

Figure 10.25
Detecting collective anomalies: an example using an upper bound.

10.4 Interactive Visual Data Analytics

Solving an urban-computing problem requires both domain knowledge and data science, which are typically obtained by two different groups of people (i.e., domain experts and data scientists). It is imperative to provide these two groups with approaches through which they can share and integrate knowledge from both sides. Interactive visual data analytics may be one of the effective approaches to achieve this goal, with the following three functions: (1) helping data analytics models to gradually incorporate complex and subtle factors of a problem through interaction and feedback, (2) guiding the parameter-tuning process, and (3) drilling down into results and gaining new insights.

10.4.1 Incorporating Multiple Complex Factors

There are usually many criteria to consider when we conduct data analytics in urban-computing projects. For example, when deploying charging stations for electrical vehicles, we need to consider the number of vehicles passing by the locations where these charging stations would be deployed, traffic conditions, and the facilities (such as shopping malls and restaurants) around these locations. It makes a problem too complex for a data analytics model to solve if all these criteria are considered at one time. In addition, there may be some subtle criteria that cannot be clearly specified or quantified by domain experts.

To address these issues, interactive visual data analytics can present some preliminary results, which are generated by considering only a few simple criteria, to domain experts. Based on their domain knowledge, which incorporates other criteria, domain experts can give feedback on these results, such as removing ineligible locations from or suggesting a few good candidate locations in the preliminary results. Using the feedback as a constraint, the data analytics model can generate another round of results. The interaction between domain experts and the data analytics model repeats multiple times until satisfactory results have been found. Through interactive visual data analytics, we are able to combine domain knowledge with data science.

10.4.2 Adjusting Parameters without Prior Knowledge

There are quite a few nontrivial parameters to tune in many data analytics models. However, we may not have prior knowledge in setting these parameters. For instance, to study the correlation between two different spatiotemporal datasets, we need to define a co-occurrence between instances from the two datasets by using a distance threshold d and a time interval threshold t. If the geographical distance between two instances is smaller than d and the time interval between them is smaller than t, they are deemed a co-occurrence. Different d and t result in different numbers of co-occurrences, which lead to different supports for correlation pattern mining. However, we do not know how to set d and t prior to seeing the mining results. To address this issue, an interactive visual data analytics system can present the preliminary results of an arbitrary d and t, allowing users to iteratively adjust the two parameters after seeing the results. Users can learn the correlation between the two parameters and the pattern-mining results as well as the trend of adjustments after seeing many rounds of results. The correlation and trend may be too complex to model by machine intelligence but can be handled by human intelligence.

10.4.3 Drilling Down into Results

Interactive visual data analytics also helps drill down results returned by an urban-computing system, gaining deeper knowledge about a problem. For example, an urban-computing system computes the traffic congestion index for a city using the GPS trajectories generated by taxicabs over the last year. Sometimes, we need to drill down into the overall results to see the traffic conditions in a specific region and over a given period (e.g., a hot spot during the holidays). Once zooming into the region, new assumptions may arise (e.g., checking the traffic conditions around a specific shopping mall on a weekend), calling for further refinements on the results through interaction.

On the one hand, the refined results may bring interesting discoveries to solve a problem or deep insights to explain a phenomenon. For instance, we may know the traffic congestion in a region is caused by a specific shopping mall that attracts many people while having insufficient parking. On the other hand, the interaction calls for a data retrieval process to find the intersecting trajectories, given spatial and temporal ranges, and then a recomputing process to generate a congestion index based on the retrieved trajectories. Thus, an interactive visual data analytics model needs to integrate data management and machine-learning algorithms as well as visualizations seamlessly in an interactive framework.

10.5 Summary

This chapter introduces a body of advanced topics for urban data analytics, consisting of strategies for selecting useful data, techniques for trajectory data mining, approaches to combining machine-learning algorithms with data management techniques, and interactive visual data analytics.

Choosing an effective and valuable dataset depends on two aspects. One is an understanding of the problem we are going to solve. The other is the insight behind a dataset.

With knowledge about these two aspects, we can further validate our assumption by using correlation analysis tools, visualizations, and experiments.

The framework that defines the scope and road map for trajectory data mining is presented, consisting of trajectory preprocessing; trajectory data management; trajectory uncertainty; trajectory pattern mining; trajectory classification; outlier detection in trajectories; and transforming trajectories into other data formats like graph, matrix, and tensor. Individual research works are well positioned, categorized, and connected in each layer of this framework. The framework provides a panorama for people who want to step into this field. Professionals can easily locate the methods they need to solve a problem or find the unsolved problems.

The data analytics models in urban-computing projects usually need to deal with a dynamic environment, complex data transformation, and citywide and multisource input. Those unique features call for the integration of data management techniques with machine-learning algorithms. Three approaches to the integration have been introduced, comprising boosting machine-learning algorithms with indexing structures, scaling down candidates for machine-learning by using pattern-mining techniques, and deriving upper and/or lower bounds to prune the search space of a learning algorithm.

Finally, interactive visual data analytics is an approach that enables domain experts and data scientists to share and integrate knowledge from both sides by providing three functions: (1) helping data analytics models to gradually incorporate the complex and subtle factors of a problem through interaction and feedback, (2) guiding the parameter-tuning process, (3) drilling down into results and gaining new insights. Interactive visual data analytics needs to integrate data management and machine-learning algorithms, as well as visualizations, seamlessly into an interactive framework.

References

[1] Abul, O., F. Bonchi, and M. Nanni. 2008. "Never Walk Alone: Uncertainty for Anonymity in Moving Objects Databases." In *Proceedings of the 24th IEEE International Conference on Data Engineering*. Washington, DC: Institute of Electrical and Electronics Engineers (IEEE) Computer Society Press, 376–385.

[2] Aggarwal, C. C., J. Han, J. Wang, and P. S. Yu. 2003. "A Framework for Clustering Evolving Data Streams." In *Proceedings of the 29th International Conference on Very Large Data Bases*. San Jose, CA: Very Large Data Bases Endowment (VLDB), 81–92.

[3] Alt, H., A. Efrat, G. Rote, and C. Wenk. 2003. "Matching Planar Maps." *Journal of Algorithms* 49 (2): 262–283.

[4] Bao, J., Y. Zheng, and M. F. Mokbel. 2012. "Location-Based and Preference-Aware Recommendation Using Sparse Geo-Social Networking Data." In *Proceedings of the 20th ACM SIGSPATIAL*

International Conference on Advances in Geographic Information Systems. New York: Association for Computing Machinery (ACM), 199–208.

[5] Bellman, R. 1961. "On the Approximation of Curves by Line Segments Using Dynamic Programming." *Communications of the ACM* 4 (6): 284.

[6] Beresford, A. R., and F. Stajano. 2003. "Location Privacy in Pervasive Computing." IEEE Pervasive Computing 2 (1): 46–55.

[7] Brakatsouls, S., D. Pfoser, R. Salas, and C. Wenk. 2005. "On Map-Matching Vehicle Tracking Data." In *Proceedings of the 31st International Conference on Very Large Data Bases*. San Jose, CA: VLDB Endowment, 853–864.

[8] Cadez, I. V., S. Gaffney, and P. Smyth. 2000. "A General Probabilistic Framework for Clustering Individuals and Objects." In *Proceedings of the 6th ACM SIGKDD Conference on Knowledge Discovery and Data Mining*. New York: ACM, 140–149.

[9] Cao, H., N. Mamoulis, and D. W. Cheung. 2005. "Mining Frequent Spatio-Temporal Sequential Patterns." In *Proceedings of the 5th IEEE International Conference on Data Mining*. Washington, DC: IEEE Computer Society Press, 82–89.

[10] Chandola, V., A. Banerjee, and V. Kumar. 2009. "Anomaly Detection: A Survey." *ACM Computing Surveys* 41 (3): 1–58.

[11] Chawathe, S. S. 2007. "Segment-Based Map Matching." *IEEE Intelligent Vehicles Symposium*. Washington, DC: IEEE Computer Society Press, 1190–1197.

[12] Chawla, S., Y. Zheng, and J. Hu. 2012. "Inferring the Root Cause in Road Traffic Anomalies." In *Proceedings of the 12th IEEE International Conference on Data Mining*. Washington, DC: IEEE Computer Society Press, 141–150.

[13] Chen, L., M. T. Ozsu, and V. Oria. 2005. "Robust and Fast Similarity Search for Moving Object Trajectories." In *Proceedings of the 24th ACM SIGMOD International Conference on Management of Data*. New York: ACM, 491–502.

[14] Chen, W., M. Yu, Z. Li, and Y. Chen. 2003. "Integrated Vehicle Navigation System for Urban Applications." In *Proceedings of the International Conference Global Navigation Satellite System*. CGNS, 15–22.

[15] Chen, Y., K. Jiang, Y. Zheng, C. Li, and N. Yu. 2009. "Trajectory Simplification Method for Location-Based Social Networking Services." In *Proceedings of the 2009 ACM SIGSPATIAL Workshop on Location-Based Social Networking Services*. New York: ACM, 33–40.

[16] Cheng, R., J. Chen, M. F. Mokbel, and C. Y. Chow. 2008. "Probabilistic Verifiers: Evaluating Constrained Nearest-Neighbor Queries over Uncertain Data." In *Proceedings of the IEEE 24th Conference on Data Engineering*. Washington, DC: IEEE Computer Society Press, 973–982.

[17] Cheng, R., D. V. Kalashnikov, and S. Prabhakar. 2004. "Querying Imprecise Data in Moving Objects Environments." *IEEE Transactions on Knowledge and Data Engineering* 16 (9): 1112–1127.

[18] Chow, C. Y., and M. F. Mokbel. 2011. "Privacy of Spatial Trajectories." In *Computing with Spatial Trajectories*, edited by Y. Zheng and X. Zhou, 109–141. Berlin: Springer.

[19] Civilis, A., C. S. Jensen, J. Nenortaite, and S. Pakalnis. 2005. "Techniques for Efficient Road-Network-Based Tracking of Moving Objects." *IEEE Transactions on Knowledge and Date Engineering* 17 (5): 698–711.

[20] Douglas, D., and T. Peucker. 1973. "Algorithms for the Reduction of the Number of Points Required to Represent a Line or Its Caricature." *Cartographica: The International Journal for Geographic Information and Geovisualization* 10 (2): 112–122.

[21] Emrich, T., H. P. Kriegel, N. Mamoulis, M. Renz, and A. Züfle. 2012. "Querying Uncertain Spatio-Temporal Data." In *Proceedings of the 28th IEEE Conference on Data Engineering*. Washington, DC: IEEE Computer Society Press, 354–365.

[22] Gaffney, S., and P. Smyth. 1999. "Trajectory Clustering with Mixtures of Regression Models." In *Proceedings of the 5th ACM SIGKDD International Conference on Knowledge Discovery and Data Mining*. New York: ACM, 63–67.

[23] Giannotti, F., M. Nanni, D. Pedreschi, and F. Pinelli. 2007. "Trajectory Pattern Mining." In *Proceedings of the 13th ACM SIGKDD International Conference on Knowledge Discovery and Data Mining*. New York: ACM, 330–339.

[24] Gidófalvi, G., X. Huang, and T. B. Pedersen. 2007. "Privacy-Preserving Data Mining on Moving Object Trajectories." In *Proceedings of the 8th IEEE International Conference on Mobile Data Management*. Washington, DC: IEEE Computer Society Press, 60–68.

[25] Greenfeld, J. S. 2002. "Matching GPS Observations to Locations on a Digital Map." In *Proceedings of the 81st Annual Meeting of the Transportation Research Board*. Washington, DC: Transportation Research Board, 576–582.

[26] Gudmundsson, J., and M. V. Kreveld. 2006. "Computing Longest Duration Flocks in Trajectory Data." In *Proceedings of the 14th Annual ACM International Symposium on Advances in Geographic Information Systems*. New York: ACM, 35–42.

[27] Gudmundsson, J., M. V. Kreveld, and B. Speckmann. 2004. "Efficient Detection of Motion Patterns in Spatio-Temporal Data Sets." In *Proceedings of the 12th Annual ACM International Symposium on Advances in Geographic Information Systems*. New York: ACM, 250–257.

[28] Hershberger, J., and J. Snoeyink. 1992. "Speeding Up the Douglas-Peucker Line Simplification Algorithm." In *Proceedings of the 5th International Symposium on Spatial Data Handling*. New York: ACM, 134–143.

[29] Hoh, B., M. Gruteser, H. Xiong, and A. Alrabady. 2010. "Achieving Guaranteed Anonymity in GPS Traces via Uncertainty-Aware Path Cloaking." *IEEE Transactions on Mobile Computing* 9 (8): 1089–1107.

[30] Jensen, C. S., D. Lin, and B. C. Ooi. 2007. "Continuous Clustering of Moving Objects." *IEEE Transactions on Knowledge and Data Engineering* 19 (9): 1161–1174.

[31] Jeung, H., H. Shen, and X. Zhou. 2008. "Convoy Queries in Spatio-Temporal Databases." In *Proceedings of the 24th IEEE International Conference on Data Engineering*. Washington, DC: IEEE Computer Society Press, 1457–1459.

[32] Jeung, H., M. Yiu, X. Zhou, C. Jensen, and H. Shen. 2008. "Discovery of Convoys in Trajectory Databases." *Proceedings of the VLDB Endowment* 1 (1): 1068–1080.

[33] Kellaris, G., N. Pelekis, and Y. Theodoridis. 2009. "Trajectory Compression under Network Constraints." In *Proceedings, Advances in Spatial and Temporal Databases*. Berlin: Springer, 392–398.

[34] Keogh, E. J., S. Chu, D. Hart, and M. J. Pazzani. 2001. "An On-Line Algorithm for Segmenting Time Series." In *Proceedings of the 2001 IEEE International Conference on Data Mining*. Washington, DC: IEEE Computer Society Press, 289–296.

[35] Kharrat, A., I. S. Popa, K. Zeitouni, and S. Faiz. 2008. "Clustering Algorithm for Network Constraint Trajectories." In *Headway in Spatial Data Handling*. Berlin: Springer, 631–647.

[36] Kido, H., Y. Yanagisawa, and T. Satoh. 2005. "An Anonymous Communication Technique Using Dummies for Location-Based Services." In *Proceedings of the 3rd International Conference on Pervasive Services*. Washington, DC: IEEE Computer Society Press, 88–97.

[37] Krumm, J. 2011. "Trajectory Analysis for Driving." In *Computing with Spatial Trajectories*, edited by Y. Zheng and X. Zhou, 213–241. Berlin: Springer.

[38] Krumm, J., and E. Horvitz. 2004. "LOCADIO: Inferring Motion and Location from Wi-Fi Signal Strengths." In *Proceedings, the First Annual International Conference on Mobile and Ubiquitous Systems*. Washington, DC: IEEE Computer Society Press, 4–13.

[39] Lee, J., J. Han, and X. Li. 2008. "Trajectory Outlier Detection: A Partition-and-Detect Framework." In *Proceedings of the 24th IEEE Conference on Data Engineering*. Washington, DC: IEEE Computer Society Press, 140–149.

[40] Lee, J. G., J. Han, and K. Y. Whang. 2007. "Trajectory Clustering: A Partition-and-Group Framework." In *Proceedings of the 2007 ACM SIGMOD Conference on Management of Data*. New York: ACM, 593–604.

[41] Lee, W.-C., and J. Krumm. 2011. "Trajectory Preprocessing." Zheng, Y. 2011. "Location-Based Social Networks: Users." In *Computing with Spatial Trajectories*, edited by Y. Zheng and X. Zhou, 1–31. Berlin: Springer.

[42] Li, Q., Y. Zheng, X. Xie, Y. Chen, W. Liu, and M. Ma. 2008. "Mining User Similarity Based on Location History." In *Proceedings of the 16th Annual ACM International Symposium on Advances in Geographic Information Systems*. New York: ACM, 34.

[43] Li, Z., B. Ding, J. Han, and R. Kays. 2010. "Swarm: Mining Relaxed Temporal Moving Object Clusters." *Proceedings of the VLDB Endowment* 3 (1–2): 723–734.

[44] Li, Z., B. Ding, J. Han, R. Kays, and P. Nye. 2010. "Mining Periodic Behaviors for Moving Objects." In *Proceedings of the 16th ACM SIGKDD International Conference on Knowledge Discovery and Data Mining*. New York: ACM, 1099–1108.

[45] Li, Z., J. Lee, X. Li, and J. Han. 2010. "Incremental Clustering for Trajectories." In *Database Systems for Advanced Applications*, edited by H. Kitagawa, Y. Ishikawa, Q. Li, and C. Watanabe, 32–46. Berlin: Springer.

[46] Li, Z., J. Wang, and J. Han. 2012. "Mining Event Periodicity from Incomplete Observations." In *Proceedings of the 18th ACM SIGKDD International Conference on Knowledge Discovery and Data Mining*. New York: ACM, 444–452.

[47] Liao, L., D. Fox, and H. Kautz. 2004. "Learning and Inferring Transportation Routines." In *Proceedings of the National Conference on Artificial Intelligence*. Palo Alto, CA: AAAI Press, 348–353.

[48] Liu, S., K. Jayarajah, A. Misra, and R. Krishnan. 2013. "TODMIS: Mining Communities from Trajectories." In *Proceedings of the 22nd ACM CIKM International Conference on Information and Knowledge Management*. New York: ACM, 2109–2118.

[49] Liu, S., L. Ni, and R. Krishnan. 2014. "Fraud Detection from Taxis' Driving Behaviors." *IEEE Transactions on Vehicular Technology* 63 (1): 464–472.

[50] Liu, W., Y. Zheng, S. Chawla, J. Yuan, and X. Xie. 2011. "Discovering Spatio-Temporal Causal Interactions in Traffic Data Streams." In *Proceedings of the 17th ACM SIGKDD International Conference on Knowledge Discovery and Data Mining*. New York: ACM, 1010–1018.

[51] Lou, Y., C. Zhang, Y. Zheng, X. Xie, Wei Wang, and Yan Huang. 2009. "Map-Matching for Low-Sampling-Rate GPS Trajectories." In *Proceedings of the 17th ACM SIGSPATIAL International Conference on Geographical Information Systems*. New York: ACM, 352–361.

[52] Maratnia, N., and R.A.D. By. 2004. "Spatio-Temporal Compression Techniques for Moving Point Objects." In *Proceedings of the 9th International Conference on Extending Database Technology*. Berlin: Springer, 765–782.

[53] McMaster, R. B. 1986. "A Statistical Analysis of Mathematical Measures of Linear Simplification." *American Cartographer* 13 (2): 103–116.

[54] Mokbel, M. F., C. Y. Chow, and W. G. Aref. 2007. "The New Casper: Query Processing for Location Services without Compromising Privacy." In *Proceedings of the 23rd IEEE Conference on Data Engineering*. Washington, DC: IEEE Computer Society Press, 1499–1500.

[55] Nergiz, M. E., M. Atzori, Y. Saygin, and B. Guc. 2009. "Towards Trajectory Anonymization: A Generalization-Based Approach." *Transactions on Data Privacy* 2 (1): 47–75.

[56] Newson, P., and J. Krumm. 2009. "Hidden Markov Map Matching through Noise and Sparseness." In *Proceedings of the 17th ACM SIGSPATIAL International Conference on Geographical Information Systems*. New York: ACM, 336–343.

[57] Nguyen, H., W. Liu, and F. Chen. 2017. "Discovering Congestion Propagation Patterns in Spatio-temporal Traffic Data." *IEEE Transactions on Big Data* 3 (2): 169–180.

[58] Niedermayer, J., A. Zufle, T. Emrich, M. Renz, N. Mamouliso, L. Chen, and H. Kriegel. 2014. "Probabilistic Nearest Neighbor Queries on Uncertain Moving Object Trajectories." *Proceedings of the VLDB Endowment* 7 (3): 205–216.

[59] Ochieng, W. Y., M. A. Quddus, and R. B. Noland. 2004. "Map-Matching in Complex Urban Road Networks." *Brazilian Journal of Cartography* 55 (2): 1–18.

[60] Pan, B., Y. Zheng, D. Wilkie, and C. Shahabi. 2013. "Crowd Sensing of Traffic Anomalies Based on Human Mobility and Social Media." In *Proceedings of the 21st Annual ACM International Conference on Advances in Geographic Information Systems*. New York: ACM, 334–343.

[61] Pang, L. X., S. Chawla, W. Liu, and Y. Zheng. 2011. "On Mining Anomalous Patterns in Road Traffic Streams." In *ADMA 2011: Advanced Data Mining and Applications*. Berlin: Springer, 237–251.

[62] Pang, L. X., S. Chawla, W. Liu, and Y. Zheng. 2013. "On Detection of Emerging Anomalous Traffic Patterns Using GPS Data." *Data and Knowledge Engineering* 87:357–373.

[63] Patterson, D. J., L. Liao, D. Fox, and H. Kaut. 2003. "Inferring High-Level Behavior from Low-Level Sensors." In *Proceedings of the 5th International Conference on Ubiquitous Computing*. New York: ACM, 73–89.

[64] Pei, J., J. Han, B. Mortazavi-Asl, and H. Pinto. 2011. "PrefixSpan: Mining Sequential Patterns Efficiently by Prefix-Projected Pattern Growth." In *Proceedings of the 29th IEEE Conference on Data Engineering*. Washington, DC: IEEE Computer Society Press, 0215.

[65] Pfoser, D., and C. S. Jensen. 1999. "Capturing the Uncertainty of Moving Objects Representation." In *Proceedings of the 6th International Symposium on Advances in Spatial Databases*. London: Springer-Verlag, 111–131.

[66] Pink, O., and B. Hummel. 2008. "A Statistical Approach to Map Matching Using Road Network Geometry, Topology and Vehicular Motion Constraints." In *Proceedings of the 11th International IEEE Conference on Intelligent Transportation Systems*. Washington, DC: IEEE Computer Society Press, 862–867.

[67] Potamias, M., K. Patroumpas, and T. Sellis. 2006. "Sampling Trajectory Streams with Spatiotemporal Criteria." In *Proceedings of the 18th International Conference on Scientific and Statistical Database Management*. Washington, DC: IEEE Computer Society Press, 275–284.

[68] Qiao, S., C. Tang, H. Jin, T. Long, S. Dai, Y. Ku, and M. Chau. 2010. "Putmode: Prediction of Uncertain Trajectories in Moving Objects Databases." *Applied Intelligence* 33 (3): 370–386.

[69] Quddus, M. A., W. Y. Ochieng, and R. B. Noland. 2006. "A High Accuracy Fuzzy Logic-Based Map-Matching Algorithm for Road Transport." *Journal of Intelligent Transportation Systems* 10 (3): 103–115.

[70] Richter, K., F. Schmid, and P. Laube. 2012. "Semantic Trajectory Compression: Representing Urban Movement in a Nutshell." *Journal of Spatial Information Science* 4:3–30.

[71] Rinzivillo, S., S. Mainardi, F. Pezzoni, M. Coscia, D. Pedreschi, and F. Giannotti. 2012. "Discovering the Geographical Borders of Human Mobility." *Künstl intell* 26 (3): 253–260.

[72] Song, R., W. Sun, B. Zheng, and Y. Zheng. 2014. "PRESS: A Novel Framework of Trajectory Compression in Road Networks." *Proceedings of the VLDB Endowment* 7 (9): 661–672.

[73] Su, H., K. Zheng, H. Wang, J. Huang, and X. Zhou. 2013. "Calibrating Trajectory Data for Similarity-Based Analysis." In *Proceedings of the 39th International Conference on Very Large Data Bases*. San Jose, CA: VLDB Endowment. 833–844.

[74] Tang, L. A., Y. Zheng, J. Yuan, J. Han, A. Leung, C. Hung, and W. Peng. 2012. "Discovery of Traveling Companions from Streaming Trajectories." In *Proceedings of the 28th IEEE International Conference on Data Engineering*. Washington, DC: IEEE Computer Society Press, 186–197.

[75] Tang, L. A., Y. Zheng, J. Yuan, J. Han, A. Leung, W. Peng, and T. L. Porta. 2012. "A Framework of Traveling Companion Discovery on Trajectory Data Streams." *ACM Transactions on Intelligent Systems and Technology* 5 (1): article no. 3.

[76] Terrovitis, M., and N. Mamoulis. 2008. "Privacy Preservation in the Publication of Trajectories." In *Proceedings of the 9th IEEE International Conference on Mobile Data Management*. Washington, DC: IEEE Computer Society Press, 65–72.

[77] Timothy, S., A. Varshavsky, A. Lamarca, M. Y. Chen, and T. Chounhury. 2006. "Mobility Detection Using Everyday GSM Traces." In *Proceedings of the 8th International Conference on Ubiquitous Computing*. New York: ACM, 212–224.

[78] Trajcevski, G., A. N. Choudhary, O. Wolfson, L. Ye, and G. Li. 2010. "Uncertain Range Queries for Necklaces." In *Proceedings of the 11th IEEE International Conference on Mobile Data Management*. Washington, DC: IEEE Computer Society Press, 199–208.

[79] Trajcevski, G., R. Tamassia, H. Ding, P. Scheuermann, and I. F. Cruz. 2009. "Continuous Probabilistic Nearest-Neighbor Queries for Uncertain Trajectories." In *Proceedings of the 12th International Conference on Extending Database Technology: Advances in Database Technology*. New York: ACM, 874–885.

[80] Trajcevski, G., O. Wolfson, K. Hinrichs, and S. Chamberlain. 2004. "Managing Uncertainty in Moving Objects Databases." *ACM Transactions on Database Systems* 29 (3): 463–507.

[81] Wang, Y., Y. Zheng, and Y. Xue. 2014. "Travel Time Estimation of a Path Using Sparse Trajectories." In *Proceedings of the 20th ACM SIGKDD International Conference on Knowledge Discovery and Data Mining*. New York: ACM, 25–34.

[82] Wei, L., Y. Zheng, and W. Peng. 2012. "Constructing Popular Routes from Uncertain Trajectories." In *Proceedings of the 18th ACM SIGKDD International Conference on Knowledge Discovery and Data Mining*. New York: ACM, 195–203.

[83] Xiao, X., Y. Zheng, Q. Luo, and X. Xie. 2010. "Finding Similar Users Using Category-Based Location History." In *Proceedings of the 18th Annual ACM International Conference on Advances in Geographic Information Systems*. New York: ACM, 442–445.

[84] Xiao, X., Y. Zheng, Q. Luo, and X. Xie. 2014. "Inferring Social Ties between Users with Human Location History." *Journal of Ambient Intelligence and Humanized Computing* 5 (1): 3–19.

[85] Xie, H., L. Kulik, and E. Tanin. 2010. "Privacy-Aware Traffic Monitoring." *IEEE Transactions on Intelligent Transportation Systems* 11 (1): 61–70.

[86] Xu, C., Y. Gu, L. Chen, J. Qiao, and G. Yu. 2013. "Interval Reverse Nearest Neighbor Queries on Uncertain Data with Markov Correlations." In *Proceedings of the 29th IEEE Conference on Data Engineering*. Washington, DC: IEEE Computer Society Press, 170–181.

[87] Xue, A. Y., R. Zhang, Y. Zheng, X. Xie, J. Huang, and Z. Xu. 2013. "Destination Prediction by Sub-Trajectory Synthesis and Privacy Protection against Such Prediction." In *Proceedings of the IEEE 29th Conference on Data Engineering*. Washington, DC: IEEE Computer Society Press, 254–265.

[88] Yan, X., J. Han, and R. Afshar. 2003. "CloSpan: Mining Closed Sequential Patterns in Large Datasets." In *Proceedings of the 3rd SIAM International Conference on Data Mining*. Washington, DC: IEEE Computer Society Press, 166–177.

[89] Yang, J., W. Wang, and S. Y. Philip. 2001. "Infominer: Mining Surprising Periodic Patterns." In *Proceedings of the 7th ACM SIGKDD International Conference on Knowledge Discovery and Data Mining*. New York: ACM, 395–400.

[90] Yang, J., W. Wang, and P. S. Yu. 2002. "Infominer+: Mining Partial Periodic Patterns with Gap Penalties." In *Proceedings, 2002 IEEE International Conference on Data Mining*. Washington, DC: IEEE Computer Society Press, 725–728.

[91] Yang, J., W. Wang, and P. S. Yu. 2003. "Mining Asynchronous Periodic Patterns in Time Series Data." *IEEE Transactions on Knowledge and Data Engineering* 15 (3): 613–628.

[92] Ye, Y., Y. Zheng, Y. Chen, J. Feng, and X. Xie. 2009. "Mining Individual Life Pattern Based on Location History." In *Proceedings of the 10th IEEE International Conference on Mobile Data Management*. Washington, DC: IEEE Computer Society Press, 1–10.

[93] Yin, H. B., and O. Wolfson. 2004. "A Weight-Based Map Matching Method in Moving Objects Databases." In *Proceedings of the 16th International Conference on Scientific and Statistical Database Management*. Washington, DC: IEEE Computer Society Press, 437–410.

[94] Yin, J., X. Chai, and Q. Yang. 2004. "High-Level Goal Recognition in a Wireless LAN." In *Proceedings of the 20th National Conference on Artificial Intelligence*. New York: AAAI Press, 578–584.

[95] Yoon, H., Y. Zheng, X. Xie, and W. Woo. 2011. "Smart Itinerary Recommendation Based on User-Generated GPS Trajectories." In *Proceedings of the 8th International Conference on Ubiquitous Intelligence and Computing*. New York: ACM, 19–34.

[96] Yoon, H., Y. Zheng, X. Xie, and W. Woo. 2012. "Social Itinerary Recommendation from User-Generated Digital Trails." *Journal on Personal and Ubiquitous Computing* 16 (5): 469–484.

[97] Yuan, J., Y. Zheng, and X. Xie. 2012. "Discovering Regions of Different Functions in a City Using Human Mobility and POIs." In *Proceedings of the 18th ACM SIGKDD International Conference on Knowledge Discovery and Data Mining*. New York: ACM, 186–194.

[98] Yuan, J., Y. Zheng, X. Xie, and G. Sun. 2011. "Driving with Knowledge from the Physical World." In *Proceedings of the 17th ACM SIGKDD International Conference on Knowledge Discovery and Data Mining*. New York: ACM, 316–324.

[99] Yuan, J., Y. Zheng, X. Xie, and G. Sun. 2013. "T-Drive: Enhancing Driving Directions with Taxi Drivers' Intelligence." *IEEE Transactions on Knowledge and Data Engineering* 25 (1): 220–232.

[100] Yuan, J., Y. Zheng, C. Zhang, X. Xie, and G. Sun. 2010. "An Interactive-Voting Based Map Matching Algorithm." In *Proceedings of the 11th IEEE International Conference on Mobile Data Management*. Washington, DC: IEEE Computer Society Press, 43–52.

[101] Yuan, J., Y. Zheng, L. Zhang, X. Xie, and G. Sun. 2011. "Where to Find My Next Passenger?" In *Proceedings of the 13th International Conference on Ubiquitous Computing*. New York: ACM, 109–118.

[102] Yuan, N. J., Y. Zheng, and X. Xie. 2012. "Segmentation of Urban Areas Using Road Networks." *Microsoft Technical Report*, MSR-TR-2012-65.

[103] Yuan, N. J., Y. Zheng, X. Xie, Y. Wang, K. Zheng, and H. Xiong. 2015. "Discovering Urban Functional Zones Using Latent Activity Trajectories." *IEEE Transactions on Knowledge and Data Engineering* 27 (3): 1041–4347.

[104] Zhang, D., N. Li, Z. Zhou, C. Chen, L. Sun, and S. Li. 2011. "iBAT: Detecting Anomalous Taxi Trajectories from GPS Traces." In *Proceedings of the 13th International Conference on Ubiquitous Computing*. New York: ACM, 99–108.

[105] Zheng, K., Y. Zheng, X. Xie, and X. Zhou. 2012. "Reducing Uncertainty of Low-Sampling-Rate Trajectories." In *Proceedings of the 28th IEEE International Conference on Data Engineering*. Washington, DC: IEEE Computer Society Press, 1144–1155.

[106] Zheng, K., Y. Zheng, N. J. Yuan, and S. Shang. 2013. "On Discovery of Gathering Patterns from Trajectories." In *Proceedings of the 28th IEEE International Conference on Data Engineering*. Washington, DC: IEEE Computer Society Press, 242–253.

[107] Zheng, K., Y. Zheng, N. J. Yuan, S. Shang, and X. Zhou. 2014. "Online Discovery of Gathering Patterns over Trajectories." *IEEE Transactions on Knowledge and Data Engineering* 26 (8): 1974–1988.

[108] Zheng, Y. 2015. "Trajectory Data Mining: An Overview." *ACM Transactions on Intelligent Systems and Technology* 6 (3): 29.

[109] Zheng, Y., Y. Chen, Q. Li, X. Xie, and W.-Y. Ma. 2010. "Understanding Transportation Modes Based on GPS Data for Web Applications." *ACM Transactions on the Web* 4 (1): 1–36.

[110] Zheng, Y., Q. Li, Y. Chen, and X. Xie. 2008. "Understanding Mobility Based on GPS Data." In *Proceedings of the 11th International Conference on Ubiquitous Computing*. New York: ACM, 312–321.

[111] Zheng, Y., F. Liu, and H. P. Hsieh. 2013. "U-Air: When Urban Air Quality Inference Meets Big Data." In *Proceedings of the 19th ACM SIGKDD International Conference on Knowledge Discovery and Data Mining*. New York: ACM, 1436–1444.

[112] Zheng, Y., L. Liu, L. Wang, and X. Xie. 2008. "Learning Transportation Mode from Raw GPS Data for Geographic Application on the Web." In *Proceedings of the 17th International Conference on the World Wide Web*. New York: ACM, 247–256.

[113] Zheng, Y., T. Liu, Y. Wang, Y. Zhu, Y. Liu, and E. Chang. 2014. "Diagnosing New York City's Noises with Ubiquitous Data." In *Proceedings of the 2014 ACM International Joint Conference on Pervasive and Ubiquitous Computing*. New York: ACM, 715–725.

[114] Zheng, Y., Y. Liu, J. Yuan, and X. Xie. 2011. "Urban Computing with Taxicabs." In *Proceedings of the 13th International Conference on Ubiquitous Computing*. New York: ACM, 89–98.

[115] Zheng, Y., and X. Xie. 2011. "Learning Travel Recommendations from User-Generated GPS Traces." *ACM Transactions on Intelligent Systems and Technology* 2 (1): 2–19.

[116] Zheng, Y., X. Xie, and W.-Y. Ma. 2010. "GeoLife: A Collaborative Social Networking Service among User, Location and Trajectory." *IEEE Data Engineering Bulletin* 33 (2): 32–39.

[117] Zheng, Y., H. Zhang, and Y. Yu. 2015. "Detecting Collective Anomalies from Multiple Spatio-Temporal Datasets across Different Domains." In *Proceedings of the 23rd SIGSPATIAL International Conference on Advances in Geographic Information Systems*. New York: ACM, 2.

[118] Zheng, Y., L. Zhang, Z. Ma, X. Xie, and W.-Y. Ma. 2011. "Recommending Friends and Locations Based on Individual Location History." *ACM Transactions on the Web* 5 (1): 5–44.

[119] Zheng, Y., L. Zhang, X. Xie, and W.-Y. Ma. 2009. "Mining Correlation between Locations Using Human Location History." In *Proceedings of the 17th Annual ACM International Conference on Advances in Geographic Information Systems*. New York: ACM, 352–361.

[120] Zheng, Y., L. Zhang, X. Xie, and W.-Y. Ma. 2009. "Mining Interesting Locations and Travel Sequences from GPS Trajectories." In *Proceedings of the 18th International Conference on the World Wide Web*. New York: ACM, 791–800.

[121] Zhu, Julie Yixuan, Chao Zhang, Huichu Zhang, Shi Zhi, Victor O. K. Li, Jiawei Han, and Yu Zheng. 2017. "pg-Causality: Identifying Spatiotemporal Causal Pathways for Air Pollutants with Urban Big Data." *IEEE Transactions on Big Data*. doi:10.1109/TBDATA.2017.2723899.

[122] Zhu, Y., Y. Zheng, L. Zhang, D. Santani, X. Xie, and Q. Yang. 2011. "Inferring Taxi Status Using GPS Trajectories." *Microsoft Technical Report*, MSR-TR-2011-144.

About the Author

Dr. Yu Zheng is the vice president and chief data scientist at JD Finance, and is passionate about using big data and artificial intelligence technology to tackle urban challenges. He is the president of the Urban Computing Business Unit and serves as director of the JD Intelligent City Research. Before joining JD Group, he was a senior research manager at Microsoft Research, with research interests across big-data analytics, spatio-temporal data mining, machine learning, and artificial intelligence.

Zheng is also a chair professor at Shanghai Jiao Tong University and an adjunct professor at the Hong Kong University of Science and Technology. He currently serves as the editor in chief of ACM *Transactions on Intelligent Systems and Technology* and is a member of the editorial advisory board of *IEEE Spectrum*. He is also an editorial board member of *IEEE Transactions on Big Data* and the founding secretary of the SIGKDD China chapter. He has served as chair of over ten prestigious international conferences, including as the program co-chair of ICDE 2014 (industrial track), CIKM 2017 (industrial track), and IJCAI (industrial track), and as the area chair of AAAI 2019.

Zheng publishes referred papers frequently as a leading author at prestigious conferences and journals, such as KDD, IJCAI, AAAI, VLDB, UbiComp, and IEEE TKDE. Those papers have been cited over twenty thousand times (Google Scholar H-Index: 63 by October 2018). He has received five best-paper awards from ICDE'13 and ACM SIGSPATIAL'10. He has been invited to give over ten keynote speeches at international conferences and forums and guest lectures at universities such as the Massachusetts Institute of Technology (MIT), Carnegie Mellon University, and Cornell University. His book *Computing with Spatial Trajectories* has been used as a textbook at universities worldwide and honored as a Top 10 Most Popular Computer Science Book published by Chinese authors at Springer.

Zheng has received three technical transfer awards from Microsoft and twenty-four granted/filed patents. His technology has been transferred to Microsoft products such as Bing Maps. One of his projects titled Urban Air has been deployed with the Chinese Ministry of Environmental Protection, predicting air quality for over three hundred Chinese cities based on big data. He also leads a Chinese pilot project on the urban big-data platform, which has been deployed in Guiyang City.

Zheng has been featured multiple times in influential journals. In 2013, he was named one of the Top Innovators under age thirty-five by the *MIT Technology Review* (TR35) and featured in *Time* magazine for his research on urban computing. In 2014, he was named one of the Top 40 Business Elites under age forty in China by *Fortune* magazine because of the business impact of urban computing he has been advocating since 2008. In 2016, Zheng was honored as an ACM Distinguished Scientist. In 2017, he was recognized as one of the Top 10 AI Innovators in China.

Index